MW00637625

Handbook of Condensation Thermoplastic Elastomer

Edited by S. Fakirov

Further Titles of Interest

Th. Meyer, J. Keurentjes (Eds.)
Handbook of Polymer Reaction Engineering
2005, ISBN 3-527-31014-2

H.-G. Elias
Macromolecules
Vols.1-4
2005, ISBN 3-527-31172-6, 3-527-31173-4, 3-527-31174-2, 3-527-31175-0

M. F. Kemmere, Th. Meyer (Eds.)
Supercritical Carbon Dioxide
in Polymer Reaction Engineering
2005, ISBN 3-527-31092-4

M. Xanthos (Ed.)
Functional Fillers for Plastics
2005, ISBN 3-527-31054-1

R. C. Advincula, W. J. Brittain, K. C. Caster, J. Rühe (Eds.)
Polymer Brushes
2004, ISBN 3-527-31033-9

S. Fakirov (Ed.)
Handbook of Thermoplastic Polyesters
2002, ISBN 3-527-30113-5

E. S. Wilks (Ed.)
Industrial Polymers Handbook
Products, Processes, Applications
2000, ISBN 3-527-30260-3

Handbook of Condensation Thermoplastic Elastomers

Edited by

Stoyko Fakirov

WILEY-VCH Verlag GmbH & Co. KGaA

Editor

Prof. Stoyko Fakirov
University of Sofia
Faculty of Chemistry
1164 Sofia
Bulgaria

Library of Congress Card No.: Applied for
British Library Cataloging-in-Publication Data:
A catalogue record for this book is available from the British Library

Bibliographic information published by Die Deutsche Bibliothek
Die Deutsche Bibliothek lists this publication in the Deutsche Nationalbibliografie; detailed bibliographic data is available in the Internet at <http://dnb.ddb.de>.

Printed in the Federal Republic of Germany
Printed on acid-free paper

Cover Design Grafik Design Schulz, Fußgönheim
Printing Strauss GmbH, Mörlenbach
Bookbinding Litges & Dopf, Heppenheim

ISBN-13 978-3-527-30976-4
ISBN-10 3-527-30976-4

Preface

Thermoplastic elastomers (TPE) belong to a relatively new and small class of engineering plastics. Nevertheless, they enjoy a steady growth because of their unusual and very important combination of properties. During service, TPE behave as elastomers (*e.g.*, as vulcanized natural rubber) but, in contrast to the classical elastomers, they can be processed by means of the conventional techniques and equipment utilized for all thermoplastics.

This peculiarity of TPE is related to their different type of crosslinking. Unlike the classical elastomers, TPE are crosslinked by thermally labile aggregates, which melt during processing and regenerate after cooling down. This is because TPE are always block copolymers comprising the so-called "hard blocks" (forming solid crystallites or glassy regions), and "soft blocks" (imparting the elastomeric properties).

TPE can be produced using all chemical approaches for polymer synthesis. By means of polycondensation and polyaddition processes the three main types of TPE can be prepared: polyester-based, polyamide-based, and polyurethanes. The polyester-based TPE were first introduced in the USA by DuPont, the polyamide-based – in France, and polyurethanes are developed and applied worldwide. All they have found new and interesting applications.

While the polymerization-based TPE and, partly, the poly(ether ester)s (PEE) are relatively well covered in the literature, this is not the case of polyamide-based TPE and polyurethanes when new and specific applications are considered.

The main goal of this project is to combine the efforts of scientists with year-long experience worldwide to fill this gap. A peculiarity of this international team of contributors is that for the first time the research results of a well known polymer institute in Ukraine will get the deserved credit that, for understandable reasons, was not possible some years ago. For instance, the application opportunities of polyurethanes were broadened significantly by preparing systems with various chemical compositions and molecular structures as (semi) interpenetrating networks. Another peculiarity is the fact that for the first time detailed information regarding the most important polyamide-based TPE will be offered by the team who created them. Last but not least, thermoplastic elastomers, being multi-block copolymers comprising entities built up of very flexible ("soft") chains and others of much less flexible ("hard") chains, are capable of displaying such unique properties as shape memory effects. For this reason, they belong to the group of intelligent ("smart") materials with interesting and completely new technical and biomedical applications.

This book covers the chemical aspects, physical structure and properties, application opportunities, life cycle assessment, recycling possibilities, and the future trends of the three classes of thermoplastic elastomers.

It should be mentioned that the Contributors, the Editor, as well as the Publisher are well aware that, strictly speaking, the title "Condensation Thermoplastic Elastomers" is chemically not sufficiently correct for the covered classes of polymers. Nevertheless, it was preferred, *e.g.*, to the chemically correct title "Thermoplastic Elastomers by Polycondensation and Polyaddition", at least for the following reasons: (i) the polyurethane group behaves chemically much more as a typical polyester or polyamide group, rather than as a polyolefin, (ii) similarly to other cases (*e.g.*, the synthesis of nylon 6) regardless of the chemical route of the synthesis, *via* polymerization or *via* polycondensation, the final polymer is distinguished by a chemical structure typical of polycondensates, and (iii) the selected title sounds more comprehensive and more attractive.

As the Editor, I have enjoyed working with the individual contributors and have greatly appreciated their support, prompt response, and patience.

By analogy with my previously edited books, this one could hardly be realized in its present shape without the generous advice and suggestions of my friend, Prof. J. Karger-Kocsis, who was involved in the design of this project from the very beginning, and to whom I would like to express my most cordial gratitude.

This book could not have been produced without the constant support of my colleague of many years, Ms. S. Petrovich, who has maintained contact with contributors, polished the English, and helped in the processing of the book. For this, I express my sincere gratitude.

The Authors, The Editor, and the Publisher would like to thank all those publishers, who generously gave their permission to reprint materials from their own publications. Details are given at the end of the book.

Last but not least, I am grateful to the Foundation of Science and Technology of New Zealand, who made possible the sabbatical visit of the Editor at the University of Auckland, Department of Mechanical Engineering, Centre for Advanced Composite Materials, where this project was finalized.

S. Fakirov

Auckland, December 2004

Contents

PART I INTRODUCTION

Chapter 1 Creation and Development of Thermoplastic Elastomers, and Their Position Among Organic Materials

E. Maréchal

Chapter 2 Polycondensation Reactions in Thermoplastic Elastomer Chemistry: State of the Art, Trends, and Future Developments

E. Maréchal

PART II POLYESTER-BASED THERMOPLASTIC ELASTOMERS

Chapter 3 Polyester Thermoplastic Elastomers: Synthesis, Properties, and Some Applications

Z. Roslaniec

Chapter 4 Terpoly(Ester-*b*-Ether-*b*-Amide) Thermoplastic Elastomers: Synthesis, Structure, and Properties

R. Ukielski

Chapter 5 High Performance Thermoplastic Aramid Elastomers: Synthesis, Properties, and Applications

H. Yamakawa, H. Miyata

Chapter 6 Poly(Ether Ester) Thermoplastic Elatomers: Phase and Deformation Behavior on the Nano- and Microlevel

S. Fakirov

Chapter 7 Condensation Thermoplastic Elastomers Under Load: Methodological Studies of Nanostructure Evolution by X-ray Scattering

N. Stribeck

Chapter 8 Dielectric Relaxation of Polyester-Based Thermoplastic Elastomers

T. A. Ezquerra

PART III POLYAMIDE-BASED THERMOPLASTIC ELASTOMERS

Chapter 9 Thermoplastic Poly(Ether-*b*-Amide) Elastomers: Synthesis

F. L. G. Malet

Chapter 10 Poly(Ether-*b*-Amide) Thermoplastic Elastomers: Structure, Properties, and Applications

R.-P. Eustache

Chapter 11 Semicrystalline Segmented Poly(Ether-*b*-Amide) Copolymers: Overview of Solid-State Structure-Property Relationships and Uniaxial Deformation Behavior

J. P. Sheth, G. L. Wilkes

PART IV POLYURETHANE-BASED THERMOPLASTIC ELASTOMERS

Chapter 12 Thermoplastic Polyurethane Elastomers in Interpenetrating Polymer Networks

O. Grigoryeva, A. Fainleib, L. Sergeeva

Chapter 13 Polyurethane Thermoplastic Elastomers Comprising Hydrazine Derivatives: Chemical Aspects

Yu. Savelyev

Chapter 14 Molecular Dynamics and Ionic Conductivity Studies in Polyurethane Thermoplastic Elastomers

P. Pissis, G. Polizos

PART V BLENDS, COMPOSITES, APPLICATIONS, AND RECYCLING OF THERMOPLASTIC ELASTOMERS

Chapter 15 Polymer Blends Containing Thermoplastic Elastomers of the Condensation and Addition Types

J. Karger-Kocsis, S. Fakirov

Chapter 16 "Nanoreinforcement" of Thermoplastic Elastomers

J. Karger-Kocsis

Chapter 17 Commercial Condensation and Addition Thermoplastic Elastomers: Composition, Properties, and Applications

O. Gryshchuk

Chapter 18 Shape Memory Effects of Multiblock Thermoplastic Elastomers

B. K. Kim, S. H. Lee, M. Furukawa

Chapter 19 Condensation and Addition Thermoplastic Elastomers: Recycling Aspects

T. Spychaj, M. Kacperski, A. Kozlowska

Contributors

Eustache, R.-P., Dr.
ARKEMA-SERDATO,
Centre d'Étude de Recherche et Développment,
27470 Serquigny, France
Fax: +33 232 46 68 89; e-mail: rene-paul.eustache@arkemagroup.com

Ezquerra, T. A., Dr.
Instituto de Estructura de la Materia, CSIC,
Serrano 119, Madrid 28006, Spain
Fax: 0034 564 2431; e-mail: imte155@iem.cfmac.csic.es

Fainleib, A., Dr.
National Academy of Sciences of Ukraine,
Institute of Macromolecular Chemistry,
Department of Interpenetrating Polymer Networks and Systems,
48, Kharkivske shose, Kyiv 02160, Ukraine
Fax: +380 44 5595372; e-mail: fainleib@i.kiev.ua

Fakirov, S., Prof. Dr.
Sofia University,
Laboratory on Structure and Properties of Polymers,
1 James Bourchier Ave, 1164 Sofia, Bulgaria
Fax: +359 2 9625438; e-mail: ohtsp@wmail.chem.uni-sofia.bg

Furukawa, M., Prof. Dr.
Nagasaki University,
Graduate School of Science and Technology,
Department of Materials Science,
Nagasaki, Japan
Fax: +81 95819 2651; e-mail: furukawa@net.nagasaki-u.ac.jp

Grigoryeva, O., Dr.
National Academy of Sciences of Ukraine,
Institute of Macromolecular Chemistry,
Department of Interpenetrating Polymer Networks and Systems,
48, Kharkivske shose, Kyiv 02160, Ukraine
Phone: +380 44 5510322; e-mail: polym@ukrpack.net

Gryshchuk, O., Dr.
National Academy of Sciences of Ukraine,
Institute of Macromolecular Chemistry,
48, Kharkivske shose, Kyiv 02160, Ukraine
Fax: +380 44 5524064; e-mail: oleg.gryshchuk@ivw.uni-kl.de

Kacperski, M., Dr.
Technical University of Szczecin,
Polymer Institute,
ul. Pulaskiego 10, 70-322 Szczecin, Poland
Fax: +48 91 449 4685; e-mail: Michal.Kacperski@ps.pl

Karger-Kocsis, J., Prof. Dr. DSc.
Institut für Verbundwerkstoffe GmbH
(Institute for Composite Materials),
Kaiserslautern University of Technology,
67663 Kaiserslautern, Germany
Fax: +49 631 2017 198; e-mail: karger@ivw.uni-kl.de

Kim, B. K., Prof. Dr.
Pusan National University,
Department of Polymer Science and Engineering,
Pusan 609-735, Korea
Fax: +82 51 514 1726; e-mail: bkkim@pusan.ac.kr

Kozlowska, A., Dr.
Technical University of Szczecin,
Polymer Institute,
ul. Pulaskiego 10, 70-322 Szczecin, Poland
Fax: +48 91 449 4098; e-mail: Agnieszka.Kozlowska@ps.pl

Lee, S. H., Dr.
Pusan National University,
Department of Polymer Science and Engineering,
Pusan 609-735, Korea
Fax: +82 51 514 1726; e-mail: hey-man2000@hanmail.net

Malet, F. L. G., Dr.
ARKEMA-CERDATO,
27470 Serquigny, France
Fax: +0033 2 32 46 69 74; e-mail: frederic.malet@atofina.com

Maréchal, E., Prof.
Université Pierre et Marie Curie,
Laboratoire de Synthese Macromoleculaire (184),
4, Place Jussieu, 75252 Paris Cedex 05, France
Fax: +33 1 44 277 054; e-mail: marechal@ccr.jussieu.fr

Miyata, H., Dr.
Tosoh Corporation,
Yokkaichi Research Laboratory,
1-8 Kasumi, Yokkaichi, Mie, 510 8540 Japan
Fax: +81 593 64 5546; e-mail: h_miyata@tosoh.co.jp

Pissis, P., Prof. Dr.
National Technical University of Athens,
Department of Physics,
Sografou Campus, 15780 Athens, Greece
Fax: +30 210 7722932; e-mail: ppissis@central.ntua.gr

Polizos, G., Dr.
Materials Science & Engineering,
Pennsylvania State University,
228 Steidle Bldg, University Park, PA 16802, USA
Fax: 001 814 865 2917; e-mail: gxp16@psu.edu

Roslaniec, Z., Prof. Dr.
Technical University of Szczecin,
Institute of Materials Science and Engineering,
Piastow Av. 19, 70-310 Szczecin, Poland
Fax: +48 91 449 43 56; e-mail: zbigniew.roslaniec@ps.pl

Savelyev, Yu., Dr.
National Academy of Sciences of Ukraine,
Institute of Macromolecular Chemistry,
48, Kharkivske shose, Kyiv 02160, Ukraine
Fax: +380 44 552 4064; e-mail: yusav@i.kiev.ua

Sergeeva, L., Prof., Dr.
National Academy of Sciences of Ukraine,
Institute of Macromolecular Chemistry,
Department of Interpenetrating Polymer Networks and Systems,
48, Kharkivske shose, Kyiv 02160, Ukraine
Phone: +380 44 554 7062; e-mail: polym@ukrpack.net

Sheth, J. P., Dr.
Xerox Corporation,
26600 SW Parkway, M/S 7060-431,
Wilsonville, OR 97070, USA
Fax: +(503) 685 3883; e-mail: jignesh.sheth@office.xerox.com

Spychaj, T., Prof., Dr.
Technical University of Szczecin,
Polymer Institute,
ul. Pulaskiego 10, 70-322 Szczecin, Poland
Fax: +48 91 449 4685; e-mail: tees@mailbox.ps.pl

Stribeck, N., Dr.
University of Hamburg,
Institute of Technical and Macromolecular Chemistry,
Bundesstr. 45, D-20146 Hamburg, Germany
Fax: +49 40 42838 6008; e-mail: Norbert.Stribeck@desy.de

Ukielski, R., Prof., Dr.
Technical University of Szczecin,
Polymer Institute,
ul. Pulaskiego 10, 70-322 Szczecin, Poland
Fax: +48 91 449 4190; e-mail: Ryszard.Ukielski@ps.pl

Wilkes, G. L., Prof. Dr.
Virginia Tech (Virginia Polytechnic Institute and State University),
Department of Chemical Engineering,
Co-Director, Polymer Materials and Interfaces Laboratory,
Blacksburg, VA 24061-0211, USA
Fax: +(540) 231 9511; e-mail: gwilkes@vt.edu

Yamakawa, H., Eng.
Tosoh Corporation,
Yokkaichi Research Laboratory,
1-8 Kasumi, Yokkaichi, Mie, 510-8540 Japan
Fax: +81 593 64 5546; e-mail: yamakawa@tosoh.co.jp

PART I
INTRODUCTION

Chapter 1

Creation and Development of Thermoplastic Elastomers, and Their Position Among Organic Materials

E. Maréchal

The thermoplastic elastomers (TPEs) concern large industrial and commercial fields, as well as academic and applied research. Often the TPEs are considered as being only an important part of the block copolymers, but they are present in many other polymeric materials, as clearly shown by Holden *et al.* [1,2] and Rader [3–5]. They are characterized by a set of properties inherent to block and graft copolymers, different blends, and some vulcanized materials. More than 7000 *Chemical Abstracts* entries directly concern TPEs and in about 12500 other publications they are closely associated to other issues (*SciFinderScholar*) [6]. Most of these references describe materials, which associate elastomeric recovery and thermoplastic properties; however some products exhibit characteristics and properties, which completely differ from those of conventional TPEs.

Historic, scientific, technical and commercial considerations [2–5] should be taken into account when outlining the TPE domain. This introductory chapter begins with historical considerations, followed by a critical evaluation of the main preparations and modern analytical techniques used in chemical, structural, and morphological studies. The TPE properties and processability, their position among organic materials, and their applications are analyzed. Finally, the most probable trends of their future development are discussed in a short conclusion.

1. Birth and development of TPEs: a brief survey

The real era of TPEs began with the advent of block and grafted copolymers. However, some blends are tacitly accepted as TPEs, even though their structure does not exhibit some of their essential characteristics, such as the separation of soft and hard phases; for instance, PVC plasticized by high-boiling liquids is often considered as one of the precursors of the TPEs [7]. The reactions between diols and diisocyanates resulting in polymeric products [8–12] were an important step in the TPE development, since these elastomers exhibit a very rapid elastic recovery and good processability. It is important to stress that these syntheses are two-step processes and their introduction on the market was the result of a new strategy. The latter was rapidly applied to polyesters when Snyder [13] polycondensed a mixture of terephthalic acid, octanedioic acid and propane-1,3-diol and, separately, terephthalic acid with ethane-1,2-diol; these two polycondensates were mixed, then reacted and the ultimate product exhibited elastomeric and plastic properties, was extrudable in the melt and spinnable from solution. In the 1950s, some other elastomers were patented, particularly polyurethanes [14] and, as Snyder's product, they behaved as vulcanized rubber even though they were not chemically crosslinked, as revealed by their complete solubility. All these elastomers exhibited properties which were not observed in natural rubber or in the first synthetic rubbers, but it is important to emphasize that they were prepared by reactions which were already classical at that time. Nevertheless, the advent of these materials generated new researches in chemistry and, perhaps more important, in structure and morphology.

When Szwarc *et al.* discovered [15,16], or rediscovered [17,18], the anionic living polymerization, a completely different preparation of these elastomers was proposed; the study of TPEs passed from infancy to maturity. These authors used sodium metal naphthalene diinitiators to prepare poly(styrene-*b*-isoprene-*b*-styrene), which was probably the first TPE with a perfectly defined structure. However, this copolymer could not be commercialized, as most of the poly-isoprene units were -3,4-, with poor elastomeric properties. It is only when the polymerization was initiated by alkyllithium that poly(styrene-*b*-isoprene-*b*-styrene) and poly(styrene-*b*-butadiene-*b*-styrene) were obtained with the classical TPE properties: very high tensile strength and elongation at break, very rapid elastic recovery, and no chemical crosslinking. Bailey *et al.* [19] announced the existence of these materials in 1966 and Holden *et al.* [20] published the corresponding theory in 1967 and extended it to other block copolymers.

In addition to their commercial success, the TPEs were the result of logical considerations and scientific effort, giving birth to a new field of science and technology. These multiphase materials stimulated many theoretical and experimental studies dealing not only with their chemistry, but also with their structure and morphology.

Later, the preparation and the characterization of new TPEs followed; once again, it is important to stress that the arrival of these materials resulted from logical considerations, as it was brilliantly confirmed by the theoretical and experimental studies by Tobolsky, predicting the existence of EPDM [21]. Based on these fundamental studies, many other TPEs were prepared by very different syntheses, commercialized or not, but with many common structural characteristics. In 1962, this new strategy was applied to prepare copolymers containing random poly(ethylene-*co*-propylene) as amorphous blocks and *linear* polyethylene or *isotactic* polypropylene as hard blocks [22]; once more, Tobolsky had predicted that such a copolymer would exhibit TPE properties.

The first polyurethane samples resulted from the pioneering work of Otto Bayer [7] and Christ [12] aiming at the preparation of new textile fibers. Coffey [23] described their elastomeric properties. However, the really scientific approach to thermoplastic polyurethane (TPU) elastomers began with the publications of Müller *et al.* [24] and Petersen *et al.* [25]. The TPU behavior was analyzed by Otto Bayer and his school [26] in a theoretical study where, for the first time, a truly linear polyurethane was prepared through a sequence of steps, announcing the classical route used to obtain thermoplastic polyurethanes: preparation of an α,β-diisocyanate prepolymer, resulting from the reaction of an α,β-dihydroxy-polyester (or polyether) with an excess of diisocyanate, then extended by water with formation of urea linkages, and then reacted with additional diisocyanate. After the importance of the addition of a short-length diol was recognized, the modern chemistry of polyurethanes was definitely born. It was the first case when polyaddition was systematically used and a new strategy was defined, opening the door to many other TPEs. The understanding of the non-chemical crosslinking was perhaps more important than the chemistry; this peculiar observation was first called "virtually crosslinked elastomer" [27]. Soon it became clear that the TPE chains were formed by a succession of long flexible blocks responsible for the elasticity and hard ones interconnecting the macromolecules. It appeared that the blocks were incompatible and localized in separate microdomains. This microphase separation was called "segregation".

The above fundamental studies were followed by the logical application of the respective strategy to polycondensation, which resulted in the preparation of some very important TPEs: poly(amide-*b*-ester) [28–30], poly(amide-*b*-ether) [31–33], and poly(ether-*b*-ester) [34–36].

Nowadays, the development of TPEs concern many branches of macromolecular chemistry: cationic and radical polymerizations, chemical modification, enzymatic catalysis or the use of microorganisms. Their respective contributions are analyzed in Section 2. The elastomers based on halogen-containing polyolefins [37] and those prepared by dynamic vulcanization [38] are also included in the TPE family. More information on these materials and techniques is given in Section 2.

2. Main routes to thermoplastic elastomer preparation

The chemistry of TPEs continuously changed with time and in this chapter the present-day situation is analyzed with emphasis on polycondensation, polyaddition, and chemical modification, which is often associated with the other processes.

An essential part of this section concerns block copolymers, but the contribution of some other processes, such as grafting or dynamic vulcanization is, by far, not negligible and their current state is discussed.

The synthesis of block copolymers follows two essential pathways: (i) a difunctional oligomer initiates the formation of two or more other blocks (Scheme 1) and (ii) two or several different difunctional oligomers react together or with a coupling agent (see Chapter 2); sometimes, the second block can be prepared in the presence of the first one.

$$\text{A–A} \xrightarrow{M_1} {}^{*}M_1(M_1)_{\overline{p}}\text{A–A–}(M_1)_q M_1^{*} \xrightarrow{M_1} {}^{*}(M_2)_r(M_1)_{\overline{p+1}}\text{A–A–}(M_1)_{q+1}(M_2)_t^{*} \quad (1)$$

A–A is an initiator

Pathway (1) is mainly encountered in chain polymerization (anionic, cationic, and controlled radical polymerizations); the second one refers essentially to polycondensation and polyaddition. There is no strict distinction between these two sets of techniques, *e.g.*, the difunctional oligoethers used in poly(ether-*b*-ester)s or poly(ether-*b*-amide)s can be prepared by ring-opening polymerization and then polycondensed with the other oligomer (Scheme 2) [39].

$$\text{X–(Block 1)–X} + \text{Y–(Block 2)–Y} \longrightarrow \text{–}[\text{(Block 1)–X}'\text{Y}'\text{–(Block 2)}]_n\text{–} \quad (2)$$

2.1. *Living anionic polymerization*

Living anionic polymerization remains an important technique for the preparation of well defined triblock copolymers, such as: poly(styrene-*b*-butadiene-*b*-styrene) and poly(styrene-*b*-isoprene-*b*-styrene) [40], and it was extended to copolymers containing polysiloxane blocks or to poly(α-methylstyrene-*b*-propylene sulfide-*b*-α-methylstyrene) [41]. In many applications, anionic polymerization no longer requires a high vacuum line and often an inert atmosphere is sufficient [42]. The ester block-containing copolymers, such as poly(styrene-*b*-butadiene-*b*-methyl methacrylate), were prepared by sequential anionic polymerization and their morphology and mechanical properties differ substantially from those of the triblock poly(styrene-*b*-butadiene-*b*-styrene). Poly(styrene-*b*-butadiene-*b*-styrene) was end-capped by alkyl methacrylate blocks, which leads to a pentablock copolymer [43]. TPEs can also be obtained by anionic ring-opening polymerization (ROP): Sipos *et al.* [44] prepared the biodegradable copolymer poly(L-lactide)-*b*-polyisobutylene-*b*-poly(L-lactide), characterized by two separated glassy phases.

2.2. *Living cationic polymerization*

The use of living cationic polymerization in the preparation of TPEs was reviewed by Kennedy [45] in relation to graft and block copolymers, but the application of cationic polymerization to TPEs began before the arrival of the living techniques. Kennedy and Maréchal [46] reviewed a large part of these elastomers in 1992. Living cationic polymerization was an essential breakthrough, as it was also the case of living anionic polymerization. The approach to this technique was vividly described by Kennedy [47], showing that it follows a three-step progression: (i) controlled initiation, (ii) reversible termination (quasiliving systems), and (iii) controlled transfer. It allows the preparation of block copolymers according to Schemes 1 or 2. The number of articles pertaining to the preparation of TPEs by living cationic polymerization is continuously increasing [48–50]. Many of them deal with styrenic TPEs, but more sophisticated architectures were also synthesized. Kwon *et al.* [49] prepared arborescent polyisobutylene-polystyrene block copolymers where the arborescent polyisobutylene was obtained by living cationic polymerization.

Anionic and cationic polymerizations are often associated. Feldthusen *et al.* [51] prepared copolymers containing linear and star-shaped blocks: a living polyisobutylene chain was prepared by cationic polymerization, its ends were converted into 2,2-diphenylvinyl groups, then metallated and used as initiators of the *tert*-butyl methacrylate anionic polymerization.

2.3. *Controlled radical polymerization*

The discovery of the controlled radical polymerization (CRP) offered additional possibilities in the chemistry of TPEs [52–54]. CRP was used in both graft and block copolymer preparation and extensively reviewed by Matyjaszewski [55] and Mayes *et al.* [56]. It allows the easy preparation of novel environmentally friendly materials, such as polar TPEs; it can be carried out in the bulk or in water and requires only a modest deoxygenation of the reaction mixture. Atom transfer polymerization (ATRP) is one of the most important aspects of CRP; it was developed by Matyjaszewski and rests on an equilibrium between active and dormant species [57]. Moineau *et al.* [58] applied ARTP to the preparation of poly(methyl methacrylate-*b*-*n*-butyl acrylate-*b*-methyl methacrylate).

2.4. *Polycondensation and polyaddition*

Strange enough, so far there are no books entirely devoted to condensation TPEs and the latter are considered only in chapters of more general works. The most important TPEs prepared by polycondensation are the subject of several chapters of this book: polyester-based TPEs, poly(amide-*b*-ethers), polyurethanes, *etc.* However, some less known condensation TPEs are described in Chapter 2: metal-containing macrocycles as monomers, liquid crystalline side chains, metallo-supramolecular block copolymers, as well as the use of enzymatic catalysis or of microorganisms.

Block copolymers can be prepared either by polycondensing a difunctional oligomer, which is often the soft block, with the precursors of the hard block, or by polycondensing or coupling two, or more, difunctional oligomers; this aspect is discussed in Chapter 2.

Many difunctional oligomers are prepared by ionic polymerization and then polycondensed with other functional species. Schmalz *et al.* [59] used a sequential preparation of a TPE with a non-polar soft segment: (i) preparation of α,ω-dihydroxy-[polyoxyethylene-*b*-(hydrogenated polybutadiene)-*b*-polyoxyethylene] (A) by anionic polymerization and (ii) polycondensation of A with dimethyl terephthalate and 1,4-butanediol. Shim *et al.* [60] associated cationic polymerization, chemical modification, and polyaddition to prepare a multiple-arm TPE: (i) preparation of a poly(styrene-*b*-isobutylene) sample (C) by living cationic polymerization, (ii) end-capping of the polyisobutylene blocks by allylic groups (D), and (iii) polyhydrosililation of D with SiH-containing cyclosiloxanes as a core. Sometimes the sequence is reversed: polycondensation, end-capping, and then anionic polymerization [61]. These examples show the importance of the association of chain polymerization with polycondensation.

Some difunctional oligomers are prepared by polycondensation and then used in chain polymerization. However, most of them are applied in block polycondensation or polyaddition: Pan *et al.* [62] prepared difunctional oligosiloxanes by polycondensation of dimethyldichlorosilane with different oligomeric diols, and the polycondensates were then reacted with diisocyanates. Yokozawa *et al.* [63] prepared a well defined poly(*p*-benzamide), with a low dispersity index, using a new polycondensation process, which is discussed in Section 6.

The contribution of the difunctional oligomers to the preparation of TPEs should enjoy an important development; their use is not limited to classical chain polymerization and polycondensation since they can be also applied to less common processes, such as the metathesis. A functional diene, for instance an α,ω-divinyl aliphatic or aromatic ester, is polyadditioned with an α,ω-divinyl-(soft-oligomer) in the presence of a metathesis catalyst, *e.g.*, a ruthenium derivative [64,65].

2.5. *Chemical modification and grafting*

Chemical modification is an important tool in the production of TPEs and the improvement of their properties [66]. It plays an important role in difunctional oligomer chemistry, particularly the modification of end-groups, which is often necessary in polycondensation or in chain polymerization where difunctional oligomers can be used as macroinitiators. Madec and Maréchal [67] prepared an α,ω-dibenzaldehyde-oligosiloxane by reacting an α,ω-dihydrogensilane-oligosiloxane with 4-allylbenzaldehyde and polycondensed it with α,ω-diamino-oligoamides.

Some functional oligomers were prepared by the degradation of a polymer; for instance, Ebdon and Flint [68] obtained α,ω-dialdehyde-[methyl oligo(methacrylate)] by oxidative cleavage of statistical methyl methacrylate-buta-1,3-diene copolymers.

The chemical modification of a block copolymer can be the essential step in TPE preparation; in this connection, the hydrogenation of unsaturated block copolymers is often an important step [69,70]. Numerous modifications aim to improve the TPE properties or to adapt TPEs to other applications, or to make them reactive to a specific compound.

The development of graft copolymers is by far less important than that of block copolymers; however, they are found in interesting patents [71–73]. Grafting is mainly used to modify the properties of a block copolymer, but it can be applied to a rubber in order to generate rigid side chains. Grafting proceeds through two different pathways: direct reaction of the backbone with a monofunctional oligomer (grafting onto) or polymerization of a monomer initiated by an active group of the polymer (grafting from). Ikeda *et al.* [74] condensed chlorinated butyl rubber with the potassium salt of an α-methyl-ω-hydroxy-polyoxyethylene, which is a "grafting onto" reaction. The resulting TPE is amphiphilic and the grafts form separate crystalline microdomains. In principle, these two techniques are applicable to any polymer, but often the first step is the modification of the chains in order to make them reactive.

Controlled chain polymerization plays an important role in "grafting from", involving cationic [47,75], anionic [76,77], and particularly radical processes [55,78] where ATRP is a powerful tool. Gaynor *et al.* [79] prepared monomers A=B–C* where A=B is a copolymerizable double bond and C* is an activated halogen, such as *p*-chloromethylstyrene or 2-(2-bromopropionyl)-ethyl acrylate; they are copolymerized with conventional vinyl monomers, leading to the formation of pendant activated groups used as macroinitiators to prepare graft copolymers. Baumert *et al.* [80] copolymerized an alkoxyamine-functionalized 1-alkene with ethylene and the resulting highly branched polyethylene initiated the controlled radical polymerization of styrene or styrene/acrylonitrile.

2.6. *Preparation by blending*

The chemistry and the structure of the elastomers prepared by blending or dynamic vulcanization greatly differ from those discussed in Sections 2.1 to 2.5, even though they have many common characteristics and properties. The elastomers based on halogen-containing polyolefins are included in very important blends, such as PVC-nitrile rubber, PVC-copolyester elastomers, and PVC-polyurethane elastomers. Their characteristics and properties were comprehensively reviewed by Hofmann [37]. As the other TPEs, these blends combine a good elastic recovery with the properties of vulcanized thermoset rubbers. They can be melt-reprocessed numerous times and exhibit good resistance to heat, oils, and many chemicals. They are single-phase polymers, in contrast to most TPEs, which are two-phase systems, and for this reason they are often defined as "processable rubbers".

2.7. *Preparation by dynamic vulcanization*

Dynamic vulcanization (DV) simultaneously performs the mixing and the cross-linking of a rubber and a thermoplastic polymer. The resulting products are called "thermoplastic vulcanizates" (TPV) and are the subject of various reviews and books [3,38,81]. TPV exhibit the two main characteristics of the

TPEs: a good elastomeric recovery and the properties of thermoplastic polymers. Their preparation, structure, and properties were carefully analyzed by Coran and Patel [38]. Their morphology has many common characteristics with the thermoplastic elastomeric polyolefins and involves a highly vulcanized elastomeric phase uniformly distributed in a melt-processable matrix; Rader [5] compares it to a "raisin pie". The vulcanization of the elastomer takes place when in a molten plastic. The crosslinking improves several TPE properties: behavior with respect to temperature, resistance to swelling in fluids, compression and tension set, creep and stress relaxation.

DV was applied to different systems, *e.g.*, a diene rubber (EPDM, butyl rubber or natural rubber) is associated with a polyolefin (polyethylene or polypropylene) or an acrylonitrile-butadiene rubber (NBR) sample is associated with a polyamide. The high incompatibility between the elastomer and the plastic may be an important obstacle in the preparation of a dynamically vulcanized material, since the properties of the latter depend on the quality of the dispersion. The dispersity of NBR in polyolefins is very low and a polymeric compatibilizer must be added, which often requires grafting and coupling processes.

3. Techniques used in the characterization of TPEs

The preparation of TPEs is closely related to the control of their structure and morphology, all the more that they are structurally complex systems, requiring accurate, efficient, and rapid analytical techniques. The characterization techniques are applied to the TPEs and their precursors. During the last thirty years, they enjoyed a fantastic development; some of them appeared during the last decade. In the following, they are grouped into analytical branches, but it is essential to keep in mind that most of them are associated with some others.

3.1. *Chromatography*

For a long time, the use of chromatography was mainly limited to the determination of molecular masses and to the qualitative estimation of the heterogeneity of the samples. At present, it is a highly efficient technique giving very accurate values [82–89]. Copolymers are complex macromolecular systems, characterized by two distributions in molecular mass and chemical composition; liquid chromatography used at the critical point of adsorption allows the determination of the molecular heterogeneities (chemical and molecular mass distributions) [83].

The association of chromatography with other techniques is successfully developing and allows, for instance, the determination of the molecular mass distribution of each block, in one run, as well as the respective chemical distribution. Kilz *et al.* [86,87] applied two-dimensional liquid chromatography to complex mixtures containing block copolymers, cycles, and functional oligomers; this technique combines the advantages of high perfoprmance liquid chromatography (HPLC) and size exclusion chromatography (SEC), and provides valuable information on the composition, the functionality, and the molecular mass distribution. Gores and Kilz [88] associated several techniques, such as multiple detection SEC, multi-angle laser light scattering, and viscometry; multiple detection allowed the determination of the chemical composition and distribution. The use of chromatography is certainly not limited to research and has become a method for routine check-up and quality control of industrial TPEs [89].

3.2. *Spectrometric techniques*

They are essential to the study of TPEs, particularly when associated with other techniques.

3.2.1. *FT-IR spectroscopy*

Although by far less used than nuclear magnetic resonance (NMR), the Fourier transform infrared (FT-IR) spectroscopy remains an efficient tool in TPE analysis, *e.g.,* in the investigations of cluster formation in thermoreversible networks [90] or in side-chain liquid crystalline TPEs [91]. FT-IR is often associated with other techniques: NMR [92,93] or X-ray diffraction [94].

Infrared dichroism is a powerful technique to study the evolution of the chain orientation in films, particularly when coupled with photoelastic modulation [95]. FT-IR is a rapid technique that is very efficient in industrial applications, such as weathering [96] or analysis of blends [97].

3.2.2. *NMR spectroscopy*

NMR is essential in TPE characterization and present in most studies; it is often associated with other techniques. This technique allows one to have a deep insight in the structure of the block copolymer chains; for instance, Boularès *et al.* [98] used ^{1}H NMR to determine the ester junctions in poly(amide-*b*-copolyether) chains, which were prepared by polycondensation of α,ω-dicarboxylic-oligoamides and α,ω-dihydroxy-copolyether.

NMR is often used as a routine technique because it is efficient in the control of the TPE purity. Frick *et al.* [99] prepared poly(lactide-*b*-isoprene-*b*-lactide)s and showed that they were not contaminated by free homopolymer and diblock copolymers using ^{1}H NMR coupled with SEC.

Some more specific aspects of NMR are necessary when more detailed studies are required. Cross polarized magic-angle ^{13}C NMR in the solid state was applied to the analysis of the microphases and their scales in TPEs [100]. Impulse NMR provided knowledge of the synergic and antagonistic deviations of the mechanical properties of TPE blends [101].

The determination of accurate molecular mass values of the copolymers and their distribution, as well as of the functionality of difunctional oligomers remains a difficult problem. It is necessary to compare the results of, at least, two different techniques, for instance SEC and titration of the end-groups but, unfortunately, the latter is inapplicable when the chain length is high. NMR is probably one of the most efficient tools in the determination of the end-group concentration, particularly when they are modified. For instance, ^{19}F NMR is used after an α,ω-dihydroxy-polymer is reacted with trifluoroacetic anhydride. In some cases, the end-units are converted into groups, which are fluorescent or adsorbing in the visible or UV light; sometimes they are converted into chemically titratable groups.

3.3. *Scattering techniques*

Lodge [102] and Norman [103] reviewed the use of scattering techniques in the characterization of polymers, particularly block copolymers. Static and dynamic light scattering, small-angle neutron and X-ray reflectivity were analyzed and emphasis was placed on their similarities and differences.

3.3.1. *Static and dynamic light scatterings*

These two techniques are coupled in many studies on block copolymers. The use of dynamic light scattering was reviewed by Stepanek and Lodge [104]; it is an efficient tool to analyze the order–disorder transition in block copolymer melts [105,106]. The static version is often used in studies dealing with aggregation and micellization in solvents [107] and is often associated with the dynamic scattering and other techniques, such as cryoscopy.

3.3.2. *Small-angle (SAXS) and wide-angle (WAXS) X-ray scattering*

SAXS and WAXS are particularly efficient in the study of amorphous polymers including microstructured materials, hence their use in block copolymers (see also Chapters 6 and 7). The advent of synchotron sources for X-ray scattering provided new information, particularly on the evolution of block copolymer microstructures with time resolution below one second. In particular, the morphology of TPEs is most often studied with these techniques; Guo *et al.* [108] applied SAXS to the analysis of the phase behavior, morphology, and interfacial structure in thermoset/thermoplastic elastomer blends. WAXS is often associated with SAXS and some other methods, such as electron microscopy, and various thermal and mechanical analyses. It is mainly used in studies of the microphase separation [109,110], deformation behavior [111], and blends [112].

3.3.3. *Small-angle neutron scattering (SANS)*

SANS is an excellent and non-destructive technique, particularly efficient in the study of order–order and disorder–order transitions. It is very useful in the characterization of the morphology of block copolymers in the ordered state and, in this case, it is complementary to electron microscopy.

Several reviews are devoted to SANS applications to TPEs and block copolymers [113,114] and almost 300 reports describe SANS studies of block copolymers. As far as TPEs are concerned, it is a very valuable technique, particularly efficient in the analysis of disorder in the melt [115] or of macroscopic phase separation [116]; it is also used in blends [117,118] and in processing operations, such as dynamic vulcanization [119] or co-molding [120].

Neutron reflection provides the atom composition of a thin film at different depths. It gives the composition profiles of block copolymers deposited as films on silicon wafers [121]. X-ray reflection is used in the same fields as neutron reflection; these are complementary techniques and often associated with some others.

3.3.4. *X-ray diffraction*

X-ray diffraction remains an important tool in morphological studies, for instance in the investigation of the semicrystalline [122] or liquid-crystalline blocks [123] and of the stretching behavior of TPEs [124,125].

3.4. *Microscopies*

Atomic force microscopy (AFM) has an important development in the structural analysis of TPEs. It was applied to different problems: thermooxidative stability and morphology [126], copolymers with arborescent blocks [127], thermoplastic vulcanizates [128], blends [129], and morphology and orientation during deformation studied both by AFM and SAXS [130].

Transmission electron microscopy (TEM) is most useful to characterize the structure and morphology of TPEs. It is almost always associated with other techniques, particularly SAXS, WAXS, SANS, and AFM. TEM is used in studies of interpenetrating networks [131], morphology and crystallinity of hard blocks [132], structural evolution of segmented copolymers under strain [111], and blends [133].

3.5. *Controlled degradation*

The control of some specific parts of a TPE may provide interesting information on its structure. Sundararajan *et al.* [134] used WCl_6-catalyzed metathesis to prepare poly(HB-*b*-butadiene-*b*-HB) where HB is a rigid block, such as polyphenylacetylene, and the blocks are linked by W atoms; their structure was analyzed by ^1H NMR, elemental analysis of tungsten and, more original, by the complete separation of the blocks when the TPE is reacted with benzaldehyde. Valiente *et al.* [135] studied the enzyme-catalyzed hydrolysis of phthalic unit-containing copolyesters, which resulted in an efficient analytical tool for the analysis of the chain structure, particularly for the determination of the block length and their distribution in the chains. Luo *et al.* [136] used pyrolysis-gas chromatography/spectrometry to analyze the microstructure of a polyester-polyether copolymer (TPE).

3.6. *Thermal techniques*

Differential scanning calorimetry (DSC) and other thermal analyses are frequently used in TPE characterization, often as routine methods. DSC is a powerful technique to improve the knowledge of the microphase structure (see also Chapter 18). For instance, Ukielski *et al.* [137] investigated the reversible endothermic processes in thermoplastic multiblock elastomers. DSC and thermo-mechanical analysis (TMA) are often used as complementary to other techniques (particularly spectroscopic analyses) or for providing additional information. Anandhan [119] studied the effect of the mixing sequence of dynamic vulcanization on the mechanical properties and DSC showed that nitrile rubber and the copolymer poly(styrene-*co*-acrylonitrile) are thermodynamically incompatible in blends.

4. Properties and processing of TPEs

The determination of the TPE properties calls for the same techniques as those used for other organic materials, though some modifications may be required; they are listed and discussed in [3]. The information they provide, whatever they are, has a sound meaning only when determined in close correlation with structural analyses. The morphology evolution, when TPEs are subjected to processing, must be carefully followed, and it is particularly important to take into account the specificity of TPE rheology in the melt. Most of the methods listed in Section 3 are essential in the understanding of the TPE properties and their changes [129,138,139].

Many of the processing techniques applied to thermoplastic polymers or elastomers hold for TPEs; a few of them are more specific to these materials. Rader [3] clearly underlined the specificity of their rheology in the melt, particularly that they are highly non-Newtonian, hence more sensitive to shear than to temperature. Nishizawa [140] discussed the recent trends in the TPE molding technology.

4.1. *Injection molding*

Injection molding is by far the most used technique in TPE processing due to its high productivity and because it is a clean process with no waste formation. It is used in a great variety of applications ranging from tubes or foams to finished articles; it can be applied to the co- or insert-injection. The use of hot runner methods in injection molding was reviewed by Lachmann [141] and this is an interesting diversification of the conventional technique; it maintains the flowability of the melt during its transportation to the individual cavities or to the individual gates. During injection molding, TPEs behave as the other thermoplastics in hot runner without major problems.

4.2. *Compression molding*

Compression molding is by far less used than injection molding. Akiba [142] reviewed this processing method concurrently to other techniques: injection molding, extrusion, transfer molding, *etc.*

4.3. *Extrusion*

The extrusion of TPEs was reviewed by Knieps [143]. This processing technique is essential in the shaping of many different profiles; the use of single-screw extruders is predominant, but some other extruders are used, such as those equipped with three-section or barrier screws. Extrusion is also applied to other shapes: foams, tubes, sheets, *etc.*

4.4. *Blow processings*

Extrusion and injection blow molding of TPEs are particularly important whatever the shape: bottles, boots, *etc.* They were reviewed by Nagaoka [144], who showed that the parameters controlling the processing are similar to those which control extrusion and injection molding. Blow processes are also used to prepare TPE foams; Brzoskowski [145] prepared low-density foamed thermoplastic vulcanizates, using a single screw.

4.5. *Thermoforming*

The number of references relative to the thermoforming of TPEs drastically increases, particularly in the last three years; most of them are patents and apparently this TPE processing technique does not enjoy so far a general appreciation.

4.6. *Reactive processings*

The corresponding literature was reviewed by Prut and Zelenetskii [146] and an important part is related to dynamic vulcanization; some of the references provide a deep insight in the chemical evolution, particularly in side reactions [147,148]. The morphology of thermoplastic vulcanizates during processing drastically depends on temperature and shear and, to a lesser extent, on the screw rate [149].

4.7. *Degradation in processing*

Endres *et al.* [150] carried out a fundamental study of the thermal decomposition of TPEs under thermoplastic processing conditions, more particularly of the respective kinetics, and an equation was given describing the thermal degradation of polyurethanes upon extrusion. The knowledge of these side phenomena during TPE processings is continuously improving and, for instance, Lee and White [151] observed that no by-products, more particularly carbon dioxide and water, are formed in the preparation of a polyetheramide triblock copolymer by reactive extrusion.

5. Position of TPEs among organic materials and their applications

TPEs take an intermediate position between rubbers which are soft, flexible and with elastic properties, and thermoplastics which are rigid; in fact they overlap both domains [152,153]. The respective positions of TPEs, thermoset rubbers, and plastics in terms of Shore A and D hardness are given in Figure 1 [5].

In organic and macromolecular chemistry, their preparation is attractive, regardless of the additional difficulties, compared to the synthesis of classical

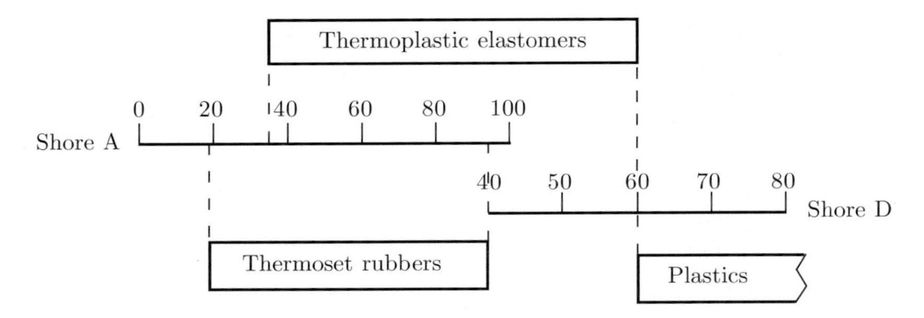

Figure 1. Respective positions of thermoset rubbers, plastics, and thermoplastic elastomers in Shore A and D hardness scales [5]

polymers, due to the incompatibility of the blocks or to the drastic control of the functionality of the precursors. On the other hand, they are processed by the same techniques and apparatuses as the thermoplastic polymers. Several authors analyzed the advantages and the drawbacks of the use of TPEs in specific fields of application. Baumann [154,155] reviewed the innovative applications of thermoplastic elastomers, showing their versatility and their advantages in design and economy.

The number of references directly related to TPE is around 7000; most of them are patents (above 95%) and many of them use the term "thermoplastic elastomer", even though the TPE structure is far from any accepted nomenclature. This is confirmed by the very limited number of publications devoted to the control of the structure. Numerous reports describe the processing of TPEs in relation to a specific application, even though some of them treat the fundamental aspect of a specific technique. Several patents describe a processing apparatus specially designed for a definite TPE [156] or the preparation of an elastomer specifically tailored to meet some processing requirements [157]. Some very few reviews analyzed the dependences of the properties and the applications on the processing conditions [142,158].

The blends of TPE with other organic materials enjoy an important development, particularly in organized morphologies, such as multilayer laminates [159]. Their final shape is obtained by different processing techniques, such as dynamic vulcanization [160]. The preparation of the blends generates interesting studies on polymer compatibility, which calls for chemical modification in the bulk or at interfaces, as well as for thermodynamic studies and careful control [161]. The preparation of a copolymer requires a long succession of steps going from the macromers to the final product; blending is the initial operation, which requires very careful structural analyses of the mixture, before and after the copolymerization [162]. The processing of blends requires coextrusion-molding, calendering, vacuum forming, *etc.*

TPE are used for the preparation of some important synthetic fibers and elastomers, and are described in many books, reviews, and articles. The thermoplastic elastomers prepared by polycondensation and polyaddition are the subjects of several chapters of this book. However, TPE are used in numerous other applications, some of them being widely commercialized. They are briefly listed below, but numerous applications cannot be classified as belonging to a specific technical field.

Many references concern layered and sandwich structures; the different layers may cooperate to provide a desired function, particularly in mechanical properties, or each layer may be responsible for a specific function such as adhesion, electric conductivity [163], specific density, [164] or noise reduction [165].

A non-negligible number of TPEs are involved in the production of foams where they provide low density, good mechanical properties and, often, advantages in processing; some of them are vulcanized [166]. The use of supercritical fluid CO_2 in the saturation/depressurization method allowed the control of TPE foaming and the material exhibited very little hysteresis [167].

The use of TPEs in films and tapes is discussed in numerous reports. However, the latter provide essentially "recipes" of mixtures. They concern many technical fields: pressure-sensitive adhesives, medical applications, barrier properties, porous films, *etc.*, and some of them describe well defined electic properties, such as an electrostrictive system formed of a conductive polymer (polypyrrole, polyaniline, polythiophene) deposited onto opposing surfaces of a TPE film, *e.g.*, a polyurethane [168]. The same comments hold for coating and painting where TPE are mainly used in the protection of metal or alloy substrates, such as electric wires; some others impart additional functions to the protection, as is the case of optical fibers or textile fibers and different fabrics. Some few patents claim that a new TPE can be used as textile fiber with interesting properties, however these patents are almost never industrialized; on the other hand, new TPE fiber processing techniques are proposed [169].

TPEs are used as gas or liquid barriers in many applications: caps, bottles, films, separators, *etc.*, where they can be responsible for the barrier effect or for the separation. Often they represent only substrates coated with an active component but imparting, for instance, the mechanical properties; the active layer can be plasma-deposited [170]. Some very few studies analyze the influence of the TPE structure on the behavior of the membranes; Ziegel [171] showed that the rigid and flexible domains of a thermoplastic polyurethane elastomer behave in different ways in gas transport.

TPE are often associated with inorganic materials, particularly metals, glass, and fillers [172] (see also Chapter 16). The adhesion of TPE films to the mineral part often requires a coupling agent or, in the case of a metal, the surface can be modified by a physical or chemical treatment. TPE found interesting applications in the pressable TPE-based explosives for metal accelerating applications [173]. A large number of references (mostly patents) are devoted to fillers. Mark [174] reviewed and analyzed the behavior of polysiloxane elastomeric composites, including TPEs, and proposed very interesting fundamental concepts. The nanocomposites are discussed in Section 6.

The applications of TPEs in the medical, surgical, and pharmaceutical activities are rapidly developing. Many references are patents, interesting reviews were also published, but apparently the most recent reviews in this field date back to 2000 [175,176]. The TPE applications are related to very different materials, such as antibacterial materials [177,178] or the emulsions used in dermatology and cosmetology [179]. There are many applications of TPEs found in pharmacology, and in cosmetology and agriculture; some references describe both fundamental and applied scientific approaches, such as the preparation and the study of the lactide block-containing copolymers [180] or poly(3-hydroxybutyric) [181].

It would be impossible to analyze all the different commercial applications of TPEs; however it is impossible not to mention their very important role in automotive industry. Many authors reviewed this field of application [154,182]. Some particular parts of the car call for specific materials, *e.g.*, poly(ether-*b*-polyesters) are particularly appraised in air bag doors and dashboards [183] (see also Chapter 17).

6. Future trends

Several reviews speculate on the future and trends of TPEs [*e.g.*, 184,185]; a considerable part of them are published in Japanese [186,187] or Chinese [188].

This book fills an important gap, since for the first time it is entirely devoted to condensation TPEs. The following considerations place landmarks of the new chemistry in this field, rather than browsing the recent improvements of the well established TPEs. Several new techniques appeared quite recently; they are already patented, if not applied, and are very promising. This is the case of the metallo-supramolecular block copolymers [189], which result from the chelation of complexing group end-capped oligomers with different metal derivatives and the homopolymers and copolymers prepared from metal-containing macrocycles [190,191].

Yokozawa and Yokoyama [192] published important reviews and articles concerning new polycondensations, which are particularly useful in condensation TPEs; they argued that the nature of these polycondensations may be regarded as living. The process mechanism is clearly described in Figure 2 and different monomers listed in [192] were polycondensed, *e.g.*, the polycondensation of phenyl *p*-amino benzoate was initiated by the system 4-nitro-benzoate/base, leading to a polymer with a polydispersity index ranging from 1.07 to 1.34. Several block copolymers were prepared according to this technique [63].

The TPEs prepared by these new chemical methods will generate novel materials, but probably on limited production scales and with high added value. Their markets will be essentially related to medical applications, cosmetology, some specific surface treatments, *etc.*

In the more distant future, some other synthetic processes will be applied to TPEs, such as the carbon-carbon polycondensations catalyzed by metal derivatives [193,194] or the use of enzymatic catalysis in organic media [195]. The bacterial polymers are of great interest as organic materials, such as the poly(3-hydroxyalkanoates), which are potential thermoplastic elastomers. The

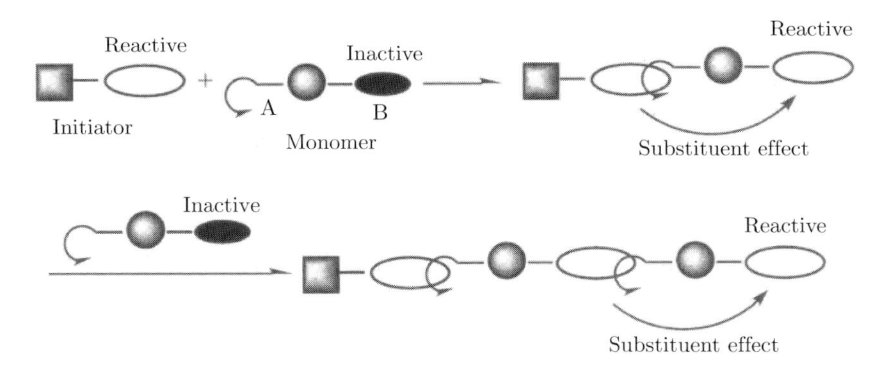

Figure 2. Effect of the substituent on the polycondensation of an A-B monomer where B becomes reactive after A has reacted with the chain [192]

respective biochemistry was reviewed by Kim and Lenz [196] and Poirier *et al.* [197–199].

The sequential reactions are rapidly developing and their contribution to the preparation of TPEs, particularly of block copolymers, will drastically increase; they were analyzed by Maréchal [39]. The sequences must not be limited to chain polymerization and polycondensation, and chemical modification will play an important role not only in grafting.

A large number of TPE syntheses depend on the existence and quality of specific functional oligomers. Their preparation and the control of their structure were reviewed by Maréchal [39] and also in Chapter 2 of this book. The oligomer chemical and structural purity, as well as functionality are responsible for the structure and morphology, hence the properties of the TPEs.

The liquid crystalline unit-containing polymers begin a timid entry in the TPE field. The liquid crystalline sequences can be part of the backbone or may be present as side chains. Nair *et al.* [200,201] carried out an abundant and very valuable work with both fundamental and applied aspects. These polymers are prepared by chain polymerization or by polycondensation and are discussed in Chapter 2. Some of them are blends and the morphology and properties of Rodrun LC3000® were the subject of two articles [202,203]. The first applications and patented products are very promising and their development should enjoy a rapid increase. This holds also for the introduction of hyperbranched and dendritic segments in block copolymers [204].

All the improvements observed in the syntheses and the advent of new products on the market result not only from the progress in organic and macromolecular chemistry, but also from the fantastic progress in structural analysis. Unfortunately, many articles and patents claim results without a deep insight in the contribution of side phenomena, whatever their nature, such as the side reactions taking place in the synthesis or the possible degradations observed during processing.

The characterization techniques will become more and more inseparable from synthesis. They were reviewed in Section 3, and the high number of techniques allow the obtaining of accurate values of the molecular masses, as well as reliable information on the nature of the blocks and that of the end-groups. Their importance will increase with the creation of new sophisticated structures and those resulting from natural products, particularly when prepared by degradation. Since the characterization of the chains remains the subject of many articles, the knowledge of their interaction and self-organization will require an increase in efficient analytical investigations; this trend is already observed in phase organization. In the same way, the morphology of the TPEs should draw increasing attention. Most of these investigations will combine theoretical and experimental studies.

Many of the improvements in TPEs and the creation of new materials result from a progressive introduction of the new achievements in chemistry, physics, and processing. However, sometimes a new technique or a new structure trigger the outburst of fundamental and applied studies and results. This is the case of the nano-technologies. Their association with TPEs is rapidly developing and several interesting reviews were published [205,206], covering an increasing number of materials: blends [207], multilayer structures [208], microemulsions [209], composites [128], stimuli-responsive polymers [210], stabilization of nanocolloidal metals [211].

The environmental protection and the recycling of TPEs are important parts of the concept and production of TPEs. Chapter 19 of this book treats these problems.

References

1. Holden G, Legge N R, Quirk R P and Schroeder H E (Eds.) (1996) *Thermoplastic Elastomers,* Hanser/Garner Publications, Munich, Vienna, New York.
2. Holden (2000) *Understanding Thermoplastic Elastomers,* Hanser/Garner Publications, Munich, Vienna, New York.
3. Rader Ch P (2003) Thermoplastic elastomers, in *Rubber Technology Special Topics* (Eds. Baranwal K and Stephens H) Rubber Div, ACS, Ch. 13, pp. 415–433.
4. Rader Ch P (2002) The rubber/thermoplastic continuum, *162nd Meet Rubber Div ACS*, Pittsburg, Paper No. 7.
5. Rader Ch P (2001) Thermoplastic elastomers, in *Elastomer Technology – Compounding and Testing for Performance* (Ed. Dick J) Hanser Publ., Ch. 10, pp. 264–283.
6. ScholarFinder (2004) *Chemical Abstracts,* Amer Chem Soc, Columbus, OH, USA.
7. Semon W L (1933) Synthetic rubber-like compositions, US patent 1,929,453, to B F Goodrich Co.
8. Bayer O, Rinke H, Siefken L, Orthner L and Schild H (1937) German patent DE 728,981, to IG Farben.
9. Bayer O (1939) Polymerization products, French patent Fr 845,917, to IG Farben.
10. Lieser T H and Marcura K (1941) Artificial organic high polymers. The mode of reaction of acyl diisocyanates with polyfunctional amino and hydroxyl compounds, *Annalen* **548**:226–254.
11. Bayer O (1941) Remarks on the work of Lieser and Marcura, *Annalen* **549**:286–287.

12. Christ R E and Hanford W E (1943) Treatment of polyesters such as those for sheets or coatings, US patent 2,333,639, to du Pont de Nemours & Co.
13. Snyder M D (1952) Elastic linear copolyesters US patent 2,623,031, to du Pont de Nemours & Co.
14. Schollenberger C S (1959) Simulated vulcanisates of polyurethane elastomers US patent 2,871,218, to B F Goodrich.
15. Szwarc M, Levy M and Milkovich R (1956) Polymerization initiated by electron transfer to monomer. A new method of formation of block polymers, *J Am Chem Soc* **78**:2656–2657.
16. Szwarc M (1956) "Living" polymers, *Nature* **178**:1168–1969.
17. Ziegler K and Bahr K (1928) Presumable mechanism of polymerization by alkali metal (preliminary communication), *Chem Ber* **61B**:253–263.
18. Ziegler K, Crossmann F, Kleiner H and Schafer O (1929) Alkali organic compounds. I. Relation between unsaturated hydrocarbons and alkali metal alkyls, *Annalen* **473**:1–35.
19. Bailey J T, Bishop E T, Hendricks W R, Holden G and Legge N R (1966) Thermoplastic elastomers. Physical properties and applications, *Rubber Age* **98**:69–74.
20. Holden G, Bishop E T and Legge N R (1969) Thermoplastic elastomers, *J Polym Sci Part C Polym Symp Ed* **26**:37–57.
21. Tobolsky A V (1959) Trends in rubber research, *Rubber world* **139**:857–862 and 868.
22. Kontos E G, Easterbrook E K and Gilbert R D (1962) Block copolymers of α-olefins prepared from macromolecules with long-chain lifetimes *J Polym Sci* **61**:69–82.
23. Coffey D H, Cook J G and Lake W H (1945) Synthetic polymeric materials, British Patent GB 574 134, to ICI Ltd.
24. Müller E, Petersen S and Bayer O (1944) German patent 76 584, to IG Farben.
25. Petersen S, Müller E, and Bayer O (12 Apr 1944) Ger Pat 77 229, to IG Farben.
26. Bayer O, Müller E, Petersen S, Piepenbrink H F and Windemuth E (1950) Polyurethanes. VI. New highly elastic synthesis. "Vulcollans", *Angew Chem* **62**:57–66.
27. Schollenberger C S, Scott H and Moore G R (1958) Polyurethan VC, a virtually crosslinked elastomer, *Rubber World* **137**:549–55.
28. Chen A T, Farrissey W J and Nelb R G II (1978) Polyester amides suitable for injection molding, US Patent 4,129,715, to Upjohn Co.
29. Chen A T, Nelb R G II and Onder K (1986) New high-temperature thermoplastic elastomers *Rubber Chem Technol* **59**:615–622.
30. Nelb R G II and Oertel, R W III (1983) Continuous, solvent-free preparation of thermoplastic polyamides, US Patent 4,420,603, to Upjohn Co.
31. Deleens G, Foy P and Maréchal E (1977) Synthèse et caractérisation de copolycondensats séquencés polyamide-*seq*-ether), *Eur Polym J* **13**:337–342; 343–351; 353–360.
32. Foy P, Jungblunt C and Deleens G (1975) Poly(ether ester amide) block mixed polymers as formable or extrudable products, German patent DE 2,523,991, to Ato Chimie.
33. Foy P, Jungblunt C and Deleens G (1982) Moldable block polyether-polyester-polyamides, US patent 4,331,786, to Ato Chimie.
34. Shivers J C, Jr (1962) Linear elastic copolymers, US patent 3,023,192, to du Pont de Nemours.
35. Nishimura A A and Komagata H (1967) Elastomers based on polyesters *J Macromol Sci–Chem* **A1**:617–625.
36. Witsiepe W K (1973) Segmented polyester thermoplastic elastomers, *Adv Chem Ser* **129**:39–60.

37. Hofmann G H (1996) Thermoplastic elastomers based on halogen-containing polyolefins, in *Thermoplastic Elastomers* (Eds. Holden G, Legge N R, Quirk R P and Schroeder H E) Hanser/Garner Publications, Munich, Vienna, New York, Ch. 6, p. 129.
38. Coran A Y and Patel R P (1996) Thermoplastic elastomers based on dynamically vulcanized elastomer-thermoplastic blends in *Thermoplastic Elastomers* (Eds. Holden G, Legge N R, Quirk R P and Schroeder H E) Hanser/Garner Publications, Munich, Vienna, New York, Ch. 7, p. 153.
39. Maréchal E (2000) Block copolymers with polyethers as flexible blocks; Future trends in block copolymers, in *Block Copolymers* (Eds. Baltá Calleja F J and Roslianec Z) Marcel Dekker, pp. 29–62; 541–572.
40. Holden G and Legge N R (1996) Styrenic thermoplastic elastomers, in *Thermoplastic Elastomers* (Eds. Holden G, Legge N R,, Quirk R P and Schroeder H E) Hanser/Garner Publications, Munich, Vienna, New York, Ch. 3, pp. 47–68.
41. Quirk R P and Morton M (1996) Research on anionic triblock copolymers, in *Thermoplastic Elastomers* (Eds. Holden G, Legge N R, Quirk R P and Schroeder H E) Hanser/Garner Publications, Munich, Vienna, New York, Ch. 4, pp. 71–499.
42. Williamson D T, Lizotte J R and Long T E (2001) Facile living anionic polymerization processes: new monomers and polymer architectures containing 1,3-cyclohexadiene without a high vacuum line, *Polym Mater Sci Eng* **84**:837–838.
43. Dubois Ph, Yu Y S, Teyssie Ph and Jerome R (1997) New polybutadiene-based thermoplastic elastomers: synthesis, morphology and mechanical properties, *Rubber Chem Tech* **70**:714–726.
44. Sipos L, Zsuga M and Deak G (1995) Synthesis of poly(L-lactide)-*block*-polyisobutylene-*block*-poly(L-lactide), a new biodegradable thermoplastic elastomer, *Macromol Chem Rapid Commun* **16**:935–940.
45. Kennedy J P (1996) Thermoplastic elastomers by carbocationic polymerization, in *Thermoplastic Elastomers* (Eds. Holden G, Legge N R, Quirk R P and Schroeder H E) Hanser/Garner Publications, Munich, Vienna, New York, Ch. 13, pp. 365–391.
46. Kennedy J P and Maréchal E (1992) Sequential (block and graft) copolymers, in *Carbocationic Polymerization,* Wiley-Interscience, New York, Ch. 8, pp. 410–441.
47. Kennedy J P (1999) Living cationic polymerization of olefins. How did the discovery come about? *J Polym Sci Part A Polym Chem* **37**:2285–2293.
48. Cao X, Sipos L and Faust R (2000) Polyisobutylene based thermoplastic elastomers: VI. Poly(α-methylstyrene-*b*-isobutylene-*b*-α-methylstyrene) triblock copolymers by coupling of living poly(α-methylstyrene-*b*-isobutylene) diblock copolymers, *Polym Bull* **45**:121–128.
49. Kwon Y, Antony P, Paulo C and Puskas J E (2002) Arborescent polyisobutylene-polystyrene block copolymers – A new class of thermoplastic elastomers, *Polym Prepr ACS* **43**:266–267.
50. Sipos L and Faust R (2004) Synthesis of poly(4-hydroxystyrene)-*b*-isobutylene-*b*-(4-hydroxystyrene) triblock copolymers via living cationic sequencial block copolymerization of isobutylene with 4-(*tert*-butyldimethylsilyloxy)styrene, *Abstr 227th ACS Natl Meet*, Anaheim, CA, USA, Poly–239.
51. Feldthusen J, Bela I and Mueller A H E (1998) Synthesis of linear and star-shaped block copolymers of isobutylene and methacrylates by combination of living cationic and anionic polymerizations, *Macromolecules* **31**:578–585.
52. Matyjaszewski K, Gaynor S, Greszta D, Mardare D and Shigemoto T (1995) Synthesis of well defined polymers by controlled radical polymerization, *Macromol Symp* **98**:73–89.

53. Matyjaszewski K, Gaynor S and Wang J S (1995) Controlled radical polymerizations: The use of alkyl iodides in degenerative transfer, *Macromolecules* **28**:2093–2095.
54. Ameduri B, Boutevin B and Gramain Ph (1997) Synthesis of block copolymers by radical polymerization and telomerization, *Adv Polym Sci* **127**:87–142.
55. Matyjaszewski K (2000) Environmental aspects of controlled radical polymerization, *Macromol Symp* **152**:29–42.
56. Mayes A M, Acar M H and Gonzalez-Leon J A (2003) Toward commodity plastics by molecular design, *Abstr 225th ACS Natl Meet*, New Orleans, LA, USA PMSE–035.
57. Gaynor S G, Balchandani P, Kulfan A, Podwika M and Matyjaszewski K (1997) Architectural control in acrylic polymers by atom transfer radical polymerization, *Polym Prepr ACS* **38**:496–497.
58. Moineau G, Minet M, Teyssie Ph and Jérome R (2000) Synthesis of fully acrylic thermoplastic elastomers by atom transfer radical polymerization (ARTP). 2. Effect of the catalyst on the molecular control and the rheological properties of the triblock copolymers, *Macromol Chem Phys* **201**:1108–1114.
59. Schmalz H, Abetz V, Lange R and Soliman M (2001) New thermoplastic elastomers by incorporation of nonpolar soft segments in PBT-based copolyesters, *Macromolecules* **34**:795–800.
60. Shim J S, Asthana S, Omura N and Kennedy J P (1998) Synthesis, characterization and properties of novel star-block polymers of PSt-b-PIB arms emanating from cyclosiloxane cores, I, demonstration of synthetic strategy, *Abstr 215th ACS Natl Meet*, Dallas, TX, USA, Poly–005.
61. Ding H and Harris F W (1995) Synthesis and characterization of a novel nylon 6-*b*-polyimide-*b*-nylon 6 copolymers, *Pure Appl Chem* **67**:1997–2004.
62. Pan M, Wang J, Zhang L and Mo L (1999) Synthesis of terminal reactive polysiloxane and study of water-soluble polysiloxane-polyurethane block copolymer, *Hebei Gongye Daxue Xuebao* **28**:26–30 (CA **132**:123254).
63. Yokozawa T, Ogawa M, Sekino A, Sugi R and Yokoyama A (2003) Synthesis of well-defined poly(*p*-benzamide) from chain-growth polycondensation and its application to block copolymers, *Macromol Symp* **199**:187–195.
64. Wagener K B, Brzezinska K, Anderson J D and Dilocker S (1997) Well phase separated segmented copolymers via acyclic diene metathesis (ADMET) polymerization, *J Polym Sci Part A Polym Chem Ed* **35**:3441–3449.
65. Tindall D and Wagener K B (1998) Using ADMET to make segmented copolymers, *Polym Prepr ACS* **39**:630.
66 Braun D and Hellmann G P (1998) Chemical modification of polymeric hydrocarbons, *Macromol Symp* **129**:43–51.
67. Madec P J and Maréchal E (1994) Block copolymers by polycondensation of α,ω-difunctional-oligomers, *Prace Naujowe Polytechniki Szczecinskiej,* **514:**11–19 (CA **123**:56784).
68. Ebdon J R and Flint N J (1996) Preparation of α,ω-aldehyde-ended telechelic methyl methacrylate oligomers by the oxidative cleavage of statistical methyl methacrylate-buta-1,3-diene copolymers, *Eur Polym J* **32**:289–294.
69 Yu J M, Yu Y, Dubois Ph and Jérome R, Teyssie Ph (1997) Synthesis and characterization of hydrogenated poly[alkyl methacrylate(-*b*-styrene)-*b*-butadiene-*b*-(styrene-*b*-)alkyl methacrylate] triblock and pentablock copolymers, *Polymer* **38**:3091–3101.
70. Sugimoto H (2004) Thermoplastic elastomer compositions and their power and moldings with good flexibility, bending, heat and cold resistance, Jpn Kokai Tokkyo Koho JP 2004 002 608, to Sumitomo Chem Co, Ltd.

71. Kichner C-P, Fritz H-G and Cai Q (2000) Thermoplastic elastomer composition and a process for making the same, European patent Appl EP 1,050,548, to Chemplast Marketing Serv Est, Liechtenstein.
72. Tasaka M and Ogawa T (2002) Thermoplastic elastomer compositions with good extrudadibility and injection moldability and adhesives containing them for tapes, Jpn Kokai Tokkyo Koho JP 2002 327 033, to Riken Technos Corp.
73. Oyama H and Nakaishi E (2003) Thermoplastic elastomer compositions, moldings and automobile interior parts therefrom with high wear resistance and good sliding property, Jpn Kokai Tokkyo Koho JP 2003,277,519, to Sumitomo Chem Co, Ltd.
74. Ikeda Y, Kodama K, Kajiwara K and Kohjiya S (1995) Chemical modification of butyl rubber. II. Structure and properties of poly(ethyleneoxide)-grafted butyl rubber, *J Polym Sci Part B Polym Phys Ed* **33**:387–394.
75. Pi Z and Kennedy J P (2002) Cationic grafting of norbornadiene, indene, and 1,3-cyclohexadiene from PVC, *Polym Bull* **48**:345–352.
76. Zheng W Y and Hammond P T (1996) Block copolymers containing smectic C* liquid crystalline segments, *Abstr 212th ACS Natl Meet*, Orlando FL, USA, Poly-204.
77. Knauss D M and Huang T (2001) Synthesis and characterization of (star polystyrene)-*bl*-polydimethylsiloxane-*bl*-(star polystyrene), *Polymer* **42**:219–220.
78. Fonagy T, Ivan B and Szesztay M (2002) New thermoplastic elastomers by quasiliving atom transfer free radical grafting, *J Reinf Plast Compos* **21**:1411–1419.
79. Gaynor S G, Balchandhani P, Kulfan A, Podwika M and Matyjaszewski K (1997) Architectural control in acrylic polymers by atom transfer radical polymerization, *Abstr 213th ACS Natl Meet*, San Francisco, CA, USA, Poly-340.
80. Baumert M, Heinemann J, Thomann R and Mulhaupt R (2000) Highly branched polyethylene graft copolymers prepared by means of migratory insertion polymerization combined with TEMPO-mediated controlled radical polymerization, *Macromol Chem Rapid Commun* **21**:271–276.
81. Prut E V (2001) Processing and properties of thermoplastic vulcanizates (TPV), *TPE Conf 2001*, Brussels, Belgium, Paper 7/1 – Paper 7/6.
82. Pasch H and Augenstein M (1993) Chromatographic investigations of macromolecules in the critical range of liquid chromatography: 5- Characterization of block copolymers of decyl and methyl methacrylate, *Makromol Chem* **194**:2533–2541.
83. Pash H (1996) Liquid chromatography at the critical point of adsorption – A new technique for polymer characterization, *Macromol Symp* **110**:107–120.
84. Tennikov M B, Nefedov P P, Lazavera M A and Frenkel S Ya (1977) Single mechanism of liquid chromatography of macromolecules on porous sorbents, *Vyssokomol Soedin Ser A* **19**:657–660 (in Russian).
85. Belenkii B G and Gankina E S (1977) Thin-layer chromatography of polymers, *J Chromatogr* **141**:13 –90.
86. Kilz P, Krüger R P, Much H and Schulz G (1995) 2-Dimensional chromatography for the deformulation of complex copolymers, *Adv Chem* **247**:223–241.
87. Pasch H and Kilz P (1999) Analysis of copolymers by online-coupled two-dimensional chromatography, *GITLabor-Fachzeitschrift* **43**:239–240, 242–244 (CA **130**:252860).
88. Gores F and Kilz P (1993) Copolymer characterization using conventional size-exclusion chromatography and molar-mass-sensitive detectors, *Chromatography of Polymers, ACS Symp Ser* **521**:122–148.
89. Aust N and Gobec G (2001) Size exclusion chromatography as a method routine check-up and quality control of industrial thermoplastic polyurethane systems – correlation between molecular weight and mechanical and thermal properties, *Macromol Mater Eng* **286**:119–125.

90. Geiger T and Stadler R (1998) Investigation of cluster formation in thermoreversible networks, *Wiley Polym Networks Group Rev Ser* **1**:129–138.
91. Nair B R, Gregoriou V G and Hammond P T (2000) FT-IR studies of side chain liquid crystalline thermoplastic elastomers, *Polymer* **41**:2961–2970.
92. Wang T-L and Huang F-J (2000) Preparation and characterization of novel thermoplastic elastomers by step/chain transformation polymerization, *Polymer* **41**:5219–5228.
93. Dardin A, Boeffel Ch, Spiess H-W and Stadler R (1994) Orienration behavior of thermoplastic elastomers studied by 2H-NMR and FT-IR spectroscopy, *Polym Mater Sci Eng* **71**:248–249.
94. Otsuka N, Yang Y, Saito H and Inoue T (1996) Structure and properties of PP/Hydrogenated SBR, *Ann Tech Conf Soc Plast Eng* **54**:2331–2333 (CA **125**:169419).
95. Wang H, Graff D K, Schoonover J R and Palmer R A (1999) Static and dynamic infrared linear dichroic study of polyester/polyurethane copolymer using step-scan FT-IR and photoelastic modulator, *Appl Spectr* **53**:687–696.
96. Nagai Y, Ogawa T, Zhen L Y, Nishimoto Y and Ohishi F (1997) Analysis of weathering of thermoplastic polyester elastomers. I. Polyether-polyester elastomers, *Polym Degr Stab* **56:**115–121.
97. Fischer W B, Poetschke P, Eichhorn K J and Siesler H W (1997) Rheo-optical FT-IR spectroscopy of polyurethane-polyolefin blends during cyclic elongation and recovery, *Mikrochim Acta Suppl* **14**:411–412.
98. Boularès A, Rozès L, Tessier M and Maréchal E (1998) Synthesis and characterization of poly(copolyethers-*b*-polyamides). I. Structural study of polyether precursors, *J Macromol Sci – Pure Appl Chem* **A35**:933–953.
99. Frick E M and Hillmeyer M A (2002) Morphological, rheological and mechanical characterization of polylactide-*b*-polyisoprene-*b*-polylactide triblock copolymers: New partially biodegradable thermoplastic elastomers, *Polym Mat Sci Eng* **87**:110.
100. Kwak S-Y and Nakajima N (1995) Magic-angle cross-polymerization ^{13}C NMR relaxation analysis of solid microstructures and their scales in a thermoplastic elastomer prepared from nitrile rubber/PVC blending, *Ann Tech Conf Soc Plast Eng* **53**:3208–3215.
101. Sukhanov P P, Musin I N and Kimel'blat V I (2002) Investigation of the polyolefins synergic blends by the impulse NMR method, *Russian Polym Rev* **7**:20–26 (CA **138**:171527).
102. Lodge T P (1994) Characterization of polymer materials by scattering techniques, with applications to block copolymers, *Mikrochim Acta* **116**:1–31.
103. Norman A I, Cabral J T, Amis E and Karim A (2004) Scattering methods applied to high throughput materials science research, *Abstr 227th ACS Natl Meet*, Anaheim, CA, USA, PMSE-207.
104. Stepanek P and Lodge T P (1997) Dynamic light scattering from block copolymers, *NATO ASI Series, Ser 3: High Technology* **40**:189–207.
105. Stepanek P, Almdal K and Lodge T P (1997) Polarized and depolarized dynamic light scattering from a block copolymer melt, *J Polym Sci Part B Polym Phys Ed* **35**:1643–1648.
106. Stepanek P and Lodge T P (1996) Dynamic light scattering from block copolymer melts near the order-disorder transition, *Macromolecules* **29**:1244–1251.
107. Hernaez E, Quintana J R and Katime I (2002) Study of the aggregation process of block copolymers in selective solvents, *Recent Res Dev in Appl Polym Sci* **1**:93–114 (AN:2003:878991).

108. Guo Q, Figueiredo P, Thomann R and Gronski W (2001) Phase behavior, morphology and interfacial structure in thermoset/thermoplastic elastomer blends of poly(propylene glycol)-type epoxy resin and polystyrene-*b*-polybutadiene, *Polymer* **42**:10101–10110.
109. Szafko J, Schulte K and Broza G (2000) Microphase separation of segmented poly(ether-ester) multiblock copolymer films crystallized under strain rates, *Fib & Text in East Eur* **8**:35–39 (CA **135**:138549).
110. Veenstra H, Hoogvliet R M, Norder B and De Boer A P (1998) Microphase separation and rheology of a semicrystalline poly(ether-ester) multiblock copolymer, *J Polym Sci Part B Polym Phys* **36**:1795–1804.
111. Muramatsu S and Lando J B (1998) Reversible crystal deformation of poly(tetramethylene terephthalate) segments in semicrystalline segmented poly(ether-ester) thermoplastic elastomers, *Macromolecules* **31**:1866–1870.
112. Chen J, Zhang J, Zhu T, Hua Z, Chen Q and Yu X (2000) Blends of thermoplastic polyurethane and polyether-polyimide: preparation and properties, *Polymer* **42**:1493–1500.
113. Wignall G D, Benoit H, Hashimoto T, Higgins J S, King S, Lodge T P, Mortensen K and Ryan A J (2002) Progress in SANS studies of polymer systems, *Macromol Symp* **190**:185–200.
114. Roestamsjah (1998) Characteristics and utilization of thermoplastic elastomers (TPE) – an overview, *JAERI-Conf 98* **015**:179–196; (CA **130**:210136).
115. Ruegg M L, Newstein M C, Balsara N P and Reynolds B J (2004) Small-angle neutron scattering from nonuniformly labeled block copolymers, *Macromolecules* **37**:1960–1968.
116. Park M J and Char K (2004) Gelation of PEO-PLGA-PEO Triblock copolymers induced by macroscopic phase separation, *Langmuir* **20**:2456–2465.
117. Balsara N P, Jonnalagadda S V, Lin C C, Han C C and Krishnamoorti R (1993) Thermodynamic interactions and correlations in mixtures of two homopolymers and a block copolymer by small angle neutron scattering, *J Chem Phys* **99**:10011–10020.
118. Jiang S, An L, Jiang B and Wolf B A (2003) Thermodynamics of phase behavior in PEO(EO-b-DMS) homopolymer and block co-oligomer mixtures under pressure, *Macromol Chem Phys* **204**:2265–2273.
119. Anandhan S, De P P, De S K, Bhowmick A K (2003) Thermoplastic elastomeric blend of nitrile rubber and poly(styrene-*co*-acrylonitrile). I. Effect of mixing sequence and dynamic vulcanization on mechanical properties, complexes *J Appl Polym Sci* **88**:1976–1987.
120. Valcarenghi A and Hill B (1998) Innovative TPE approaches in co-molding, *TPE'98*, Westminster, UK, Conf Proc Pap 18/1–Pap 18/5.
121. Foster M D, Sikka M, Singh N, Bates F S, Satija S K and Majkrzak C F (1992) Structure of symmetric polyolefin block copolymer thin films, *J Chem Phys* **96:**8605–8615.
122. Inoue T and Svoboda P (2000) Structure-properties of PP-EPDP thermoplastic elastomer: Origin of strain recovery, *58th Ann Tech Conf Soc Plast Eng* **2**:1676–1679.
123. Huang F-j and Wang T-l (2003) Synthesis and characterization of new segmented polyurethanes with side-chain, liquid-crystalline chain extenders, *J Polym Sci Part A Polym Chem Ed* **42**:290–302.
124. Li H and White J L (2000) Preparation and characterization of biaxially oriented films from polybutylene terephthalate based thermoplastic elastomer block copolymers, *Polym Eng Sci* **40**:2299–2310.
125. Volegova I A, Godovsky Y K and Soliman M (2003) Glass transition of undrawn and drawn copolyetherester thermoplastic elastomers, *Int J Polym Mater* **52**:549–564.

126. Skachkova V K, Erina N A, Chepel L M and Prut E V (2003) Thermooxidative stability and morphology of polypropylene-rubber-paraffin oil blends, *Vyssokomol Soedin Ser A i Ser B* **45**:2040–2046 (CA **140**:254336) (in Russian).
127. Antony P, Kwon Y, Puskas J E, Kovar M and Norton P R (2004) Atomic force microscopic studies of novel arborescent block and linear triblock polystyrene-polyisobutylene copolymers, *Eur Polym J* **40**:149–157.
128. Thakkar H and Goettler L A (2003) The effects of dynamic vulcanization on the morphology and rheology of TPV's and their nanocomposites, *Tech Papers ACS Rubber Div* **163**:482–491.
129. Pham L P and Sung C (2002) Effects of blending SIBS and SMA on morphology and mechanical properties, *Mat Res Soc Symp Proceed* **734**:391–395.
130. Sauer B B, McLean R S and Brill D J Londono J D (2001) Morphology and orientation during deformation of segmented elastomers studied by SAXS and atomic force microscopy, *Abstr 222nd ACS Meet*, Chicago III, USA, PMSE–163.
131. Burford R P, Markotsis M G and Knott R B (2003) Small angle neutron scattering and transmission electron microscopy studies of interpenetrating polymers networks from thermoplastic elastomers, *Nucl Instrum & Meth in Phys Res Sect B* **208**:58–65.
132. Schmalz H, Abetz V and Lange R (2003) Thermoplastic elastomers based on semicrystalline block copolymers, *Compos Sci Technol* **63**:1179–1186.
133. Svodoba P, Saito H, Chiba T, Inoue T and Takemura Y, (2000) Morphology and elastomeric properties of isotactic polypropylene/hydrogenated polybutadiene blends, *Polym J* **32**:915–920.
134. Sundararajan G, Vasudevan V and Reddy K A (1998) Synthesis of triblock copolymers-(polyA-polybutadiene-polyA) via metathesis polymerization, *J Polym Sci Part A Polym Chem Ed* **36**:2601–2610.
135. Valiente N, Lalot T, Brigodiot M and Marechal E (1998) Enzymatic hydrolysis of phthalic unit containing copolyesters as a potential tool for block length determination, *Polym Degr Stab* **61**:409–415.
136. Luo A-Q, Ling Y E, Fu R-N, Xie G-Y and Wang X-L (2001) Pyrolysis-gas chromatography/mass spectrometry for microstructure and pyrolysis pathway of polyester-polyether multiblock copolymer, *J Beijing Inst Techn* **10**:45–50 (CA **136**:103641).
137. Ukielski R and Kozlowska A (2003) Reversible endothermic process observed by DSC in thermoplastic multiblock elastomers at middle temperature range, *Polimery* **48**:809–815 (AN 2004:3830).
138. Hild S, Schroth K and Doering A (2002) Investigating the microscopic properties of thermoplastic-elastic polymers using pulsed force mode SFM, *Techn Papers 161st ACS Rub Div Spring Tehn Prog*, Savannah, GA, USA, **161**:467–486.
139. Pesestkii S S, Jurkowski B, Olkhov Y A , Olkhova O M, Storozhuk I P and Mozheiko U M (2001) Molecular and topological structures in polyester block copolymers, *Eur Polym J* **37**:2187–2199.
140 Nishizawa H (2000) Recent trends of thermoplastic elastomer molding technology, *Porima Daijesuto* **52**:17–36 (CA **133**:178563).
141. Lachmann M (1997) Use of hot runner method in the injection molding of TPE, *Thermoplastische Elastomere – Herausforderung an die Elastomerverarbeiter* 141–161 (CA **127**:347437).
142. Akiba M (2003) Progress in molding of rubber materials, *Porima Daijesuto* **55**:23–37 (CA **139**:246807).
143. Knieps H (1997) Extrusion of thermoplastic elastomers, *Thermoplastische Elastomere – Herausforderung an die Elastomerverarbeiter* 103–115 (CA **127**:332617).

144. Nagaoka T (1999) Blow molding of thermoplastic elastomers (TPE), *Porima Daijesuto* **51**:42–54 (CA **132**:309538).
145. Brzoskowski R, Wang Y, La Tulippe C, Dion B, Cai H and Sadhegi R (1998) Extrusion of low density chemically foamed thermoplastic vulcanizates, *56th Annual Techn Conf Soc Plast Eng* **3**:3204–3208.
146. Prut E V and Zelenetskii A N (2001) Chemical modification and blending of polymers in an extruder reactor, *Russian Chem Rev* **70**:65–79 (CA **135**:138226).
147. Martin P P C G, Van Duin M, Van Gurp M (2003) Process for making thermoplastic composition comprising dynamic crosslinking, PCT Int Appl WO 2003 099 927, to DSM Ip Assets B V Neth.
148. Kojina J (2003) New thermoplastic elastomers by dynamic vulcanization, *Nippon Gomu Kyokaishi* **76**:310–315 (AN:2003:812376).
149. Pesneau I, Champagne M F and Huneault M A (2000) Influence of processing parameters on dynamic vulcanization, *Polym Mater Sci Eng* **83**:384–385.
150. Endres W, Lechner M D and Steinberger R (2003) The kinetics of the thermal decomposition of thermoplastic polyurethane elastomers under thermoplastic processing conditions, *Macromol Mater Eng* **288**:525–530.
151. Lee B H and White J L (2002) Formation of a polyetheramide triblock copolymer by reactive extrusion; process and properties, *Polym Eng Sci* **42**:1710–1723.
152. O'Connor G E and Fath M A (1981) Thermoplastic elastomers, Part I: Can TPEs compete against thermoset rubbers? *Rubber World* **185**:25–29.
153. O'Connor G E and Fath M A (1982) Thermoplastic elastomers, Part II: A new thermoplastic rubber, *Rubber World* **185**:26–32.
154. Baumann M (2001) Application design advances through TPEs, *Thermopl Elast Topical Conf Publisher: Soc Plast Eng,* Houston, Tx, pp. 99–106.
155. Baumann M (2002) Application design advances through plastics, *Prepr 60th Ann Tech Conf Soc Plast Eng* **3**:2742–2744.
156. Sutton S P (2003) Molding of thermoplastic elastomer compositions and heating apparatus for the process, US patent 2003 193 116.
157. Tasaka M and Suka T (2003) Thermoplastic elastomer compositions with good flowability for powder molding and their powders, Jpn Kokai Tokkyo Koho JP 2003 246 887, to Riken Technos Corporation, Japan.
158. Sengupta P and Noordermeer J W M (2003) Effects of composition and processing conditions on morphology and properties of thermoplastic elastomer blends of SEBS/PP/oil and of dynamically vulcanized EPDM/PP/oil, *163rd Spring ACS Rub Div Tech Meet*, San Francisco, Ca, USA, 149–176.
159. Omori M (2004) Ethylene-vinyl alcohol copolymer/polyester rubber blend and multilayer laminate, US patent 2004 006 182.
160. Murakami S (2004) Dynamically crosslinked thermoplastic elastomer compositions and their moldings with good flexibility and low compression set, Jpn Kokai Tokkyo Koho JP 2004 010 770, to Mitsui Chem Inc.
161. Ou Y, Lei Y, Fang X and Yang G (2004) Maleic anhydride grafted thermoplastic elastomer as an interfacial modifier for polypropylene/polyamide 6 blends, *J Appl Polym Sci* **91**:1806–1815.
162. Nele M and Soares J B P (2003) Molecular weight and long chain branch distributions of branch-block olefinic thermoplastic elastomers, *Macromol Theor & Simul* **12**:386–400.
163. Akari H (2003) Layered organic silicate and quaternary ammonium salt-containing electrification components, Jpn Kokai Tokkyo Koho JP 2003 223 038, to Sekisui Chem Co Ltd.

164. Takeshima K (2003) Laminated polyolefin sheets with low density and excellent heat resistance and elongation at high temperatures and moldings using them, Jpn Kokai Tokkyo Koho 2003 260 766, to Sikisui Chem Co Ltd.
165. Utazaki K and Kumaki J (2003) Noise-reducing lightweight polyester multilayer articles concisely manufactured by injection molding for electric apparatus, Jpn Kokai Tokkyo Koho 2003 260 759, to Toray Ind Inc.
166. Morino K and Furuichi M (2003) Vulcanization-expandable thermoplastic elastomer composition and manufacture of lightweight foam moldings with excellent conformity to mold cavities from them, Jpn Kokai Tokkyo Koho JP 2003 292 667, to JSR Ltd.
167. Indukuri K K and Lesser A J (2003) Controlled solid-state foaming of elastomeric systems, *Polym Mater Sci Eng* **89**:767–768.
168. Zhang Q, MacDiarmid A G, Su J, Wang P-c and Wynne K J (1999) Polymeric electrostrictive systems, PCT Int Appl WO 9 917 929, to the trustees of Uni Pennsylvania.
169. Hoffmann M, Beyreuther R, Vogel R and Taendler B (2003) Melt spinning of cross-linkable thermoplastic elastomers by means of twin screw extrusion, *Chem Fib Int* **53**:442–444.
170. Cuomo J J, Sakhrani V G and Vernon P M Jr (2002) Plasma-deposited coatings for elastomeric materials, deposition method, and articles, PCT Int Appl WO 2002 100 928, to the Uni North Carolina and the authors.
171. Ziegel K D (1971) Gas transport in segmented block copolymers, *J Macromol Sci-Phys* **B5**:11–21.
172. Hordijk A C, Schoolderman C, Van der Heijden A E D M (2001) Highly filled thermoplastic elastomers (TPEs) – processing and rheology, *Polym Rheol Conf*, Shrewsbury, IXT13/1-P13/9 (CA **140**:376339).
173. Hollands R E, Jordan T H, Leach C J and Murray I E P (1996) Pressable TPE-based explosives for metal accelerating applications, *27th Int Ann Conf of ICT* (Energetic Mater) 65.1–65.12 (CA **125**:172513).
174. Mark J E (2004) Polysiloxane elastomers, *Abstr 227th ACS Natl Meet*, Anaheim, CA, USA, Poly-308.
175. Severyns K (1999) Engineered thermoplastic elastomers (TPEs). New materials, *Collected Papers 13th Int Conf Medical Plastics'99*, Copenhagen, Denmark, 15.1–15.13.
176. Mochizuki A (2000) Application of thermoplastic elastomer to medical device, *Seikei Kako* **12**:770–774 (CA **135**:262059).
177. Karasawa Y and Sato H (2004) Antibacterial adhesive film containing polyolefin elastomer components and polylysine, Jpn Kokai Tokkyo Koho JP 2004 049 540, to Chisso Corp Petroch Corp.
178. Fujishima A, Chih C-F and Kubota Y (2003) Antibacterial medical materials, their manufacture, and regeneration method, Jpn Kokai Tokkyo Koho JP 2003 260 126, to Kanagawa Acad Sci Tech, Japan, Japan Sci and Tech Corp.
179. Auguste S and Desmaison N (2003) Solid emulsions based on thermoplastic elastomers for medical goods and cosmetics, PCT Int Appl WO 2003 087 220, to Laboratoires Ugo.
180. Frick E M and Hillmyer M A (2001) Polylactide-b-polyisoprene-b-polylactide triblock copolymers: New thermoplastic elastomers containing biodegradable segments, *Abstr 221st ACS Meeting Poly–077*.
181. Steinbuchel A and Fuchtenbusch B (1998) Bacterial and other biological systems for polyester production, *Trends Biotech* **16**:419–427.
182. Klingensmith W, Dendringer T, McConnell A, Standley P, Beckett J, Eller R, Otterstedt C and Walker J (2003) Elastomers for automotive applications, in *Elastomer Technology: Special Topics* (Eds. Baranwal K C and Stephens H L) ACS Rubber Division, Akron, OH, pp. 455–491.

183. Rutgers G and Lange R F M (2001) The use of co-poly(ether esters) in automotive applications, *TPE Conf*, Brussels, Belgium, Paper 20/1–20/4 (CA **138**:171511).
184. Holden G (2000) Thermoplastic elastomers, *219th ACS Natl Meet*, San Francisco, CA, USA MACR–050 (AN 2000:331599).
185. Spontak R J and Patel N P (2000) Thermoplastic elastomers: fundamentals and applications, *Current and Opinion in Colloid Interf Sci* **5**:334–341 (CA **134**:223865).
186. Akiba M (2002) Recent status and task of thermoplastic elastomers, *Parasuchikkusu* **53**:9–16 (CA **137**:264211).
187. Akiba M (2003) TPE, technology of the 21st century, *Porima Daijesuto* **55**:17–30 (CA **139**:23115).
188. Yu Q (2002) Development of thermoplastic elastomers, *Hecheng Xiangjiao Gongye* **25**:202–206 (CA **140**:95356).
189. Lohmeijer B G G and Schubert U S (2003) Playing LEGO with macromolecules: Design, synthesis, and self-organization with metal complexes, *J Polym Sci Part A Polym Chem Ed* **41**:1413–1427.
190. Kricheldorf H R and Langanke D (1999) Macrocycles 7. Cyclization of oligo- and poly(ethyleneglycol)s with dibutyltin dimethoxide – a new approach to (super)macrocycles, *Macromol Chem Phys* **200**:1174–1182.
191. Kricheldorf H R and Langanke D (1999) Macrocycles 8. Multiblock copoly(etheresters) of poly(THF) and ε-caprolactone via macrocyclic polymerization, *Macromol Chem Phys* **200**:1183–1190.
192. Yokozawa T and Yokoyama A (2004) Chain-growth polycondensation: polymerization nature in polycondensation and approach to condensation polymer architecture, *Polym J* **36**:65–83.
193. Heitz W (1995) Metal catalyzed polycondensation reactions, *Pure Appl Chem* **67**:1951–1964.
194. Maréchal E (2002) New aspects of catalysis in polycondensation, *Current Org Chem* **6**:177–208.
195. Lalot T and Marechal E (2001) Enzyme-catalyzed polyester synthesis *Int J Polym Mater* **50**:267–286.
196. Kim Y B and Lenz R W (2001) Polyesters from microorganisms, *Adv Biochem Eng* **71**:51–79.
197. Poirier Y and Nawrath C (1998) Transgenic plants for the production of polyhydroxyalkanoates, a family of biodegradable thermoplastics and elastomers, in *Transgenic Plants Research* (Ed. Keith L) Harwood Publisher Amsterdam, Ch. 12, pp. 201–218.
198. Nawrath C and Poirier Y (1996) Review on polyhydroxyalkanoate formation in the model plant *Arabidopsis thaliana, Publ Natl Res Council of Canada Int Symp Bact Polyhydroxyalkanoates*, pp. 119–126.
199. Poirier Y and Gruys K J (2002) Production of polyhydroxyalkanoates in transgenic plants, *Biopolymers* **3a**:401–435.
200. Nair B R and Hammond P T (1998) Synthesis and characterization of novel side chain liquid crystalline polyurethanes, *ACS Polym Prepr* **39**:1077–1078.
201. Nair B R, Gregoriou V G and Hammond P T (2000) FT-IR studies of side chain liquid crystalline elastomers, *Polymer* **41**:2961–2970.
202. Saikrasun S, Bualek-Limcharoen S, Kohjiya K and Urayama K (2003) Thermotropic liquid crystalline/thermoplastic elastomer in situ composites. I. Rheology, morphology, and mechanical properties of extruded strands, *J Appl Polym Sci* **89**:2676–2685.
203. Saikrasun S, Bualek-Limcharoen S, Kohjiya S and Urayama K (2003) Thermotropic liquid copolyester (Rodrun LC3000)/thermoplastic SEBS elastomer in situ composites:

II. Mechanical properties and morphology of monofilaments in comparison with extruded strands *J Appl Polym Sci* **90**:518–524.

204. Antony P, Puskas J E and Paulo C (2002) Arborescent polyolefinic thermoplastic elastomers and products therefrom, PCT Int WO 2002 096 967, to the authors and the Uni Western Ontario, Canada.

205. Krueger P (2003) Nanoparticles as filler in polymers. Prospects and challenges of a new technology rich in tradition, *VDI-Berichte* **1772**:21–37 (CA **140**:94680).

206. Imaizumi K (2003) Nano-structures of the novel cyclic conjugated diene block copolymers and their application to thermoplastic elastomers, *Nippon Gomu Kyokaishi* **76**:281–284 (AN 2003:812370).

207. Dong W-F, Zhang S-J, Liu Y-Q, Zhang X-H, Huang F, Gao J-M, Wei G-S and Qiao J-L (2003) Development progress of morphology control of rubber phase in rubber-plastics blending – new technology for preparation of rubber-plastics blends *Zhongguo Suliao* **17**:1–16. (AN 2003:858846).

208. Wang X and Folz V J (2003) Multilayer polymeric nanoparticle preparation and applications in rubbers, US Patent Appl 2003149185, to Bridgestone Corp.

209. El-Aasser M S, Li M, Jeong P, Daniels E S, Dimonie V L and Sudol E D (2001) Nano-hybrid polymer particles via thee mini emulsion process *DECHEMA Monographien* **137**:1–13 (CA **136**:387428).

210. Koerner H, Price G, Pearce N A, Alexander M and Vaia R A (2004) Remotely actuated polymer nanocomposites-stress-recovery of carbon-nanotube-filled thermoplastic elastomers, *Nature Materials* **3**:115–120.

211. Nandi A, Dutta G M and Banthia A K (2002) Sulfonated polybutadiene random ionomer as stabilizer for colloidal copper nanoparticles, *Colloids & Surf, A* **197**:119–124.

Chapter 2

Polycondensation Reactions in Thermoplastic Elastomer Chemistry: State of the Art, Trends, and Future Developments

E. Maréchal

1. Introduction

The great majority of thermoplastic elastomers (TPE) are copolymers associating soft and hard blocks, such as polyamide-*b*-polyethers, polyurethane-*b*-polyethers, or polyether-*b*-polyesters. However, polymers with less classical structures, such as liquid crystalline copolymers or block copolymers based on supramolecular segments, are the subject of extensive research and will probably be commercialized in the near future.

Lodge [1] reviewed the synthesis, structure, morphology, properties, and processing of block copolymers in a well documented article. The part "synthesis" is mainly devoted to living anionic polymerization, radical polymerization, and chemical modification, but polycondensation and polyaddition reactions are not mentioned, despite that the above examples show their importance in the production of organic materials.

The first part of this chapter is a critical review of the various techniques of polycondensation used in the preparation of block copolymers [2]. It is followed by an analysis of the research trends and future, and the developments of polycondensation reactions leading to the obtaining of thermoplastic elastomers.

2. Preparation of block copolymers by polycondensation. A critical review

2.1. *General considerations*

Block copolymers can be prepared by ionic and radical polymerizations, including ring-opening polymerization, and by polyaddition and polycondensation. The chemical modification is often associated to other techniques used, *e.g.*, in the preparation of the oligomers (such as end-capping) or the copolymer chains. Chemical modification frequently takes place in one or several steps of a process (see Section 4).

IUPAC recommends different definitions for polycondensation and polyaddition [3], but in the following we will use the word polycondensation as a generic term, except for the cases when it can be confusing.

Block copolymers can be prepared by polycondensation of several α,ω-difunctional oligomers or by polycondensation of α,ω-difunctional oligomers with the precursors of the other blocks. They can be also prepared by the coupling of different difunctional oligomers. Most often, the number of the different oligomers does not exceed 2.

2.2. *Direct polycondensation of α,ω-difunctional oligomers*

The corresponding pattern is given in Scheme 1:

X—[oligomer 1]—X + Y—[oligomer 2]—Y ⟶

~~~X'—[block 1]—X'Y'—[block 2]—Y'~~~ (1)

The direct polycondensation of α,ω-difunctional oligomers leads to well defined block copolymers. The oligomers must be soluble in the reaction mixture or molten and stable when the reaction is carried out in the melt, which is rarely the case of aromatic polymers. Moreover, the direct polycondensation requires that the functionality of the oligomers is exactly 2.

The polycondensation of α,ω-diacid (or diester) polyamides with α,ω-dihydroxy polyethers leads to thermoplastic elastomers with well defined structures and valuable properties. For instance, the polycondensation of α,ω-dicarboxylic polydodecanamides with different α,ω-dihydroxy (polyoxyethylene-*b*-polyoxypropylenes) led to polydodecanamide-*b*-copolyethers [4,5] (Scheme 2).

$$\text{HOOC—polyamide—COOH} + \text{HO—polyether—OH} \xrightarrow[\text{Zr(OBu)}_4]{240^\circ\text{C}} \quad (2)$$

$$\sim\!\sim\text{polyamide—}\underset{\underset{\text{O}}{\|}}{\text{C}}\text{O—polyether}\sim\!\sim$$

The knowledge of the kinetics and the mechanisms of such reactions is poor for both experimental and theoretical reasons. Tetrabutoxytitanium and tetrabutoxyzirconium, which are the most frequently used catalysts in these reactions, are very sensitive to hydrolysis. Traces of water change their structure into condensed species and agglomerates showing only a surface activity; this effect reaches its maximum when the polyether block is oxyethylene. On the
~~~

other hand, the reaction mixture is often inhomogeneous, at least on the molecular level, and the location of the catalyst is almost unknown; in any case, its distribution in the medium is certainly not homogeneous. Laporte *et al.* [6,7] studied the polycondensation of α,ω-dicarboxylic polyundecanamide or of 11-dodecylamideundecanoic acid with α,ω-dihydroxy-polyoxyethylene, or their condensation with 1-dodecanol, or 2-tridecanol, showing that the kinetics drastically depends on the reaction medium; the reaction rate greatly decreases when the monoalcohols are replaced by α,ω-dihydroxy-polyoxyethylene.

The elimination of water is a major problem even when no catalyst is added. Michot *et al.* [8] prepared poly(imide-*seq*-polyethers) by polycondensing different anhydrides deriving from benzene, naphthalene, perylene, and benzophenone with α,ω-di(amino-2-propyl)-oligooxyethylenes (Scheme 3).

(cyclic anhydride) + H–N(H)~~ → (amic acid: –COOH / –CO–NH~~) → (cyclic imide N~~) + H_2O (3)

The medium is dehydrated by anhydrous argon bubbling for 2 h while kept at 80°C. The cyclization is carried out at 200°C in the presence of an acetic anhydride/pyridine mixture; the use of solvents, such as dimethylacetamide at 120–140°C, allows a one-pot synthesis from the starting ingredients. The determination of the molar mass and cyclization extent is difficult because of the insolubility of aromatic polyimides, and the accuracy of the values obtained is poor. Cyclization is often incomplete due to the effect of polycondensation water, which is difficult to eliminate when the medium becomes quite viscous at high conversions.

Einsenbach and Heinlein [9] prepared thermotropic block copolymers by polycondensing α,ω-dichloroformate oligomers (Scheme 4) with α,ω-dipiperazinyl-oligooxyethylenes (or oligooxytetramethylenes). Protection of the functional groups is required for the obtaining of rigorously bifunctional oligomers; several chemical steps should be carried out, showing the difficulties related to the direct polycondensation.

$$ClCO-R-OC(=O)-N\langle\rangle N-CO-(CH_2CH_2O)_3-C(=O)-N\langle\rangle N-CO-R-OCCl(=O) \;+$$

Hard liquid crystalline block (LCB)

$$HN\langle\rangle N-C(=O)-(OCH_2CH_2CH_2CH_2)_n-OC(=O)-N\langle\rangle NH \;\longrightarrow \qquad (4)$$

Soft block

$$\left[-O-LCB-C(=O)-N\langle\rangle N-C(=O)-(OCH_2CH_2CH_2CH_2)_n-OC(=O)-N\langle\rangle N-C(=O)-\right]_n$$

R is C_6H_4–CO–C_6H_4

More recently, Pospiech *et al.* [10] prepared block copolymers by the direct polyesterification of an α,ω-diester-polysulfone with different α,ω-dicarboxylic oligomers (Scheme 5).

$$m\ H_3C\text{-}C(=O)\text{-}O\text{-}C_6H_4\text{-}C(CH_3)_2\text{-}C_6H_4\text{-}\left(O\text{-}C_6H_4\text{-}SO_2\text{-}C_6H_4\text{-}O\text{-}C_6H_4\text{-}C(CH_3)_2\text{-}C_6H_4\right)_n\text{-}O\text{-}C(=O)CH_3 \ (\text{PES}) + m\ HOOC\text{-}[\text{block 2}]\text{-}COOH \xrightarrow{-\,CH_3COOH}$$

$$\left[O\text{-}C_6H_4\text{-}C(CH_3)_2\text{-}C_6H_4\text{-}\left(O\text{-}C_6H_4\text{-}SO_2\text{-}C_6H_4\text{-}O\text{-}C_6H_4\text{-}C(CH_3)_2\text{-}C_6H_4\right)_n\text{-}OC(=O)\text{-}[\text{block 2}]\text{-}C(=O)\right] \quad (5)$$

Blocks 2 can be poly(oxytetramethylene), a semifluorinated polyester, poly-[(ethylene terephthalate)-*co*-oxybenzoate], a polyesterimide and a poly(ethylene terephthalate)-*co*-polyesterimide. The structure of the copolymer polysulfone-*b*-polyoxytetramethylene is given as an example in Scheme 6.

$$\left[O\text{-}C_6H_4\text{-}C(CH_3)_2\text{-}C_6H_4\text{-}PES\text{-}OC(=O)\text{-}(\text{phthalimide})N\text{-}(CH_2)_3\text{-}[O(CH_2)_4]_p\text{-}O(CH_2)_3\text{-}N(\text{phthalimide})\text{-}C(=O)\right]_x \quad (6)$$

PES is $\left[O\text{-}C_6H_4\text{-}SO_2\text{-}C_6H_4\text{-}O\text{-}C_6H_4\text{-}C(CH_3)_2\text{-}C_6H_4\right]_n$

Bloom and Sheares [11] prepared 4-fluorobenzophenone end-capped poly(4'-methyl-2,5-benzophenone) (I, Scheme 7) and polycondensed it with bisphenol A end-capped polyaryletherketone in the presence of potassium carbonate (Scheme 8).

$$H_3C\text{-}C_6H_4\text{-}C(=O)\text{-}C_6H_3Cl_2 + Cl\text{-}C_6H_4\text{-}C(=O)\text{-}C_6H_4\text{-}F \xrightarrow[\text{NMP, 60°C, 2 h}]{NiCl_2,\ Zn,\ PPh_3,\ BiPy}$$

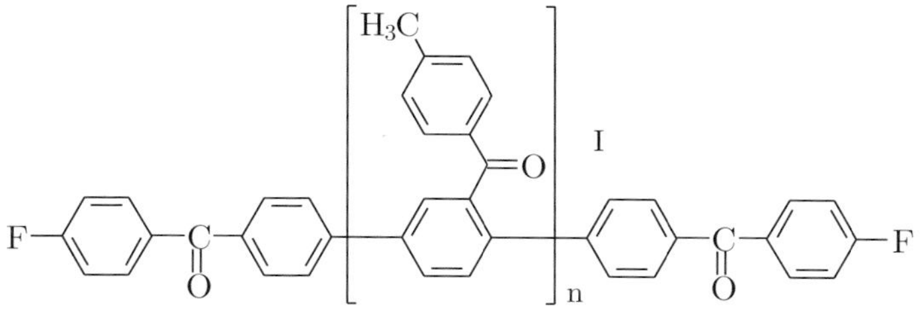

(7)

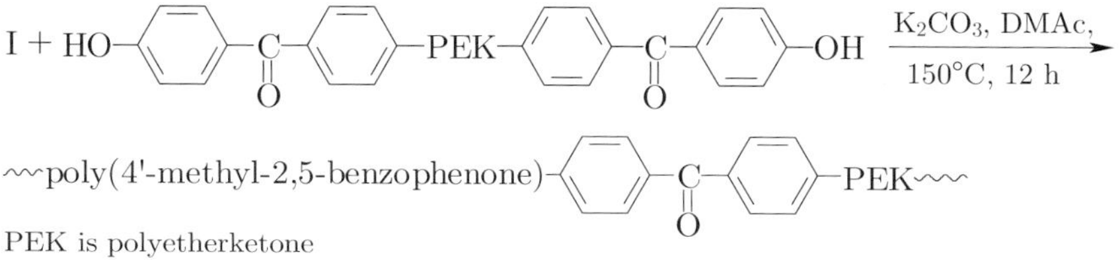

I + HO–C₆H₄–C(=O)–C₆H₄–PEK–C₆H₄–C(=O)–C₆H₄–OH $\xrightarrow[150°C,\ 12\ h]{K_2CO_3,\ DMAc,}$

~~poly(4'-methyl-2,5-benzophenone)–C₆H₄–C(=O)–C₆H₄–PEK~~ (8)

PEK is polyetherketone

It is important to emphasize the high quality of the control of Structure I, more particularly, its molar mass and functionality. The authors determined the characteristics obtained by chromatography and chemical techniques, and compared the end-groups of I to those of models resulting from the reaction of K_2CO_3 with monofunctional compounds, such as 4-isopropanyl-phenol.

Rodewald and Ritter [12] polycondensed α,ω-di(4-hydroxyphenyl) oligoethersulfones with α,ω-di(methanesulfonyl)-polyoxyalkylenes (Scheme 9). Unfortunately, this article does not propose a convincing control of the oligo(ethersulfones) and oligoether functionality.

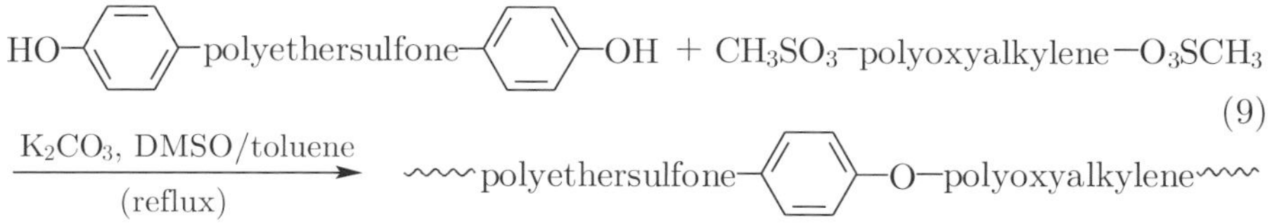

HO–C₆H₄–polyethersulfone–C₆H₄–OH + CH_3SO_3–polyoxyalkylene–O_3SCH_3

$\xrightarrow[(reflux)]{K_2CO_3,\ DMSO/toluene}$ ~~polyethersulfone–C₆H₄–O–polyoxyalkylene~~ (9)

Nuyken [13] used a "criss-cross" cycloaddition to prepare the hard block of a copolymer (Scheme 10) followed by a polyaddition (Scheme 11).

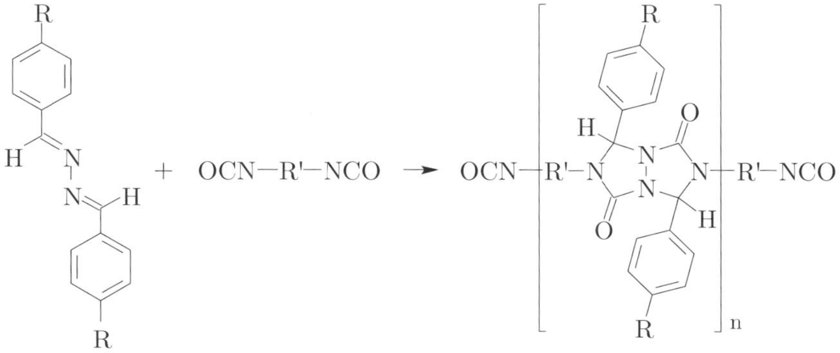

R–C₆H₄–CH=N–N=CH–C₆H₄–R + OCN–R'–NCO → OCN–[R'–(bicyclic triazolidinedione ring bearing two R–C₆H₄ groups)–]ₙR'–NCO (10)

$$OCN-[R'-N\cdots N]_n-R'-NCO \; + \; H-[O-CH_2CH_2]_n-OH \; \rightarrow \qquad (11)$$

$$\left(-\underset{\underset{O}{\|}}{C}-NH-[R'-N\cdots N]_n-R'-NH-\underset{\underset{O}{\|}}{C}-[O-CH_2CH_2]_n-O- \right)_p$$

(R, H, N, O: substituents of the repeating unit in brackets $[\ ]_n$, drawn as structure in the original)

2.3. *Polycondensation of an α,ω-difunctional oligomer with the precursors of another block*

The general process is described by Scheme 12:

X—[oligomer 1]—X + A—R—A + B—R'—B ⟶

~~X'—[block 1]—X'Y'—[block 2]—Y"~~~ (12)

where A–R–A and B–R'–B are the precursors of block 2
and Y'–[block 2]–Y" is A'–R–A'–B'–R'–B'–[A'RA'B'R'B']$_n$–A'–R–A'

The direct method (Section 2.2) is the best way to prepare block copolymers with well controlled structures. However, it cannot be used when one of the initial oligomers is fragile or its melting temperature is too high, or when its structure is ill-defined. In this case, it is possible to prepare this block by polycondensation of its precursors in the presence of the telechelic oligomer, which generates the second block. Many block copolymers were prepared in this way; however their structures are often ill-defined, since it is difficult to control the distribution of the units in the chains. This method requires a very careful identification of the product.

Thuillier *et al.*[14,15] prepared thermotropic polyester-*b*-polyethers by polycondensing α,ω-dihydroxy-polyoxytetramethylene with a mixture of 1,5-

pentanediol and 4,4'-dicarboxylic-*trans*-stilbene acid or 4,4'-dicarboxylic-biphenyl acid in the bulk. Scheme 13 relates to the preparation of poly(1,5-pentanediyl-4,4'-*trans*-stilbene dicarboxylate)-*b*-polyoxytetramethylenes.

$$\text{HOOC}-C_6H_4-\underset{trans}{\text{CH=CH}}-C_6H_4-\text{COOH} + \text{HO(CH}_2)_5\text{OH} + \text{H}-[\text{O(CH}_2)_4]_p-\text{OH}$$

$$\downarrow$$

$$\sim\!\!\left[-\underset{\|}{\overset{}{C}}(=O)-C_6H_4-\underset{trans}{\text{CH=CH}}-C_6H_4-C(=O)-O-(CH_2)_5O\right]_x-C(=O)-//\tag{13}$$

$$-C_6H_4-\underset{trans}{\text{CH=CH}}-C_6H_4-C(=O)-[\text{O(CH}_2)_4]_p\sim$$

The presence of double bonds in the chains allows a thorough analysis of the polymer structure by 500 MHz nuclear magnetic resonance spectroscopy (^{1}H NMR), showing that the product contains a cycle and linear chains built up of three different kinds of units (Scheme 14).

$$-O-(CH_2)_5-O-C(=O)-C_6H_4-CH=CH-C_6H_4-C(=O)-O-(CH_2)_5-O-$$

$$\left[O-(CH_2)_4\right]_p-O-C(=O)-C_6H_4-CH=CH-C_6H_4-C(=O)-O-\left[(CH_2)_4-O\right]_p$$

$$\left[O-(CH_2)_4\right]_p-O-C(=O)-C_6H_4-CH=CH-C_6H_4-C(=O)-O-(CH_2)_5-O-\tag{14}$$

$$\text{cyclo}\left[C(=O)-C_6H_4-CH=CH-C_6H_4-C(=O)-O-(CH_2)_5-O\right]_2$$

Gaymans *et al.* [16] prepared poly(tetramethylene-1,6-hexanediamide)-*b*-polyoxytetramethylene by polycondensation of α,ω-diamino-polyoxytetramethylene with the salt of hexanedioic acid and 1,4-diamino-butane (Scheme 15).

$$H_2N-\left[(CH_2)_4O\right]_n-(CH_2)_4-NH_2 + HOOC(CH_2)_4COOH, H_2N(CH_2)_4NH_2 \tag{15}$$

The diacid and the diamine remain associated in the polycondensation mixture as their salt is independently prepared, which leads to well defined block copolymers and the dispersity index of the rigid block is probably close to 2.

When the copolycondensation is carried out in the melt, very often phase segregation takes place between the oligomer and the precursors of the other block, which may result in quite heterogeneous copolymers. This difficulty can be overcome in different ways. Gaymans *et al.* [16] carried out a pre-polycondensation in a solvent (pyrrolidone) which resulted in a homogeneous mixture of oligomers which was post-polycondensed in the solid state (250–260 °C). Deleens *et al.* [17–19] observed phase segregation when polycondensing various α,ω-diamino-polyamides with different α,ω-dihydroxy-polyethers; however, when diblock or triblock prepolymers were added to the reaction mixture, it became homogeneous. When no pre-polymers were added, the kinetics was erratic, whereas it obeyed a well defined second order law when the medium was homogeneous.

More recently, McCarthy [20] prepared polyoxyalkylene-*b*-polyester TPEs by polycondensing dimethyl terephthalate with a 50/50 mixture of an α,ω-dihydroxy-polyether and 1,4-butanediol (Scheme 16).

$$CH_3OOC{-}C_6H_4{-}COOCH_3 + HO{-}(CH_2)_4{-}OH + H{-}[O(CH_2)_m]_{\overline{n}}{-}OH \rightarrow$$

$$\sim\!\!\left(\underset{\displaystyle O}{\overset{\|}{C}}{-}C_6H_4{-}\underset{\displaystyle O}{\overset{\|}{C}}O{-}(CH_2)_4{-}O\right)_p{-}\underset{\displaystyle O}{\overset{\|}{C}}{-}C_6H_4{-}\underset{\displaystyle O}{\overset{\|}{C}}O{-}(CH_2)_{\overline{m}}{-}[O(CH_2)_m]_{n-2}{-}O(CH_2)_{\overline{m}}{-}O\sim \qquad (16)$$

m values: 4, 6, and 10

The transesterification and polycondensation steps were carried out in the bulk at 200°C and 250°C, respectively, in the presence of transition metal derivatives. The molar masses were determined by size exclusion chromatography (SEC) and ^1H NMR, and the correlation was good.

These techniques can be extended to more complex systems. Kozlowska and Slonecki [21] prepared polyether-*b*-polyamide-*b*-polyesters by polycondensing a dimerized fatty acid ($\overline{M}_n = 570$), an α,ω-diamino-polyether, dimethyl-terephthalate, and butane-1,4-diol. The soft segments consisted of the polyether and the fatty acid moiety. Their general structure is given in Scheme 17.

$$\left[O{-}(CH_2)_4{-}O\left[\underset{\displaystyle O}{\overset{\|}{C}}{-}C_6H_4{-}\underset{\displaystyle O}{\overset{\|}{C}}O{-}(CH_2)_4{-}O\right]_p{-}C{-}DFA{-}\underset{\displaystyle O}{\overset{\|}{C}}{-}NH{-}R{-}NH{-}\underset{\displaystyle O}{\overset{\|}{C}}{-}DFA{-}\underset{\displaystyle O}{\overset{\|}{C}}\right] \qquad (17)$$

R = $(CH)_6$ or $(CH_2)_3[O(CH_2)_4]_3O(CH_2)_3$;
DFA = dimerized fatty acid

In the same way, El Fray and Slonecki [22] prepared (polyether-*b*-polyester)s by polycondensation of a diol, an α,ω-dihydroxy-polyether, a dimerized fatty acid, and dimethyl terephthalate.

These TPEs have a high degree of phase separation; differential scanning calorimetry (DSC) shows a single glass transition temperature of the soft segment, indicating that the polyether and the aliphatic units formed from DFA are associated in a homogeneous amorphous phase. Such syntheses include both polycondensation and inter-exchange reactions.

In some cases, one of the precursors contributes to the formation of both the hard and the soft blocks. Aleksandrovic and Djonlagic [23] copolycondensed a mixture of diesters of terephthalic and fumaric acids, 1,4-butanediol and α,ω-dihydroxy-poly(oxytetramethylene) in the presence of tetrabutoxytitanium

(Scheme 18). The comments on References [21] and [22] hold for [23] and [24], as well: the polycondensation is accompanied by reorganization of the growing chain units.

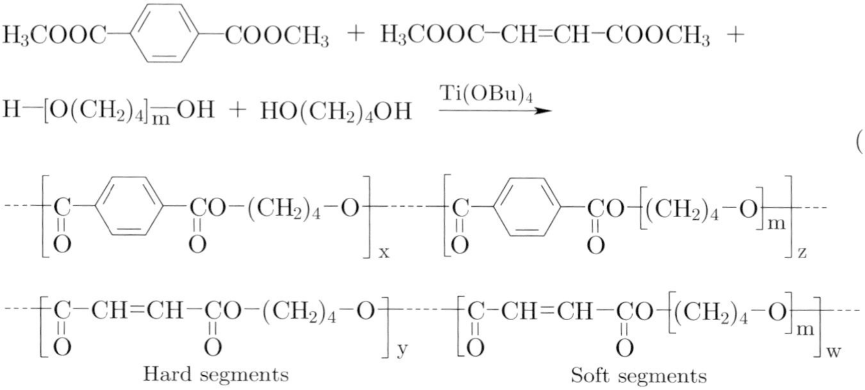

$$H_3COOC-C_6H_4-COOCH_3 + H_3COOC-CH{=}CH-COOCH_3 + H-[O(CH_2)_4]_m-OH + HO(CH_2)_4OH \xrightarrow{Ti(OBu)_4} \quad (18)$$

$$\cdots\left[\underset{\|}{\overset{}{C}}(O)-C_6H_4-\underset{\|}{C}(O)O-(CH_2)_4-O\right]_x\cdots\left[C(O)-C_6H_4-C(O)O-[(CH_2)_4-O]_m\right]_z\cdots$$

$$\cdots\left[C(O)-CH{=}CH-C(O)O-(CH_2)_4-O\right]_y\cdots\left[C(O)-CH{=}CH-C(O)O-[(CH_2)_4-O]_m\right]_w\cdots$$

Hard segments — Soft segments

Jeong *et al.* [25] studied the phase structure and the properties of some polyesteramide TPEs prepared by polycondensation of acid-terminated poly(butylene hexanedioate) with a mixture of an aliphatic diacid and methylenebis(4-phenylisocyanate) (Scheme 19).

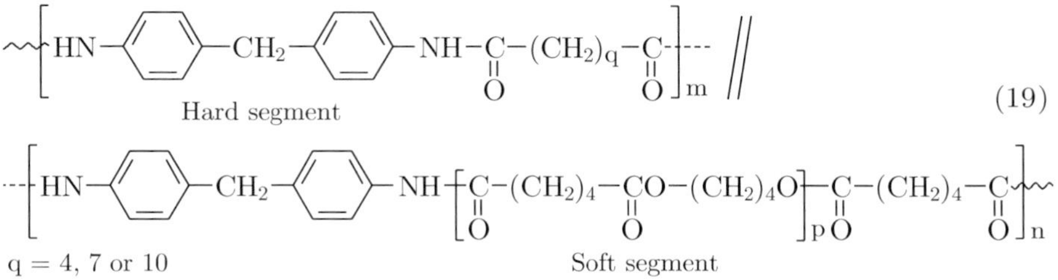

$$\sim\left[HN-C_6H_4-CH_2-C_6H_4-NH-C(O)-(CH_2)_q-C(O)\right]_m\cdots \;/\!/ \quad (19)$$

Hard segment

$$\cdots\left[HN-C_6H_4-CH_2-C_6H_4-NH-\left[C(O)-(CH_2)_4-C(O)O-(CH_2)_4O\right]_p-C(O)-(CH_2)_4-C(O)\right]_n\sim$$

$q = 4$, 7 or 10 — Soft segment

The aliphatic diacid was present in both hard and soft blocks and its distribution was determined by ^{1}H NMR; it influences the tensile properties of the material. The mixing of hard and soft phases depends on the chain length of the aliphatic diacid; it decreases in the order hexanedioic $<$ dodecanedioic, nonanedioic.

Kang *et al.* [26] pointed out the influence of the order of introduction of the reactants when studying the preparation of multiblock poly(dimethylsiloxane)-*b*-polyamides. Three different strategies were used (Schemes 20–22):

(i) (polysiloxane + diamine)s reacted with the dichloride

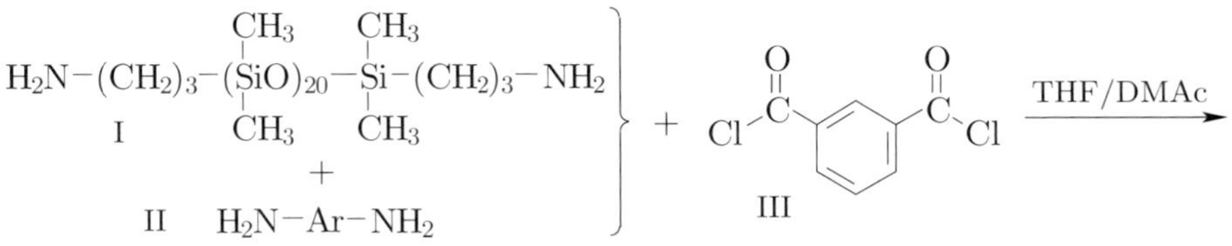

$$\left.\begin{array}{l} \text{I} \quad H_2N-(CH_2)_3-(Si(CH_3)_2O)_{20}-Si(CH_3)_2-(CH_2)_3-NH_2 \\ \quad + \\ \text{II} \quad H_2N-Ar-NH_2 \end{array}\right\} + \underset{\text{III}}{Cl-C(O)-C_6H_4-C(O)-Cl} \xrightarrow{THF/DMAc}$$

$$\left\{\left[-\overset{O}{\overset{\|}{C}}-C_6H_4-\overset{O}{\overset{\|}{C}}NH-Ar-NH\right]_p-\overset{O}{\overset{\|}{C}}-C_6H_4-\overset{O}{\overset{\|}{C}}NH-(CH_2)_3-(\underset{CH_3}{\overset{CH_3}{Si}}O)_{20}-\underset{CH_3}{\overset{CH_3}{Si}}-(CH_2)_3-NH-\right\}_n \quad \text{IV} \qquad (20)$$

Ar = $-C_6H_4-O-C_6H_4-$ (meta, para) or $-C_6H_4-O-C_6H_4-$ (para, para)

(ii) reaction of II with chloride end-capped I

$$\text{II} + Cl-\overset{O}{\overset{\|}{C}}-C_6H_4-\overset{O}{\overset{\|}{C}}NH-(CH_2)_3-(\underset{CH_3}{\overset{CH_3}{Si}}O)_{20}-\underset{CH_3}{\overset{CH_3}{Si}}-(CH_2)_3-NH-\overset{O}{\overset{\|}{C}}-C_6H_4-\overset{O}{\overset{\|}{C}}-Cl \quad \text{V}$$

$$\downarrow \text{THF/DMAc} \qquad (21)$$

$$\left[-\overset{O}{\overset{\|}{C}}-C_6H_4-\overset{O}{\overset{\|}{C}}NH-Ar-NH-\overset{O}{\overset{\|}{C}}-C_6H_4-\overset{O}{\overset{\|}{C}}NH-(CH_2)_3-(\underset{CH_3}{\overset{CH_3}{Si}}O)_{20}-\underset{CH_3}{\overset{CH_3}{Si}}-(CH_2)_3-NH-\right]_n \quad \text{VI}$$

(iii) reaction of V with II and the product is reacted with III

$$\text{II (excess)} + \text{V} \xrightarrow{\text{THF/DMAc}}$$

$$H_2NArNH\overset{O}{\overset{\|}{C}}-C_6H_4-\overset{O}{\overset{\|}{C}}NH-(CH_2)_3-(\underset{CH_3}{\overset{CH_3}{Si}}O)_{20}-\underset{CH_3}{\overset{CH_3}{Si}}-(CH_2)_3-NH\overset{O}{\overset{\|}{C}}-C_6H_4-\overset{O}{\overset{\|}{C}}NHArNH_2 \quad \text{VII}$$

$$\text{VII} + \text{III} \xrightarrow{\text{THF/DMAc}} \qquad (22)$$

$$\left\{\left[-\overset{O}{\overset{\|}{C}}-C_6H_4-\overset{O}{\overset{\|}{C}}NH-Ar-NH\right]_2-\overset{O}{\overset{\|}{C}}-C_6H_4-\overset{O}{\overset{\|}{C}}NH-(CH_2)_3-(\underset{CH_3}{\overset{CH_3}{Si}}O)_{20}-\underset{CH_3}{\overset{CH_3}{Si}}-(CH_2)_3-NH-\right\}_n \quad \text{VI}$$

Copolymer IV has a random distribution of the sequences and it is a rubber-toughened material at low content of I, whereas it behaves as a TPE when the I content increases. On the other hand, copolymer VI is characterized by alternating distributions of the sequences and behaves like very soft rubbers showing a high value of elongation at break.

2.4. *Oligomer-coupling reactions*

Numerous coupling systems may be used to prepare block copolymers (Scheme 23).

$$\text{X—[oligomer 1]—X} + \text{Y—[oligomer 2]—Y} + \text{C—C} \rightarrow$$

$$\sim\sim\text{X'—[block 1]—X'C'C'Y'—[block 2]—Y'}\sim\sim \qquad (23)$$

Oligomer-coupling reactions were reviewed by Fradet [27]. This is the only way to obtain block copolymers prepared from oligomers with identical end-groups or with different end-groups, which are unable to interact. In contrast to the second method (Section 2.3), coupling leads to well defined blocks since the functional oligomers are prepared and characterized before the polycondensation. On the other hand, the distribution of the blocks could be statistical, since one of the blocks can react several times before the other does.

2.4.1. *Coupling of oligomers with hydroxy end-groups*

The most frequently used coupling agents are diisocyanates (Scheme 24).

$$\text{HO}{-}\!\left[\boxed{\text{polyether}}\right]\!{-}\text{OH} + \text{HOCH}_2{-}\!\left[\boxed{\text{polysiloxane}}\right]\!{-}\text{CH}_2\text{OH} + \text{OCN}{-}\text{X}{-}\text{NCO} \longrightarrow$$

$$\sim\!\left[\boxed{\text{polyether}}\right]\!{-}\text{O}{-}\underset{\underset{\text{O}}{\|}}{\text{C}}{-}\text{NH}{-}\text{X}{-}\text{NH}{-}\underset{\underset{\text{O}}{\|}}{\text{C}}{-}\text{O}{-}\text{CH}_2{-}\!\left[\boxed{\text{polysiloxane}}\right]\!\sim \qquad (24)$$

Tsai *et al.* [28] polycondensed methylenebis(4-phenylisocyanate) with a mixture of α,ω-dihydroxy-poly(pentamethylene-4,4'-dibenzoate) and α,ω-dihydroxy-poly(tetramethylene butane-1,4-dicarboxylate) oligomers (Scheme 25).

$$\text{OCN-DPM-NCO} + \text{HO}{-}(\text{CH}_2)_4{-}\text{O}\!\left[\underset{\underset{\text{O}}{\|}}{\text{C}}{-}\text{DPh}{-}\underset{\underset{\text{O}}{\|}}{\text{C}}\text{O}{-}(\text{CH}_2)_5\text{O}\right]_n\!\text{H} +$$

$$\text{HO}{-}(\text{CH}_2)_4{-}\text{O}\!\left[\underset{\underset{\text{O}}{\|}}{\text{C}}{-}(\text{CH}_2)_4{-}\underset{\underset{\text{O}}{\|}}{\text{C}}\text{O}{-}(\text{CH}_2)_4\text{O}\right]_p\!\text{H} \xrightarrow[\text{DMF}]{\text{Bu}_2\text{Sn dodecanoate}}$$

$$\sim\underset{\underset{\text{O}}{\|}}{\text{C}}{-}\text{NH}{-}\text{DPM}{-}\text{NH}{-}\underset{\underset{\text{O}}{\|}}{\text{C}}{-}\text{O}{-}(\text{CH}_2)_4{-}\text{O}\!\left[\underset{\underset{\text{O}}{\|}}{\text{C}}{-}\text{DPh}{-}\underset{\underset{\text{O}}{\|}}{\text{C}}\text{O}{-}(\text{CH}_2)_5\text{O}\right]_n\!/\!/ \qquad (25)$$

$$-\underset{\underset{\text{O}}{\|}}{\text{C}}{-}\text{NH}{-}\text{DPM}{-}\text{NH}{-}\underset{\underset{\text{O}}{\|}}{\text{C}}{-}\text{O}{-}(\text{CH}_2)_4{-}\text{O}\!\left[\underset{\underset{\text{O}}{\|}}{\text{C}}{-}(\text{CH}_2)_4{-}\underset{\underset{\text{O}}{\|}}{\text{C}}\text{O}{-}(\text{CH}_2)_4\text{O}\right]_p\!\sim$$

where DPM = $-C_6H_4-CH_2-C_6H_4-$ and DPh = $-C_6H_4-C_6H_4-$

Isocyanates are efficient coupling agents, but when the temperature is too high they can trimerize. Diacid chlorides can be used instead of diisocyanates, but the released hydrogen chloride often induces side reactions.

Bis(cyclic iminoesters) are very efficient coupling agents, much less sensitive to temperature than isocyanates, their polycondensation with molten poly(alkylene terephthalate)s being thus effected at 280°C (Scheme 26) [29].

$$2\ \text{HO}{-}(\text{Al}){-}\text{OH} + \text{[2,2'-bis(4H-3,1-benzoxazin-4-one)]} \xrightarrow[\text{5 min}]{\text{bulk, 280°C}}$$

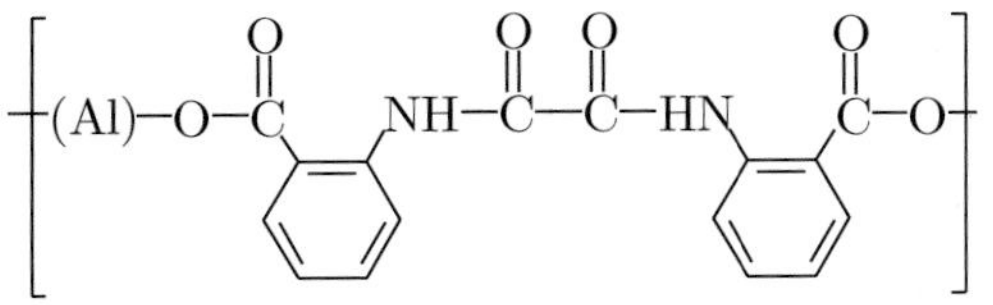

$$\left[(Al)-O-\overset{O}{\overset{\|}{C}}-C_6H_4-NH-\overset{O}{\overset{\|}{C}}-\overset{O}{\overset{\|}{C}}-HN-C_6H_4-\overset{O}{\overset{\|}{C}}-O \right] \quad (26)$$

(Al) = poly(alkylene terephthalate)

The formulas of several very efficient bis(cyclic iminoester)s are given in Scheme 27 [27].

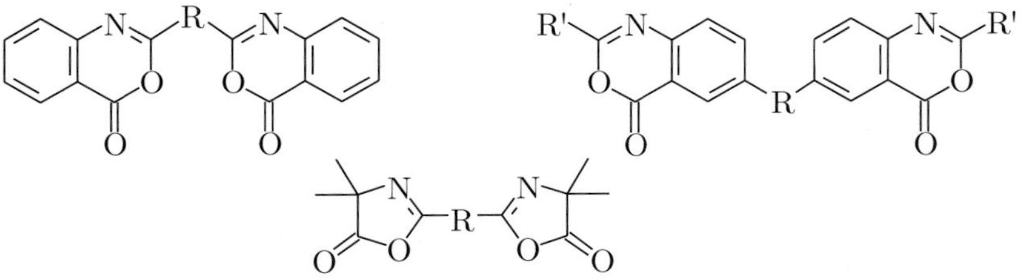

(27)

R and R' are aliphatic and aromatic hydrocarbon groups

Acevedo and Fradet [30] studied the polyaddition of bisoxazolones with several dihydroxy polyethers, leading to diblock polyethers (Scheme 28).

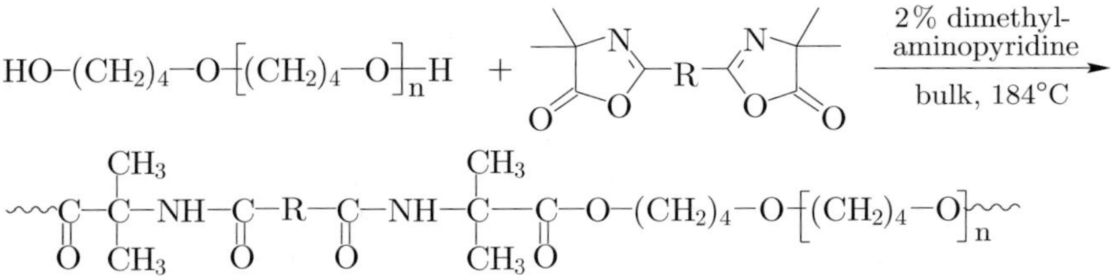

$$HO-(CH_2)_4-O-[(CH_2)_4-O]_n-H \; + \; \text{bisoxazolone} \xrightarrow[\text{bulk, 184°C}]{\text{2\% dimethylaminopyridine}}$$

$$\sim\sim\overset{O}{\overset{\|}{C}}-C(CH_3)_2-NH-\overset{O}{\overset{\|}{C}}-R-\overset{O}{\overset{\|}{C}}-NH-C(CH_3)_2-\overset{O}{\overset{\|}{C}}-O-(CH_2)_4-O-[(CH_2)_4-O]_n\sim\sim \quad (28)$$

4-Dimethylaminopyridine is an efficient catalyst. When the system consists of α,ω-dihydroxy-polyoxytetramethylene and 2,2'-bis[4,4'-dimethyl-5(4H)-oxazolinone], the conversion is 100% after 30 min; $\overline{M}_w$ ranges from 7800 to 19000. The reaction is particularly fast when carried out in a double-screw extruder.

Pollock *et al.* [31] designed and prepared segmented polyurethanes mimicking the spider silk morphology which is characterized by large-scale crystalline ordered domains within a continuous elastomeric matrix. They were prepared by polyadditioning a macromolecular diisocyanate (I) with hexamethylenebisisocyanate and 1,4-butanediol (Scheme 29). II is a physical crosslinked network; it is important to stress that its structure is characterized by self-ordering groups between the hard and soft segments of the polymer backbone.

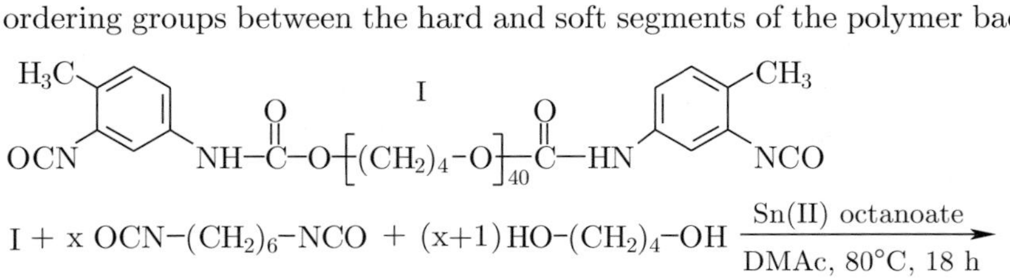

I

$$OCN(H_3C)C_6H_3-NH-\overset{O}{\overset{\|}{C}}-O-[(CH_2)_4-O]_{40}-\overset{O}{\overset{\|}{C}}-HN-C_6H_3(CH_3)NCO$$

$$I + x\ OCN-(CH_2)_6-NCO + (x+1)\,HO-(CH_2)_4-OH \xrightarrow[\text{DMAc, 80°C, 18 h}]{\text{Sn(II) octanoate}}$$

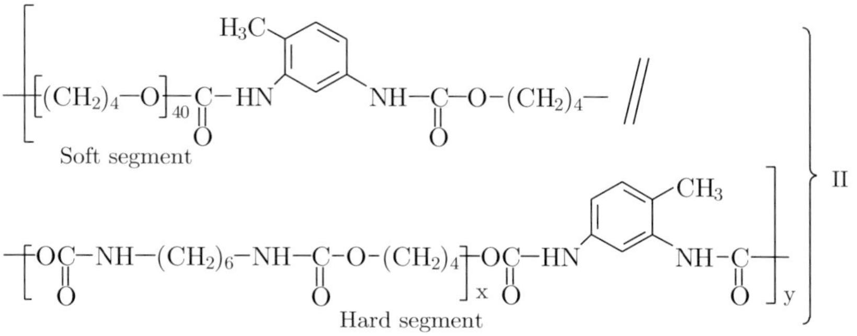

$$\left[\left[(CH_2)_4-O \right]_{40} \underset{\underset{O}{\|}}{C}-HN-C_6H_3(CH_3)-NH-\underset{\underset{O}{\|}}{C}-O-(CH_2)_4- \; // \right.$$

Soft segment

$$\left. - \left[O\underset{\underset{O}{\|}}{C}-NH-(CH_2)_6-NH-\underset{\underset{O}{\|}}{C}-O-(CH_2)_4 \right]_x O\underset{\underset{O}{\|}}{C}-HN-C_6H_3(CH_3)-NH-\underset{\underset{O}{\|}}{C} \right]_y \quad \text{II} \tag{29}$$

Hard segment

2.4.2. *Coupling of oligomers with amino end-groups*

Diisocyanates are efficient coupling agents for the systems shown in Scheme 30.

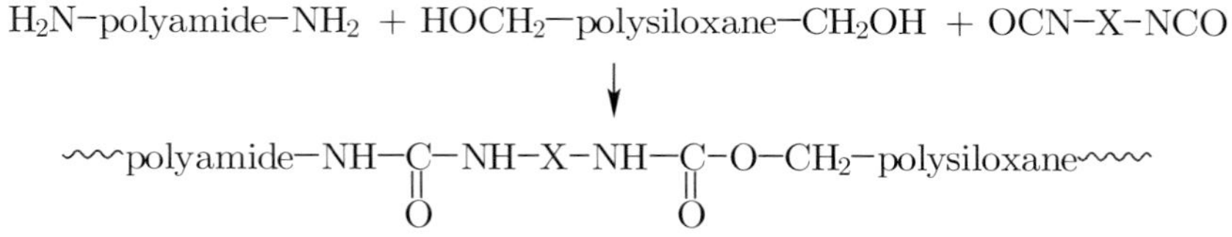

$$H_2N\text{-polyamide-}NH_2 + HOCH_2\text{-polysiloxane-}CH_2OH + OCN\text{-}X\text{-}NCO$$
$$\downarrow$$
$$\sim\sim\text{polyamide-}NH-\underset{\underset{O}{\|}}{C}-NH-X-NH-\underset{\underset{O}{\|}}{C}-O-CH_2\text{-polysiloxane}\sim\sim \tag{30}$$

This holds also for bis(cyclic iminoesters); bisoxazolones are particularly efficient because they are stable and lead to linear polymers when reacted with diamines in solution [32].

Acevedo and Fradet [33] prepared polyether-*b*-polyamides by polyadditioning bisoxazolones with the mixture of an α,ω-diamino-polyether and an α,ω-diamino-polyamide (Scheme 31).

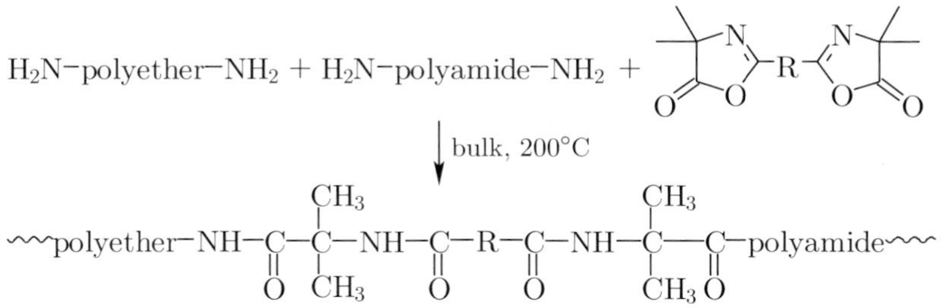

$$H_2N\text{-polyether-}NH_2 + H_2N\text{-polyamide-}NH_2 + \text{(bisoxazolone: 4,4-dimethyl-oxazol-5-one rings linked by R)}$$
$$\downarrow \text{bulk, 200°C}$$
$$\sim\sim\text{polyether-}NH-\underset{\underset{O}{\|}}{C}-\underset{CH_3}{\overset{CH_3}{C}}-NH-\underset{\underset{O}{\|}}{C}-R-\underset{\underset{O}{\|}}{C}-NH-\underset{CH_3}{\overset{CH_3}{C}}-\underset{\underset{O}{\|}}{C}\text{-polyamide}\sim\sim \tag{31}$$

No side reaction accompanied this block copolycondensation, except for the formation of a very small amount of 2-imidazolin-5-one.

Jan *et al.* [34] took advantage of the different reactivities of the chlorine atoms of 2,4,6-trichloro-1,3,5-triazine (cyanuric chloride) to prepare polyoxyethylene-*b*-polyoxypropylenes. The reactivity of the chloride functions drastically depends on the temperature (Scheme 32).

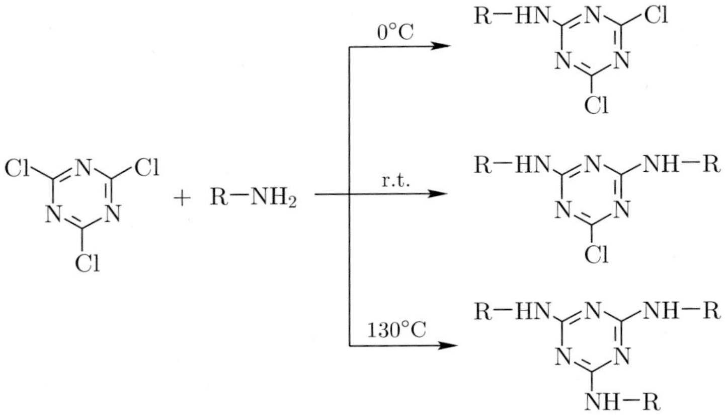

$$\text{Cl}_3\text{C}_3\text{N}_3 + \text{R–NH}_2 \xrightarrow{0°\text{C}} \text{R–HN–C}_3\text{N}_3\text{Cl}_2;\quad \xrightarrow{\text{r.t.}} (\text{R–HN})_2\text{C}_3\text{N}_3\text{Cl};\quad \xrightarrow{130°\text{C}} (\text{R–HN})_3\text{C}_3\text{N}_3 \tag{32}$$

This allows the obtaining of the multiblock copolymer (Scheme 33).

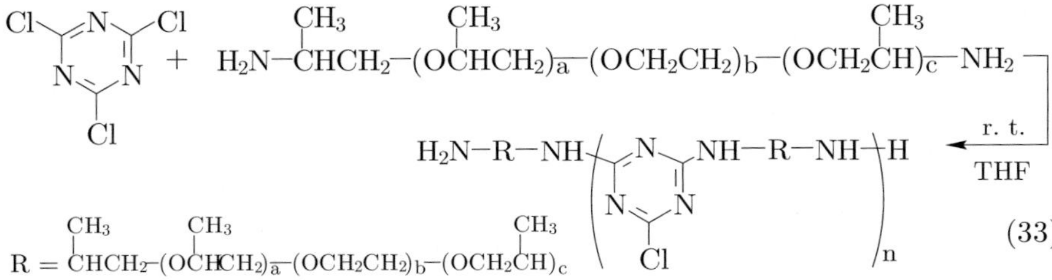

$$\text{Cl}_3\text{C}_3\text{N}_3 + \text{H}_2\text{N–}\underset{|}{\overset{\text{CH}_3}{\text{C}}}\text{HCH}_2\text{–(O}\overset{\text{CH}_3}{\text{C}}\text{HCH}_2)_a\text{–(OCH}_2\text{CH}_2)_b\text{–(OCH}_2\overset{\text{CH}_3}{\text{C}}\text{H)}_c\text{–NH}_2 \xrightarrow[\text{THF}]{\text{r. t.}} \text{H}_2\text{N–R–NH}\left(\text{C}_3\text{N}_3(\text{Cl})\text{–NH–R–NH}\right)_n\text{H} \tag{33}$$

$$\text{R} = \overset{\text{CH}_3}{\text{C}}\text{HCH}_2\text{–(O}\overset{\text{CH}_3}{\text{C}}\text{HCH}_2)_a\text{–(OCH}_2\text{CH}_2)_b\text{–(OCH}_2\overset{\text{CH}_3}{\text{C}}\text{H)}_c$$

2.4.3. *Coupling of oligomers with carboxylic end-groups*

Fischer *et al.* [35] prepared multiblock polyether-*b*-amides by reacting a triblock α,ω-dicarboxylic-poly(ether-*b*-amide) with bisoxazoline (Scheme 34)

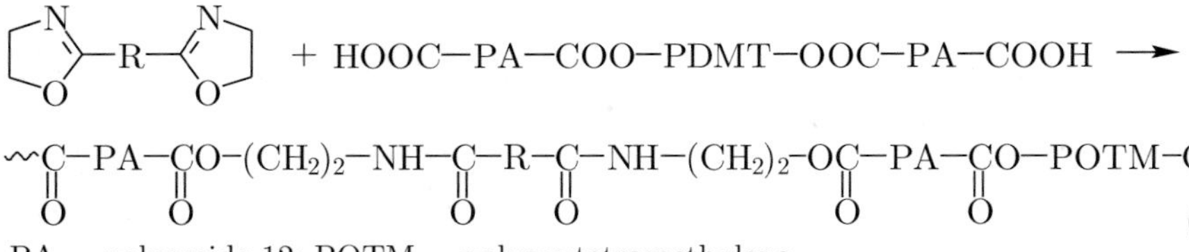

$$\text{bisoxazoline (oxazoline–R–oxazoline)} + \text{HOOC–PA–COO–PDMT–OOC–PA–COOH} \longrightarrow$$

$$\sim\text{C(=O)–PA–C(=O)O–(CH}_2)_2\text{–NH–C(=O)–R–C(=O)–NH–(CH}_2)_2\text{–OC(=O)–PA–C(=O)O–POTM–O}\sim \tag{34}$$

PA = polyamide 12; POTM = polyoxytetramethylene

The structure of the chains was carefully characterized by ^{13}C NMR and SEC; the separation of the phases was studied by DSC.

2.5. *Characterization techniques. Side reactions*

2.5.1. *Functional oligomers*

The preparation of block copolymers by polycondensation requires very well defined oligomers. More particularly, it is important to control: (i) their exact composition because many commercial oligomers are mixtures of homopolymers and copolymers, (ii) their functionality, which is an essential parameter determining the molar mass of the polymers prepared by polycondensation, and (iii) their stability under the experimental conditions chosen for the blocking reaction.

In many reports, it is mentioned that "the telechelic oligomers were used as received". Unfortunately, the actual structure can be far from the composition given in the analytical card.

Boularès [4,5] studied the polycondensation of an α,ω-dicarboxylic-polydodecanamide with different commercial triblock polyethers having a structure supposed to obey the formula given in Scheme 35 (data card):

polyoxyethylene-*b*-polyoxypropylene-*b*-polyoxyethylene (35)

Unfortunately, their actual structures are far from Scheme 35, as it was shown by SEC, mass spectrometry, and ^{1}H NMR. The nature and the location of the reactive end-groups are essential and those of the oligomer given in Scheme 35 are primary alcohols. However, NMR studies show that part of them are secondary groups, which drastically changes the kinetics of the polycondensation and the distribution of the blocks in the chain. Moreover, the ratio [oxyethylene]/[oxypropylene] determines the hydrophilicity/hydrophobicity balance and the water content of the oligomer which certainly modifies the activity of organometallic catalysts. The location of the end-group is sometimes erratic due to side reactions, such as transfer; it can be controlled by the fractionation of a sample and the determination of the functionality of each fraction. It is not rare to find values ranging from 1 to 3 or even 4, and average functionality has little meaning in polycondensation.

Heatley [36] provided very valuable ^{1}H NMR spectroscopic information on polyethers. The ^{1}H NMR spectra of the copolyethers after trifluorination allowed the determination of the oxyethylene-to-oxypropylene unit ratio in the chain and the concentrations of primary (triplet at 4.42 ppm) and secondary (sextet at 5.16 ppm) hydroxyl groups. Their ^{13}C NMR spectra were compared to those of α,ω-dihydroxy-polyoxyethylene and α,ω-dihydroxy-polyoxypropylene, which confirmed the ^{1}H NMR data and permitted the identification of each configuration in the chain.

The presence of low molecular mass impurities in the initial oligomers changes the stoichiometric balance and these impurities often behave as chain-termination agents. For instance, Thuillier [14,15] showed that many telechelic oligoethers contain significant amounts of 1-butanol and butyrolactone.

The formation of macrocycles may also concern the rigid block. Rozès [37] prepared liquid crystalline copolymers with flexible blocks of aliphatic polyethers and rigid blocks of aromatic polyesters. They were prepared by polytransesterification of aliphatic diols with aromatic diesters derived from biphenyl or *trans*-stilbene. Scheme 36 shows the formation of cycles when the diester is dimethyl-2,2'-biphenyldicarboxylate. In this case, the respective location of the ester groups is responsible for the formation of the cyclic compound.

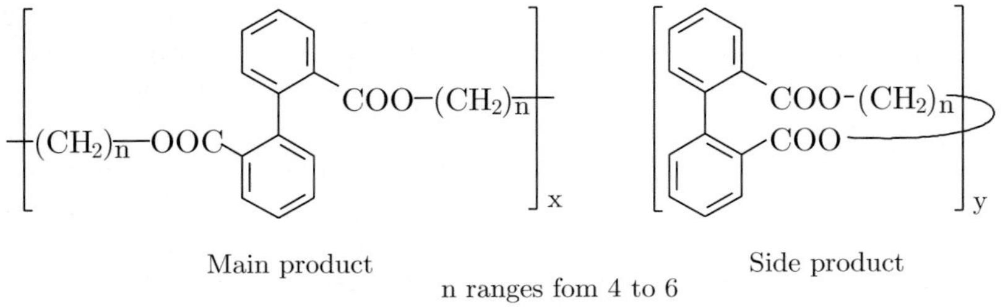

$$\left[-(CH_2)_n-OOC-C_6H_4-C_6H_4-COO-(CH_2)_n- \right]_x \quad \left[C_6H_4(COO-(CH_2)_n)-C_6H_4(COO) \text{ (cyclic)} \right]_y \tag{36}$$

Main product Side product

n ranges fom 4 to 6

The determination of the functionality is an essential parameter in block polycondensation. Unfortunately, many of the values given in the literature are doubtful because they are obtained by only one analytical technique or they are implicitly assumed. Only the comparison of the molar masses obtained by direct determination (chromatography, tonometry, mass spectrometry, osmometry) and those calculated from the concentration of the end-groups is relevant. The latter is determined by *e.g.,* chemical titration, ^{1}H NMR, or infrared spectroscopy; the use of several methods increases the accuracy of the values. Reliable values are obtained when both direct and indirect determinations are taken into account.

2.5.2. *Block copolymers*

The preparation of block copolymers is often accompanied by different side phenomena.

Degradation of the oligomers. Many telechelic oligomers contain light components; this is the case of α,ω-dihydroxy-oligoethers. These impurities are either eliminated during the polycondensation or react with the end-groups, which may stop the chain growth.

On the other hand, the thermal stability of the oligomers in the reaction medium can be very high. When studying the polycondensation of α,ω-dihydroxy-oligooxyethylenes with aliphatic α,ω-dicarboxylic-polyamides at different temperatures, Deleens [17–19] did not observe significant degradation of the polyether, despite that some of the reactions were carried out at 260°C. Thuillier [38] showed that α,ω-dihydroxy-oligooxytetramethylenes contain light fractions, the amount of which increases during the polycondensation.

Homogeneity of the reaction medium. The reaction medium is often heterogeneous, at least in the first step of the reaction. This may result in an ill-defined distribution of the end-groups and the location of the catalyst is often based on pure assumptions. The case of transition metal derivatives is particularly critical because their unoccupied orbitals accept the doublets of donating atoms, such as oxygen atoms in aliphatic polyethers.

Formation of macrocycles and homopolymers. The formation of macrocycles was analyzed in Sections 2.3 and 2.5.1. On the other hand, there are only a few unambiguous proofs of the formation of block copolymers free of homopolymers.

The values obtained by elemental analysis and spectroscopy are common to the block copolymer and to a mixture of the corresponding homopolymers. In some very rare cases, it is possible to characterize the junction between the blocks. Boularès [4,5] prepared polyamide-*b*-copolyethers by polycondensation of an α,ω-dicarboxylic-oligoamide (PA, $\overline{M}_n = 1200$) with different α,ω-dihydroxy-copolyethers (PE, $\overline{M}_n = 1000$); their ^{1}H NMR spectra clearly show the signals (in ppm) relative to the ester junction between the blocks (Scheme 37), even though the molecular mass of the block copolymer is close to 20 000. SEC often gives an unambigous proof of the copolymer formation.

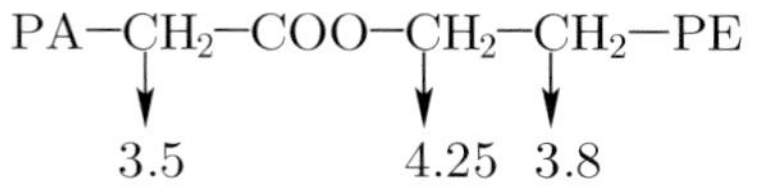

$$\underset{}{PA-}\underset{\overset{\downarrow}{3.5}}{CH_2}-COO-\underset{\overset{\downarrow}{4.25}}{CH_2}-\underset{\overset{\downarrow}{3.8}}{CH_2}-PE \qquad (37)$$

Otsuki *et al.* [39] prepared multiblock copolymers by reacting a mixture of α,ω-dicarboxylic oligooxyethylenes, isophthalic acid, and nonanedioic acid with methylenebis(4-phenylisocyanate). Two techniques were used: a one-step process where all the components are mixed (I) or a two-step one (II) where the diacid mixture is reacted with the diisocyanate and the resulting α,ω-diisocyanate-polyamide A is reacted with B (Scheme 38).

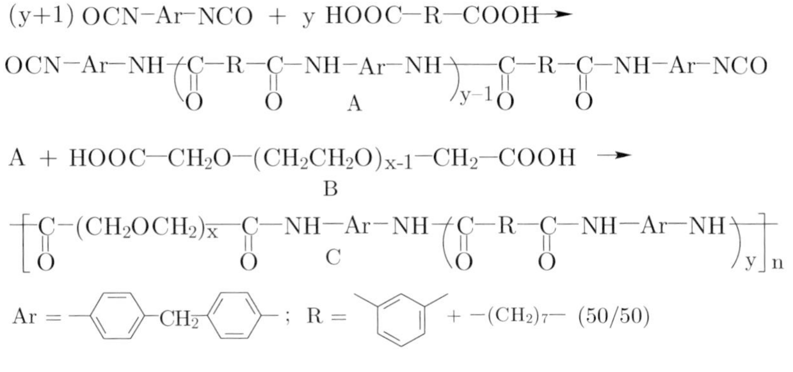

$$(y+1)\ OCN-Ar-NCO + y\ HOOC-R-COOH \rightarrow$$

$$OCN-Ar-NH\left(\underset{\underset{O}{\|}}{C}-R-\underset{\underset{O}{\|}}{C}-NH-Ar-NH\right)_{y-1}\underset{\underset{O}{\|}}{C}-R-\underset{\underset{O}{\|}}{C}-NH-Ar-NCO$$
$$\text{A}$$

$$A + \underset{B}{HOOC-CH_2O-(CH_2CH_2O)_{x-1}-CH_2-COOH} \rightarrow$$

$$\left[\underset{\underset{O}{\|}}{C}-(CH_2OCH_2)_x-\underset{\underset{O}{\|}}{C}-NH-Ar-NH\left(\underset{\underset{O}{\|}}{C}-R-\underset{\underset{O}{\|}}{C}-NH-Ar-NH\right)_y\right]_n \qquad (38)$$
$$\text{C}$$

$$Ar = -C_6H_4-CH_2-C_6H_4-\ ;\quad R = \text{1,3-}C_6H_4 + -(CH_2)_7- \ (50/50)$$

The elemental analysis of C gave little information, since the compositions of the copolymer and the reactant mixture are identical. Its infrared spectrum showed that the reaction took place, since new absorption groups (amide) were observed. On the other hand, the SEC traces of the products obtained by processes I and II are almost identical, suggesting that they probably do not contain homopolymer because its content has little chance to be the same. The solubility of the block copolymers, of the initial oligomers, and of the corresponding homopolymers provides useful information, despite that it is mainly a qualitative property (Table 1).

Table 1 shows that the copolymers are soluble in dimethylsulfoxide even at room temperature, whereas PA is soluble only on heating. C_1 is soluble in pyridine on heating, but PA is not. The products are copolymers, since

Table 1. Solubility of polyethylene-*b*-polyamide [39]. The molar masses of the oligoethers (PE) and oligoamides (PA) are 3900 and 3200 (copolymer C_1) and 7400 and 8300 (copolymer C_2), respectively. Processes I and II are described in the text. Solubility: ++ soluble at room temperature, + soluble on heating, ± swelling, – insoluble

Solvent	POE	PA	C_1(I)	C_1(II)	C_2(I)	C_2(II)
N-methyl-2-pyrrolidone	++	++	++	++	++	++
N,N'-dimethylacetamide	++	++	++	++	++	++
N,N'-dimethylformamide	++	++	++	++	++	++
Dimethylsulfoxide	++	+	++	++	++	++
m-Cresol	++	++	++	++	++	++
Pyridine	++	–	++	–	±	±
Methanol	++	–	–	–	–	–

otherwise the polyamide would probably separate from the solution. C_1 is more soluble than C_2, which could be expected since the ratio [POE]/[PA] is 1.2 in the case of C_1 and 0.9 for C_2. C_1(I) is soluble in *m*-cresol, whereas C_1(II) is not; this probably results from different block distributions — statistical in the case of the copolymers prepared by process I and more regular when process II is used.

The comparison of the solid state ^{13}C NMR spectra of the block copolymers and of the initial oligomers [37,40,41] provides interesting information. Boularès *et al.* [4,5] analyzed the crystalline structure of different polydodecanamide-*b*-copolyethers using the results of Mathias and Johnson [42] relative to different crystalline forms of polydodecanamide: α-form (precipitated from phenol-ethanol solution), γ-form (by thermal annealing), and γ'-form (quenched from the melt). Comparison of the spectra of the telechelic polydodecanamide precursor and the block copolymer shows that the crystalline parts of the copolymer are γ, which is confirmed by X-ray diffractometry.

This morphology is preserved in the block copolymers even though they were prepared by polycondensation in the melt. The analysis provides also interesting information on the flexible block.

3. New structures

3.1. *Block copolymers containing liquid crystalline structures*

Numerous copolymers contain liquid crystalline segments either as the main chain [43–47] or as side chains [48–55]. They are prepared by chain polymerization [48–50] or by polycondensation [43,44,46,47,51–57]. In the same way, the introduction of hyperbranched and dendritic segments in block copolymers has a rapid development [54–56].

3.2. *Liquid crystalline sequences as part of the backbone*

This is the case of thermoplastic polyurethane elastomers [45,46]. Hsu and Lee [45] prepared a segmented copolymer containing both urethane and biphenyl units by polyaddition of 4,4'-bis(n-hydroxyalkyloxy)biphenyl with 1,6-hexane-diisocyanate or 2,4-toluene-diisocyanate. Scheme 39 relates to the aliphatic diisocyanate.

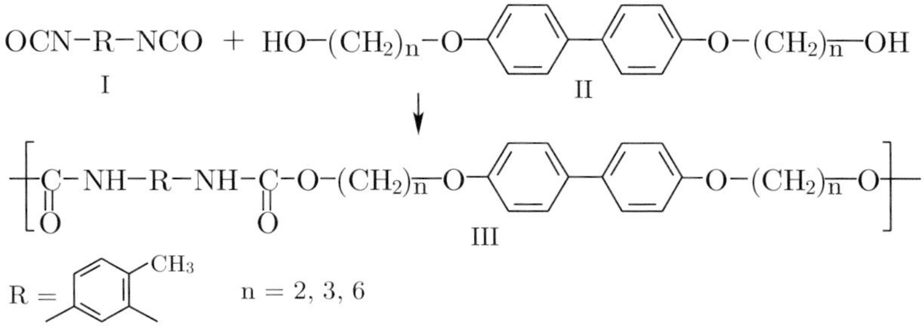

$$\underset{\text{I}}{OCN-R-NCO} + \underset{\text{II}}{HO-(CH_2)_n-O-C_6H_4-C_6H_4-O-(CH_2)_n-OH}$$
$$\downarrow$$
$$\underset{\text{III}}{\left[-\underset{\|}{\overset{}{C}}(O)-NH-R-NH-\underset{\|}{C}(O)-O-(CH_2)_n-O-C_6H_4-C_6H_4-O-(CH_2)_n-O-\right]} \qquad (39)$$

$R = $ (2,4-tolylene, CH_3) $\quad n = 2, 3, 6$

DSC and polarized light microscopy showed that III exhibits a smectic type mesophase and the phase transition temperature increases with increasing n.

Jeong *et al.* [46] prepared thermoplastic elastomers containing polycaprolactone (PCL) as soft block, the hard block resulting from the reaction of hexamethylenediisocyanate (HDI) with 4,4'-dihydroxy-biphenyl (DHB) (Scheme 40).

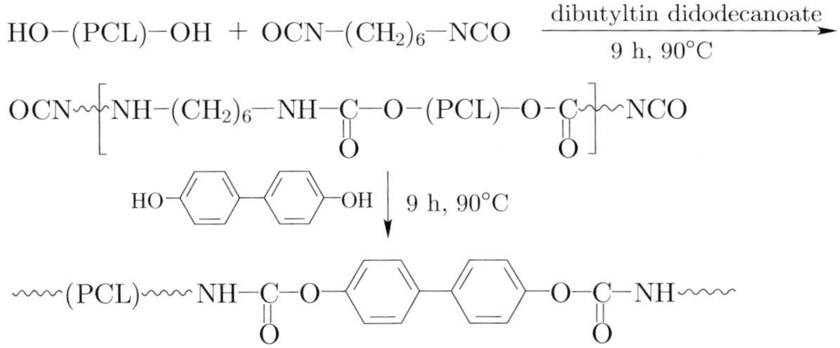

$$HO-(PCL)-OH + OCN-(CH_2)_6-NCO \xrightarrow[\text{9 h, 90°C}]{\text{dibutyltin didodecanoate}}$$
$$OCN\sim\left[NH-(CH_2)_6-NH-\underset{\underset{O}{\|}}{C}-O-(PCL)-O-\underset{\underset{O}{\|}}{C}\right]\sim NCO$$
$$\downarrow HO-C_6H_4-C_6H_4-OH,\ \text{9 h, 90°C} \qquad (40)$$
$$\sim(PCL)\sim NH-\underset{\underset{O}{\|}}{C}-O-C_6H_4-C_6H_4-O-\underset{\underset{O}{\|}}{C}-NH\sim$$

These segmented polymers are both thermoplastic and liquid crystalline elastomers. When the molar mass of PCL is 4000 and the hard domain content is 40 wt%, an enantiotropic mesophase is formed. The tensile storage moduli values of these TPEs, both in the crystalline state below the melting temperature of the PCL block and in the rubbery state above its T_m, are much higher than those of the TPEs prepared from the systems PCL/HDI.

Inspired by the spider silk, which is among the toughest natural materials, Pollock and Hammond [47] prepared an interesting silk-like thermoplastic polyurethane elastomer with main-chain liquid crystalline soft segments (Schemes 41–43).

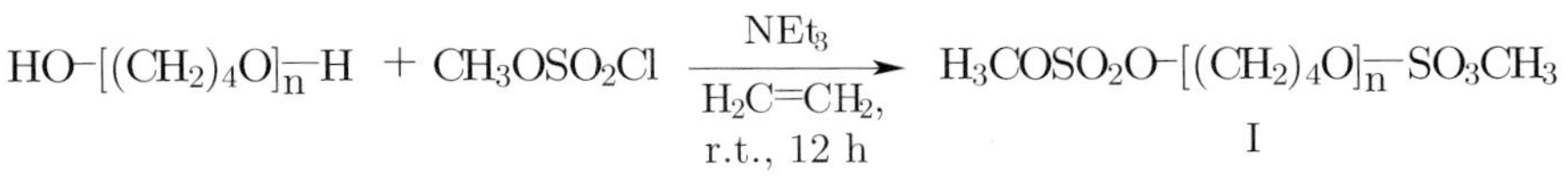

$$HO-[(CH_2)_4O]_n-H + CH_3OSO_2Cl \xrightarrow[\text{H}_2\text{C=CH}_2,\ \text{r.t., 12 h}]{NEt_3} \underset{\text{I}}{H_3COSO_2O-[(CH_2)_4O]_n-SO_3CH_3}$$

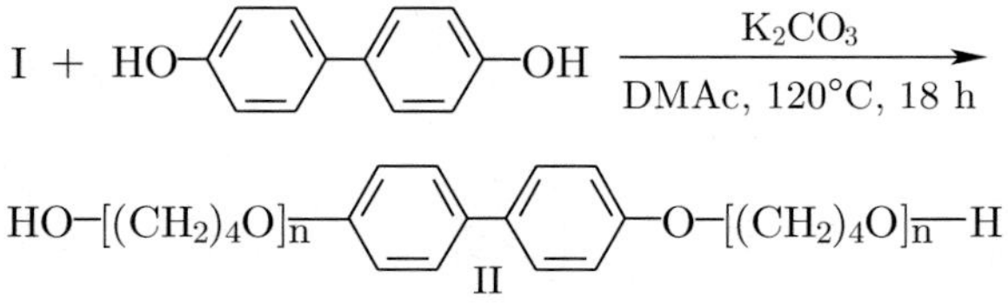

$$\mathrm{I} + \mathrm{HO{-}C_6H_4{-}C_6H_4{-}OH} \xrightarrow[\text{DMAc, 120°C, 18 h}]{K_2CO_3} \tag{41}$$

$$\mathrm{HO{-}[(CH_2)_4O]_n{-}C_6H_4{-}C_6H_4{-}O{-}[(CH_2)_4O]_n{-}H} \quad \mathrm{II}$$

In the following, II is symbolized by HO〰▬〰OH

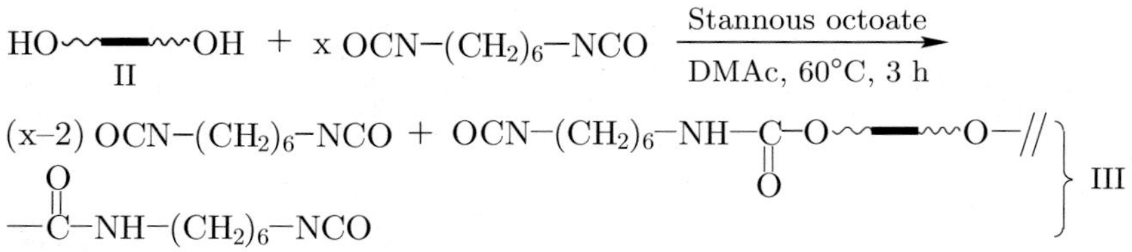

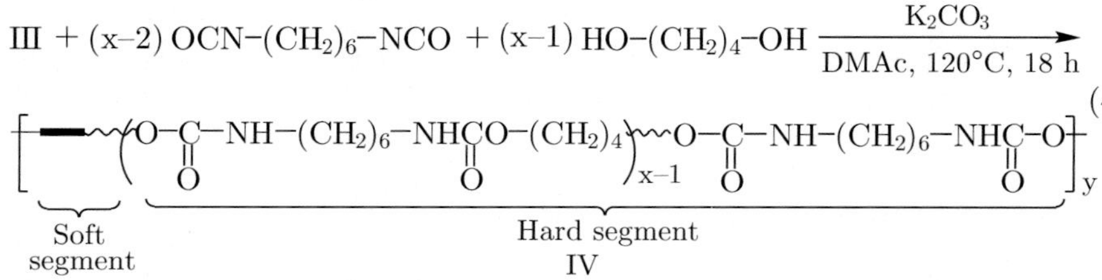

$$\underset{\mathrm{II}}{\mathrm{HO〰▬〰OH}} + \mathrm{x\ OCN{-}(CH_2)_6{-}NCO} \xrightarrow[\text{DMAc, 60°C, 3 h}]{\text{Stannous octoate}} \tag{42}$$

$$\left.\begin{array}{l}\mathrm{(x{-}2)\ OCN{-}(CH_2)_6{-}NCO + OCN{-}(CH_2)_6{-}NH{-}C(=O){-}O〰▬〰O{-}//} \\ \mathrm{{-}C(=O){-}NH{-}(CH_2)_6{-}NCO}\end{array}\right\} \mathrm{III}$$

$$\mathrm{III + (x{-}2)\ OCN{-}(CH_2)_6{-}NCO + (x{-}1)\ HO{-}(CH_2)_4{-}OH} \xrightarrow[\text{DMAc, 120°C, 18 h}]{K_2CO_3} \tag{43}$$

$$\Big[\underbrace{▬〰}_{\text{Soft segment}}\underbrace{\big(O{-}C(=O){-}NH{-}(CH_2)_6{-}NHC(=O)O{-}(CH_2)_4\big)_{x-1}〰O{-}C(=O){-}NH{-}(CH_2)_6{-}NHC(=O){-}O}_{\text{Hard segment}}\Big]_y \quad \mathrm{IV}$$

The DSC thermogram of IV shows a much higher segregation between the hard and soft segments than in a polyurethane, which does not contain liquid crystalline segments.

3.3. *Liquid crystalline sequences as side chains*

Nair *et al.* [51–53] prepared TPEs containing liquid crystalline side chains. The liquid crystalline blocks are formed of polysiloxanes substitued by 4-cyano-biphenylalkanes. Their preparation [51] includes polycondensation and chemical modification steps (Scheme 44). The first one is the polycondensation of a mix-

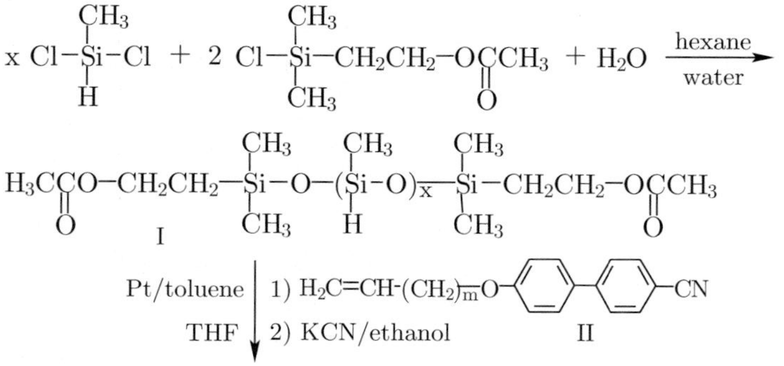

$$\mathrm{x\ Cl{-}Si(CH_3)(H){-}Cl + 2\ Cl{-}Si(CH_3)_2{-}CH_2CH_2{-}OC(=O)CH_3 + H_2O} \xrightarrow[\text{water}]{\text{hexane}}$$

$$\mathrm{H_3CC(=O)O{-}CH_2CH_2{-}Si(CH_3)_2{-}O{-}(Si(CH_3)(H){-}O)_x{-}Si(CH_3)_2{-}CH_2CH_2{-}OC(=O)CH_3} \quad \mathrm{I}$$

$$\Big\downarrow \text{Pt/toluene, THF: 1) } \mathrm{H_2C{=}CH{-}(CH_2)_m{-}O{-}C_6H_4{-}C_6H_4{-}CN}\ (\mathrm{II}) \text{; 2) KCN/ethanol}$$

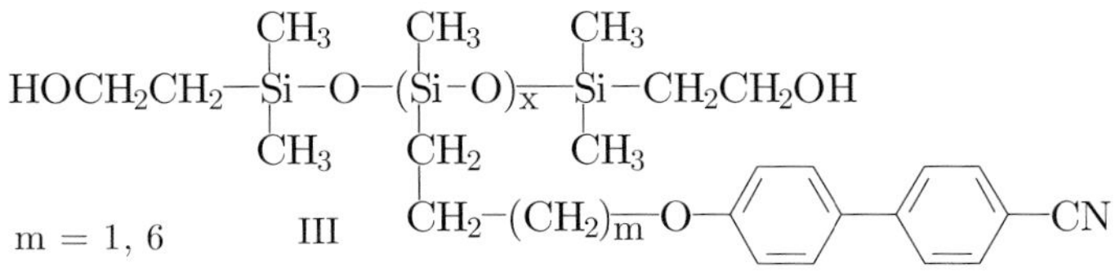

$$HOCH_2CH_2{-}\underset{CH_3}{\overset{CH_3}{Si}}{-}O{-}(\underset{CH_2{-}CH_2{-}(CH_2)_m{-}O{-}C_6H_4{-}C_6H_4{-}CN}{\overset{CH_3}{Si}}{-}O)_x{-}\underset{CH_3}{\overset{CH_3}{Si}}{-}CH_2CH_2OH \tag{44}$$

m = 1, 6 III

ture of dichloromethylsilane and acetoxyethyldimethylchlorosilane in the presence of water, which results in α,ω-acetoxyethyl-poly(hydrogenomethylsiloxane) (I). The second step is the hydrogenosilylation between I and a vinylmesogen (II), catalyzed by hexachloroplatinic acid and then the carbinol end-groups are deprotected by transesterification of the acetoxyethyl groups with ethanol in the presence of KCl and tetrahydrofuran.

The formation of the TPE follows the classical polyurethane preparation (Scheme 45): end-capping of I by methylene-bis(4-phenylisocyanate) followed by the polyaddition of the mixture of II, 1,4-butanediol, and the diisocyanate.

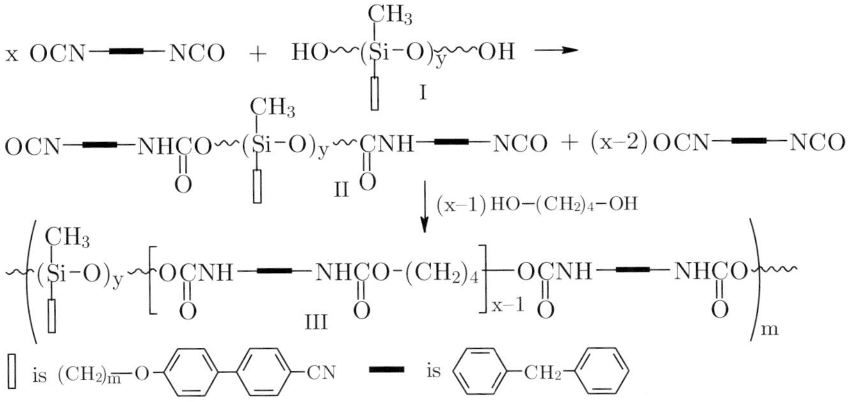

$$x\ OCN{-}\blacksquare{-}NCO + HO{\sim}(\underset{\square}{\overset{CH_3}{Si}}{-}O)_y{\sim}OH \longrightarrow \quad \text{I}$$

$$OCN{-}\blacksquare{-}NH\underset{O}{\overset{}{C}}O{\sim}(\underset{\square}{\overset{CH_3}{Si}}{-}O)_y{\sim}\underset{O}{C}NH{-}\blacksquare{-}NCO + (x{-}2)\,OCN{-}\blacksquare{-}NCO \quad \text{II}$$

$$\downarrow (x{-}1)\,HO{-}(CH_2)_4{-}OH$$

$${\sim}\left((\underset{\square}{\overset{CH_3}{Si}}{-}O)_y{\sim}\left[O\underset{O}{C}NH{-}\blacksquare{-}NH\underset{O}{C}O{-}(CH_2)_4\right]_{x-1}O\underset{O}{C}NH{-}\blacksquare{-}NH\underset{O}{C}O\right)_m{\sim} \quad \text{III} \tag{45}$$

$\square$ is $(CH_2)_m{-}O{-}C_6H_4{-}C_6H_4{-}CN$ $\quad\blacksquare$ is $C_6H_4{-}CH_2{-}C_6H_4$

Nair *et al.* [51] studied the structure of the block copolymer given in Scheme 46.

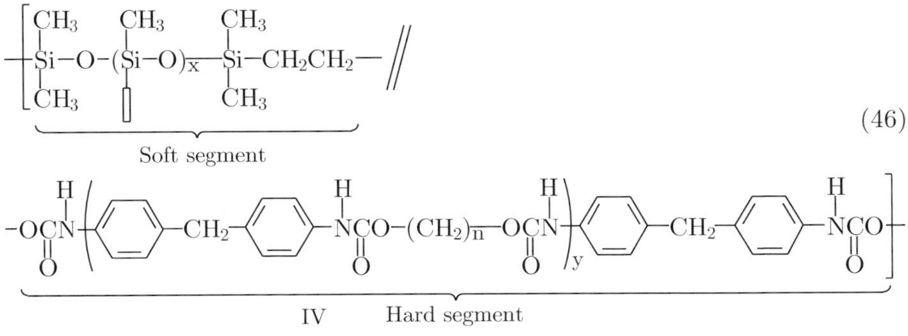

$$\left[\underset{CH_3}{\overset{CH_3}{Si}}{-}O{-}(\underset{\square}{\overset{CH_3}{Si}}{-}O)_x{-}\underset{CH_3}{\overset{CH_3}{Si}}{-}CH_2CH_2{-}\right] /\!/ \tag{46}$$

Soft segment

$$\left.{-}O\underset{O}{C}\overset{H}{N}\left(C_6H_4{-}CH_2{-}C_6H_4{-}\overset{H}{N}\underset{O}{C}O{-}(CH_2)_n{-}O\underset{O}{C}\overset{H}{N}\right)_y C_6H_4{-}CH_2{-}C_6H_4{-}\overset{H}{N}\underset{O}{C}O\right]$$

IV Hard segment

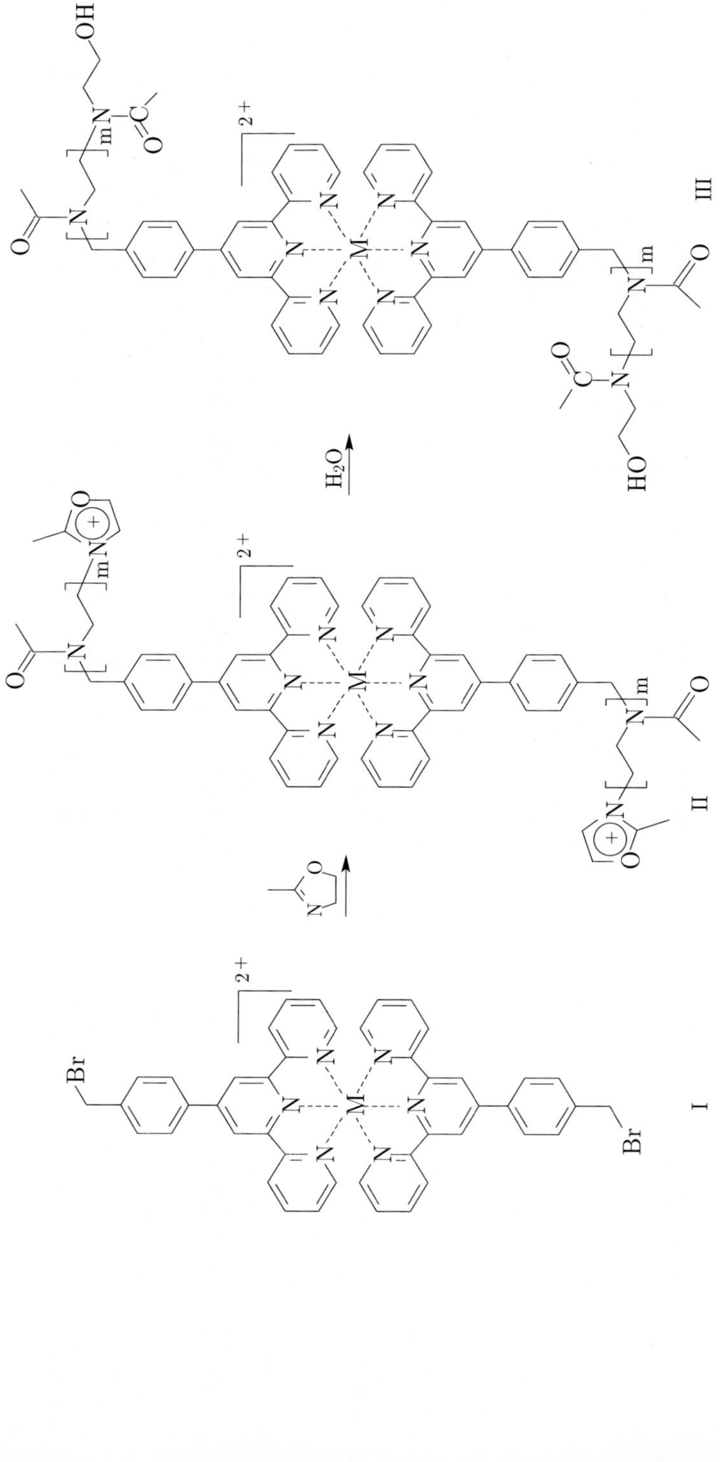

Scheme 47

Copolymers IV exhibit a liquid crystalline phase behavior over a broad temperature range and its nature depends on both the spacer length (n = 3, 6, 8) and the temperature. When n = 6 or 8, the thermoplastic polyurethane elastomer exhibits a smectic phase; on the other hand, when n = 3, it shows a nematic phase at high temperatures. Even though accurate values of the molecular mass could not be obtained due to solubility difficulties, their estimation from viscosity determinations shows that they are low. The sample with n = 3 does not indicate any segregation; the latter was observed in the samples with longer alkyl spacers (n = 6 or 8) as shown by SAXS studies. The morphology of these thermoplastic elastomers was analyzed by FTIR dichroism, which showed the relative orientations of the hard segments and their changes when the samples were subjected to deformation. The re-orientations of the hard domains and of the liquid crystalline mesogens are coupled. These studies are reported in [54–56], and Anthamatten and Hammond [57] have published a simple free energy model, which predicts phase ordering in liquid crystalline block copolymers. Even though the materials studied were different, Abeysekera *et al.* [58] carried out interesting studies of the organization of the blocks in discotic liquid crystalline triblock copolymers and its shear-induced change.

3.4. *Metallo-supramolecular block copolymers*

New polymeric architectures were prepared by polyaddition of mono- or difunctional oligomers having complexing groups as terminal units, such as terpyridine ligands, which form chelates with different metals; they can be block or graft copolymers. These new techniques were particularly developed by Schubert and his group and reviewed in a recent publication [59]. Newkome *et al.* [60] reported tailored architectures through the formation of metal complexes, mainly used to prepare metallodendrimers. Brunsveld *et al.* [61] prepared supramolecular polymers based on hydrogen bonding or discotic molecules, and Schubert and Heller used metallo-supramolecular initiators to obtain star (block co)polymers [62].

Several different strategies can be used, *e.g.*, diblock copolymers are first prepared by ring-opening polymerization using a difunctional metallic complex and the second step is a coupling polyaddition of these diblock copolymers leading to a multiblock structure.

For instance, Schubert and Heller polymerized methyloxazoline using iron(II) or ruthenium(II) bi- and terpyridine complexes (Scheme 47) [62].

A multiblock copolymer can be obtained when III is reacted with a coupling agent.

Lohmeijer and Schubert [63] prepared a copolymer with two polyoxyethylene blocks and a hard block made of a metallic complex (Scheme 48).

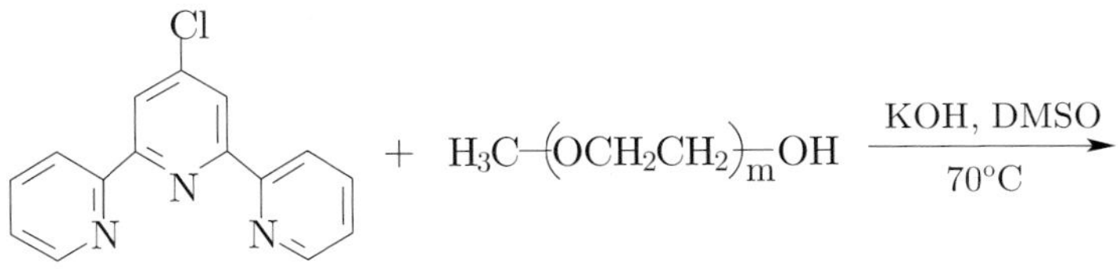

Cl
N N N

$+ \ H_3C{-}(OCH_2CH_2)_m{-}OH \xrightarrow[70°C]{KOH,\ DMSO}$

$$H_3C{-}OCH_2CH_2{-}(OCH_2CH_2)_m{-}O{-}CH_2CH_2O{-}\text{terpyridine} \xrightarrow[\text{2) } NH_4PF_6]{\text{1) } M(OAc)_2,\ MeOH,\ reflux}$$

m = 72, 122, 227, 377

$$[CH_3O{-}(CH_2CH_2O)_{m+1}{-}\text{terpy}{-}M{-}\text{terpy}{-}(OCH_2CH_2)_{m+1}{-}OCH_3]^{2+}\ 2\,PF_6^-$$

m = 72
M = Fe, Ru, Co, Ni, Cu, Zn, Cd

(48)

The ^{1}H NMR spectrum of the triblock copolymer clearly shows the formation of the complex by the shift of several signals, particularly those of protons 5':3' (8.4 to 8.57 ppm), 6:6" (6.68 to 7.20 ppm), and the methylene group (arrow, 4.40 to 3.83 ppm). These supramolecular structures are sensitive to pH changes: Fe-, Co-, Zn-, and Cd-complexes are destroyed and the initial compounds are recovered when pH is high (13) or low (1), whereas the Cu-complex is destroyed only when maintained at low pH for an extended time; Ru- and Ni-polymers are insensitive to pH changes.

These techniques were also used to prepare triblock copolymers with two different soft blocks [64,65] in a three-step process: (i) preparation of a trichlororuthenium complex of α-methyl, ω-terpyridinyl-polyoxyethylene, (ii) preparation of α-terpyridinyl-polystyrene, and (iii) association of these monofunctional oligomers (Scheme 49).

$$CH_3O{-}(CH_2CH_2O)_m{-}\text{terpy} + RuCl_3 \xrightarrow[60°C,\ 3\ h]{MeOH} CH_3O{-}(CH_2CH_2O)_m{-}\text{terpy}{-}RuCl_3\ \ (I)$$

$$I + \text{polystyrene}_n{-}O{-}\text{terpy} \longrightarrow$$

(49)

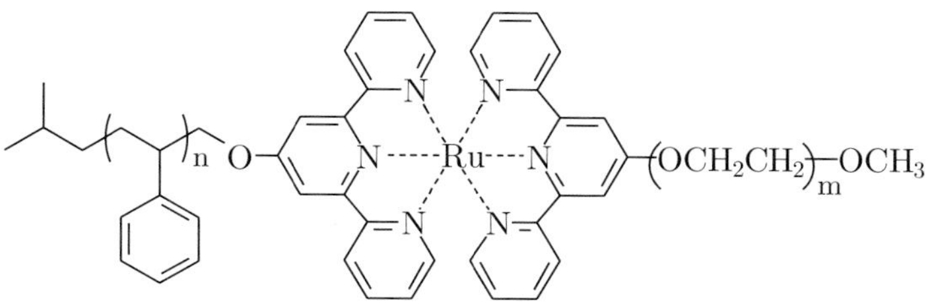

A completely different synthetic strategy can be used to prepare multiblock copolymers. For instance, a copolymer containing an oligoether and two complex-forming blocks gives a multiblock copolymer with double helical supramolecular segments when its solution in chloroform is added with Cu(I)-trifluoromethanesulfonate (Schemes 50 and 51) [66,67].

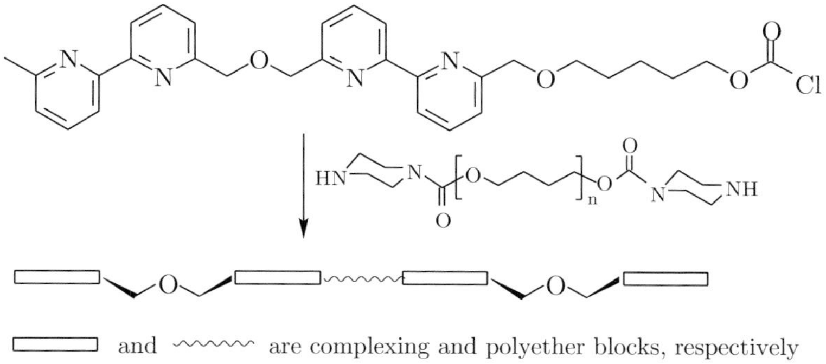

(50)

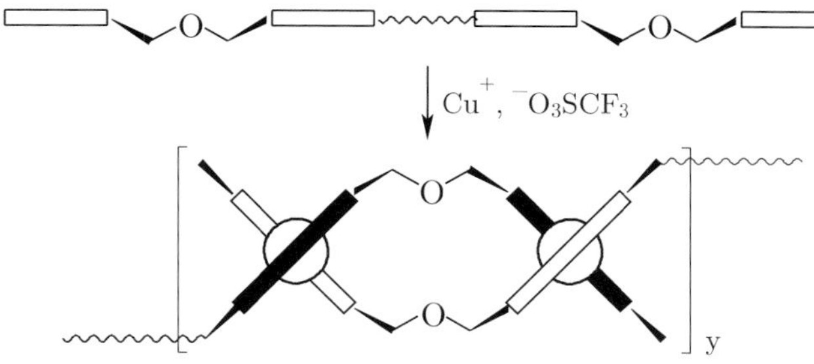

and are complexing and polyether blocks, respectively

(51)

The formation of this supramolecular structure affects substantially the crystallization behavior of the triblock copolymer; in the absence of the complex, it exhibits distinct melting endotherms, but after complex formation with Cu(I), only a weak melting endotherm is observed.

Schubert *et al.* [68,69] prepared copolymers with polyether soft blocks, the hard blocks being complexes of transition metal and terpyridine groups, by non-covalent polyaddition of terpyridine end-capped poly(oxyethylene) or poly(oxytetramethylene) with a transition metal derivative, such as Co(II), Fe(II), or Ni(II) diacetates (Schemes 52 and 53).

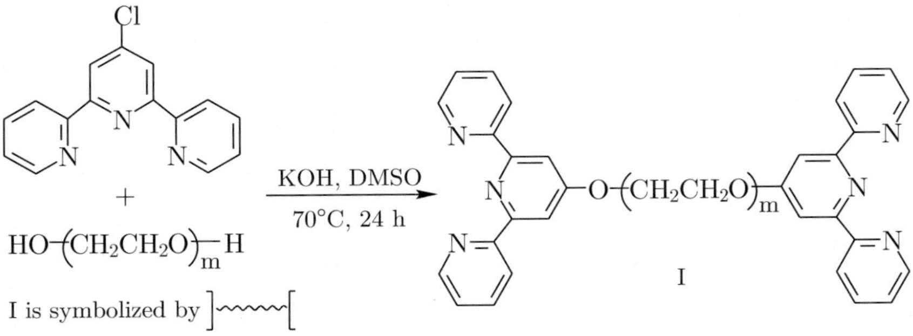

$$\text{4'-chloro-2,2':6',2''-terpyridine} + HO\text{+}CH_2CH_2O\text{+}_m H \xrightarrow[70°C,\ 24\ h]{KOH,\ DMSO} \text{tpy}-O\text{+}CH_2CH_2O\text{+}_m-\text{tpy} \quad (I) \tag{52}$$

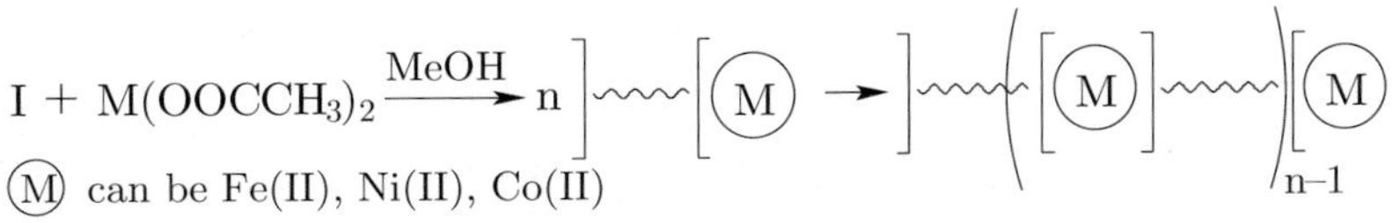

I is symbolized by]~~~~~[

$$I + M(OOCCH_3)_2 \xrightarrow{MeOH} n\ ]\sim\sim[(M) \rightarrow]\sim\sim\left([(M)]\sim\sim\right)_{n-1}[(M) \tag{53}$$

(M) can be Fe(II), Ni(II), Co(II)

Reaction (53) can be reversed by modifying pH, or by electrochemical or thermal changes.

3.5. *Block copolymers prepared from metal-containing macrocycles*

Kricheldorf *et al.* [70–76] studied the use of metal-containing cycles and macrocycles to prepare homopolymers and copolymers.

Cyclic or non-cyclic tin alkoxides tend to associate through donor-acceptor interactions between oxygen and tin atoms. Most of the cyclic tin alkoxides give dimers both in the solid state and in solution (Scheme 54) [72].

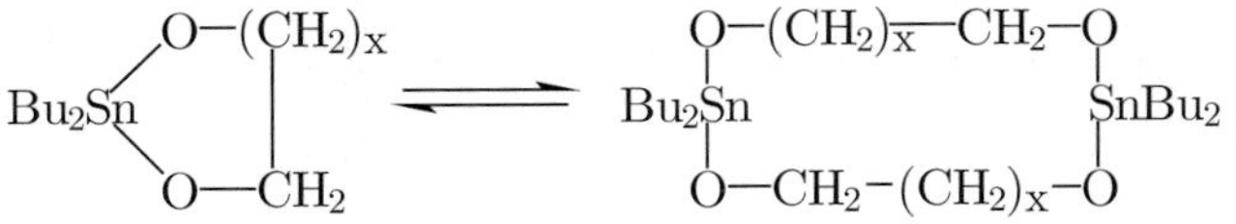

$$Bu_2Sn\langle{}^{O-(CH_2)_x}_{O-CH_2}\rangle \rightleftharpoons Bu_2Sn\langle{}^{O-(CH_2)_x-CH_2-O}_{O-CH_2-(CH_2)_x-O}\rangle SnBu_2 \tag{54}$$

These dimerizations are equilibria, but when x = 3, the reaction mixture mainly contains the monomeric species because the dimerization is probably not favored by thermodynamic parameters. These cyclic tin alkoxides are efficient initiators of lactones [70,73] and morpholines [76].

The initiation of ε-caprolactone by 2-dibutylstanna-1,3-dioxepane leads to the segmented macrocycle I (Scheme 55) [71].

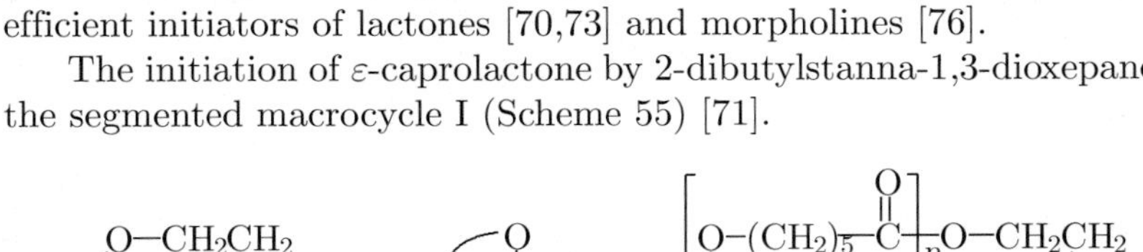

$$Bu_2Sn\langle{}^{O-CH_2CH_2}_{O-CH_2CH_2}\rangle + (H_2C)_5\langle{}^{O}_{C=O}\rangle \rightarrow Bu_2Sn\langle{}^{[O-(CH_2)_5-C(=O)]_n-O-CH_2CH_2}_{[O-(CH_2)_5-C(=O)]_m-O-CH_2CH_2}\rangle \quad I \tag{55}$$

However, the macrocycle I can be also obtained from the corresponding linear segmented copolymer II [72] (Scheme 56):

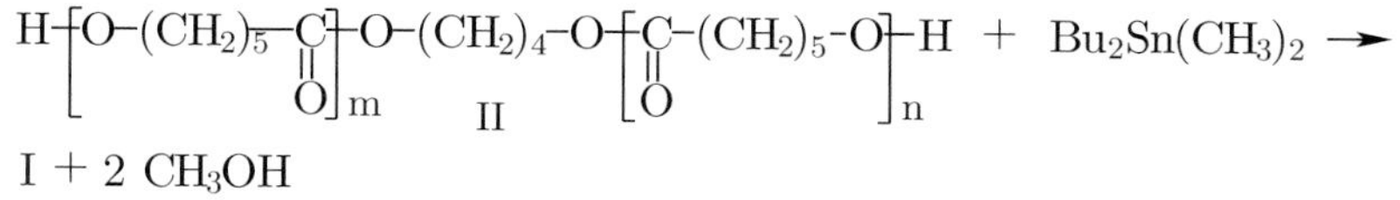

$$H{-}\left[O{-}(CH_2)_5{-}\underset{\Vert}{C}\right]_m{-}O{-}(CH_2)_4{-}O{-}\left[\underset{\Vert}{C}{-}(CH_2)_5{-}O\right]_n{-}H \;(\text{II}) + Bu_2Sn(CH_3)_2 \longrightarrow I + 2\,CH_3OH \quad (56)$$

(the C atoms in brackets bear =O)

The structures of the macrocycles and linear polymers were analyzed by matrix-assisted laser desorption ionization with time of flight detection (MALDI-TOF) [71].

Triblock-copolymers are obtained when the macrocycle I is reacted with a carboxylic diacid dichloride giving either a macrocycle (II) or a linear multiblock copolymer (III) (Scheme 57) [70,73].

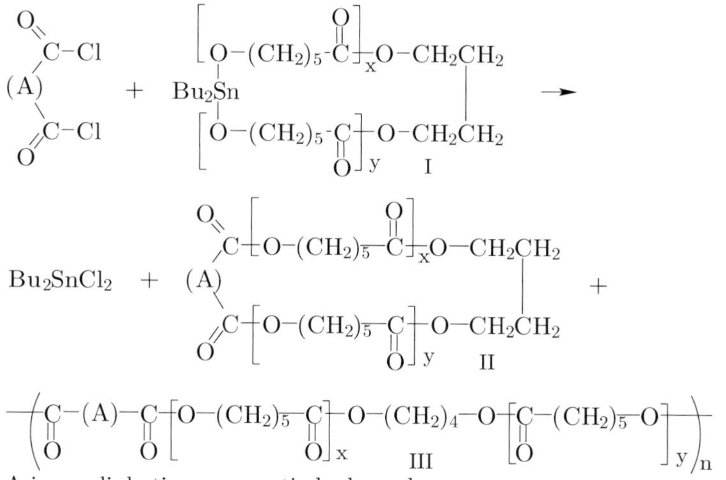

$$ClOC{-}(A){-}COCl + \text{I: } Bu_2Sn\begin{cases}[O{-}(CH_2)_5{-}C(=O)]_x{-}O{-}CH_2CH_2\\ [O{-}(CH_2)_5{-}C(=O)]_y{-}O{-}CH_2CH_2\end{cases}\text{(ring closed via } CH_2CH_2{-}CH_2CH_2) \longrightarrow$$

$$Bu_2SnCl_2 + \text{II: } (A)\begin{cases}C(=O){-}[O{-}(CH_2)_5{-}C(=O)]_x{-}O{-}CH_2CH_2\\ C(=O){-}[O{-}(CH_2)_5{-}C(=O)]_y{-}O{-}CH_2CH_2\end{cases} + \quad (57)$$

$$\text{III: } {-}\left(C(=O){-}(A){-}C(=O){-}[O{-}(CH_2)_5{-}C(=O)]_x{-}O{-}(CH_2)_4{-}O{-}[C(=O){-}(CH_2)_5{-}O]_y\right)_n{-}$$

A is an aliphatic or aromatic hydrocarbon

When terephthaloyl dichloride is used, an aromatic multiblock copolymer is formed [73] (Schemes 58–60).

First step:

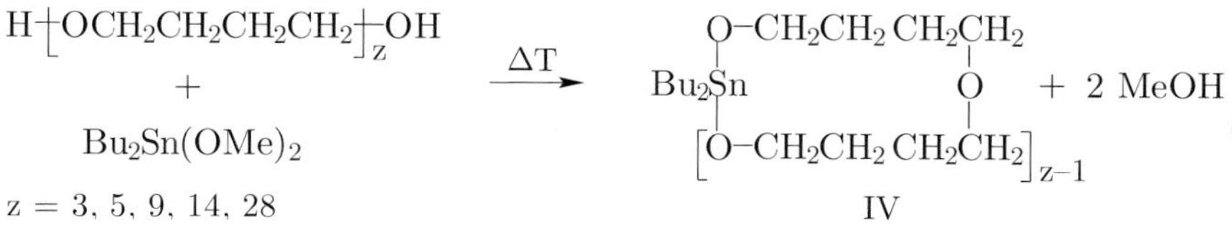

$$H{-}[OCH_2CH_2CH_2CH_2]_z{-}OH + Bu_2Sn(OMe)_2 \xrightarrow{\Delta T} \text{IV: } Bu_2Sn\begin{cases}O{-}CH_2CH_2\,CH_2CH_2\\ [O{-}CH_2CH_2\,CH_2CH_2]_{z-1}\end{cases}\text{(ring closed via O)} + 2\,MeOH \quad (58)$$

$z = 3, 5, 9, 14, 28$

Second step: IV reacts with ε-caprolactone ($z = 28$) to give macrocycle V

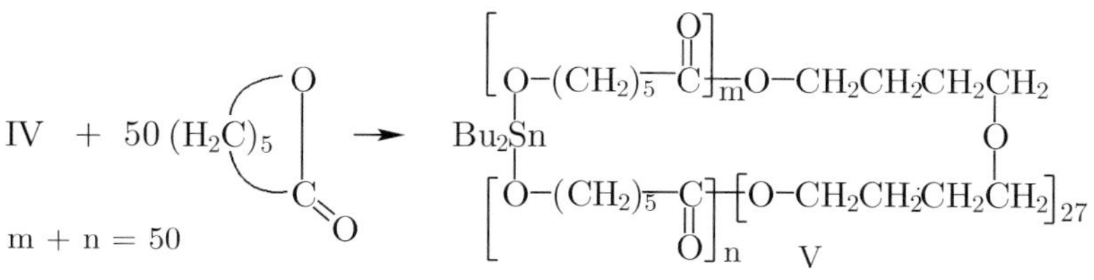

$$\text{IV} + 50\,(H_2C)_5\!\left<\begin{matrix}O\\ C{=}O\end{matrix}\right. \longrightarrow \text{V: } Bu_2Sn\begin{cases}[O{-}(CH_2)_5{-}C(=O)]_m{-}O{-}CH_2CH_2CH_2CH_2\\ [O{-}(CH_2)_5{-}C(=O)]_n{-}[O{-}CH_2CH_2CH_2CH_2]_{27}\end{cases}\text{(ring closed via O)} \quad (59)$$

$m + n = 50$

Third step: V reacts with terephthaloyl dichloride

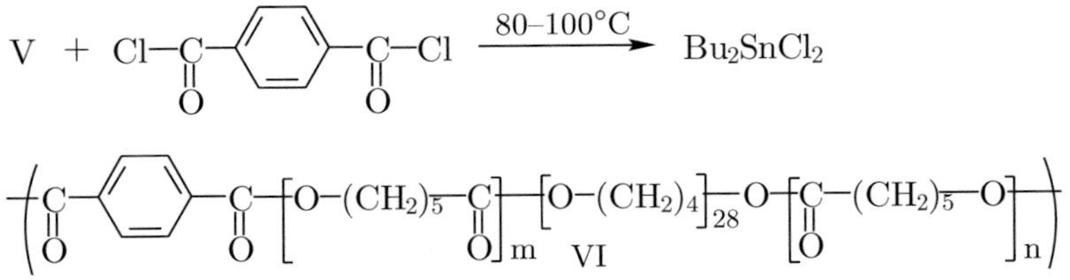

$$\mathrm{V} + \mathrm{Cl{-}C(=O){-}C_6H_4{-}C(=O){-}Cl} \xrightarrow[\quad]{80\text{–}100^\circ\mathrm{C}} \mathrm{Bu_2SnCl_2}$$

$$\left(\!\!-\mathrm{C(=O){-}C_6H_4{-}C(=O)}\left[\mathrm{O{-}(CH_2)_5{-}C(=O)}\right]_m\left[\mathrm{O{-}(CH_2)_4}\right]_{28}\mathrm{O}\left[\mathrm{C(=O){-}(CH_2)_5{-}O}\right]_n\right) \quad \mathrm{VI} \tag{60}$$

The characteristics of VI depend on the reaction time and temperature; after 6 h at 200°C, $\overline{M}_n = 56\ 000$ and $\overline{M}_w/\overline{M}_n = 1.8$; at 220°C, $\overline{M}_n = 42\ 000$ and $\overline{M}_w/\overline{M}_n = 1.6$.

Reaction (60) was also carried out with 1,4-butanedioyl dichloride (Scheme 61) or 1,10-decanedioyl dichloride (Scheme 62).

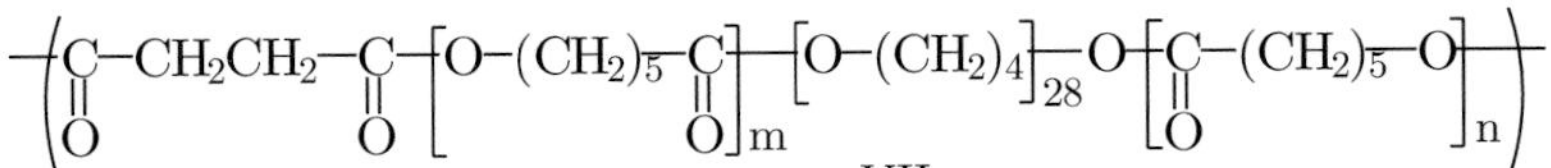

$$\left(\!\!-\mathrm{C(=O){-}CH_2CH_2{-}C(=O)}\left[\mathrm{O{-}(CH_2)_5{-}C(=O)}\right]_m\left[\mathrm{O{-}(CH_2)_4}\right]_{28}\mathrm{O}\left[\mathrm{C(=O){-}(CH_2)_5{-}O}\right]_n\right) \quad \mathrm{VII} \tag{61}$$

$$\left(\!\!-\mathrm{C(=O){-}(CH_2)_8{-}C(=O)}\left[\mathrm{O{-}(CH_2)_5{-}C(=O)}\right]_m\left[\mathrm{O{-}(CH_2)_4}\right]_{28}\mathrm{O}\left[\mathrm{C(=O){-}(CH_2)_5{-}O}\right]_n\right) \quad \mathrm{VIII} \tag{62}$$

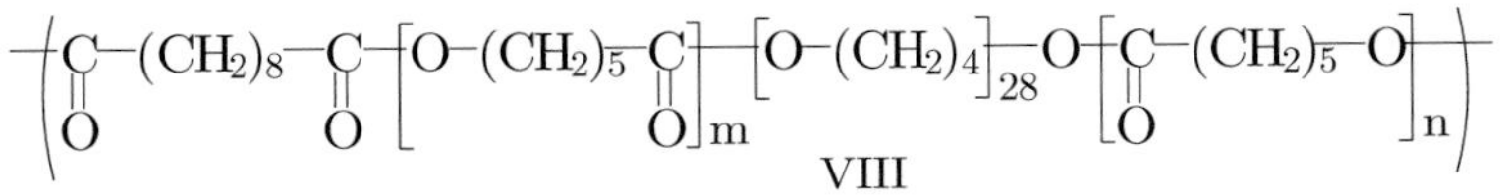

As in the case of VI, $\overline{M}_n$ and $\overline{M}_w/\overline{M}_n$ are temperature-dependent; their values are 35 000–50 000 and 1.55-1.65 for VII, and 42 000–63 000 and 1.6–1.7 for VIII, respectively.

It is worth noting that dibutyl dimethoxide can react with a polymer, such as α,ω-dihydroxy-poly(oxyethylene) to form a macrocycle, rather than a linear polymer (Scheme 63) [73].

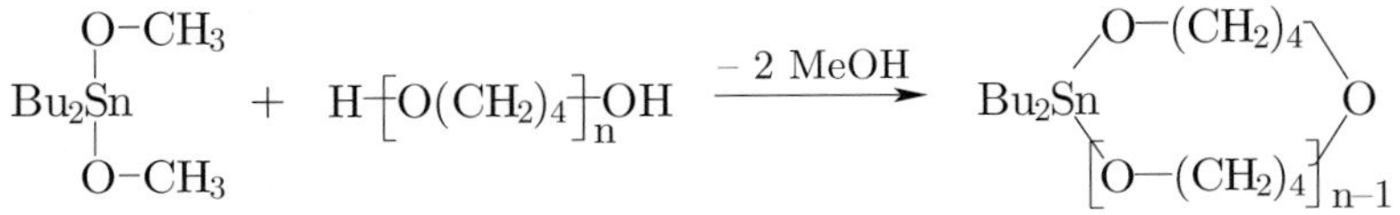

$$\mathrm{Bu_2Sn(O{-}CH_3)_2} + \mathrm{H}\left[\mathrm{O(CH_2)_4}\right]_n\mathrm{OH} \xrightarrow{-\,2\,\mathrm{MeOH}} \mathrm{Bu_2Sn}\!\!\left\langle\begin{matrix}\mathrm{O{-}(CH_2)_4{-}O}\\ \left[\mathrm{O{-}(CH_2)_4}\right]_{n-1}\end{matrix}\right. \tag{63}$$

Many authors [76–80] studied the ring-opening polymerization of morpholine-2,5-diones initiated by various species, *e.g.*, 2,2-dibutyl-2-stanna-1,3-dioxacyclobutane (Scheme 64) [76]. This holds for both homo- and copolymerization.

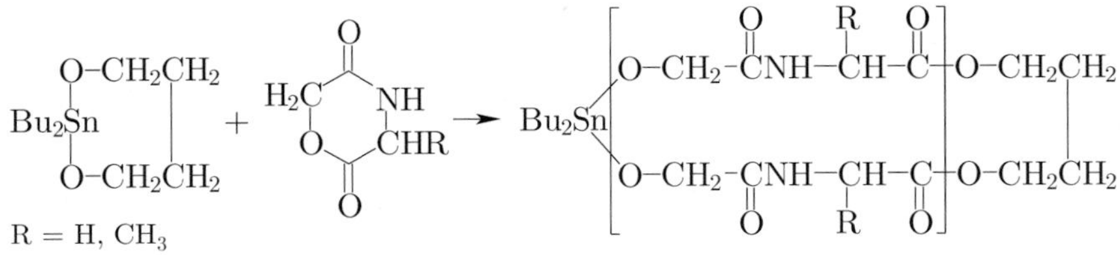

$$\mathrm{Bu_2Sn}\!\left\langle\begin{matrix}\mathrm{O{-}CH_2CH_2}\\ \mathrm{O{-}CH_2CH_2}\end{matrix}\right| + \text{morpholine-2,5-dione } (\mathrm{H_2C,\ NH,\ CHR,\ O}) \longrightarrow \mathrm{Bu_2Sn}\!\left\langle\begin{matrix}\left[\mathrm{O{-}CH_2{-}C(=O)NH{-}CH(R){-}C(=O)}\right]\mathrm{O{-}CH_2CH_2}\\ \left[\mathrm{O{-}CH_2{-}C(=O)NH{-}CH(R){-}C(=O)}\right]\mathrm{O{-}CH_2CH_2}\end{matrix}\right| \tag{64}$$

R = H, CH_3

Thiolactones can be quantitatively inserted in the Sn–O bond of Sn-containing macrocycles [71,75,81,82] since the Sn–S bond is more stable than Sn–O (Scheme 65) [75].

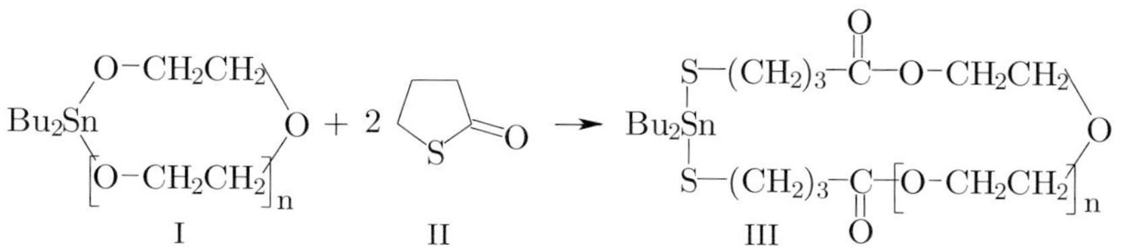

$$\mathrm{Bu_2Sn}\begin{matrix}\mathrm{O-CH_2CH_2} \\ [\mathrm{O-CH_2CH_2}]_n\end{matrix}\!\!\!>\mathrm{O} + 2\ \text{(thiobutyrolactone)} \longrightarrow \mathrm{Bu_2Sn}\begin{matrix}\mathrm{S-(CH_2)_3-C(=O)-O-CH_2CH_2} \\ \mathrm{S-(CH_2)_3-C(=O)-[O-CH_2CH_2]_n}\end{matrix}\!\!\!>\mathrm{O} \qquad (65)$$

I II III

The reaction of 2,2-dibutyl-2-stanna-l,3-dioxycycloalkanes with thiobutyrolactone leads to a macrocycle; its formula fits the reactant stoichiometric balance ([VI] = 2 [IV]) (Scheme 66) [71,81].

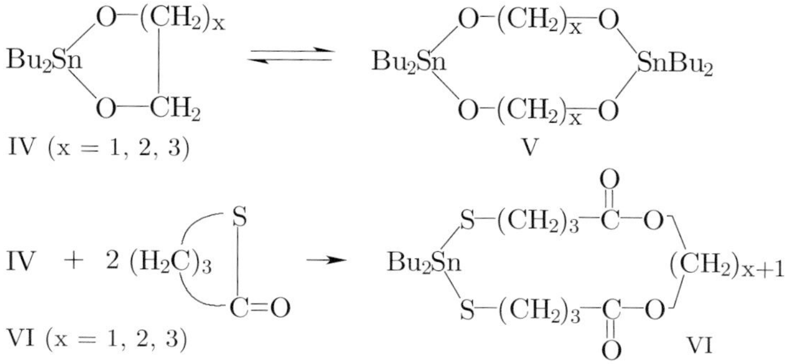

$$\mathrm{Bu_2Sn}\begin{matrix}\mathrm{O-(CH_2)_x} \\ \mathrm{O-CH_2}\end{matrix} \rightleftharpoons \mathrm{Bu_2Sn}\begin{matrix}\mathrm{O-(CH_2)_x-O} \\ \mathrm{O-(CH_2)_x-O}\end{matrix}\mathrm{SnBu_2}$$

IV (x = 1, 2, 3) V

$$\mathrm{IV} + 2\,\mathrm{(H_2C)_3}\begin{matrix}\mathrm{S} \\ \mathrm{C=O}\end{matrix} \longrightarrow \mathrm{Bu_2Sn}\begin{matrix}\mathrm{S-(CH_2)_3-C(=O)-O} \\ \mathrm{S-(CH_2)_3-C(=O)-O}\end{matrix}\mathrm{(CH_2)_{x+1}} \qquad (66)$$

VI (x = 1, 2, 3) VI

On the other hand, when the same reaction is carried out in thiocaprolactone in excess, an insertion takes place (Scheme 67) [81].

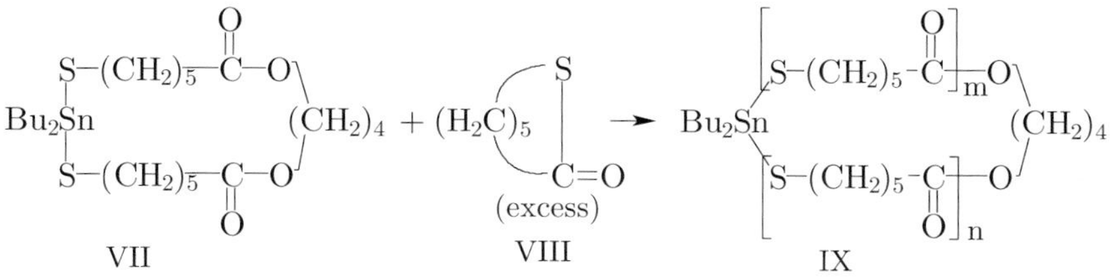

$$\mathrm{Bu_2Sn}\begin{matrix}\mathrm{S-(CH_2)_5-C(=O)-O} \\ \mathrm{S-(CH_2)_5-C(=O)-O}\end{matrix}\mathrm{(CH_2)_4} + \mathrm{(H_2C)_5}\begin{matrix}\mathrm{S} \\ \mathrm{C=O}\end{matrix}\ \text{(excess)} \longrightarrow \mathrm{Bu_2Sn}\begin{matrix}\mathrm{[S-(CH_2)_5-C(=O)]_m-O} \\ \mathrm{[S-(CH_2)_5-C(=O)]_n-O}\end{matrix}\mathrm{(CH_2)_4} \qquad (67)$$

VII VIII IX

When I reacts with the cyclic diesters of dimercaptoethane, the products are not stable, they split into dioxalane (X) and poly(ether-ester)s are formed (Scheme 68) [75].

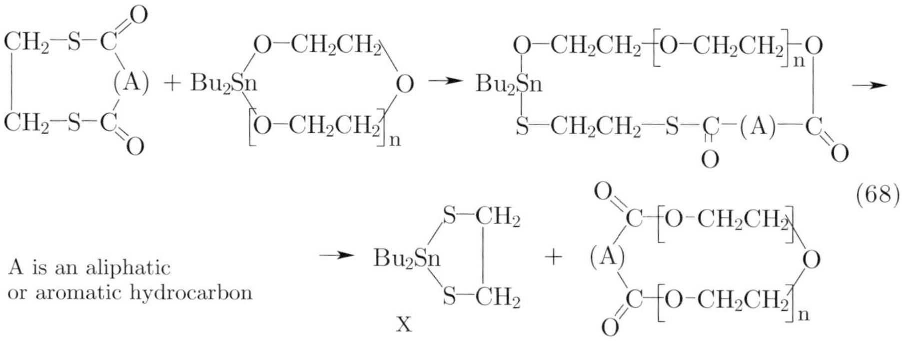

$$\begin{matrix}\mathrm{CH_2-S-C(=O)} \\ \mathrm{CH_2-S-C(=O)}\end{matrix}\!\!>\mathrm{(A)} + \mathrm{Bu_2Sn}\begin{matrix}\mathrm{O-CH_2CH_2} \\ [\mathrm{O-CH_2CH_2}]_n\end{matrix}\!\!>\mathrm{O} \longrightarrow \mathrm{Bu_2Sn}\begin{matrix}\mathrm{O-CH_2CH_2-[O-CH_2CH_2]_n-O} \\ \mathrm{S-CH_2CH_2-S-C(=O)-(A)-C(=O)}\end{matrix} \longrightarrow$$

$$\longrightarrow \mathrm{Bu_2Sn}\begin{matrix}\mathrm{S-CH_2} \\ \mathrm{S-CH_2}\end{matrix} + \mathrm{(A)}\begin{matrix}\mathrm{(O=)C-[O-CH_2CH_2]} \\ \mathrm{(O=)C-[O-CH_2CH_2]_n}\end{matrix}\!\!>\mathrm{O} \qquad (68)$$

X

A is an aliphatic or aromatic hydrocarbon

In the same way, when I reacts with a thiospirophthalide, the insertion product is unstable, the dioxalane X is eliminated and a cyclic polyoxyethylene phthalate is formed (Scheme 69) [75].

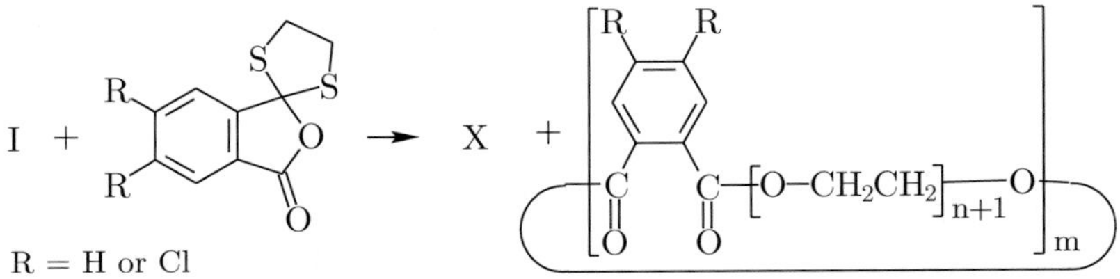

$$\text{I} + \text{(R,R-thiospirophthalide)} \rightarrow \text{X} + \left[\text{C(=O)-C}_6\text{H}_2\text{R}_2\text{-C(=O)}-\left[\text{O-CH}_2\text{CH}_2\right]_{n+1}\text{O} \right]_m \quad (69)$$

R = H or Cl

All these reactions were also applied to α,ω-dihydroxypropyl-polysiloxanes [82,83] (Scheme 70).

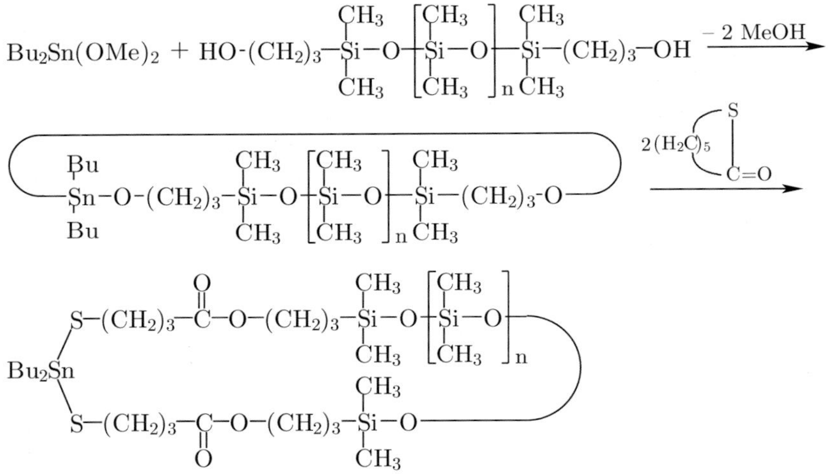

$$\text{Bu}_2\text{Sn(OMe)}_2 + \text{HO-(CH}_2)_3\text{-Si(CH}_3)_2\text{-O}\left[\text{Si(CH}_3)_2\text{-O}\right]_n\text{Si(CH}_3)_2\text{-(CH}_2)_3\text{-OH} \xrightarrow{-2\,\text{MeOH}}$$

$$\left(\text{-Sn(Bu)}_2\text{-O-(CH}_2)_3\text{-Si(CH}_3)_2\text{-O}\left[\text{Si(CH}_3)_2\text{-O}\right]_n\text{Si(CH}_3)_2\text{-(CH}_2)_3\text{-O-}\right) \xrightarrow{2\,(\text{H}_2\text{C})_5\text{(S)(C=O)}}$$

$$\text{Bu}_2\text{Sn}\left(\text{S-(CH}_2)_3\text{-C(=O)-O-(CH}_2)_3\text{-Si(CH}_3)_2\text{-O}\left[\text{Si(CH}_3)_2\text{-O}\right]_n \ \ldots \ \text{S-(CH}_2)_3\text{-C(=O)-O-(CH}_2)_3\text{-Si(CH}_3)_2\text{-O}\right) \quad (70)$$

3.6. *The use of microorganisms*

The bacterial polymers raise increased interest due to their applications in thermoplastics, elastomers, adhesives, and in biocompatible and biodegradable materials. Poly(3-hydroxyalkanoate)s are extensively studied and some of them have properties, which make them potential TPEs. Their biochemistry was reviewed by Fuller, more particularly the use of photosynthetic bacteria [84], and more recently Kim and Lenz [85] classified poly(hydroxyalkanoate)s according to the length of their macromolecules. Poly(3-hydroxybutyrate), poly(3-hydroxypentanoate), and poly(3-hydroxybutyrate-*co*-3-hydroxypentanoate) are short-length polyesters. The chains of the medium-length poly(3-hydroxyalkanoate)s contain either higher 3-hydroxy-*n*-alkanoate units or unsaturated, aromatic or halogenated constitutive units; some of these units can be present in the same chain.

Poirier and Nawrath [86,87] and Poirier and Gruys [88] provided major information on the production of these polymers. Most poly(hydroxyalkanoate)s are built up of constitutive units deriving from R(−)-3-hydroxyalkanoic acids. The structure of poly(3-hydroxybutyrate) is given in Scheme 71.

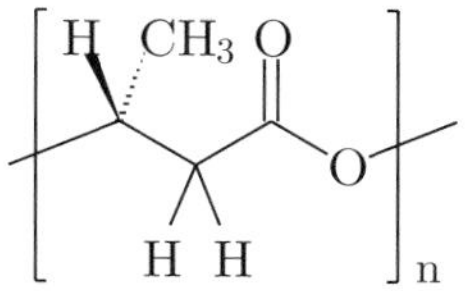

$$\left[-CH(CH_3)-CH_2-C(=O)-O- \right]_n \qquad (71)$$

Poly(hydroxyalkanoate)s can be prepared from a large range of bacteria, *e.g.*, the system poly(hydroxybutanoate)/*Alcaligene eutrophus* (*A. eutrophus*) was extensively studied. On the other hand, medium-sized poly(hydroxyalkanoate)s were essentially prepared by *Pseudomonas oleovorans* (*P. oleovorans*).

When *A. eutrophus* is used to prepare poly(hydroxybutanoate), the bacteria grow in media containing excess carbon, such as glucose, but limited in one essential nutrient, such as phosphate. Under these experimental conditions, the poly(hydroxybutanoate) inclusions can accumulate up to 80–90% dry mass of the bacteria with a molar mass of 10^3 to 10^4.

When *P. oleovorans* is used, the type of poly (hydroxyalkanoate)s essentially depends on the nature of the carbon source. For instance, when the substrate is octanoic acid, the product is a random copolymer containing 89 mole% of C8 units and 11 mole% of C6 units. On the other hand, dodecanoic acid generates copolymers containing C12 (31 mole%), C10 (36 mole%), C8 (31 mole%), and C6 (2 mole%) units [89].

Kim and Lenz [85] proposed a classification of the homopolymers and copolymers prepared from hydroxyalkanoic acids (Figure 1).

The poly(HASCL) containing units other than 3HB and 3HV include different tercopolymers, such as poly(3HB-*co*-3HV-*co*-5HV) *(R. eutropha)* [90] and poly(3HB-*co*-4HB) (*R. eutropha*) [91]. Poly(3HB-*co*-3HV-*co*-5HV) and poly-(3HB-*co*-4HB) are less crystalline and more biodegradable than poly(3HB-*co*-4HB). 3-Hydroxy-2,2-dimethyl-propanoic acid was polycondensed *(Rhodocodus ruber)* [92], as well as 3-pentenoic acid *(Burkolderia)* [93] and 4-pentanoic acid *(Rhodospirilbum)* [94]; these acids were also copolycondensed with 3-hydroxy-butanoate, 3-hydroxy-pentanoate, and 3-hydroxy-2,2-dimethyl-propanoic acid.

Most poly(HAMCL) are prepared from *P. oleovorans* and *P. putida* which are grown from alkanes, alkenes, and carboxylic acids. The structures of these polymers were studied by different techniques, such as ^{13}C labeling [95] and X-ray diffraction. The latter showed that the PHAs prepared from octanoic, nonaoic, and decanoic acids have comb-like structures [96]. *P. putida* can grow not only on alkyl substances, but also on compounds such as glucose; the polymers contain both saturated and unsaturated units [97–99]. Substituted PHAs were prepared from *P. oleovorans* grown on different substrates. Polymers with unsaturated side groups were obtained from double or triple carbon–carbon bond-containing substrates (alkenes or unsaturated acids, such as undecylenic or 10-undecynoic acids) [100]. PHAs with aromatic side groups were obtained by growth of *P. oleovorans* on 5-phenylpentanoic acid [101], many

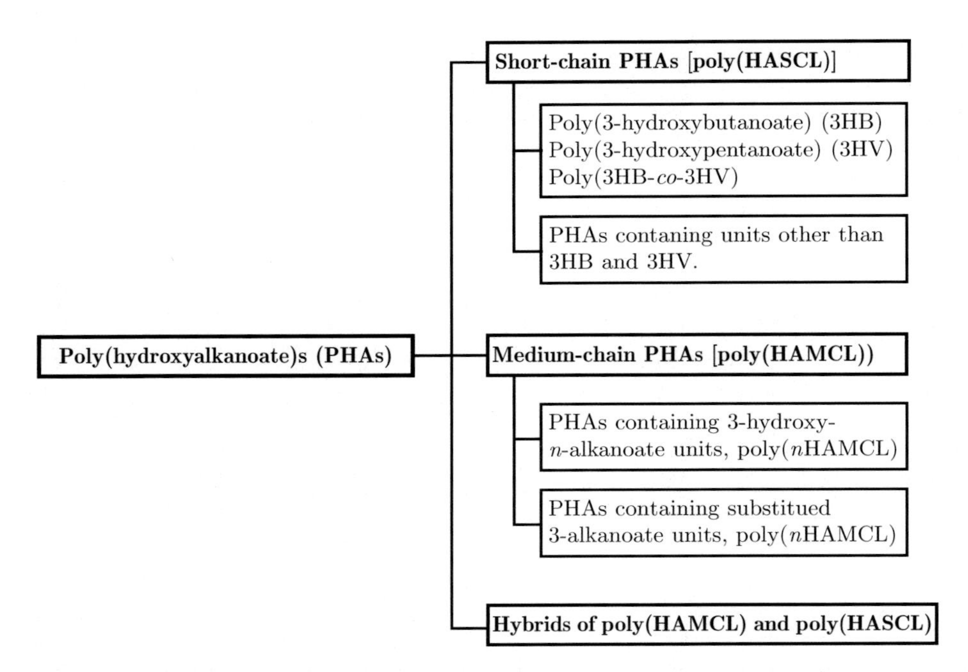

Figure 1. Classification of poly(hydroxyalkanoate)s [85]

other aromatic PHAs were prepared with phenoxy, methylphenoxy, nitroxyphenoxy, or cyanophenoxy groups as substituents. Halogenated PHAs were obtained from *P. putida* growing on either mixtures of alkanes and chloro- or fluoroalkanes or mixtures of alkanoic acids and halogenated alkanoic acids.

These polyesters have a high molar mass and are isotactic, they exhibit crystalline properties and are biodegradable, they can be either thermoplastics or elastomers.

Copolymers containing poly(3-hydroxyalkanoate) and polyether blocks were prepared. Marchessault and Yu [102] prepared block copolymers of poly(3-hydroxybutyrate) and polyoxyethylene by coupling α-methoxy-oligooxyethylenes with α-hydroxy-ω-carboxylic-oligo(3-hydroxybutyrates) using 1,3-dicyclohexylcarbodiimide as a coupling agent. The telechelic poly(hydroxybutyrate)s were obtained by non-random hydrolysis of poly(hydroxybutyrate) in a heteregeneous medium, which generates well defined difunctional oligomers (Scheme 72).

$$\mathrm{HO{-}\overset{\overset{\large CH_3}{|}}{C}H{-}CH_2{-}\overset{\overset{\large O}{\|}}{C}{-}\left[O{-}\overset{\overset{\large CH_3}{|}}{C}H{-}CH_2{-}\overset{\overset{\large O}{\|}}{C}\right]_n{-}O{-}\overset{\overset{\large CH_3}{|}}{C}H{-}CH_2{-}\overset{\overset{\large O}{\|}}{C}{-}OH} \qquad (72)$$

4. Conclusions

Over the past few years, new routes were proposed to prepare block copolymers, such as the carbon–carbon polycondensations catalyzed by metal derivatives

[103–105] or the use of enzymatic catalysis in organic media [106]. However, for the time being, the sequential reactions are the more promising techniques and some of them are either already commercialized or under development. The use of sequential reactions in the synthesis of block copolymers was analyzed by Maréchal [2]; they create the basis of important synthetic strategies, such as anionic-to-radical polymerization, radical-to-cationic polymerization, metasynthesis-to-radical polymerization, *etc.* The polycondensation in sequential reactions is often associated to chemical modification.

The characterization of the block copolymers has changed significantly, especially by the association of several techniques [107]. Many of them were frequently mentioned throughout this chapter since they are inseparable from synthesis. The techniques used to determine the molar masses, the nature of the blocks, and that of the end-groups were critically analyzed and NMR, mass spectrometry, SEC, and selective chemical degradation led to important breakthroughs in the knowledge of the structure of the chains. From this point of view, enzymatic catalysis would be useful as it was shown by Valiente *et al.* [108], who succeeded in degrading a block without modifying the other one. Murgasova *et al.* [109] characterized the soft and hard blocks of a polyesterurethane (I) by a combination of MALDI, SEC, and chemical degradation. They used isocyanatolysis as the degradation step where I is reacted with phenylisocyanate, which led to a mixture of different oligomers mainly formed of the soft blocks (Scheme 73).

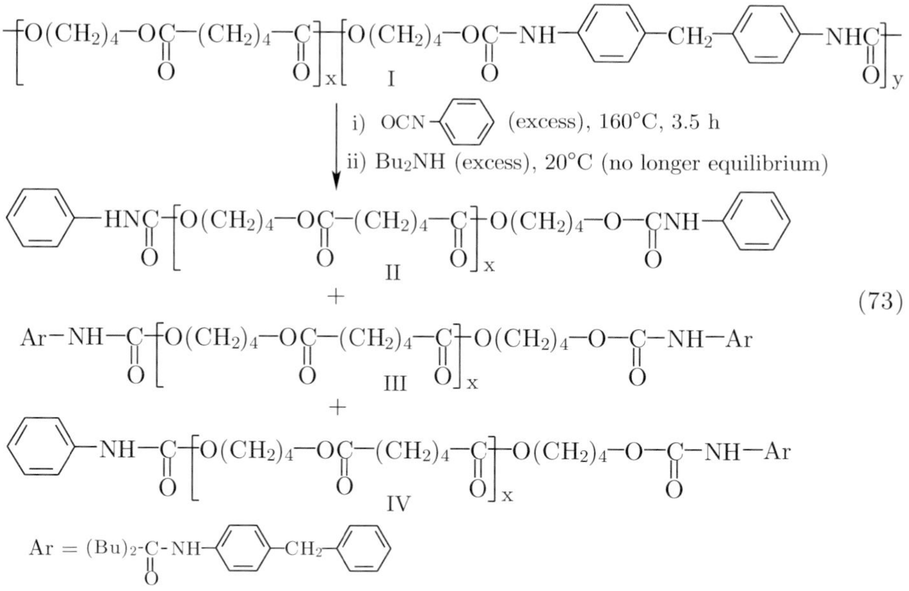

$$\left[O(CH_2)_4-O\overset{O}{\overset{\|}{C}}-(CH_2)_4-\overset{O}{\overset{\|}{C}}\right]_x\left[O(CH_2)_4-O\overset{O}{\overset{\|}{C}}-NH-C_6H_4-CH_2-C_6H_4-NH\overset{O}{\overset{\|}{C}}\right]_y \quad \text{I}$$

$$\downarrow \text{i) } OCN-C_6H_5 \text{ (excess), 160°C, 3.5 h} \quad \text{ii) } Bu_2NH \text{ (excess), 20°C (no longer equilibrium)}$$

$$C_6H_5-HN\overset{O}{\overset{\|}{C}}\left[O(CH_2)_4-O\overset{O}{\overset{\|}{C}}-(CH_2)_4-\overset{O}{\overset{\|}{C}}\right]_x O(CH_2)_4-O-\overset{O}{\overset{\|}{C}}NH-C_6H_5 \quad \text{II}$$

$$+ \qquad\qquad (73)$$

$$Ar-NH-\overset{O}{\overset{\|}{C}}\left[O(CH_2)_4-O\overset{O}{\overset{\|}{C}}-(CH_2)_4-\overset{O}{\overset{\|}{C}}\right]_x O(CH_2)_4-O-\overset{O}{\overset{\|}{C}}-NH-Ar \quad \text{III}$$

$$+$$

$$C_6H_5-NH-\overset{O}{\overset{\|}{C}}\left[O(CH_2)_4-O\overset{O}{\overset{\|}{C}}-(CH_2)_4-\overset{O}{\overset{\|}{C}}\right]_x O(CH_2)_4-O-\overset{O}{\overset{\|}{C}}-NH-Ar \quad \text{IV}$$

$$Ar = (Bu)_2\text{-}\overset{O}{\overset{\|}{C}}\text{-}NH-C_6H_4-CH_2-C_6H_5$$

The oligomers I, II, and III were characterized by SEC and MALDI mass spectrometry. Hard and soft blocks were isolated by acid hydrolysis and then identified by MALDI and SEC analysis.

The characterization of the chains is essential, but it must be complemented by knowledge of their interaction and self-organization. This is the subject of numerous publications, especially the segregation of the blocks into different phases [110–112]. The number of articles dealing with the determination of the phase organization is rapidly increasing [113].

The knowledge of the morphology of the block copolymers makes a great progress. Numerous studies are theoretical, but they are often coupled with one or several experimental techniques. Hashimoto *et al.* [114] used SAXS to investigate the hexagonally packed cylindrical particles with paracrystalline distortion by comparing experimental and theoretical values. Stocker *et al.* [115] associated transmission electron microscopy to atomic force microscopy or SAXS and theoretical calculations or electron microscopy to treat surface problems [116] (see also Chapters 6 and 7).

The deep insight in the structure of the chains is often accompanied by little attention to their association in the solid state; on the other hand, some authors are only concerned by the material without paying great attention to the macromolecule itself. However, some articles associate these two aspects, as did Furukawa *et al.* [117], who prepared polysiloxane-*b*-polyamide by solution polycondensation (xylene/NMP) of 3,3',4,4'-diphenylsulfonetetracarboxylic dianhydride, 2,2-bis[4-(4-aminophenoxy)phenyl]propane, and α,ω-diamino-polysiloxane (the synthesis belongs to the techniques listed in Section 2.3). The chains were analyzed by ^{1}H, ^{13}C, and ^{29}Si NMR, as well as by infrared spectroscopy, whereas the spin-lattice relaxation time was measured by solid-state NMR. A kinetic study of the imidization was carried out in NMP solution; it showed that the value of the activation energy is partly determined by the solvation of the amide groups and polyamic acids.

The contribution of theoretical studies is rapidly growing, especially the use of self-consistent schemes associated with experimental studies [118–121]. Many of the calculations are relative to the dynamics of the copolymer chains, to order-disorder problems and to diffusion problems [122,123]. The simulation in block copolymer studies will rapidly develop in the near future, both in static and dynamic analyses [124,125].

References

1. Lodge T P (2003) Block copolymers: past successes and future challenges, *Macromol Chem Phys* **204**:265–273.
2. Maréchal E (2000) Block copolymers with polyethers as flexible blocks, Future trends in block copolymers, in *Block Copolymers* (Eds. Baltá Calleja F J and Roslaniec Z) M. Dekker, pp. 29–63, 531–572.
3. Jenkins A D, Kratochvil P, Stepto R F T and Suter U W (1996) Glossary of basic terms in polymer science, *Pure Appl Chem* **68**:2288–2311.
4. Boularès A, Rozès M, Tessier M and Maréchal E (1997) Structure and properties of block copolymers with polyethers as flexible blocks, *Europhysics Conf Abstr* **21A**:24.

5. Boularès A, Tessier M and Maréchal E (1998) Synthesis and characterization of poly-(copolyether-*block*-polyamides); I. Structural study of polyether precursors, *J Macromol Chem* **A35**:933–953.
6. Laporte P, Fradet A and Maréchal E (1987) Kinetics study on models of the noncatalyzed reaction between α,ω-dicarboxy-polyamide-11 and α,ω-dihydroxy-polyoxyalkylenes, *J Macromol Sci–Chem* **A24**:1269–1287.
7. P. Laporte, A. Fradet and E. Maréchal (1987) Kinetics study on models of the tetrabutoxyzirconium-catalyzed reaction between α,β-dicarboxy-polyamide-11 and α,ω-dihydroxy-polyoxyalkylenes, *J Macromol Sci–Chem* **A24**:1289–1302.
8. Michot C, Baril D and Armand M (1995) Polyimide-polyether mixed conductors as switchable materials for electrochromic devices, *Solar Energy Mater and Solar Cells* **39**:289–299.
9 Einsenbach C D and Heinlein J (1996) Synthesis and properties of block copolymers consisting of liquid crystalline/amorphous segment, *ACS Polym Prepr* **37**(1):75–76.
10. Pospiech D, Häussler L, Eckstein K, Voigt D, Jehnichen D, Gottwald A, Kollig W, Janke A, Grundke K, Werner C and Kricheldorf H R (2001) Tailoring of polymer properties in segmented block copolymers, *Macromol Symp* **163**:113–126.
11. Bloom P D and Sheares V V (2001) Synthesis of poly(*p*-phenylene) macromonomers and multiblock copolymers, *J Polym Sci A Polym Chem* **39**:3505–3512.
12. Rodewald B and Ritter H (1999) Oligo(ethersulfones). 3 Block copolymers *via* condensation reactions of telechelic oligo(ethersulfones) bearing phenolic end-groups and oligomeric α,ω-diols, *Macromolecules* **32**:1697–1700.
13. Nuyken O (1994) Massgeschneiderte Polymere, *Angew Makromol Chem* **223**:29–46.
14. Thuillier P, Tessier M and Maréchal E (1993) Synthesis and characterization of thermotropic polyester-*block*-polyethers, *Makromol Chem Macromol Symp* **70/71**:37–45.
15. Thuillier P, Tessier M and Maréchal E (1994) Synthesis and characterization of thermotropic polyester-*block*-polyethers, *Molec Cryst Liquid Cryst A* **254**:1–16.
16. Gaymans R J, Schwering P and de Haan J L (1989) Nylon 46-polytetramethylene oxide segmented block copolymers, *Polymer* **30**:974–977.
17. Deleens G, Foy P and Maréchal E (1977) Synthèse et caractérisation de copolycondensats séquencés (polyamide-*seq*-ether) I Synthese et étude de divers oligomères ω,ω'-difonctionnels du poly(amide-11), *Eur Polym J* **13**:337–342.
18. Deleens G, Foy P and Maréchal E (1977) Synthèse et caractérisation de copolycondensats séquencés (polyamide-*seq*-ether) II Polycondensation d'oligomeres polyamides-11 ω,ω'-diacides ou diesters avec des oligomères polyethers ω,ω'-dihydroxy, *Eur Polym J* **13**:343–351.
19. Deleens G, Foy P and Maréchal E (1977) Synthèse et caractérisation de copolycondensats séquencés (polyamide-*seq*-ether) III Etude de la réaction de polycondensation du polyamide-11 ω,ω'-dicarboxylique et du polyoxyethylene ω,ω'-dihydroxy. Détermination des constantes de vitesse et de l'énergie d'activation, *Eur Polym J* **13**:353–360.
20. McCarthy S J, Meijs G F and Gunatillake P (1997) Synthesis, characterization and stability of poly[(alkylene oxide) ester] thermoplastic elastomers, *J Appl Polym Sci* **65**:1319–1332.
21. Kozlowska A and Slonecki J (1998) Segmented block copolymers based on oligoamides of dimerized fatty acid and poly(butylene terephthalate), *Polymery* **43:**188–191.
22. El Fray M and Slonecki J (1999) Dimer fatty acid-modified poly [ester-*b*-ether]s. Synthesis and properties, *Polym-Plast Technol Eng* **38**:51–69.
23. Aleksandrovic V and Djonlagic J (2001) Synthesis and characterization of thermoplastic copolyester elastomers modified with fumaric moieties, *J Serb Chem Soc* **66**:139–152.

24. Antic V V, Balaban M R and Djonlagic J (2001) Synthesis and characterization of thermoplastic poly(ester-siloxane)s, *Polym Int* **50**:1201–1208.
25. Jeong H M, Moon S W, Jho J Y and Ann T O (1998) Phase structure and properties of some thermoplastic polyesteramide elastomers, *Polymer* **39**:459–465.
26. Kang E-C, Kaneko T, Shiino D and Akashi M (2003) Novel functional polymers: poly(dimethylsiloxane)-polyamide multiblock copolymers. XI. The effects of sequence regularity on the thermal and mechanical properties, *J Polym Sci A Polym Chem* **41**:841–852.
27. Fradet A (1996) Coupling reactions in polymer synthesis, in *Comprehensive Polymer Science* (Eds. Aggarwal S L and Russo S) Oxford: Pergamon 2nd Suppl. pp. 133–162.
28. Tsai R-S, Lee Y-D and Tsai H-B (2002) Block copolyesters of poly(pentamethylene-*p,p*'-bibenzoate) and poly(tetramethylene adipate), *J Polym Sci A Polym Chem* **40**:2626–2636.
29. Inata H and Matsumura S (1987) Chain extenders for polyesters. V. Reactivities of hydroxyl-addition-type chain extender; 2,2'-bis(4H-3,l-benzoxazin-4-one), *J Appl Polym Sci* **34**:2609–2617.
30. Acevedo M and Fradet A (1993) A chain coupling reaction of α,ω-dihydroxy-poly(oxy-tetramethylene) by bisoxazolones, *ACS Polym Prepr* **34**(1):457–458.
31. Pollock G S, Ko M S, Kojic N, James-Korley L T, McKinley G H and Hammond P (2002) Synthesis and characterization and resin spinning of segmented polyurethanes with copolymerized hard segments, *ACS Polym Prepr* **43**(2):397–398.
32. Rasmussen J K, Krepski L R, Heilmann S M, Sakizadeh K, Smith II H K and Katrizky A R (1986) Polyimidazolinones *via* thermal cyclodehydration of polyamides containing α-aminoacids, *ACS Polym Prepr* **27**(2):17–20.
33. Acevedo M and Fradet A (1993) Study of bulk chain coupling reactions. II. Reaction between bisoxazolones and amine-terminated polyether: synthesis of polyether-*block*-polyamides, *J Polym Sci A Polym Chem* **31**:1579–1588.
34. Jan J Z, Huang B H and Lin J J (2003) Facile preparation of amphiphilic oxyethylene-oxypropylene block copolymers by selective triazine coupling, *Polymer* **44**:1003–1011.
35. Fischer T, Lefèbvre H and Fradet A (1997) Synthesis of block copolymers by oligomer-coupling reactions in the bulk: bisimidazoline/COOH and bisoxazolinone/NH_2 reactions, *Macromol Symp* **118**:79–87.
36. Heatley F, Luo Y Z, Ding J-F, Mobbs R H and Booth C (1988) A ^{13}C nuclear magnetic resonance study of the triad sequence structure of block and statistical copolymers of ethylene oxide and propylene oxide, *Macromolecules* **21**:2713–2721.
37. Rozès L (1996) Synthesis of thermotropic poly(polyester-*block*-polyether)s, Thesis, University of Paris.
38. Thuillier P (1992) Synthesis and characterization of mesogenic groups containing polyesters-*block*-polyethers, Thesis, University of Paris.
39. Otsuki T, Kakimoto M-A and Imai Y (1990) Synthesis and properties of multiblock copolymers based on polyoxyethylene and polyamides by diisocyanate method, *J Appl Polym Sci* **40**:1433–1443.
40. Hatfield G R, Guo Y, Killinger W E, Andrejack R A and Roubicek P M (1993) Characterization of structure and morphology in two poly(ether-*block*-amide) copolymers, *Macromolecules* **26**:6350–6353.
41. Hatfield G R, Bush R W, Killinger W E and Roubicek P M (1994) Phase modification and polymorphism in two poly(ether-*block*-amide) copolymers, *Polymer* **35**:3943–3947.
42. Mathias L J and Johnson C G (1991) Solid-state NMR investigation of Nylon 12, *Macromolecules* **24**:6114–6122.

43. Leblanc J-P, Tessier M, Judas D, Friedrich C, Noël C and Maréchal E (1993) Aromatic copolyesters with stilbene mesogenic groups. 1. Liquid crystalline properties of compounds containing a stilbene, terephthaloyl, or hydroquinone central group, *Macromolecules* **26**:4391–4399.
44. Leblanc J-P, Tessier M, Judas D, Friedrich C, Noël C and Maréchal E (1995) Aromatic copolyesters with stilbene mesogenic groups, 2. Synthesis and thermal behavior, *Macromolecules* **28**:4837–4850.
45. Hsu T-F and Lee Y-D (1999) Properties of thermoplastic polyurethane elastomers containing liquid crystalline chain extender (I) Synthesis and properties of hard segments, *Polymer* **40**:577–587.
46. Jeong H M, Kim B K and Choi Y J (2000) Synthesis and properties of thermotropic liquid crystalline polyurethane elastomers, *Polymer* **41**:1849–1855.
47. Pollock G S and Hammond P T (2002) Synthesis and characterization of silk-like polyurethanes with main-chain liquid crystalline soft segments, *ACS Polym Prepr* **43**(1):478–479.
48. Bai S and Zhao Y (2002) Azobenzene elastomers for mechanically tunable diffraction gratings, *Macromolecules* **35**:9657–9664.
49. Sänger J and Gronski W (1998) Thermoplastic liquid crystalline triblock elastomers, 1. Synthesis and phase behaviour, *Macromol Chem Phys* **199**:555–561.
50. Gronski W and Sänger J (2001) Liquid crystalline triblock copolymers. Mechanical behaviour and orientation under uniaxial strain, *Macromol Symp* **163**:127–134.
51. Nair B R and Hammond P T (1998) Synthesis and characterization of novel side chain liquid crystalline polyurethanes, *ACS Polym Prepr* **39**(2):1077–1078.
52. Nair B R, Osbourne M A R and Hammond P T (1998) Synthesis and characterization of new segmented copolymers with side chain liquid crystalline soft segments, *Macromolecules* **31**:8749–8756.
53. Nair B R, Gregoriou V G and Hammond P T (2000) FT-IR studies of side chain liquid crystalline elastomers, *Polymer* **41**:2961–2970.
54. Hammond P T (2002) Designing for supramolecular assembly in new architecturally nonsymmetric diblock copolymers, *Abstr Papers 224th ACS Natl Meet*, Boston, Poly–313.
55. Hammond P T (2002) Designing for supramolcular assembly in architecturally nonsymmetric diblock copolymers, *ACS Polym Prepr* **43**(2):370.
56. Antony P, Puskas J and Paulo C (2002) Arborescent polyolefinic thermoplastic elastomers and products therefrom, WO 02 096967; Cl. JPC C0 F 297/00.
57. Anthamatten M and Hammond P T (2001) Free-energy model asymmetry in side chain liquid crystalline block copolymers, *J Polym Sci B Polym Phys* **39**:2671–2691.
58. Abeysekera R, Bushby R J, Caillet C, Hamley I W, Lozman O R, Lu Z and Robards A W (2003) Discotic liquid crystalline triblock copolymers: interplay of liquid crystal architecture microphase separation, *Macromolecules* **36**:1526–1533.
59. Lohmeijer B G G and Schubert U S (2003) Playing LEGO with macromolecules: design, synthesis, and self-organization with metal complexes *J Polym Sci A Polym Chem* **41**:1413–1427.
60. Newkome G R, He E and Moorefield C N [1999] Suprasupermolecules with novel properties: metallodendrimers, *Chem Rev* **99**:1689–1746.
61. Brunsveld L, Folmer B J B, Meijer E W and Sijbesma R P [2001] Supramolecular polymers, *Chem Rev* **101**:4071–4097.
62. Schubert U S and Heller M (2001) Metallo-supramolecular initiators for the preparation of novel functional architectures, *Chemistry: A Eur J* **7**:5252–5259.

63. Lohmeijer B G G and Schubert U S (2003) Water-soluble building blocks for terpyridine-containing supramolecular polymers: Synthesis, complexation and pH stability studies of poly(ethylene oxide) moieties, *Macromol Chem Phys* **204**:1072–1078.
64. Lohmeijer B G G and Schubert U S (2001) Synthesis of metallo-supramolecular block-copolymers, *Polym Mater Sci Eng* **85**:460–461.
65. Lohmeijer B G G and Schubert U S (2002) Supramolecular engineering with macromolecules: an alternative concept for block copolymers, *Angew Chem Int Ed* **41**:3825–3829.
66. Einsenbach C D and Schubert U S (1993) Synthesis and chain extension of bipyridine-terminated polyethers with copper(I) ions, *Macromolecules* **26**:7372–7374.
67. Schubert U S and Einsenbach C D [1999] Block copolymers with double helical supramolecular segments, *ACS Polym Prepr* **40**(1):222–223.
68. Schubert U S and Eschbaumer C (2000) Functional (block) copolymers with metal complexing segments, *ACS Polym Prepr* **41**(1):542–543.
69. Schubert U S and Schmatloch S (2001) Approach towards high molecular mass polymers *via* metal complexing oligomers, *ACS Polym Prepr* **42**(2):395–396.
70. Kricheldorf H R and Eggerstedt S (1998) Macrocycles 4. Macrocyclic polylactones as bifunctional monomers for polycondensations, *J Polym Sci A Polym Chem* **36**:1373–1378.
71. Kricheldorf H R and Eggerstedt S ((1999) Macrocycles 6. MALDI-TOF mass spectroscopy of tin-initiated macrocyclic polylactones in comparison to classical mass-spectroscopic methods, *Macromol Chem Phys* **200**:1284–1291.
72. Kricheldorf H R and Langanke D (1999) Macrocycles 7. Cyclization of oligo- and poly(ethyleneglycol)s with dibutyltin dimethoxide — a new approach to (super)macrocycles, *Macromol Chem Phys* **200**:1174–1182.
73. Kricheldorf H R and Langanke D (1999) Macrocycles 8. Multiblock copoly(ether-esters) of poly(THF) and ε-caprolactone *via* macrocyclic polymerization, *Macromol Chem Phys* **200**:1183–1190.
74. Ryner M, Finne A, Albertson A-C and Kricheldorf H R (2001) L-Lactide macromonomer synthesis initiated by new cyclic tin alkoxides functionalized for brushlike structures, *Macromolecules* **34**:7281–7287.
75. Kricheldorf H R, Al-Masri M and Schwarz G (2002) Macrocycles 20. Cyclic poly(ethylene glycol) phthalates *via* ring-exchange substitution, *Macromolecules* **35**:8936–8942.
76. Kricheldorf H R and Hauser K (2001) Macrocycles 45. Homo- and copolymerizations of 3-methylmorpholine-2,5-dione initiated with a cyclic tin alkoxide, *Macromol Chem Phys* **202**:1219–1226.
77. Ouchi T, Nozaki T, Okamoto Y, Shiratani M and Ohya Y (1996) Synthesis and enzymatic hydrolysis of polydepsipeptides with functionalized groups, *Macromol Chem Phys* **197**:1823–1833.
78. Ouchi T, Nozaki T, Ishikawa A, Fujimoto I and Ohya Y (1997) Synthesis and enzymatic hydrolysis of lactic acid-depsipeptide copolymers with functionalized pendant groups, *J Polym Sci A Polym Chem* **35**:377–383.
79. Jorres V, Keul H and Hocker H (1998) Polymerization of (3S, 6S)-3-isopropyl-6-methyl-2,5-morpholinedione with tin octoate and tin acetylacetonate, *Macromol Chem Phys* **199**:835–843.
80. John G and Morita M (1999) Biodegradable cross-linked microspheres from poly(ε-caprolactone)-co-glycolic acid-co-L-serine) based polydepsipeptides, *Macromol Chem Rapid Commun* **20**:265–268.
81. Kricheldorf H R, Lee S-R and Schittenhelm N (1998) Macrocycles 1. Macrocyclic polymerizations of (thio)lactones — stepwise ring expansion and ring contraction, *Macromol Chem Phys* **199**:273–282.

82. Kricheldorf H R (2003) Macrocycles 21. Role of ring-ring equilibria in thermodynamically controlled polycondensations, *Macromolecules* **36**:2302–2308.
83. Kricheldorf H R and Langanke D (2001) Polylactones 56. ABA copolymers derived from ε-caprolactone or L-lactide and a central polysiloxane block, *Macromol Biosci* **1**:364–369.
84. Fuller R C (1995) Polyesters and photosynthetic bacteria from lipid cellular inclusions to microbial thermoplastics, in *Anoxygenic Photosynthetic Bacteria* (Eds. Blankenship R E, Madigan M T and Bauer C E) Kluwer Academic Publishers, The Netherlands, Ch. 60, pp. 1245–1256.
85. Kim Y B and Lenz R W (2001) Polyesters from microorganisms, *Adv Biochem Eng* **71**:51–79.
86. Poirier Y and Nawrath C (1998) Transgenic plants for the production of polyhydroxyalkanoates, a family of biodegradable thermoplastics and elastomers, in *Transgenic Plants Research* (Ed. Keith L) Harwood Publisher, Amsterdam, Ch. 12, pp. 201–218.
87. Nawrath C and Poirier Y (1996) Review on polyhydroxyalkanoate formation in the model plant *Arabidopsis thaliana, Publ Natl Res Council of Canada Int Symp Bact Polyhydroxyalkanoates*, pp. 119–126.
88. Poirier Y and Gruys K J (2002) Production of polyhydroxyalkanoates in transgenic plants, *Biopolymers* **3a**:401–435.
89. Lageveen R G, Huisman G W, Preusting H, Ketelaar P, Eggink G and Withold B (1988) Formation of polyesters by *Pseudomonas oleovorans:* effect of substrates on formation and composition of poly[(-R)-3-hydroxyalkanoate]s and poly[(-R)-3-hydroxyalkenoate]s, *Appl Environ Microbiol* **54**:2924–2932.
90. Doi Y, Tamaki A, Kunioka M and Soga K (1987) Biosynthesis of an unusual copolyester (10 mol% 3-hydroxybutyrate and 90 mol% 3-hydroxyvalerate units) in *Alcaligenes eutrophus* from pentanoic acid, *J Chem Soc Chem Commun* **21**:1635–1636.
91. Doi Y, Tamaki A, Kunioka M and Soga K (1988) Production of copolyesters of 3-hydroxybutyrate and 3-hydroxyvalerate by *Alcaligenes eutrophus* from butyric and pentanoic acids, *Appl Microbiol Biotech* **28**:330–334.
92. Fuchtenbusch B, Fabritius D, Waltermann M and Steinbuchel A (1998) Biosynthesis of novel copolyesters containing 3-hydroxypivalic acid by *Rhodococcus ruber* NCIMB 40126 and related bacteria, *FEMS Microbiol Lett* **159**:85–92.
93. Rodrigues M F A, da Silva L F, Gomez J G C, Valentin H E and Steinbuchel A (1995) Biosynthesis of poly(3-hydroxybutyric acid-co-3-hydroxy-4-pentenoic acid) from unrelated substrates by *Burkholderia sp, Appl Microbiol Biotech* **43**:880–886.
94. Ballistreri A, Montaudo G, Impallomeni G, Lenz R W, Ulmer H W and Fuller R C (1995) Synthesis and characterization of polyesters produced by *Rhodospirillum rubrum* from pentenoic acid, *Macromolecules* **28**:3664–3671.
95. Huijberts G N M, De Rijk T C, de Waard P and Eggink G (1994) ^{13}C nuclear magnetic resonance studies of *Pseudomonas putida* fatty acid metabolic routes involved in poly(3-hydroxyalkanoate) synthesis, *J Bacterol* **176**:1661–1666.
96. Marchessault R H, Monasterios C J, Morin F G and Sundararajan P R (1990) Chiral poly(3-hydroxyalkanoates): an adaptable helix influenced by the alkane side-chain, *Intern J Biol Macromol* **12**:158–165.
97. Haywood G W, Anderson A J, Ewing D F and Dawes E A (1990) Accumulation of a polyhydroxyalkanoate containing primarily 3-hydroxydecanoate from simple carbohydrate substrates by *Pseudomonas* sp. Strain NCIMB 40135, *Appl Environ Microbiol* **56**:3354–3359.
98. Huijberts G N M, Eggink G, de Waard, Huisman G W and Witholt B (1992) *Pseudomonas putida* KT2442 cultivated on glucose accumulates poly(3-hydroxyalkanoates)

consisting of saturated and unsaturated monomers, *Appl Environ Microbiol* **58**:536–544.
99. Timm A and Steinbiichel A (1990) Formation of polyesters consisting of medium-chain-length 3-hydroxyalkanoic acids from glutanate by *Pseudomonas aeruginosa* and other fluorescent pseudomanads, *Appl Environ Microbiol* **56**:3360–3367.
100. Kim D Y, Kim Y B and Rhee Y H (1998) Bacterial poly(3-hydroxyalkanoates) bearing carbon-carbon triple bonds, *Macromolecules* **31**:4760–4763.
101. Fritzsche K, Lenz R W and Fuller R C (1990) An unusual bacterial polyester with a phenyl pendant group, *Makromol Chem* **191**:1957–1965.
102. Marchessault R H and Yu G (1999) Preparation and characterization of low molecular weight poly(3-hydroxybutyrate)s and their block copolymers with poly(oxyethylene)s, *ACS Polym Prepr* **40**(1):527–528.
103. Heitz W (1995) Metal catalyzed polycondensation reactions, *Pure Appl Chem* **67**:1951–1964.
104. Maréchal E (2002) New aspects of catalysis in polycondensation, *Current Org Chem* **6**:177–208.
105. Takagi T, Tomita I and Endo J (1997) Block copolymerization of alkoxyallenes with phenylallene by living coordination system with *p*-allylnickel catalyst, *Polym Bull* **39**:685–692.
106. Lalot T and Maréchal E (2001) Enzyme-catalyzed polyester synthesis, *Int J Polym Mater* **50**:267–286.
107. Pash H (2002) Analytical techniques for polymers with complex architectures, *Macromol Symp* **178**:25–37.
108. Valiente N, Lalot T, Brigodiot M and Maréchal E (1998) Enzymatic hydrolysis of phthalic unit containing copolyesters as a potential tool for block length determination, *Polym Degrad Stabil* **61**:409–415.
109. Murgasova R, Brantley E L, Hercules D M and Nefzger H (2002) Characterization of a polyester-polyurethane soft and hard blocks by a combination of MALDI, SEC and chemical degradation, *Macromolecules* **35**:8338–8345.
110. Olvera de la Cruz M, Mayes A M and Swift B W (1992) Transition to lamellar-catenoid structure in block-copolymer melts, *Macromolecules* **25**:944–948.
111. Erukhimovich I, Abetz V and Stadler R (1997) Microphase separation in ternary ABC block copolymers: ordering control in molten diblock AB copolymers by attaching a short strongly interacting C block, *Macromolecules* **30**:7435–7443.
112. de Luca Freitas L, Jacobi M M, Gonçalves G and Stadler G (1998) Microphase separation induced by hydrogen bonding in poly(1,4-butadiene)-*block*-poly(1,4-isoprene) diblock copolymer — an example of supramolecular organization *via* tandem interactions, *Macromolecules* **31**:3379–3382.
113. Stadler R, Krappe U and Voigt-Martin I (1994) Chiral nanostrutures based on amorphous ABC-triblock copolymers. *ACS Polym Prepr* **35**(2):542–543.
114. Hashimoto T, Kawamura T, Harada M and Tanaka H (1994) Small-angle scattering from hexagonally packed cylindrical particles with paracrystalline distortion, *Macromolecules* **27**:3063–3072.
115. Stocker W, Beckmann J, Stadler R and Rabe J P (1996) Surface reconstruction of the lamellar morphology in a symmetric poly(styrene-*block*-butadiene-*block*-methyl methacrylate) triblock copolymer: a tapping mode scanning force microscopy study, *Macromolecules* **29**:7502–7507.
116. Breiner U, Krappe U, Jakob T, Abetz V and Stadler R (1998) Spheres on spheres — a novel spherical multiphase morphology in symmetric polystyrene-*block*-polybutadiene-*block*-poly(methyl methacrylate) triblock copolymers, *Polym Bull* **40**:219–226.

117. Furukawa N, Yuasa M and Kimura Y (1998) Structure analysis of a soluble polysiloxane-*block*-polyamide and kinetic analysis of the solution imidization of the relevant polyamide acid, *J Polym Sci A Polym Chem* **36**:2237–2245.
118. Benmouna M, Maschke U and Ewen B (1996) Properties of polymer blends and triblock copolymers obtained by neutron scattering, *J Polym Sci B Polym Phys* **34**:2161–2168.
119. Matsen M W and Bates F S (1997) Conformationally asymmetric block copolymers, *J Polym Sci B Polym Phys* **35**:945–952.
120. Netz R R and Schick M (1996) Liquid-crystalline phases of semiflexible diblock copolymer melts, *Phys Rev Lett* **77**:302–305.
121. Pesetskii S S, Jurkowski B, Olkhov Y A, Olkhova O M, Storozhuk I P and Mozheiko U M (2001) Molecular and topological structures in polyester block copolymers, *Eur Polym J* **37**:2187–2199.
122. Guenza M, Tang H and Schweizer K S (1998) Mode-coupling theory of self-diffusion in diblock copolymers I. General derivation and qualitative predictions, *J Chem Phys* **108**:1257–1270.
123. Guenza M, Tang H and Schweizer K S (1998) Mode-coupling theory of self-diffusion in diblock copolymers II. Model calculations and experimental comparisons, *J Chem Phys* **108**:1271–1283.
124. Pakula T (1996) Simulation of copolymers by means of the cooperative-motion algorithm, *J Computer-Aided Mater Design* **3**:329–340.
125. Smith A P, Douglas J F, Meredith J C, Amis E C and Karim A (2001) High-throughput characterization of pattern formation in symmetric diblock copolymer films, *J Polym Sci B Polym Phys* **39**:2141–2158.

PART II

POLYESTER-BASED THERMOPLASTIC ELASTOMERS

Chapter 3

Polyester Thermoplastic Elastomers: Synthesis, Properties, and Some Applications

Z. Roslaniec

1. Introduction

The classical vulcanized rubbers and the relatively new thermoplastic elastomers (TPEs) are technical elastomeric materials. At present, TPEs are becoming increasingly important in various industrial applications. They combine the end-use physical properties of vulcanized rubbers with the easy processing of thermoplastics [1–6].

The thermoplastic elastomers are phase-separated systems comprising a *hard* and solid (crystalline or glassy) phase and another phase, which is *soft* (liquid and rubbery) at room temperature. Strength is provided by the hard phase, while the elastomeric behavior arises from the soft phase. The TPE properties depend on the nature and amount of the hard phase present. The elastomers are processed at a high temperature when the hard phase melts and flows. Upon cooling, the hard phase solidifies and acts as physical crosslinks, which behave like the sulphur crosslinks in conventional vulcanized rubbers. The glass transition temperatures, T_{g1}, T_{g2}, and the melting point of crystallites, T_m, of the soft and hard phases determine the temperature range of their application.

The TPE polyolefin copolymers and blends (including stereo-copolymers [7]), as well as the reactive rubber-plastic blends, dynamically vulcanized TPEs, and melt-processible rubbers are beyond the scope of this chapter.

The term *polyester thermoplastic elastomers* is widely used for segmented poly(ether-ester) block copolymers with alternating/random length sequences (segments) of poly(alkylene oxide) joined by ester bonds from butylene terephthalate polyester segments. The great differences in the molecular mobilities of these segments and particularly their immiscibility cause phase separation and formation of an amorphous soft phase of polyether-rich segments ($T_g < -40°C$) and hard crystalline polyester domains of high melting point.

Poly(ether-ester) (PEE) copolymers obtained by modification of poly(ethylene terephthalate) with up to 20 wt% of poly(ethylene ether) glycol was first described by Coleman [8]. Subsequently, the DuPont Co. developed poly(ether-ester) elastomers, which were commercially introduced in 1972 under the trade name Hytrel® [4,9]. The polyester thermoplastic elastomers are nowadays produced by several companies. Apart from DuPont, these are DSM, The Netherlands (Arnitel®), General Electric, USA (Lomed®), Hoechst Celanese, USA (Retiflex®), Toyobo, Japan (Pelprene®), Elana, Poland (Elitel®) [2,10]. The synthesis, chemical structure, physical properties, and some new applications of polyester TPE are discussed in this chapter (about the development of TPE, see also Chapter 1, while details on some commercial TPE products can be found in Chapter 17).

2. Chemical structure of polyester elastomers

The polyester elastomers were so far defined as multiblock (segmented) poly(ether-ester) copolymers of the general formula (Scheme 1):

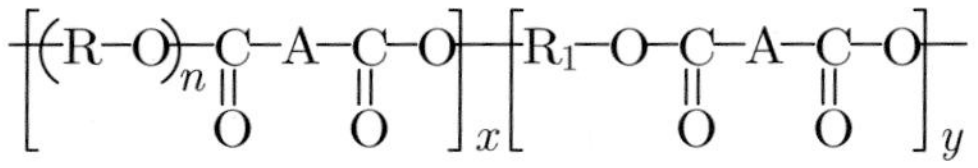

$$\left[(R{-}O)_n{-}\underset{\underset{O}{\|}}{C}{-}A{-}\underset{\underset{O}{\|}}{C}{-}O \right]_x \left[R_1{-}O{-}\underset{\underset{O}{\|}}{C}{-}A{-}\underset{\underset{O}{\|}}{C}{-}O \right]_y$$

where R is tetramethylene, isopropylene, ethylene or their mixtures (copolymers), R_1 is tetramethylene, and A is terephthalate or a terephthalate/isophthalate mixture

Scheme 1

The commercial polyester elastomers are mostly based on poly(tetramethylene oxide) (PTMO) as flexible segment and poly(butylene terephthalate) (PBT) as rigid segment. Yet, many chemical modifications of some particular segments and also segments of totally different chemical structure have been examined and applied. Thus, in addition to PTMO [1,2,11–16], poly(ethylene oxide) (PEO) [17–22], poly(aliphatic oxide) (C_2–C_4) copolymers [23–30], poly(butylene succinate), and other aliphatic polyesters [31–35], polycaprolactone (PCL), polypivalolactone (PVL) [36,37], aliphatic polycarbonates (PC) [38–41], dimerized fatty acid (DFA)-based polyesters [42–50], polyamide 66 and derivatives [47–57], polyolefins [58–60], rubbers [61–63], and polydimethylsiloxane [64,65] are used as flexible segments of polyester elastomers.

In addition to PBT [1,2,4,5,18,19,66,67], poly(butylene terephthalate-*co*-isophthalate) [68–70], PBT copolymers with unsaturated alkylene dicarboxylates or unsaturated glycols [20,71–74], poly(ethylene terephthalate) (PET)

[75–85], poly(tetramethylene 2,6-naphthalene dicarboxylate) [86,87], caprolactone [88], D,L-lactide [89], as well as other copolyesters (*e.g.*, with liquid crystalline [90–102] or ionic units [103,104]), and aromatic polycarbonate [105] are used as rigid segments.

Branched PEE [106,107] and ABA-type triblock copolymers [102] were also synthesized. Polyalcohols (butanetriol, pentaerythritol), aliphatic polyacids and branched polyether polyols are used as branching agents [108–111].

From the point of view of the synthesis course, one can refer to random copolyesters prepared from terephthalic acid and two types of glycols. Phase separation occurs if the molecular weight of the polyether glycol is high enough (1000 g/mole and higher). The soft phase of low glass transition temperature (similar to that of the homopolyether) and the hard phase with high melting point are formed. Numerous publications report on various segment modifications of typical copoly(ether-ester)s and synthesis methods other than polyesterification (*e.g.*, in an end-linking reaction). Thus, the following groups are used as the elements joining polyether and polyester segments: ester (as in commercial PEEs), amide (*e.g.*, using amine alcohols, diamines or bisoxasolines) [112], urethane (using diisocyanates) [73,74], and ether groups (with the use of diepoxides) [73] (see also Chapters 2 and 9).

The number of reports on the obtaining of polyesters (mainly PBT-based) and on rubber blends for dynamic vulcanization is growing. The authors point to the multiphase structure (in which the partially vulcanized rubber is the soft phase) and to the properties of the TPEs or the high impact thermoplastics. PEE/nitryl rubber blends [113–115], PBT or PET/acrylic rubber blends [116,117], PBT/rubbery polymer blends [118,119], PBT/ethylene-propylene rubber blends, unsaturated PEE/EPDM rubber blends [73,120,121], and PBT/rubber blends [122,123] are reported.

Block terpolymers of poly(hydroxyoctanoic acid), poly(hydroxyhexanoic acid) and poly(hydroxydecanoic acid), and the so-called “bacterial elastomers”, produced by bacteria [124] also belong to the polyester thermoplastic elastomer family (see also Chapter 2).

3. Poly(alkylene oxide) flexible segment-based polyester elastomers

PTMO, PEO, poly(propylene oxide) (PPrO), and their random or block copolymers are most often used as flexible segments. Poly(alkylene ether) glycols are synthesized mainly by ring-opening polymerization and radical polymerization. Extensive reviews on the chemical structure and synthesis of oligo(alkylene ether)s with functional end-groups and their copolymers were published by Maréchal [11] and Xie [17].

PTMO is used as a flexible segment in most of the commercial PEE elastomers (Hytrel®, DuPont; Elitel®, Elana, *etc.*) [1,2]. The PTMO segments of molecular weight higher than 1500 g/mole tend to crystallize causing a decrease in the rubbery phase content in the material. The use of copolymers, *e.g.*, PTMO with PEO and PTMO with PPrO, allows to preserve the amorphous polyether phase structure [23]. Since the copolymerization of PTMO with PPrO [24] or

PEO [25] is difficult to perform, mixtures of poly(alkylene ether) glycols are used [26]. The physical properties and structure of polyether segments in PEE differ from those of the mixture of oligoether glycols [27,28].

Higashiyama *et al.* [12] synthesized PEE with different distributions and lengths of PBT and PTMO segments. The chemical structure of PEE has been examined by means of nuclear magnetic resonance (^{1}H NMR and ^{13}C NMR). It has been confirmed that the probability of long sequence formation (Scheme 1) in the flexible segments in PBT-*b*-PTMO copolymers of lower PBT content is greater than in the case of higher PBT content.

Interesting results are reported on the basis of studies on the influence of the distribution and molecular weight of the PTMO segments on the PEE mechanical characteristic [13,14]. For instance, the hardness of PTMO-PBT based PEEs as a function of temperature and PTMO molecular weight is reported to decrease in the temperature range of 0–50°C for PEEs with flexible segment contents of more than 60 wt%. The tensile strength of PEE depends mainly on the rigid segment content and temperature and less on PTMO segment molecular weight within the range M_n = 650 to 2000 g/mole.

It is known that sample annealing causes the reorganization of the copolymer phase structure. Yet, there are very interesting results suggesting structural changes after annealing of highly oriented PEE (based on PTMO-PBT) samples [15,16]. The increase in the rubber elasticity and particularly in the resilience (>98%) is probably caused by an additional phase separation and reorganization of the PEE morphology.

Over the last decade, Fakirov *et al.* carried out extensive studies on the synthesis, structure, and physical properties of PEE based on PBT and PEO [18–21]. The investigations of the oriented PEE structure are particularly interesting [18,19,21,29,30,125–128,130,131]. Textile fibers have been obtained with the use of these copolymers [19]. The hardness of PEO-PBT based PEE changes with temperature and PEO molecular weight similarly to the PTMO-PBT block copolymers [129]

Slonecki described the synthesis and characteristics of PEE with PPrO flexible segment [29,30]. The copolymers were synthesized by two-stage polycondensation in the molten state with $Zn(Ac)_2$ and a TiMg organometallic complex as catalysts. Oxyethylated poly(propylene ether) glycol (M_n= 1200 g/mole) was used to increase the reactivity. The T_g changes from –15 to –35°C and T_m changes from 227 to 132°C depending on the PPrO content in the copolymer (Figure 1). The difference between T_g and T_m is the “plateau” of low modulus characteristic of the thermoplastic elastomers. The tensile strength changes from 10 to 50 MPa.

4. Modified poly(butylene terephthalate) rigid segment-based polyester elastomers

To decrease the crystallization ability of the PEE hard phase, some polymer manufacturers apply isophthalic acid (Scheme 2) [68] or unsaturated glycols [71] in the modification of the PBT rigid segments.

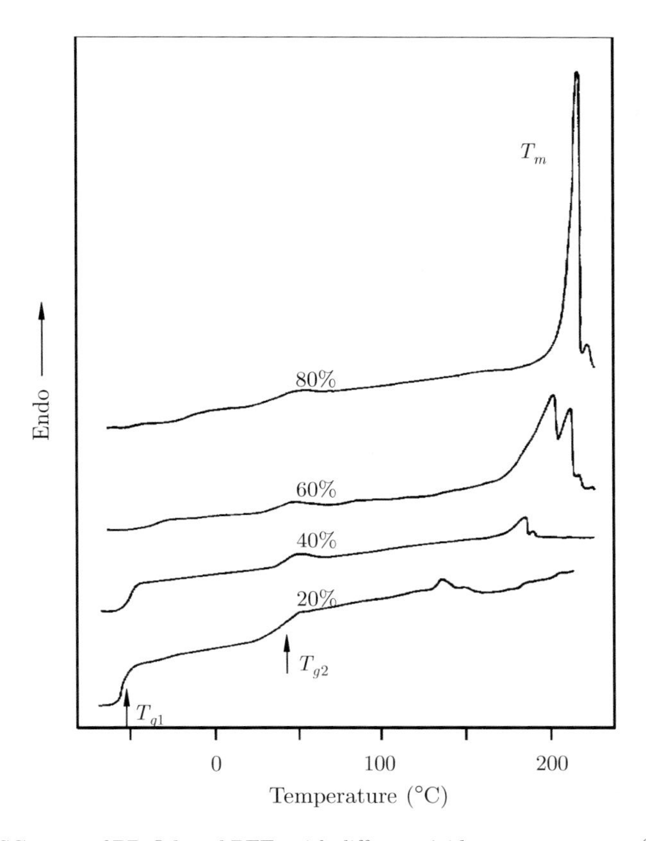

Figure 1. DSC traces of PPrO-based PEEs with different rigid segment contents (20–80 wt%) [29]

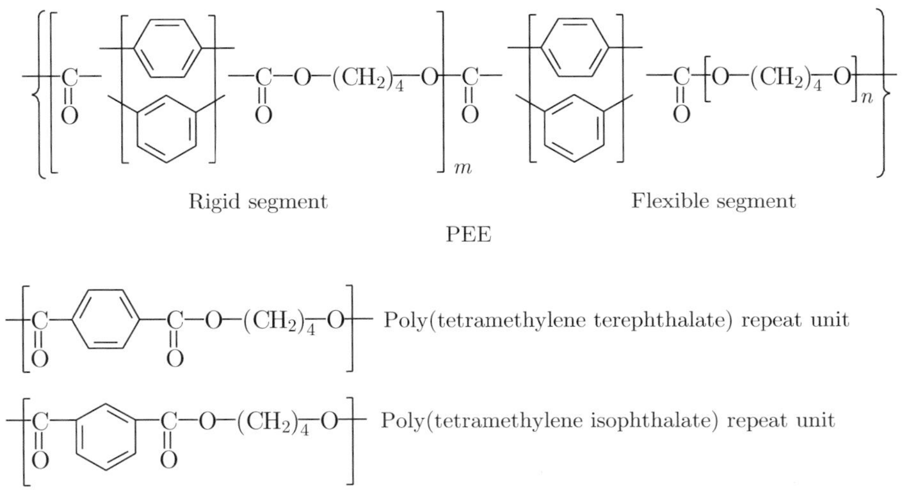

$$\left\{ \left[-\underset{\displaystyle O}{\overset{}{C}}- (C_6H_4) -\underset{\displaystyle O}{\overset{}{C}}-O-(CH_2)_4-O- \right]_m -\underset{\displaystyle O}{\overset{}{C}}- (C_6H_4) -\underset{\displaystyle O}{\overset{}{C}}-\left[O-(CH_2)_4-O \right]_n \right\}$$

Rigid segment

Flexible segment

PEE

$$\left[-\underset{\displaystyle O}{C}- (C_6H_4) -\underset{\displaystyle O}{C}-O-(CH_2)_4-O- \right]$$ Poly(tetramethylene terephthalate) repeat unit

$$\left[-\underset{\displaystyle O}{C}- (C_6H_4) -\underset{\displaystyle O}{C}-O-(CH_2)_4-O- \right]$$ Poly(tetramethylene isophthalate) repeat unit

Scheme 2

Cooper *et al.* [68–70] investigated the effect of butylene terephthalate and butylene isophthalate unit content in the rigid segment on the physical properties and structure of PEE. The isophthalate influences the crystallization and melting conditions of the PEE hard phase. This is directly related to the material mechanical properties.

Gogeva and Fakirov [20] synthesized polyoxides and PBT-based PEE with 1,4-butyn-2-diol (B3D) (Scheme 3) using tetrabutyl titanate as a catalyst. The chemical structure of the PEE was determined using ^{1}H NMR. The content

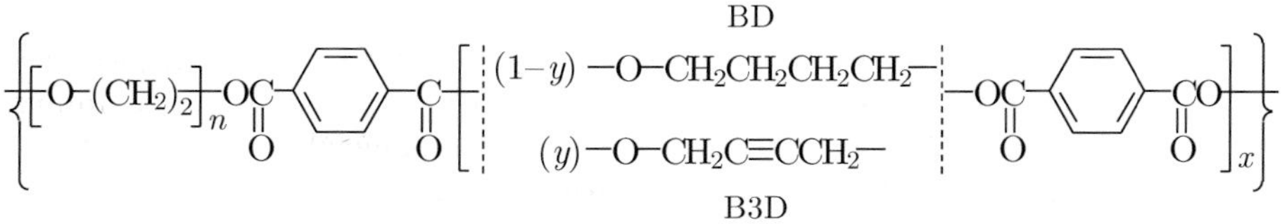

$$\left\{ \left[O-(CH_2)_2 \right]_n \underset{\|}{\overset{}{OC}}\!\!-C_6H_4-\underset{\|}{C}- \left[\begin{array}{l} \overset{BD}{(1-y)\ -O-CH_2CH_2CH_2CH_2-} \\ \underset{B3D}{(y)-O-CH_2C\equiv CCH_2-} \end{array} \right] -\underset{\|}{OC}-C_6H_4-\underset{\|}{CO} \right]_x \Bigg\}$$

(carbonyl groups: $C=O$ on each OC, C and CO)

Scheme 3

of B3D in the reaction mixture did not exceed 20 wt% of low molecular weight glycols (BD+B3D). Probably the reactivity of B3D is lower than that of BD. The unsaturated PEE obtained according to this method showed good properties for the B3D maximum content of 10 wt%.

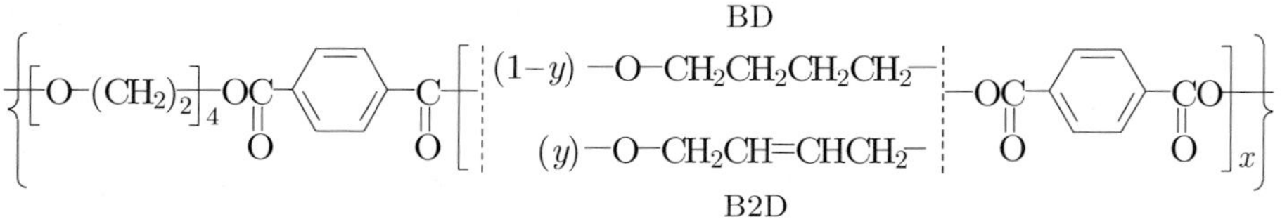

$$\left\{ \left[O-(CH_2)_2 \right]_4 \underset{\|}{OC}-C_6H_4-\underset{\|}{C}- \left[\begin{array}{l} \overset{BD}{(1-y)\ -O-CH_2CH_2CH_2CH_2-} \\ \underset{B2D}{(y)-O-CH_2CH=CHCH_2-} \end{array} \right] -\underset{\|}{OC}-C_6H_4-\underset{\|}{CO} \right]_x \Bigg\}$$

Scheme 4

PEE was also synthesized with the mixture of BD and 1,4-butene-2-diol (B2D) [71,73,74] (Scheme 4). Nelsen *et al.* [71] proved that the addition of up to 15 wt% of B2D instead of BD in the PEE synthesis improves the physical and processing properties without changing the polymer color.

Siemiński and Roslaniec [73,74] synthesized an unsaturated PEE with PTMO ($M_n = 1000$ g/mole) as the flexible segment. The BD-to-B2D ratio in the reaction mixture was varied from 0 to 100 mole% of B2D. The content of the soft segment was constant (50 wt%). A TiMg organometallic complex was used as catalyst.

The content of unsaturated bonds in copolymers, determined by ^{1}H NMR, is quite different from that obtained in the stoichiometric calculation. It points to the occurrence of unidentified side reactions during the polycondensation in which the unsaturated bonds decrease. Differential scanning calorimetry (DSC) and dynamic mechanical thermal analysis (DMTA) have shown a higher glass transition temperature and a lower melting point of the unsaturated PEE. A

Table 1. Mechanical properties of unsaturated copoly(ether-ester)s [73]

Unsaturated glycol content (mole/mole)	Tensile strength (MPa)	Elongation at break (%)	Shore D hardness (MPa)	Young's modulus (MPa)
0	21.38	680	52	104.8
0.5	14.50	540	47	88.3
1.0	13.99	500	46	87.2

comparison of some physical properties of unsaturated copoly(ether-ester)s (PEE2) is shown in Table 1. The stress-strain curves of PEE2 have a similar trend as those of saturated PEE, but PEE2 show somewhat lower strength parameters (tensile strength and elongation at break) as a result of lower crystallinity. Copoly(ether-ester)s containing unsaturated bonds in the polyester segment are thermoplastic elastomers.

5. Branched polyester elastomers

By introducing a small amount of pentaerythritol or some other polyfunctional monomer in the PEE synthesis [106–108], it is possible to obtain a defined content of branched chains. Ukielski [106] studied the influence of pentaerythritol on the physical properties of PEE based on PTMO and PEO. The pentaerythritol content was changed from 0.1 to 0.44 phr with respect to dimethyl terephthalate (DMT). Branching causes an increase in the PEE viscosity, but it changes only insignificantly the strength characteristics of the copolymers (Table 2). It was reported [111] that the considerable content of branchings in PEE distinctly influences the flow rate of the molten polymer. These copolymers are suitable for extrusion and blow molding.

Table 2. Physical properties of PEO-*b*-PBT and PTMO-*b*-PBT copolymers with various degrees of branching [106]

Property	PEO-*b*-PBT					PTMO-*b*-PBT				
	1A	2A	3A	4A	5A	1B	2B	3B	4B	5B
Limiting viscosity number (dL/g)	1.46	1.42	1.58	1.58	1.70	1.25	1.42	1.44	1.61	2.0
Melting point temperature (°C)	205	205	202	202	199	214	216	217	218	214
Swelling in benzene (wt%)	143	142	146	148	153	93	87	95	100	83
Resilience (%)	57	58	59	58	59	44	42	42	40	40
Shore A hardness	90	90	89	88	86	84	81	80	80	78
Tensile strength (MPa)	25.2	25.0	24.8	24.3	24.1	24.1	23.2	23.5	22.8	22.9
Elongation at break (%)	860	830	750	680	660	840	800	750	680	580

Pentaerythritol content: 1 – 0; 2 – 0.11; 3 – 0.22; 4 – 0.33; 5 – 0.43 phr, based on DMT

6. Synthesis of poly(ether ester) block copolymers

PEE are mostly synthesized in the same way as typical polyesters (PBT, PET) by a step-growth condensation polymerisation in the molten state with BD, poly(tetramethylene ether) glycol (PTMEG) and DMT or terephthalic acid in two or three stages [132–135]: (i) transesterification of DMT or condensation of terephthalic acid with the hydroxyl groups of both polyester glycol and low molecular weight diol, (ii) low pressure melt polycondensation at high temperatures (250–350°C), and (iii) post-polycondensation in the solid state (if a polymer of higher molecular weight is required). As catalysts, mainly salts and metallic oxides, as well as organometallic complex compounds containing sodium, potassium, titanium, zirconium, magnesium, cadmium, cobalt, manganese, antimony, tin, germanium, selenium, *etc.*, are used [123,132–141]. The transesterification of DMT with PTMEG and BD takes place in the presence of metallic catalysts according to the reactions in Scheme 5.

Monohydroxybutyl terephthalate, bishydroxybutyl terephthalate, and their mixture are formed.

During the transesterification, oligomerization is possible. The oligomer content does not increase before BD is removed from the reaction medium. Transesterification, similar to PET and PBT homopolymer synthesis, is carried out in excess of glycol, and the conversion is most often 80–90%. Thus, the reactions are carried out in the low pressure melt polycondensation stage [143–147] and in the post-polycondensation process [112,148–151].

Hsu and Choi [132,146] pointed out that the reactivity of PTMEG with DMT depends on the temperature, catalyst concentration, and molecular weight of PTMEG, whereas the reaction progress is less dependent on the PTMEG excess in the system. The conversion levels off at about 50% for the highest catalyst concentration of 3.1×10^{-4} mole/L. It has been observed during the experiments [132] that, when a higher catalyst concentration was employed, an undesirable entrainment of DMT to the distillation column and condenser occurred. The equilibrium conversion of the methyl ester groups of DMT at 200°C is only about 40%. It is much lower than the final conversion obtainable in the transesterification of DMT with BD (~90%).

Some authors suggest that transesterification should be run in two stages [66,152,153]. In the first stage of PEE synthesis, only the reaction of DMT with BD is carried out to the conversion level of over 90%. In the next stage, PTMEG is introduced in the reactor, together with a catalyst and an antioxidant, and the reaction mixture is heated slowly under reduced pressure. Under these conditions, the transesterification of DMT with BD ceases, further oligomerization occurs, and ester exchange takes place.

The removal of BD from the reaction medium by distillation under reduced pressure brings the reaction to completion. At the same time, the polyether segment thermal degradation, caused by the long heating during the transesterification of DMT-BD-PTMEG, is eliminated. The action of the titanium catalyst can be explained by forming temporary adducts of titanium and the carbonyl group [139,142].

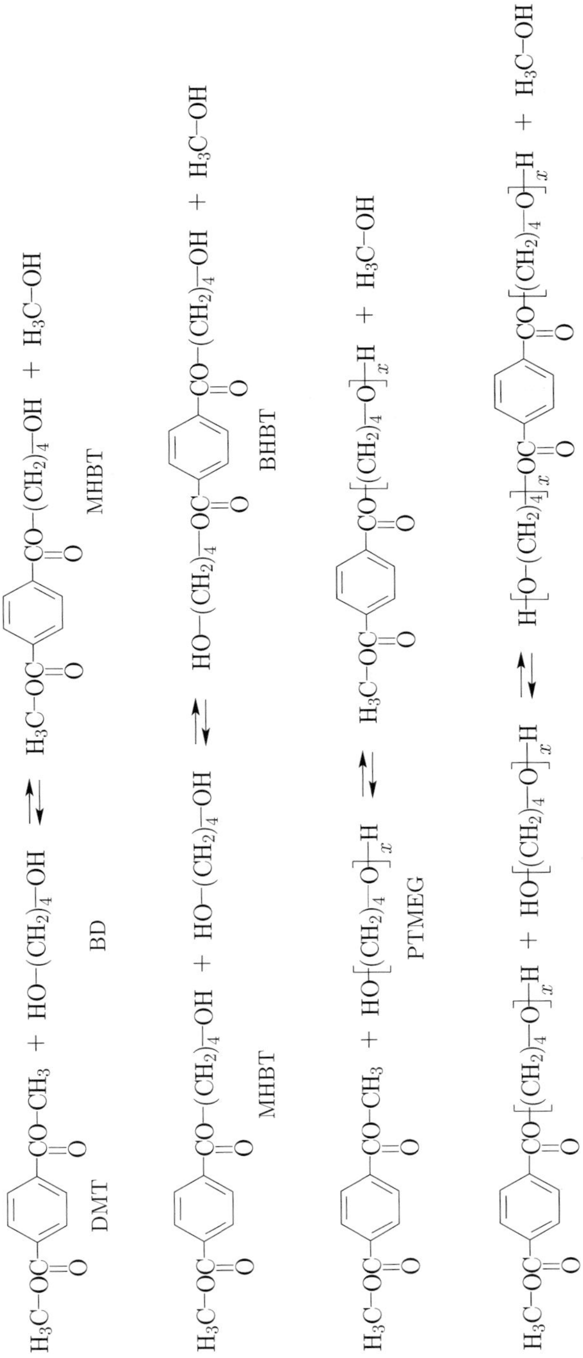

$$\underset{\text{DMT}}{H_3C{-}O\overset{O}{\overset{\|}{C}}{-}C_6H_4{-}\overset{O}{\overset{\|}{C}}O{-}CH_3} + \underset{\text{BD}}{HO{-}(CH_2)_4{-}OH} \rightleftharpoons \underset{\text{MHBT}}{H_3C{-}O\overset{O}{\overset{\|}{C}}{-}C_6H_4{-}\overset{O}{\overset{\|}{C}}O{-}(CH_2)_4{-}OH} + H_3C{-}OH$$

$$\underset{\text{MHBT}}{H_3C{-}O\overset{O}{\overset{\|}{C}}{-}C_6H_4{-}\overset{O}{\overset{\|}{C}}O{-}(CH_2)_4{-}OH} + HO{-}(CH_2)_4{-}OH \rightleftharpoons \underset{\text{BHBT}}{HO{-}(CH_2)_4{-}O\overset{O}{\overset{\|}{C}}{-}C_6H_4{-}\overset{O}{\overset{\|}{C}}O{-}(CH_2)_4{-}OH} + H_3C{-}OH$$

$$H_3C{-}O\overset{O}{\overset{\|}{C}}{-}C_6H_4{-}\overset{O}{\overset{\|}{C}}O{-}CH_3 + \underset{\text{PTMEG}}{HO{-}\left[(CH_2)_4{-}O\right]_x{-}H} \rightleftharpoons H_3C{-}O\overset{O}{\overset{\|}{C}}{-}C_6H_4{-}\overset{O}{\overset{\|}{C}}O{-}\left[(CH_2)_4{-}O\right]_x{-}H + H_3C{-}OH$$

$$H_3C{-}O\overset{O}{\overset{\|}{C}}{-}C_6H_4{-}\overset{O}{\overset{\|}{C}}O{-}\left[(CH_2)_4{-}O\right]_x{-}H + HO{-}\left[(CH_2)_4{-}O\right]_x{-}H \rightleftharpoons H{-}\left[O{-}(CH_2)_4\right]_x{-}O\overset{O}{\overset{\|}{C}}{-}C_6H_4{-}\overset{O}{\overset{\|}{C}}O{-}\left[(CH_2)_4{-}O\right]_x{-}H + H_3C{-}OH$$

Scheme 5

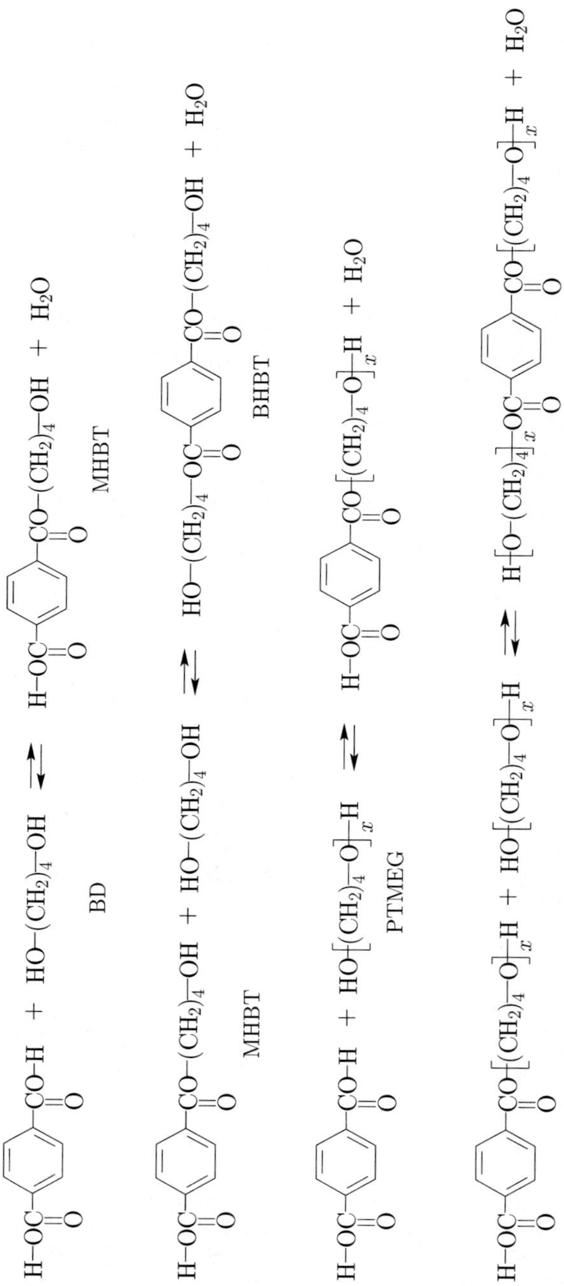

$$\text{H–OC(=O)–C}_6\text{H}_4\text{–C(=O)O–H} + \underset{\text{BD}}{\text{HO–(CH}_2)_4\text{–OH}} \rightleftharpoons \underset{\text{MHBT}}{\text{H–OC(=O)–C}_6\text{H}_4\text{–C(=O)O–(CH}_2)_4\text{–OH}} + \text{H}_2\text{O}$$

$$\underset{\text{MHBT}}{\text{H–OC(=O)–C}_6\text{H}_4\text{–C(=O)O–(CH}_2)_4\text{–OH}} + \text{HO–(CH}_2)_4\text{–OH} \rightleftharpoons \underset{\text{BHBT}}{\text{HO–(CH}_2)_4\text{–OC(=O)–C}_6\text{H}_4\text{–C(=O)O–(CH}_2)_4\text{–OH}} + \text{H}_2\text{O}$$

$$\text{H–OC(=O)–C}_6\text{H}_4\text{–C(=O)O–H} + \underset{\text{PTMEG}}{\text{HO}\left[(\text{CH}_2)_4\text{–O}\right]_x\text{H}} \rightleftharpoons \text{H–OC(=O)–C}_6\text{H}_4\text{–C(=O)O}\left[(\text{CH}_2)_4\text{–O}\right]_x\text{H} + \text{H}_2\text{O}$$

$$\text{H–OC(=O)–C}_6\text{H}_4\text{–C(=O)O}\left[(\text{CH}_2)_4\text{–O}\right]_x\text{H} + \text{HO}\left[(\text{CH}_2)_4\text{–O}\right]_x\text{H} \rightleftharpoons \text{H}\left[\text{O–(CH}_2)_4\right]_x\text{OC(=O)–C}_6\text{H}_4\text{–C(=O)O}\left[(\text{CH}_2)_4\text{–O}\right]_x\text{H} + \text{H}_2\text{O}$$

etc.; similar to DMT transesterification

Scheme 6

The PEE preparation by polycondensation of terephthalic acid (TA), BD, and PTMEG has been given little attention by the industry. The situation is similar to the case of the preparation of PBT by a TA process in which the formation of tetrahydrofuran (THF) by cyclization is a serious problem [133,155]. This situation can be schematically summarized as follows (Schemes 6–8):

(i) The esterification is similar to DMT transesterification (Scheme 6).

(ii) Then the polycondensation reaction takes place (Scheme 7).

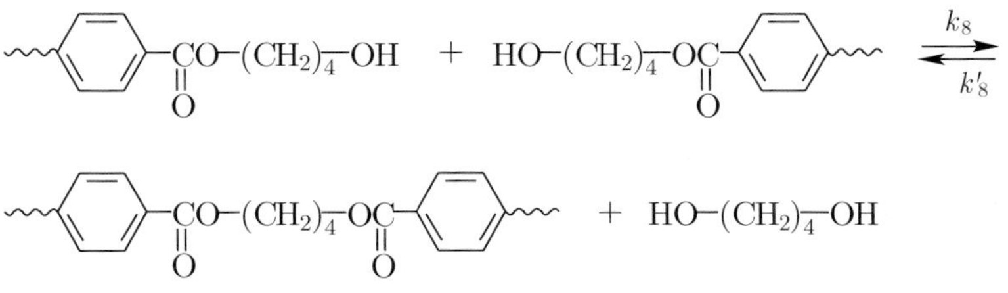

$$\sim C_6H_4{-}\underset{\|}{\overset{}{C}}O{-}(CH_2)_4{-}OH + HO{-}(CH_2)_4{-}O\underset{\|}{C}{-}C_6H_4\sim \ \underset{k'_8}{\overset{k_8}{\rightleftarrows}}$$

$$\sim C_6H_4{-}\underset{\|}{C}O{-}(CH_2)_4{-}O\underset{\|}{C}{-}C_6H_4\sim + HO{-}(CH_2)_4{-}OH$$

(in each structure, the ester C carries a =O)

Scheme 7

(iii) The side reaction (cyclization) proceeds according to Scheme 8.

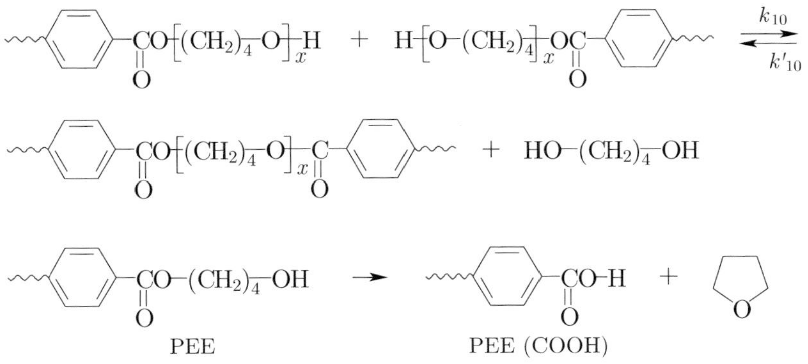

$$\sim C_6H_4{-}C(=O)O{-}\left[(CH_2)_4{-}O\right]_x{-}H + H{-}\left[O{-}(CH_2)_4\right]_x{-}OC(=O){-}C_6H_4\sim \ \underset{k'_{10}}{\overset{k_{10}}{\rightleftarrows}}$$

$$\sim C_6H_4{-}C(=O)O{-}\left[(CH_2)_4{-}O\right]_x{-}C(=O){-}C_6H_4\sim + HO{-}(CH_2)_4{-}OH$$

$$\underset{\text{PEE}}{\sim C_6H_4{-}C(=O)O{-}(CH_2)_4{-}OH} \longrightarrow \underset{\text{PEE (COOH)}}{\sim C_6H_4{-}C(=O)O{-}H} + \text{THF (tetrahydrofuran ring, O)}$$

Scheme 8

The cyclization reaction can be reduced by using salts, such as phosphates, terephthalates, citrates and carbonates. The addition of 0.00115 wt% sodium phosphate or 0.005 wt% potassium terephthalate shortens the polycondensation time and lowers the amount of THF formed [133].

It is well known that the reaction temperature for PEE and PBT synthesis is about 250–260°C, but after the reaction temperature reaches 260°C, the cyclization and degradation reactions become considerable. As the synthesis temperature increases, the degree of the polycondensation decreases and the acid group concentration increases [144]. The kinetic polyesterification model and the effect of various physico-chemical factors on the reaction of dicarboxylic acids with glycols are discussed in [134,136,156].

The increase in the molecular weight and the limiting viscosity number, $[\eta]$, is controlled by the progress of polycondensation. The carboxyl group content is controlled by the side reactions. The lower the pressure in the reactor (Figure 2), the easier is the removal of BD from the reaction medium. As the temperature increases, the –COOH group content increases. The increase in

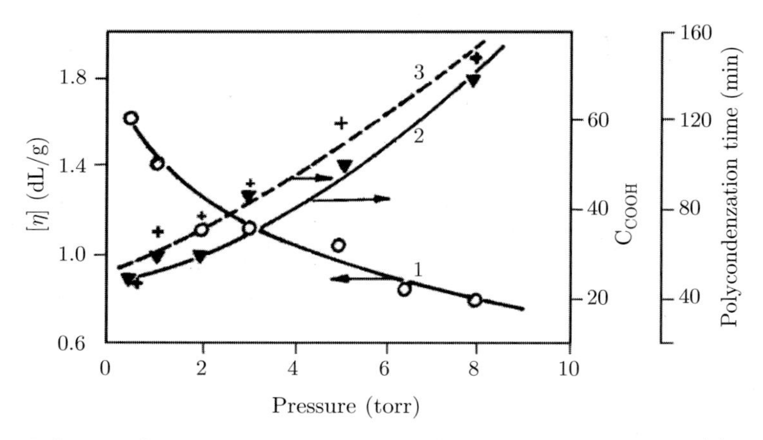

Figure 2. Influence of the reactor pressure on the limiting viscosity number (1), carboxyl group content (2) and polycondensation time (3) for a PBT-PTMO based PEE (60/40 wt%). PTMO $M_n = 1000$ [66]

the melt viscosity or, indirectly, the stirrer torque depends on the reaction progress (Figure 3). The real polycondensation time in the molten state can be determined in practice as the time required to obtain the maximum viscosity or slowly increasing stirrer torque. This allows the estimation of the influence of the chemical structure, content, and molecular weight of polyether glycols on the PEE polycondensation time under comparable conditions [66].

Solid-state post-polycondensation is widely used in order to increase the average molecular weight of polyesters [112,149–151]. The process is particularly important for achieving high values of M_n in the case of crystalline condensation polymers of high melting point, when thermal degradation takes place in the melt. Polycondensation in the solid state is carried out mainly in vacuum or in an inert gas stream. Typically, the polymer is heated to temperatures, which enhance chain propagation reactions, usually up to 20–50°C below T_m. PEE of higher molecular weight can be used for articles made by blow molding and for shock absorber elements.

The influence of physical conditions on the PEE post-polycondensation efficiency has been studied [148]. Investigations of the post-polycondensation reaction of typical engineering polyesters (PBT, PET) have been carried out for a long time. The kinetics and process mechanisms have been identified [149–151,157–160]. The PEE post-polycondensation proceeds according to similar mechanisms [112].

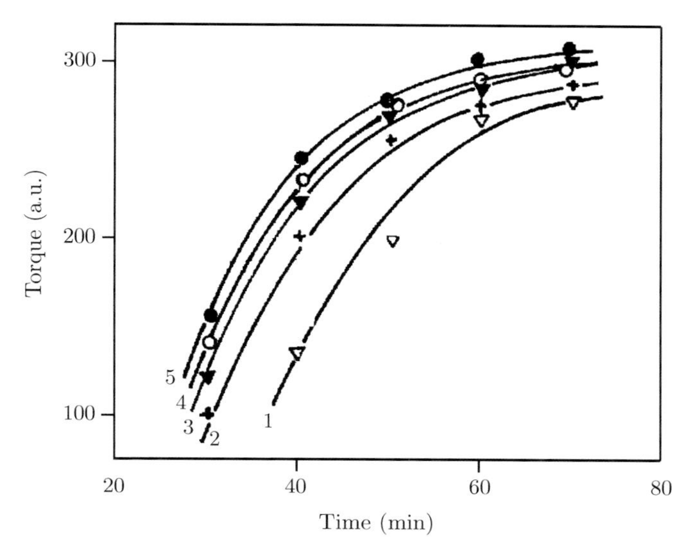

Figure 3. Influence of the polycondensaton time and the MgTi catalyst content on the degree of polycondensation of PEE expressed by the stirrer torque. Catalyst content (g/kg): (1) – 0; (2) – 0.5; (3) – 2; (4) – 4; (5) – 5 [66]

Very interesting methods of molecular weight upgrading are the swollen-state post-polycondensation [161–164] and the chain extension (end-linking) [165–172]. Bisozoxalines, diepoxides and diisocyanates are used as chain extenders. The last reaction was successfully used to upgrade the molecular weight of PEE [73,74,112]. The high reactivity of isocyanate groups with carboxyl and hydroxyl end-groups is known. Theoretically, an equimolar ratio of PEE and diisocyanate should be used, but possible errors in the determination of the concentration of end-groups in the polymer, as well as the reaction of isocyanate with moisture and differences in isocyanate functionality can influence the efficiency of chain extension. The appropriate ratios were determined experimentally. For the PEE with 50 wt% PBT segments and 50 wt% PTMO segments, extended with methylene-diphenylene-diisocyanate (MDI), a direct dependence of the limiting viscosity number of the final product on the ratio of the reacting substances was established. The result was to almost double the average molecular weight of PEE (for the synthesis of PEE, see also Chapter 2).

7. Other multiblock polyester elastomers

Some interesting attempts at synthesizing multiblock copolymers with aliphatic polyester or aliphatic polycarbonate flexible segments are described in [31,35, 38–41]. Such copolymers (Schemes 9–11) were synthesized by polycondensation and transesterification. Two polymers or oligomers were prepared separately and then blended in the molten state. During heating of this blend, partial macromolecular fragmentation and segment exchange occurred [35]. These

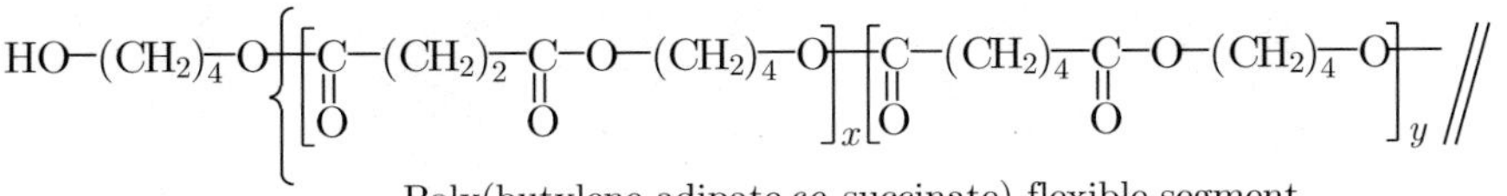

$$\text{HO}-(\text{CH}_2)_4-\text{O}\left\{\left[\underset{\|}{\overset{}{\text{C}}}-(\text{CH}_2)_2-\text{C}(=\text{O})-\text{O}-(\text{CH}_2)_4-\text{O}\right]_x\left[\text{C}(=\text{O})-(\text{CH}_2)_4-\text{C}(=\text{O})-\text{O}-(\text{CH}_2)_4-\text{O}\right]_y\right.$$

Poly(butylene adipate-*co*-succinate) flexible segment

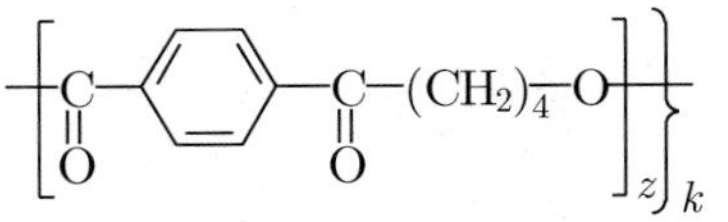

$$\left.\left[\text{C}(=\text{O})-\text{C}_6\text{H}_4-\text{C}(=\text{O})-(\text{CH}_2)_4-\text{O}\right]_z\right\}_k$$

Poly(butylene terephthalate) rigid segment

PBAS-*b*-PBT

Scheme 9

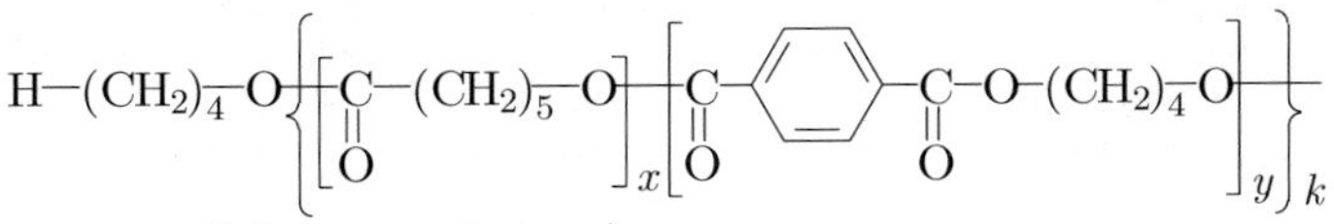

$$\text{H}-(\text{CH}_2)_4-\text{O}\left\{\left[\text{C}(=\text{O})-(\text{CH}_2)_5-\text{O}\right]_x\left[\text{C}(=\text{O})-\text{C}_6\text{H}_4-\text{C}(=\text{O})-\text{O}-(\text{CH}_2)_4-\text{O}\right]_y\right\}_k$$

Poly(ε-caprolactone) flexible segment

PBT rigid segment

PCL-*b*-PBT

Scheme 10

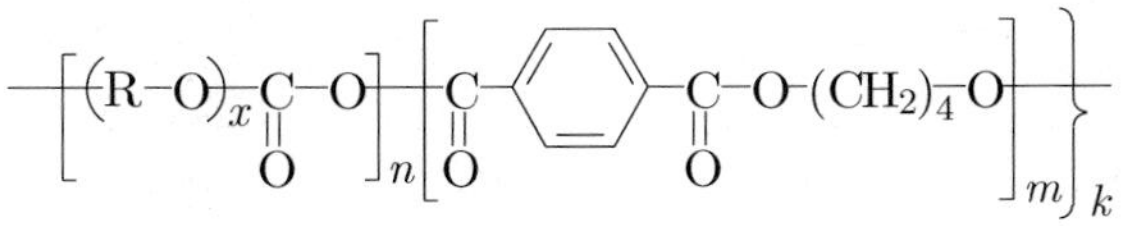

$$\left\{\left[(\text{R}-\text{O})_x-\text{C}(=\text{O})-\text{O}\right]_n\left[\text{C}(=\text{O})-\text{C}_6\text{H}_4-\text{C}(=\text{O})-\text{O}-(\text{CH}_2)_4-\text{O}\right]_m\right\}_k$$

Polyethercarbonate flexible segment; R = C_{6-8}

PBT rigid segment

PC-*b*-PBT

Scheme 11

syntheses can also take place during polycondensation [34,41], similar to that described in Section 6.

Using NMR and DSC methods, Kang and Park [35], and Cheng *et al.* [34] proved unequivocally the blocky structure of poly(ester-ester) copolymers. Prolonged ester exchange decreases the copolymer randomness (Figures 4–6).

The synthesis and characteristics of poly(ester-ester) multiblock copolymers from DMT, alkyl glycols (C_2–C_{10}), and pivalolactone have also been described [36,37]. More details about the exchange of units between chains or segments in polymer systems, *i.e.*, the type of the sequential order in condensation copolymers, were given by Fakirov and Denchev [185]. Roslaniec *et al.* described the synthesis and physical properties of PEE containing aliphatic polycarbonate soft segments [38,40,41]. Multiblock copolyesters can also be obtained by an

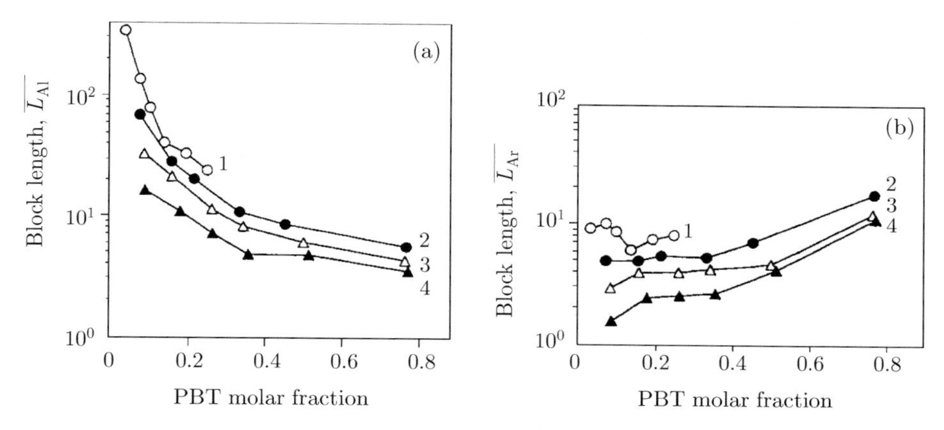

Figure 4. Dependence of the block length on the PBT molar fraction for PBAS-PBT copolymers. Mixing times at 290°C: (1) – 0 min, (2) – 10 min, (3) – 20 min, (4) – 30 min; (a) aliphatic units, (b) aromatic units [35]

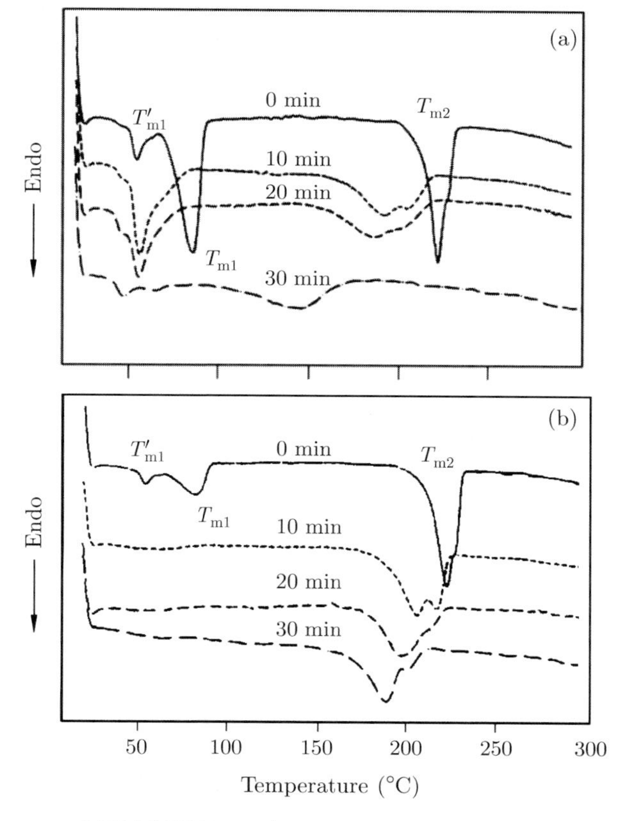

Figure 5. DSC traces of PBAS-PBT copolymers with mixing times of 0, 10, 20, and 30 min at 290°C: (a) 40 mole% terephthalate units, (b) 60 mole% terephthalate units [35]

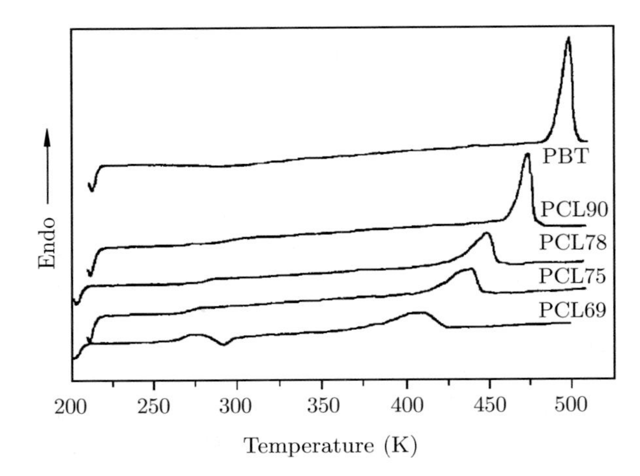

Figure 6. DSC thermograms of PCL-PBT copolymers (numbers after PCL indicate the flexible segment contents by wt) [34]

end-linking method. Depending on the content of flexible and rigid segments along the chains, some of them reveal thermoplastic elastomer properties and better resistance to thermooxidative aging than copoly(ether-ester)s. Poly(ester-ester) elastomers are produced by DSM (The Netherlands, Arnitel® U) [186,187] (see also Chapter 17).

The application of PET as a rigid segment is interesting because this polymer does not crystallize under the usual processing conditions, *i.e.*, the phase separation conditions for PET-based PEE differ from those for PBT-based PEE. The synthesis and properties of PET-PEO [76–82], PET-PTMO [75,76], and PET-PCL [83–85] block copolymers have been described. The addition of 10 wt% of polyether segments in PET-based PEE hampers the so-called "cold crystallization" of PET, and these copolymers can be considered semicrystalline, with a very poor crystallinity. It is worth noting the melting point depression caused by the polyether segments (Figure 7) [76]. The miscibility in these systems is better than that of PBT and PTMO segments.

The PEEs with poly(tetramethylene 2,6-naphthalenedicarboxylate) (PBN) hard segment (Scheme 12) were prepared by melt polycondensation. PTMEG of various molecular weights was introduced as the flexible segment. Tetrabutyl orthotitanate (0.1 wt%) and lead acetate (0.1 wt%) were used as catalysts [86].

The glass transition temperature of copolymers containing PTMO (molecular weight of 1000 and 2000 g/mole) changed from –46 to –68°C with the rise in the flexible segment content. The melting points changed from 234 to 169°C [87]. A relatively high decomposition temperature was observed.

Kricheldorf and Langarke [142] synthesized a poly(ether-ester) block copolymer, using poly(ε-caprolactone) (PCL) as the rigid segment. The PTMO-PCL block copolymer was prepared by ring-opening polymerization (Scheme 13) of tin-containing PTMO macrocycles and caprolactone, with dibutyltin dimethoxide

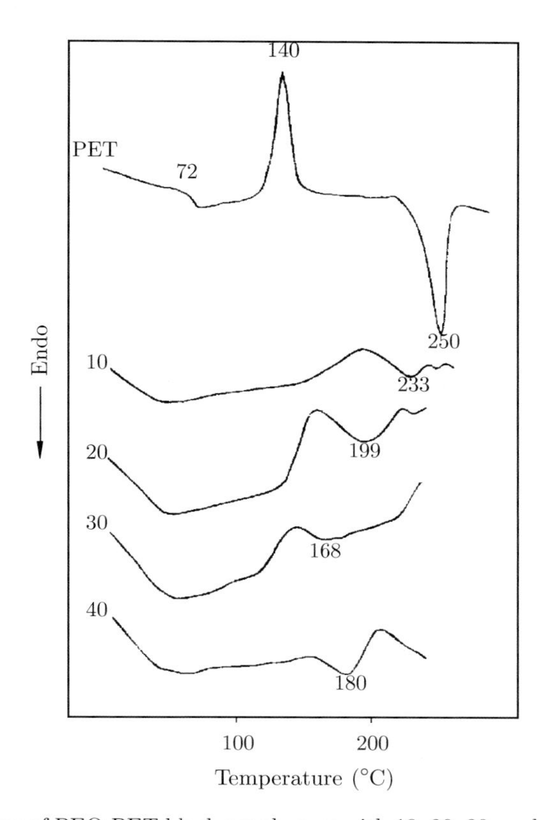

Figure 7. DSC traces of PEO-PET block copolymers with 10, 20, 30, and 40 wt% PEO [76]

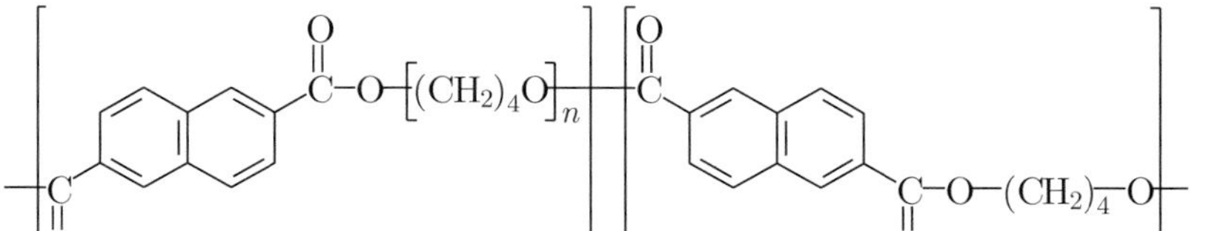

PTMO flexible segment

PBN rigid segment

PTMO–*b*–PBN

Scheme 12

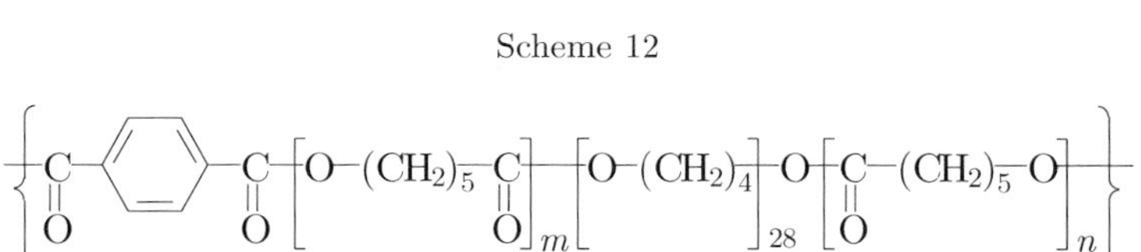

$m + n = 50$

PTMO–*b*–PCL

Scheme 13

as catalyst. A broad glass-transition step of the PTMO-PCL block copolymers is detected between –70 and –55°C. It combines the glass transitions of the PTMO blocks and of the amorphous part of the PCL blocks. The PTMO-PCL multiblock copolymers show a high deformation (Figure 8) and are thermoplastic elastomers. A PEO-PCL block copolymer was also synthesized and characterized [88].

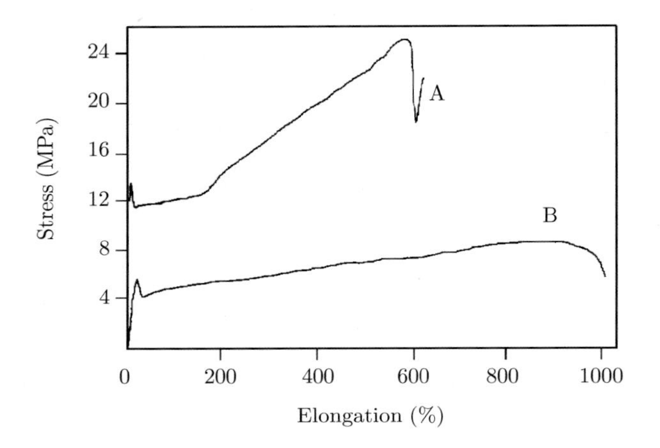

Figure 8. Stress-strain curves of commercial PCL (A) and a PTMO-PCL based multiblock copolymer (B) [88]

PTMO-D,L-lactide block copolymers have similar physical properties [89]. Very interesting are the chemical structure and the synthesis of a poly(ester-carbonate) multiblock copolymer based on poly(lactic-glycolic acid) and PCL segments [188]. The carbonate group was used as the end-linking element of two different polyester segments.

Dimerized fatty acids (DFA) can be used as an alternative to polyether segments in TPE. The polyester elastomers based on DFA were first prepared by Hoeschele [42]. The properties affecting the degradation stability were improved, but the properties at lower temperatures were unsatisfactory, because the higher glass transition temperature was influenced by the lengths of the amorphous segments. DFA can be coupled with diols and diamines to synthesize telechelic polyesters (Scheme 14) and polyamides, respectively.

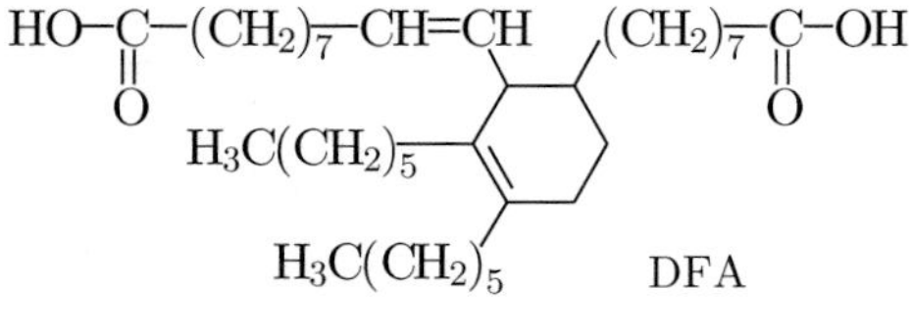

HO–C(=O)–$(CH_2)_7$–CH=CH $(CH_2)_7$–C(=O)–OH

$H_3C(CH_2)_5$

$H_3C(CH_2)_5$ DFA

Scheme 14

Gaymans and Manuel [51,52], and Slonecki *et al.* [43–50,53–57] carried out extensive research on this subject. Soybean oil-based dimer acids can be prepared by condensation of unsaturated fatty acids, such as oleic and linolic acid, in the presence of a catalyst (Scheme 15). Hydrogenated dimer acids (HDFA) were also examined (Scheme 16) [189].

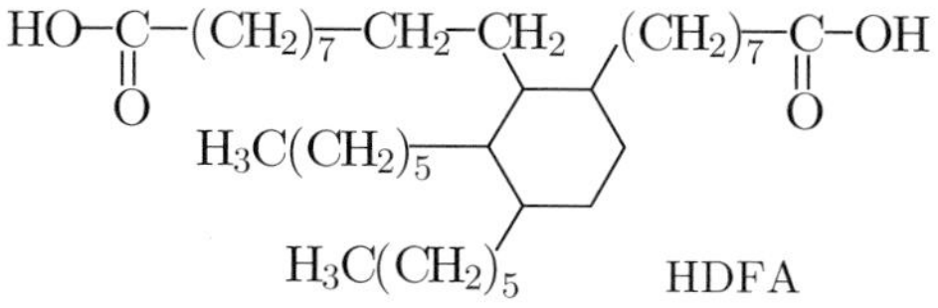

$HO-C(=O)-(CH_2)_7-CH_2-CH_2$ $(CH_2)_7-C(=O)-OH$
$H_3C(CH_2)_5$
$H_3C(CH_2)_5$ HDFA

Scheme 15

$\left[-C(=O)-C_6H_4-C(=O)-(O-CH_2)_4-O- \right]_x \left[-C(=O)-R^3-CHR^1-CHR^2-CH_2-R^4-C(=O)-O-(CH_2)_4-O- \right]_y$

Rigid segment Flexible segment

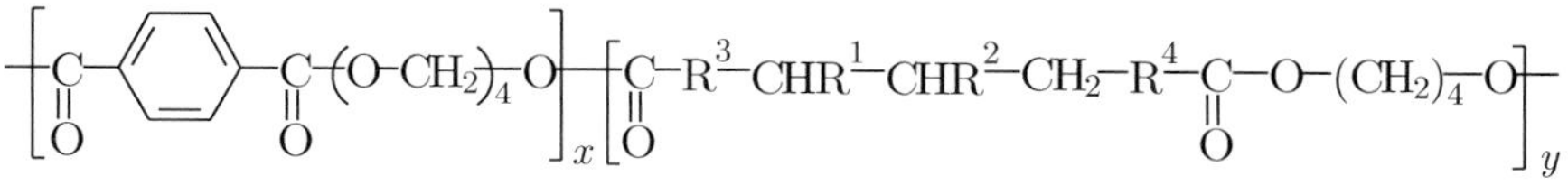

x, y – degree of polycondensation of rigid and flexible segments, respectively
$R^1 + R^2 + R^3 + R^4 = C_{31}$

Scheme 16

The melting and crystallization temperatures of the PBT-rich hard phase are controlled by the content of DFA segments in the copolymers (Figure 9). The glass transition temperature of the soft phase also strongly depends on the DFA content (Figure 10) [55]. Carboxyl-terminated oligoamides of DFA ($M_n = 575$ g/mole) and diamines, *e.g.*, 1,6-hexamethylene diamine, 1,4-cyclohexamethylene

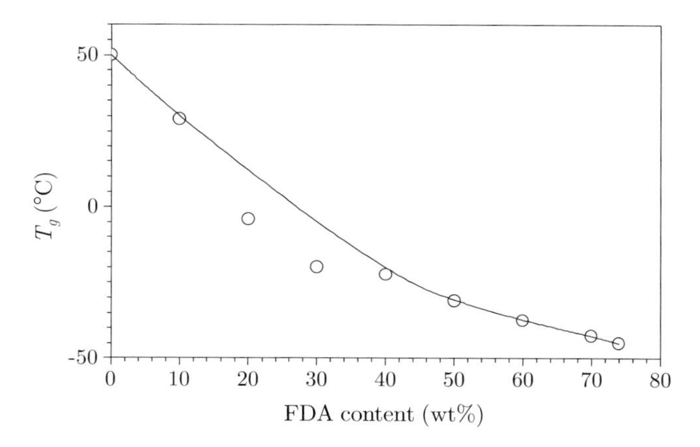

Figure 9. Dependence of the glass transition temperature on the FDA soft segment content in copoly(ester-ester) multiblock copolymers [55]

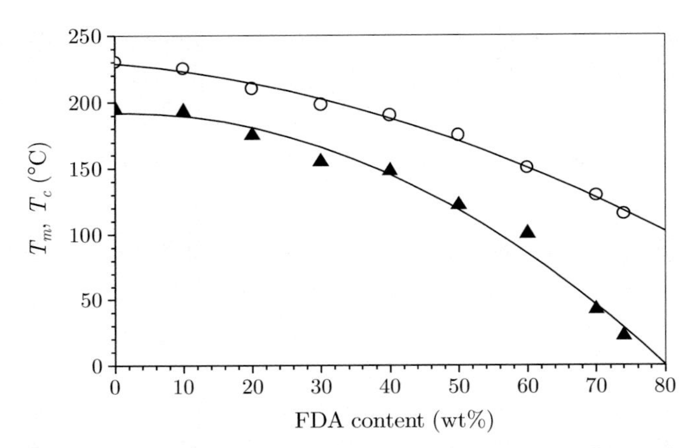

Figure 10. Melting (○) and crystallization (▲) temperature of a DFA-based copoly(ester-ester) multiblock copolymer *vs.* flexible segment content [55]

diamine, piperazine, and α,ω-di(3-aminopropyl)-poly(oxytetramethylene) were synthesized and used to obtain a new poly(amide-ester) multiblock copolymer.

Methyloxylithium and a MgTi organometallic complex were used as catalysts [56]. The physical properties of the poly(amide ester) (PAE) are similar to those of the poly(ester-ester) copolymers mentioned above (Figure 11) [57].

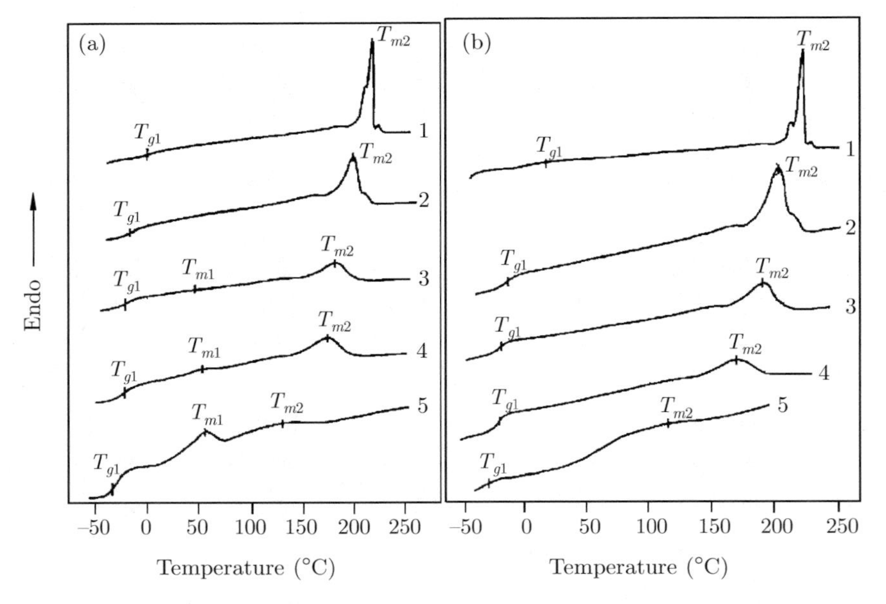

Figure 11. DSC plots of the PAE series with various contents of DFA/diamine-based flexible segments: (a) hexamethylene diamine, (b) piperazine; (1) – 20, (2) – 40, (3) – 50, (4) – 60, and (5) – 80 wt% PA segment content [57]

PAE reveals a stress-strain dependence typical of thermoplastic elastomers, as shown in Figure 12 [57].

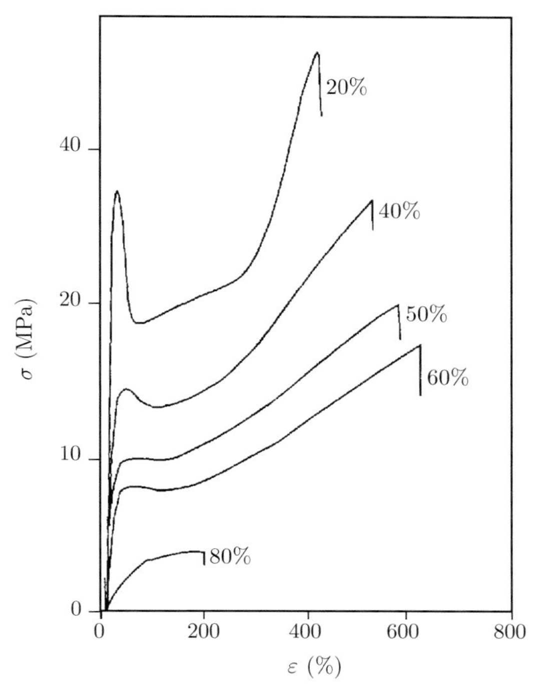

Figure 12. Stress-strain curves of PAE with DFA/hexametylene diamine-based flexible segments and various PA segment contents (wt%) [57]

Poly(olefin-ester) multiblock copolymers belong to a quite new and fast developing class of polyester elastomers [58–60]. The copolymer with PBT rigid segments and flexible polyisobutylene (PIB) segments was prepared by polycondensation in the molten state with DMT, BD, and functionalized PIB, in the presence of titanium tetrabutoxide as catalyst. α,ω-Dianhydride [58] and dihydroxyl-terminated PIB of molecular weights 1000, 2200, 4800, and 10 000 g/mole were used (Schemes 17 and 18). A multiblock copolymer with PBT and

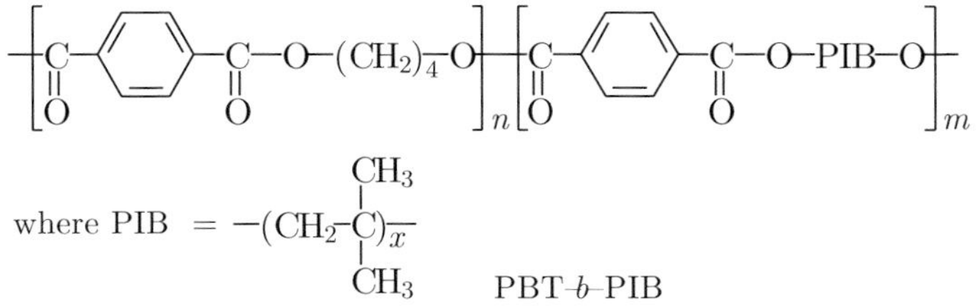

$$\left[\!\!-\overset{\displaystyle O}{\overset{\|}{C}}-C_6H_4-\overset{\displaystyle O}{\overset{\|}{C}}-O-(CH_2)_4-O-\right]_n\left[-\overset{\displaystyle O}{\overset{\|}{C}}-C_6H_4-\overset{\displaystyle O}{\overset{\|}{C}}-O-PIB-O-\right]_m$$

where PIB $= -(CH_2-C(CH_3)_2)_x-$ PBT-*b*-PIB

Scheme 17

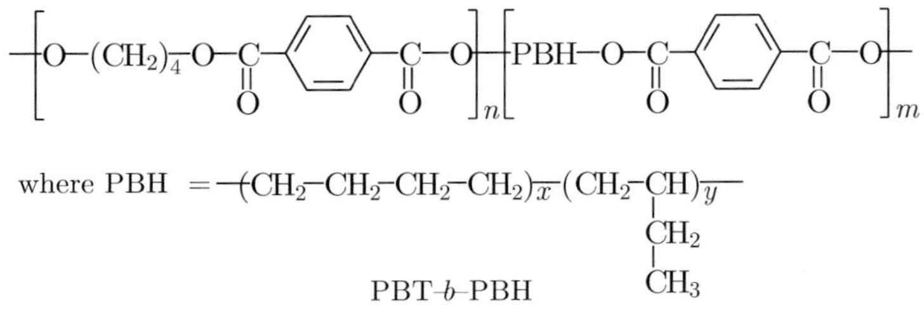

$$\left[O-(CH_2)_4-O-\underset{\|}{\overset{}{C}}-C_6H_4-\underset{\|}{C}-O \right]_n \left[PBH-O-\underset{\|}{C}-C_6H_4-\underset{\|}{C}-O \right]_m$$
(each C=O)

where PBH $= -(CH_2-CH_2-CH_2-CH_2)_x-(CH_2-CH(CH_2-CH_3))_y-$

PBT-*b*-PBH

Scheme 18

hydrogenated polybutadiene (PBH, $M_n = 2000$) as flexible segment was also synthesized and characterized.

The process was carried out in the molten state, using tetrapropoxy titanium as catalyst or *via* terephthaloyl chloride in solution. The copolymers show the properties of thermoplastic elastomers. All PBT-PIB multiblock copolymers have a low glass transition temperature of about −65°C, *i.e.*, much lower than that of the common commercial TPEs with polyether or aliphatic polyester as flexible segment. The PIB segment does not crystallize. The melting point of the copolymers depends on the PBT block length (Figure 13). The replacement of the polyether flexible segment by PIB allowed the synthesis of TPE with a lower modulus (Figure 14) than it was possible for PEE. PIB-based copolymers show also better thermal and thermooxidation stability.

The multiblock copolymers of poly(alkylene terephthalate) with telechelic amine-terminated poly(butadiene-*co*-acrylonitryle) rubber (ATBN) are also very interesting [61–63]. The copolymer based on poly(pentamethylene tere-

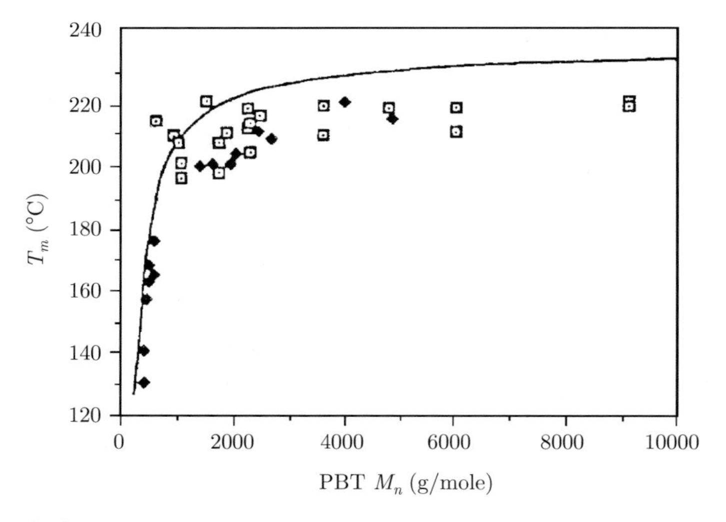

Figure 13. Melting temperature *vs.* the calculated PBT block length. ▣ PBT-*b*-PIB copolymer; ■ PBT-*b*-PTMO copolymer; the theoretical curve for extended chains is also given [58]

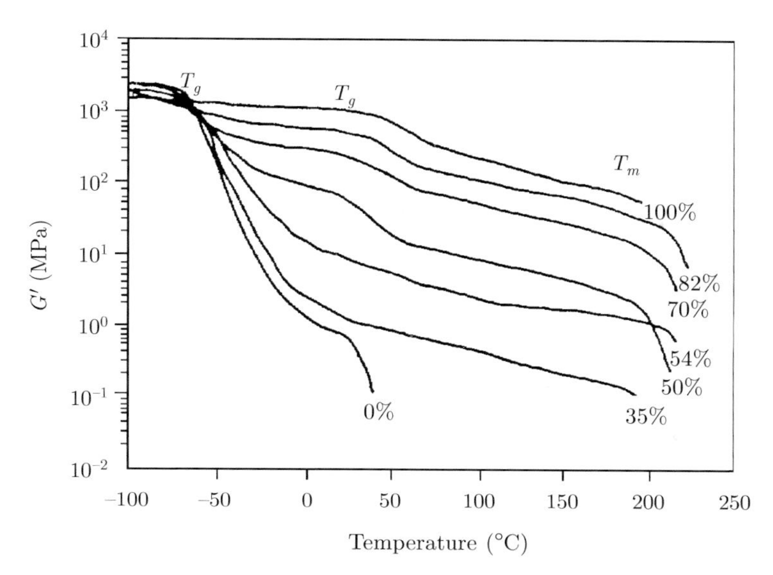

Figure 14. Torsional modulus *vs.* temperature dependence of some PBT-*b*-PIB copolymers with a PIB M_n = 1500 and of the respective homopolymers. PBT content in wt% [58]

phthalate) and poly(butadiene-*co*-acrylonitrile) (PHT-*b*-ATBN) is an example of this structure (Scheme 19) [63].

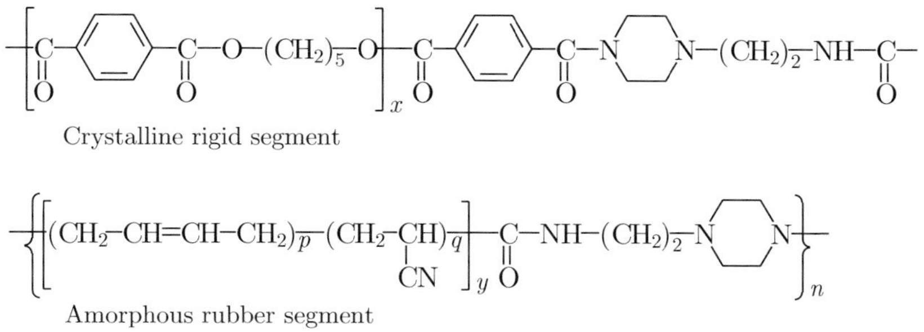

$$\left[-\overset{\overset{\displaystyle O}{\|}}{C}-C_6H_4-\overset{\overset{\displaystyle O}{\|}}{C}-O-(CH_2)_5-O- \right]_x -\overset{\overset{\displaystyle O}{\|}}{C}-C_6H_4-\overset{\overset{\displaystyle O}{\|}}{C}-N(C_4H_8)N-(CH_2)_2-NH-\overset{\overset{\displaystyle O}{\|}}{C}-$$

Crystalline rigid segment

$$\left\{ \left[(CH_2-CH=CH-CH_2)_p-(CH_2-\underset{CN}{CH})_q \right]_y -\overset{\overset{\displaystyle O}{\|}}{C}-NH-(CH_2)_2-N(C_4H_8)N- \right\}_n$$

Amorphous rubber segment

PHT-*b*-ATBN

Scheme 19

8. Polyester thermoplastic elastomers from blends

Reactive blends of rubber and plastics are well known thermoplastic elastomers [132]. However, only a few blends with PBT are described. Cai and Isayev [113,114] reported the obtaining of new thermoplastic elastomers from blends of a copoly(ether-ester) (PBT-*b*-PTMO) and poly(acrylonitrile-butadiene) rubber (NBR) by the dynamic vulcanization process. These types of TPE show very good mechanical and rheological properties, and can be used in the production of shock absorber elements [115].

The blends of PBT or PET with poly(ethylene-acrylate) rubber [116,117], PBT with polybutadiene rubber [121], and PBT with EPDM [122,123] obtained by *in situ* reactive blending were also examined. TPEs by grafting of PBT on the acryl units of the poly(ethylene-acrylate) rubber or by covulcanization of unsaturated PBT with polybutadiene rubber were also reported. The covulcanization of unsaturated PEE with EPDM was studied by Siemiński [73]. Some authors modified PBT with small amounts of rubber in order to improve its impact resistance [190]. Manas-Zloczower *et al.* [191] and Utracki [192] reviewed the procedures of preparation of such blends.

Thermoplastic elastomers containing PET (50 wt%), compatibilizer (glycidyl methacrylate grafted rubber or glycidyl methacrylate containing copolymer, 30 wt%) and various rubbers (20 wt%) were produced by melt blending with and without dynamic curing (dicumyl peroxide-initiated) [118].

The preparation of a new material by means of reactive blending of PET bottle waste and poly(tetramethyleneoxide) (PTMO) has been described. The diphase structure of this material has been shown by DSC (Figure 15). The stress-strain characteristics of the prepared materials are typical of thermoplastic elastomers (Figure 16) [119].

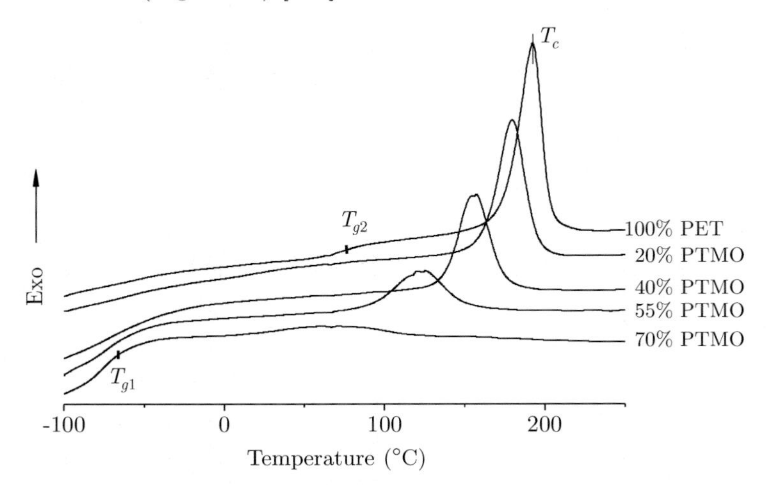

Figure 15. DSC plots of PET and PET/PTMO reactive blends with various PTMO contents. Cooling rate 10°C/min, T_g and T_c – glass transition and crystallization temperature, respectively [119]

9. A new processing aspect: weldability of polyester elastomers

Polyester elastomers have good processability by extrusion, injection molding, blow molding, and casting molding (hoses, pipes, technical goods, fire hose films, see also Chapter 17). Poly(ether-ester) elastomers containing predominantly

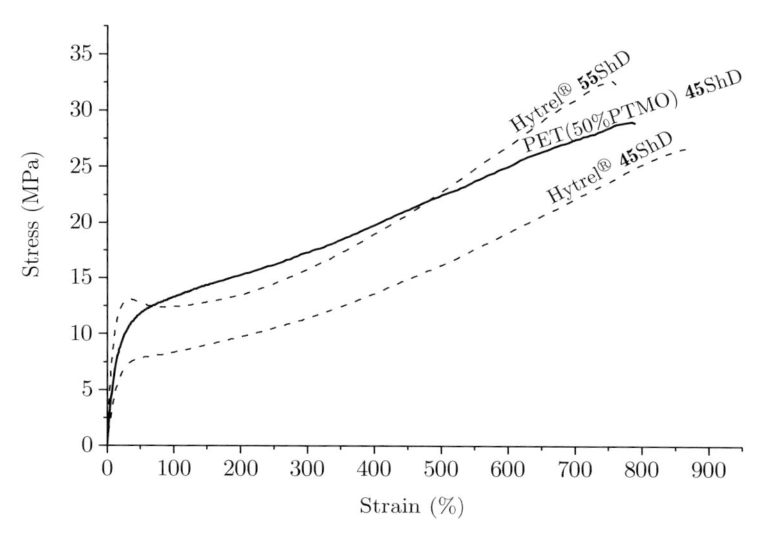

Figure 16. Stress-strain curves of a PET waste-based elastomer and Hytrel® (DuPont, USA) elastomers [119]

the PTMO soft phase are weldable [193]. SEM micrographs of the welded fracture surface of these specimens show a stable joint between the weld and the original material. The mechanical properties of this weld are similar to those of the original material, except for the elongation at break, which is considerably lower.

The mechanical properties deteriorate with increasing rigid segment content. The ester bonds in the rigid segments are susceptible to hydrolytic degradation, which occurs in seams produced when a non-dried material is used. SEM micrographs of the weld cross-sections show defects in the form of bubbles. The mechanical properties of extruded welds are generally slightly better than those of the wire welds. When a polymer is thermally sensitive, wire welding gives better results, since the weld is not melted. The attempts to produce seams using materials with hardness differing from that of the materials being joined were promising and allowed to influence the seam properties. This is important, because almost always the same material has to be used both for the seam and the welded parts. It is useful to investigate the welding of PEE elastomers with materials dried immediately before processing. Stabilizing agents should be developed in parallel, in order to prevent hydrolytic degradation and to improve the weldability of elastomers with higher contents of ester rigid segments.

10. Polyester elastomers for biomedical application

Synthetic commodity polymers are generally hardly degradable in nature and cause ecological problems when disposed. Aromatic polyesters (PBT, PET, PEN) do not show any sign of degradation by microorganisms [173–175], but their copolymers with aliphatic esters and ethers can be biodegradable [154,176–184]. The prefix "bio" is considered to reflect phenomena resulting from the contact with living tissues, cells, body fluids, or microorganisms [174]. Accordingly, water, oxygen, and enzymes are also regarded as biological elements, although recent trends tend to limit the meaning of biodegradation to enzymatic attack only. Sometimes, the biodegradability of polymers is related to bioabsorbability and biocompatibility [82,176]. A bioabsorbable (biocompatible) polymer is a polymer that can be assimilated by a biological system. Terephthalic acid polyester-based copolymers with polycaprolactone, adipic acid, sebacic acid, poly(glycolic acid) and poly(L-lactic acid) appeared to be biodegradable [35,177–179]. The biodegradability of PET-based copolyesters has been reviewed by Kint and Muñoz-Guerra [174].

Nagata *et al.* [181] studied the biodegradability of terephthalic acid-based block copolymers with PEO and PTMO as polyether segments. The biodegradation mechanism of PEO-PET block copolymers was studied *in vitro* by Reed and Gilding [82]. These materials are promising as biodegradable elastomers, similar to poly(ethylene oxide-hydantoin)-poly(butylene terephthalate) block copolymers [176]. Zhou *et al.* [182] have found that aliphatic polycarbonates, which can be used as flexible segments of polyester block copolymers, are degradable by *Rhizopus delemalippase*. The biocompability of PEO, PPrO, and PTMO segment-based block copolymers has been investigated by Mochizuki *et al.* [183]. The results indicate that these copolymers have good blood compatibility.

The biocompatible dimerized fatty acid (DFA)-based poly(aliphatic-aromatic ester) elastomers (PED) have been synthesized and studied for biomedical applications by El Fray *et al.* [194–200]. The design of nanostructured elastomeric biomaterials (mimicking biological materials) has been realized by using renewable resources, *i.e.*, DFA. They are prepared by transesterification and polycondensation from the melt (see Section 7). The exceptional properties of DFA, *e.g.*, excellent resistance to oxidative and thermal degradation, allow the preparation of PEDs without the use of thermal (often irritating) stabilizers. This is a particularly important feature making these polymers environmentally friendly and additive-free. What is equally important, by the use of the same method and stabilizer-free conditions, it was possible to prepare specially modified PED copolymers with an increased surface hydrophobicity.

Differences in morphology and nanostructure formation result in different mechanical properties, placing these new PED materials between commercially available thermoplastic poly(urethane-ether) and poly(ether-ester) elastomers. The advantages offered by PED, however, are not only related to their higher hydrophobicity, but also to their fatigue properties. These properties are particularly important in terms of mimicking natural biostructural materials, such as tendons, which experience cyclic loading patterns *in vivo*. By means of the hysteresis measurement method, it was possible to evaluate the load-carrying properties of polymers containing high amounts of the hard phase. PED copolymers show much better creep resistance when compared to thermoplastic biomedical polyurethanes [198,199]. The increasing concentration of poly(dimethyl siloxane) (second flexible segment) improves the creep resistance at ambient and elevated (37°C) temperature in the presence of a simulated physiological fluid. PED materials show good dynamic relaxation during stepwise increasing strain, particularly in the region of small strains (3–10%), which are critical for natural tendon tissues.

The evaluation of the biocompatibility of these new materials was an important achievement [199]. *In vitro* screening tests revealed controlled biodegradation susceptibility, cell proliferation and viability comparable to medical grade polyurethane elastomers. The amount of extractables, mainly terephthalate derivatives, has been found below the GC/MS assay sensitivity limit (0.02 ppm). The identified low molecular weight extractables from saline extracts of PED copolymers did not show pyrogenic activity on rabbits, thus confirming their non-toxicity. Finally, a long term *in vivo* implantation test (6 months) confirmed the good biocompatibility of PED copolymers. The animal implantation test has been used and documented. Polymer rods were implanted in the soft tissue of the abdominal wall and into muscles of Wistar rats weighing 200–250 g and into muscles of rabbits. A promising issue is the impregnation of PED materials with antibiotics or silver salts to gain antibacterial properties.

PED copolymers show the stress-strain curves typical of thermoplastic elastomers [194–197] (Figure 17). Their shapes change with composition; at low concentrations of the hard phase, the stress-strain curves are typical of elastomers while an increase of the hard (semicrystalline) phase imparts a higher toughness to the material. The values of the ultimate tensile strength vary between 3 and 45 MPa (the elastic-plastic deformation point, the so-called yield strength varies from 5 to 17 MPa), while the elongation at break is between

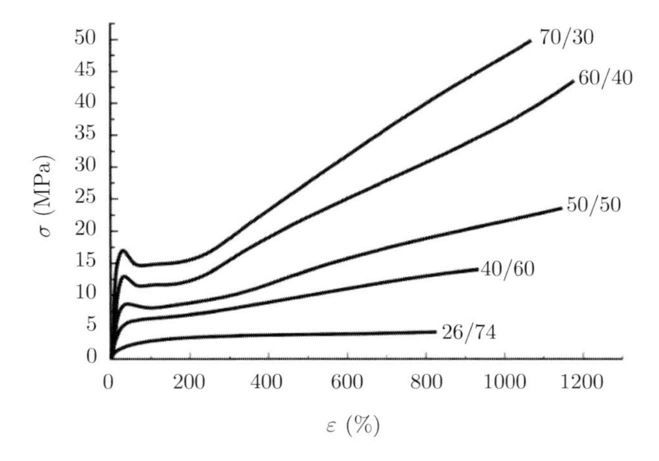

Figure 17. Stress-strain curves of PED copolymers (measurements performed on 0.5 mm thick films; crosshead speed of 100 mm/min). The PBT-to-DFA segment ratio in the samples varies from 70/30 to 26/74 [194]

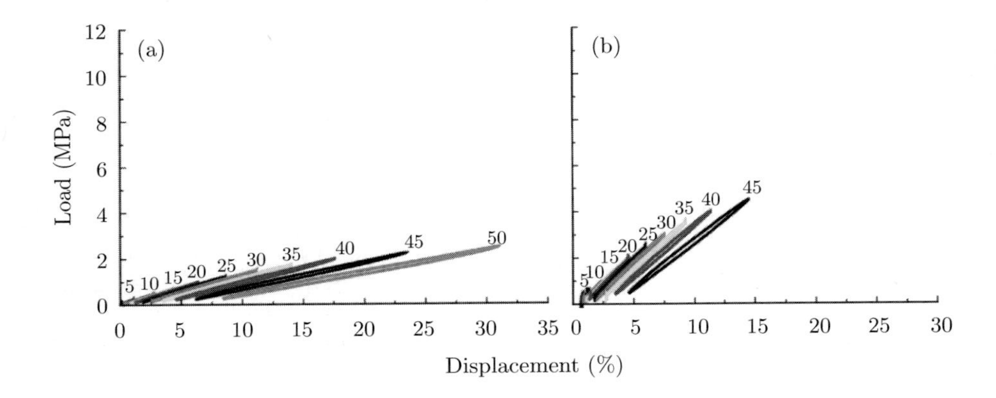

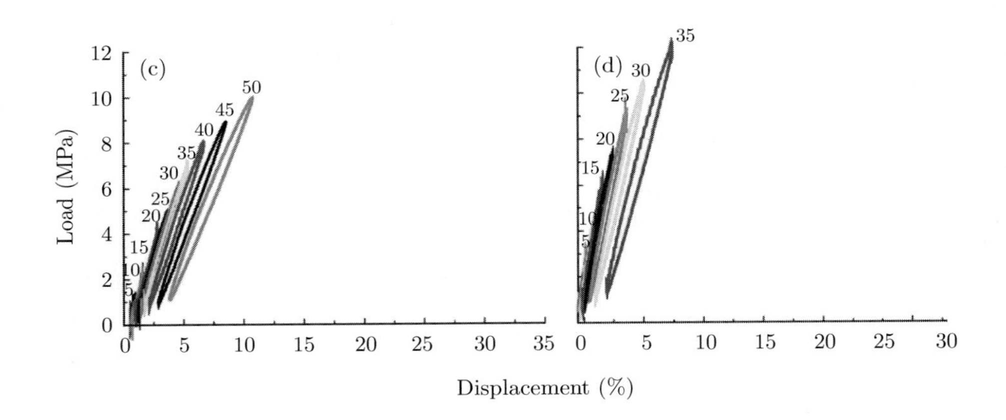

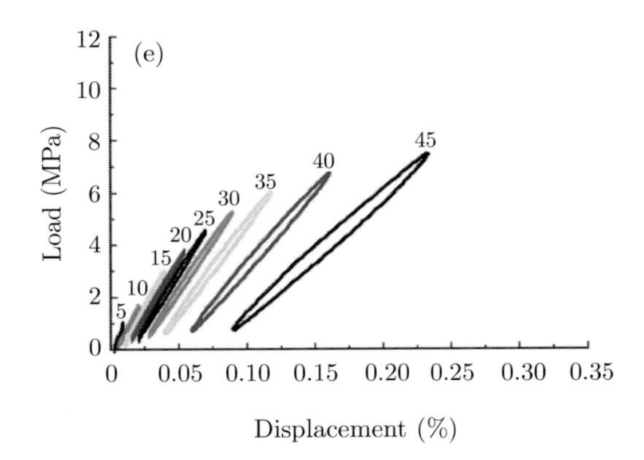

Figure 18. Representative hysteresis loops for each load level (numbers of loops refers to percent of ultimate tensile strength): (a)-(d) PED copolymers containing 26, 40, 60 and 70 wt% rigid segments, respectively, testing frequency of 4 Hz; (e) sample 60/40 tested at 1 Hz; T = 24°C [194]

800 and 1200%. These mechanical characteristics of PED copolymers match the mechanical properties of flexor tendons and have much better tensile properties than the silicone elastomers presently used in flexor tendon grafting (tensile strength of 3–11 MPa). In addition to their good mechanical properties, PED copolymers show excellent rubber elasticity. At higher rigid segment contents (higher crystallinity), a lower displacement and a larger area of the hysteresis loop are observed. The area of the hysteresis loop increases with increasing load level for each polymer. It has also been found that the displacement of the hysteresis loop depends on the test frequency, as demonstrated in Figure 18e, representing a copolymer tested at 1 Hz. It shows much lower displacement values than the same material tested at 4 Hz.

PED copolymers tested at different temperatures reveal a significant effect at 37°C on the creep behavior. No significant influence, however, of the tested environment (air or liquid) was observed at 24°C. The nanostructured elastomeric biomaterials can be used for soft tissue reconstruction.

11. Conclusions and outlook

The world demands for thermoplastic elastomers are forecast to expand by 7.5% per year to 2.6 million metric tons in 2006. TPEs will continue to find the majority of their applications as replacements for natural and synthetic rubbers, as well as rigid thermoplastics and metals. The global TPE industry will remain heavily concentrated in the USA, Western Europe and Japan, particularly for specialty materials, such as polyester elastomers. From the two trends of application of polyester elastomers as engineering plastics and functional materials, the second one is prioritized. In terms of engineering applications, the PEE production will remain closely related to the motor vehicle industry, sporting goods, hoses, and small household goods.

Extensive investigations on the synthesis of block copolymers based on polyester (not only terephthalate) rigid segments and various flexible segments (polyether, aliphatic polyester, dimerized fatty acid, polyolefin, and even polyamide segments) resulted in enlarging the group defined as polyester thermoplastic elastomers. The modifications of the synthesis aimed both at the establishment of the effect of the chemical structure on the synthesis conditions and the copolymer physical properties and at the improvement of some of their useful properties, such as the thermal stability [201], processability, and service temperature range. At present, attention should be focused on polyester elastomers synthesized from monomers obtained from renewable resources. Another future trend in TPE development should be related to the creation of biodegradable copolymers. In this way, the solution of the environmental problems created by petrochemical synthetic polymers could be facilitated.

Another important trend in TPE developments refer to biomedical applications. In addition to the chemistry of the polyester-based TPE and some of their physical properties reviewed in this chapter, specific applications are also considered, such as, *e.g.*, in biomedical practice. In this respect, the TPE are very promising materials since, due to their peculiar physical structure, they reveal properties and behavior typical of the intelligent ("smart") materials, *e.g.*, shape memory effects.

References

1. *Handbook of Thermoplasic Elastomers* (1988) (Eds. Walker B M and Rader Ch P) Van Nostrand Reinhold, New York.
2. *Thermoplastic Elastomers* (1996) (Eds. Holden G, Legge N R, Quirk R and Schroeder H E) Hanser Publishers, Munich, Vienna, New York.
3. *Thermoplastic Elastomers from Rubber Plastic Blends* (1990) (Eds. De S K and Bhowmick A K) Ellis Horwood, New York, London, Toronto, Sydney, Tokyo, Singapore.
4. Hofmann W (1987) Thermoplastic elastomers, *Kunststoffe* **77**:767–776 (in German).
5. Domininghaus H (1989) Thermoplastic elastomers: Development, structure, types – a review, *Plastverarbeiter* **40**(1):39–46; (2):42–47; (3):30–38 (in German).
6. Lauhus W P, Haberstroh F and Ehrig F (1997) Technical elastomers, *Kunststoffe* **87**:706–716.
7. Rajan G S, Vu Y T, Mark J E and Myers Ch L (2004) Thermal and mechanical properties of polypropylene in the thermoplastic elastomeric state, *Eur Polym J* **40**:63–71.
8. Coleman D (1954) Block copolymers: copolymerization of ethylene terephthalate and polyoxyethylene glycols, *J Polym Sci* **14**:15–28.
9. Hoeschele G K and Witsiepe W K (1973) Polyetherester-block-copolymers – a new group of thermoplastic elastomers, *Angew Makromol Chem* **29/30**:267–289 (in German).
10. Ash M and Ash I (1992) *Handbook of Plastic Compounds, Elastomers and Resins*, VCH Publ. Inc., New York.
11. Maréchal E (2000) Block copolymers with polyethers as flexible blocks, in *Block Copolymers* (Eds. Baltá Calleja F J and Roslaniec Z) Marcel Dekker Inc., New York, Basel, pp. 29–63.
12. Higashiyama A, Yamamoto Y, Chujo R and Wu M (1992) NMR Characterization of segment sequence in polyester-polyether copolymers, *Polymer J* **24**:1345–1349.
13. Slonecki J (1992) The effect of mass fraction and molecular weight of segments on the conditions of preparation, structure and properties of thermoplastic copolyester-ethers (KPEE). Part III. The effect of mass fraction and molecular weight of segments on some mechanical properties of KPEE, *Polimery* **37**:19–24 (in Polish).
14. Slonecki J (1991) The effect of mass fraction and molecular weight of segments on the conditions of preparation, structure, and properties of thermoplastic copolyester-ethers (KPEE). Part II. The effect of KPEE structure on its thermal and relaxation properties, *Polimery* **36**:422–427 (in Polish).
15. Broza G (1995) Thermoplastc poly(etherester) block copolymers, PhD Thesis, Technical University of Szczecin, Szczecin (in German).
16. Szafko J, Roslaniec Z, Schulte K, Broza G and Petermann J (1994) Morphology studies of high oriented poly(ether-ester) block copolymer, in 3rd Intl Symp Thermoplastic Elastomers, Kolobrzeg (Poland), *Sci Papers Tech Uni Szczecin* **514**:151–159.
17. Xie H Q and Xie D (1999) Molecular design, synthesis and properties of block and graft copolymers containing polyoxyethylene segments, *Progr Polym Sci* **24**:275–313.
18. Fakirov S and Gogeva T (1990) Poly(ether esters) based on poly(butylene terephthalate) and polyethylene glycol. 1. Poly(ether esters) with various polyether:polyester ratios. 2. Effect of polyether segment length. 3. Effect of thermal treatment and

drawing on the structure of the poly(ether esters), *Makromol Chem* **191:**603–614, 615–624, 2341–2354.

19. Bogdanov B, Mikhailov M, Gavrilova G and Panev P. (1991) Melting of oligo(ethylene terephthalate) oligo(oxyethylene) block copolymers, *Acta Polymerica* **42**:255–260.
20. Gogeva T and Fakirov S (1990) Poly(ether/ester)s based on poly(tetramethylene terephthalate) and poly(ethylene terephthalate) and poly(ethylene glycol) 4. Modification with 2-butyne-1,4-diol, *Makromol Chem* **191**:2355–2365.
21. Gogeva T, Fakirov S, Mishinev J and Sarkisova L (1990) Poly(ether ester) fibres, *Acta Polymerica* **41**:31–36.
22. Slonecki J (1995) The effect of chain length and soft oligo(oxyethylene) segments content on some properties of copoly(ester-ethers). Part I. Thermal properties, *Polimery* **40**:572–577 (in Polish).
23. Chu B and Li Y (1993) Synchrotron SAXS studies of segmented polyurethanes, *Progr Colloid Polym Sci* **91**:51–54.
24. Bednarek M, Kubisa P and Penczek P (1999) Coexistence of activated monomer and active chain end mechanisms in cationic copolymerization of tetrahydrofuran with ethylene oxide, *Macromolecules* **32**:5257–5263.
25. Killmann E, Cordt F, Möller F and Zellner H (1995) Thermodynamics of mixing of propylene oxide oligomers with different end groups and of statistical and block cooligomers of ethylene oxide and propylene oxide in tetrachloromethane, *Macromol Chem Phys* **196**:47–62.
26. Yokoyama M, Anazawa H, Takahashi A, Inoue S, Kataoka K, Yui N and Sakurai Y (1990) Synthesis and permeation behavior of membranes from segmented multiblock copolymers containing poly(ethylene oxide) and poly(β-benzyl L-aspartate) blocks, *Makromol Chem* **191**:301–311.
27. Nicholas Ch V, Luo Y Z, Deng N J, Attword D, Collett J H, Prince C and Booth C (1993) Effect of chain lenght on the micellization and gelation of block copoly(oxyethylene/oxybutylene/oxyethylene) $E_mB_nE_m$, *Polymer* **34**:138–144.
28. Higgins J S and Carter A J (1984) Molecular conformation and interactions in oligomeric mixtures of poly(ethylene glycol) and poly(propylene glycol) methyl ethers: A small-angle neutron scattering study, *Macromolecules* **17**:2197–2202.
29. Slonecki J (1991) Block copoly(ether-ester)s with oligo(1,2-oxypropylene) soft segments, *Acta Polymerica* **42**:655–660.
30. Lembicz F and Slonecki J (1991) Thermomechanical properties of poly(1,2-oxypropylene)-poly(tetramethylene terephthalate) block copolymers, *Kaut Gummi Kunstst* **44**: 668–670.
31. Wu C, Woo K F, Luo X L and Ma D Z (1994) A modified light-scattering method for the characterization of the segmented copolymer poly(ethylene terephthalate-co-caprolactone), *Macromolecules* **27**:6055–6060.
32. Ma D Z and Prud'home R E (1990) Miscibility of caprolactone/ethylene terephthalate copolymers with chlorinated polymers: a differential scanning calorimetry and Fourier-transform infrared study, *Polymer* **31**:917–923.
33. Zhang R Y, Luo X L and Ma D Z (1995) Miscibility of polyhydroxy ether of bisphenol A with ethylene terephthalate-caprolactone copolyesters, *Eur Polym J* **31**:1011–1014.
34. Cheng X, Luo X L and Ma D Z (1999) Proton NMR characterization of chain structure in butylene terephthalate-ε-caprolactone copolymers, *J Polym Sci A Polym Chem Ed* **37**: 3770–3777.
35. Kang H J and Park S S (1999) Characterization and biodegradability of poly(butylene adipate-*co*-succinate/poly(butylene terephthalate) copolyester, *J Appl Polym Sci* **72**:593–608.

36. Tijsma E J, van der Does L, Bantjes A and Vulic I (1994) Machanical properties and chemical stability of pivalolactone-based poly(ether esters)s, *Polymer* **35**:5483–5490.
37. Tijsma EJ, van der Does L, Banties A, Vulic I and Werumeus Buning G H (1994) Poly(ether ester)s from pivalolactone, alkanediols, and dimethyl terephthalate. 2. Synthesis and characterization, *Macromolecules* **27**:179–186.
38. Roslaniec Z and Wojcikiewicz H (1988) Synthesis and characterization of elastomers based on poly(terephthalate-co-carbonate)s, *Polimery* **33**:360–363.
39. Roslaniec Z (1992) Dynamic mechanical and microcalorimetric investigations of the phase structure of multiblock elastomers, *Polimery* **37**:328–335.
40. Ezquerra T A, Roslaniec Z, Lopez-Cabarcos E and Baltá Calleja F J (1995) Phase separation and crystallization phenomena in a poly(ester-carbonate) block copolymer: A real-time dielectric spectroscopic and X-ray scattering study, *Macromolecules* **28**:4516–4522.
41. Roslaniec Z (1997) Poly(carbonate-ester) block copolymers. Part. I. Synthesis and characterization, *Polimery* **42**:465–469 (in Polish).
42. Hoeschele G K (1997) Thermostable polyester-block-copolymers, *Angew Makromol Chem* **58/59**:299–319 (in German).
43. El Fray M and Slonecki J (1996) Multiblock copolymers consisting of polyester and polyaliphatic blocks, *Angew Macromol Chem* **234**:103–117.
44. El Fray M, Slonecki J and Broza G (1997) Melt-crystallized segmented aromatic-aliphatic copoly(ester-ester)s based on poly(butylene terephthalate) and a dimerized fatty acid, *Polimery* **42:**35–39.
45. El Fray M and Slonecki J (1999) Dimer fatty acid-modified poly(ester-*b*-ether)s: Synthesis and properties, *Polym Plast Technol Eng* **38**:51–69.
46. Lembicz F, Majszczyk J, Slonecki J and El Fray M (1998) Ultrasonic characterization of thermoplastic multiblock elastomers, *J Macromol Sci Phys* **B37**:161–170.
47. El Fray M and Slonecki J (1997) Synthesis and structure of elastomeric poly(ester-*block*-ether)s, *Macromol Symp* **122**:335–342.
48. El Fray M and Slonecki J (1996) Synthesis and properties of multiblock ester-aliphatic-ether terpolymers, *Kautch Gummi Kunstst* **49**:692–697.
49. El Fray M (1996) Influence of chemical structure of flexible segments on some properties of segmented multiblock copolyesters, PhD Thesis, Technical University of Szczecin, Szczecin (in Polish).
50. El Fray M and Slonecki J (1996) The influnce of the soft segment chemical structure on some selected properties of thermoplastic multiblock terephthalate copolymers, *Polimery* **41**:214–221.
51. Manuel H J and Gaymans R J (1993) Segmented block copolymers based on dimerized fatty acids and poly(butylene terephthalate), *Polymer* **34**:636–641.
52. Manuel H J and Gaymans R J (1993) Segmented block copolymers based on poly(butylene terephthalate) and telechelic polyesters and polyamides of dimerized fatty acids, *Polymer* **34**:4325–4329.
53. Kozlowska A and Slonecki J (1998) Segmented block copolymers based on ologoamides of dimerized fatty acid and poly(butylene terephthalate), *Polimery* **43**:188–191.
54. Kozlowska A and Slonecki J (1998) Synthesis and some properties of multiblock poly(ester-block-amide)s, *Elastomery* **2**:3–11 (in Polish).
55. El Fray M, Kozlowska A and Slonecki J (1997) Influence of the oligoamide's soft segments mass concentration on some selected properties of copoly(ester-amide)s, *Elastomery* **1**:12–20.
56. Slonecki J and Woropaj A (1996) Preparation and some properties of multiblock copoly(ester-amide)s (KPEA) containing oligo(tetramethylene terephthalate) hard segments and oligoamide soft segments, *Polimery* **41**:344–349.

57. Kozlowska A (2000) Influence of chemical structure of flexible segments on some properties of multiblock copolymers, PhD Thesis, Technical University of Szczecin, Szczecin.
58. Walch E and Gaymans R J (1994) Synthesis and properties of poly(butylene terephthalate)-*b*-polyisobutylene segmented block copolymers, *Polymer* **35**:636–641.
59. Marossy K, Deak G, Keki S and Zsuga M (1999) Thermally stimulated discharge current and dynamic mechanical investigation of polyisobutylene-polybutylene terephthalate thermoplastic multiblock copolymers, *Macromolecules* **32**:814–818.
60. Boutevin B, Khamlichi M, Pietrasanta Y and Robin J J (1995) Synthesis and characterization of a new block copolymer: poly(butylene terephthalate-co-olefin) application on PP/PBT blend and PBT homopolymer, *J Appl Polym Sci* **55**:191–199.
61. Mahato B M and Maiti S (1988) New block copolymers. V. Synthesis, characterization and morphological studies of poly(pentamethylene terephthalate)-b-(acrylonitrile butadiene rubber), *Colloid Polym Sci* **266**:601–607.
62. Mahato B M, Shit S C, Maiti M M and Maiti S (1985) New block copolymers 1. Synthesis and characterization of an $(A\text{-}B)_n$ type block copolymer, *Angew Makromol Chem* **134**:113–123.
63. Mahato B M, Shit S C and Maiti S (1985) New block copolymers III. Synthesis, characterization and crosslinking studies of an $(A\text{-}B)_n$ type block copolymer, *Eur Polym J* **21**:925–32.
64. Roslaniec Z (1992) Characteristics of multiblock terpoly(ester-ether-siloxane) elastomers, *Polymer* **33**:1717–1723.
65. Roslaniec Z (1997) Synthesis and characteristics of multiblock terpolymers with two flexible segments. Part I. Synthesis of poly(siloxane-ether-ester) terpolymers, *Polimery* **42**:367–372 (in Polish).
66. Slonecki J (1992) Morphology and some properties of copoly(ether-ester)s, *Sci Papers Tech Uni Szczecin*, Szczecin, **479.**
67. Roslaniec Z (1993) Polymer systems with elastothermoplastic properties, *Sci Papers Tech Uni Szczecin*, Szczecin, **503.**
68. Stevenson J C and Cooper S L (1988) Multiple endothermic melting behavior in poly(tetramethylene terephthalate)-containing polyesters and block copolyether-esters, *J Polym Sci B Polym Phys Ed* **26**:953–966.
69. Castles J L, Vallance M A, Mckenna J M and Cooper S L (1985) Thermal and mechanical properties of short-segment block copolyesters and copolyether-esters, *J Polym Sci Polym Phys Ed* **23**:2119–2147.
70. Stevenson J C and Cooper S L (1988) Microstructure and property changes accompanying hard-segment crystallization in block copoly(ether-ester) elastomers, *Macromolecules* **21**:1309–1316.
71. Nelsen S B, Gromelski S J and Charles J J (1983) New segmented thermoplastic polyester elastomers, *J Elast Plast* **15**:256–264.
72. Case L C and Case L K (1968) Vinyl-crosslinked unsaturated polyester-polyether copolymers, US Patent 3,375,301.
73. Siemiński J (1996) Reactive polymer blends, PhD Thesis,Technical University of Szczecin, Szczecin.
74. Siemiński J and Roslaniec Z (1994) Synthesis and characteristics of multiblock poly-(ether-(urethane)ester) elastomers, *3rd Int Symp on Thermoplastic Elastomers*, Kolobrzeg, Poland, *Sci Papers Tech Uni Szczecin* **514**:125–130.
75. Zeilstra J J (1986) Influencing the crystallization behavior of PET-based segmented copoly(ether ester), *J Appl Polym Sci* **31**:1977–1997.

76. Misra A and Garg S N (1986) Block copolymers of poly(ethylene terephthalate)-poly(butylene terephthalate). II. Small-angle light scattering studies, *J Polym Sci B Polymer Phys* **24**:999–1008.
77. Ma D, Wang M, Wang M, Zhang X and Luo X (1998) Compositional heterogeneity, thermostable and shape memory properties of ethylene oxide-ethylene terephthalate segmented copolymer with long soft segment, *J Appl Polym Sci* **69**:947–955.
78. Bogdanov B, Michajlov M and Gavrailova G (1991) Infrared spectroscopy of oligo-ethyleneterephthalate-oligooxyethylene block copolymers, *Acta Polymerica* **47**:560–565.
79. Bogdanov B, Michajlov M, Popov A and Uzov Ch (1988) X-ray study of block copolymers of oligo(ethylene terephthalate) and oligo(oxyethylene), *Acta Polymerica* **39**:385–389.
80. Kong X, Yang X, Zhou E and Ma D (2000) Nonisothermal crystallization kinetics of ethylene terephthalate-ethylene oxide segmented copolymers with two crystallizing segments, *Eur Polym J* **36**:1085–1090.
81. Wang M T, Zhang L D, Ma D Z (1999) Degree of microphase separation in segmented copolymers based on poly(ethylene oxide) and poly(ethylene terephthalate), *Eur Polym J* **35**:1335–1343.
82. Reed M A and Gilding D K (1981) Biodegradable polymers for use in surgery poly(ethylene oxide)/poly(ethylene terephthalate) (PEO/PET) copolymers: 2. *In vitro* degradation, *Polymer* **22**:499–504.
83. Tang W, Sanjeeva Murthy N, Mares F, Mcdonnell M E and Curran S A (1999) Poly(ethylene terephthalate)-poly(caprolactone) block copolymer. I. Synthesis, reactive extrusion, and fiber morphology, *J Appl Polym Sci* **74**:1858–1867.
84. Fai W K, Chi W, Zhu M D and Lie L X (1995) Molecular-weight distribution of segmented copolymer of PET-PC, *J Appl Polym Sci* **57**:1285–1290.
85. Zhang R, Luo X and Ma D (1995) Multiple melting endotherms from ethylene terephthalate-caprolactone copolyesters, *Polymer* **36**:4361–4364.
86. Tsai R-S and Lee Y-D (1997) Synthesis and characterization of block copolyetheresters with poly(tetramethylene 2,6–naphthalenedicarboxylate) segments, *J Appl Polym Sci* **66**:1411–1418.
87. Tsai R-S, Lee Y-D (1999) Effect of thermal history on properties of *block*-copolyether-esters with poly(tetramethylene 2,6-naphthalenedicarboxylate) segments, *J Appl Polym Sci* **73**: 1441–1449.
88. Nojima S, Ono M and Ashida T (1992) Crystallization of block copolymers II. Morphological study of poly(ethylene glycol)-(poly(ε-caprolactone) block copolymers, *Polymer J* **24**:1271–1280.
89. Xiong C D, Cheng L M, Xu R P and Deng X M (1995) Synthesis and characterization of block copolymers from D,L-lactide and poly(tetramethylene ether glycol), *J Appl Polym Sci* **55**:865–869.
90. Mitrach K, Pospiech D, Haubler L, Voigt D, Jehnichen D and Ratzsch M (1993) Thermotropic block copolymers: polyesters with flexible poly(tetramethylene glycol) units in the main chain, *Polymer* **34**:3469–3474.
91. Bilibin A Yu, Tenkovtsev A V, Piraner O N, Pashkovsky E E and Skorokhodov S S (1985) Thermoplastic polyesters 2. Synthesis of regular polyesters from aromatic dicarboxylic acids and phenols or aliphatic diols, and study of their mesomorphic properties, *Makromol Chem* **186**:1575–1591.
92. Bilibin A Yu, Tenkovtsev A V and Skorokhodov S S (1985) Thermoplastic polyesters 1. Synthesis of complex monomers for polycondensations, *Makromol Chem Rapid Commun* **6**:209–213.

93. Bilibin A Yu and Piraner O N (1991) Thermotropic polyesters. 6. New approach to the synthesis of segmented liquid-crystalline polymers with an increased length of mesogenic and flexible elements, *Makromol Chem* **192**:201–214.
94. Sonpatki M M, Ravindranath K and Ponrathnam S (1994) Segmented copoly(ether-ester) elastomers. Influence of hard segment lenght and substitution on mesophase formation, *Polymer J* **26**:804–815.
95. Roslaniec Z (2000) Poly(ether-ester) block copolymers with LC segments, in *Block Copolymers* (Eds. Baltá Calleja F J and Roslaniec Z) Marcel Dekker, Inc., New York, Basel, pp. 451–478.
96. Pietkiewicz D and Roslaniec Z (2000) Liquid crystalline block copoly(ether-ester)s, *Polimery* **45**:69–78.
97. Tsai H B, Lee C and Chang N S (1992) Synthesis and thermal properties of thermotropic block copolyetheresters, *Polymer J* **24**:157–164.
98. Tsai H B, Lee D K, Liu J L, Tsao Y S and Tsai R S (1996) Block copolyetheresters. V. Low-temperature properties of thermotropic block copolyetheresters, *J Appl Polym Sci* **59**:1027–1031.
99. Tsai H B (1997) Characterization of thermotropic block copolyetheresters by X-ray diffraction and DSC, *J Macromol Sci-Phys* **B36**:175–185.
100. Rozes L, Tessier M and Maréchal E (1996) Synthesis and characterization of thermotropic poly(polyesters-block-polyethers, *Proc Int Symp Polycondensation'96*, Paris, p. 334.
101. Kawatsuki N, Kikai A, Fukae R, Yamamoto T and Sangen O (1997) Synthesis, characterization, and photoreaction of photoreactive liquid crystalline block copolyetheresters, *J Polym Sci A Polym Chem Ed* **35**:1849–1855.
102. Schulze U and Schmidt H W (1998) Synthesis and characterization of ABA-triblock copolymers with poly(ethylene glycol) segments and LC-segments, *Polym Bull* **40**:159–166.
103. Szymczyk A (1999) Synthesis and physical properties of ionic poly(ether-ester) elastomers, PhD Thesis, Technical University of Szczecin, Szczecin.
104. Szymczyk A and Roslaniec Z (1999) Sulfonated poly(ether-block-ester) ionomers with anions in the polyester hard segments, *Polym Adv Technol* **10**:579–587.
105. Perry K P, Jackson W J and Caldwell J R (1965) Elastomers based on polycyclic bisphenol polycarbonates, *J Appl Polym Sci* **9**:3451–3463.
106. Ukielski R and Wojcikiewicz H (1988) Poly(ether-ester) block copolymers modified by pentaerythritol, *Polimery* **33**:9–12.
107. Magryta J, Pyskło L, Roslaniec Z and Kapko E (1988) Influence of modification on physical properties of copoly(ether-ester)s, *Polimery* **33**:464–466.
108. Roslaniec Z, Slonecki J and Wojcikiewicz H (1991) Manufacture of copoly(ether ester) elastomers for electrical cable coatings, Polish patent 158339, to Technical University of Szczecin.
109. Matsuki T, Kuwata J and Takayama H (1989) Catalysts for the preparation of polyester-polyethers with high degree of polymerisation, Japanese patent 01095127, to DuPont-Toray Co., Ltd.
110. Tamura S, Matsuki T, Kuwata J and Ishii H (1990) Flexible polyester-polyether elastomers, Japanese patent 02269118, to DuPont-Toray Co., Ltd.
111. Pozdzal R and Roslaniec Z (1999) Rheological properties of multiblock ether-ester copolymers, *Kaut Gummi Kunstst* **52**:656–662.
112. Roslaniec Z (1995) Increasing methods of molecular weight of thermoplastic polyesters, Proc 7th *Sci Conf New Aspects of Modification and Use of Plastics*, Rydzyna, Poland, pp. 12–21.

113. Cai F and Isayev A I (1993) Dynamic vulcanization of thermoplastic copolyester elastomer/nitrile rubber alloys: I. Various mixing methods, *J Elast Plast* **25**:74–89.
114. Cai F and Isayev A I (1993), Dynamic vulcanization of thermoplastic copolyester elastomer/nitrile rubber alloys: II. Rheology, morphology and properties, *J Elast Plast* **25**:249–265.
115. Roslaniec Z and Poślednik S (1988) Rubber modification of poly(ether-ester) elastomer, *Sci Papers Tech Uni Szczecin* **14**:217–225.
116. Gravalos K G, Kallitsis J K and Kalfoglou N K (1995) *In situ* compatibilization of poly(ethylene terephthalate)/(poly(ethylene–*co*-ethyl acrylate) blends, *Polymer* **36**:1393–1399.
117. Okamoto M and Shiomi K (1994) Structure and mechanical properties of poly(butylene terephthalate)/rubber blends prepared by dynamic vulcanization, *Polymer* **35**:4618–4622.
118. Papke N and Karger-Kocsis J (2001) Thermoplastic elastomers based on compatibilized poly(ethylene terephthalate) blends: effect of rubber type and dynamic curing, *Polymer* **42**:1109–1120.
119. Kwiatkowski K and Roslaniec Z (2003) Preparation of thermoplastic elastomers by the reactive modification of poly(ethylene terephthalate), *Sci Papers Inst Org Polym Technol Wroclaw Uni Technol* **52**:500–504.
120. Coran A Y, Chung O and Laokijcharoen P (1998) The phase morphology of rubber/plastic blends, *Kaut Gummi Kunstst* **51**:342–347.
121. Hourston D J, Lane S, Zhang H X, Bootsma J P C and Koetsier D W (1991) Toughened thermoplastics: 1. The synthesis and use of maleate-modified poly(butylene terephthalate) in polyester/rubber blends, *Polymer* **32**:1140–1145.
122. Hourston D J, Lane S and Zhang H X (1991) Toughened thermoplastics: 2. Impact properties and fracture mechanisms of rubber modified poly(butylene terephthalates), *Polymer* **32**:2215–2220.
123. Laurienzo P, Malinconico M, Martuscelli E and Volpe M G (1989) Rubber modification of polybutyleneterephthalate by reactive blending concurrently with polymerization reaction, *Polymer* **30**:835–841.
124. Gagnon K D, Lenz R W, Farris R J and Fuller R C (1992) Crystallization behavior and its influence on the mechanical properties of a thermoplastic elastomer produced by *Pseudomonas oleovorans*, *Macromolecules* **25**:3723–3728.
125. Stribeck N, Fakirov S and Sapoundijeva D (1999) Deformation behavior of a poly(ether ester) copolymer. Quantitative analysis of SAXS fiber patterns, *Macromolecules* **32**:3368–3378.
126. Stribeck N, Sapoundijewa D, Denchev Z, Apostolov A A, Zachman H G, Stamm M and Fakirov S (1997) Deformation behavior of poly(ether ester) copolymer as revealed by small-and wide-angle scattering of X-ray radiation from synchrotron, *Macromolecules* **30**:1329–1339.
127. Apostolov A A and Fakirov S (1992) Effect of the block length on the deformation behavior of polyetheresters as revealed by small-angle X-Ray scattering, *J Macromol Sci-Phys* **B31**: 329–355.
128. Apostolov A A, Fakirov S and Mark J F (1998) Mechanical properties in torsion for poly(butylene terephthalate) and a poly(ether ester) based on poly(ethylene glycol) and poly(butylene terephthalate), *J Appl Polym Sci* **69**:495–502.
129. Slonecki J (1990) Investigations of the hardness and thermal properties of copoly(ether-ester)s containing segments of different molecular weights, *Polymer* **31**:1464–1466.

130. Fakirov S, Fakirov C, Fischer E W and Stamm M (1991) Deformation behaviour of poly(ether ester) thermoplastic elastomers as revealed by small-angle X-ray scattering, *Polymer* **32**:1173–1180.
131. Fakirov S, Apostolov A A, Boeseke P and Zachmann H G (1990) Structure of segmented poly(ether ester)s as revealed by synchrotron radiation, *J Macromol Sci-Phys* **B29**:379–395.
132. Hsu J and Choi K Y (1987) Kinetics of transesterification of dimethyl terephthalate with poly(tetramethylene ether) glycol and 1,4-butanediol catalyzed by tetrabutyl titanate, *J Appl Polym Sci* **33**:329–351.
133. Chang S J, Chang F Ch and Tsai H B (1995) Block copolyetheresters. Part 3: Preparation of block copolyetheresters by a terephthalic acid process in the presence of salts, *Polym Eng Sci* **35**:190–194.
134. Paatero E, Närhi K, Salmi T, Still M, Nyholm P and Immonen K (1994) Kinetic model for main and side reactions in the polyesterification of dicarboxylic acids with diols, *Chem Eng Sci* **49**:3601–3616.
135. Dairanieh I S and Khraish N (1991) Empirical modelling and scale-up of a semi-batch polyestrification process, *J Appl Polym Sci* **42**:1631–1637.
136. Leverd F, Fradet A and Maréchal E (1987) Study of model esterifications and of polyesterifications catalyzed by various organometallic derivatives, *Eur Polym J* **23**:695–698, 699–704, 705–709.
137. Kaźmierczak A, Leo J and Paryjczak T (1983) Synthesis of poly(ethylene terephthalate) with high limiting viscosity number, *Polimery* **28**:227–229.
138. Ukielski R and Wojcikiewicz H (1978) Preparation of thermoplastic copoly(ether ester) elastomers, *Polimery* **23**:48–51.
139. Pilati F, Manaresi P, Fortunato B, Munari A and Monari P (1983) Models for the formation of poly(butylene terephthalate): kinetics of the titanium tetrabutylate-catalyzed reactions, *Polymer* **24**:1479–1483.
140. Tomita K and Ida H (1975) Formation of poly(ethylene terephthalate). 3. Catalytic activity of metal compounds in transesterification of dimethyl terephthalate with ethylene glycol, *Polymer* **16**:185–190.
141. Tomita K (1976) Formation of poly(ethylene terephthalate). 6. Catalytic activity of metal compounds in polycondensation of bis(2-hydroxyethyl) terephthalate, *Polymer* **17**:221–224.
142. Kricheldorf H R and Langarke D (1999) Multiblock copoly(ether-esters) of poly(THF) and poly(ε-caprolactone) via macrocyclic polymerization, *Macromol Chem Phys* **200**:1183–1190.
143. Yoon K H and Park O O (1995) Diffusion of butanediol in poly(butylene terephthalate) (PBT) melt and analysis of PBT polymerization reactor with surface renewal, *Polym Eng Sci* **35**:703–708.
144. Yoon K H and Park O O (1994) Analysis of reactor with surface renewal for poly(ethylene terephthalate) synthesis, *Polym Eng Sci* **34**:190–200.
145. Tomita K and Ida H (1973) Formation of poly(ethylene terephthalate). 2. Role of transesterification of dimethyl terephthalate with ethyleneglycol, *Polymer* **14**:55–60.
146. Hsu J and Choi K Y (1986) Kinetics of transesterification of dimethyl terephthalate with 1,4-butanediol catalyzed by tetrabutyl titanate, *J Appl Polym Sci* **32**:3117–3132.
147. Vaidya U R and Nadkarni V M (1989) Polyester polyols from glycolyzed PET waste: Effect of glycol type on kinetics of polyestrification, *J Appl Polym Sci* **38**:1179–1190.
148. Pawlaczyk K and Slonecki J (1988) Post-polycondensation of copoly(ether-ester)s in solid state, *Sci Papers Tech Uni Szczecin* **330**:43–47.

149. Kokkalas D E, Bikiaris D N and Karayannidis G P (1995) Effect of the Sb_2O_3 catalyst of the solid-state postpolycondensation of poly(ethylene terephthalate), *J Appl Polym Sci* **55**:787–791.
150. Kang Ch-K (1998) Modeling of solid-state polymerization of poly(ethylene terephthalate), *J Appl Polym Sci* **68**:837–846.
151. Gostoli C, Pilati F, Sarti G and Giacomo B (1984) Chemical kinetics and diffusion in poly(butylene terephthalate) solid-state polycondensation: Experiments and theory, *J Appl Polym Sci* **29**:2873–2887.
152. Slonecki J (1991) The effect of mass fraction and molecular weight of segments on the conditions of preparation, structure and properties of thermoplastic copolyester-ethers (KPEE). Part I. Synthesis and physicochemical characterization of KPEE differing in mass fraction and molecular weight of segments, *Polimery* **36**:225–229.
153. Slonecki J, Wojcikiewicz H, Kapelanski A, Kurek P, Ukielski R, Roslaniec Z and Mackow Z (1990) Manufacture of thermoplastic block copoly(ether ester) elastomers, Polish patent 150278, to Technical University of Szczecin.
154. Deng Y, Price C and Booth C (1994) Preparation and properties of block copolymers with two statistical copoly(oxyethylene/oxypropylene) blocks, *Eur Polym J* **30**:103–111.
155. Chang S J and Tsai H B (1992) Effect of salts on the formation of THF in preparation of PBT by TPA process, *J Appl Polym Sci* **45**:371–373.
156. Maréchal E (1987) Activation in polycondensation, *Bull Soc Chim France* **4**:713–723 (in French).
157. Lee K J, Moon D Y, Park O O and Kang Y S (1992) Diffusion of ethylene glycol accompanied by reactions in poly(ethylene terephthalate) melts, *J Polym Sci B Polym Phys Ed* **30**:707–716.
158. Wagener K B (1989) Polymerization and bulk crystallization behavior of poly(ethylene terephthalate) (PET), *J Appl Polym Sci* **75**:78–86.
159. Yoon K H, Kwon M H, Jeon M H and Park O O (1993) Diffusion of ethylene glycol in solid state poly(ethylene terephthalate), *Polymer J* **25**:219–226.
160. Chen L-W, Chen J-W (1999) Kinetics of diethylene glycol formation from bishydroxyethyl terephthalate with zinc catalyst in the preparation of poly(ethylene terephthalate), *J Appl Polym Sci* **75**:1229–1234.
161. Tate S, Watanabe Y and Chiba A (1993) Synthesis of ultra-high molecular weight poly(ethylene terephthalate) by swollen-state polymerization, *Polymer* **34**:4974–4977.
162. Tate S and Ishimaru F (1995) Swollen-state polymerization of poly(ethylene terephthalate): kinetic analysis of reaction rate and polymerization conditions, *Polymer* **36**:353–356.
163. Tate S and Watanabe Y (1995) Swollen-state polymerization of poly(ethylene terephthalate) in fibre form, *Polymer* **36**:4991–4995.
164. Parashar M K, Gupta R P, Jain A and Agarwal U S (1998) Reaction rate enhancement during swollen-state polymerization of poly(ethylene terephthalate), *J Appl Polym Sci* **67**:1589–1595.
165. Cardi N, Po R, Giannotta G, Occhiello E, Garbassi F and Messina G (1993) Chain extension of recyclated poly(ethylene terephthalate) with 2,2'-bis(2–oxazoline), *J Appl Polym Sci* **50**:1501–1509.
166. Wörner Ch, Müller P and Mülhaupt R (1997) Toughened poly(butylene terephthalate)s and blends prepared by simultaneous chain extension, interfacial coupling, and dynamic vulcanization using oxazoline intermediates, *J Appl Polym Sci* **66**:633–642.
167. John J, Tang J and Bhattacharya M (1998) Grafting of oxazoline functional group to polycaprolactone, *J Appl Polym Sci* **67**:1947–1955.

168. Böhme F, Leistner D and Baier A (1995) Comparison of the coupling behaviour of diisocyanate and bisoxazolines in modification of liquid crystalline poly(ethylene terephthalate-*co*-oxybenzoate)s, *Angew Makromol Chem* **224**:167–178.
169. Lefebvre H and Fradet A (1998) Chain-coupling reaction of amine-terminated oligomers by bis(4-monosubstituted-5(4H)oxazolinones), *Macromol Chem Phys* **199**:815–824.
170. Kim H S and Ma P (1998) Mode II fracture mechanisms of PET-modified brittle epoxies, *J Appl Polym Sci* **69**:405–415.
171. Bikiaris D N and Karayannidis G P (1996) Thermomechanical analysis of chain-extended PET and PBT, *J Appl Polym Sci* **60**:55–61.
172. Loontjens T, Belt W, Stanssens D and Weerts P (1993) Synthesis of bisoxazolines and their application as chain extender for poly(ethylene terephthalate), *Makromol Chem Macromol Symp* **75**:211–216.
173. Lefevre C, Mathieu C, Tidjani A, Dupret I, Vander Wauven C, De Winter W and David C (1999) Comparative degradation by micro-organisms of terephthalic acid, 2,6-naphthalene dicarboxylic acid, their esters and polyesters, *Polym Degrad Stabil* **64**:9–16.
174. Kint D and Muñoz-Guerra S (1999) A review on the potential biodegradability of poly(ethylene terephthalate), *Polym Int* **48**:346–352.
175. Wagener K B (1989) Physical constans of poly(oxytetramethylene-oxyterephthaloyl) and copolymers with tetramethylene oxide thermoplastic elastomers, in *Polymer Handbook* (Eds. Brandup J and Immergut E H) Wiley-Interscience, New York, V/107–108.
176. Bakker D, Van Blitterswijk C A, Deams W T and Grote J J (1988) Biocompatibility of six elastomers *in vitro*, *J Biomed Mat Res* **22**:423–439.
177. Witt U, Müller R J and Deckwer W D (1996) Studies on sequence distribution of aliphatic/aromatic copolyesters by high-resolution C-13 nuclear magnetic resonsnce spectroscopy for evaluation of biodegradability, *Macromol Chem Phys* **197**:1525–1535.
178. Tokiwa V, Ando T, Suzuki T and Takedo T (1990) Biodegradation of synthetic polymers containing ester bonds, *Polym Mater Sci Eng* **62**:988–992.
179. Lee S H, Lim S W and Lee K H (1999) Properties of potentially biodegradable copolyesters of (succinic acid-1,4–butanediol)/(dimethyl terphthalate-1,4–butanediol), *Polym Int* **48**:861–867.
180. Baltá Calleja F J and Fakirov S (2000) Structure-microhardness correlations in condensation copolymers, in *Block Copolymers* (Eds. Baltá Calleja F J and Roslaniec Z) Marcel Dekker Inc., New York, Basel, pp. 179–213.
181. Nagata M, Kiyotsukuri T, Minami S, Tsutsumi N and Sakai W (1996) Biodegradability of poly(ethylene terephthalate) copolymers with poly(ethylene glycol)s and poly(tetramethylene glycol), *Polym Int* **39**:83–89.
182. Zhou M, Takayanagi M, Yoshida Y, Ishi S and Noguchi H (1999) Enzyme-catalyzed degradation of aliphatic polycarbonates prepared from epoxides and carbon dioxide, *Polym Bull* **42**:419–424.
183. Mochizuki A, Seita Y, Nakashima T, Endo F and Yamashita S (1998) Polyether-segmented nylon hemodialysis membrane. V. Evaluation of blood compatibility of polyether-segmented nylons, *J Appl Polym Sci* **67**:1253–1257.
184. Gagnon K D, Lenz R W, Farris R J and Fuller R C (1994) Chemical modification of bacterial elastomers: 1. Peroxide crosslinking, *Polymer* **35**:4358–4367.
185. Fakirov S and Denchev Z (1999) Sequential reordering in condensation copolymers, in *Transreactions in Condensation Polymers* (Ed. Fakirov S) Wiley-VCH, Weinheim, Ch. 8, pp. 319–389.

186. Alma D H (1986) Thermoplastic polyester elastomers, *Synthetic* **6**:47–50 (in German).
187. DSM Engineering Plastics (1995) Arnitel Copolyester Elastomers: DSM Technical Informations, Evansville.
188. Penco M, Donetti R, Mendichi R and Feruti P (1998) New poly(ester-carbonate) multi-block copolymers based on poly(lactic-glycolic acid) and poly(ε-caprolactone) segments, *Macromol Chem Phys* **199**:1737–1745.
189. Fan X D, Deng J, Waterhouse J and Peromm P (1998) Synthesis and characterization of polyamide resins from soy-based dimer acids and different amides, *J Appl Polym Sci* **68**:305–314.
190. Neuray D and Off K-H (1981) New rubber-modified thermoplastics, *Angew Makromol Chem* **98**:213–224 (in German).
191. Manas-Zloczower I, Nir A and Tadmor Z (1984) Dispersive mixing in rubber and plastics, *Rubber Chem Technol* **57**:583–620.
192. Utracki L A (1991) *Two-Phase Polymer Systems*, Hanser Publishers, Munich, Vienna, New York, Barcelona.
193. Kwiatkowski K and Roslaniec Z (2004) Study of welding of poly(ether-ester) elastomers. I. Influence of contents of flexible segments on morphology and mechanical properties, *Polimery* **49**:268–274.
194. El Fray M (2003) *Nanostructured Elastomeric Biomaterials for Soft Tissue Reconstruction*, Publishing House of the Warsaw University of Technology, Warsaw.
195. El Fray M and Altstadt V (2003) Fatigue behaviour of multiblock thermoplastic elastomers. 1. Stepwise increasing load testing of poly(aliphatic/aromatic-ester) copolymers, *Polymer* **44**:4635–4642.
196. El Fray M and Altstadt V (2003) Fatigue behaviour of multiblock thermoplastic elastomers. 2. Dynamic creep of poly(aliphatic/aromatic-ester) copolymers, *Polymer* **44**:4643–4650.
197. El Fray M and Altstadt V (2003) Fatigue behaviour of multiblock thermoplastic elastomers. 3. Stepwise increasing strain test of poly(aliphatic/aromatic-ester) copolymers, *Polymer* **45**:263–273.
198. El Fray M (2002) Multiblock alicyclic/aliphatic polyesters. Evaluation of the polymer's susceptibility to enzymatic degradation, *Polimery* **47**:191–195.
199. Prowans P, El Fray M and Slonecki J (2002) Biocompatibility studies of new multiblock poly(ester-ester)s composed of poly(butylenes terephthalate) and dimerized fatty acid, *Biomaterials* **23**:2973–2978.
200. El Fray M, Bartkowiak A, Prowans P and Slonecki J (2000) Physical and mechanical behavior of electron-beam irradiated and ethylene oxide sterilized multiblock polyester, *J Mater Sci Mater Medicine* **11**:757–762.
201. Roslaniec Z and Pietkiewicz D (2002) Synthesis and characteristics of polyester-based thermoplastic elastomers: chemical aspects, in *Handbook of Thermoplastic Polyesters* (Ed. Fakirov S) Wiley-VCH, Weinheim, Ch.13, pp. 581–658.

Chapter 4

Terpoly(Ester-*b*-Ether-*b*-Amide) Thermoplastic Elastomers: Synthesis, Structure, and Properties

R. Ukielski

1. Introduction

Multiblock thermoplastic elastomers (TPE) are multiphase systems with a blurred separation surface (large interfacial area). The properties of these polymers result from their specific multiphase morphology (*physical microstructure*) and molecular structure. The microstructure is a result of the microphase separation of blocks (*local segregation*), which is a consequence of the differences in the chemical structure of the respective blocks [1–7]. The microphase separation in multiblock polymers depends on the change in Gibbs' free energy of mixing, ΔG_s, of the block system. When $\Delta G_s > 0$, the change in the entropy of mixing approaches zero (in the case of macromolecules with a high degree of polymerization) and only the change in the enthalpy of mixing, ΔH_s, is responsible for the solubility. The mixing enthalpy change can be characterized by the difference in the Hildebrand solubility parameters, $\Delta\delta$, of the components (blocks) [8]. Thus, the microstructure of multiblock TPE is affected by the mutual solubility of the blocks, the latter being the decisive factor in the selection of blocks used for the manufacture of block TPE. The microstructure and phase separation mechanisms of multiblock copolymers $-(A_xB_y)_n-$ were established and described in [7,9–12].

In multiblock terpolymers, the macromolecules are composed of multiple repeat units of three chemically different blocks $-(A_xB_yC_z)_n-$ [13,14]. Because of the incorporation into the macromolecule of a third block, the system becomes complex and, depending on the location, the molecular weight, and the contribution of this block, various combinations of interactions and mutual solubility of blocks are possible. In copolymers, these are the interactions of A with A, B with B, and A with B, and the mutual solubility concerns only the pair A and B. In terpolymers $-(A_xB_yC_z)_n-$, six interactions between the blocks are possible, and three block pairs can be soluble in each other.

For instance, when block A is: (i) insoluble in the phases of blocks B and C, then the polymer structure will change and a new phase morphology will be formed, (ii) partly soluble in the phase of B (low molecular fractions of block A) and insoluble in the phase of blocks C, then the properties of the B phase will change (increase of the glass transition temperature or decrease in the melting point), the interphase area may be enlarged and a new phase morphology will probably be formed, (iii) completely soluble in the B phase and insoluble in C, then the properties of the B phase will change, (iv) partly soluble in the phases B and C, then the interphase area may be enlarged, the properties of the phases can change, and a new phase morphology can be formed, (v) completely soluble in the B phase and partly soluble in C, then both the properties of phases B and C and the morphology of the system as a whole will change, (vi) completely soluble in B and C, then the properties of these two phases will change, and the diphase morphology could vanish.

From the above considerations, it follows that by the proper choice of the chemical structure of the blocks and by predetermining their location in the macrochain, as well as their composition and fraction, interactions may occur in the terpolymers, causing the formation of a nanophase morphology, which is beneficial from the point of view of their elastic properties [1,4,12].

In this chapter, the synthesis, the chemical and the physical structure of terpoly(ester-*b*-ether-*b*-amide)s, as well as their influence on the final material properties are considered.

These multiblock terpolymers are laboratory products. They were prepared in order to evaluate the influence of the incorporation of a third block into the macromolecule on the copolymer phase structure. An attempt was made to change the morphology (from microphase to nanophase) in the Hytrel®-type elastomers; thereby, the interphase could be expanded. The elastomers having a nanophase structure are polymers with a better shape memory than those with coarse morphology. By the incorporation of a third block, the properties of multiblock polymers can be modified so that different shape memory effects can be achieved (for shape memory effects, see also Chapter 18).

2. Chemical structure of terpoly(ester-*b*-ether-*b*-amide)s

The starting materials used in the preparation of terpoly(ester-*b*-ether-*b*-amide)s were polyetherdiols, lactams, diamines, diacids or their diesters, and glycols [15,16]. The general chemical composition of their macromolecule, PE-*b*-PO-*b*-PA, is represented in Scheme 1. Terpoly[(tetramethylene terephtalate)-*b*-(oxy-

*a*G T *a*G T PO*b* PA SBA

$H-\{[O-(CH_2)_a-O-C(=O)-C_6H_4-C(=O)]_x-O-(CH_2)_a-O-C(=O)-C_6H_4-C(=O)-[O-(CH_2)_b]_y-O-[C(=O)-(CH_2)_c-NH]_z-C(=O)-(CH_2)_8-C(=O)\}_n-OH$

$A_x = a$GT $B_y =$ POb $C_z =$ PA$(c+1)$

$A_x = a$GT,	oligoester block	a=2–6	x=1–14	M_w ~220–3080 g/mole
$B_y =$ POb,	oligoether block	b=2, 4	y=1–14	M_w ~650–2900 g/mole
$C_z =$ PA$(c+1)$,	oligoamide block	c=5, 11	z=1–14	M_w ~1000–4000 g/mole

B_y/C_z = 1–5 and when B_y/C_z = 2, the most probable arrangement is the following:

Scheme 1

tetramethylene)-*b*-laurolactam)], $-(4GT\text{-}b\text{-}PO4\text{-}b\text{-}PA12)_n-$ is a representative of this group of polymers.

The structure of the $-(GT\text{-}b\text{-}PO4\text{-}b\text{-}PA)_n-$ macromolecule can be regarded as a copoly(ether-*b*-ester) chain, $-(GT\text{-}b\text{-}PO4)_n-$, in which some of the terephthalate groups (T) are replaced by PA blocks. A characteristic feature of the preparation of $-(GT\text{-}b\text{-}PO4\text{-}b\text{-}PA)_n-$ (similarly to the synthesis of $-(GT\text{-}b\text{-}PO4)_n-$) is the formation of the ester block during the polycondensation step. This enables the control of the polymerization degree of this block through the change in the dimethyl terephthalate (DMT) mole fraction with respect to the other reactants. The PA/PO weight ratio can be used to affect the distribution of these blocks in the macromolecule. However, one should bear in mind that the terpoly(ester-*b*-ether-*b*-amide)s are random block polymers.

3. Synthesis of triblock $-(GT\text{-}b\text{-}PO4\text{-}b\text{-}PA)_n-$ polymers

The synthesis of terpoly(ester-*b*-ether-*b*-amide)s is a four-step process (Scheme 2) [12,17–20]. The second, third and fourth step take place in the presence of a magnesium titanate organometallic complex as a catalyst (10 wt% solution in *n*-butanol). The first step involves a hydrolytic condensation polymerization (the pressure decreases from 0.8–1.5 MPa to 0.1 MPa, at 280–320°C for 6–10 h) in the presence of sebacic acid as a stabilizer of the molecular weight.

The second step involves the transesterification reaction of DMT (2 moles) and the glycol (G, 3 moles), leading to the formation of GTGTG and the release of CH_3OH. The third step, taking place simultaneously in another reactor, is the esterification reaction of α,ω-dicarboxylic-oligoamide (PA) with poly(oxytetramethylene)diol. The product obtained is PO4-*b*-PA12-*b*-PO4 and water is formed. The fourth step is the polycondensation reaction of these two previously prepared intermediate compounds:

$$\text{GTGTG} + \text{PO4-}b\text{-PA12-}b\text{-PO4} \rightarrow [\text{PO4-}b\text{-PA12-}b\text{-PO4-}b\text{-(GT)}_x]_n + \text{G}\uparrow.$$

From the respective amounts of methanol and water collected from these reactions, it was concluded that the conversion in the transesterification step

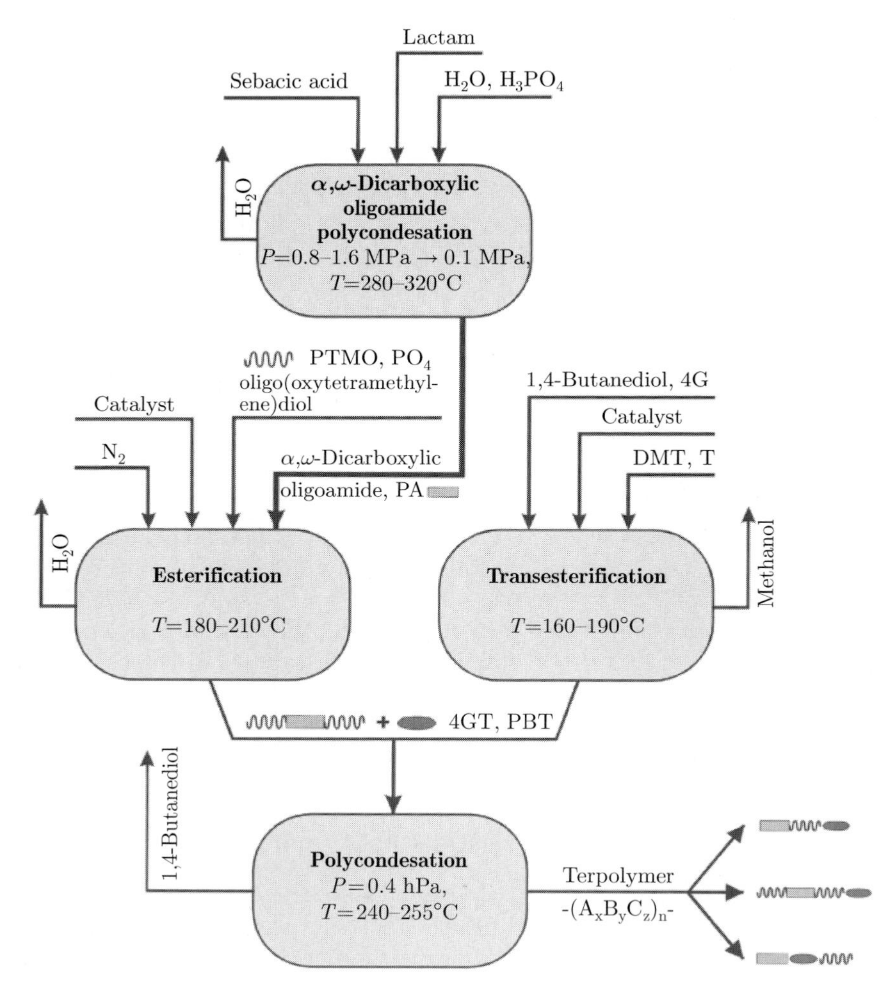

Scheme 2. Production flow chart

amounts to 98%, and the degree of esterification is 88%. The degrees of conversion in the esterification and transesterification reactions were expressed as the weight ratios (in percent) of the released water or methanol to the respective stoichiometric amounts of these products.

The boundary conditions of polycondensation, resulting in the obtaining of terpolymers with good mechanical properties, are represented by broken lines in Figure 1.

The compositions of the multiblock polymers described in this chapter are given in Table 1.

The obtained products are light cream-colored; by their appearance and on touch they resemble the polyurethane elastomers. A detailed description of the apparatus and synthesis is given in [12,17–20].

Table 1. Compositions of the different terpoly(ester-*b*-ether-*b*-amide) samples, a copoly(ester-*b*-amide), a copoly(ether-*b*-amide), and a copoly(ether-*b*-ester)

Sample	M_{PO4} (g/mole)	M_{PA12} (g/mole)	M^*_{PA12} (g/mole)	DP_{4GT}	Molar ratio: Esterification PO4 (mole)	Esterification PA12 (mole)	Transesterification DMT (mole)	Transesterification 4G (mole)	Series
1 M_{PA12}	1000	960	1000	4	2	1	5	9	
2 M_{PA12} Standard	1000	2250	2000	4	2	1	5	9	I
3 M_{PA12}	1000	4050	4000	4	2	1	5	9	
4 M_{PO4}	650	2250	2000	4	2	1	5	9	
5 M_{PO4}	1400	2250	2000	4	2	1	5	9	
6 M_{PO4}	2000	2250	2000	4	2	1	5	9	II
7 M_{PO4}	2900	2250	2000	4	2	1	5	9	
8 M_{4GT}	1000	2250	2000	0.25	2	1	1.25	3	
9 M_{4GT}	1000	2250	2000	2	2	1	3	5	
10 M_{4GT}	1000	2250	2000	7	2	1	8	15	III
11 M_{4GT}	1000	2250	2000	9	2	1	10	19	
12 M_{4GT}	1000	2250	2000	14	2	1	15	29	
PE-b-PA	-	2250	2000	4	-	1	4	8	
PO-b-PA	1000	2250	2000	-	1	1	-	-	
PO-b-PE	1000	-	-	4	1	-	5	9	

M_{PA12} – molecular weight calculated from the end-groups;

M^*_{PA12} - molecular weight from the stoichiometric ratio;

2 M_{PA12} – standard terpolymer included in each series

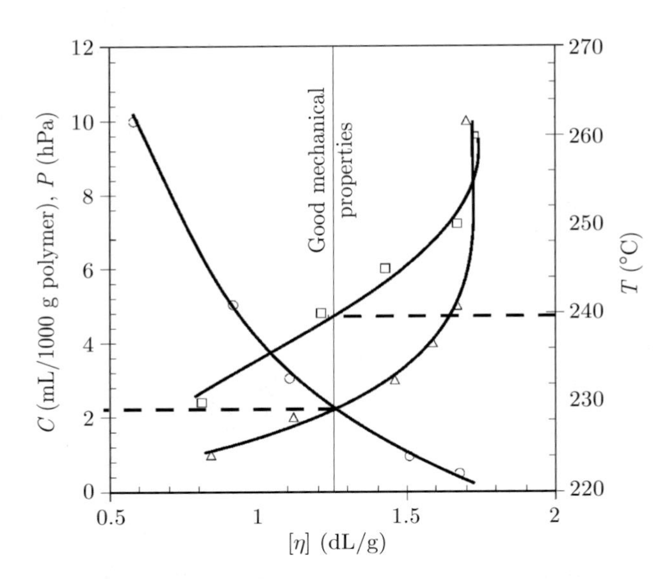

Figure 1. Effect of catalyst concentration, C (△), pressure, P (○), and polycondensation temperature, T (□), on the intrinsic viscosity, $[\eta]$, of terpoly(tetramethylene terephthalate-*b*-oxytetramethylene-*b*-laurolactam), $-(4GT\text{-}b\text{-}PO4\text{-}b\text{-}PA12)_n-$

4. Solubility of the blocks

It is well known that the addition of a component soluble in each of the phases of a polymer blend has a stabilizing effect. Hence, the modification of a copolymer chain by an additional block can produce a similar effect, *i.e.*, stabilization of the biphase supermolecular structure on the nano-level. The knowledge of the solubility parameters of the blocks allows the prediction of a critical degree of polymerization (DP) of the respective blocks (the maximum DP, at which the blocks dissolve in each other). The data presented in Table 2 show that the blocks PBT and PA12 are mutually soluble within the DP range of 1÷12. The PA12 block with the molecular weight of 2000 g/mole (DP = 1) is insoluble in the PO4 phase, whereas the PBT block dissolves in this phase when its DP ≤ 3.

5. Structure-property relationships

The effects of block lengths of the hard and the soft segments of terpoly[(tetramethylene terephthalate)-*b*-(oxytetramethylene)-*b*-(laurolactam)] are shown below. An appropriate way of investigation of the phase ratios is to combine differential scanning calorimetry (DSC) with dynamic mechanical thermal analysis (DMTA) and wide-angle X-ray scattering (WAXS).

The DSC and WAXS data of the respective blocks A_x, B_y, and C_z, and of the copolymers $(A_xB_y)_n$, $(A_xC_z)_n$, and $(B_yC_z)_n$ combining blocks included in the terpolymer compositions are summarized in Table 3 and Figure 2.

Table 2. Studies of the solubility parameters of the three blocks and the diblock copolymers incorporated into the composition of the terpolymers

Solubility parameters, δ (MPa$^{0.5}$)		Solubility parameters of the blocks, δ (MPa$^{0.5}$)			Difference in the solubility parameters of block pairs, $\Delta\delta$ (MPa$^{0.5}$)		
		$PO4_{1000}$	$(4GT)_4$	$PA12_{2000}$	PE/PA	PO/PA	PO/PE
Calculated by Hoy's method [8]	δ_{calc}	19.9	21.8	21.3	3.9*	2.2*	5.7*
	δ_d	17.3	15.6	17.7	2.1	0.4	1.7
	δ_p	7.5	11.9	9.2	2.7	1.7	4.4
	δ_H	6.2	9.4	7.5	1.9	1.3	3.2
Empirical [5,8,21]	δ	18.5	21.6	22.1	0.5	4.6	4.1
$M_{cr,calc}$					356	1150	168
$DP_{cr,calc}$					1.8	5.8	0.8
$M_{cr,emp}$					21800	421	567
$DP_{cr,emp}$					111	2.1	2.6

δ_d, δ_p, δ_H – resulting from dispersion, polar interaction, and hydrogen bonding, respectively; M_{cr}, DP_{cr} – critical values of the molecular weight and the degree of polymerization (when $M \leq M_{cr}$, a real solution can be achieved);
* – $\Delta\delta = [(\delta_{d1}\text{-}\delta_{d2})^2+(\delta_{p1}\text{-}\delta_{p2})^2+(\delta_{h1}\text{-}\delta_{h2})^2]^{0.5}$

Table 3. DSC data for the three blocks and the diblock copolymers incorporated into the composition of the terpolymers

Property		$PO4_{1000}$	$(4GT)_4$	$PA12_{2000}$	PE-*b*-PA	PO-*b*-PA	PO-*b*-PE
T_{g1} (°C)	Soft phase	–88	-	-	-	–77	–65
ΔC_{p1} (J/g K)		0.34	-	-	-	0.09	0.21
T_{c1} (°C)		2	-	-	-	–3.5	-
ΔH_{c1} (J/g)		91.3	-	-	-	23	-
T_{m1} (°C)		19	-	-	-	20	-
ΔH_{m1} (J/g)		111	-	-	-	22	-
w_c^s (%)		56.5	-	-	-	32	-
T_{g2} (°C)	Hard phase	-	60	10	16	-	40–65
ΔC_{p2} (J/g K)		-	0.18	0.17	0.18	-	~0.07
T_{c2} (°C)		-	127	134	94	134	158
ΔH_{c2} (J/g)		-	83.5	75.5	38	37	26
T_{m2} (°C)		-	193	163	135	163	181
ΔH_{m2} (J/g)		-	90.5	81	45	40	26.5
w_c^h (%)		-	63	84.5	67.5	62	42.5

T_{g1}, T_{c1}, T_{m1} - glass transition, crystallization, and melting temperature, respectively, in the low-temperature region; ΔC_{p1} – heat capacity change at T_{g1}; ΔC_{p2} – heat capacity change at T_{g2}; ΔH_{m1} – heat of fusion at T_{m1}; ΔH_{c1} - heat of crystallization at T_{c1}; T_{g2}, T_{c2}, T_{m2} – glass transition, crystallization, and melting temperature, respectively, in the high-temperature region; ΔH_{c2} – heat of crystallization at T_{c2}; ΔH_{m2} – heat of fusion at T_{m2}; w_c – degree of crystallinity, w_c^s – soft phase, w_c^h – hard phase

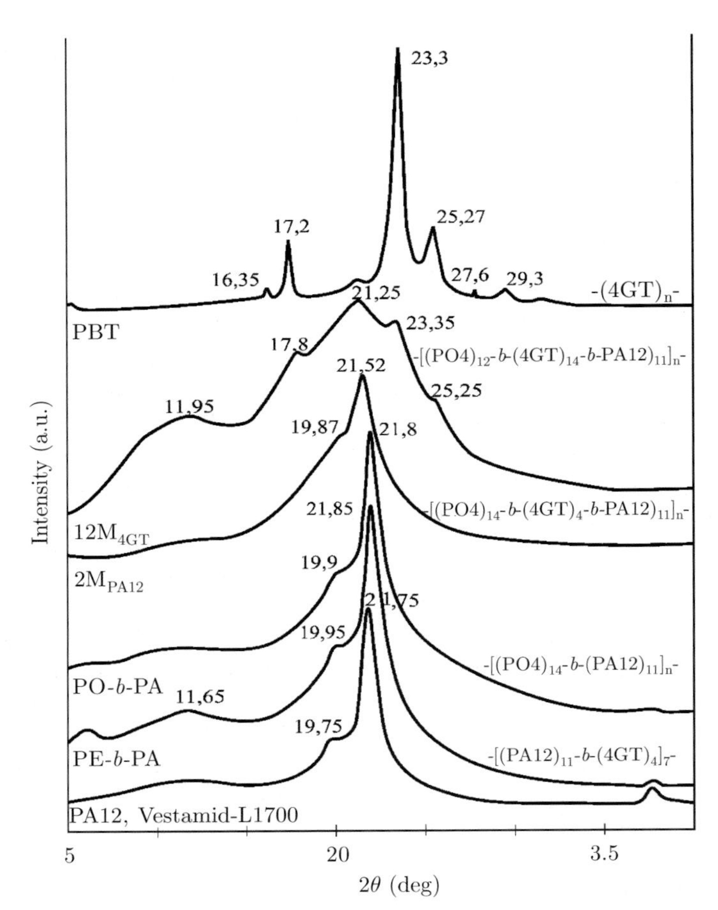

Figure 2. WAXS diffractograms of PBT, Vestamid-L1700 (PA12), PE-*b*-PA and PO-*b*-PA copolymers, and multiblock terpolymers

These studies confirm the assumptions concerning the solubilities of the PE/PA, PO/PA, and PO/PE systems, based on the analysis of the solubility parameters given in Table 2.

The glass transition temperature of PO4 is $T_{g1} = -88°C$ and slightly differs from $T_{g1,PO\text{-}b\text{-}PA}$ (by 11°C). It is believed that the immobilization of the chain-ends of the flexible block by the chemical bond enhances its T_g by 10°C. However, the interactions of the hard phase with this block can be responsible for an increase of this temperature by maximum 5°C [3,21–24]. Hence, it can be assumed that $T_{g1,PO\text{-}b\text{-}PA}$ is the glass transition temperature of PO4, *i.e.*, that PA12 with molecular weight of 2000 g/mole does not dissolve in the PO4 phase.

The PO-*b*-PE copolymer shows a completely different behavior, since its $T_{g1,PO\text{-}b\text{-}PE}$ differs from $T_{g1,PO4}$ by 23°C. Such an increase in T_g cannot be explained by the immobilization of the chain-ends, or by interphase interactions. In this case, dissolution of 4GT short segments in the PO4 soft phase is responsible for the T_g increase. The melting temperature decrease of the PE-*b*-PA copolymer $T_{m2,PE\text{-}b\text{-}PA} = 135°C$ ($T_{m2,(4GT)4} = 193°C$, $T_{m2,PA12} = 163°C$) is caused by dissolution of the $(4GT)_4$ blocks in the amorphous phase of PA12 blocks.

The qualitative evaluation of the diffractograms (Figure 2) suggests that in the copolymers and terpolymers composed of the hard blocks 4GT and PA12, the latter block is responsible for the formation of the hard crystalline phase.

The diffraction patterns of these polymers reveal only reflections with the same 2Θ angle as in the diffraction pattern of PA12. This is valid for the weight ratios PA12/4GT > 1 and 4GT/polymer ≤ 0.35. In the case of multiblock terpolymers, these values also depend on the macromolecular architecture (distribution of blocks with respect to each other). Exceeding the above mentioned boundary values causes the formation of opaque products, as observed with the $[(PO4)_{14}\text{-}b\text{-}(4GT)_{14}\text{-}b\text{-}(PA12)_{11}]_n$ terpolymer. Its diffraction pattern differs from the others (Figure 2) and represents a superposition of PA12 (Vestamid) and PBT. This is confirmed by the values of the 2Θ angles determined by the peaks of the respective reflections. In this case, two different crystalline phases PA and 4GT are formed in the polymer.

In summary, three very important conclusions should be emphasized: (i) the oligomers forming the PA12 blocks with molecular weight of 2000 g/mole and higher are insoluble in the phase formed by the PO4 blocks, (ii) the oligomers forming the 4GT blocks, under the defined conditions of composition and molecular weight, will be completely dissolved (provided that they do not crystallize) or partly dissolved (only the amorphous part) in the phases formed by the PO4 or PA12 blocks, and (iii) in the presence of the 4GT blocks, the hard PA12 blocks with molecular weight of 2000 g/mole always form the crystalline phase, whereas the crystalline phase of the 4GT blocks is only formed when the weight ratio of the blocks PA12/4GT is smaller than unit. In the copolymers and terpolymers formed by these blocks after exceeding this ratio, two crystalline phases appear. In practice, this means that the terpolymer composed of the blocks PA12, PO4, and 4GT is the system in which the 4GT amorphous block could be partly soluble in PO4 and partly or completely in PA12, and good phase separation can only take place at the PO4/PA12 phase boundary (see examples (iv) and (v), page 118).

5.1. *Thermal properties of* $-(4GT\text{-}b\text{-}PO4\text{-}b\text{-}PA12)_n-$

The DSC curves of multiblock elastomers can be divided into two parts. The trend of the first part, which is in the low-temperature range of $-100°C < T < 30°C$, characterizes the processes caused by the changes in the soft phase. The trend of the second part of these curves (above 30°C) characterizes the thermal properties of the hard phase. The influence of the PA12 block molecular weight on the thermal transitions taking place in the $-(4GT\text{-}b\text{-}PO4\text{-}b\text{-}PA12)_n-$ terpolymers of Series I (see Table 1) is shown in Figure 3.

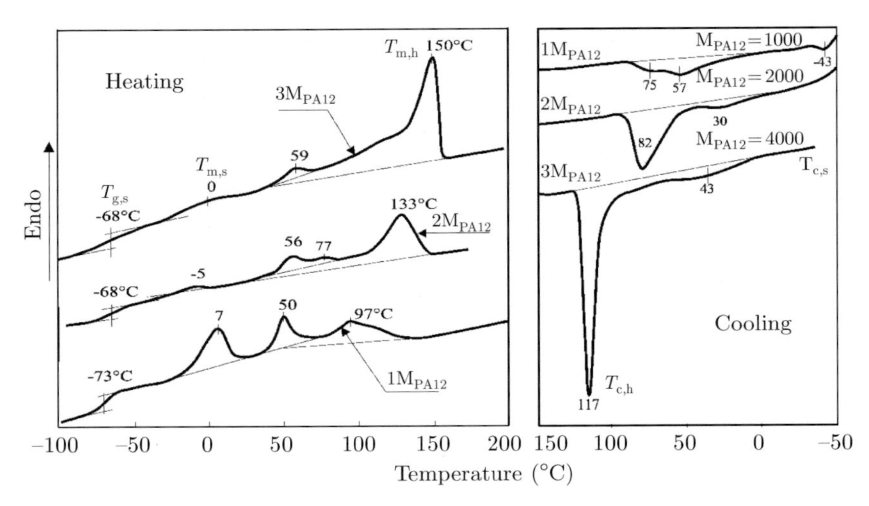

Figure 3. DSC heating and cooling scans of the terpolymers of Series I with varied molecular weight of PA12 and constant $DP_{4GT} = 4$ and PO4 molecular weight of 1000 g/mole

An increase in the amide block molecular weight extends the temperature range of the glass transition and increases the glass transition temperature $T_{g,s}$ of the soft phase. Moreover, the heat of fusion of this phase also decreases to the complete disappearance of the crystallization extremum at $T_{m,s}$. This means that the better the separation of the PA12 hard phase, the more readily soluble are the amorphous 4GT blocks in the PO4 phase.

Two melting endotherms are observed in the high-temperature part of the DSC curves. The first thermal effect (annealing endoterm) was observed in many polymers crystallized from the melt. It occurs during the first heating, whereas the same effect is frequently observed in elastomers during the second heating. In multiblock copolymers with crystallizable hard blocks, this thermal effect is believed to be responsible for the melting of short branches and of the microcrystallites formed between the boundary layers of larger crystallites [4,23,24].

This endothermal effect shields the glass transition taking place in the same temperature range. The second endothermal transition determines the heat of fusion $\Delta H_{m,h}$ and the glass transition temperature $T_{g,h}$ of the hard crystalline phase. $T_{g,h}$ and $T_{m,h}$ shift to higher values with increasing the fraction and the molecular weight of the PA12 block. The melting endoterm in the high-temperature region of the $1M_{PA12}$ terpolymer is small, poor-shaped and reaches a maximum at $T_{m,h} = 97$°C (a very large depression between the melting temperatures of $(4GT)_4$ and $PA12_{1000}$). In this case, one can assume a paracrystalline phase and dissolution of the $(4GT)_4$ blocks in $PA12_{1000}$. The endotherms in the curves $2M_{PA12}$ and $3M_{PA12}$ are well-shaped, and the shape of the $3M_{PA12}$ endoterm and the value of its maximum are associated with a slightly imperfect crystalline phase.

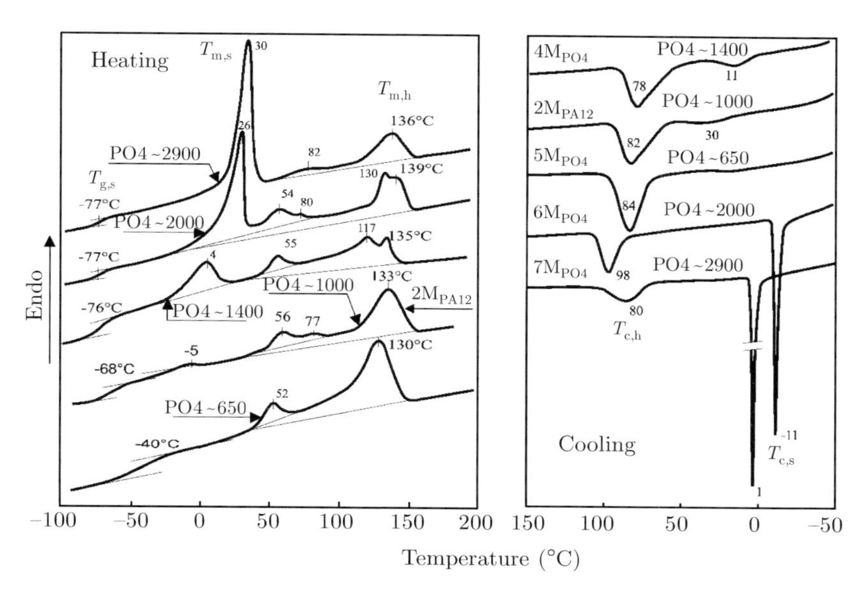

Figure 4. DSC heating and cooling scans of the terpolymers of Series II with varied PO4 molecular weight and constant DP_{4GT} and PA12 molecular weight of 2000 g/mole

The influence of the PO4 block molecular weight on the thermal transitions taking place in the $-(4GT\text{-}b\text{-}PO4\text{-}b\text{-}PA12)_n-$ terpolymers of Series II (see Table 1) is shown in Figure 4.

With the increase in the ether block (M_{PO4}) molecular weight, the temperature range of the glass transition becomes narrower and $T_{g,h}$ in the low-temperature region decreases and approaches a constant value of –77°C. The PO4 blocks with molecular weight $M_{PO4} > 1400$ g/mole cause the appearance of the melting transition and crystallization. The $T_{m,s}$ and $T_{c,s}$ of the soft phase increase, *i.e.*, with the rise in M_{PO4} the separation conditions between the phases PO/PA and PO/4GT are improved, thereby the "purity" of the PO4 soft phase increases. The two endothermal processes can be distinguished in the high-temperature region. It can be seen that the melting temperatures T_m increase with increasing molecular weight of the PO4 block, whereas the melting enthalpies decrease and it can be concluded that the PA12 crystalline structure is more perfect in the presence of PO4 having higher molecular weights. The confirmation of this conclusion can be found in the cooling cycle. The DSC curves at cooling have broad crystallization exotherms of the PA12 phase and perfectly shaped exotherms of the PO4 phase, when the PO4 block molecular weight is 2000 g/mole and higher.

The DSC thermograms of terpolymers with variable degree of polymerization of the $(4GT)_x$ block (Series III, Table 1) are shown in Figure 5. Along with an increase in DP of the ester block, the temperature range of glass transition becomes broader, T_g of the soft phase increases and the crystallization enthalpies become weaker. This behavior suggests a deterioration of the phase

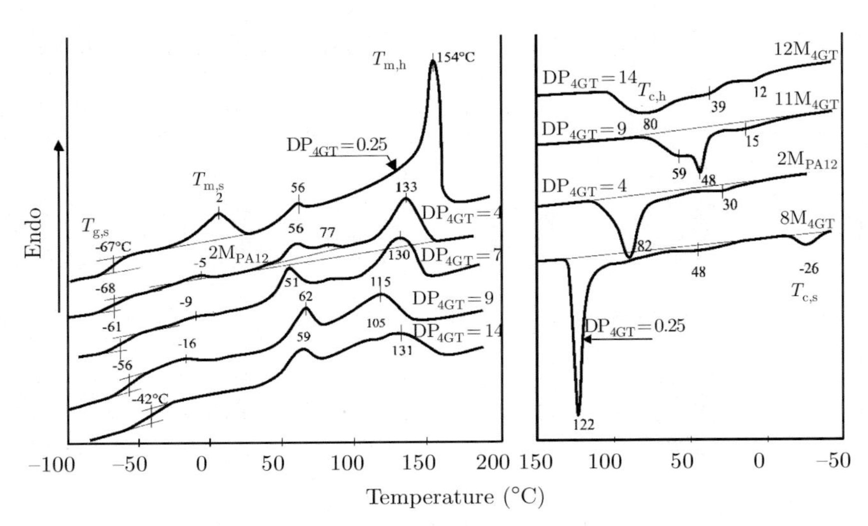

Figure 5. DSC heating and cooling scans of the terpolymers of Series III with varied DP_{4GT} and constant molecular weights of PA12 = 2000 g/mole and PO4 = 1000 g/mole

separation and "contamination" of the PO4 phase with other blocks. The endoterm determining the heat of fusion $\Delta H_{m,h}$ and the melting temperature

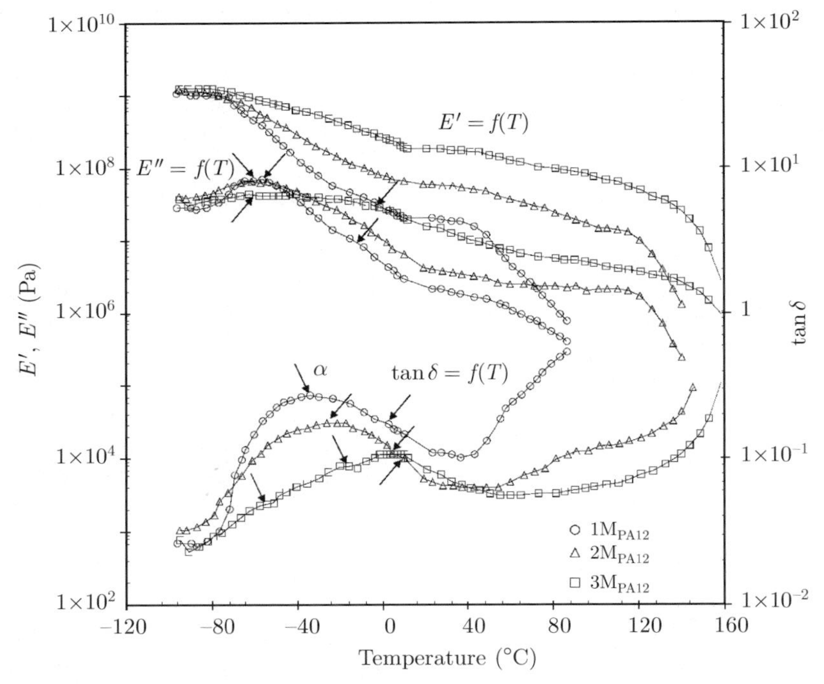

Figure 6. DMTA analysis of the terpolymers of Series I with varied length of the PA12 block and constant $DP_{4GT} = 4$, PO4 molecular weight of 1000 g/mole

$T_{m,n}$ of the hard crystalline phase decreases and flattens with increasing DP_{4GT}, its maximum shifting to lower temperatures [25].

The observed phenomena result from dissolution of 4GT in PA12 and deterioration of the crystalline phase perfection. Analyzing the cooling scans, it can be noticed that the 4GT paracrystalline phase begins to form when the $(4GT)_x$ block has DP = 9, *i.e.*, when the molecular weight of this block becomes equal to that of the PA12 block; two exothermal effects are observed on the thermogram (about the thermal properties of poly(ether ester) thermoplastic elastomers, see also Chapter 6).

The influence of the molecular weight of the PA12, PO4, and 4GT blocks on the relaxation processes associated with the motion of macromolecular aggregates, macrochains, or their fragments, or with the side groups or the chain-ends of the terpolymers is illustrated in Figures 6, 7, and 8 [12]. Each figure shows the temperature dependences of the storage modulus E', the loss modulus E'', and the loss $\tan\delta$. In the curves of the storage modulus, five temperature ranges can be distinguished in which the E' values are essentially different. In the temperature range from −100°C to −80°C, the E' curves are flat, the modulus does not change, and the terpolymers are in the glassy state. In the temperature range from −80°C to −15°C, the modulus decreases, the macromolecular segments activate, and the polymer passes into the viscoelastic state.

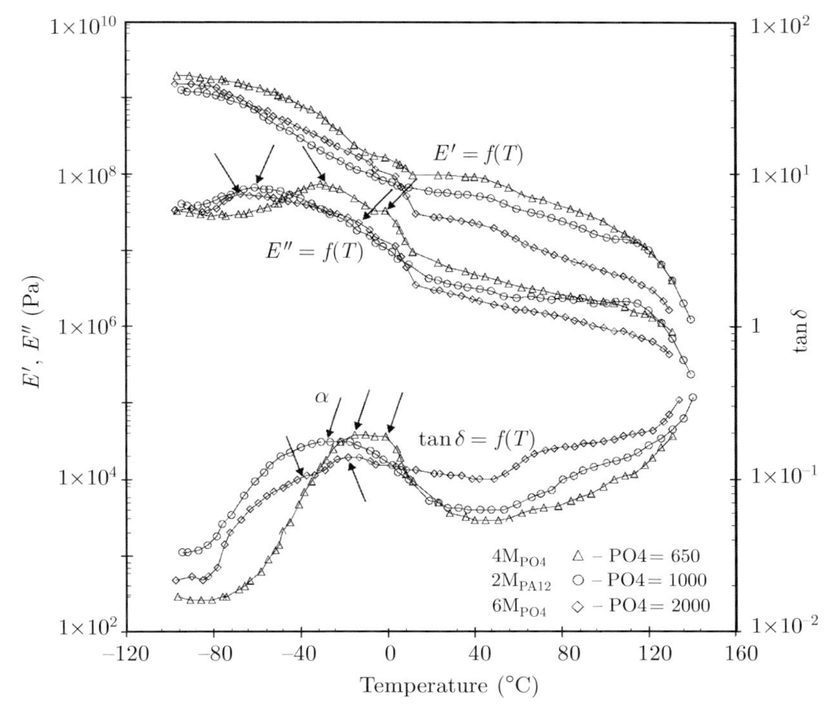

Figure 7. DMTA analysis of the terpolymers of Series II with varied length of the PO4 block and constant $DP_{4GT} = 4$, PA12 molecular weight of 2000 g/mole

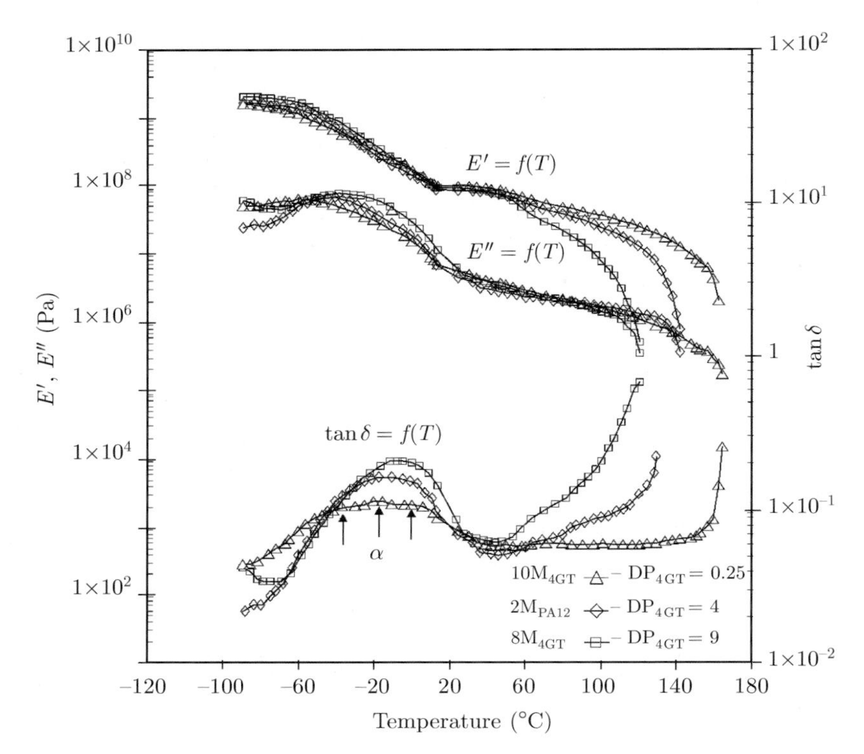

Figure 8. DMTA analysis of the terpolymers of Series III with varied DP_{4GT} and constant molecular weights of PA12 = 2000 g/mole and PO4 = 1000 g/mole

In the third range from –15°C to 15°C (or to 25°C), relaxation phenomena occur, which are associated with the PO4 crystallization and/or the glass transition of the interphase being a PO/4GT/PA12 solution. The fourth range from 15°C to 120°C comprises "a plateau of elasticity", the modulus is constant, and the terpolymer is in the high-elastic state. Above the softening point $T_s = 120°C$ (the fifth range), the elastomers soften and their modulus rapidly decreases; the polymers pass into the plastic state.

The temperature dependences of E'' and $\tan\delta$ have maxima attributed to the respective relaxation processes, as well as a broad maximum resulting from the plasticizing effect of 4GT. The main peak corresponds to the glass transition and is in agreement with the DSC data. The maxima of damping ($\tan\delta$) are very broad and have many additional inflections (marked with arrows).

The temperature dependence of E'' also shows additional inflections as a result of superposition of the relaxation processes associated with the glass transitions of the soft phase, the interphase, and the amorphous phase of the hard blocks, as well as with melting of the PO4 block crystalline phase. The inflections of all curves at high temperatures, associated with the rapid decrease of modulus or increase of damping, characterize the softening point (plasticity),

T_s, of the polymer (temperature at which viscosity factors suppress the elasticity). When this temperature significantly deviates from the melting points of the crystalline phases, the existence of an expanded interphase is assumed [5].

The detailed analysis of the thermal properties based on DSC and DMTA data shows that the terpoly(ester-*b*-ether-*b*-amide)s are capable to form a nanophase structure when their blocks satisfy the following conditions:

$$\bar{M}_{n,PA12} \geq 2000 \text{ g/mole; } 1000 \leq \bar{M}_{n,PO4} \leq 1500 \text{ g/mole; DP}_{(4GT)} \leq 7.$$

The temperature dependences of the imaginary part of the dielectric losses, ε'', of a standard terpolymer are shown in Figure 9. They characterize the energy losses in the polymer caused by a dipole reorientation and an increase in the conductivity. The $\varepsilon'' = f(\mathrm{T})$ curves exhibit four maxima (even five at the frequency of 1 kHz), which should be attributed to the respective relaxation process in the elastomer. These maxima, occurring at –90, –50, 10, and 60°C (β and α-processes) shift to higher temperatures with increasing frequency. The relaxation effect at *ca.* –100°C is assigned to a restricted motion of short segments in the PO4 chain composed of at least four $-CH_2-$ groups, that is undoubtedly the β-relaxation type. The maximum at *ca.* –50°C overlaps with

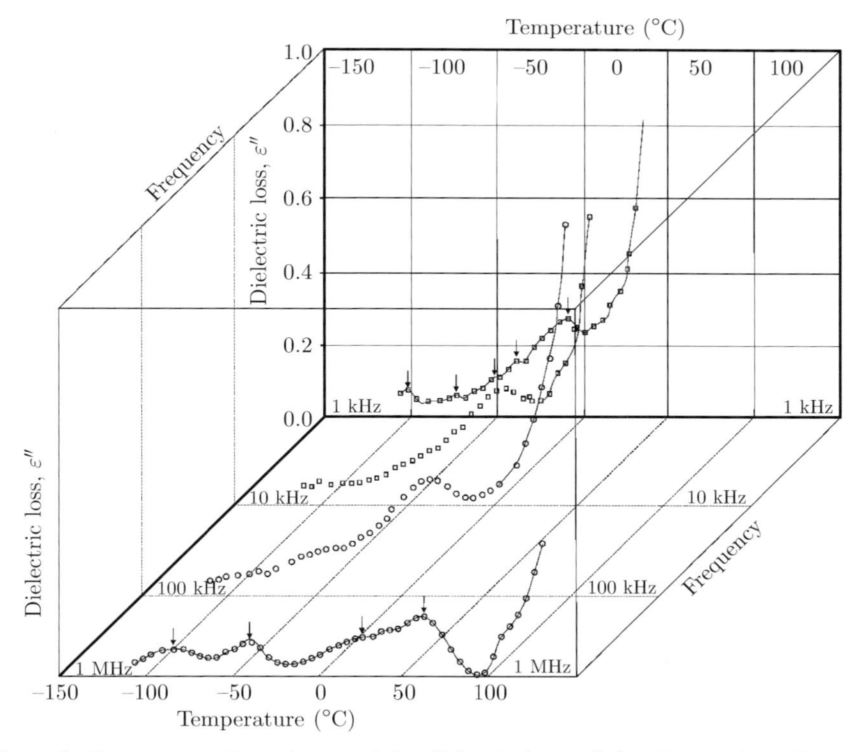

Figure 9. Temperature dependences of the dielectric loss, ε'', for the standard elastomer $(PO4_{1000}\text{-}b\text{-}PA12_{2000}\text{-}b\text{-}(4GT)_4)_n$ at various frequencies [26]

the α-maximum of mechanical damping ($\tan\delta = f(\mathrm{T})$, Figure 6), and is an α-relaxation process of the PO4 block. The relaxation occurring at 10°C has its counterpart in the mechanical damping curve at *ca.* 0°C. Because the PO4 phase of the standard terpolymer does not crystallize (see the DSC traces in Figure 3), it can be concluded that this is the glass transition α' of the intermediate phase (solutions of PO/PA/4GT). The maximum of the curve of dielectric losses at 60°C is also observed as a small peak in the DMTA curve of mechanical damping. These peaks can be assigned to the annealing endoterm.

The processes taking place in the standard terpolymer at various temperatures are shown in Figure 10. This is a diagram of the relaxation processes of the homopolymers PO4, 4GT, and PA12 (bold lines) investigated by various methods onto which relaxation data for the standard terpolymer were superimposed. The α-relaxation processes are associated with the glass transitions of

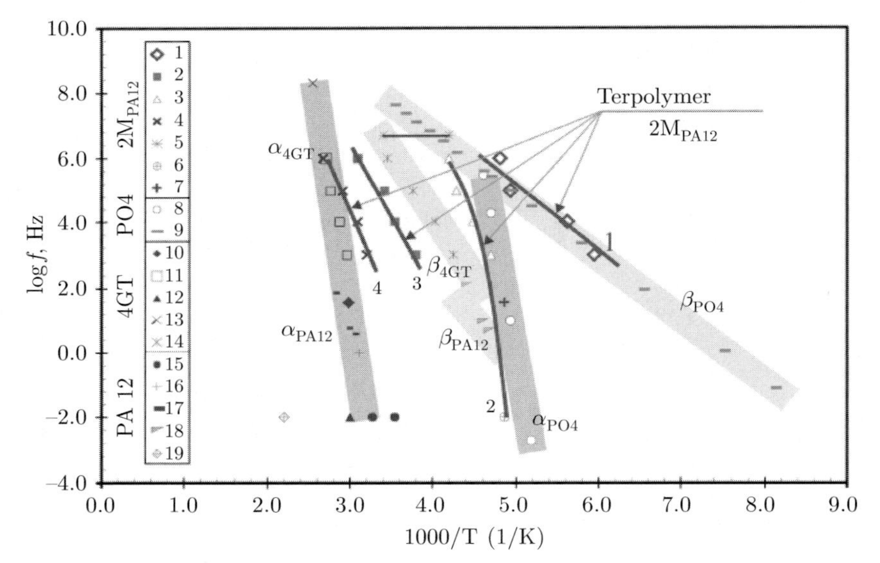

Figure 10. Relaxation diagram of terpolymer $2M_{PA12}$ (1–7), PO4 (8,9), 4GT (10–14), and PA12 (15–19); 1–4, 11, 14, – dielectric data, 5 – ultrasonic data, 6, 12, 15, 19 – DSC data, 7, 10, 16–18 – DMTA data, 8, 9 – ESR, dielectric and DMTA data, 13 – NMR data [26]

blocks or their homopolymers. When no side groups are present in the chain, the β-processes are motions of short segments of this chain; the displacement of these segments requires only a small free volume. Additionally, in polyamides, one should take into account the presence and motion of "bonded water" by the hydrogen bond with the carbonyl groups. As it is constructed, the diagram allows the attribution of the observed relaxation effects to the appropriate phase transitions in the terpolymer:

(i) The points forming the straight line 1 overlap with the β-relaxation of the PO4 homopolymer, *i.e.,* this is the β-relaxation of PO4 blocks. In the case of the PO4 chain, owing to a lack of side groups, it can be assumed that these are the so-called "crank" motions between the ether bonds.

(ii) The points forming curve 2 overlap with the α-relaxation of the PO4 homopolymer. Therefore, they describe the glass transition of the soft phase, which is rich in PO4 blocks. This can be interpreted as the main glass transition of the terpolymer.

(iii) The points forming the straight line 3 are located between the β- and α-relaxation of the homopolymers PBT and PA12. One should conclude that this is the glass transition of a solution of PO4/4GT/PA12 blocks (interphase).

(iv) The points forming the straight line 4 are located in the region of α-relaxation of the homopolymers 4GT and PA12. Hence, they could be the glass transition of insoluble amorphous fractions of 4GT or PA12 (α'') or, more probably, they could be the effect of dissipation of the paracrystalline aggregates associated with the annealing endoterm.

The thermal transitions and the relaxation processes observed in multiblock terpolymers allow to evaluate their phase morphology. At room temperature, these polymers are composed of three phases: hard, soft, and strongly expanded interphase. The two latter phases are amorphous and form a matrix (continuous phase), whereas the hard (crystalline) phase is the dispersed phase. The thermal transition and relaxation processes occurring in the interphase of the multiblock copolymers are not detected by the DSC and DMTA methods. The incorporation of the third short block into the copolymer chain causes an increase in the volume of the interphase. This facilitates the establishment of the processes occurring in this phase at various temperatures. Moreover, it enables the evaluation of the influence of the dimension and composition of this phase on the polymer properties. (About the number of phases in poly(ether ester) thermoplastic elastomers, see also Chapter 6.)

5.2. *Phase composition of terpoly(ester-b-ether-b-amide)s*

A DSC analysis combined with the dielectric, DMTA and WAXS studies enables the quantitative evaluation of the phase composition in terpoly(ester-*b*-ether-*b*-amide)s. The phase composition is calculated by known methods [2,5,27,28] from the following data:

- Degree of phase separation (SR_s) of the soft phase (Eq. (1)):

$$SR_s = [(\Delta c_{p,exp}/\Delta c_{p,a}) + (\Delta H_{exp}/\Delta H_{t,c})]^{*}1/w^s \qquad (1)$$

where $\Delta c_{p,exp}$ is the experimental heat capacity increase of the polymer, $\Delta c_{p,a}$ is the heat capacity increase of the amorphous homopolymer of a given block, $\Delta c_{p,a,PO4} = 0.72$ J/g K* [5]), and w^s is the weight fraction of the soft phase (real solution of 4GT in PO4) in the polymer ($SR_s + w_{x4GT}$)

* Heat capacity change of amorphous PO4 calculated on the basis of data taken from Table 3 amounts to $\Delta c_{p,a,PO4} = 0.34$ J/g K/(1 − 0.556) = 0.766 J/g K.

– Degree of crystallinity, w_c^h, of the hard phase (Eq. (2)):

$$w_c^h = (\Delta H_{m,exp}/\Delta H_{m,c})*1/w^h \quad (2)$$

where $\Delta H_{m,exp}$ is the heat of fusion of the polymer sample, $\Delta H_{m,c}$ is the heat of fusion of the crystalline homopolymer in a given block ($\Delta H_{t,c4GT} = 144.5$ J/g [8,29], $\Delta H_{t,cPO4} = 199.7$ J/g [8], $\Delta H_{t,cPA12} = 96$ J/g [5]), and w^h is the weight fraction of the crystallizing hard block in the polymer.

By the following assumptions: (i) the PA_{2000} blocks are insoluble in the phase of PO4 blocks (they do affect $T_{g,PO4}$) and the remaining PO4/4GT fraction can be treated as a separate system, (ii) the glass transition temperature of the PO4 phase, due to the immobilization of the PO4 chain-ends by chemical bonds, is higher by 5°C than that determined for PO4 homopolymer, and (iii) the 4GT blocks with $DP_{4GT} < 7$ do not crystallize in the presence of $PA12_{2000}$, the fraction of short 4GT segments dissolved in PO4 ($w_{x,4GT}$) was calculated from the Gordon-Taylor equation (3). This enables the calculation of the weight fraction of the PBT/PO4 real solution and the degree of soft block phase separation in the terpolymer.

$$w_{x,4GT} = w_{PO4}(T_{g,PO4} - T_{g,polym})/k(T_{g,polym} - T_{g,4GT}) \quad (3)$$

where w_{PO4} and w_{4GT} are the weight fractions of PO4 and 4GT in the amorphous phase, $T_{g,polym}$ is the glass transition temperature of PO4-*b*-$(4GT)_4$, T_{gPO4} is the glass transition temperature of PO4, T_{g4GT} is the glass transition temperature of 4GT, and k is a coefficient of intermolecular interactions, which reflects the magnitude of deviation from the additive change of T_g as a function of the solution composition. The deviation coefficient k was calculated assuming $\Delta c_{p,4GT} = 0.484$ J/g K [30]. For the 4GT/PO4 system, this coefficient is k = $\Delta c_{p2}/\Delta c_{p1} = 0.484$ J/g K:0.72 J/g K= 0.672.

The knowledge of SR_s and $w_{c,h}$ allows to calculate the interphase composition. This is possible by the assumption that the interphase comprises a solution of PO/4GT/PA, and the sum of the fractions of the PO, PA, and 4GT blocks, which are not included in the soft and in the hard phase. The composition of the standard terpolymer calculated by this method is presented in Figure 11.

5.3. *Mechanical properties of* $-(4GT\text{-}b\text{-}PO4\text{-}b\text{-}PA12)_n-$

The elastic properties of the terpolymers studied are characterized by the mechanical hysteresis loops shown in Figures 12–14. The ability of an instant recovery after deformation is expressed by the elastic elongation, ε_s, and the area A, which is proportional to the dissipated elastic energy. These magnitudes provide information about the capability of the spatial network formed in the elastomer to transfer large stresses. The ability to a delayed recovery after the deformation is characterized by the high-elasticity elongation, ε_{hs}, and the area B, which is proportional to the dissipated high-elastic energy. These magnitudes are the measures of the capability of the continuous phase to show a viscoelastic response. The accumulated energy (area C) and the permanent set, ε_{ps}, are

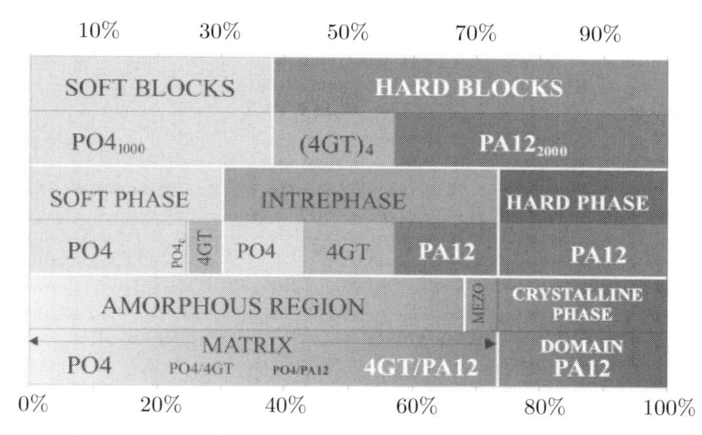

Figure 11. Composition of blocks and phases in the standard elastomer $2M_{PA12}$

the measures of the irreversible changes in the material caused by the stress. A large energy accumulation during the first cycle of elongation is assigned to a change in the spatial distribution of domains (disperse phase), *i.e.*, in their orientation and reorientation. The energy accumulation in the first cycle, accompanied by small accumulations in the consecutive cycles, proves the validity of this assumption.

The trends of the curves obtained during the mechanical cyclic test of various types of elastomers are compared in Figure 12. The PUE elastomers and terpolymers are characterized by a small energy accumulation in the first cycle,

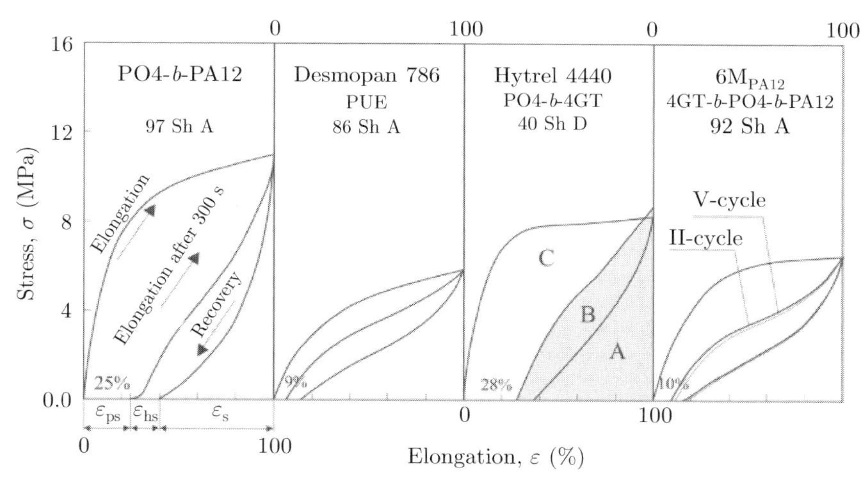

Figure 12. Mechanical hysteresis loops at a constant elongation of 100% (room temperature) of copoly(ether-*b*-amide), copoly(ether-*b*-urethane) (Desmopan® 786), copoly(ester-*b*-ether) (Hytrel® 4440), and the terpolymer $((GT)_4$-*b*-$PO4_{1400}$-*b*-$PA12_{2000})_n$ ($5M_{PO4}$): ε_s – elastic elongation, ε_{hs} – delayed high-elastic elongation, ε_{ps} – permanent set, A – area proportional to the dissipated elastic energy, B – area proportional to the high-elastic dissipated energy, and C – area proportional to the accumulated energy

whereas PEE and PEA accumulate a larger part of energy supplied during this very cycle of elongation. Hence, it can be assumed that in PUE elastomers and in –(GT-*b*-PO4-*b*-PA)$_n$–, the hard domains are in the matrix, which exhibits an ability to recover almost completely. This in turn is associated with the effect of stronger interactions at the domain-matrix contact in PEE and PEA, *e.g.*, through chain folds, hydrogen bonds, or other Van der Waals forces. It seems highly probable that the interphase interactions are caused by an appropriately distributed and adequately large intermediate phase. The studies by Camberlin and Pascault [2,5] have proved the existence of a widespread interphase in PUE, whereas PEE are characterized by a small interphase [4,29].

The terpolymers and PUE are elastomers with a better mechanical shape memory than PEE and PEA. This as a result of the greater amount of the interphase which is a part of the matrix. The mechanical hysteresis loops shown

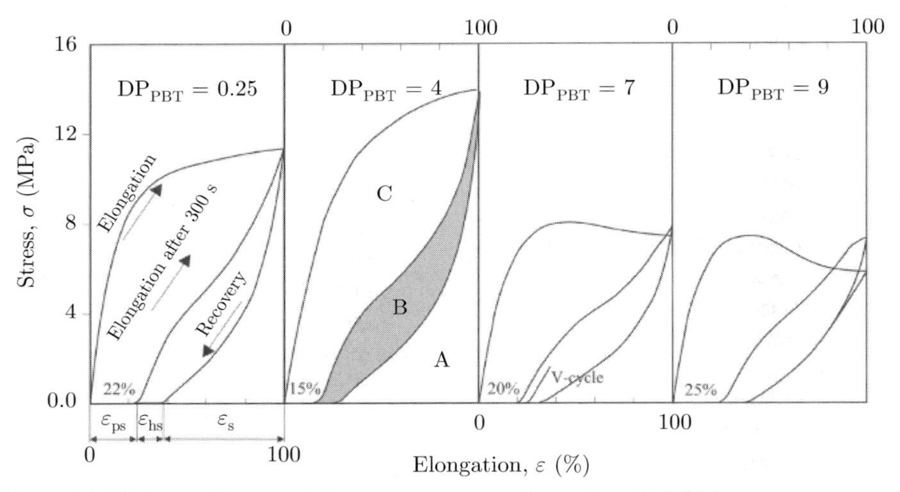

Figure 13. Mechanical hysteresis loops at a constant elongation of 100% (room temperature) of the terpolymers ((GT)$_x$-*b*-PO4$_{1000}$-*b*-PA12$_{2000}$)$_n$

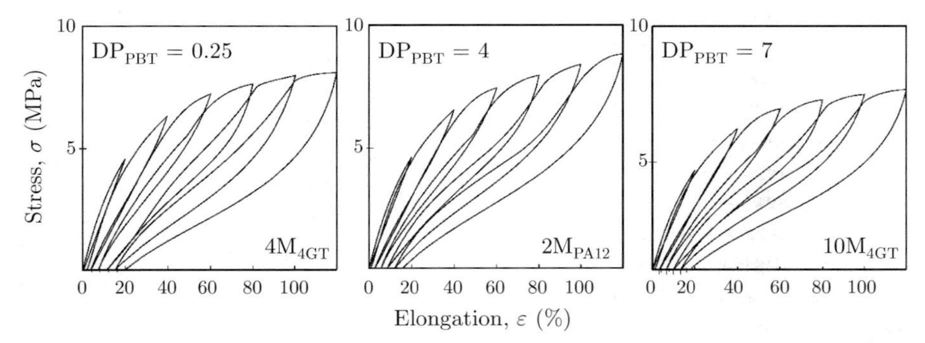

Figure 14. Mechanical hysteresis loops at elongation growing by 20% (room temperature) of the terpolymers ((GT)$_x$-*b*-PO4$_{1000}$-*b*-PA12$_{2000}$)$_n$

in Figures 13 and 14 demonstrate that the terpolymers in which the 4GT block has a DP = 4 show the best recovery. This is the block which readily dissolves in the amorphous phases of the remaining blocks and causes an increase in the interphase dimensions and a decrease in crystalline domain sizes.

The basic properties of the terpolymers studied are summarized in Table 4.

Table 4. Properties of the $-(4GT\text{-}b\text{-}PO4\text{-}b\text{-}PA12)_n-$ terpolymers

Sample	$[\eta]$ (dL/g)	H (Shore D)	σ_r (MPa)	ε (%)	T_m (°C)
1 M_{PA12}	0.62	25	9.2	165	97
2 M_{PA12}	1.70	31	19.4	340	133
3 M_{PA12}	1.85	54	19.4	250	150
4 M_{PO4}	1.25	37	13.3	210	130
5 M_{PO4}	1.52	32	10.2	310	135
6 M_{PO4}	1.06	31	9.8	295	139
7 M_{PO4}	0.92	23	7.5	295	136
8 M_{4GT}	1.71	36	21.8	400	154
9 M_{4GT}	1.75	35	18.7	380	138
10 M_{4GT}	1.58	28	13.3	310	130
11 M_{4GT}	1.30	27	9.4	275	115
12 M_{4GT}	0.82	26	6.1	210	105,131

$[\eta]$ – Limiting viscosity number, H – Shore D hardness, σ_r – tensile strength, ε – elongation at break, T_m – melting point

6. Conclusions and outlook

The incorporation of a third block into the main chain of a multiblock copolymer not only changes its chemical structure, but also influences its morphology [31]. The changes in the supermolecular structure depend on the solubility of the additional block in the amorphous phases of the two other blocks. This solubility is affected by the chemical structure, molecular weight, polydispersity and the contribution of each of the blocks in the terpolymer (Figure 15). The thermoplastic elastomers of the $-(4GT\text{-}b\text{-}PO4\text{-}b\text{-}PA12)_n-$ type have a nanostructure and possess useful properties when the molecular weights of the blocks are appropriately chosen: $1500 \leq M_{n,PA12} \leq 2500$ g/mole, $1000 \leq M_{n,PO4} \leq 1500$ g/mole, $400 \leq M_{n,4GT} \leq 1100$ g/mole. The terpolymers with the above composition of blocks are elastomers containing: (i) from 30 to 40 wt% of the intermediate phase and it would be beneficial if this phase is uniformly composed of all the three or of two blocks, provided that one of them is a soft block, (ii) the soft phase in amounts $\geq$ 40 wt%, and (iii) the hard phase in an amount varying from 20 to 25 wt%.

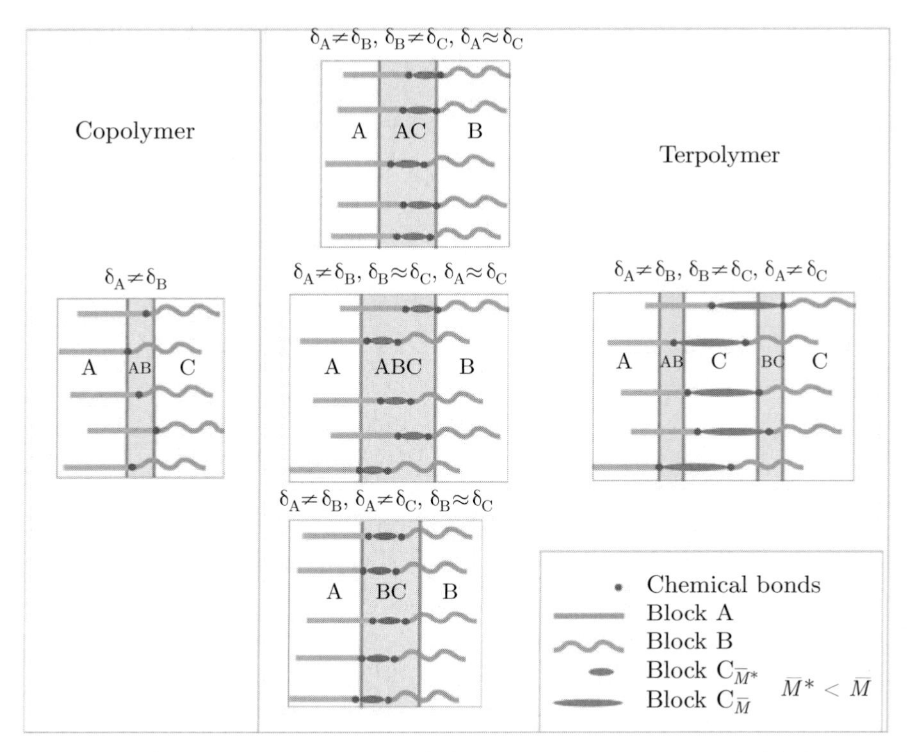

Figure 15. Schematic representation of the phase separation in multiblock co- and terpolymers depending on the solubility parameters of the blocks (A, B, C - phases; AB, AC, BC, ABC - interphases)

It was established that a properly selected third block enlarges the interphase and the increase of the interphase volume (but only to same extent) enhances the ability of the material to recover its original shape after mechanical deformations. The interphase volume enlargement is a result of the change in the structure from microphase to nanophase (see also Chapter 7).

The interphase (intermediate phase) is a result of an imperfect separation of the soft and hard phases; it can be in the high-elastic, viscoelastic, or glassy state, and these physical states determine the material flexibility.

Under normal conditions, the interphase is in the high-elastic state only in the case when its glass transition is below room temperature. The dimensions and composition of the interphase influence the glass transition temperature of the soft phase and of the interphase itself, as well as the melting point of the hard phase. These parameters are in turn decisive as far as the flexibility plateau is concerned. Hence, they influence the range of applicability of this elastomer and its mechanical memory (recovery).

The quantitative and qualitative evaluation of the phase composition led to the conclusion that the third block with appropriately selected chemical and physical parameters, incorporated into the multiblock copolymer, plays the role of a plasticizer and stabilizer of the structure, as well as of a compatibilizer at the matrix-domain boundary. An advantage of the modification of a multiblock copolymer chain by the third block is the possibility of preparation of materials showing a shape memory effect, *i.e.*, having the ability to transform a temporary shape into a permanent shape when triggered by some external factor, such as temperature, light, radiation, ultrasound, humidity, electric and magnetic field, pH variation, and others. Hence, this method leads to the preparation of polymers with various types of shape memory (about shape memory, see also Chapter 18).

References

1. Bates F S and Fredrickson G H (February 1999) Block copolymers – designer soft materials, *Physics Today*, American Institute of Physics, pp. 32–38.
2. Camberlin Y and Pascault J P (1983) Quantitative DSC evaluation of phase segregation rate in linear segmented polyurethanes and polyurethaneureas, *J Polym Sci* **21**:415–423.
3. Lilaonitkul A, West J C and Cooper S L (1976) Properties of poly(tetramethylene oxide)-poly(tetramethylene terephthalate) block polymers, *J Macromol Sci–Phys* **B12**:563–597.
4. Vallance M A and Cooper S L (1984) Microstructure in linear condensation block copolymers: a modeling approach, *Macromolecules* **17**:1208–1219.
5. Xie M and Camberlin Y (1986) Morphological study of block copoly(ether-amide)s, *Makromol Chem* **187**:383–400 (in French).
6. Bornschlegl E, Goldbach G and Meyer K (1985) Structure and properties of segmented polyamides, *Progr Colloid Polym Sci* **71**:119–124.
7. *Processing, Structure and Properties of Block Copolymers* (1985) (Ed. Folkes M J) Elsevier Applied Science Pubublications, London.
8. Van Krevelen D W (1990) *Properties of Polymers, their Estimation and Correlation with Chemical Structure*, Elsevier Scientific Pubublications Co., Amsterdam-Lausanne-New York-Oxford-Shannon-Tokyo.
9. *Handbook of Thermoplasic Elastomers* (2001) (Eds. Bhowmick A K and Stephens H L) Marcel Dekker, Inc., New York-Basel.
10. *Thermoplastic Elastomers* (1996) (Eds. Holden G, Legge N R, Quirk R and Schroeder H E) Hanser, Munich, Vienna, New York.
11. *Block Copolymers* (2000) (Eds. Baltá Calleja F J and Roslaniec Z) Marcel Dekker, Inc., New York-Basel.
12. Ukielski R (2000) Multiblock terpoly(ester-*b*-ether-*b*-amide) elastomers: synthesis, structure, properties, Habilitation Thesis, *Sci Papers Techn Uni Szczecin* **556:**1–150 (in Polish).
13. Baltá Calleja F J, Fakirov S, Roslaniec Z, Krumova M, Ezquerra T A and Rueda D R (1998) Microhardness of condensation polymers and copolymers. 2. Poly(ester ether carbonate) thermoplastic terpolymers, *J Macromol Sci–Phys* **B37**:219–237.
14. Roslaniec Z (1999) Block copolymers, terpolymers and poly(ether-ester-amide) blends, *Polimery* **44**:483–491 (in Polish).
15. Polish Patent 165 712 (1993).
16. Polish Patent 162 304 (1991).
17. Ukielski R (2000) New multiblock terpoly(ester-ether-amide) with various chemically constitutive amide blocks, in *Block Copolymers* (Eds. Baltá Calleja F J and Roslaniec Z) Marcel Dekker, Inc., New York-Basel, pp. 65–91.

18. Pawlaczyk K and Ukielski R (1997) Synthesis of α,ω-dicarboxyl oligoamides *Polimery* **42**:680–684 (in Polish).
19. Ukielski R (1995) Synthesis of block polyesteretheramide terpolymers, *Polimery* **40**:160–163 (in English).
20. Ukielski R (1996) Synthesis of block terpolymer of the polyoxytetramethylene-*block*-(tetramethylene terephthalate)-*block*-laurolactam type, *Polimery* **41**:286–289 (in English).
21. Stevenson J C and Cooper S L (1988) Microstructure and property changes accompanying hard-segment crystallization in block copoly(ether-ester) elastomers, *Macromolecules* **21**:1309–1316.
22. Ukielski R (2000) New multiblock terpoly(ester-ether-amide) thermoplastic elastomers with various chemical compositions of ester block, *Polymer* **41**:1893–1904.
23. Ukielski R (2001) Low temperature endotherm in multiblock terpoly(ester-*b*-ether-*b*-amide) thermoplastic elastomers, *Macromol Mater Eng* **286**:337–341.
24. Castles J L, Vallance M A, Mckenna J M and Cooper S L (1985) Thermal and mechanical properties of short-segment block copolyesters and copolyether-esters, *J Polym Sci B Polym Phys* **23**:2119–2147.
25. Ukielski R and Pietkiewicz D (1998) Influence of changes in the degree of polycondensation of polyester block on properties of poly(ether-*block*-ester-*block*-amide) terpolymers, *J Macromol Sci–Phys* **B37**:255–264.
26. Ukielski R, Lembicz F and Majszczyk J (1999) New multiblock terpoly(ester-ether-amide) thermoplastic elastomers, *Angew Makromol Chem* **271**:53–60.
27. Khanna Y P and Kuhn W O (1997) Measurement of crystalline index in nylons by DSC: complexities and recommendations, *J Polym Sci B Polym Phys* **35**:2219–2231.
28. Fakirov S and Gogeva T (1990) Poly(ether/esters) based on poly(butylene terephthalate) and poly(ethylene glycol). 1. Poly(ether/esters) with various polyether: polyester ratios. 2. Effect of polyether segment length. 3. Effect of thermal treatment and drawing on the structure of the poly(ether/esters), *Makromol Chem* **191:**603–614, 615–624, 2341–2354.
29. Gilbert M and Hybart F J (1972) Effect of chemical structure on crystallisation rates and melting of polymers, *Polymer* **13**:327–332.
30. Cheng S Z D, Pan R and Wunderlich B (1988) Thermal analysis of poly(butylene terephthalate) for heat capacity, rigid-amorphous content, and transition behavior, *Makromol Chem* **189**:2443–2458.
31. Ukielski R and Piatek M (2003) The influence of chemical composition of amide block on the thermal properties and structure of terpoly(ester-*b*-ether-*b*-amide) elastomers, *Polimery* **48:**690–695 (in English).

Chapter 5

High Performance Thermoplastic Aramid Elastomers: Synthesis, Properties, and Applications

H. Yamakawa, H. Miyata

1. Introduction

Thermoplastic aramid elastomers have alternating sequences of highly crystalline *aramid hard segments* and *rubbery soft segments*. The aramid segments form crystalline hard domains acting as physical crosslinks of the soft segments. The two segments are thermodynamically immiscible, which therefore results in a two-phase structure composed of crystalline aramid phase and semicrystalline or amorphous phase. Figure 1 is a schematic representation

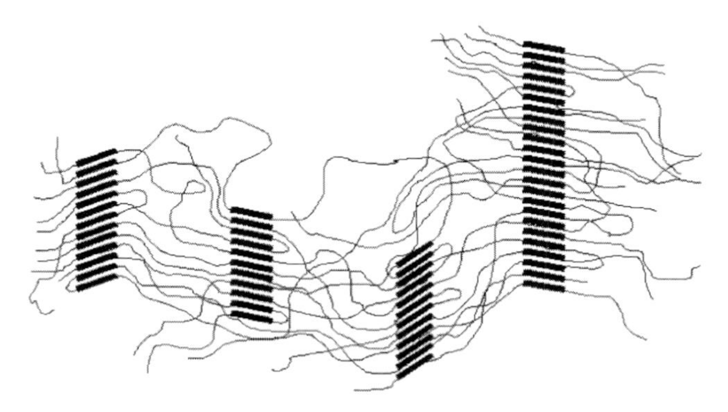

Figure 1. Schematic representation of the phase structure of aramid elastomers; bold lines: aramid hard segments; thin lines: soft segments

of the phase structure of the aramid elastomers. The crystalline phase melts upon heating to disrupt the physical crosslinks. The melt processability however is only achieved for aramid structures with short and uniform segment length. The incorporation of aramid structures and the uniformity of the segment length are therefore structural features of great importance to obtain high melting elastomers with excellent melt processability. For this reason, the length of the aramid segments has to be strictly controlled [1–3]. The uniformity of hard segments is a remarkable structural difference, compared to those of other thermoplastic elastomers with segmented block structures [4–7], *e.g.*, copolyester [8–17], copolyamide [18–24], and polyurethane elastomers [25–30].

2. Development of thermoplastic aramid elastomers

Two types of aramid elastomers are distinguished according to the chemical structure of the aramid segments. Scheme 1 shows the typical chemical structures of aramid elastomers of the types I and II, the aramid segments differring in the arrangement of amide groups.

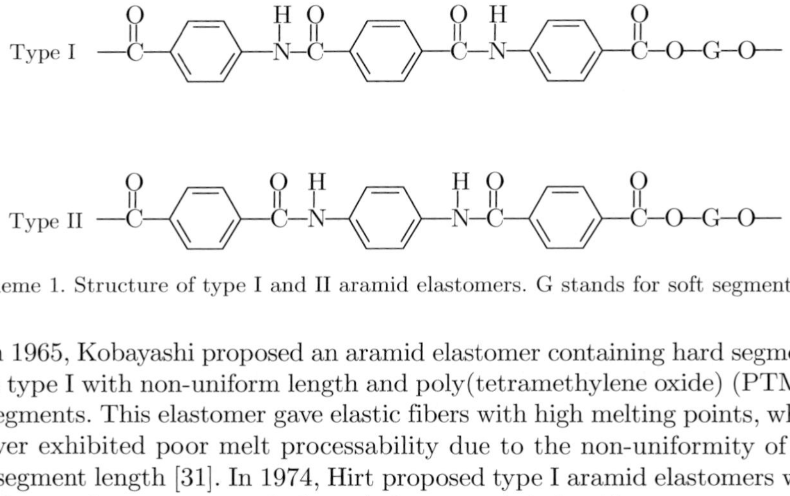

Type I —C(=O)—C₆H₄—N(H)—C(=O)—C₆H₄—C(=O)—N(H)—C₆H₄—C(=O)—O—G—O—

Type II —C(=O)—C₆H₄—C(=O)—N(H)—C₆H₄—N(H)—C(=O)—C₆H₄—C(=O)—O—G—O—

Scheme 1. Structure of type I and II aramid elastomers. G stands for soft segments

In 1965, Kobayashi proposed an aramid elastomer containing hard segments of the type I with non-uniform length and poly(tetramethylene oxide) (PTMO) soft segments. This elastomer gave elastic fibers with high melting points, which however exhibited poor melt processability due to the non-uniformity of the hard segment length [31]. In 1974, Hirt proposed type I aramid elastomers with PTMO as soft segments and showed that terephthalamide segments act as thermally stable physical crosslinks of the PTMO segments [32]. In 1994, type I aramid elastomers were synthesized by direct polycondensation of a terephthalamide compound and poly(tetramethylene oxide) glycol (PTMG); they had a controlled segment length distribution. The low molecular weight polymers obtained exhibited poor mechanical strength [33]. Later, in 1995, high molecular weight aramid elastomers were obtained by using a transesterification reaction in the presence of an amide solvent [34–42].

Various kinds of soft segments have been incorporated into aramid elastomers to improve anti-oxidative degradation and low temperature flexibility, including poly(oxyethylene)s, poly(oxypropylene)s [43], poly(dimethylsiloxane)s (PDMS) [42,44], hydrogenated poly(butadiene)s [43], and poly(butylene-adipate)s (PBA) [42,44]. The aramid-*b*-PBA elastomer was synthesized according to a novel synthetic route using a terephthalamide compound with activated carbonyl end-groups. The terephthalamide compound gave high molecular weight aramid elastomers simply by heating with various glycols under reduced pressure [42,44].

In 1999, Niesten and Gaymans proposed type II aramid elastomers [45], which exhibited excellent thermal and mechanical properties similar to those of type I aramid elastomers. Niesten synthesized a series of aramid elastomers consisting of terephthalamide hard segments and PTMO soft segments with various molecular weights, and studied the mechanical, chemical and thermal properties, as well as the phase structures [46–50].

3. Type I poly(aramid-*b*-polyether) elastomers

This section overviews the synthesis of "reactive aramid compounds" and various synthetic routes leading to type I aramid elastomers with PTMO soft segments.

3.1. *Synthesis of reactive aramid compounds*

A reactive aramid compound 4,4′-(terephthaloyldiimino)diethylbenzoate (Scheme 2, I) was synthesized by the condensation reaction of terephthaloyl chloride with 4-aminoethylbenzoate in an aprotic organic solvent, *e.g.*, N-methyl-2-pyrrolidone (NMP) at room temperature. The reaction proceeded

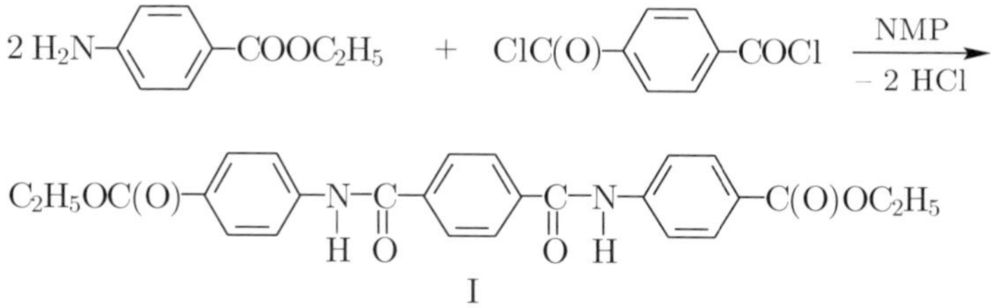

$$2\,H_2N{-}C_6H_4{-}COOC_2H_5 \;+\; ClC(O){-}C_6H_4{-}COCl \xrightarrow[-\,2\,HCl]{NMP}$$

$$C_2H_5OC(O){-}C_6H_4{-}NH{-}C(O){-}C_6H_4{-}C(O){-}NH{-}C_6H_4{-}C(O)OC_2H_5$$

I

Scheme 2

quantitatively without catalysts in the solvent. Due to the limited solubility of compound I in NMP, crystallites of I precipitated during the reaction. Recrystallization of the precipitates from NMP yielded the compound I with purity sufficient for polycondensation reactions [51,52]. This compound was also used as a monomer of copolyesteramide resins [53–55].

3.2. *Synthesis of aramid elastomers*

Several methods for the synthesis of type I aramid elastomers have been proposed: (i) transesterification reaction between the aramid compound I and PTMG (Route A) [33–35], (ii) polycondensation between PTMG end-capped with *p*-aminobenzoate and terephthaloyl chloride (Route B) [56], (iii) polycondensation of PTMG end-capped with *p*-aminobenzoate with diiodobenzene (Route C) [57,58], (iv) polycondensation of an aromatic amide having activated carbonyl end-groups with glycols (Route D). Elastomers with the highest molecular weight have been obtained by Route A.

3.2.1. *Route A*

The titanate-catalyzed transesterification reaction between compound I and PTMG (Scheme 3) gave only low molecular weight elastomers, because of the insolubility of compound I in PTMG, a poor solvent for I. The solubility is quite limited even at elevated temperatures, preventing I from reacting with PTMG [44].

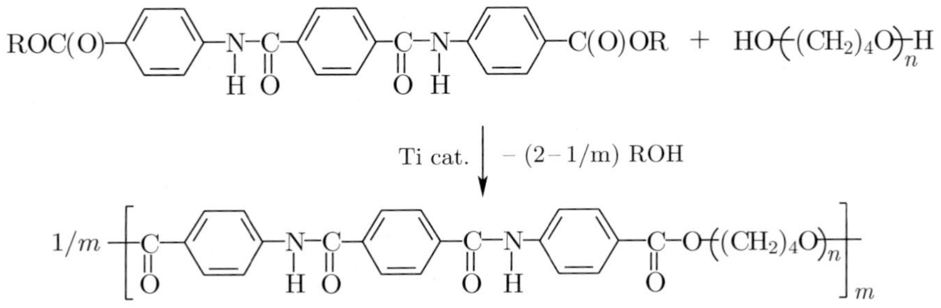

$$\mathrm{ROC(O)}-C_6H_4-\underset{H}{N}-\underset{O}{\overset{\|}{C}}-C_6H_4-\underset{O}{\overset{\|}{C}}-\underset{H}{N}-C_6H_4-\mathrm{C(O)OR} + \mathrm{HO}\!\left(\mathrm{(CH_2)_4O}\right)_n\!\mathrm{H}$$

$$\xrightarrow[\text{Ti cat.}]{-(2-1/m)\,\mathrm{ROH}}$$

$$1/m\left[\underset{O}{\overset{\|}{C}}-C_6H_4-\underset{H}{N}-\underset{O}{\overset{\|}{C}}-C_6H_4-\underset{O}{\overset{\|}{C}}-\underset{H}{N}-C_6H_4-\underset{O}{\overset{\|}{C}}-O\left(\mathrm{(CH_2)_4O}\right)_n\right]_m$$

Scheme 3

The obtaining of high molecular weight polymers was achieved by the addition of an appropriate amount of NMP into the reaction system [34–41]. In this route, the molecular weight depended on both the amount of NMP and the catalyst, titanium tetrabutoxide. Figure 2 shows a plot of the molecular weight of elastomers *vs.* the amount of NMP. Each curve reaches a maximum at a specific amount of NMP, depending on the molecular weight of PTMG. The addition of NMP increased the solubility of compound I, thus facilitating the reaction. Increasing amounts of NMP decreased the reaction rate of compound I, because of the reduces concentration of the reaction species [44].

The effect of the catalyst amount on the molecular weight was also studied and it was shown that the molecular weight of the elastomers reached a maximum at a specific amount of catalyst (Figure 3). The catalysts SnO, Sb_2O_3, $Zn(OCOCH_3)_2$, and $Ca(OCOCH_3)_2$ showed a low activity [44], although they are known to have high activities in the ordinary transesterification reaction between carboxylic acid alkyl esters and alkylene glycols [59,60].

3.2.2. *Route B*

The reactions in the bulk of PTMG and 4-aminoethylbenzoate with transesterification catalysts at *ca.* 180°C lead to an NH_2 group-terminated PTMG, which yielded type I aramid elastomers by reacting with terephthaloyl chloride [56].

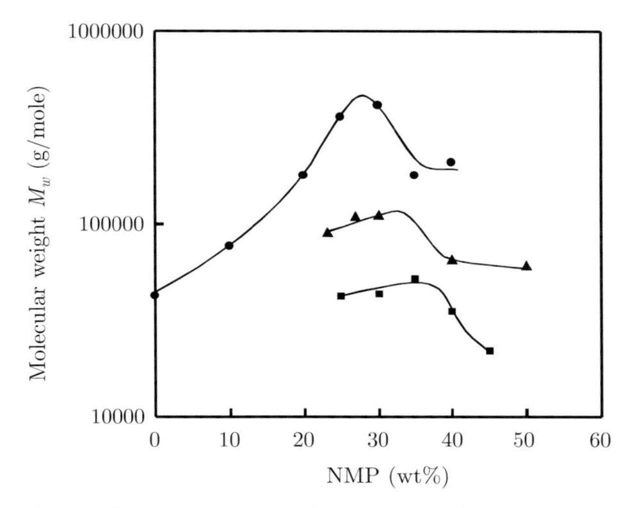

Figure 2. Effect of the NMP amount on the molecular weight of aramid elastomers: (●) PTMG $M_n = 1000$; (▲) PTMG $M_n = 1500$; (■) PTMG $M_n = 2100$; polymerization temperature: 210°C; aramid compound I/PTMG molar ratio = 1/1; 0.73 mole% $Ti(OBu)_4$ [44]

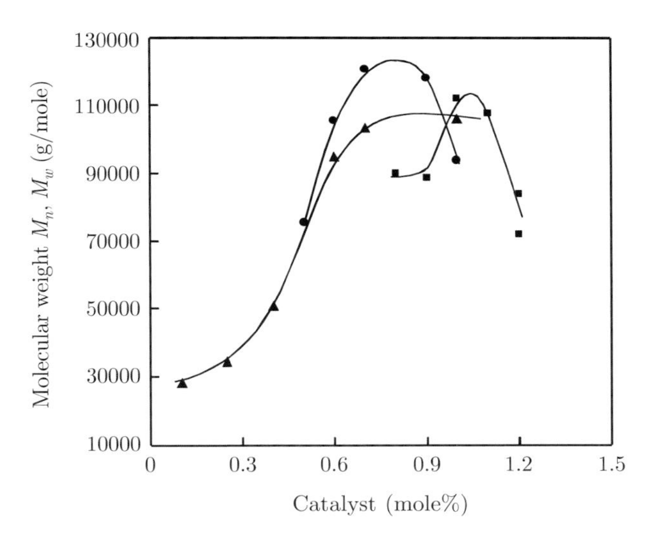

Figure 3. Effect of the Ti catalyst amount on the molecular weight of the aramid elastomers: (●) PTMG: $M_n = 1000$; (▲) PTMG: $M_n = 1500$; (■) PTMG: $M_n = 2100$; polymerization temperature: 210°C; aramid compound I/PTMG molar ratio = 1/1; weight ratio NMP/(I + PTMG)= 0.3 (wt/wt) [44]

3.2.3. *Route C*

Kraft synthesized type I aramid elastomers by a novel route, a Pd(0) catalyzed carbonylation of amino group-terminated telechelic PTMO and aromatic

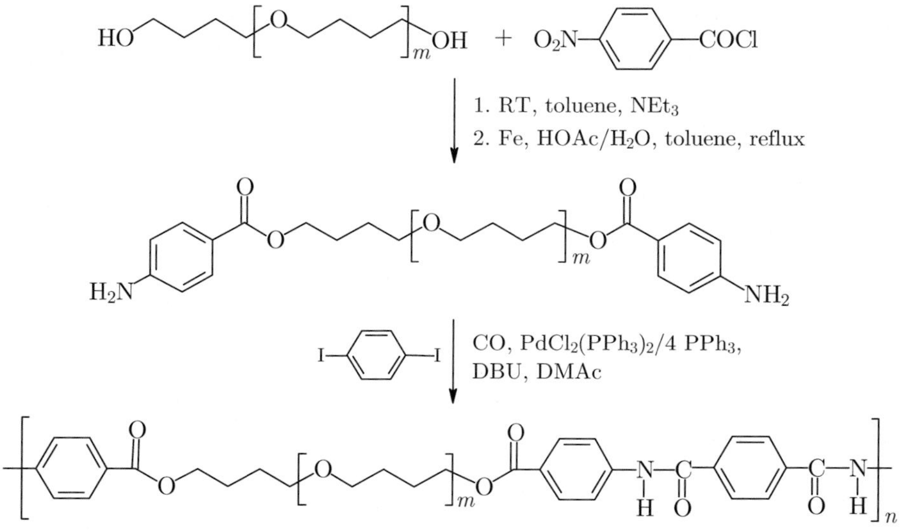

1. RT, toluene, NEt_3
2. Fe, $HOAc/H_2O$, toluene, reflux

CO, $PdCl_2(PPh_3)_2$/4 PPh_3, DBU, DMAc

Scheme 4

diiodide. The telechelic PTMO was obtained by the end capping of PTMG with 4-nitrobenzoylchloride, followed by hydrogenation of the nitro groups (Scheme 4). The method provided low temperature polymerization routes to the preparation of type I aramid elastomers. Polymerization at 115°C in dimethylacetamide (DMAc) yielded polymers with a molecular weight ranging from 5.9 to 6.9 kg/mole. Starting from other diiodoaryl compounds, such as 1,3-diiodobenzene, 4,4′-diiododiphenyl or 4,4′-diiododiphenylether, new types of aramid elastomers were synthesized [57,58].

3.2.4. *Route D*

An aromatic amide compound II with activated carbonyl groups at both ends was synthesized (Scheme 5). This type of groups easily reacted with glycols simply on heating by eliminating caprolactam [61]. Using Route D, aramid elastomers with various kinds of soft segments were synthesized: polybutylene-adipate, poly(ε-caprolactam) and polydimethylsiloxane [42,44,45], and a detailed description of this procedure is given in Section 5. The reaction of acid chlorides of terephthalamide compounds [53] and caprolactam also leads to the obtaining of compound II.

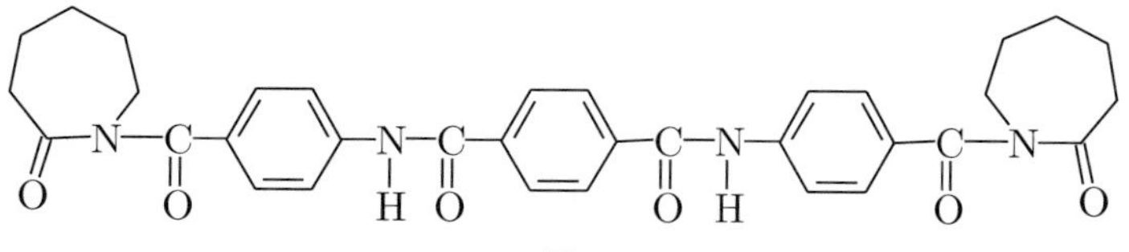

II

Scheme 5

3.3. *Thermal properties*

Table 1 lists the thermal properties of a series of type II aramid elastomers. Differential scanning calorimetry (DSC) measurements revealed that the aramid elastomers exhibited three transition points: a glass transition, T_g, below room temperature (RT), and melting transitions, T_m, below RT and around 200°C. The DSC thermograms are typical of block copolymers consisting of crystalline

Table 1. Thermal properties of aramid elastomers [39]

Polymer	ω_h (wt%)	PTMO M_n (g/mole)	T_g (°C)	T_m (°C)	T_s (°C)	T_b (°C)
TPEA1	28.2	1000	−78	−37, 225	200	<−70
TPEA2	20.7	1500	−79	−23, 210	187	<−70
TPEA3	15.6	2100	−81	0, 190	171	<−70
TPEA4	8.6	4100	−81	20, 173	–	<−70

ω_h: weight fraction of aramid segments; T_s: softening temperature; T_b: brittleness temperature

hard segments and semicrystalline soft segments (Figure 4). Sharp melting peaks at around 200°C were observed for the elastomers synthesized from PTMG with an M_n below 3000 g/mole due to the uniformity of the aramid segment [44]. The DSC measurements also confirmed that the PTMO phase involved crystallized oxytetramethylene (OTM) units.

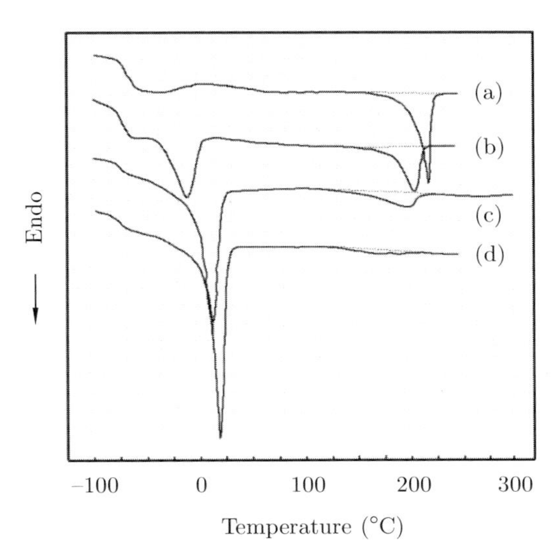

Figure 4. DSC thermograms of aramid elastomers of various molecular weights of the PTMO soft segments: (a) M_n = 1000; (b) M_n = 1900; (c) M_n = 2900; (d) M_n = 4100 [44]

The melting point of the aramid elastomer varied from 173 to 225°C with decreasing molecular weight of the soft segments. The melting point depression was successfully explained by the solvent effect [62] defined by

$$1/T_m - 1/T_m^0 = -(\mathrm{R}/\Delta H_u^0)\ln X, \tag{1}$$

where X is the molar fraction of the aramid segment, R the gas constant, ΔH_u^0 the enthalpy of fusion, and T_m and T_m^0 are the melting points of aramid elastomers and of aramid segments, respectively.

Figure 5 shows a plot of melting points of type I aramid elastomers as dependent on the molar fraction of aramid hard segments, X. Using this plot, the melting point of the aramid segments, T_m^0, and the equilibrium heat on fusion, ΔH_u^0, were determined to be 638 K and 53.1 kJ/mole, respectively. An equilibrium melting point of 638 K is close to the value of 633 K reported for type II aramid elastomers [43]. The observed enthalpy of fusion $\Delta H_u^0 = 53.1$ kJ/mole

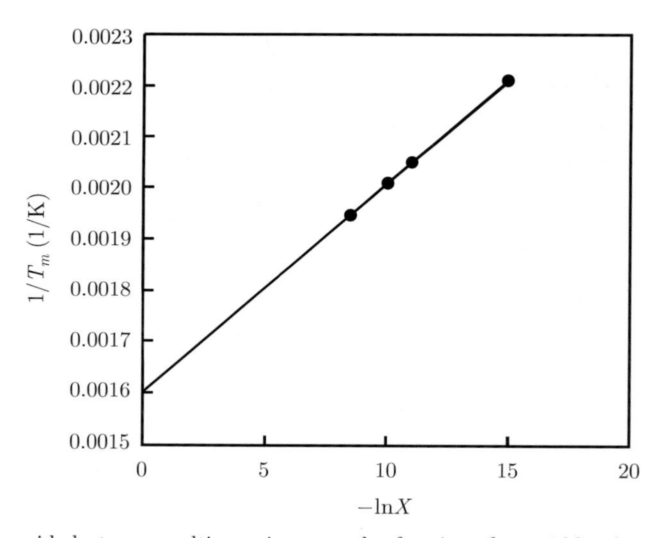

Figure 5. Aramid elastomer melting point *vs.* molar fraction of aramid hard segments, X [44]

was equivalent to that of the PBT perfect crystal [63], indicating that aramid segments form thermally stable crystalline domains. A small change in the entropy of fusion $\Delta S_u^0 = 0.82$ J/K·mole, derived from the relation $\Delta H_u^0/\Delta T_m^0$, suggests that some ordering exists even in the molten state. The ordered structure is probably formed by hydrogen bonding between amide groups of the aramid segments.

Figure 6 shows a plot of melting points of PTMO segments *vs.* the molar fraction of OTM units, X_s. From this plot, the equilibrium melting temperature, T_m^0, of the PTMO and the heat of fusion, ΔH_u^0, were determined to be 323 K and 0.486 kJ/mole, respectively. The heat of fusion value was dramatically lower than that reported for homo-PTMO, 6.8 kJ/mole [64]. Furthermore, the change

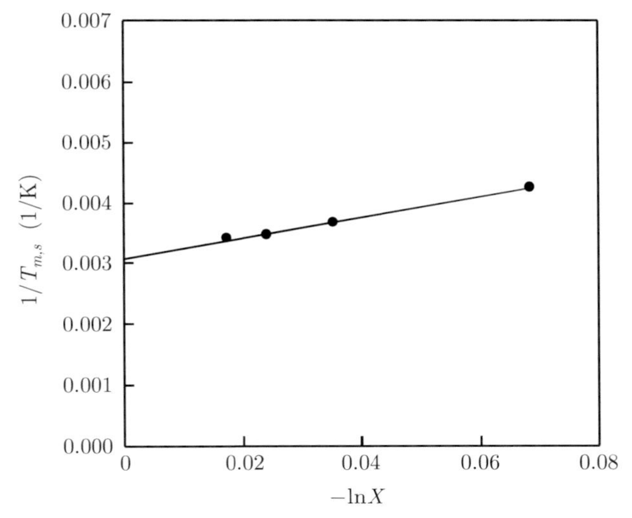

Figure 6. Melting point of crystallized PTMO segments $T_{m,s}$ *vs.* molar fraction of oxytetramethylene (OTM) units, X_s [44]

of the entropy of fusion ΔS_u^0 obtained (0.82 J/K·mole) was extremely low, compared to a reported value of 21.4 J/K·mole for homo-PTMO [64]. This reduced value indicates that the aramid segments effectively reduce the crystallization of the PTMO segments [64]. Table 2 lists other thermodynamic parameters

Table 2. Thermodynamic parameters [44]

Segment	Density (g/cm^3)	T_m^0 (K)	V_m (cm^3/mole)	ΔH_u^0 (kJ/mole)	ΔS_u^0 (J/K•mole)
Aramid	1.405	638	274.2	53.1	83.1
PTMO	0.989 [64]	323	74.3 [64]	0.486	0.82

V_m: molar volume of aramid and OTM units

obtained. Figure 7 shows a Hoffman-Weeks plot [65] of the aramid elastomers. The gradient of the line directly gave the stabilization parameter, f, defined by

$$T_m = T_m^0(1-f) + T_c f, \tag{2}$$

where T_m and T_c are the melting and crystallization temperatures, and T_m^0 is the equilibrium melting temperature of the aramid elastomer. The stabilization parameter obtained was 0.49, suggesting the aramid segments formed needle-like crystals [44].

Kraft analyzed the crystalline structure of several kinds of aramid compounds, including compound I [58]. Figure 8 shows the crystal structures of I and another aramid compound having a naphthalene group. The analysis revealed that the hydrogen atom of the N–H group and the oxygen atom of the C=O

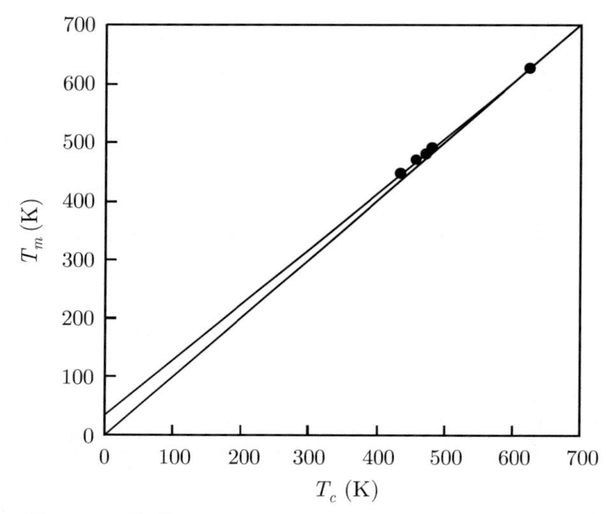

Figure 7. Hoffman-Weeks plot for aramid elastomers [44]

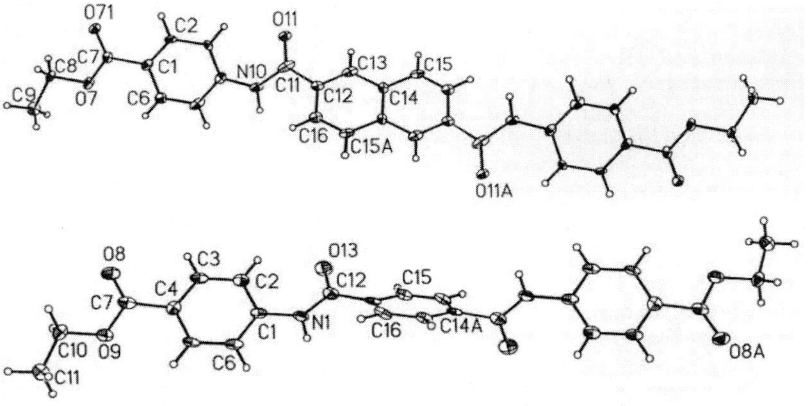

Figure 8. Crystal structures of the aramid compounds naphthalene-2,6-dicarboxylic acid bis[4-(ethoxycarbonyl)phenylamide (top) and benzene-1,4-dicarboxylic acid bis(4-ethoxycarbonyl)phenylamide (bottom) [58]

group form hydrogen bonds. The distances between the hydrogen and ester carbonyl oxygen were determined to be 2.32 Å and 2.15 Å for the terephthalate- and the naphthalate-based compound, respectively.

3.4. *Mechanical properties*

Table 3 summerizes the mechanical properties of the aramid elastomers listed in Table 1. Hardness decreases with increasing molecular weight of PTMO segments. The tensile strength at break, σ_b, strongly depends on the M_n of the PTMO segments. The elongation at break, ε_b, ranges from 810 to 900% indepen-

Table 3. Mechanical properties of aramid elastomers [39]

Polymer designation	ω_h (wt%)	PTMG M_n (g/mole)	Shore A hardness	M_{100} (MPa)	T_r (kN/m)	σ_b (MPa)	ε_b (%)
TPEA1	28.2	1000	93	12	93	18	810
TPEA2	20.7	1500	88	9	88	25	900
TPEA3	15.6	2100	83	8	83	23	850
TPEA4	8.6	4100	66	–	66	31	850

M_{100}: modulus at 100% strain; T_r: tear strength

dently of the weight fraction of aramid segments, ω_h. The compression set at 70°C varies from 59 to 66% depending on the M_n of PTMO segments [44]. Figure 9 shows stress-strain curves at various temperatures from 25°C to 100°C, for TPEA1, an elastomer synthesized from PTMG with an M_n of 1000 g/mole.

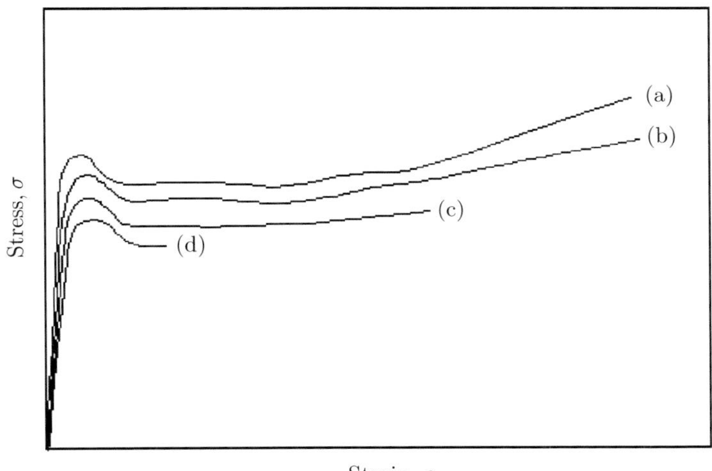

Figure 9. Strain-stress curves of an aramid elastomer synthesized from PTMG with M_n 1000 g/mole at different temperatures: (a) 25°C, (b) 50°C, (c) 80°C, (d) 100°C; draw rate: 200 mm/min [44]

The elongation at break reduces with increasing temperature, but the modulus is temperature independent. This indicates that aramid domains retain a crystal state and act as stable physical crosslinks even at elevated temperatures.

3.5. *Dynamic-mechanical properties*

Figure 10 shows storage moduli and loss tangents of the samples with 83 and 93 Shore A hardness *vs.* temperature, as found by dynamic-mechanical thermal analysis (DMTA). The storage modulus has two transition points typical of

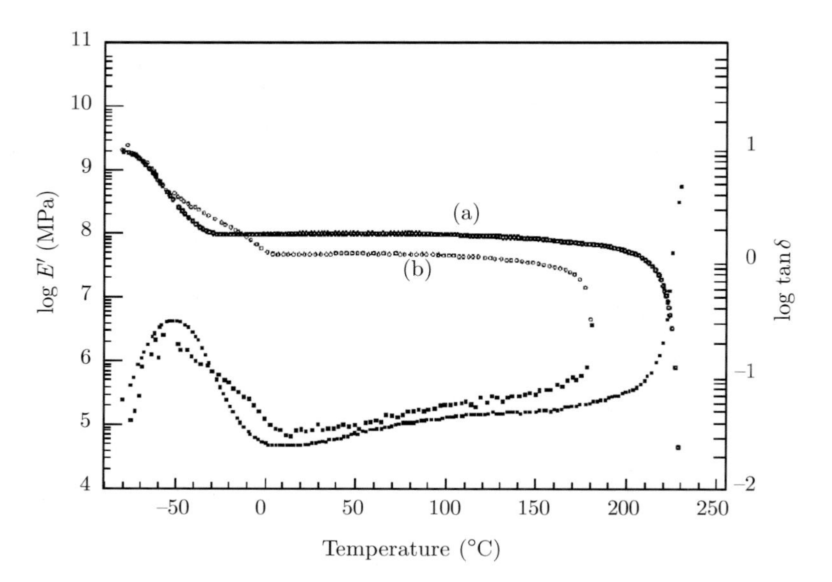

Figure 10. Effect of hard segment content on the storage modulus and loss tangent of samples TPEA1 (a) and TPEA2 (b); frequency 11 Hz [44]

block copolymers ranging from *ca.* −80 to −60°C, and from *ca.* 170 to 220°C. The former is the glass transition of PTMO segments, and the latter is the melting transition of aramid segments. The modulus of each sample shows a sharp drop at the melting point [39].

3.6. *Chemical properties*

Table 4 summarizes the solvent resistance of sample TPEA1 having PTMO segments with an M_n of 1000 g/mole. TPEA1 is soluble in THF, chloroform, aniline, NMP, hot DMAc, and dimethylsulfoxide (DMSO) [44]. It absorbs chlorinated hydrocarbons to swell extensively, which largely reduces the

Table 4. Resistance to organic solvents at room temperature [44]

Polymer	Solvent	ΔV (%)	ΔW (%)	H_s
TPEA1	2-aminoethanol	0.9	20.9	94
	ethylene glycol	2.8	1.3	93
	ethanol	23.7	17.9	92
	tetrachloroethane	61.1	130.4	89
	acetone	92.3	26.9	88
	1,2-dichloroethane	94.4	104.1	77

ΔV: volume change; ΔW: weight change; H_s: Shore A hardness after immersion for 200 h at RT

Table 5. Oil resistance of aramid elastomers [39]

Polymer	JIS #3 oil, 100°C, 70 h				Morilex© No.2 Grease 120°C, 168 h			
	H_s	$\Delta\sigma_b$ (%)	$\Delta\varepsilon_b$ (%)	ΔV (%)	H_s	$\Delta\sigma_b$ (%)	$\Delta\varepsilon_b$ (%)	ΔV (%)
TPEA1	92	+28	+11	20	93	−28	−24	12
TPEA2	80	+3	+1	34	85	−35	17	17
TPEA3	61	−22	+1	57	73	−25	30	25

H_s: Shore A hardness after immersion; $\Delta\sigma_b$: change of tensile strength at break; $\Delta\varepsilon_b$: change of elongation at break; ΔV: volume change

hardness. TPEA1 shows an excellent resistance to JIS #3 oil and to the lithium-containing Morilex© No.2 grease (Table 5).

An aramid elastomer sample immersed in water at 23°C showed a weight gain of 1.19% for 25 h, and of 1.40% for 200 h. No weight gain after the 200 h immersion has been observed, indicating that 1.40% is the maximum value of water absorption [44].

4. Type II poly(aramid-*b*-polyether) elastomers

This section overviews the synthetic routes and the thermal, mechanical, and chemical properties of type II aramid elastomers with PTMO soft segments.

4.1. *Synthesis of reactive aramid compounds*

The reactive aramid compound III, bisesteramide, was synthesized by reacting *p*-phenylenediamine with terephthaloyl chloride, and then with alcohols, *e.g.*, methanol (Scheme 6) [66,67] in the absence of catalysts and acid scavengers. Since side reactions were involved, generating oligomeric products, the removal

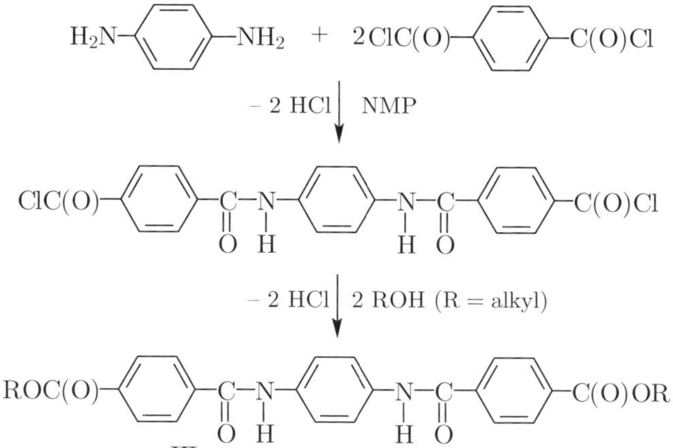

H_2N–C₆H₄–NH_2 + 2ClC(O)–C₆H₄–C(O)Cl

↓ − 2 HCl, NMP

ClC(O)–C₆H₄–C(O)–NH–C₆H₄–NH–C(O)–C₆H₄–C(O)Cl

↓ − 2 HCl, 2 ROH (R = alkyl)

ROC(O)–C₆H₄–C(O)–NH–C₆H₄–NH–C(O)–C₆H₄–C(O)OR

III

Scheme 6

of higher oligomers by filtration and recrystallization from NMP was required. Compound III exhibited a sharp melting point at 371°C and its purity varied from 96 to 98% [43]. An alternative route was reported, reacting *p*-phenylenediamine with methyl (4-chlorocarbonyl) benzoate to avoid the formation of higher oligomers [43].

4.2. *Synthesis of aramid elastomers*

The polymerization Route A was successfully applied to the synthesis of type II aramid elastomers: titanate-catalyzed transesterification between the aramid compound III and PTMG (Scheme 7). Using Route A, a high molecular weight elastomer with inherent viscosity of 2.8 dL/g was obtained [43].

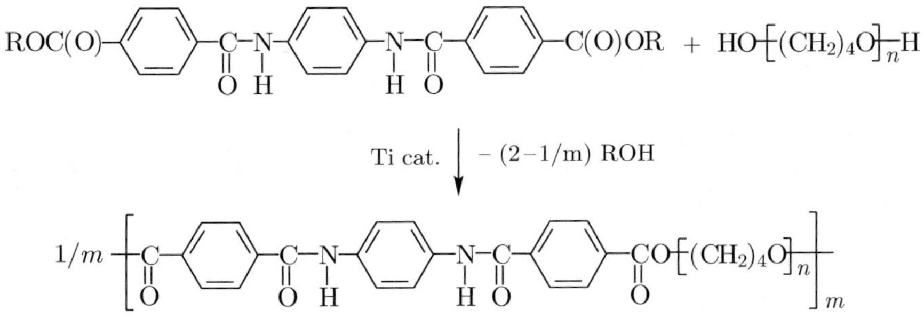

ROC(O)–C₆H₄–C(=O)–NH–C₆H₄–NH–C(=O)–C₆H₄–C(O)OR + HO–[(CH₂)₄O]ₙ–H

Ti cat. ↓ – (2 – 1/m) ROH

1/m –[C(=O)–C₆H₄–C(=O)–NH–C₆H₄–NH–C(=O)–C₆H₄–C(=O)O–[(CH₂)₄O]ₙ]ₘ–

Scheme 7. Synthesis of type II aramid elastomers

Melt polycondensation at 275°C in an extruder was tested by extruding an aramid compound, dry blended with a catalyst solution and PTMG. The extrudate was further treated in a reactor at 250°C, which yielded polymers with an inherent viscosity ranging from 0.7 to 1.1 dL/g [43].

4.3. *Thermal properties*

The thermal properties of a series of type II aramid elastomers (Table 6) with varying molecular weight of the PTMO soft segments, M_s, have been reported [43]. The elastomers exhibit high melting and crystallization points. The melting point depression of the segments is to be explained by the solvent effect [62]. The equilibrium melting point, T_m^0, of the aramid segments was determined to be 360°C [43]. The measurements also revealed that the PTMO segments have partially crystallized when the samples were synthesized from PTMG with an M_n higher than 1400 g/mole. The melting point of the soft segments, $T_{m,s}$, ranged from –8 to 9°C and shifted to higher temperatures with increasing M_s.

4.4. *Mechanical properties*

Table 7 lists the mechanical properties of the type II aramid elastomers, which include yield stress, σ_y, stress at break, σ_b, fracture strain, λ_b, stress relaxation

Table 6. DSC and DMTA data of type II aramid elastomers [43]

M_s (g/mole)	$T_{m,s}$ (°C)	$\Delta H_{m,s}$ (°C)	$T_{c,s}$ (°C)	$\Delta H_{c,s}$ (J/g)	$T_{m,h}$ (°C)	$\Delta H_{m,h}$ (J/g)	$T_{c,h}$ (°C)	T_g (°C)	T_m (°C)
650	–	–	–	–	266	12	238	–58	247
1000	–	–	–	–	222	10	203	–69	216
1400	–8	12	–46	–6	–	–	–	–70	198
2000	0	17	–39	–13	–	–	–	–65	191
2900	9	30	–31	–22	–	–	–	–70	170

$T_{m,h}$: melting temperature of aramid segments; $\Delta H_{m,s}$: melting enthalpy of crystallized PTMO segments; $\Delta H_{m,h}$: enthalpy of melting of aramid segments; $T_{c,s}$: crystallization temperature of PTMO segments; $T_{c,h}$: crystallization temperature of aramid segments; $\Delta H_{c,s}$: enthalpy of melting of crystallized PTMO segments; T_g: glass transition temperature

Table 7. Mechanical properties of type II aramid elastomers [43]

Polymer	ω_h (wt%)	PTMG M_s (g/mole)	G′ (MPa)	σ_y (MPa)	σ_b (MPa)	λ_b (–)	SR25 (–)	SR100 (–)
$PTMO_{650}$	29	650	118	14.0	–	–	–	–
$PTMO_{1000}$	22	1000	44	9.8	38.1	14.7	0.68	0.82
$PTMO_{1400}$	17	1400	15	–	–	–	0.37	0.70
$PTMO_{2000}$	13	2000	10	5.7	51.9	17.6	0.21	0.48
$PTMO_{2900}$	9	2900	6	3.4	58.8	16.6	0.11	0.37

(SR) at a strain of 0.25 (SR25) and 1.00 (SR100) [43]. The storage modulus at room temperature, G', linearly decreases with decreasing weight content of aramid segments, ω_h. A high strength elastomer with $\sigma_b = 51.9$ MPa was obtained using PTMG with an M_s of 2000 g/mole. Compression sets, at 55% strain and 20°C, of 19 and 42% have been reported for the sample $PTMO_{2000}$. The tension sets, at a strain of 3.0, of the elastomer ranged from 39 to 46% and reduced with increasing molecular weight of the PTMO segments, M_s [50].

4.5. *Chemical properties*

Chemical properties of type II aramid elastomers have been reported in [43]. Water absorption tests were conducted at RT and at 100% relative humidity. The water absorption of a PTMO-containing aramid elastomer was only 1.0 wt%, while that of a poly(ethyleneoxide) (PEO)-containing elastomer was 80.5 wt% [43].

4.6. *Aramid elastomers with other polyether soft segments*

Aramid elastomers containing poly(2-methyltetramethyleneoxide) (PMTMO) and PEO as soft segments have been synthesized [43]. The copolymer with

PMTMO exhibited superior low temperature flexibility, compared to those with PTMO soft segments. By using PMTMO, an elastomer with a T_g of –75°C was synthesized. The study also showed that methyl side groups incorporated into the PTMO soft segments effectively reduced the crystallinity of the PTMO segments. Poly(ethylene oxide) glycol (PEG), however, yielded only low molecular weight elastomers due to thermal degradation of PEO or phase separation during polymerization.

5. Poly(aramid-*b*-polyester) elastomers

This section summerizes syntheses and properties of aramid elastomers with polyester soft segments.

Aramid elastomers consist of thermally stable aramid segments, as well as of crystalline and thermally less stable rubbery segments, *i.e.*, the soft segments determine the service temperature of the material. In fact, aramid elastomers with PTMO soft segments degrade at elevated temperatures above 120°C by oxidative degradation of the PTMO segments [68–74]. Aliphatic polyesters are promising soft segments in improving the thermal stability of aramid elastomers, because the resistance of polyester soft segments against oxidative degradation has been already validated by copolyester elastomers with polyester soft segments, which exhibit excellent thermal stability, compared to those with polyether soft segments [75–82].

5.1. *Synthesis and properties of aramid-b-polyester elastomers*

As mentioned in Section 3, typical aramid-*b*-polyether elastomers are synthesized by the polycondensation reaction of polyether diol with the aramid compound I in the presence of transesterification catalysts. Under these conditions, the synthesis of aramid-*b*-polyester elastomers gave only low molecular weight elastomers with a broad segment length distribution due to transesterification reactions of the polyester segments. This result inidicated that in the obtaining of aramid-*b*-polyester elastomers, transesterification catalysts should be avoided. Later, a method for the obtaing of this type of elastomers was developed, which consisted in the copolymerization of an activated acyl lactam-terminated aramid compound II with polyester diols in the molten state, in the absence of transesterification catalysts [40,42]. Compound II was obtained by the reaction of N-(*p*-aminobenzoyl) caprolactam with terephthaloyl chloride, as shown in Scheme 8 [61].

Copolymerization of the aramid compound II with the polyester diols, poly(butylene-1,4-adipate) glycol and poly(ε-caprolactone) glycol, yielded aramid-*b*-polyester elastomers (Scheme 9) [40,42]. The reaction of the polyester diols with compound II resulted in high molecular weight elastomers with a well defined structure by eliminating caprolactam [83]. However, this reaction involves a possible side reaction, the ring opening of lactam rings [61,83].

Table 8 summarizes the melting points, T_m, and stresses at break, σ_b, of the elastomers obtained. All elastomers exhibit high melting temperatures above

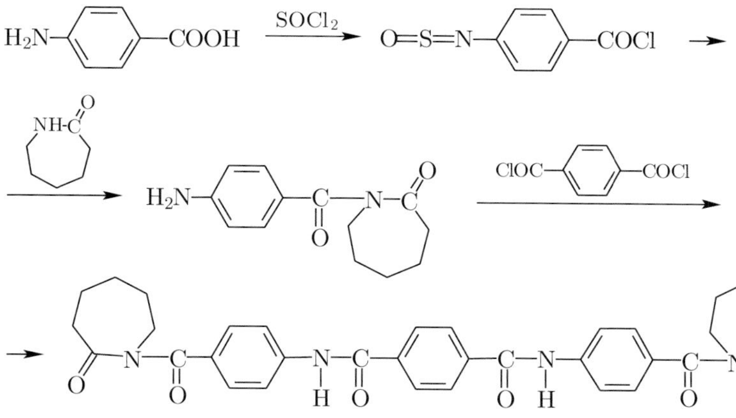

Scheme 8. Acyl lactam-terminated aramid compound

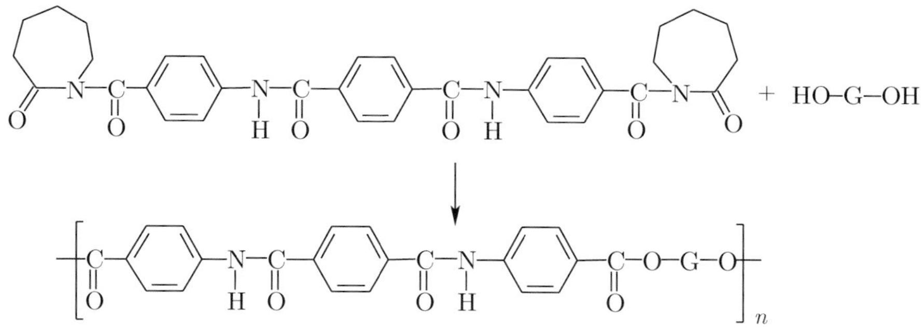

G = poly(butylene-1,4-adipate) or poly(ε-caprolactone)

Scheme 9. Aramid-*b*-polyester elastomers

Table 8. Properties of aramid elastomers [42]

Elastomer	Soft segment	M_w (g/mole)	T_m (°C)	σ_b (MPa)
TPAR1	Poly(butylene-1,4-adipate)	104 000	215	33.6
TPAR2	Poly(ε-caprolactone)	102 000	201	24.5
TPAR3	Poly(tetramethyleneoxide)	100 000	220	25.0

200°C. Figure 11 shows the molecular weight change during thermal aging at 120°C in air; samples TPAR1 and TPAR2 having polyester soft segments are considerably less susceptible to oxidative degradation than sample TPAR3 having PTMO soft segments [42].

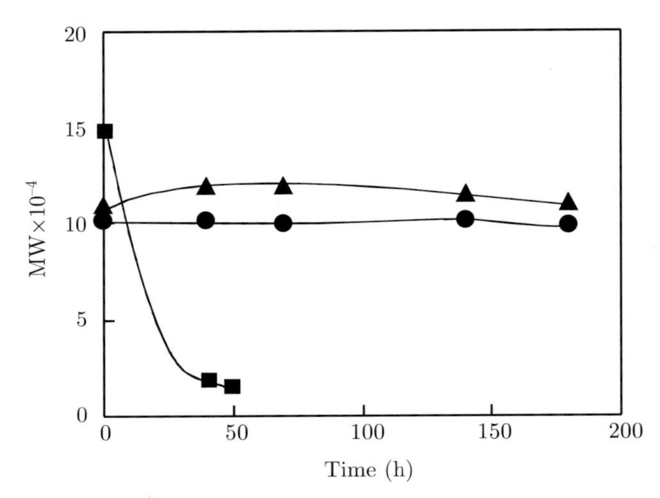

Figure 11. Thermal aging tests at 120°C for the aramid elastomer samples: (▲) TPAR1; (●) TPAR2; (■) TPAR3 [42]

5.2. *A novel synthetic route to aramid-b-polyester elastomers*

Recently, aramid-*b*-polyester elastomers (TPARs) of a new type were synthesized using a hydroxy-terminated aramid compound IV, as shown in Scheme 10 [84,85]. Compound IV was synthesized in a two-step reaction. First, N,N′-bis(4-hydroxyphenyl)terephthalanilide was synthesized by condensation of terephthaloyl chloride with *p*-aminophenol in polar organic solvents and then the anilide was reacted with ethylene carbonate.

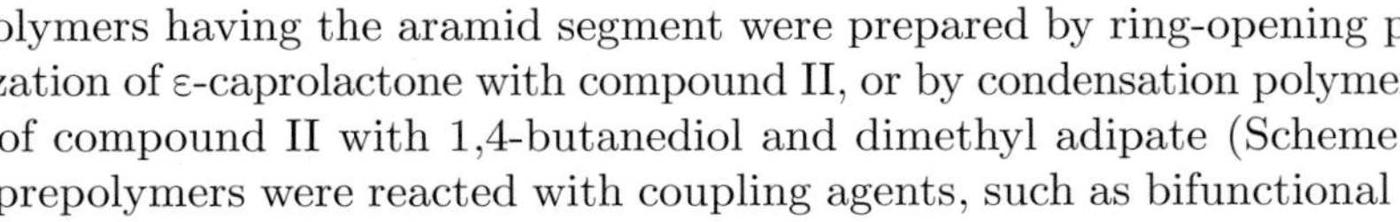

$HOCH_2CH_2O-C_6H_4-N(H)-C(=O)-C_6H_4-C(=O)-N(H)-C_6H_4-OCH_2CH_2OH$

IV

Scheme 10

TPARs were synthesized in the following way. First, hydroxy-terminated prepolymers having the aramid segment were prepared by ring-opening polymerization of ε-caprolactone with compound II, or by condensation polymerization of compound II with 1,4-butanediol and dimethyl adipate (Scheme 11). The prepolymers were reacted with coupling agents, such as bifunctional acyl halide compounds or bisacyllactam compounds.

Table 9 summarizes the thermal properties of a typical TPAR sample with poly(ε-caprolactone) soft segments together with those of commercial thermoplastic copolyester elastomers TPEE1 and thermoplastic urethane elastomers TPU1 with Shore A hardness 95. TPAR4 exhibits the highest melting and softening points among these samples, and a low glass transition temperature close to that of TPEE1.

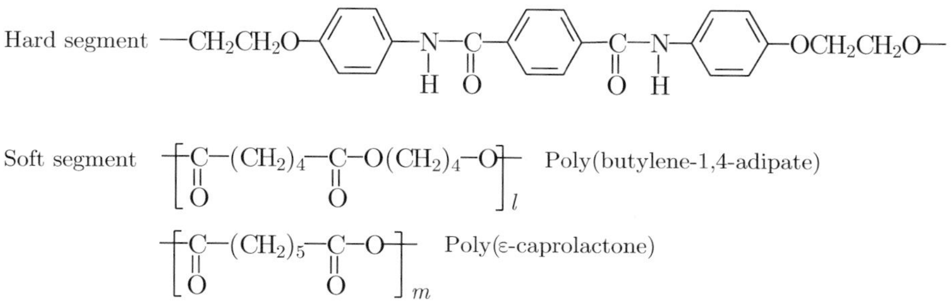

Hard segment $—CH_2CH_2O—C_6H_4—\underset{H}{N}—\underset{O}{\overset{\|}{C}}—C_6H_4—\underset{O}{\overset{\|}{C}}—\underset{H}{N}—C_6H_4—OCH_2CH_2O—$

Soft segment $\left[\underset{O}{\overset{\|}{C}}—(CH_2)_4—\underset{O}{\overset{\|}{C}}—O(CH_2)_4—O\right]_l$ Poly(butylene-1,4-adipate)

$\left[\underset{O}{\overset{\|}{C}}—(CH_2)_5—\underset{O}{\overset{\|}{C}}—O\right]_m$ Poly(ε-caprolactone)

Scheme 11

Table 9. Thermal properties of TPEs [85]

Elastomer	T_m (°C)	T_g (°C)	T_s (°C)
TPAR4	219	–56	187
TPEE1	204	–58	171
TPU1	182	–31	140

The temperature dependence of the storage modulus, E', of TPAR4 at 11 Hz shows a rubbery plateau region over a wide temperature range from –20 to 180°C (Figure 12), *i.e.*, E' in the rubbery region is temperature independent,

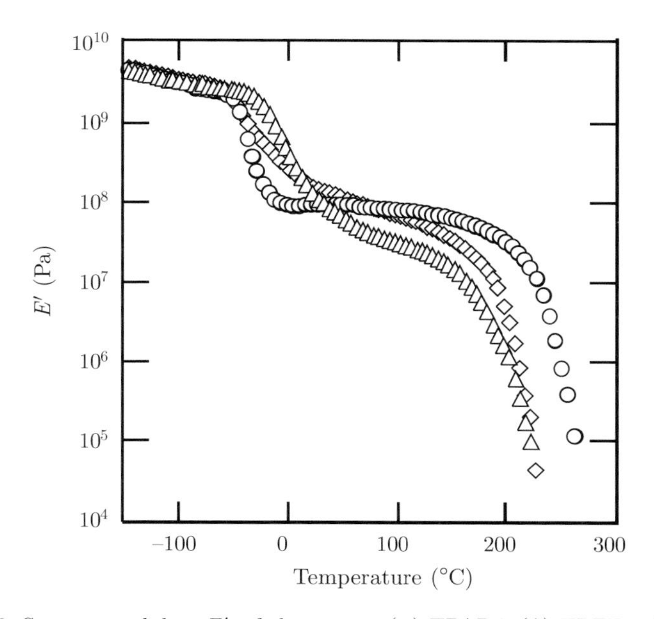

Figure 12. Storage modulus, E', of elastomers: (○) TPAR4; (◇) TPEE1; (△) TPU1 [85]

whereas E' of TPEE1 and TPU1 decreases with increasing temperature. The results suggest that TPAR4 has a well-controlled structure consisting of hard segments with uniform length and soft segments with a narrow length distribution. Thus, TPAR4 markedly differs from TPEE1 and TPU1, which have broad molecular weight distributions of both hard and soft segments. Figure 13 shows the temperature dependence of the loss tangent, $\tan\delta$, for all three samples. The $\tan\delta$ peak of TPAR4 is narrower than those of TPU1 and TPEE1. Furthermore, Figure 13 reveals that TPAR4 has the lowest $\tan\delta$ value at room temperature, suggesting that TPAR4 would have a high rebound resilience. This is because TPAR4 has a narrow segment length distribution.

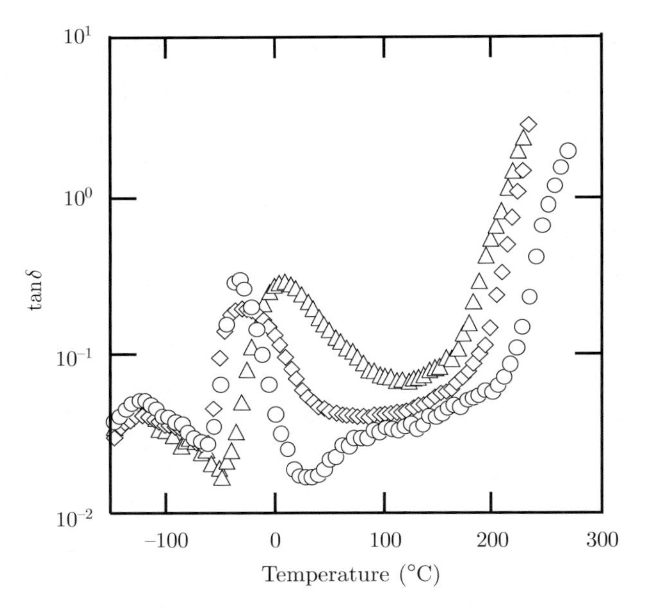

Figure 13. Temperature dependence of $\tan\delta$ for TPEs: (○) TPAR4; (◇) TPEE1; (△) TPU1 [85]

Tensile tests of the TPEs revealed that all samples except for TPU1 had yield points, as shown in Figure 14. The tensile strength at break of TPAR4 is 50 MPa and is comparable to that of the thermoplastic polyurethane elastomer having the highest mechanical strength among the TPE samples. Moreover, TPAR4 shows the highest ultimate elongation among these samples. Table 10 summarizes the mechanical properties of TPAR4, TPEE1 and TPU1. It is seen that TPAR4 exhibits an excellent permanent set in between those of TPU1 and TPEE1[85].

6. Applications

One of the possible applications of the thermoplastic aramid elastomers is in the production of melt spinnable elastic fibers [41,48,86]. Strong elastic fibers

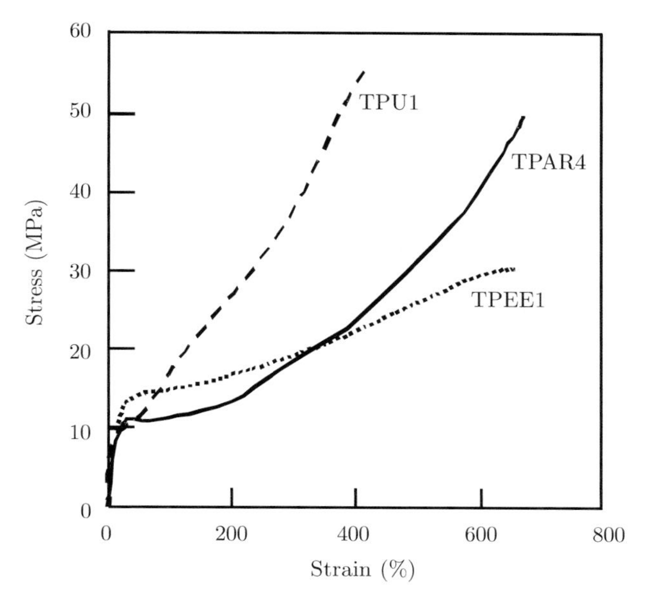

Figure 14. Stress-strain curves of TPEs

Table 10. Mechanical properties of TPAR, TPEE, and TPU samples [85]

Elastomer	Shore A hardness (point)	Shore D hardness (point)	Tensile strength (MPa)	Elongation at break (%)	Permanent set (%)
TPAR4	95	48	50	700	30
TPEE1	96	48	28	680	39
TPU1	95	52	54	440	20

from 70 to 305 dtex were spun from the melt and exhibited a tensile set superior by 13% than those of commercial copolyetherester elastic fibers [41]. Other applications proposed are semi-permeable clothes [41,47], automotive boots [87], tubing [88], and injection molding products [39].

7. Conclusions

Thermoplastic aramid elastomers consisting of aramid hard segments with uniform length and rubbery soft segments are novel thermoplastic elastomers exhibiting unique properties: high melting points, fast crystallization, temperature independent moduli and high mechanical strength. The elastomers exhibit temperature independent rubbery plateau moduli, providing a dimensionally stable elastic material retaining its mechanical strength over a wide temperature range. Furthermore, aramid elastomers with polyester soft segments are high strength materials with excellent thermal aging resistance.

Therefore, these materials are attracting research interest, especially in designing novel elastomers with unique combinations of aramid and soft segments, and in improving the synthetic routes and the structure analyses. So far, the phase and crystal structures of these materials have not been fully described and understood. Consequently, the understanding of the microstructures would be a key of great importance to enhance the unique properties of thermoplastic aramid elastomers.

References

1. Ng H N, Allegrazza A E, Seymour R W and Cooper S L (1973) Effect of segment size and polydispersity on the properties of polyurethane block polymers, *Polymer* **14**:255–261.
2. Harrell L L J (1969) Segmented polyurethans. Properties as a function of segment size and distribution, *Macromolecules* **2**:607–612.
3. Miller J A, Lin S B, Hwang K K S, Wu K S, Gibson P E and Cooper S L (1985) Properties of polyether-polyurethane block copolymers: Effects of hard segment length distribution, *Macromolecules* **18**:32–44.
4. Roderick P Q and Alsamarraie M A (1988) Polyester thermoplastic elastomers, in *Handbook of Elastomers, New Developments and Technology* (Eds. Bhowmick A K and Stephens H L) Marcel Dekker, New York, Ch. 10, pp. 341–372.
5. Estes G M, Cooper S L and Tobolsky A V (1970) Block polymers and related heterophase elastomers, *J Macromol Sci Macromol Chem* **4**:313–366.
6. Bonart R (1979) Thermoplastic elastomers, *Polymer* **20**:1389–1403.
7. Mangaraj D and Markle R (1991) New generation of thermoplastic elastomers, *Proc TPE Europe'91*, pp. 299–310.
8. Witsiepe W K (1972) Segmented thermoplastic copolyester elastomers, US Patent 3,651,014, to E.I. du Pont de Numours and Co.
9. Wolfe J R Jr (1973) Segmented thermoplastic copolyesters, US Patent 3,775,373, to E.I. du Pont de Numours and Co.
10. Brown M, Hoeschele G K and Witsiepe W K (1974) Process for improving thermoplastic elastomeric compounds, US Patent 3,835,098, to E.I. du Pont de Numours and Co.
11. Shen M, Mehra U, Niinomi M, Koberstein J T and Cooper S L (1974) Morphological, rheo-optical, and dynamic mechanical studies of a semicrystalline block copolymer, *J Appl Phys* **45**:4182–4189.
12. Seymour R W, Overton J R and Corley L S (1975) Morphological characterization of polyester-based elastoplastics, *Macromolecules* **8**:331–335.
13. Wolfe J R Jr (1977) Elastomeric polyether-ester block copolymers. I. Structure-property relationships of tetramethylene terephthalate/polyether terephthalate copolymers, *Rubber Chem Technol* **50**:688–703.
14. Lilaonitkul A, West J C and Cooper S L (1976) Properties of polytetramethylene oxide-polytetramethylene terephthalate block polymers, *J Macromol Sci–Phys* **B12**:563–597.
15. Lilaonitkul A and Cooper S L (1977) Properties of polyether-polyester thermoplastic elastomers, *Rubber Chem Technol* **50**:1–23.

16. Kobayashi T, Kitagawa H and Kaji A(1992) Influence of raw material molecular weight on the block length of polyester-polyester block copolymer, *Kobunshi Ronbunshu*, **49**:625–633 (in Japanese).
17. Kobayashi T, Kaji A, Kitagawa H and Takemoto K (1992) Polymerization behavior of ε-caprolactone for the formation of polyester-polyester block polymer, *Kobunshi Ronbunshu*, **49**:561–568 (in Japanese).
18. Deleens G, Foy P and Maréchal E (1977) Synthèse et caractérisation de copolycondensates séquencés poly(amide-seq-éther)–I. Synthese et etude de divers oligomers ω,ω'-difonctionnels du poly(amide-11), *Eur Polym J* **13**:337–342.
19. Deleens G, Foy P and Maréchal E (1977) Synthèse et caractérisation de copolycondensates séquencés poly(amide-seq-éther)–II. Polycondensation d'oligomères polyamides-11 ω,ω'-diacides ou diesters avec des oligomères polyéthers ω,ω'- dihydroxy, *Eur Polym J* **13**:343–351.
20. Deleens G, Foy P and Maréchal E (1977) Synthèse et caractérisation de copolycondensates séquencés poly(amide-seq-éther)–III. Etude de la réaction de polycondensation du polyamide-11 ω,ω'-dicarboxylicque et du polyoxyéthylène ω,ω'-dihydroxy. Determination des constantes de vitesse et de l'energie d'activation, *Eur Polym J* **13**:353–360.
21. Mumcu S, Burzin K, Felsmann R and Feinauer R (1978) Copolyesteramide aus laurinlactam, 1,10-decandicarbonsäure und α,ω- dihydroxy-(polytetrahydrofuran), *Angew Makromol Chem* **74**:49–60.
22. Muncu S (1980) Preparation of polyether ester amides, US Patent 4,345,064, to Chemische Werke Huls AG.
23. Isler W and Schmid E (1980) Polyetherpolyamide copolymers, US Patent 4,356,300, to Inventa AG für Forschung und Patentverwertung.
24. Bornschlegl E, Goldbach G and Meyer K (1985) Structure and properties of segmented polyamides, *Progr Colloid Polym Sci* **71**:119–124.
25. Ma E C (1989), TPUs – the first TPEs, *Rubber World* **199**:30–35.
26. Estes G M, Seymour R W, Huh D S, and Cooper S L (1969) Mechanical and optical properties of block copolymers. I. Polyester-urethanes, *Polym Eng Sci* **9**: 383–387.
27. Dietrich D, Grigat E and Hahn W (1985) Effects of hard segment domains on mechanical and thermal properties, in *Polyurethane Handbook* (Ed. Oertel G) Karl Hanser Verlag, Munich, Ch. 2.5.2.4, pp. 35–36.
28. Huh D S and Cooper S L (1971) Dynamic mechanical properties of polyurethane block polymers, *Polym Eng Sci* **11**:369–376.
29. Seymour R W and Cooper S L (1971) DSC studies of polyurethane block polymers, *J Polym Sci B Polym Lett Ed* **9**:689–694.
30. Wang C B and Cooper S L (1983) Morphology and properties of segmented polyether polyurethaneureas, *Macromolecules* **16**:775–786.
31. Kobayashi H, Sasaguri K and Makita M (1969) Thermostable and light-resistant elastomers, Japanese Patent 44030751, to Asahi Chemical Ind Co Ltd.
32. Hirt V P and Herlinger H (1974) Neue elastomerfasersysteme, *Angew Makromol Chem* **40**:71–88.
33. Kubo Y, Yamakawa H, Kirikihira I and Shimosato S (1996) Ester/amide block copolymer, Japanese Patent 8134210, to Tosoh Corporation.
34. Yamakawa H, Kirikihira I and Kubo Y (1994) Polyetheresteramide elastomer and production thereof, Japanese Patent 6279583, to Tosoh Corporation.
35. Kirikihira I, Yamakawa H, Kubo Y and Shimosato S (1996) Production of high-molecular-weight ester amide copolymer, Japanese Patent 8253580, to Tosoh Corporation.

36. Kirikihira I, Yamakawa H and Kubo Y (1998) Esteramide copolymers and production thereof, US Patent 5,811,495, to Tosoh Corporation.
37. Kubo Y, Yamakawa H, and Simosato S (1997) High-viscosity modified polyesteramide elastomer composition, its production and molded form made therefrom, Japanese Patent 9324121, to Tosoh Corporation.
38. Duchácek V (1998) Recent developments for thermoplastic elastomers, *J Macromol Sci–Phys* **B37**:275–282.
39. Yamakawa H, Kirikihira I, Kubo Y and Shimosato S (1995) High performance TPE with terephthalamide hard segment of uniform length, *Proc Int Rubber Conf*, Kobe, pp. 523–526.
40. Kubo Y, Yamakawa H, Kirikihira I and Shimosato S (1996) Aromatic amide having active end and production of esteramide block copolymer using the same, Japanese Patent 8143549, to Tosoh Corporation.
41. Yamakawa H, Kubo Y and Shimosato S (1997) Production of polyetheresteramide elastomer, Japanese Patent 9316194, to Tosoh Corporation.
42. Kubo Y, Yamakawa H, Kirikihira I and Shimosato S (1998) Ester-amide block copolymer and process for producing same, US Patent 5,760,143, to Tosoh Corporation.
43. Niesten M C E J (2000) Polyether based segmented copolymers with unform aramid units, www.ub.utwente.nl/webdocs/ct/1/t0000018.pdf, Thesis, University of Twente, Enschede, the Netherlands.
44. Yamakawa H, Kubo Y and Simosato S, unpublished data
45. Niesten M C E J, Gaymans R J and ten Brinke A (1999) Thermoplastic elastomers from segmented copolyetheresteramides with self assembling aramid units, *Polym Prepr ACS Div Polym Chem* **40**:1012–1013.
46. Niesten M C E J and Gaymans R J (2000) Elastic behavior of segmented copolymers having crystallizable segments of uniform length, www.huntsman.com/pu/Media/annapolis.
47. Niesten M C E J, Feijen J and Gaymans R J (2000) Synthesis and properties of segmented copolymers having aramid units of uniform length, *Polymer* **41**:8487–8500.
48. Niesten M C E J, Krijgsman J and Gaymans R J (2000) Melt spinnable elastane (spandex) fibers from segmented copolyetheresteramids, *Chem Fibers Int* **50**:256, 258–260.
49. Niesten M C E J, ten Brinkel J W and Gaymans R J (2001) Segmented copolyetheresteraramids with extended poly(tetramethyleneoxide) segments, *Polymer* **42**:1461–1469.
50. Niesten M C E J and Gaymans R J (2001) Comparison of properties of segmented copolyetheresteramides containing unform aramid segments with commercial segmented copolymers, *J Appl Polym Sci* **81**:1372–1381.
51. Kubo Y, Yamakawa H and Kirikihira I (1995) Aromatic ester amide and its production, Japanese Patent 7112961, to Tosoh Corporation.
52. Shimosato S, Yamakawa H and Kubo Y (1997) Production of aromatic ester amide, Japanese Patent 9295961, to Tosoh Corporation.
53. Sek D, Wolinska A and Janeczek H (1986) Structure-liquid crystalline properties relationship of polyesteramides, *J Polym Mater* **3**:225–233.
54. Minami S, Saito S and Itoga M (1974) Linear copoly(ester amides), Japanese Patent 49063785, to Toray Industries Inc.
55. Aharoni S M and Edwards S F (1994) Polyamide networks with stiff "liquid-crystalline" segments and trifurcated flexible junctions, *Polymer* **35**:1298–1305.
56. Kirikihira I, Yamakawa H and Kubo Y (1994) Production of esteramide copolymer, Japanese Patent 6207005, to Tosoh Corporation.

57. Rabani G and Kraft A (2002) Synthesis of poly(ether-esteramide) elastomers by a palladium-catalyzed polycondensation of aromatic diiodides with terechelic diamines and carbon monoxide, *Macromol Chem Rapid Commun*, **23**:375–379.
58. Rabani G, Rosair G M and Kraft A (2004) Low-temperature route to thermoplastic polyamide elastomers, *J Polym Sci A Polym Chem Ed* **42**:1449–1460.
59. Wilfong R E (1961) Linear polyester, *J Polym Sci* **54**:385–410.
60. Griehl W and Schnock G (1958) Zur kinetik der polyesterbildung durch umesterung, *J Polym Sci* **30**:413–422 (in German).
61. Mathias L J, Moore D R and Smith A S (1987) Polymerization of N-(*p*-aminobenzoyl)caprolactam: Block and alternating copolymers of aromatic and aliphatic polyamides, *J Polym Sci A Polym Chem Ed* **25**:2699–2709.
62. Flory, P J (1955) Theory of crystallization in copolymers, *Trans Faraday Soc* **51**:848–857.
63. Antic V V, Balaban M R and Djonlagic J (2001) Synthesis and characterization of thermoplastic poly(ester-siloxanes)s, *Polym Int* **50**:1201–1208.
64. Dreyfuss P (1982) *Poly(tetrahydrofuran),* Gordon and Breach, New York.
65. Hoffman J D and Weeks J J (1962) Melting process and equilibrium melting temperature of poly(chlorotrifluoroethylene), *J Res Nat Bureau Standards* **66A**:13–28.
66. Magnus R (1992) Terephthalamide-PTMO segmented copolymers with a unform block length, Twaio Report, University of Twente, Enschede, The Netherlands.
67. Den Breejen C (1994) Segmented block copolymers based on uniform aramid blocks, Twaio Report, University of Twente, Enschede, The Netherlands (in Dutch).
68. Grosborne P and de Roch I S (1967) Étude cinétique de l'oxydation des éthers en phase liquide, *Bull Soc Chim France*, **7**:2260–2267.
69. Thomassin C and Marchal J (1977) Réactions primaires de dégradation oxydante au cours de l'autoxydation du poly(oxytétraméthylène) à 25°C, 1, *Makromol Chem* **178**:981–1003.
70. Thomassin C and Marchal J (1977) Réactions primaires de dégradation oxydante au cours de l'autoxydation du poly(oxytétraméthylène) à 25°C, 2, *Makromol Chem* **178**:1295–1312.
71. Thomassin C and Marchal J (1977) Réactions primaires de dégradation oxydante au cours de l'autoxydation du poly(oxytétraméthylène) à 25°C, 3, *Makromol Chem* **178**:1313–1325.
72. Thomassin C and Marchal J (1977) Réactions primaires de dégradation oxydante au cours de l'autoxydation du poly(oxytétraméthylène) à 25°C, 4, *Makromol Chem* **178**:1327–1339.
73. Li C X and Aklonis J J (1988) Thermal degradation of poly(tetrahydrofuran), *Polym Prepr ACS Div Polym Chem* **29**:393–394.
74. Kumar N S (1991) Oxidative degradation of poly(tetramethylene glycol), *Acta Polymerica* **42**:535–536.
75. Roslaniec Z and Pietkiewicz D (2002) Synthesis and characteristics of polyester-based thermoplastic elastomers: chemical aspects, in *Handbook of Thermoplastic Polyesters* (Ed. Fakirov S) Wiley-VCH, Weinheim, Vol. 1, Ch. 13, pp. 581–658.
76. Hsu J and Choi K Y (1987) Kinetics of transesterification of dimethyl terephthalate with poly(tetramethylene ether)glycol and 1,4-butanediol characterized by tetrabutyl titanate, *J Appl Polym Sci* **33**:329–351.
77. Dairanieh I S and Khraish N (1991) Empirical modelling and scale-up of a semi-batch polyesterification process, *J Appl Polym Sci* **42**:1631–1637.
78. Higashiyama A, Yamamoto Y, Chujo R and Wu M (1992) NMR characterization of segment sequence in polyester-polyether copolymers, *Polymer J* **24**:1345–1349.

79. Chang S J, Cheng F C and Tsai H B (1995) Block copolyetheresters. Part 3: Preparation of block copolyetherester by a terephthalic acid and processes in the presence of salts, *Polym Eng Sci* **35**:190–194.
80. Wu C, Woo K F, Luo X L and Ma D Z (1994) A modified light-scattering method for the characterization of the segmented copolymer poly(ethylene terephthalate-co-caprolactone), *Macromolecules* **27**:6055–6060.
81. Cheng X, Luo X L and Ma D Z (1999) Proton NMR characterization of chain structure in butylene terephthalate-ε-caprolactone copolymers, *J Polym Sci A Polym Chem Ed* **37**:3770–3777.
82. Kang H J and Park S S (1999) Characterization and biodegradability of poly(butylene adipate-co-succinate/poly(butylene terephthalate) copolyester, *J Appl Polym Sci* **72**:593–608.
83. Demharter S, Rösch J and Mülhaupt R (1993) Anisotoropic polyamide/polyol dispersions – new components for the preparation of polyurethane microcomposites, *Polym Bull* **31**:421–428.
84. Miyata H, Mori K, Shimosato S, Kondo S and Yamakawa H (2003) Development of thermoplastic aramid elastomers, *Proc 12th Polym Mater Forum*, Osaka, p. 173 (in Japanese).
85. Miyata H, Kondo S, Shimosato S, Mori K and Yamakawa H (2004) Development of thermoplastic aramid elastomers, *Proc Int Rubber Conf'04*, Moscow, pp. 292–293.
86. Shimosato S, Yamakawa H, and Kubo Y (1998) Elastic yarn of polyether ester amide elastomer, Japanese Patent 10008324, to Tosoh Corporation.
87. Yamakawa H, Kirikihira I, Kubo Y and Shimosato S (1995) Boots of thermoplastic elastomer, Japanese Patent 7313201, to Tosoh Corporation.
88. Yamakawa H, Kirikihira I, Kubo Y and Shimosato S (1996) Tubular hollow molding, Japanese Patent 8336912, to Tosoh Corporation.

Chapter 6

Poly(Ether Ester) Thermoplastic Elatomers: Phase and Deformation Behavior on the Nano- and Microlevel

S. Fakirov

1. Introduction

Thermoplastic elastomers (TPE) represent a relatively new class of engineering plastics. They are distinguished by some peculiarities, making them quite interesting from both the scientific and the industrial point of view. Due to their extraordinary combination of elasticity, toughness, low-temperature flexibility, and strength at 150°C, they are now also of great commercial importance as engineering type TPE. This set of properties is related mainly to the existence of crosslinks tying the array of macromolecules into an infinite network. In natural rubber and synthetic elastomers, these crosslinks represent chemical bonds. In thermoplastic elastomers, they are replaced by thermally labile tie points held together by physical forces. These junctions may be glassy, crystalline, or even hydrogen bonded molecules or ionic associations [1].

In order to behave as a thermoplastic elastomer, the molecules must contain two types of units, blocks, or chain segments: amorphous type (above the glass transition temperature, T_g) referred to as *soft segments* or blocks, and *hard segments* or blocks, the latter being usually of the crystalline type. The soft segments impart elastomeric character to the copolymer while the hard blocks are capable of intermolecular association with other hard blocks; they should form a solid phase within a desired temperature range in order to impart dimensional stability to the array of molecules. At high temperatures, dissociation of the physical bonds occurs. In order to ensure the formation of a three-dimensional network, each molecule should contain, on the average, at least two hard blocks. The soft and hard blocks may be arranged in various ways, randomized or ordered. Their way of ordering affects the physical and mechanical properties of the material [2].

The basic difference in the nature of network formation in classical rubbers and thermoplastic elastomers influences not only their processing conditions, but also such fundamental and characteristic properties as their high elasticity and the extent of its deformation reversibility. Since elastomers of the condensation type combine strength and flexibility, they are quite suitable for applications requiring spring action, recovery after deformation, impact resistance, and capability to withstand repeated flexure. They are now being used as heavy duty shock absorbers, actually replacing hydraulic systems with a completely new concept in shock absorber design. Due to their ability to demonstrate shape memory effects, TPE behave as intelligent ("smart") materials, which is related to the polyblock nature of this class of polymers. These properties result in completely new, unexpected technical and biomedical applications (see also Chapter 18).

Among the variety of thermoplastic elastomers synthesized and studied so far, those based on segmented polyblock poly(ether ester)s have attracted special attention. They comprise repeating higher-melting, crystallizable blocks and noncrystallizable blocks having a relatively low glass transition temperature, T_g. Typically, the hard segments are composed of multiple short-chain ester units (usually tetramethylene terephthalate, commonly called poly(butylene terephthalate) (PBT)), whereas the soft segments are derived from aliphatic polyether or polyester glycols (see Chapters 2 and 3).

A great variety of starting materials can be used for the preparation of segmented polyblock elastomers. By varying the hard-to-soft segment ratio, materials ranging from soft elastomers to relatively hard plastics can be obtained. For the better understanding of the material properties, it is thus necessary to accumulate data on the properties of the crystallizable and noncrystallizable segments. This means to determine the morphology of these polymers under various conditions including application of stress and different thermal treatments.

The present chapter summarizes our results on deformation, reflecting the behavior of the morphological elements on the nano-level, obtained by means of X-ray scattering with samples of poly(ether ester) (PEE) thermoplastic elastomers, under or in the absence of stress during the measurements. Further, results of two methods for the quantitative analysis of two-dimensional (2D) small-angle X-ray scattering (SAXS) patterns with fiber symmetry are described (for the 2D and 3D analysis, see Chapter 7).

All studies are carried out on polyblock PEEs consisting of PBT as hard segments and poly(ethylene glycol)s (PEG) as soft segments in different ratios. It should be noted that the starting PEGs are characterized (according to size exclusion chromatographic (SEC) analysis) [3] by a very narrow molecular weight distribution ($\bar{M}_w/\bar{M}_n = 1.30$). They are denoted PEG (600), PEG (1000), and PEG (2000), although their molecular weights are actually lower, as shown in Table 1.

Table 1. Some characteristics of PEE samples

Sample designation	Composition (from the starting ratio) (wt%)		Degree of polymerization		Average mol. wt. of PEG, $\bar{M}_n^s$ (by SEC)	Segment length (from chem. composition) (Å)		Maximal melting temp. (°C)	Degree of crystallinity (from DSC) (%)	
	PBT	PEG	PBT	PEG		l^h	l^s		w_c	w_c^{PBT}
100/0	100	0	–	–	–	–	–	230	-	50
75/25(1000)	75	25	13.8	20.2	890	160	71	212	30	40
49/51(1000)	49	51	4.4	20.2	890	51	71	188	13	27
24/76(1000)	24	76	1.4	20.2	890	16	71	none	0	0
67/33(600)	67	33	4.1	7.3	320	48	26	184	22	33
57/43(1000)	57	43	6.1	20.2	890	71	71	196	19	33
41/59(2000)	41	59	3.2	8.6/32	380/1400	37	30/112	198	13	31

For the sake of comparison, commercial engineering grade poly(ethylene terephthalate) (PET) (Yambolen, Bulgaria) and PBT (Aldrich Chemical Company, melt index = 50) were used as homopolymers. In addition, two PEEs (Arnitel®, DSM, The Netherlands), designated as EM400 and EM550, with molecular weight/weight content of the poly(tetramethylene glycol) (PTMG) soft segments 2000/60 and 1000/35, respectively, were used independently and in the blend with PBT.

Calculations based on normal kinetics of polyester formation and ^{1}H NMR as well as ^{13}C NMR measurements were used for the determination of the average degree of polymerization of the hard segments (Table 1). Assuming a total molecular weight of the copolymer of 20 000 (for products of technical interest it is about 2-4.10^4 [2]), these data suggest an average total number of blocks in each macromolecule of about 20. For an extended chain conformation, the average soft segment length, l^s, and the average length of the hard segments, l^h, were calculated (Table 1). For details on the synthesis of this sample series (Table 1) see [4] and for the more general case, see Chapters 2 and 3.

Bristles with diameter of about 1 mm were prepared from all materials by means of melt extrusion. These isotropic (according to X-ray tests) bristles were drawn at room temperature until the entire sample underwent neck formation, corresponding to a draw ratio λ = 3.5–5. The drawn material exhibited a reversible deformation at room temperature of about 50%. The drawn and undrawn bristles were annealed with fixed ends in vacuum at various temperatures, T_a, for 6 h.

Some measurements were performed using an X-ray source with rotating anode and a pinhole collimation [5]. The rest of the measurements were carried out using the synchrotron radiation generated at the beamline A2 of HASYLAB in Hamburg, Germany. The intensity was measured by means of one- and two-dimensional position-sensitive detectors, respectively.

SAXS measurements under stress were performed using a frame allowing controlled variation of the sample length. Each measurement under stress was followed by one in the absence of stress (in the "relaxed state") before a larger deformation was applied in the subsequent cycle. Thus, the samples were studied in various deformation ranges, depending on their extensibility. The elongation ε in percent is defined as: $\varepsilon = 100 \times (l - l_0)/l_0$, where l_0 and l are the initial and the actual bristle length, respectively, as measured between two marks close to the irradiated part of the bristle. In the case of measurements in the relaxed state, the same equation was used to compute the residual elongation, ε_r.

2. Phase behavior of PEEs

2.1. *Number of phases present and their miscibility*

More than two phases were observed with poly(ether ester) TPE mostly based on PTMG as soft segments. Decreasing miscibility between hard and soft segments has been reported with the rise of the segment length. Increasing PTMG molecular weight resulted in a transition from a single phase of soluble PBT and PTMG segments into a two-phase melt of a soft segment-rich phase and a hard segment-rich one. The dephasing tendency has been evaluated as the lowest in the case of PEG as soft segments, as compared to PTMG and poly(propylene oxide) [2,6]. These relationships were checked more precisely in the case of PEE based on PBT and PEG.

In general, one can expect four phases in a multiblock PEE, two crystalline and two amorphous. The actual phase structure will depend on the chemical composition and the thermal prehistory of the sample, as well as on the temperature at which the analysis is carried out. The observation of two transition temperatures, T_g^{PEG} and T_g^{PBT}, at about −46°C and 55°C, respectively (Figure 1), proves the presence of the two different amorphous phases, each one consisting of PEG or PBT segments. Evidence of the existence of two crystalline phases comprising either soft (PEG) or hard (PBT) segments can be found in the thermograms, where different melting temperatures (T_m^{PEG} and T_m^{PBT}) are observed. Homo-PEG crystallites melt at about 35°C (Figure 1E), while the PBT homopolymer melts at 231°C (Figure 1A).

These conclusions are supported also by SAXS measurements performed at various temperatures using synchrotron radiation. The changes in the scattering intensity and the long spacing, L, with temperature and composition were followed [7].

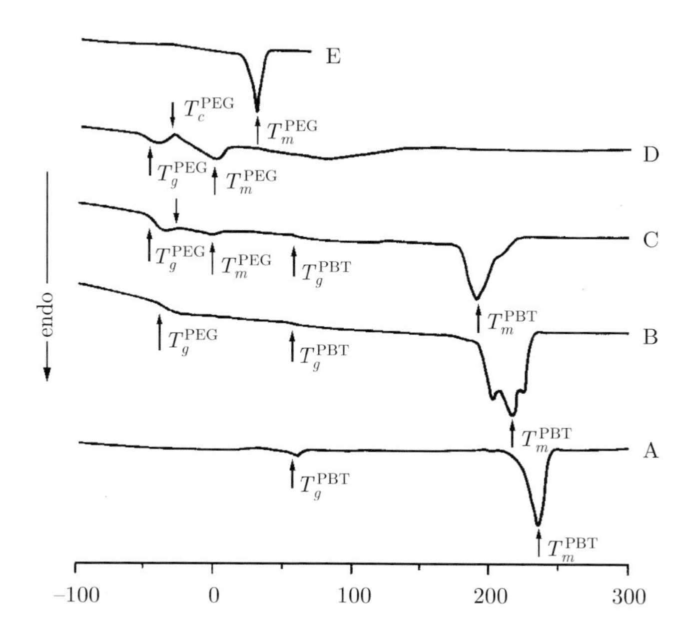

Figure 1. DSC curves of PEE annealed for 6 h at T_a, having the following PBT/PEG(1000) ratios (by wt): A – 100/0 (PBT), $T_a = 205°C$; B – 75/25, $T_a = 190°C$; C – 49/51, $T_a = 170°C$; D – 24/76, unannealed; E – 0/100 (PEG), unannealed [7]

For homopolymers having a well-defined chemical composition (*e.g.*, polyethylene (PE)), a strong increase in the SAXS intensity with the temperature of measurement has been observed [8]. The rise of scattering intensity originates mainly from the increase in the density difference between the crystalline, ρ_c, and amorphous, ρ_a, regions with temperature, due to different thermal expansion coefficients. The dependence of the square root of the normalized integral SAXS intensity on the temperature of measurement, T, for homopolyesters, *e.g.*, PBT, consists of two linear parts, as shown in Figure 2 (dashed line). Rather different are these curves for PEE. In contrast to PBT, the curves for copolyesters are not reversible in the heating and cooling mode, and some hysteresis is observed (Figure 2). The common trend of these curves suggests that the higher the PEG content, the stronger is the deviation from linearity [7].

An attempt was made to explain the results obtained (Figure 2) assuming a three-phase model and in this way to prove the existence of the three phases, *i.e.*, amorphous PEG, amorphous PBT and crystalline PBT [7]. One of the theoretical curves for the scattering intensity is shown in Figure 2 by the solid line; a relatively good agreement is achieved.

These data are supported by the changes in the long spacing with temperature of measurements and composition, too [7]. The observed linearity and reversibility of the L value of homo-PBT [7] means that the changes observed are related solely to the thermal expansion of the samples during the measure-

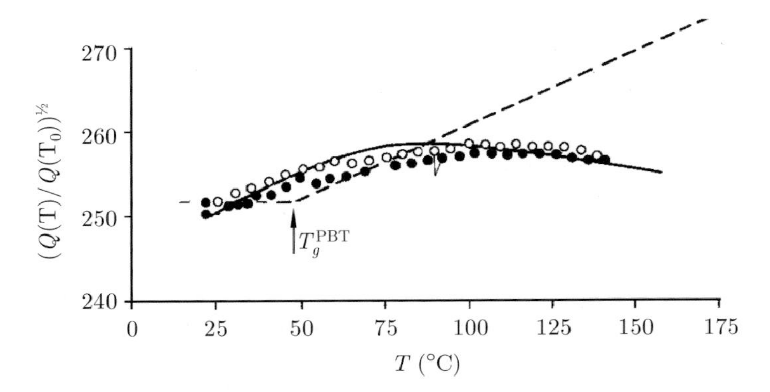

Figure 2. Dependence of the SAXS intensity $(Q(T)/Q(T_0))^{1/2}$ on the temperature of measurement T for annealed homo-PBT (dashed line) and PEE copolyester 49/51 (by wt): ○ – heating, ● – cooling. The solid line represents the theoretically derived scattering curve for the same copolyester assuming the presence of three phases [7]

ments. On the contrary, the same relationship is not linear and reversible for the PEE copolymers.

In order to explain the curved rise of L with temperature in the case of copolyesters, one has to take into account the existence of amorphous domains consisting of sub-domains of PBT and PEG, each forming a single phase, indicated by circles in Figure 3. Thus the long period may represent the spacing of PBT and PEG domains as well as the spacing of PBT crystals. Therefore, the rise of L can be attributed to the growth of the amorphous domains and/or sub-domains. This growth may have different reasons: (i) Small and/or imperfect PBT crystallites may melt when the temperature during the measurements rises, thus increasing the average distance between the crystals. Similar processes are also possible in samples with higher PBT content. The effect on SAXS, however, is expected to be small because of the small degree of crystallinity. (ii) Growth by coalescence of amorphous PBT and/or PEG sub-domains; calorimetric studies [4] support the PEG sub-domain growth. It was found that samples with intermediate compositions have two peculiarities. First, the unannealed samples have very high T_g^{PEG} values (–15°C for the undrawn samples and –25°C for the drawn ones), which drop with the rise of the annealing temperature, almost reaching the value of pure PEG. Second, no crystallization and melting of PEG segments can be observed before reaching the annealing temperature of 150°C. These two facts clearly indicate that phase separation in the amorphous domains does actually occur upon thermal treatment at elevated temperatures.

This tendency of the amorphous sub-domains to grow is also supported by a decrease in the miscibility of the two amorphous phases (PBT and PEG) with temperature, as suggested by a consideration of the solubility parameters [7].

Concerning the number of phases, their transitions and miscibility, as derived on the basis of X-ray and calorimetric studies, it could be concluded that domains of four types exist in PBT/PEG copolyesters: crystalline PBT, amorphous PBT, amorphous PEG, and amorphous PBT/PEG intermixed (as shown schematically in Figure 3 by dashed line) at least before annealing. Crystalline PEG is present only at temperatures below 0°C. A partially irreversible growth of domains is observed with the rise of the temperature. Furthermore, partial melting of PBT crystallites occurs to a much greater extent than in the homopolymer heated under the same conditions.

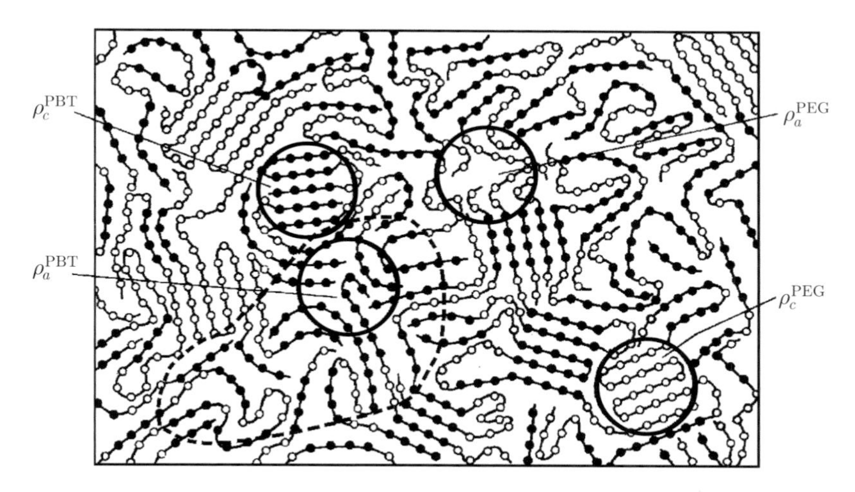

Figure 3. Model of the PEE copolyester structure with hard (-●-●-●-) PBT segments and soft (-○-○-○-) PEG ones in equimolar amounts [7]

2.2. *Amorphous phase distribution in PEE copolymers*

The main characteristic feature of highly oriented semicrystalline homopolymers is the existence of a regular sequence of crystalline and amorphous layers placed normally to the orientation axis and giving rise to the long spacing in SAXS. They are linked by a large number of *tie molecules.* Another peculiarity of oriented polymers is the existence of *microfibrils* aligned parallel to the orientation axis, as concluded mainly from microscopic observations. Further, it is known that the amorphous phase is situated mainly inside the fibrils between the crystalline lamellae (*intrafibrillar amorphous phase*), whereas a relatively small amount exists between the fibrils (*interfibrillar amorphous phase*). The latter equals 10-15% for poly(ethylene terephthalate) [9] and PBT [10].

The structure of oriented multi-block crystallizable copolymers should be basically the same. However, some peculiarities are caused by the limited length of the crystallizable blocks affecting the interfibrillar and intrafililllar amorphous phase distribution, as well as the number of tie molecules, discussed below.

In the present case of multi-block $(AB)_n$ poly(ether ester)s, the PBT segments are in the crystalline state at room temperature and their length, l^h, as seen in Table 1, is of the order of magnitude of the lamella thickness, l_c. Consequently, they crystallize either in the extended state or slightly folded (1 to 3 folds). In this way, the chains leave the crystalline lamella much more frequently and longer parts of the molecules remain in the amorphous regions, in contrast to the case of homopolymers, where essentially longer parts of the molecules form multiple folds and remain in the same crystallite. It follows that the amount of tie molecules in the case of copolymers should be higher than that in homopolymers. This peculiarity of the structure could result in a different amorphous phase distribution, *i.e.*, one can expect a larger amount of interfibrillar amorphous phase than in homopolymers. This assumption is supported by the calculation of the lamella thickness, l_c, according to the common approach

$$l_c = \varphi_c^{DSC} L \tag{1}$$

where the superscript DSC means that the DSC volume degree of crystallinity φ_c was used.

Equation (1) implicitly assumes that all the amorphous material is situated inside the fibrils between the crystalline lamellae, *i.e.*, it is of interfibrillar type only. The value of l_c thus obtained turned out to be twofold smaller than the lamella thickness, l_c^{WAXS}, obtained by the Scherrer equation [3]. For this reason, an attempt was made to calculate the amount of the intra- and interfibrillar amorphous material, as well as the relative cross-section of a fibril core.

For the sake of simplicity, let the total cross-section of a fibril (including the interfibrillar amorphous material) be unit and let $S_f < 1$ be the relative cross-section area (dimensionless) of the fibril crystalline core. It follows from Figure 4 that

$$\varphi_c = S_f l_c / L \tag{2}$$

$$\varphi_{intra} = S_f (L - l_c) / L \tag{3}$$

where φ_{intra} is the intrafibrillar amorphous material volume fraction. It is related to the interfibrillar amorphous material volume fraction φ_{inter} through the equation $\varphi_a = \varphi_{inter} + \varphi_{intra}$. From Eqs. (2) and (3) one obtains:

$$\varphi_{intra} = \varphi_c (L - l_c) / l_c \tag{4}$$

The calculation performed using the model of Figure 4 [3] shows that the relative fraction (in %) of the amorphous interfibrillar material strongly depends on the sample type and is between 45% and 80% in five of the samples and 35% in the sample 75/25(1000). This last sample is distinguished by the highest PBT content and the greatest l^h, *i.e.*, among all six samples it is the closest to the case of a homopolymer, which explains the relatively low value of the fraction of interfibrillar amorphous material. The values of the relative cross-section area of the fibril core are between 0.3 and 0.8.

It can be concluded that, in contrast to homopolymers, the interfibrillar amorphous material fraction in the case of poly(ether ester)s is large (30–80%).

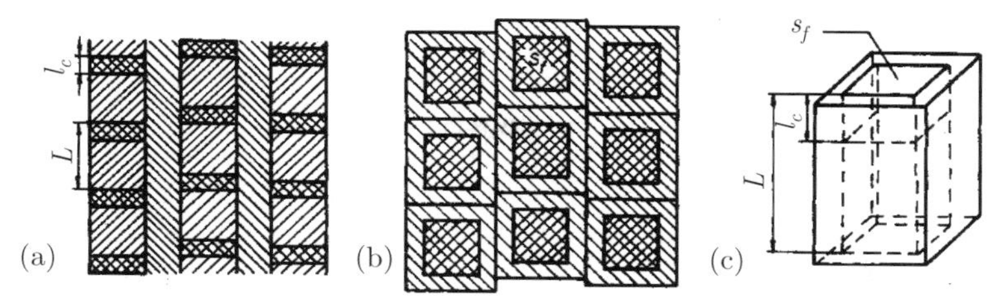

Figure 4. Schematic representation of the lateral (a) and cross section (b) of the polymer structures, as well as a 3D view of a microfibril fragment (c): double shaded, crystallites; rightward shaded and leftward shaded, intra- and interfibrilar amorphous material, respectively. s_f is the relative cross-section area of the fibril crystalline core; L is the long spacing and l_c is the lamella thickness [3]

2.3. *Does crystal thickening exist in segmented and multiblock copolymers?*

The phenomenon of *crystal thickening* in semicrystalline homopolymers can explain the strong increase of the long spacing, L, with crystallization temperature in segmented or multiblock copolymers [11]. At the same time, indications exist [12] that this explanation is not a universal one. Some previous measurements demonstrated that blocks with very short length restrict crystal thickening at least in the c-direction in drawn materials [3,13]. For this reason, the change in the long spacing and crystallite size was followed directly by SAXS and WAXS as dependent on the crystallization temperature in the case of PEE differing in their PBT/PEG ratio, as shown in Figure 5. Further, in order to demonstrate the dominating role played by the intercrystalline amorphous regions in the

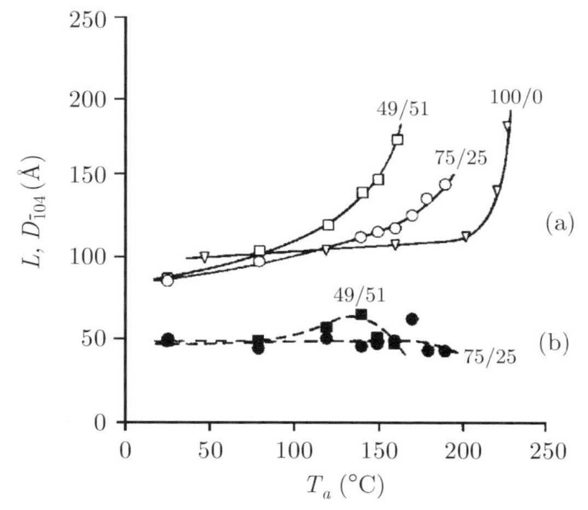

Figure 5. Effect of the annealing temperature T_a on the long spacing L (a) and crystallite size $D_{\bar{1}04}$ (b) for drawn PBT and PEE with various PBT/PEG(1000) ratios (by wt): (□) – L of 49/51, (○) – L of 75/25, (▽) – L of 100/0, (■) – $D_{\bar{1}04}$ of 49/51, (●) – $D_{\bar{1}04}$ of 75/25 [13]

formation of the long spacing, the effect of the soft segment length, l^s, on the long spacing was studied for various annealing temperatures [13].

The results obtained lead to several conclusions. The situation with oriented crystals is much clearer (Figure 5). Direct measurements of the lamellae thickness by WAXS for two samples (75/25(1000) and 49/51(1000) show that l_c varies between 50 and 60 Å within the entire annealing range, in agreement with previous measurements of samples with a larger variety in their PBT/PEG ratios [3]. This rather constant value of l_c suggests the lack of crystal thickening in the samples studied (Figure 5b [3]). It should be noted that in the case of sample 49/51(1000), this could be due to the very short hard segments (l^h = 51 Å, Table 1). However, sample 75/25(1000) has l^h = 160 Å (Table 1); obviously, in this second case the absence of crystal thickening is related to the copolymer nature of the macrochains – the presence of non-crystallizable blocks restricts the recrystallization ability of the hard segments due to the large number of inter- and intrafibrillar tie molecules, as shown below.

The observed increase in the long spacing by 80% at higher annealing temperatures (Figure 5a) can be thus explained solely by the volume changes in the non-crystalline interlamellar regions consisting of soft and hard segments [3,13,14]. This conclusion is supported also by the fact that the longer the soft segments, the larger and more temperature-sensitive are the long spacings, as reported by Wegner [12] and observed in [7].

The same general conclusion concerning the nature and temperature dependence of the long spacing in PEE can be drawn from the measurements on undrawn samples, too [13]. In this particular case, crystal thickening of about 20–30% was observed in the (100) and (010) directions; it is by far not enough so as to explain the rise in L by 80% for sample 49/51(1000).

The best proof in favor of this statement is the appearance of long spacing even in PEE free of crystalline phase (sample 24/76(1000)). This interesting observation has been reported earlier [15] for a similar PEE system, with the indication that the peak in the SAXS curves can have a completely different origin. As shown by our earlier studies [3,7], the existence of long spacing in the absence of crystalline phase is related to the multiphase character of the systems under investigation. The rise in L with the annealing temperature, T_a, for such samples is due to dephasing processes, as proved by means of synchrotron radiation and discussed above.

The measurements on isotropic samples support earlier results that the higher the soft segment content, the larger L is [4,12].

The predominant contribution of the amorphous intercrystalline regions to the formation of the long spacing and the strong temperature dependence of the latter are best illustrated by the measurements on samples containing soft segments of different length. It has been found [13] that the longest soft segments result in the highest and most temperature-dependent long spacing. The variation in the L values up to 150% resulting from the PBT/PEG ratio, from the PEG molecular weight, and the thermal treatment when the crystallite size remains almost constant (Figure 5b) is related solely to a volume increase of the amorphous regions. Usually, this volume change is accompanied by dilution effects leading to a rise in the density difference between ordered and disordered regions, as demonstrated by the SAXS curves [13].

It should be emphasized again that the well known strong dependence of L on T_a is confirmed, but it cannot be explained by the phenomenon of crystal thickening, generally accepted for homopolymers [16], for the following reasons: (i) the lamellae thickness, l_c (for drawn samples) and the crystallite size (for undrawn samples) remain almost constant with the rise of T_a, (ii) L drops with the increase of the crystallizable segment content in PEE, and (iii) L is most sensitive to T_a in samples with the highest PEG molecular weight. The repeatedly observed strong increase of the long spacing with temperature for segmented and multiblock copolymers originates from the expansion of the intercrystalline amorphous regions caused by various factors.

3. Deformation behavior of PEE as revealed by small-angle X-ray scattering

Thermoplastic elastomers are attractive subjects for structural investigations not only because of their peculiar mechanical properties. In addition, they offer many modeling opportunities due to their crystallization ability, multiblock character, and possibility of varying both the block flexibility and length. There are numerous unanswered questions concerning the deformation mechanism of this class of polymers, as compared to classical rubbers. For this reason, the relationship between macro- and microdeformation (the latter reflecting the behavior of the morphological elements on the nano-level) was studied by means of SAXS. Only deformation in a tensile mode is considered. Deformations in bending [17] or torsion [18] modes are beyond the scope of this chapter.

3.1. *Effect of chain flexibility on the deformation of PET, PBT, PEE, and PBT/PEE blend*

The main purpose of this subsection is, by applying the same approach, namely by following the relationship between macro- and microdeformation on the nano-level by SAXS recorded during the measurement under and without stress, (i) to study the behavior of the neat PET and PBT, as well as the blend PBT/PEE and to compare them with the results obtained earlier on PEE [3,5,14,19–21], and (ii) to study the behavior of commercial PEE with PTMG as soft segment. Such a comparison will shed light on the validity of the models derived from PEE with PEG-based soft segments [3–9] and will help to better understand the contribution of the PBT hard segment to the overall behavior of PEE.

The patterns from the relaxed samples are denoted, for instance, as $\varepsilon_r = 4.7(8)$, where the first number represents the tensile set (*i.e.*, the residual plastic deformation), ε_r, after straining to a tensile deformation, ε, given by the number in brackets.

The next figures display the results obtained with the homopolymers PET and PBT, the commercial PEE with PTMG as soft segments, the PBT/PEE blend, and two types of the same PEE differing in their hard-to-soft segment ratios. The results are represented as 2D scattering patterns and as graphs showing the long spacings, L, as a function of the elongation (ε or ε_r). The L values were calculated by means of Bragg's law from the positions of the peak maxima of both two- and four-point scattering patterns. The results shown in Figures 6–9 represent the two basic states of the samples, both under stress as a function of the tensile deformation, ε, and after relaxation as a function of the residual deformation, ε_r.

The results are presented in the sequence defined by the increasing flexibility of the "soft" chain constituents of the materials studied, *i.e.*, PET, PBT, PBT/PEE, PEE EM550, and PEE EM400. Such a presentation allows one to follow the effect of the chain flexibility on the relationship between macro- and

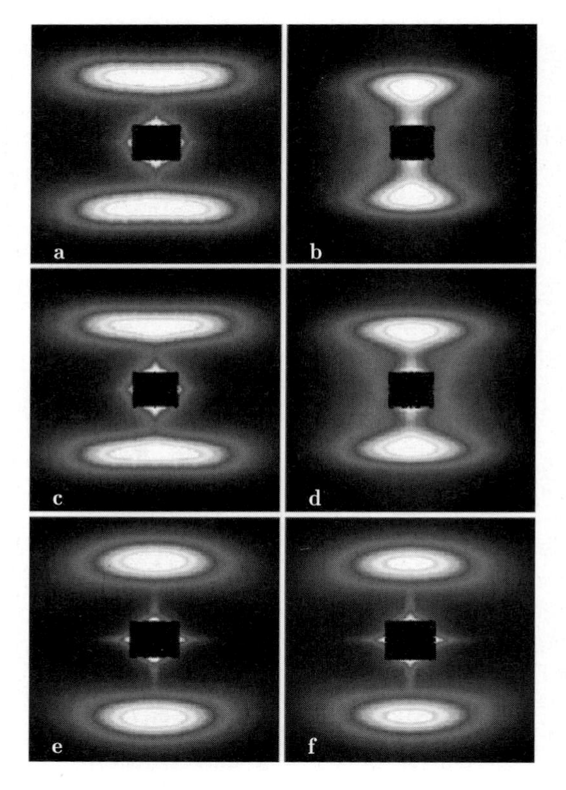

Figure 6. 2D SAXS patterns of PET and PBT bristles, cold drawn, $\lambda = 3.5$ ($\lambda = 2.3$) and annealed with fixed ends for 6 h at 240°C (180°C), recorded at room temperature at a forced tensile deformation ε or tensile set (residual elongation) ε_r in percent: (a) PET, $\varepsilon = 0$, (b) PET, $\varepsilon = 8$, (c) PET, $\varepsilon_r = 0(5.5)$, (d) PET, $\varepsilon_r = 4.7(8)$, (e) PBT, $\varepsilon = 0$, and (f) PBT, $\varepsilon_r = 6.7(16.7)$. The value in brackets is the forced elongation in percent during the previous measurement under stress. Each square covers the range $-0.15\ \text{nm}^{-1} < s_{12}, s_3 < 0.15\ \text{nm}^{-1}$ with the modulus of the scattering vector defined by $\mathbf{s} = (s_{12}^2 + s_3^2)^{0.5} = (2/\lambda)\sin\theta$. Vertical straining direction [22]

microdeformation on the nano-level, the latter being expressed by the variations of L in the stressed ($\sigma \neq 0$) and relaxed ($\sigma = 0$) state.

Figure 6 shows selected SAXS patterns from the deformation-relaxation cycles at different forced (ε) or residual (ε_r) deformations for bristles of PET and PBT homopolymers, cold drawn and annealed with fixed ends. It is important to note that the type of the scattering patterns is the same for the two homopolymers, regardless of the extent and character of deformation (ε or ε_r); all patterns are of the two-point type. What is changed during these measurements is the shape of the peaks and the distance between the two reflections, which indicates variation in the long spacing with the progress of deformation. These changes are better expressed in the next two figures showing the effect of the level and character of deformation on L. Before discussing these results, let us point out an important difference in the behavior of the two homopolymers. While PET breaks after the measurement at $\varepsilon = 8\%$ ($\varepsilon_b > 8\%$), PBT displays twice the elongation at break.

Figure 7a shows the dependence of L on the forced ($\sigma > 0$) elongation ε and residual ($\sigma = 0$) elongation ε_r in the case of PET. Figure 7b represents the same dependence for PBT homopolymer. Obviously, the experimental points

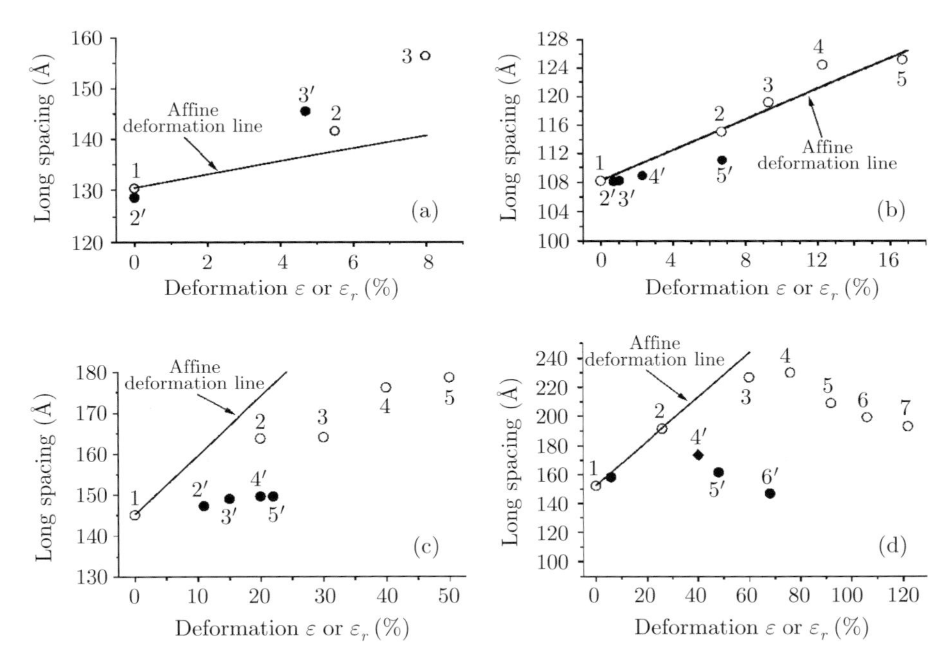

Figure 7. Long spacing L calculated from two-point patterns, *vs.* elongation ε or ε_r for homopolymers, blends and copolymers: ○ – samples under stress ($L(\varepsilon)$), ● – sample in relaxed state ($L(\varepsilon_r)$). The digits show the sequence of the measurements with the progress of the external elongation ε. The "primed" digits depict measurements without stress at $\sigma = 0$ at the respective residual elongation ε_r: (a) homo-PET, (b)-homo PBT, (c) the PBT/PEE blend (40:60 by wt, PEE is Arnitel® EM550), (d) PEE (Arnitel® EM550) [22]

for PET (Figure 7a) are rather scattered for the two kinds of measurements ($\sigma > 0$ and $\sigma = 0$). For this reason, it is impossible to derive any well-defined relationship between deformation and long spacing value. Nevertheless, there are two noteworthy results. First, a completely reversible response of L (ε, ε_r) for ε up to 5.5% (Figure 7a, points 2, 2′) is observed, whereas there is only partial reversibility at higher deformations. Second, comparing the increase of the long spacing for PET and PBT (Figure 7a,b), one can see that, for the same elongation interval ε, the increase of L for PET is approximately twice that for PBT.

The behavior of PBT displayed in Figure 7b is quite different; the dependence of L on the external macrodeformation can be followed in a much broader deformation interval (up to $\varepsilon_b > 18\%$) and a very well expressed linear relationship is observed between the macro- (ε) and microdeformation (L). The respective line coincides with that of the affine deformation model, and the deformation is completely reversible up to $\varepsilon = 12\%$. The similarity of the PET and PBT scattering patterns on the one hand, and the differences in their deformation behavior on the other hand (Figure 7a and b, respectively), *i.e.*, their macrodeformation limits, the reversibility of deformation and its character (affine or non-affine) lead to a preliminary conclusion regarding the reason for the differences observed. They can be attributed to the different chemical composition — replacement of the ethylene glycol (EG) moieties of PET by the longer and more flexible tetramethylene glycol (TMG) units of PBT.

Let us now examine the deformation of the TPE samples of the poly(ether ester) type containing more flexible segments, PTMG. Before considering the neat PEE, the behavior of a PBT/PEE blend (40:60 by wt) comprising both TMG and PTMG will be discussed. The 2D SAXS scattering patterns (not shown) are similar to those presented in Figure 6, *i.e.*, all of them are of the two-point type, regardless of the extent and kind of deformation (ε or ε_r). The dependence of the long spacing L on the deformation for the same PBT/PEE blend is shown in Figure 7c. It is clearly seen that the elongation at break of the blend, $\varepsilon_b = 50\%$ is much higher than those of the neat PET and PBT samples (Figure 7a,b). Another difference, especially as far as PBT is concerned, is in the character of macrodeformation under stress. Although L increases with ε more of or less linearly, the increase is significantly lower than the prediction of an affine microdeformation.

A common feature of the PBT/PEE blend (Figure 7c) and the neat PBT (Figure 7b) is the reversibility of the L values measured under and without stress. The corresponding values for the measurements without stress are close to the initial one, although the tensile set, ε_r, increases with the progress of the forced elongation, ε, and becomes almost half the latter ($\varepsilon_r \approx \varepsilon/2$), similar to earlier studied PEE samples with PEG as the soft segment [3,5,19,20].

The next sample of the series is the commercial thermoplastic elastomer Arnitel® EM550 comprising 35 wt% of PTMG. The dependence of L on ε and ε_r for this sample is plotted in Figure 7d. The 2D SAXS patterns obtained with this sample are again of the two-point type, as those shown in Figure 6, except for the pattern at $\varepsilon_r = 40\%$ where one observes a trend toward the formation of a four-point diagram, similarly to the next sample (Figure 8) and in accordance with the reports for PEE based on PBT and PEG [3,5,14,19–21].

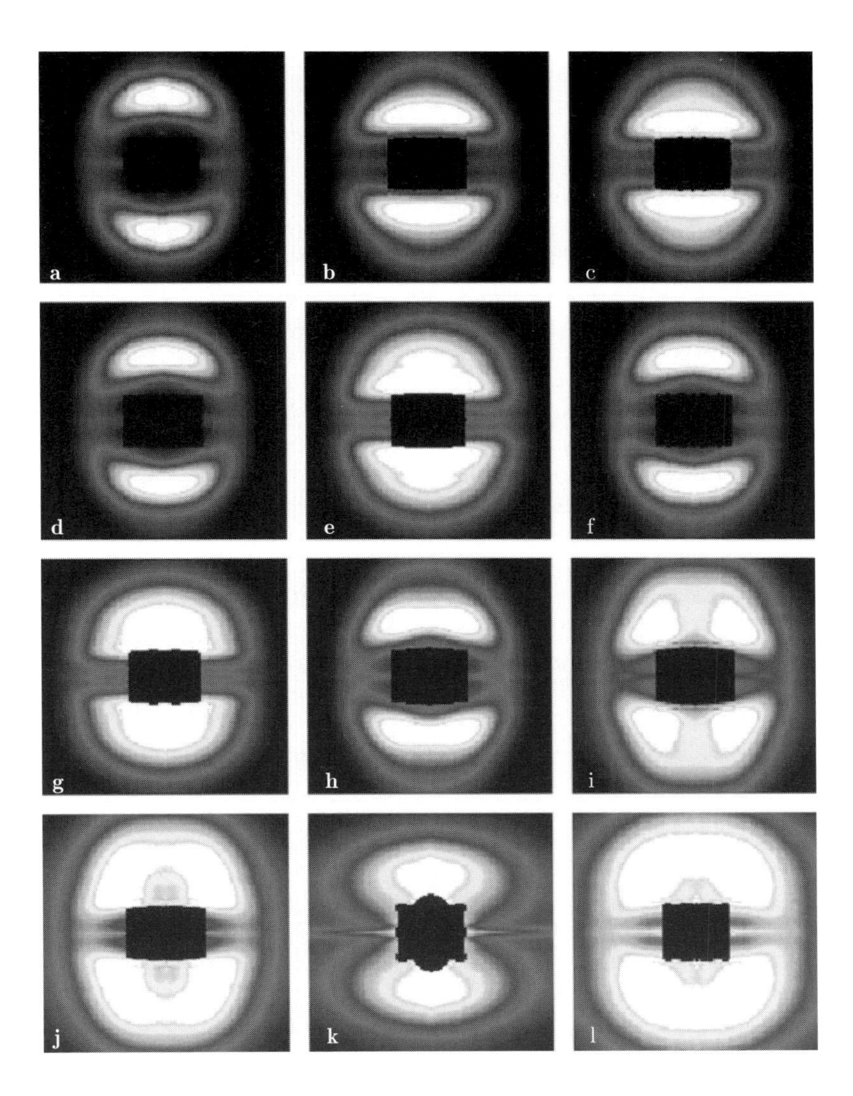

Figure 8. 2D SAXS patterns of a PEE (Arnitel® EM400) bristle cold drawn ($\lambda = 2.3$) and annealed with fixed ends for 6 h at 180°C, collected in the strained (elongation ε in percent) or the relaxed (tensile set ε_r, in percent) state, respectively: (a) $\varepsilon = 0$, (b) $\varepsilon_r = 0(50)$, (c) $\varepsilon = 80$, (d) $\varepsilon_r = 25(80)$, (e) $\varepsilon = 100$, (f) $\varepsilon_r = 25(100)$, (g) $\varepsilon = 140$, (h) $\varepsilon_r = 25(140)$, (i) $\varepsilon_r = 51(250)$, (j) $\varepsilon_r = 106(360)$, (k) $\varepsilon = 510$, and (l) $\varepsilon_r = 205(510)$. The values in brackets are the elongations in percent during the previous measurements under stress. Each square covers the range $-0.1\ \text{nm}^{-1} < s_{12}, s_3 < 0.1\ \text{nm}^{-1}$ with the modulus of the scattering vector defined by $\mathbf{s} = (s_{12}{}^2 + s_3{}^2)^{0.5} = (2/\lambda)\sin\theta$. Vertical straining direction [22]

What is interesting with this neat PEE is its much higher extensibility; this material breaks at $\varepsilon_b = 120\%$. Furthermore, L (ε) shows an almost linear increase up to $\varepsilon = 50\%$ and obeys the affine model of deformation. At $\varepsilon = 70\%$, the L value starts to drop significantly, again following an almost linear dependence. Concerning the L values measured without stress, they are close to the initial one at $\varepsilon = 0\%$ (Figure 7d). What is more, the value obtained at $\varepsilon_r = 70\%$ is slightly below the initial one, as found in earlier studies for PEE with PEG as soft segment [3,5,19].

The last sample also represents a PEE (Arnitel® EM400) with PTMG as soft segment (PBT/PTMG 40:60 by wt). It should be stressed immediately that its deformation behavior differs completely from those of all previous samples. First, its extensibility is extremely high, reaching $\varepsilon_b = 510\%$. Second, the scattering patterns in this broad deformation range vary in type; both two-point and four-point diagrams are observed, as shown in Figure 8.

With respect to their chemical composition, the materials subjected to this investigation have a common characteristic feature. All of them can be viewed as being built of a rigid and a flexible part. For the two homopolymers PET and PBT, the rigid part is the terephthalic acid residue, and in PEE it comprises the PBT hard segments. As to the flexible parts, these are ethylene glycol (EG for PET), TMG (for PBT) and PTMG (for the PEE). Obviously, the chain flexibility increases from EG to PTMG. This fact is reflected by the values of their glass transition temperatures, T_g, which are 70, 40 and –50°C, respectively (the last value is for a PEE with PEG as the soft segment). Thus, it is not surprising that the three polymers exhibit different limits of external deformations, including the PBT/PEE blend (Figures 7 and 9).

The different chain flexibility, or more precisely the flexibility of the "soft" part of the molecules, is responsible for the quite different fractions of the residual deformation, ε_r, from the total one, as well as the limit of ε below which the change of L is more or less completely reversible (Figures 7 and 9). The flexibility of the "soft" parts strongly affects the character of microdeformation, making it to obey or disobey the affine model of deformation. The latter can only be applied for the homo PBT and PEE Arnitel® EM400 (Figure 9).

3.2. *Relationship between macro- and nano-deformation in PEE*

Selected 2D scattering patterns from PEE bristles (Arnitel® EM400) are displayed in Figure 8.

Analyzing all these 12 patterns taken in the range of ε between 0 and 510% under stress, one can define three deformation intervals, characterized by different types of scattering. In the first interval ($\varepsilon = 0$–80%), the scattering patterns are of the two-point type, regardless of whether they are taken under or in the absence of stress (Figure 8, $\varepsilon = 0$ up to 80%). In the second interval ($\varepsilon = 80$–140%), the scattering patterns are again of the two-point type, but a second long spacing is observed when the sample is under stress (Figure 8, $\varepsilon = 100$ and 140%). Such a coexistence of two long periods has been found

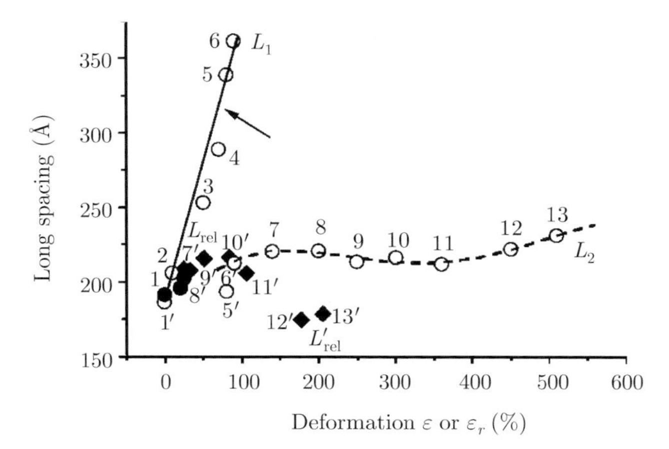

Figure 9. Long spacing L *vs.* deformation ε or ε_r for a PEE sample (Arnitel® EM400): ○ – sample under stress; ● – sample in relaxed state; ◆ – sample in relaxed state (but L calculated from four-point pattern). The meaning of the digits is the same as in Figure 7 [22]

earlier in similar investigations of PEE with PEG as the soft segment [3,5,14,19–21].

In the third interval (ε = 140–510%), the two-point scattering patterns are present only in the measurements carried out under stress. Those taken in the relaxed state are distinguished by a perfect four-point shape (Figure 8, ε_r = 25 (140%) up to 205 (510%)).

The results of the quantitative determinations of the long spacings of all patterns taken under stress ($\sigma \neq 0$) or in the absence of stress ($\sigma = 0$) are presented in Figure 9. The majority of L values are obtained from two-point patterns (Figure 9, ○, ●), and the rest from four-point patterns (Figure 9, ◆). The data in Figure 9 demonstrate also a rather different deformation behavior of this sample, compared to those of the other ones (Figure 7). The very high elongation at break has already been discussed. Further, the L values measured as a function of the forced (ε) or residual deformation (ε_r) show at least three well expressed tendencies. Up to ε = 80–100%, L increases linearly with ε, the straight line being quite close to that reflecting the affine deformation model. In this interval of ε, the long spacing measured under stress is almost doubling, whereas the corresponding values in the absence of stress ($\sigma = 0$) are very close to the initial one (Figure 9, points 2′–5′). This means that the macrodeformational changes caused by the external deformation ε are completely reversible. This conclusion is supported by the observation that up to a total deformation of ε = 50%, ε_r is close to zero (Figure 9).

In the intermediate deformation interval (ε = 80–140%), a tendency to the simultaneous formation of two long spacings is observed. This was already demonstrated when discussing the 2D SAXS patterns (Figure 8) and is not surprising. It has been observed and interpreted also in the case of PEE with PEG as the soft segment [3,5,14,19–21].

With the further progress of the external deformation (ε up to 510%), only one long spacing arising from the two-point patterns ($\sigma \neq 0$) or from the four-point patterns ($\sigma = 0$) can be observed. They are much smaller than that registered at a lower deformation, L_1, in the vicinity of the initial L value, L_0. Nevertheless, they differ in value while those measured under stress (L_2) remain constant and slightly higher than L_0. For $\varepsilon_r \leq 100\%$, the measured relaxed long period, L_{rel}, shows values close to those of L_2. For the highest forced ($\varepsilon = 450$ and 510%) and residual ($\varepsilon_r = 177$ and 205%) deformations, the values of L_{rel} are close to the initial value L_0 (Figure 9, ◆).

The relationship between macro- and microdeformation on the nano-level is reflected in the relationship $L(\varepsilon)$. For PET (Figure 7a), there is no well-defined relationship. For the other materials, one observes a linear relation (Figures 7b-d and 9) that obeys the affine deformation model, except for the blend (Figure 7c). At low elongations, this increase is completely reversible, *i.e.*, the L values measured in the corresponding deformation range in the absence of stress, are very close to the initial value L_0 (best seen in Figure 9).

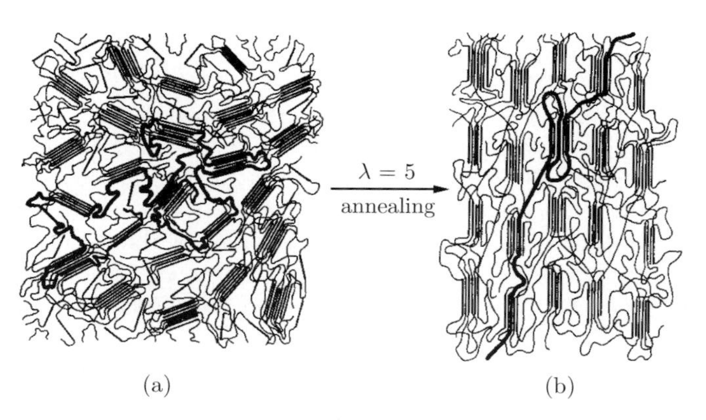

Figure 10. Schematic model of the structure and chain conformation of the thermoplastic elastomer (proposed by Fischer *et al.* for polyurethane block copolymers [23]) in isotropic (a) and oriented (b) ($\lambda = 5$) state. The thick line depicts a tie molecule [24]

On the basis of the models describing the structure and chain conformation of homo- and copolymers (Figure 10), as well as amorphous and crystalline regions organized in alternating layers in the oriented state, giving rise to the long spacing, L, one can derive further models, reflecting the structural and conformational changes occurring during deformation, in order to explain the scattering behavior of these copolymers under stress. These models are shown schematically in Figure 11.

Figure 11a shows the structure of the starting material characterized by a long spacing, L_1^0. Deformation of the sample within the range of $\varepsilon = 10$–20% leads to stretching of the chains in the amorphous regions, resulting in a decrease of their density, hence in the rise of the scattering intensity and the

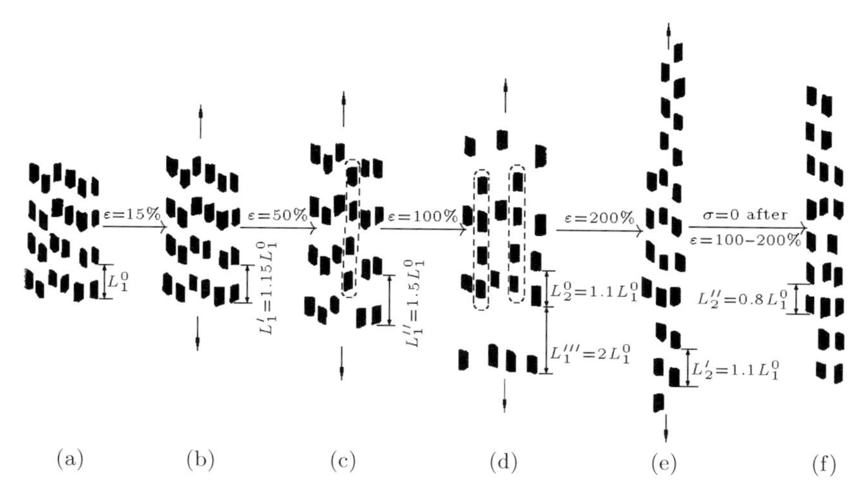

Figure 11. Schematic model of the structural changes in a drawn semicrystalline TPE at different stages of deformation: (a) no stress applied, (b)–(e) under stress, (f) in the absence of stress after deformation (only the crystalline regions from Figure 10 are depicted). The dashed lines show relaxed microfbrils after elimination of the interfibrillar tie-molecular contacts [24]

long spacing by the same percentage [24]. If the inter- and intrafibrillar contacts remain unaffected by these deformations, it is to be expected that the changes described should be reversible, as actually observed in the present case (Figures 7b-d and 9). It is reasonable to expect that with the further progress of the deformation, many tie molecules will pull out from the neighboring microfibrils and at a given stage of deformation (ε = 75–100%), their relaxation will be sufficient to give rise to a second long spacing, L_2^0. This situation is visualized in Figure 11d. The samples subjected to the highest deformations (ε = 150–200%) and measured thereafter in the absence of stress exhibit an L value smaller by 20% than the initial one (Figure 9, steps 12′ and 13′). This peculiar result can be interpreted only in terms of the structural changes described above and presented in Figure 11. The microfibrils free of interfibrillar contacts have more freedom to relax than those in the starting material (Figure 10b). The higher extent of relaxation results in shrinkage of the microfibrils, and the long spacing decreases ($L_2'' = 0.8\ L_1^0$, Figure 11f).

The results obtained suggest that the progressive increase of the residual plastic deformation is a result of structural changes taking place at very high relative deformations — continuous loss of interfibrillar contacts by pullout of tie molecules and subsequent relaxation and shrinkage of the microfibrils. The lack of interfibrillar links required for the transfer of the mechanical forces hampers the regeneration of the initial dimensions. For the same reason, the samples with destroyed interfibrillar bonds are unable to exhibit a high elastic deformation since during extension, instead of stretching, slippage of the microfibrils occurs. In order to restore the elastic properties of the material, one has first to regenerate the interfibrillar links [24]. Recall that the models described (Figure 11) are derived from PEE samples based on PBT and PEG only [24].

In this context, it is noteworthy to mention an early observation by Zhurkov *et al.* [25] on PA6 films. It has been demonstrated that the SAXS intensity increases continuously with drawing ($\varepsilon = 12$, 14, and 31%) during the measurement. Performing similar experiments, Fischer and Fakirov [26] showed that the scattering intensity is very sensitive to strain or stress also in the case of PET. For instance, measurements carried out on 1 mm thick PET sheets drawn 5× and annealed at 260°C for 6 h have shown that the SAXS intensity increases sharply and reversibly when the sample is under strain (5–7%) at the time of measurement. If the strain is larger, the scattering intensity increases further, but is no longer fully reversible [26]. Such a reversibility (up to $\varepsilon = 13\%$) was recently established for PET in a PET/PE blend (50:50 by wt) from both the changes in L^{PET} and the variation of the scattering intensity [27].

The observed SAXS behavior of PET, either in the blend with PE [27] or as homopolymer [26], as well as the changes of L for PET and, particularly, for PBT in the present case, *i.e.*, the increase of the scattering intensity and/or the long spacing with extensional deformation, can be properly explained by the model proposed by Zhurkov *et al.* [25] assuming a density decrease of the soft (non-crystalline) phase due to its elongation. Such an explanation is supported by the reversible changes of both the intensity and L values in the respective deformation range; the observed external elongation is related to conformational changes only in the amorphous domains, the crystallites remaining unaffected at these deformation levels. This mechanism is proved in the case of PEE [3,5,19,20,24].

It should be stressed here that such an interpretation of the SAXS behavior (for both intensity and L) cannot be applied, *e.g.*, to PE in the above-mentioned PET/PE (50:50 by wt) blend. The possible reason for this is that the PE crystallites melt at much lower temperatures (around 120, compared to 255°C for PET). In other words, their mechanical resistance during deformation is lower and during straining defects are introduced in crystallites leading to a decrease of the density difference between the hard and the soft phase that explains the observed continuous decrease in SAXS intensity of PE [27].

Coming back to the polymers studied here (Figures 6–9), one has to stress that the reversibility of the long spacing L in a certain range of lower macrodeformation is expressed in the best way by PBT (Figure 7b) and by PEE Arnitel® EM400 (Figure 9). As discussed above, it is related to conformational changes of the chains in the amorphous intercrystalline regions, as repeatedly reported for similar [28] and other polymer systems [29,30]. This conclusion is supported by the observed changes in the shape of the 2D SAXS patterns themselves, as shown in Figures 6 and 8 (up to $\varepsilon = 80\%$). The application of stress in limited deformation ranges (ε up to 18% for homopolymers and up to 80% for PEE) causes the two-point reflections to get closer to one another and to move apart when the stress is removed (Figures 6 and 8). These results suggest that at low deformation levels no substantial persistent morphological changes occur — some of the morphological elements (ensembles of crystalline domains) remain the same as entire entities and preserve their internal structure.

The last sample, the PEE Arnitel® EM400, is distinguished by properties that are most typical of poly(ether ester)s (Figures 8 and 9). It is interesting to see to what extent the model derived from the behavior of PEE with PEG as the soft segment (Figure 11) is applicable to the case when PEG is replaced by PTMG. In addition to the behavior up to $\varepsilon = 80\%$ discussed above, in the range of $\varepsilon = 80$–140%, in contrast to the rest of the samples, a secondary periodicity (L_2) appears and coexists with the primary long spacing L_1 (Figures 8 and 9, $\varepsilon = 100$ and 140%, respectively). Such a second long spacing L_2 is reported for PEE with PEG as the soft segment [3,5] and explained by the loss of interfibrillar contacts under such deformations. The result is relaxation of some microfibrils and the observation of a corresponding shorter long period L_2. With the further increase of the external deformation ε from 140 to 500%, the first periodicity L_1 disappears while L_2 remains constant. The nanostructure evolution as a function of elongation is further elucidated by the description of the features of the corresponding chord distribution function that adds quantitative data concerning the domain shapes and orientations. It is noteworthy to mention that both a microfibrillar component and a lamellar component are existing in the staring material, and that partition of the inclined double lamellae into fragments is already noted during the initial process of elongation.

3.3. *Chord distribution of a neat EM400 bristle*

In this subsection, we present the multidimensional chord distribution functions (CDF) computed from the scattering patterns (Figure 8) as described below. It seems noteworthy that while for the determination of long periods from peak maxima this is not a problem, for the quantitative analysis of SAXS intensity it is. Nevertheless, by automatic iterative spatial frequency filtering it is still possible to extract the nanostructure information from the SAXS, after a suitable transformation in order to display it as a multidimensional chord distribution function, and finally to discuss it [31–34]. The discussion is straightforward, since the CDF has been defined [31] by the Laplacian of Vonk's multidimensional correlation function [35]. Thus, it presents the autocorrelation of the surfaces from the (nano-size) domains in space in a similar manner as Ruland's interface distribution function does [36–38] for one-dimensional structures as a function of distance. For samples with fiber symmetry, the CDF $z(r_{12}, r_3)$ is a function of two coordinates only (transverse direction r_{12} and the fiber direction r_3). Therefore, it can be displayed by means of contours in a plane. Positive peaks found in the vicinity of the origin are size distributions of the primary domains and their size, shape, and orientation in space are depicted. The next negative peaks exhibit "long periods", *i.e.*, the distance between two adjacent domains. The further positive peaks describe the size and orientation of superdomains (*i.e.*, assemblies made of two primitive domains separated by a rather probable distance and measured from the beginning of the first to the end of the second domain), and correlations among more distant domains are manifested in consecutive peaks at longer distances. As expatiated in several papers [31,33,34], the CDF is computed from the scattering intensity by

$$z(r_1, r_3) = -\mathcal{F}^2 \left(4\pi^2(s_1^2 + s_3^2)\{I(\mathbf{s})\}_2(s_1,s_3) - B(s_1,s_3)\right) \qquad (5)$$

with the projection of the scattering intensity on the representative plane

$$\{I(\mathbf{s})\}_2(s_1,s_3) = \int I(\mathbf{s})\, ds_2 \qquad (6)$$

the factor $-4\pi^2(s_1^2 + s_3^2)$ being equivalent to the Laplacian in the physical space, $B(s_1, s_3)$ describing a background determined by spatial frequency filtering of $4\pi^2(s_1^2 + s_3^2)\{I(\mathbf{s})\}_2(s_1,s_3)$, and $\mathcal{F}^2$ denoting a 2D Fourier transformation (for more details see Chapter 7).

Figure 12 exhibits the nanostructure of the neat Arnitel® EM400 bristle before straining. The fiber direction is vertical (r_3) and r_{12} is the transversal direction in cylindrical coordinates. Figure 12a shows the elevated contours from the CDF that exhibit the domain nanostructure. The central pair of peaks (1) describes the basic domain structure. The elongated shape parallel to the equator indicates lamellae with an average diameter of 20 nm.

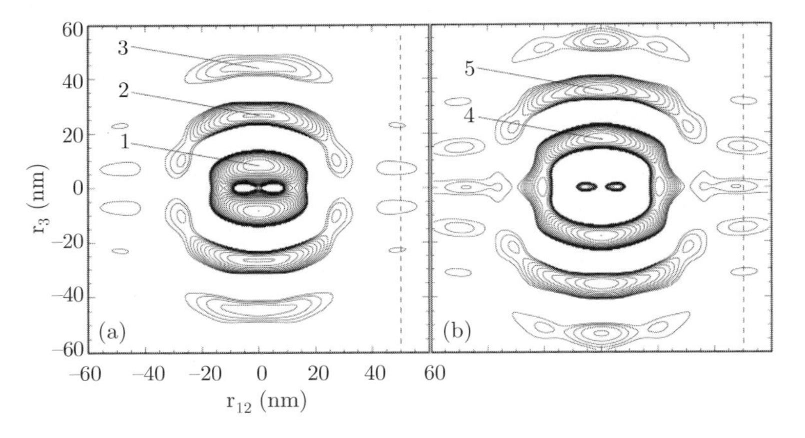

Figure 12. Bristle of neat EM400. Chord distribution function (CDF, logarithmic scale) computed from 2D SAXS pattern. Fiber direction is vertical. (a) Positive contours describing the domains from the nanostructure. (b) Negative contours describing the lattice properties [22]

The average layer thickness, as determined from the maximum on the meridian, is $d = 8.3$ nm. Both the soft- and the hard-domain layers contribute to this peak. The second peak (2) describes sandwiches made of two hard domains with a soft layer in between, called here "bi-domains". The average bi-domain thickness is 26.6 nm, and "tri-domains" (3) with an average thickness of 44.3 nm are observed as well. There are fewer bi-domains than single domains, and even less tri-domains, although in the contour plots they appear emphasized due to a logarithmic scaling of the peak heights that was chosen for the sake of clarity. The diameter of those lamellae, which are members of a stack and form bi- and tri-domains is higher (30 nm) than that of the average monolayers. The number of stacks containing more than 3 lamellae is negligible. A dashed vertical line indicates a very weak lateral correlation among adjacent stacks that is also present in the contours describing the lattice properties (Figure 12b). The most common lattice property is the long period peak (4). From the position of the maximum we determine $L = 18.0$ nm, which is in good agreement with the value directly derived from the SAXS pattern (Figure 9). A second (5) and even a third long period peak are observed at 35.2 nm and 53.4 nm in the fiber direction and describe correlations among phase boundaries that are placed at twice and trice the "lattice constant", L. In general, there is no long ranging lattice in the studied material, and consequently the negative peaks in the CDF that are related to the lattice properties are weaker by one order of magnitude than the positive peaks that describe the ensemble of nano-size domains.

In summary, the neat bristle comprises lamellae of hard and soft domains that are oriented perpendicular to the fiber direction. The average layer thickness is 8 nm. The diameter of the lamellae is varying considerably. Extended layers with a diameter of 30 nm have formed stacks comprising no more than 3 members. Imperfect layers with diameters of less than 20 nm are more frequent than the extended ones. Their correlation with neighboring domains is low [22] (see also Chapter 7).

4. Nanostructure evolution during the straining cycle

Figure 13 shows the nanostructure evolution during the straining cycle as revealed in the CDF for low and medium elongation ($50\% \leq \varepsilon \leq 140\%$). The left column shows the chord distributions of the domains in the strained state. For all elongations, the strong central domain peak exhibits a diamond shape that can be explained by the superposition of lamellae and other, rather compact single domains with a low aspect ratio. A separation by hand is impossible. Nevertheless, the bi-domain peak outside the central diamond is a set of four straight and elongated lines. As a first approximation, it can be described by a (sub-)stack of two inclined lamellae. The increase of the inclination angle with increasing elongation from 51° to 69° is indicated in the plots. At 140% strain, the linear peak begins to disintegrate into two parts indicating the formation of small domains by disruption of the lamellae. Beginning at an elongation of 100%, two well-separated peaks of cylindrical soft and hard domains located in microfibrils are observed. The peak closer to the center is still constant and thus assigned to the hard domain cylinders with a height of 8.9 nm. The average height of the soft domains from the microfibrils is 21.3 nm at 100% and 25.1 nm at 140%. During the further straining cycle, this soft domain height will not increase beyond 28 nm. Instead, the corresponding component will decrease and vanish, indicating the maximum extensibility of the soft phase between two hard domains.

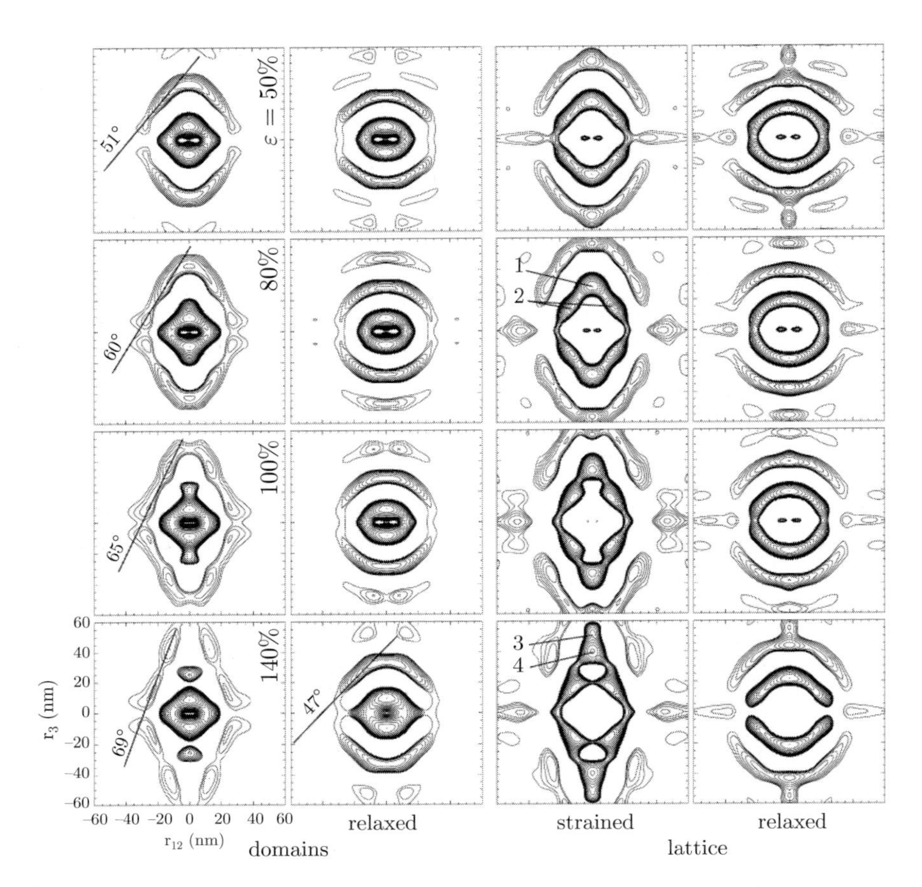

Figure 13. Bristle of neat EM400 in a straining cycle. Lower elongations. Chord distribution function (CDF, logarithmic scale) computed from 2D SAXS pattern. Fiber direction is vertical. Elongation $50\% \le \varepsilon \le 140\%$ as indicated in the left column. The contour plots of this column show the domain contours of the nanostructure in the strained states. The meaning of the other columns is given in the labels underneath [22]

The second column shows the bristle in the relaxed state after the previous elongation. We observe a system of lamellar stacks with some orientation distribution about the meridian. Released from 140%, for the first time this orientation distribution of the bi-domains appears to be bimodal, with maxima of the preferential layer orientation at the meridian (0°) and at ≈47°. There are no tri-domains with layer shape any more.

The third column shows the lattice peaks in the strained state. The long periods of the microfibrillar component (1) and of the tilted layers (2) are the most prominent features. Several higher orders are present. At $\varepsilon = 140\%$, we observe both a long period of microfibrils under tension (3) and a long period of relaxed microfibrils (4).

The fourth column shows the arrangement of long periods in the relaxed state. Up to the second order, the long periods are clear and continuous, indicating an orientation distribution of lamellar stacks that contain no more than three layers.

Figure 14 presents nanostructure evolution during the straining cycle as revealed in the CDF for high elongation ($250\% \leq \varepsilon \leq 510\%$). For the strained state, the first column reveals the transition from a macrolattice with hexagonal

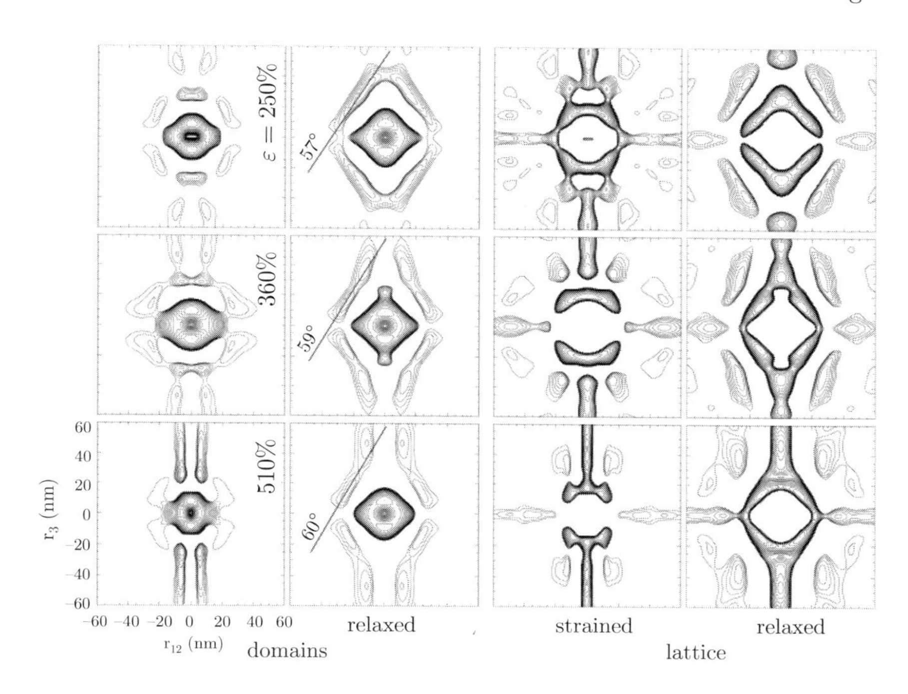

Figure 14. EM400 bristle in straining cycle. High elongations. Chord distribution function (CDF, logarithmic scale) computed from 2D SAXS pattern. Fiber direction is vertical. Elongation $250\% \leq \varepsilon \leq 510\%$ as indicated in the left column. The contour plots of this column show the domain contours of the nanostructure in the strained states. The meaning of the other columns is given in the labels underneath [22]

arrangement of small and compact domains into a purely microfibrillar system with perfect orientation.

The second column, again, shows the relaxed state. It is interesting to note that, when releasing the stress, the disrupted domains merge again to form stacks of tilted lamellae. The inclination angle now is rather high and almost constant at 60°. Again, the diameter of these lamellae is 30 nm. None of these stacks comprises more than 3 domains. The last two columns show the peaks related to the long periods in the strained and in the relaxed state. In the strained state (third column) the emerging microfibrillar component is clearly detected by the row of peaks on the meridian. Their strength is growing as the elongation increases, whereas a long period from some remnant lamellae oriented perpendicular to the fiber axis is vanishing.

All the discussed changes in L (Figure 11) concern only measurements under stress. From $\varepsilon = 140\%$, the scattering patterns taken in the absence of stress are of the four-point type and the question arises whether the material is built up of tilted lamellae or of a macrolattice of microfibrils. The CDFs (Figures 13 and 14) exhibit the presence of bi-domains from tilted lamellae that support the model of zigzag lamellae [19]. With increasing deformation ε, the inclination angle is growing to 60° at $\varepsilon = 510\%$. The trend toward the formation of zigzag assemblies of parallel lamellae is strongly enhanced when the sample is not loaded. It increases significantly as the residual deformation in the straining cycle ($e = 200$–510%) is increased (Figure 8, after $\varepsilon = 200$–510%, $\sigma = 0$). Simultaneously, the shape of the scattering patterns taken under stress at $\varepsilon = 200$–500% is closer to the lobe-type than to the four-point type. Such a difference between the scattering patterns suggests substantial morphological changes. These changes are clearly identified in the CDF (Figure 14). In the strained state, it shows a disruption of tilted lamellae into smaller pieces. At $\varepsilon = 250\%$, the fragments have formed a macrolattice with hexagonal order. When the elongation is increased beyond this value, the lattice is transformed into an ensemble of microfibrils with little correlation in the lateral direction. From the CDF, it is clearly deduced that the samples subjected to the highest deformations and thereafter unloaded consist of parallel crystalline lamellae tilted with respect to the fiber axis. A zigzag-shaped arrangement of these layers in the lateral direction is, on the other hand, not discernible from the CDF. A simplified sketch of the basic states of the nanostructure, as extracted from the SAXS data, is shown in Figure 15.

Whether really filled lamellae or only assemblies of blocks are observed cannot be deduced from the presented scattering data. In the latter case, a rotation of a lamella could be performed by a more probable process of progressive displacement of adjacent blocks in the straining direction.

It is important to emphasize here that for the nanostructure at low and medium elongation: (i) the proximate morphology and the described morphological transition between inclined lamellar assemblies and flat ones is completely reversible and (ii) the trend to the formation of zigzag morphologies (or the increase of the inclination to the fiber axis in the case of the unloaded samples) becomes much stronger with the progress of the macrodeformation ε. It can be assumed that such a reversible morphological transition contributes to the formation of the elastic properties of thermoplastic elastomers, as suggested previously [19].

One can assume that the considerable number of molecules connecting parallel lamellae causes a better-coordinated response to the changes of stress and/or strain. The interfibrillar tie molecules holding together the crystalline domains of neighboring microfibrils (Figure 10b) are responsible for the more or less smooth distribution of the external stress in the sample. Taking into

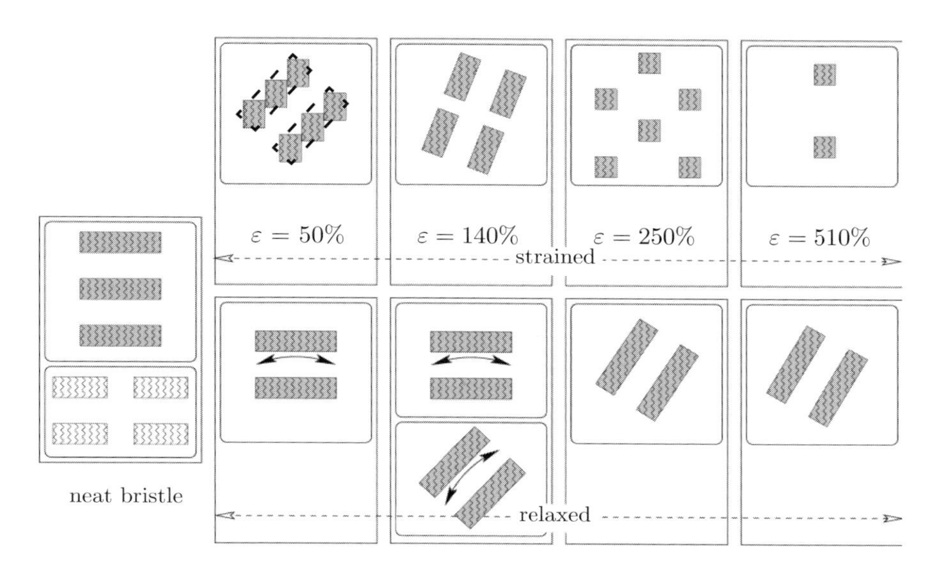

Figure 15. Nanostructure of a neat EM400 bristle as a function of elongation in the strained (top) and relaxed (bottom) state, respectively. A microfibrillar structure always present (like the one outlined in the top right-hand part) is never sketched. A less simplified and more probable view of the tilted lamellae is indicated at $\varepsilon = 50\%$ [22]

account the larger amount of amorphous material in the interfibrillar regions than in the fibrils (Figure 4, [3]), the creation of new tie molecules appears more probable in the phase surrounding the microfibrils than in the amorphous regions inside the microfibrils [22].

5. Conclusions and outlook

The scattering behavior under stress of annealed drawn and undrawn poly(ether ester)s based on poly(butylene terephthalate) as the hard segment was studied. Two glass and two melting transitions were observed by DSC, suggesting the existence of at least four phases. The long period increases with the temperature of measurement, which is attributed to growth of domains due to dephasing of both amorphous phases.

By SAXS measurements during the progressive elongation of some polyester and poly(ether ester) samples containing glycols of varying length, it was found that the extensibility strongly depends on the flexibility of the glycol residues. It increases from 10% (PET) to 510% (commercial PEE Arnitel® EM400).

Studying the relationship between the macro- (or the external deformation denoted by ε) and the microdeformation on a nano-level, expressed by the change of the long spacing L, in accordance with previous observations, the existence of three typical deformation ranges was established. Within the lowest range (up to 10% for PBT and 80% for PEE), the deformation obeys the affine model and is due to conformational changes occurring only in the amorphous intercrystalline areas. In the next range (80–140%), observable for the PEE samples only, a second periodicity appears as a result of relaxation of a part of the microfibrils owing to a loss of microfibrillar contacts at this level of deformation. Finally, within the third range for PEE (between 140 and 510%), four-point scattering patterns appear when the measurements are performed after removing the stress. Two-point diagrams characterize all measurements under stress. Using the multidimensional chord distribution function (CDF) analysis, it was verified that the two-point diagram results from sets of parallel lamellar stacks, perpendicular to the stretching direction, the four-point diagram being related to stacks of layers tilted with respect to the longitudinal direction (fiber axis). The transition between these two types of lamellar arrangement is reversible. It is believed that this reversibility on the morphological level contributes to the deformation reversibility of the thermoplastic materials.

Acknowledgement

The author appreciates the hospitality of the Centre for Advanced Composite Materials at the University of Auckland, New Zealand, where this chapter was prepared.

References

1. Noshay A and McGrath J E (1977) *Block Copolymers – Overview and Critical Survey*, Academic Press, New York, p. 278.
2. Schröder H and Cella R J (1988) Polyesters, elastomeric, in *Encyclopedia of Polymer Science and Engineering* (Eds. Mark H F, Bikales N M, Overberger C G and Menges G) John Wiley & Sons, New York, Chichester, Brisbane, Toronto, Singapore, Vol. 12, pp. 75–117.
3. Apostolov A A and Fakirov S (1992) Effect of the block length on the deformation behaviour of poly(ether ester)s as revealed by SAXS, *J Macromol Sci–Phys* **B31**:329–355.
4. Fakirov S and Gogeva T (1990) Poly(ether ester)s based on poly(butylene terephthalate) and poly(ethylene oxide)glycols. Parts 1-3, *Makromol Chem* **191**:603–614, 615–624, 2341–2354.
5. Fakirov S, Fakirov C, Fischer E W and Stamm M (1991) Deformation behaviour of poly(ether ester) thermoplastic elastomers as revealed by SAXS, *Polymer* **32**:1173–1180.
6. Legge R N, Holden G, Schröder H and Quirk R P, (Eds.) (1996) *Thermoplastic Elastomers*, Hanser Publishers, Munich, Vienna, New York, 2nd edition.
7. Fakirov S, Apostolov A A, Boesecke P and Zachmann H G (1990) Structure of segmented poly(ether ester)s as revealed by synchrotron radiation, *J Macromol Sci–Phys* **B29**:379–395.
8. Fischer E W (1978) Studies of structure and dynamics of solid polymers by elastic and inelastic neutron-scattering, *Pure Appl Chem* **50**:1319–1341.
9. Fakirov S, Fischer E W, Hoffman R and Schimdt G F (1977) Structure and properties of poly(ethylene terephthalate) crystallized by annealing in the highly oriented state.

Part 2. Melting behavior and the mosaic block structure of the crystalline layers, *Polymer* **18**:1121–1129.

10. Fakirov S and Wendorff J (1979) On the anisotropy of thermal expansion in drawn and annealed poly(butylene terephthalate), *Angew Makromol Chem* **81**:217–228.
11. Adams R K and Hoeschele G K (1987), in *Thermoplastic Elastomers – a Comprehensive Review* (Eds. Legge R N, Holden G and Schröder H) Hanser Publishers, Munich, Vienna, New York, pp. 163–196.
12. Wegner G (1987) Model studies toward a molecular understanding of the properties of segmented block copolyetheresters, in *Thermoplastic Elastomers, Research and Development* (Eds. Legge R N, Holden G and Schröder H) Carl Hanser Verlag, Munich.
13. Fakirov S, Apostolov A A and Fakirov C (1992) Long spacing in segmented block copoly(ether ester)s – origin and features, *Int J Polym Mater* **18**:51–70.
14. Fakirov S, Fakirov C, Fischer E W and Stamm M (1992) Deformation behaviour of poly(ether ester) thermoplastic elastomers with destroyed and regenerated structure as revealed by SAXS, *Polymer* **33**:3818–3827.
15. Perego G, Cesari M and Vitali R (1984) Structural investigations of segmented block copolymers. II. The morphology of poly(ether ester)s based on poly(tetramethylene oxide) "soft" segments: comparison between poly(ether ester)s and poly(ether esteramide)s, *J Appl Polym Sci* **29**:1157–1169.
16. Wunderlich B (1980) *Macromolecular Physics*, Vol. 3 *Crystal Melting*, Academic Press, New York, London, Toronto, Sydney, San Francisco.
17. Shuhong W, Mark J E, Erman B and Fakirov S (1994) Mechanical properties of thermoplastic elastomers of poly(butylene terephthalate) and poly(ethylene glycol) in a bending deformation, *J Appl Polym Sci* **51**:145–151.
18. Apostolov A A, Fakirov S and Mark J E (1998) Mechanical properties in torsion for poly(butylene terephthalate) and a poly(ether ester) based on poly(ethylene glycol) and poly(butylenes terephthalate), *J Appl Polym Sci* **69**:495–502.
19. Fakirov S, Fakirov C, Fischer E W, Stamm M and Apostolov A A (1993) Reversible morphological changes in poly(ether ester) thermoplastic elastomers during deformation as revealed by SAXS, *Colloid Polym Sci* **271**:811–823.
20. Fakirov S, Denchev Z, Apostolov A A, Stamm M and Fakirov C (1994) Morphological characterization during deformation of a poly(ether ester) thermoplastic elastomers by SAXS, *Colloid Polym Sci* **272**:1363–1372.
21. Stribeck N, Sapoundjieva D, Denchev Z, Apostolov A A, Zachmann H G, Stamm M and Fakirov S (1997) Deformation behaviour of poly(ether ester) copolymer as revealed by small- and wide-angle scattering of X-ray radiation from synchrotron, *Macromolecules* **30**:1329–1339.
22. Stribeck N, Fakirov S, Apostolov A A, Denchev Z and Gehrke R (2003) Deformation behavior of PET, PBT and PBT-based thermoplastic elastomers as revealed by SAXS from Synchrotron, *Macromol Chem Phys* **204**:1000–1013.
23. Fischer E W and Struth U, Unpublished data; Struth U (1986) PhD Thesis, University of Mainz, Germany.
24. Fakirov S and Stribeck N (2002) Flexible copolyesters involving PBT: strain-induced structural changes in thermoplastic elastomers, in *Handbook of Thermoplastic Polyesters* (Ed. Fakirov S) Wiley-VCH, Weinheim, Ch. 15, pp. 672-716.
25. Zhurkov S A, Slutsker A I and Yastrebinski A A (1963) Influence of load on the supermolecular structure of oriented polymers, *Dokl Acad Nauk SSSR* **153**:303–314 (in Russian).
26. Fischer E W and Fakirov S (1976) Structure and properties of poly(ethylene terephthalate) crystallized by annealing in the highly oriented state. Part 1. Morphological structure as revealed by SAXS, *J Mater Sci* **11**:1041–1065.

27. Fakirov S, Samokovlijsky O, Stribeck N, Apostolov A A, Denchev Z, Sapoundjieva D, Evstatiev M, Meyer A and Stamm M (2001) Nanostructural Deformation Behavior in Poly(ethylene terephthalate)/Polyethylene Drawn Blend as revealed by Small Angle Scattering of X-Ray Radiation from Synchrotron *Macromolecules* **34**:3314–3317.
28. Pakula T, Saijo K and Hashimoto T (1985) Structural changes in polystyrene-polybutadiene-polystyrene block polymers caused by annealing in highly oriented state, *Macromolecules* **18**:2037–2044.
29. Gerasimov V I, Zanegin V D and Tsvankin D Ya (1978) Influence of neighboring fibrils and shear deforamations of crystals on the small-angle X-ray scattering, *Vyssokomol Soedin* **A20**:846–853 (in Russian).
30. Gerasimov V I, Zanegin V D and Smirnov V D (1979) Shear strain of oriented polyethylene, *Vyssokomol Soedin* **A21**:765–776 (in Russian).
31. Stribeck N (2001) Extraction of the domain structure information from small angle X-ray patterns, *J Appl Cryst* **34**:496–503.
32. Stribeck N and Fakirov S (2001) Three-dimensional chord distribution function SAXS analysis of the strained domain structure of a poly(ether ester) thermoplastic elastomer, *Macromolecules* **34**:7758–7761.
33. Stribeck N, Buzdugan E, Ghioca P, Serban S and Gehrke R (2002) Nanostructure evolution of SIS thermoplastic elastomers during straining as revealed by USAXS and multi-dimensional chord distribution analysis, *Macromol Chem Phys* **203**:636–644.
34. Stribeck N, Bayer R, von Krosigk G and Gehrke R (2002) Nanostructure evolution of oriented high-pressure injection-molded polyethylene during heating, *Polymer* **43**:3779–3784.
35. Vonk C G (1979) A SAXS study of PE fibers, using the two-dimensional correlation function, *Colloid Polym Sci* **257**:1021–1032.
36. Ruland W (1977) The evaluation of the small-angle scattering of lamellar two-phase systems by means of interface distribution functions, *Colloid Polym Sci* **255**: 417-427.
37. Ruland W (1978) The evaluation of the small-angle scattering of anisotropic lamellar two-phase systems by means of interface distribution functions, *Colloid Polym Sci* **256**:932–936.
38. Stribeck N and Ruland W (1978) Determination of the interface distribution function of lamellar two-phase systems, *J Appl Cryst* **11**:535–539.

Chapter 7

Condensation Thermoplastic Elastomers Under Load: Methodological Studies of Nanostructure Evolution by X-ray Scattering

N. Stribeck

1. Introduction

1.1. *Phase separation*

Elastic at room temperature and fusible: these two properties characterizing thermoplastic elastomers (TPEs) result from phase separation, taking place whenever the polymer is cooled from the melt. In this process, parts of the copolymer chains arrange into hard domains, whereas other parts of the same chains form soft domains. The final material is elastic when hard domains act as nodes in a network [1].

1.2. *"Living" nanostructure*

The domain size is of the order of several nanometers. Variations in the processing parameters affect the domain size, shape, and perfection of arrangement. However, the nanostructure is altered not only during manufacturing. Even in service, it responds to applied mechanical and/or thermal load, as well as to the absorption of low molecular mass compounds [2] or, *e.g.*, to the chemical crosslinking of a printing plate. Whenever the nanostructure is modified, the material properties change. In principle, the nanoscopic processes have been noticed several decades ago (*e.g.*, void formation [3,4], microfibrillation [5], fine chain slip [6], strain-induced crystallization [7,8]). The understanding of the effects of these processes and of their intertwinning would enable one to better assess the material capacity and to tailor material properties by a proper processing.

1.3. *X-ray techniques in process monitoring*

Numerous methods can be used to investigate the nanostructure, but few are as suitable for integration into manufacturing processes, as the X-ray techniques. In the X-ray beam, polymers can be drawn, spun, molten, or swollen, while scattering patterns are recorded. However, the major problem to be solved is that the structure evolution has to be monitored with sufficient time resolution. For this reason, strong X-ray sources, such as synchrotrons are required. Since several years, powerful and stable sources are available to materials science, and presently we are witnessing promising developments as far as detectors and special beam optics are concerned. Even the project of an X-ray free electron laser (XFEL) source has been lined up in the frame of a European effort.

1.4. *Progress in technology and methodology*

Scattering studies have, however, a fundamental drawback. Unlike microscopic studies, in this case the information concerning the nanostructure is not suggestive. Scattering patterns dwell in the reciprocal space. In order to draw conclusions concerning the physical space, inversion is necessary either by analysis or by interpretation.

The fact that the methodology of the 1970s is not well adapted to to-date high-resolution multidimensional scattering data is not the only problem. The flood of data emerging from a modern synchrotron beamline is equally problematic; automatic algorithms must be developed and the instrumentation of a beamline has to be designed in such a way that automated processing is not obstructed [9].

Published papers indicate the rapid progress of radiation sources, but also a considerable backlog as far as the advance of data analysis in the field of polymer nanostructure studies by small-angle X-ray scattering (SAXS) is concerned. In several fields of materials science, the importance of method developments has been early recognized and supported. For instance, the great success of protein crystallography is mainly based on the development and standardization of methods for the evaluation of synchrotron radiation data [10], which evidenced an unparalleled precision in crystal structure studies. A similar awareness is still insufficient in the field of polymer materials science.

This chapter aims not only at the illustration of X-ray methods, which may be suitable for the investigation of a special class of TPEs, but also at drawing the reader's attention to the perspectives of an advanced analysis of scattering data.

2. Materials

In this chapter, a poly(ether ester) (PEE) synthesized in the laboratory of Fakirov and characterized elsewhere [11], as well as a commercial material (Arnitel® by DSM) are denoted by two numbers, *e.g.*, PEE 1000/43. The first number indicates the molecular mass of the soft blocks, while the second one gives their weight fraction (in percent).

The poly(ether amides) (commercial PEBAX® by Elf Atochem) are listed in Table 1.

Table 1. PEBAX® grades and their designation. Soft block weight fraction indicated by the first two digits. Soft and hard block polymers: PTMO – poly(tetramethylene oxide); PEO – poly(ethylene oxide); PA6 – polyamide-6; PA12 – polyamide-12

PEBAX® grade	Material designation	PEBAX® grade	Material designation
2533	80PTMO/PA12	1074	55PEO/PA12
3533	70PTMO/PA12	4011	57PEO/PA6
4033	53PTMO/PA12		

3. Basic notions

3.1. *Nanostructure topology*

The question of how the hard and soft domains share space is the topological aspect of the nanostructure. Figure 1 illustrates the principle topology of a TPE at room temperature. Hard domains form the nodes of the network. Single hard segments may be dissolved in the soft phase. Hard domains of more or less granular (spherical) shape are shown. Nevertheless, cylinders and lamellae, or even networks of interpenetrating phases have also been found [12]. Domains of varying shape and size can be included into different kinds of aggregates and, at the same time, they can dwell in the material. This means that the two phase system is, in general, profoundly distorted and is only of some short-range order.

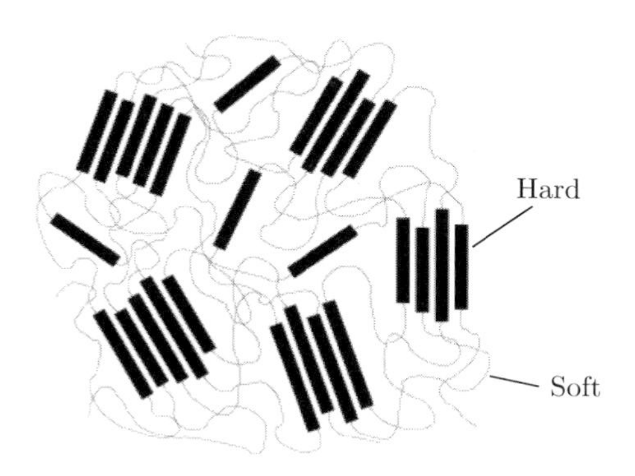

Figure 1. Principle scheme of the nanostructure of a TPE at ambient temperature

3.2. *Non–topological parameters*

Furthermore, the SAXS is also related to some non-topological parameters of the nanostructure [13,14]. They describe the contrast between the phases, $\Delta\rho_{el}$, (*i.e.*, the perfection of phase separation), the amount of the internal surface per unit volume generated by the phase separation, S/V, the width of the phase transition zone at the phase boundary, t_i, and the homogeneity of each phase.

3.3. *Isotropic vs. anisotropic SAXS patterns*

After manufacturing from the melt, TPEs often exhibit isotropic scattering patterns (Figure 2) containing quite limited information. Only a broad circular reflection (long period) is observed, and it is related to the distance between the hard domains. The long period is the most basic topological feature of the scattering pattern. When the investigation aims at the elucidation of the topology, the analysis of the isotropic pattern is only possible if the hard domains are assumed to be stack-forming lamellae. On the other hand, the study of the non-topological parameters of isotropic materials is especially simple. In this case, the study of the perfection of phase separation as a function of different manufacturing processes is a promising issue.

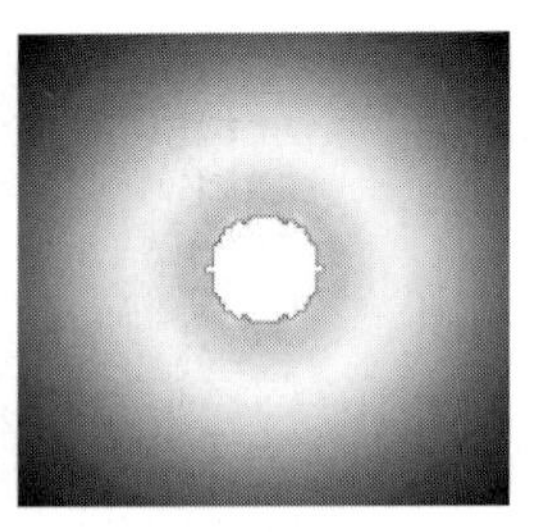

Figure 2. Isotropic SAXS from a TPE showing a long period peak

As known from wide-angle X-ray diffraction studies of polymers, the topological information obtained from SAXS can be considerably increased by straining the specimen (Figure 3, top row). Nevertheless, in SAXS studies, the anisotropic patterns are generally subject to little analysis. Based on simplified models, the patterns are mentally inverted, resulting in sketches, as shown in the second row of Figure 3. This can result in a misinterpretation, as demonstrated in the second column of the figure. In addition, the positions of the peak maxima are frequently identified and from these values long periods are determined, which then are attributed to distances between domains. The considerable disorder of the polymer nanostructure is not taken into account and, consequently, it frequently remains unmentioned that the determined long period values do not represent an ordinary average. Instead, they generally mark the upper limit of a long period distribution [15]. In the case of a microfibrillar system under load, the SAXS reflection marks the maximum length between crosslinks (before failure). Thus, it is not astonishing that during straining the long period peak of many condensation TPEs is changing little (see also Chapter 6). Nevertheless, the shape of the long period distribution, as well as the shapes and size distributions of the domains are altered. In order to extract this information from the high-precision scattering data that have become available, a profound data analysis is required.

3.4. *Multidimensional chord distributions*

The most recently developed method for the analysis of SAXS patterns of TPEs in the strained state is utilizing the multidimensional chord distribution function (CDF) [16]. The CDF is defined in the physical space, rather than in the reciprocal space. It has peaks with both positive and negative values. In the third row of Figure 3, the elevations of the CDF are shown; the depressions are depicted in the fourth row. If the hard domains were distributed at random, no negative values in the CDF would be observed, because depressions describe

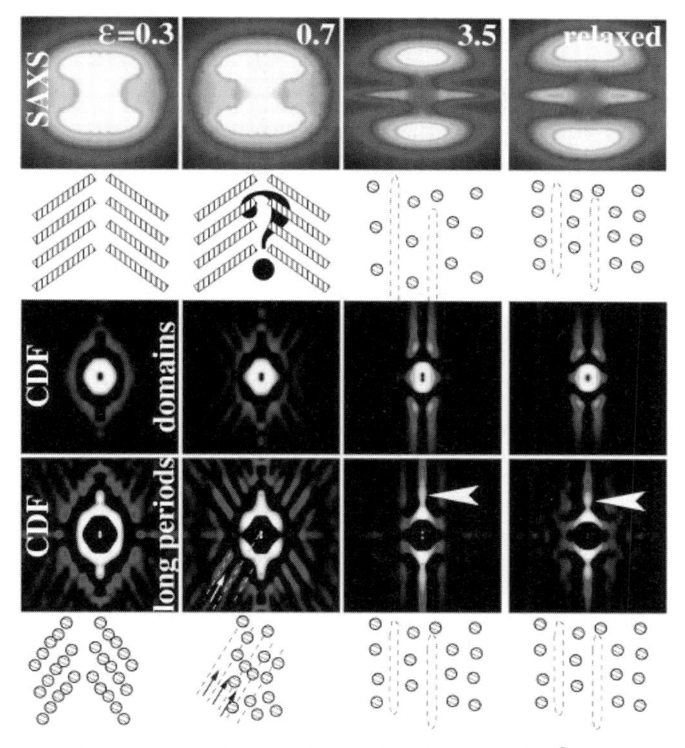

Figure 3. SAXS and related nanostructure during drawing of Arnitel® E1500/50. Rows from top to bottom: recorded SAXS patterns as a function of elongation $\varepsilon = (\ell - \ell_0)/\ell_0$ with ℓ_0 and ℓ the initial and the actual sample length, respectively; sketches of related domain structure; CDF displaying the topological information from SAXS in the physical space; sketches of the structure derived from CDF interpretation

the arrangement (long periods, lattice properties) from the nanostructure topology. The SAXS at $\varepsilon = 0.7$ is a four-point pattern but, compared to the pattern at $\varepsilon = 0.3$, it is no longer the typical "butterfly pattern" commonly related to a nanostructure made from stacks of tilted lamellae [17], nor is it looking like a pattern related to a so-called macrolattice [18]. This different structure revealed in the CDF is sketched in the bottom row and is discussed at the end of Section 5.1.3.

The interpretation of this function is straightforward, since the CDF has been defined [16] by the Laplacian of Vonk's multidimensional correlation function [19]. For this reason, it presents the autocorrelation in space of the surfaces from the domains in a similar manner as Ruland's interface distribution function does [20–22] for one-dimensional (1D) structures as a function of distance. For samples with uniaxial symmetry, the CDF $z(r_{12}, r_3)$ is a function of two coordinates only (transverse direction r_{12} and draw direction r_3). Therefore, it can be displayed by means of contours or density plots in a plane. Positive peaks found in the vicinity of the origin are size distributions of the primary domains. In this way, their size, shape, and orientation in space are depicted. The farther negative peaks exhibit "long periods", *i.e.*, the distance between two adjacent domains. The next positive peaks describe the size and orientation of super domains (*i.e.*, clusters made from two adjacent domains), and correlations among more distant domains are manifested by consecutive peaks at even longer distances.

The contributions to the value of the CDF are arising from correlations between domain surfaces. For instance, a cylindrical domain is characterized by two sharp peaks on the meridian, their distance from the origin denoting the cylinder height. In the equatorial direction, the peak is not sharp, falls off almost linearly, and becomes zero at the diameter of the cylinder.

The CDF interpretation results in a detailed, yet only qualitative description of the complex nanostructure at each stage of the straining experiment. Several features are frequently superimposed, and the quantitative analysis requires a complex three-dimensional (3D) adapted model to be fitted to the CDF in order to retrieve precise data concerning the nanostructure evolution. For this reason, it may be reasonable to quantitatively study only a partial aspect of the nanostructure topology [23].

3.5. *Longitudinal and transverse structure*

Considering TPEs under uniaxial load, Bonart [23] has proposed the study of two aspects of the nanostructure, called longitudinal and transverse structure. They can readily be extracted from the scattering pattern by projections [24]. In both cases, the result is a curve and, obviously, curves are analyzed with less computational effort than 2D scattering patterns.

The longitudinal structure evaluates the (chord) lengths of soft and hard domains, only along lines running parallel to the draw direction. Thus a quantitative representation is obtained, describing both the average extensions of the domains (Figure 4a, $\bar{d}_h$ and $\bar{d}_s$) and of their statistics represented by the standard deviations (Figure 4a, σ_h and σ_s) in the draw direction, r_3. This information can be retrieved *in situ* during deformation in the synchrotron beam. Thereafter, the curves can be fitted using one of the well known models of 1D disorder (paracrystal [25–27], stacking statistics [28], homogeneous long period distribution [22,29–31]) or a unified model [32,33].

The transverse structure (Figure 4b) retrieves the information on the extension of hard and soft domains in the transverse direction ($r_{12} = \sqrt{r_1^2 + r_2^2}$) only. The extracted curve intrinsically describes a two-dimensional function with radial symmetry. Therefore, in contrast to the longitudinal structure, no fundamental method for data analysis can be outlined. The analysis becomes simple, if the correlations among domains are negligible, by the assumption that the domain cross sections are represented by circular discs [34]. In this case, the distribution of the domain cross sections can readily be determined as a function of elongation.

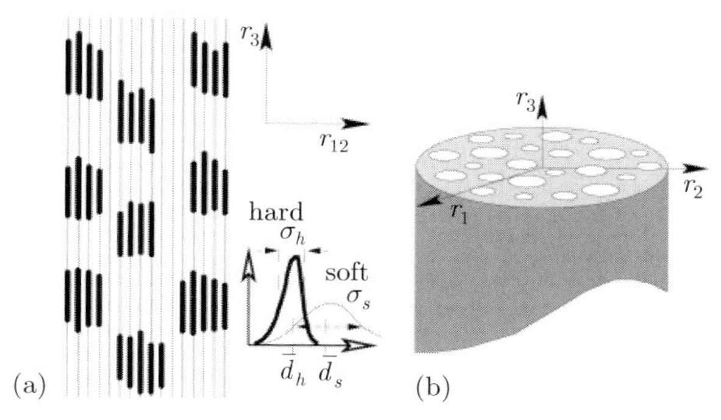

Figure 4. The information contents of (a) longitudinal and (b) transverse structure that can be extracted from SAXS patterns of drawn TPEs

3.6. *Void scattering*

At high elongations, an equatorial streak [3] (Figure 3, top right) extending transverse to the draw direction is occasionally observed. It is caused by high and thin (*i.e.*, needle-shaped, rodlike, *etc.*) domains extending in the draw direction. Such domains are frequently voids, but even microfibrils, after the complete destruction of hard domains or needle-shaped crystalline domains [35,36], can reveal such a scattering feature. The height, width, and orientation distribution of the needles can be determined from SAXS studies [37–42].

4. Theoretical

4.1. *Basic definitions in SAXS*

Figure 5a shows the scattering pattern of a poly(ether ester) bristle (Arnitel® E2000/60) recorded after relaxation from an elongation $\varepsilon = 3.5$.

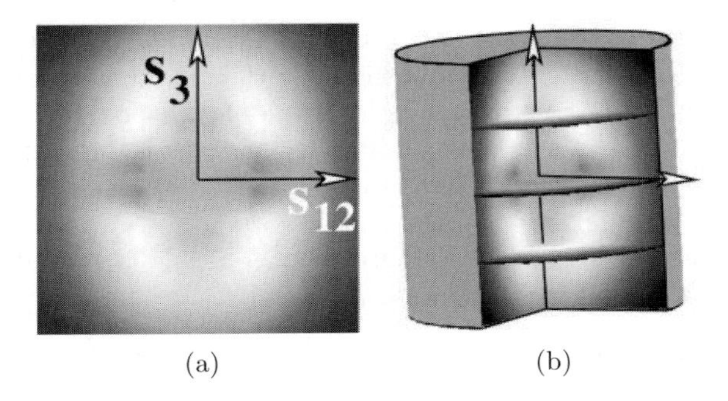

(a) (b)

Figure 5. SAXS of Arnitel® E2000/60 bristle after relaxation from $\varepsilon = 3.5$. (a) Recorded pattern (draw direction vertical). The components of the scattering vector $\mathbf{s} = (s_{12}, s_3)$ in cylindrical coordinates are indicated. (b) Sketch of the reciprocal space (cylindrical symmetry). Displayed range $-0.1\ \mathrm{nm}^{-1} < s_{12}, s_3 < 0.1\ \mathrm{nm}^{-1}$

The directions of the principal components of the scattering vector $\mathbf{s} = (s_{12}, s_3)$ are indicated. The modulus of the scattering vector, $s = (2/\lambda)\sin\theta$, is defined by the wavelength of the X-rays, λ, and the scattering angle, 2θ. The scattering vector spans in the reciprocal space. Since the scattering pattern does not change when the fiber is rotated about its axis, the total information concerning the reciprocal space can be obtained from a single 2D scattering pattern (Figure 5b).

The fundamental relation between the scattering intensity, $I(\mathbf{s})$, and the corresponding nanostructure is given by Eq. (1)

$$I(\mathbf{s}) = \mathscr{F}^3(\Delta\rho_{\mathrm{el}}^{*2}(\mathbf{r})) = QV\mathscr{F}^3(\gamma(\mathbf{r})) \tag{1}$$

with the nanostructure in physical space represented either by the correlation function, $\gamma(\mathbf{r})$ [29,43], or by the electron density $\rho_{el}(\mathbf{r})$. In the ideal two-phase system of a TPE, the electron density (in units of electrons/nm^3, *i.e.*, e.u./nm^3) would be a function, which can take only two values

$$\rho_{el}(\mathbf{r}) = \begin{cases} \rho_h & \text{for } \mathbf{r} \text{ in the hard phase} \\ \rho_s & \text{for } \mathbf{r} \text{ in the soft phase} \end{cases}. \tag{2}$$

$\Delta\rho_{el}(\mathbf{r}) = \rho_{el}(\mathbf{r}) - \langle\rho_{el}\rangle_V$ is the variation of the electron density with respect to the irradiated volume V, and the operator ^{*2} designates the autocorrelation (self-convolution) of a function defined by

$$f^{*2}(\mathbf{r}) = \iiint f(\mathbf{r}')f(\mathbf{r}' + \mathbf{r})d^3\mathbf{r}'. \tag{3}$$

From this equation, it is obvious that $\mathbf{r}$ is a displacement of (the nanostructure) $f(\mathbf{r})$ with respect to itself. Moreover, in Eq. (1), $\mathscr{F}^3(\,)$ stands for the 3D Fourier transformation and the scattering power (or invariant) Q is defined

$$Q = \iiint I(\mathbf{s})/V d^3s = \Delta\rho_{el}^{*2}(0)/V \tag{4}$$

in a way that it normalizes the correlation function $\gamma(0) = 1$. By definition, Q is the total intensity per irradiated volume, which is scattered into the reciprocal space.

Finally, the n-dimensional Fourier transformation of a function $f(\mathbf{r})$ is defined

$$F(\mathbf{s}) = \mathscr{F}^n(f(\mathbf{r})) = \int f(\mathbf{r}) \exp(2\pi i \mathbf{r}\mathbf{s})\, d^n r \quad \mathbf{r},\mathbf{s} \in \mathbb{R}^n \tag{5}$$

with $\mathbb{R}^n$ denoting the n-dimensional real vector space. The chosen definition of $\mathbf{s}$ and the symmetry of the scattering functions guarantee that this equation is valid for reverse transformations as well.

4.2. *Projections and sections*

The invariant Q is defined by an integral operation from the class of projections, and its relation to the nanostructure is frequently utilized. For an ideal two-phase system,

$$Q = (\rho_h - \rho_s)^2\, \phi_s(1-\phi_s), \tag{6}$$

it is related to the contrast, $\rho_h - \rho_s$, and to the volume fraction, ϕ_s, occupied by the soft phase.

SAXS data analysis gives an insight in the nanostructure in the physical space ($\mathbf{r}$), but the SAXS intensity is gathered in the reciprocal space. Because both ($\mathbf{r}$) and $I(\mathbf{s})$ are present in the Fourier relation (Eq. (1)), projections should play the major role in SAXS analysis [23], despite that numerous examples are found in the literature, where nanostructure parameters are determined from scattering intensity curves cut from the SAXS pattern along a straight line. From the mathematical point of view, this cut is a section, and the informational content of sections and projections is readily established by means of mathematical reasoning.

The interchange theorem. The interest in the longitudinal structure in the physical space means that one intends to gather information on the shape of the correlation function $\gamma(\mathbf{r})$ in the direction r_3 in physical space. Since this is the definition of a section, one can write

$$\gamma(0,0,r_3) = \lceil \gamma(\mathbf{r}) \rceil_1 (r_3)$$

for γ restricted to a 1D subspace in the draw direction, r_3. Using Eqs. (1) and (5), it is readily established that

$$QV \lceil \gamma(\mathbf{r}) \rceil_1 (r_3) = \mathscr{F}^1 \left(\iint \mathrm{I}(s_1, s_2, s_3,)\, ds_1\, ds_2 \right),$$

with

$$\iint \mathrm{I}(s_1, s_2, s_3,)\, ds_1\, ds_2 = \{I(\mathbf{s})\}_1 (s_3)$$

yielding the definition of the projection of $I(\mathbf{s})$ onto a 1D subspace in the $\mathbf{s}_3$ direction. Projections and sections interchange under Fourier transformation.

Thus, mathematical reasoning evidences that information concerning sections in the physical space can be obtained only after the respective projection of the intensity in the reciprocal space. It is frequently argued that projection "smears the details" of the scattering intensity and that this is felt to be incorrect. In order to disprove this intuititive reasoning, let us respond by an intuitive argument. Yes, the peaks of the SAXS intensity get blurred by projection and concentrate in the center of the pattern – and, because of Fourier transformation, the opposite is true in the physical space. Projection sharpens and extends the correlations in the real space. Moreover, because of its definition, by an integration it increases the signal-to-noise ratio of the recorded pattern, so that the exposure time can be reduced.

4.3. *Projections useful for TPEs studied under uniaxial load*

Four kinds of projections are of particular importance for the study of the nanostructure from SAXS patterns obtained in drawing experiments using thermoplastic elastomers:

Invariant Q	$\{I(\mathbf{s})\}_0 / V$,
longitudinal scattering	$\{I(\mathbf{s})\}_1(s_3)$,
transverse scattering	$\{I(\mathbf{s})\}_2(s_{12})$,
representative fiber scattering	$\{I(\mathbf{s})\}_2(s_1,s_3)$.

The information content of most of these projections was already described in this chapter, except for the representative fiber scattering,

$$\{I(\mathbf{s})\}_2(s_1,s_3) = \int I(\mathbf{s})\, ds_2. \tag{7}$$

This is the first step towards a multidimensional real-space nanostructure analysis of SAXS data with fiber symmetry, because it is related to the 2D correlation function in the representative plane of the real space,

$$\frac{1}{QV}\mathscr{F}^2(\{I(\mathbf{s})\}_2(s_1,s_3)) = \lceil\gamma(\mathbf{r})\rceil_2(r_1,0,r_3),$$

a function which has first been proposed and studied by Vonk in 1979 [19]. Correlation function analysis is frequently undertaken in SAXS analysis of two-phase systems and, in particular, the 1D correlation function, $\lceil\gamma(\mathbf{r})\rceil_1(r_3)$, [44–48], is often utilized. The analysis is coupled to the concept of a multiphase system. Thus, it is not astonishing that deviations from an ideal multiphase system [13] have to be considered. In general, these deviations contribute to a slowly varying background intensity in the SAXS, but if only the correlation function is to be computed, it is generally sufficient to subtract a constant background determined from the lowest valid intensity reading in the pattern. Then, from the correlation function, information concerning the average cross sections of both hard and soft domains can be extracted. Nevertheless, the important statistical distribution of the domain sizes is not retrieved.

4.4. *Correlation functions and their derivatives*

4.4.1. *The 1D problem*

A solution of the problem of extracting information concerning the imperfection and statistical distributions from scattering data has first been proposed by Guinier and Fournet [49] using the concept of chord length distributions [50,51]. In fact, the isotropic chord length distribution

$$g(r) = \frac{d^2}{dr^2}\gamma(r) \tag{8}$$

was defined as the second derivative of the correlation function of an isotropic scattering sample, the derivative being taken with respect to the radial component of the vector in the physical space, r. The physical meaning and the advantage of the chord distribution, as compared to the correlation function, become clear in a 1D sketch (Figure 6). In the correlation function (Figure 6a), the overlapping of the density function with its displaced ghost generates the value of the function, whereas in the chord distribution (Figure 6b), the signal is generated from the autocorrelation of the first derivative of the density function $\Delta\rho'_{el}$, which is zero everywhere, except for the edges of the nanostructure domains. Thus, the chord length distribution directly reflects all the distances between surfaces and their statistics. The 1D case sketched in Figure 6 has frequently and successfully been applied to practical problems. It covers both the cases of longitudinal scattering and of the scattering from lamellar systems, and was developed and called interface distribution function (IDF), $g_1(r)$, by Ruland [20] for application to semicrystalline polymers containing lamellar crystallites. In publications by Strobl, it is denoted as $K''(z)$ [52].

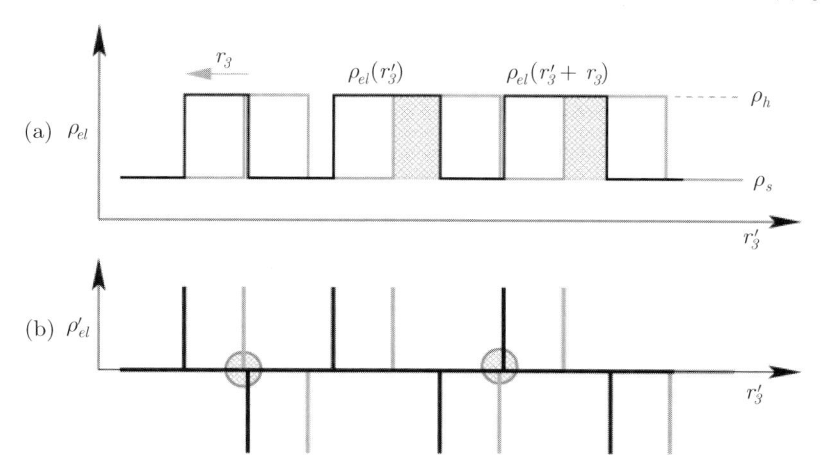

Figure 6. Autocorrelation of (a) the density, ρ_{el}, and (b) the density derivative, ρ'_{el}, generates contributions (hatch-marked) to (a) the correlation function and (b) the chord distribution, respectively. ρ_h, ρ_s — electron densities of hard and soft phase

The computation of chord distributions or IDFs is not as simple as the computation of correlation functions, because of considerable difficulties to correctly determine the different contributions to the scattering background, which cause deviations from Porod's law [29,53,54], a law predicting the fall-off of the scattering intensity with $A_P\, s^{-4}$ and being valid for every multiphase topology (A_P is Porod's second invariant).

4.4.2. *The multidimensional case*

For anisotropic samples, such as thermoplastic elastomers showing uniaxial orientation, the analysis of the longitudinal structure gives only a fraction of the total information concerning the nanostructure topology. By analogy with the IDF concept, the complete information should be displayed in a multidimensional function that maps the distances between all the domain surfaces.

A solution proposed by the author was the multidimensional CDF [16]. The extension of the background subtraction and the derivative to the multidimensional case created the main problems in this approach.

The first problem was solved by using (recursive) spatial frequency filtering to construct the background from the scattering data themselves. The method does not describe the background by physical quantities, but simply extracts the topological information from the scattering pattern, whereas the background is eliminated. The algorithm is independent of the dimensionality of the data. Thus, it can also be used to automatize background elimination in the cases of, *e.g.*, longitudinal scattering or isotropic data [55].

The second problem was solved by replacing the second derivative of a 1D correlation function by the Laplacian operator. It was deduced that, by analogy with the 1D case, the Laplacian of the correlation function

$$QV\,\Delta\gamma(\mathbf{r}) = (\nabla\rho_{el})^{*2}(\mathbf{r})$$

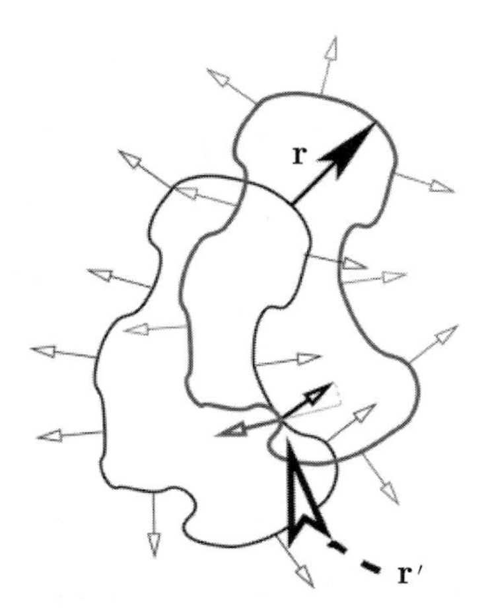

Figure 7. Particle-ghost autocorrelation of the gradient vectors generates the multidimensional CDF. The vectors emerge from the surfaces of domains and their displaced (by **r**) ghosts

is the autocorrelation of the gradient vector field, $\nabla\rho_{el}(\mathbf{r})$, computed from the multidimensional density function, with the gradient vanishing everywhere, except for the domain surfaces (Figure 7). When both the gradient of the electron density, $\nabla\rho_{el}(\mathbf{r}')$, and its displaced ghost, $\nabla\rho_{el}(\mathbf{r}'+\mathbf{r})$, do not vanish at $\mathbf{r}'$, the scalar product of the two gradient vectors contributes to the CDF, $z(\mathbf{r})$. Its positive and negative values depend on the relative surface orientation.

4.5. *Data processing*

The analysis is generally performed in three steps: (i) image preprocessing [56], (ii) extraction of topological information, and (iii) extraction of structure parameters. A uniform concept is not available. Thus programming skills and a toolkit, such as *pv–wave*® [57] holding the fundamental image processing algorithms [57–59] are recommended, but at least the commercial toolkit may become some times dispensable because of strong efforts in the open domain [60]. The sources of the procedures developed by the author in Hamburg for *pv–wave* are freely accessible [61].

4.5.1. *Image preprocessing*

Image preprocessing is closely related to the experiment, and the choice of an inappropriate instrumental setup or design for data collection may obstruct the automatization of data analysis [9].

Sample scattering and blind scattering are corrected for the detector characteristics, aligned, normalized, and subtracted. Valid data are separated from invalid regions (blind spots, penumbras) by defining masks (or regions of interest). The scattering pattern is centered and aligned so that the straining direction becomes vertical. Then part of the invalid regions can be filled from symmetry considerations. Finally, it may be reasonable to fill the central blind spot and to extend the data into an apron zone by a multidimensional extrapolation procedure [62]. The artificial data supplied in this way can help to protect measured data from being distorted in the following steps.

4.5.2. *Extraction of nanostructure topology*

The extraction of topological information is independent of the experiment and, after adjustment, it can be carried out automatically for a complete series of data. Figure 8 shows the processing steps. The preprocessed (raw) pattern is projected according to Eq. (7), the Laplacian in real space is considered by a multiplication by $4\pi s^2$ in the reciprocal space, a background is constructed from the low spatial frequencies in the scattering pattern, and after its subtraction an interference function, $G(\mathbf{s})$, is obtained. Finally, the requested CDF is computed by 2D Fourier transformation

$$z(\mathbf{r}) = -\mathscr{F}^2(G(\mathbf{s})). \tag{9}$$

If the longitudinal structure or the transverse structure is to be studied, similar steps are carried out, resulting in an interface distribution, $g_1(r_3)$ [34,63], or in a 2D chord distribution, $g_2(r_{12})$ [64], respectively.

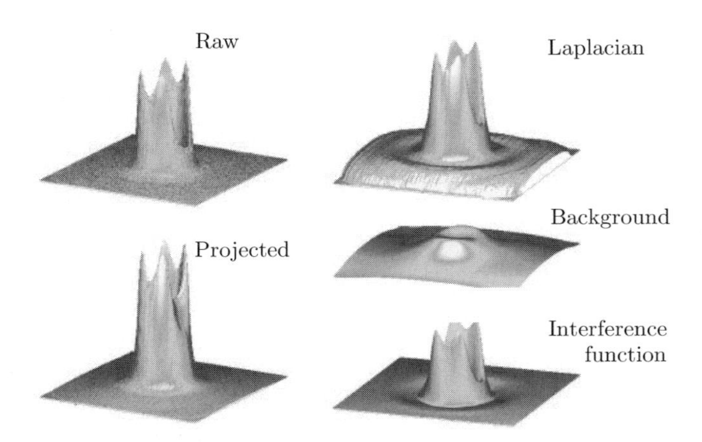

Figure 8. SAXS pattern of a thermoplastic elastomer and the evaluation steps leading to an interference function $G(\mathbf{s})$ of the ideal domain nanostructure. The 2D Fourier transformation of $G(\mathbf{s})$ is the sought CDF $z(\mathbf{r})$

4.5.3. *Structure parameter determination*

The first step in structure parameter determination is the interpretation of the obvious features in the real-space representation of the nanostructure topology. This interpretation leads to a qualitative model describing the nanostructure. Based on this description, a mathematical model may be set up and fitted to the data. Such fitting can be sometimes replaced by the direct determination of parameters of physical meaning. An example concerning the analysis of the transverse structure is presented in the sequel.

5. Nanostructure evolution and processes observed with condensation TPEs

5.1. *Poly(ether ester)s during straining*

5.1.1. *Transverse contraction of destroyed microfibrils*

From a drawing series of the material Arnitel® E2000/60, the equatorial scattering was extracted and projected onto the transverse plane, as discussed in the theoretical section. The resulting curves of $\{I\}_2(s_{12})$ are presented in Figure 9a. After background correction, the 2D chord distribution $g_2(r_{12})$ was computed [64]. The respective curves are shown in Figure 9b. When such equatorial scattering was first observed [3], it was attributed to the existence of elongated microvoids. In our PEE samples, the formation of these elongated domains is preceded by a destruction of hard domains and growth of a long period reflection, which merges with the primary beam before the equatorial scattering emerges [64]. This peculiarity led to the assumption that the observed equatorial scattering is most likely originating from elongated soft domains ("soft

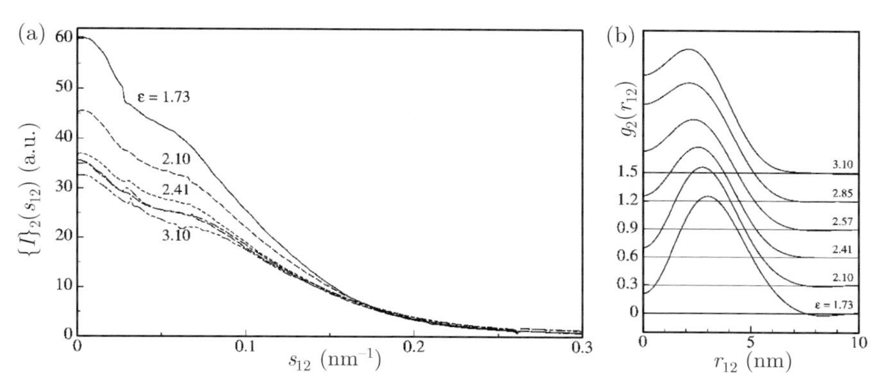

Figure 9. (a) Transverse scattering from Arnitel® E2000/60; (b) chord distributions of Arnitel® E2000/60 computed from the curves in (a) [64]

needles"), which are the product of destruction of hard domains. These elongated soft domains should thus contain almost equal parts of soft and hard segments, according to the chemical composition of the poly(ether ester).

In good approximation, the 2D chord distributions are positive everywhere (Figure 9b), *i.e.,* the correlation among needle cross sections is negligible. Assuming additionally circular cross sections, it follows that $g_2(r_{12})$ is represented by the chord distribution of the unit circle, Mellin-convoluted [32] with a domain diameter distribution $h_D(D)$, the latter describing the nanostructure as a function of the macroscopic load of the specimen. Because of the favorable mathematical properties of Mellin convolution, parameters can be computed directly from $g_2(r_{12})$ by moment arithmetics. In this way, the average needle diameter, $\bar{D}$, the relative width of the needle diameter distribution, $\sigma/\bar{D}$, and the total cross section of the needles with respect to the total cross section of the fiber were computed [64].

From the data shown in Figure 10a, it is obvious that the mean diameter of the needle-shaped domains decreases almost linearly with increasing elongation, whereas for rubber elastic behavior one should expect a decrease according to $\bar{D}(\varepsilon) = \bar{D}_0/\sqrt{\varepsilon+1}$. As depicted in Figure 10b, this is because the disk diameter distribution changes its shape.With increasing elongation, more and more thin needles are emerging and causing the average diameter to decrease considerably. Extrapolating linearly towards $\varepsilon = 0$, one finds a hypothetic average initial diameter $\bar{D}_0 = 4.8$ nm of the soft domain needles. The relative width parameter of the disk diameter distribution, $\sigma/\bar{D}(\varepsilon)$, hardly increases. The total needle cross section per fiber cross section becomes constant for elongations $\varepsilon > 2.5$, after a considerable decrease in the range $1.7 < \varepsilon < 2.5$. This decrease indicates strain hardening of the soft needles; during the straining process, the soft material of the needles is compressed in transverse direction with respect to the surrounding matrix material. A more comprehensive discussion of the theoretical background and the results can be found in the original paper [64].

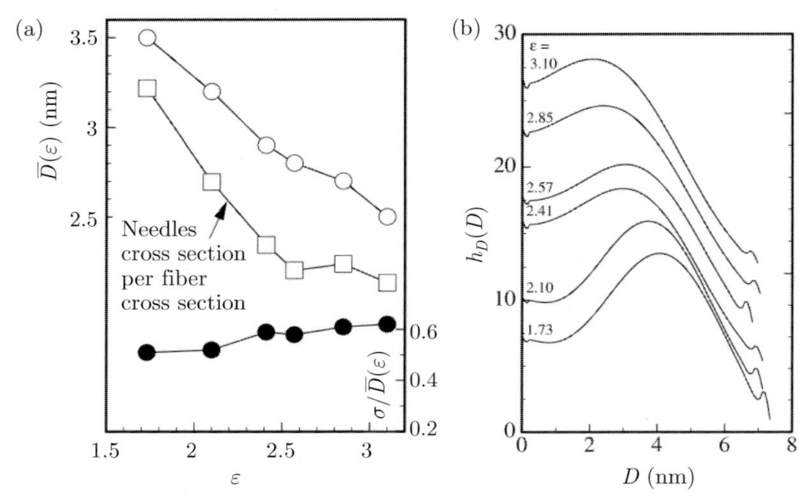

Figure 10. (a) Characterization of the ensemble of needle-shaped soft domains in Arnitel® E2000/60 as a function of elongation, ε; (b) diameter distributions, $h_D(D)$, of soft-domain needles computed by numerical Mellin deconvolution of the curves shown in Figure 9b

5.1.2. *Evolution of hard and soft domain heights*

In an experiment, PEE samples underwent a series of draw/relaxation cycles with increasing final elongation [34]. SAXS longitudinal projections from sample PEE 1000/43 as dependent of elongation are shown in Figure 11a, together

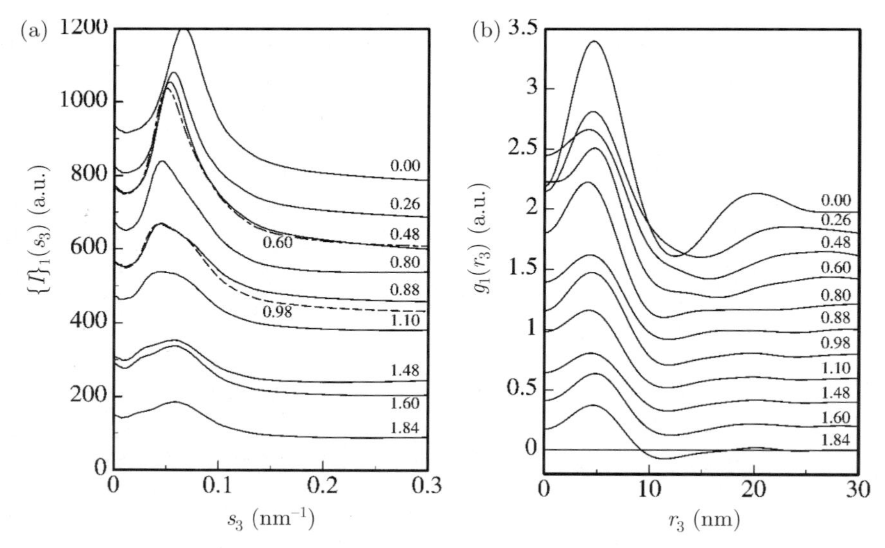

Figure 11. PEE 1000/43: (a) 1D scattering curves $\{I\}_1(s_3)$ obtained by projection of scattering patterns onto the fiber direction; the elongations, ε, are given to the curves; (b) IDFs $g_1(r_3)$ from longitudinal scattering curves $\{I\}_1(s_3)$

with the interface distributions in the real space, $g_1(r_3)$ (Figure 11b). For quantitative data analysis, these curves are fitted to a nanostructure model [33], which unifies the two most frequently discussed models for 1D statistics, *i.e.*, stacking statistics [28] and "homogeneous long period distribution" [30,65]. It turns out that whenever a second long period is observed in the original scattering patterns (Figure 13a), the fit is possible only if a two-component model is used.

In each component of the model, the nanostructure is described by two domain height distributions $h_h(r_3)$ and $h_s(r_3)$, characterized by their centers of gravity ($\overline{d}_h$ and $\overline{d}_s$, respectively) and their widths ($\sigma_h/\overline{d}_h$ and $\sigma_s/\overline{d}_s$, respectively). The skew of the distributions or the heterogeneity of the stacked domains is considered by a global parameter σ_H. Obviously, $\overline{d}_h$ and $\overline{d}_s$ are the respective average heights of the hard and soft domains, and suitable visualization of the parameters as a function of elongation, ε, should help to gain a better understanding of the structural changes during sample straining.

A plot of the average domain heights *vs.* elongation is shown in Figure 12a. Although the determination of soft and hard domain heights is ambiguous in principle, the interaction of the different domain height distributions with external strain provides sufficient information to identify each distribution. Then, as a first result, the volume fraction of soft domains in the original sample,

$$\phi_s = \frac{\overline{d}_s}{\overline{d}_h + \overline{d}_s}, \tag{10}$$

can be computed from its average hard domain height, $\overline{d}_h = 7.4$ nm, and its average soft domain height, $\overline{d}_s = 5.2$ nm. The computed value of $\phi_s = 0.41$ is close to the hard-to-soft segment ratio of the polymer. Such an agreement cannot be expected since there are several structural features discussed in the literature in order to predict or to explain deviations [66].

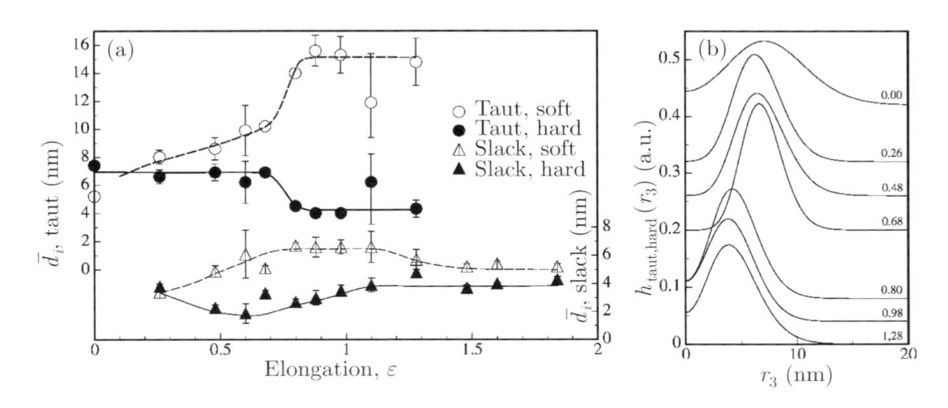

Figure 12. PEE 1000/43: (a) average domain heights of taut (top) and slack (bottom) microfibrillar components as determined from fits of $\boldsymbol{g}(r_3)$; (b) height distributions of the hard domains in the taut microfibrillar component

At medium elongation, one observes two kinds of microfibrils. The taut component (Figure 12a, top) interacts elastically with the external strain while the slack component (Figure 12a, bottom) collects "garbage" from microfibrils, which are no longer connected to the surrounding elastic network and remain in the relaxed state, regardless of whether mechanical load is applied or not.

At low elongations in the taut microfibrillar component, soft domain heights start to grow continuously as a function of external elongation. The internal elongation of these soft domains is much higher than the externally applied elongation because of the rigid nature of the hard domains filling the elastic network. Nevertheless, the average long period

$$L = \overline{d}_h + \overline{d}_s \tag{11}$$

of the taut component increases slower than the external elongation, ε. This finding reflects the known fact that PEE polymers are far from an ideal elastic material. Pullout of taut tie molecules from hard domains and a collapse of the hard domains observable at $\varepsilon \approx 0.8$ are characteristic of the straining process in such polymers.

Figure 12b shows that not only the average domain heights, but also the domain height distributions are considered in the data evaluation. Here, the height distributions $h_h(r_3)$ of the hard domains in the taut microfibrillar component are presented. In the plot of the average domain heights, one could hardly observe any change during the initial stage of straining. Considering the height distributions as a whole, one observes a narrowing, indicating that the domains of medium height are the most stable ones. By the loss of high and tiny hard domains included into the slack component, the height distribution narrows considerably. Thus, a material with a narrow distribution of hard domain heights would probably be a more perfect elastomer.

5.1.3. *CDF analysis: the multidimensional view of nanostructure*

Figure 13 shows two original scattering patterns of PEE 1000/43 in the strained and in the relaxed state. In the strained state (Figure 13a), a "six-point pattern" is seen. During relaxation (Figure 13b), a well separated four-point pattern is observed. After considering the evolution of domain thickness in the straining direction only [34], now the multidimensional CDF analysis is demonstrated [67].

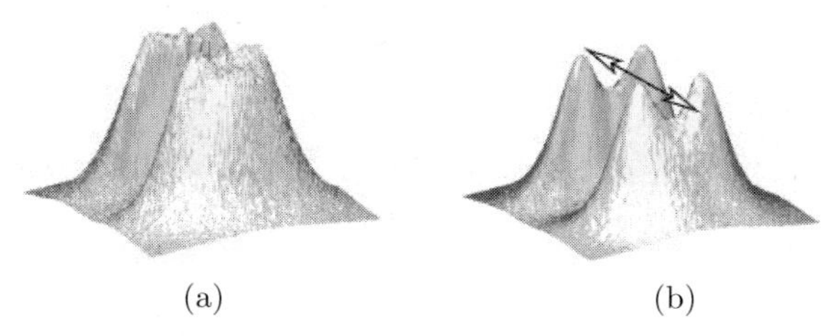

Figure 13. 2D scattering patterns of PEE 1000/43: (a) at $\varepsilon = 0.88$; (b) relaxing from $\varepsilon = 0.88$. The arrow indicates the draw direction. Range $-0.15\ \text{nm}^{-1} \leq s_{12}, s_3 \leq 0.15\ \text{nm}^{-1}$

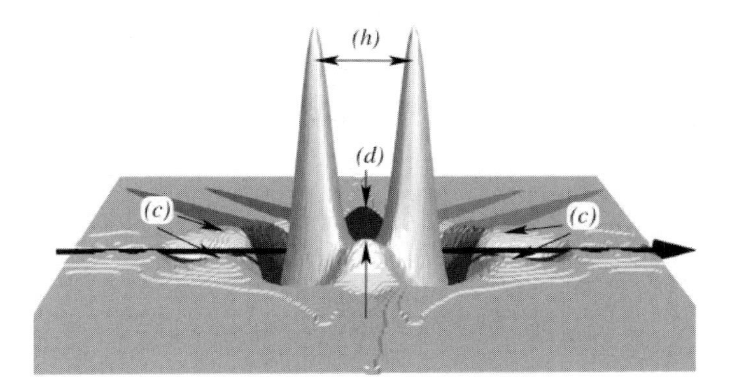

Figure 14. 3D CDF of PEE 1000/43 at $\varepsilon = 0.88$. The "domain peaks" are visible: (h) domain height peaks; (d) domain diameter peaks; (c) interdomain correlation peaks. Range $|r_{12}, r_3| \leq 40$ nm

Figure 14 shows the 3D CDF computed directly from the scattering pattern in Figure 13a by an automated procedure. The draw direction r_3 is indicated by the bold arrow crossing the basal plane. The observer is facing the side from which the domain peaks are turning outward.

The strongest peaks (h) show the correlations between the upper and lower faces of the basic domains. Two meridional peaks (d) indicate the average diameter of the domains. The aspect ratio computed from the positions of the height and diameter peaks indicates that the hard domains are cylinders. Under oblique angle with respect to the fiber axis, four interdomain correlation peaks (c) show that the closest neighbors are not in line with the fiber axis; they form a macrolattice [18] with short-range correlation. These peaks carry a positive sign, because they collect chords starting from the front surface of a cylinder and ending at the back surface of its closest neighbor. The long periods (measured from front surface to front surface of neighboring domains) are indicated by the deepest indentations of the function and are best viewed after turning it upside down (Figure 15).

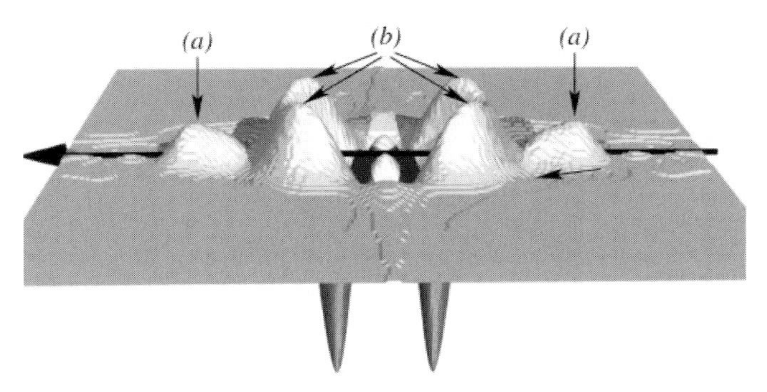

Figure 15. 3D CDF of PEE 1000/43 at $\varepsilon = 0.88$: (a) long period to the next neighbor in the draw direction; (b) stronger long period under oblique angle

It is obvious that the long periods in the draw direction (a) are less pronounced than the long periods observed under an oblique angle (b). The topology shows only short-range correlations among the domains. In the draw direction, there is a long period of 25 nm (a), but the size of the neighboring domain behind its front surface is already undetermined. On the other hand, the movement of the domain in the oblique direction (b) reveals a better defined correlation, because the size of the neighboring domain in the oblique direction is rather well defined (Figure 14, (c)).

After releasing the strain, another scattering pattern was taken. The corresponding 3D CDF is shown in Figure 16. Compared to the strained state, the positions of the long period under oblique angle (b) remain constant, suggesting that the cylindrical domains are surrounded by domains, which are firmly coupled to their center. Such an ensemble has some of the properties of tilted

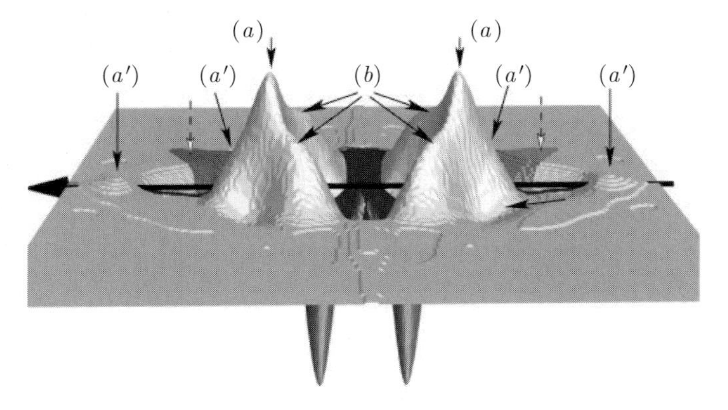

Figure 16. 3D CDF of PEE 1000/43 relaxing from $\varepsilon = 0.88$: (a) the strongest long period in the draw direction (13 nm) (dashed arrow indicates the former position of this peak in the strained state); (a') best correlated long period (17 nm) and its second order; (b) constant long period to the next neighbor under oblique angle

lamellae except for a missing intradomain bridge. If, on the other hand, the hard domains were formed by a composite material comprising both crystalline and amorphous zones, then the contrast between these zones could make them look like different domains firmly coupled to each other, as it is actually observed.

Let us compare the nanostructure in the draw direction after releasing the strain to the morphology in the strained state. The most prominent change is in the long period (a), which has relaxed to half its value in the strained state. Whereas this long period, again, describes a next-neighbor correlation only, now there is a clear set of peaks (a') revealing a microfibrillar component with medium range of order in the draw direction. Thus, this component is an ensemble of several well defined hard domain cylinders arranged in a row and oriented parallel to the draw direction. This is the topological definition of a microfibril.

Discussion of Figure 3. If condensation TPEs are drawn to a sufficient extent, a microfibrillar nanostructure is frequently observed. In Figure 3, this is clearly depicted in the third column at $\varepsilon = 0.88$ and after relaxation. In the respective experiment, films from Arnitel E1500/50 were drawn and relaxed in the synchrotron beam. In principle, the interpretation of the SAXS patterns would be similar to that of the CDFs. Only from the CDF central intense ring (third row) it becomes clear that the shape of the hard domains is a granular one. The arrow tips in the fourth row indicate the second-order microfibrillar long period. In the drawn state, not only the average value is higher, but also the long period distribution is extending far out along the meridian. The formation process of the microfibrils can be deduced from the first and second column. At $\varepsilon = 0.3$, a typical four-point "butterfly" pattern [17] related to stacks from tilted lamellae is observed, whereas the interpretation of the structure from the pattern at $\varepsilon = 0.7$ is ambiguous. In the CDF, granular domains are observed at any elongation. At $\varepsilon = 0.3$, the long periods from the CDF show that lamellae are formed from clusters of these granular domains. At $\varepsilon = 0.7$, we were lucky to "catch" the critical elongation at which the lamellae are about to fail. One observes that there is still a rather well defined distance between the granules, which is related (up to this point) to the elastic response of lamellar stacks. However, there are already many granules torn out from their assemblies, displaced by widely varying amounts (dashed lines) in an oblique direction with respect to the applied force.

Thus, in the TPE, lamellae are assembled from granules. At a critical elongation, granules are separated from each other in a peculiar manner. At higher elongations, the granules form a microfibrillar nanostructure. Finally, more and more of the hard domain granules are completely decomposed and soft domain needles show up.

5.2. *Swollen and drawn poly(ether amide)s*

Gas permeation through a polymer membrane is not only a function of the chemical structure of the polymer chains, but is also determined by a morphology inside the film with typical domain dimensions of several nanometers. Membranes from commercial polyether-*b*-polyamide (PEBA) polymers with varying chemical composition, cast from both *n*-butanol and cyclohexanol are studied by SAXS, in dry form and water-swollen, and as a function of strain. The nanostructure from soft and hard domains is determined [68].

5.2.1. *Dry and water-swollen isotropic materials*

Figure 17 shows some of the measured SAXS curves of unstrained, isotropic films. The curves for water-swollen samples are indicated by dashed lines. Most of them exhibit a long period reflection, which is characteristic of correlation among domains in a multiphase system. Water-swollen materials show a more pronounced long period peak. A considerable shift of the peak position is observed only with the 57PEO/PA6 material.

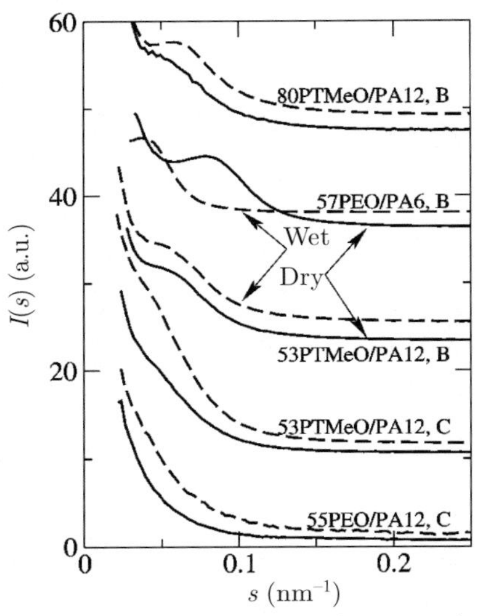

Figure 17. SAXS of neat, isotropic PEBA films, dry (solid lines) and water-swollen for 12 h (dashed lines). Different PEBA grades solvent-cast from n-butanol (B) and cyclohexanol (C), respectively. The curves are shifted for the sake of clarity

In order to quantitatively analyze the nanostructure of isotropic materials, the morphology of a layer system is assumed, the measured data are transformed into $g_1(r)$, and finally fitted by our 1D statistical model [32,33].

The fit results are parameter values describing the lamellar nanostructure in terms of the average layer thicknesses and the widths of the layer thickness distributions in the stacks (Table 2). The association of layer thicknesses to the two phases has been deduced from the known composition of the block copolymer, taking into account that hard domains (polyamide) should be somewhat more perfect in order to keep them stable.

Layer thicknesses, $\bar{t}_E$ and $\bar{t}_A$, of the polyether and polyamide domains, respectively, of the order of 6 nm and average long periods, $\bar{L} = \bar{t}_E + \bar{t}_A$, ranging from 12 nm to 18 nm are determined. For dry films having a balanced composition of hard and soft blocks cast from n-butanol, polyether domain volume fractions, $\phi_E = \bar{t}_E/\bar{L}$, very close to the polyether weight fractions in the block copolymers are computed. After saturation of the materials with water, the average layer thicknesses of 53PTMO/PA12 remain unchanged. Only a tendency to a narrow polyamide layer distribution can be observed. On the contrary, 57PEO/PA6 exhibits a considerable swelling of the lamellar nanostructure. Both soft and hard layers swell by the same factor. Thus, the volume fraction of soft domains is slightly changing.

53PTMO/PA12 cast from cyclohexanol exhibits a nanostructure differing from that observed with the same material when cast from n-butanol. The long period is slightly enlarged by 1 nm, but the volume fraction of the soft phase has increased from $\phi_E = 0.53$ to $\phi_E = 0.60$. The polyamide hard domains

Table 2. Lamellar nanostructure parameters of isotropic PEBA membranes as determined by SAXS IDF analysis

	53PTMO/PA12		57PEO/PA6		70PTMO/PA12	
	Dry	Wet	Dry	Wet	Dry	Wet
			Cast from *n*-butanol			
$\bar{t}_E$ [nm]	6.9(5)	7.0(5)	5.7(2)	10.5(4)	7.2(2)	7.1(2)
$\bar{t}_A$ [nm]	6.0(4)	6.0(3)	4.6(2)	7.5(2)	5.1(1)	5.2(1)
$\sigma_E/\bar{t}_E$	0.59(3)	0.58(4)	0.46(5)	0.44(4)	0.52(3)	0.52(3)
$\sigma_A/\bar{t}_A$	0.44(5)	0.35(4)	0.46(4)	0.47(2)	0.42(2)	0.41(2)
			Cast from cyclohexanol			
$\bar{t}_E$ [nm]	8.2(3)	8.3(3)	5.8(2)	9.1(4)	7.0(2)	7.1(2)
$\bar{t}_A$ [nm]	5.5(1)	5.3(1)	4.1(1)	6.0(2)	5.1(1)	4.8(1)
$\sigma_E/\bar{t}_E$	0.65(8)	0.51(2)	0.51(2)	0.50(4)	0.51(2)	0.51(3)
$\sigma_A/\bar{t}_A$	0.42(7)	0.31(4)	0.45(2)	0.44(4)	0.42(3)	0.42(3)

$\bar{t}_E$, $\bar{t}_A$ – average layer thicknesses of polyether and polyamide phase, respectively; $\sigma_E/\bar{t}_E$, $\sigma_A/\bar{t}_A$ – observed relative widths of the layer thickness distributions of polyether and polyamide phase, respectively. IUCR nomeclature is used for the error of determination, *e.g.*, 0.65(8) means 0.65 ± 0.08

have shrinked. Again, swelling narrows the layer thickness distributions and both the polyether and the polyamide thickness distributions are affected.

70PTMO/PA12 is not a lamellar system. This conclusion can be drawn from several consistency checks [68]. Thus, the "layer thicknesses" reported for this material should not be compared with other materials or with the results of other methods. Their constancy only confirms that the method of preparation does not change the size and arrangement of the domains in this material.

5.2.2. *PEBA nanostructure as a function of strain*

Figure 18 shows the SAXS from the material 80PTMO/PA12 as a function of elongation. Data of the neat material (before straining) are already presented

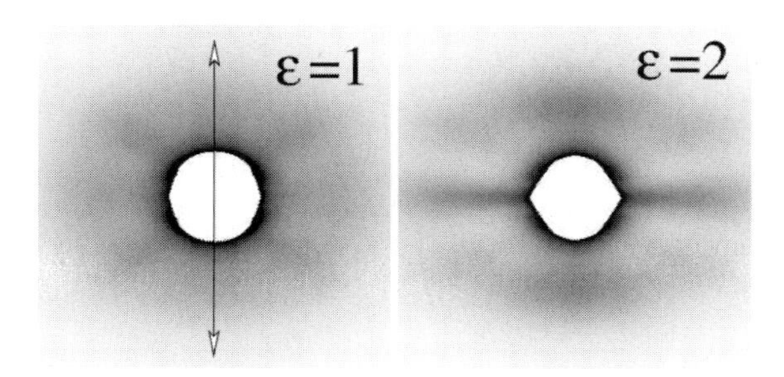

Figure 18. SAXS intensity, $I(s_{12}, s_3)$ ($-0.1\ \mathrm{nm}^{-1} \leq s_{12}, s_3 \leq 0.1\ \mathrm{nm}^{-1}$) as a function of elongation, ε. Water-swollen 80PTMO/PA12 film cast from *n*-butanol. Draw direction (s_3) is indicated

in Figure 17. The two patterns are representative of all samples of this material when the obvious features of the SAXS patterns are concerned. Nanostructure topology is revealed in the CDF $z(r_{12},r_3)$ (Figure 19). Figure 19a shows the CDF viewed from the top. The strong ring-shaped peak ("a-peak") in the center indicates that the basic domain is an almost spherical ellipsoid. The four peaks surrounding the central ring indicate the distributions of chords made from a domain diameter, a gap formed from matrix material and another domain diameter ("aba-peaks"; the different letters in the term "aba" indicate each one of the two phases, respectively). Each pair of these peaks is separated by a deep valley at the meridian caused by a strong and narrow, ridge-shaped long period peak on the meridian of the CDF. Because of the fact that the lateral extension of such a pair does not extend beyond the lateral extension of the central ring-shaped peak, we hesitate to describe this morphology by the

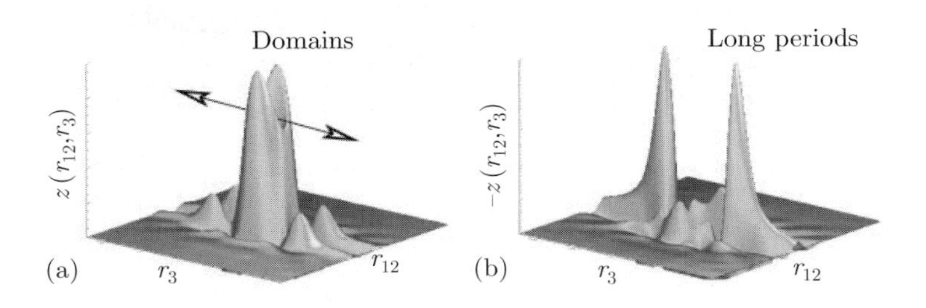

Figure 19. Nanostructure revealed in the multidimensional CDF, $z(r_{12},r_3)$ (-50 nm $\leq r_{12},r_3 \leq$ 50 nm), as computed from SAXS. Water-swollen 80PTMO/PA12 film cast from n-butanol, at elongation $\varepsilon = 2$. The draw direction is indicated. (a) CDF viewed from the top; (b) CDF turned upside down

concept of a central sphere surrounded by four neighbors, as in Section 5.1.3.

Instead, it appears more probable that we observe a structure comprising two components with different ranges of order. A macrolattice of ellipsoids in a matrix forms the component with a longer range of order. The component with a shorter range of order is a distorted nanofibrillar system extending along the straining direction. The lateral extension of its correlation volume [69] is rather narrow and causes the broad aba-peaks to split about the meridian.

The plot on the right hand side of Figure 19 shows the CDF viewed from the bottom; long period peaks are observed and a considerable skew of the long period distribution is found. Most of the chords made from one domain thickness plus one matrix thickness (ab-peaks) are rather short, but there are considerable amounts of such distances, which are much longer. Further long period peaks are observed, which are not placed on the meridian and indicate a macrolattice [18]. Their shape is less skewed.

Figure 20 represents elevations of $z(r_{12},r_3)$ with respect to its zero level as a function of elongation and water content for the samples cast from n-butanol. The respective plots for films cast from cyclohexanol look similar. The plots

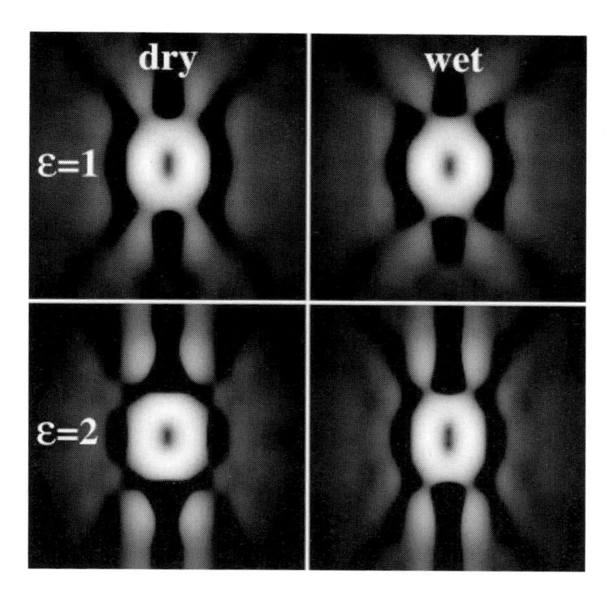

Figure 20. Positive peaks ("domains", logarithmic scale) of the multidimensional CDF $z(r_{12},r_3)$ (-40 nm $\leq r_{12},r_3 \leq 40$ nm) as a function of elongation and water content of 80PTMO/PA12

at $\varepsilon = 1$ suggest that water-swollen samples show a higher lateral correlation among the domains than the dry materials. At $\varepsilon = 2$, the lateral correlation for both samples is low and the contour plots are similar for both the dry and the water-swollen sample. More significant differences are found in the corresponding "negative" contours, marking depressions in $z(r_{12},r_3)$ below its zero level, as depicted in Figure 21. At $\varepsilon = 2$, one still observes some lateral correlation among the ellipsoids, although the macrolattice now appears to be somewhat distorted as well. A detailed discussion of the nanostructure can be found in the original paper [68].

According to the results of nanostructure analysis, the strained material 80PTMO/PA12 contains slightly anisotropic hard domains of approximately 8 nm height and 6 nm diameter. The variation of the SAXS as a function of elongation indicates a moderate destruction of hard domains as a result of straining. There is a strongly microfibrillar component, in which domains are aligned in the draw direction. Its most probable long period is 17 nm and does not change as a function of strain. Instead, the shape of the distribution is changed, indicating that a majority of the mesh lengths resist, whereas an increasing minority is extended by varying amounts. Additionally, a considerable amount of hard domains shows correlation with domains perpendicular to the direction of strain. These correlated assemblies form a distorted macrolattice with a lattice constant of *ca.* 13 nm in the equatorial direction at $\varepsilon = 1$, which is decreasing to 11 nm upon straining to $\varepsilon = 2$. Samples cast from both *n*-butanol and cyclohexanol show almost the same two-phase topology. Swelling in

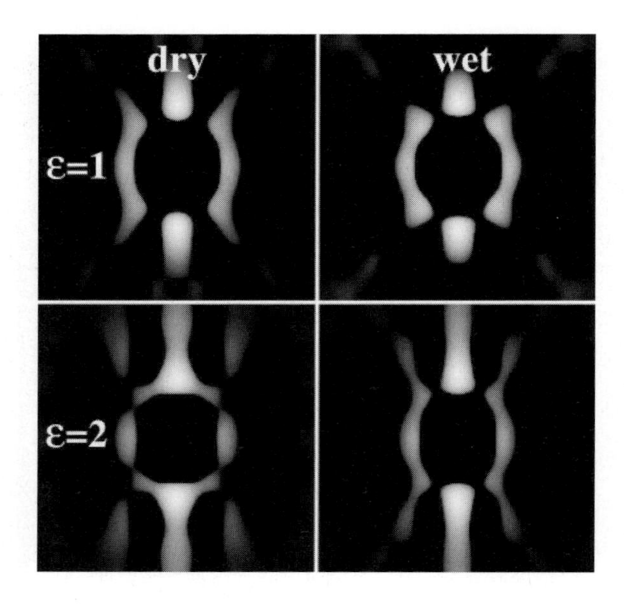

Figure 21. Negative peaks ("long periods", logarithmic scale) of the multidimensional CDF $z(r_{12},r_3)$ (-40 nm $\leq r_{12},r_3 \leq 40$ nm) as a function of elongation and water content of 80PTMO/PA12

water causes small changes, but only as far as the hard domain size is concerned. The inherent inhomogeneity of the strained nanostructure is decreased in the water-swollen samples (commercial poly(ether-*b*-amides) (PEBAX®) are discussed also in Chapter 11).

Acknowledgement

The support of HASYLAB, Hamburg, within the frame of Project II-01-041 and of GKSS, Geesthacht, within the frame of Cooperation Project 6-T3-00-G-01–HS 1 is gratefully acknowledged.

References

1. Spontak R J and Patel N P (2000) Thermoplastic elastomers: fundamentals and applications, *Curr Opin Colloid Interface Sci* **5**:334–341.
2. Polizzi S, Stribeck N, Zachmann H G and Bordeianu R (1989) Morphological changes in SBS block copolymers caused by oil extension as determined by absolute small angle x-ray scattering, *Colloid Polym Sci* **267**:281–291.
3. Statton W O (1962) Microvoids in fibers as studied by small-angle scattering of X-rays, *J Polym Sci* **58**:205–220.
4. Statton W O (1968) Crystallographic studies of synthetic fibers, *Z Kristallogr* **127**:229–260.
5. Peterlin A (1972) Morphology and properties of crystalline polymers with fiber structure, *Text Res J* **42**:20–30.

6. Peterlin A (1971) Molecular model of drawing polyethylene and polypropylene, *J Mater Sci* **6**:490–508.
7. Flory P J (1947) Thermodynamics of crystallization in high polymers. I. Crystallization induced by stretching, *J Chem Phys* **15**:397–408.
8. Andrews E H (1966) Microfibrillar textures in polymer fibers, *J Polym Sci A2 Polym Phys* **4**:668–672.
9. Stribeck N (2003) Nanostructure evolution studies of bulk polymer materials with synchrotron radiation. Progress in method development, *Anal Bioanal Chem*, in press.
10. Grosse-Kunstleve R W, Sauter N K, Moriarty N W and Adams P D (2002) The computational crystallography toolbox: crystallographic algorithms in a reusable software framework, *J Appl Cryst* **35**:126–136.
11. Stribeck N, Sapoundjieva D, Denchev Z, Apostolov A A, Zachmann H G, Stamm M and Fakirov S (1997) Deformation behavior of poly(ether ester) copolymer as revealed by small- and wide-angle scattering of X-ray radiation from synchrotron, *Macromolecules* **30**:1329–1339.
12. Wohlgemuth M, Yufa N, Hoffman J and Thomas E L (2001) Triply periodic bicontinuous cubic microdomain morphologies by symmetries, *Macromolecules* **34**:6083–6089.
13. Ruland W (1971) Small-angle scattering of two-phase systems: determination and significance of systematic deviations from Porod's law, *J Appl Cryst* **4**:70–73.
14. Wolff T, Burger C and Ruland W (1994) Small-angle X-ray scattering study of lamellar microdomains in diblock copolymers in the weak segregation regime, *Macromolecules* **27**:3301–3309.
15. Santa Cruz C, Stribeck N, Zachmann H G and Bálta Calleja F J (1991) Novel aspects in the structure of poly(ethylene terephthalate) as revealed by means of small-angle X-ray scattering, *Macromolecules* **24**:5980–5990.
16. Stribeck N (2001) Extraction of domain structure information from small-angle X-ray patterns of bulk materials, *J Appl Cryst* **34**:496–503.
17. Bonart R, Bötzl F and Schmid J (1987) Cross interferences in the small-angle X-ray patterns of strained segmented PU elastomers, *Makromol Chem* **188**:907–919.
18. Fronk W and Wilke W (1985) SAXS of partially oriented polymers: model calculations with monoclinic macrolattice, *Colloid Polym Sci* **263**:97–108.
19. Vonk C G (1979) A SAXS study of PE fibers, using the two-dimensional correlation function, *Colloid Polym Sci* **257**:1021–1032.
20. Ruland W (1977) The evaluation of the small-angle scattering of lamellar two-phase systems by means of interface distribution functions, *Colloid Polym Sci* **255**:417–427.
21. Ruland W (1978) The evaluation of the small-angle scattering of anisotropic lamellar two-phase systems by means of interface distribution functions, *Colloid Polym Sci* **256**:932–936.
22. Stribeck N and Ruland W (1978) Determination of the interface distribution function of lamellar two-phase systems, *J Appl Cryst* **11**:535–539.
23. Bonart R (1966) Colloidal structures in strained polymers, *Kolloid Z Z Polymere* **211**:14–33.
24. Stribeck N (2000) Analysis of SAXS fiber patterns by means of projections, *ACS Symp Ser* **739**:41–56.
25. Hosemann R (1962) Crystallinity in high polymers, especially fibres, *Polymer* **3**:349–392.
26. Porod G (1961) Application and results of small-angle X-ray scattering from solid polymers, *Fortschr Hochpolym Forsch* **2**:363–400.
27. Hosemann R and Bagchi S N (1962) *Direct Analysis of Diffraction by Matter*, North-Holland, Amsterdam.

28. Hermans J J (1944) On the influence of lattice distortions on the X-ray diagram, particularly of gels, *Rec Trav Chim Pays Bas* **63**:211–218.
29. Porod G (1951) The small-angle X-ray scattering from densely packed colloidal systems, *Kolloid Z* **124**:83–114.
30. Strobl G R and Müller N (1973) Small-angle X-ray scattering experiments for investigating the validity of the two-phase model, *J Polym Sci B Polym Phys* **11**:1219–1233.
31. Blundell D J (1978) Models for small-angle X-ray scattering from highly dispersed lamellae, *Polymer* **19**:1258–1265.
32. Stribeck N (1993) SAXS data analysis of a lamellar two-phase system. Layer statistics and compansion, *Colloid Polym Sci* **271**:1007–1023.
33. Stribeck N (1993) Complete SAXS data analysis and synthesis of lamellar two-phase systems. Deduction of a simple model for the layer statistics, *J Phys IV* **3**:507–510.
34. Stribeck N, Fakirov S and Sapoundjieva D (1999) Deformation behavior of a poly(ether ester) copolymer. Quantitative analysis of SAXS fiber patterns, *Macromolecules* **32**:3368–3378.
35. Gent A N (1954) Crystallization and the relaxation of stress in stretched natural rubber vulcanizates, *Trans Faraday Soc* **15**:521–533.
36. Grubb D T, Keshav P and Adams W (1991) SAXS of Kevlar using synchrotron radiation, *Polymer* **32**:1167–1172.
37. Hermans P H, Heikens D and Weidinger A (1959) A quantitative investigation on the X-Ray small-angle scattering of cellulose fibers. Part II. The scattering power of various cellulose fibers, *J Polym Sci* **35**:145–165.
38. Ruland W (1969) Small-angle scattering studies on carbonized cellulose fibers, *J Polym Sci B Polym Phys* **28**:143–151.
39. Schurz J, Janosi A, Wrentschur E, Krässig H and Schmidt H (1982) A small-angle X-ray analysis of polyacrylonitrile (PAN) fibers, *Colloid Polym Sci* **260**:205–211.
40. Hentschel M P, Hosemann R, Lange A and Uther B (1987) Small-angle X-ray refraction from metal wires, glass fibers and hard elastic poly(propylene), *Acta Cryst* **A43**:506–513.
41. Stribeck N (1996) Small-angle X-ray scattering functions in the vicinity of zero scattering angle with an application to polymer blends, *Macromolecules* **29**:7217–7220.
42. Thünemann A F and Ruland W (2000) Microvoids in polyacrylonitrile fibers: a small-angle X-ray scattering study, *Macromolecules* **33**:1848–1852.
43. Debye P and Bueche A M (1949) Scattering by an inhomogeneous solid, *J Appl Phys* **20**:518–525.
44. Vonk C G and Kortleve G (1967) X-ray small-angle scattering of bulk polyethylene, *Kolloid Z Z Polymere* **220**:19–24.
45. Kortleve G and Vonk C G (1968) X-ray small-angle scattering of bulk polyethylene. III. Results, *Kolloid Z Z Polymere* **225**:124–131.
46. Vonk C G (1973) Investigation of non-ideal two-phase polymer structures by small-angle X-ray scattering, *J Appl Cryst* **6**:81–86.
47. Strobl G R and Schneider M (1980) Direct evaluation of the electron density correlation function of partially crystalline polymers, *J Polym Sci B Polym Phys* **B18**:1343–1359.
48. Vonk C G and Pijpers A P (1985) An X-ray diffraction study of nonlinear polyethylene. I. Room-temperature observations, *J Polym Sci B Polym Phys* **23**:2517–2537.
49. Guinier A and Fournet G (1955) *Small-Angle Scattering of X-Rays*, Chapman and Hall, London.

50. Méring J and Tchoubar D (1968) Interpretation of the SAXS from porous systems. Part I., *J Appl Cryst* **1**:153–165.
51. Tchoubar D and Méring J (1969) Interpretation of the SAXS from porous systems. Part II., *J Appl Cryst* **2**:128–138.
52. Hauser G, Schmidtke J and Strobl G (1998) The role of co-units in polymer crystallization and melting: new insights from studies on syndiotactic poly(propene-co-octene), *Macromolecules* **31**:6250–6258.
53. Glatter O and Kratky O, Eds. (1982) *Small Angle X-ray Scattering*, Academic Press, London.
54. Baltá Calleja F J and Vonk C G (1989) *X-Ray Scattering of Synthetic Polymers*, Elsevier, Amsterdam.
55. Stribeck N (2002) Utilising spatial frequency filtering to extract nanoscale layer structure information from isotropic small-angle X-ray scattering data, *Colloid Polym Sci* **280**:254–259.
56. Stribeck N (1997) Data analysis of 2D-SAXS patterns with fibre symmetry from some elastomers, *Fibre Diffraction Rev* **6**:20–24.
57. pv-wave Version 7.5 (2001), Visual Numerics Inc., Boulder, Colorado.
58. Haberäcker P (1989) *Digitale Bildverarbeitung*, Hanser, München.
59. Rosenfeld A and Kak A C (1982) *Digital Picture Processing*, Academic Press, London, Vol. 1.
60. Rasband W, Image J – Image processing and analysis in Java, http://rsb.info.nih.gov/ij.
61. Stribeck N, Web Page, http://www.chemie.uni-hamburg.de/tmc/stribeck/.
62. Buhmann M D (2000) Radial basis functions, *Acta Numerica* **9**:1–38.
63. Wang Z, Hsiao B S, Stribeck N and Gehrke R (2002) Nanostructure evolution of isotropic high-pressure injection-molded UHMWPE during heating, *Macromolecules* **35**:2200–2206.
64. Stribeck N (1999) The equatorial small-angle scattering during the straining of poly(ether ester) and its analysis, *J Polym Sci B Polym Phys* **37**:975–981.
65. Stribeck N and Ruland W (1978) Determination of the interface distribution function of lamellar two-phase systems, *J Appl Cryst* **11**:535–539.
66. Zachmann H G and Wutz C (1993) Studies of the mechanism of crystallization by means of WAXS and SAXS employing synchrotron radiation, in *Crystallization of Polymers* (Ed. Dosière M) NATO ASI Ser. C, Vol. 405, pp. 403–414.
67. Stribeck N and Fakirov S (2001) Three-dimensional chord distribution function SAXS analysis of the strained domain structure of a poly(ether ester) thermoplastic elastomer, *Macromolecules* **34**:7758–7761.
68. Barbi V, Funari S S, Gehrke R, Scharnagl N and Stribeck N (2003) SAXS and the gas transport in polyether-block-polyamide copolymer membranes, *Macromolecules* **38**:749–758.
69. Stribeck N and Wutz C (2002) Layer morphology of a poly(ester imide) LCP in different solid states, *Macromol Chem Phys* **203**:328–335.

Chapter 8

Dielectric Relaxation of Polyester-Based Thermoplastic Elastomers

T. A. Ezquerra

1. Introduction

Thermplastic elastomers are a class of block copolymers, in which one segment, usually referred to as "soft", is in the rubbery state and imparts flexibility to the material, while the other, "hard" segment is either semicrystalline or in the glassy state and imparts high elastic modulus, strength and mechanical stability to the material [1,2]. Thermoplastic elastomers are of interest for a great variety of industrial processes, such as extrusion, injection molding, and composite preparation. Generally, the chemical and physical characteristics of an individual segment in a block copolymer have profound effects on both the phase behavior and the structure of the resulting microphases (phase behaviour is discussed in Chapter 6). For this reason, it is important to well understand the relationships between physical properties, chemical composition, and microstructure. Various mechanisms, taking place in the amorphous phase, can lead to energy dissipation in polymeric materials [3]. At temperatures below the glass transition, T_g, the mechanical energy can be dissipated by the so-called *secondary relaxation*, which involves local motions of the macromolecular chain [3,4]. Additionally, polymers exhibit a *primary relaxation* above T_g, involving segmental motions [4–6]. Typically for the thermoplastic elastomers in the temperature region of technological interest, both relaxation processes coexist or eventually merge, thus making the study of the relaxation behavior relatively complex. About the synthesis of these materials, see Chapters 2 and 3.

2. Study of the relaxation behavior by means of dielectric spectroscopy

Dielectric spectroscopy is a technique developed to measure the complex dielectric permittivity ($\varepsilon^* = \varepsilon' - i\varepsilon''$, where $i = \sqrt{-1}$) as a function of both the frequency of the exciting field and the temperature [4–6]. The principles of application of this technique have been recently reviewed [4]. The real part of ε^* (ε') corresponds to the dielectric constant and is associated to the energy stored in the material through polarization. The imaginary part, ε'', is related to the energy dissipated in the medium and therefore is frequently referred to as dielectric loss [4–6]. When dielectric spectroscopy is used to study the molecular motion in polymeric materials, the frequency of interest typically covers the range from 10^{-2} to 10^9 Hz [4–6]. Within the $10^{-2} < F/\text{Hz} < 10^6$ range, ε^* measurements can be performed by using impedance or frequency response analyzers and lock-in amplifiers. In this case, thin films with circular gold metallic electrodes on both free surfaces (typically 3 to 4 cm in diameter) are prepared and placed between two metallic electrodes, building up a capacitor. For frequencies between 10^6 and 10^9 Hz, this "sandwich geometry" is inappropriate and reflectometer techniques are required. Here, ε^* can be obtained from reflection coefficient measurements [4–7]. To obtain the temperature dependence of ε^*, the dielectric cell including the sample and electrodes is introduced in a cryostat operating under controlled temperature conditions. Polymers may exhibit a great variety of molecular motions, which can be investigated by dielectric spectroscopy (DS), provided that dipolar groups are involved. In general, below T_g, homopolymers exhibit the occurrence of local motions, giving rise to relaxations. Above T_g, segmental mobility extended to several molecular units appears, causing the primary relaxation associated to the glass transition temperature. Block copolymers present a more complicated dynamic scenario because relaxations above and below T_g associated to each block building up the copolymer may occur. Dielectric spectroscopy is also considered in Chapter 14.

3. Dielectric spectroscopy of poly(ether ester) thermoplastic elastomers

Among poly(ether ester) (PEE) thermoplastic elastomers, special attention has been devoted to the block copolymers with poly(butylene terephthalate) (PBT) as hard segments and poly(tetramethylene oxide) (PO4) as soft segments [1,2,8–18]. Such materials are, *e.g.*, Hytrel® (DuPont), Arnitel® (DSM) or Elitel® (Elana). The ability of the PBT segments to crystallize controls the appearance of microphase-separated crystalline hard segment domains. The resulting semicrystalline copolymers may combine the properties of vulcanized rubbers and thermoplastics [1,2]. A general scheme of these systems is given in Figure 1a. A typical molecular weight of the tetramethylene oxide moiety is 1000 g/mole. The dielectric behavior of these block copolymers has been the subject of several investigations [8–18].

In order to illustrate the general relaxation behavior, Figure 2 shows isochronal plots of ε'' and ε' as a function of temperature at different frequencies for a PBT-PO4 copolymer [12] with 50 wt% PBT and a polymerization degree

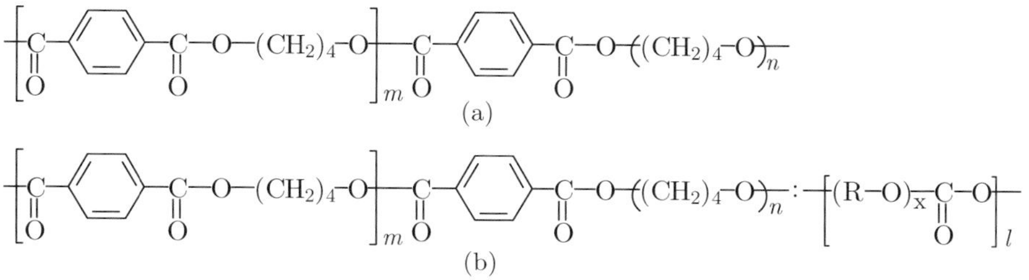

Figure 1. (a) Scheme of a poly(ether ester) thermoplastic elastomer based on PBT (hard block) and PO4 (soft block); (b) scheme of a poly(ester ether carbonate) terpolymer based on PBT, PO4, and aliphatic polycarbonate

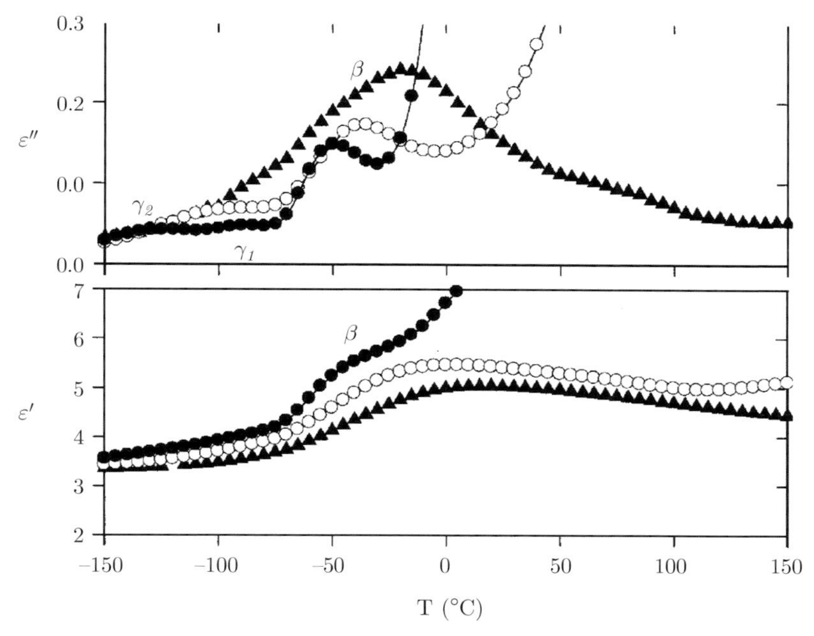

Figure 2. Temperature dependences of (a) ε'' and (b) ε' for a PBT-PO4 block copolymer (50/50 by wt) at various frequencies: ● – 10 Hz, ○ – 10^3 Hz, and ▲ – 10^5 Hz

of $m = 5$. The most intense relaxation (observed at –50°C and 10 Hz) appears as a broad maximum in the dielectric loss, ε'', measurements and as a step in the dielectric constant, ε', measurements. The process corresponds to large-scale motions taking place above the glass transition temperature in the amorphous phase of the copolymer [9–13]; this relaxation is currently referred to in the literature as β. The β relaxation depends on the hard segment concentration, as it becomes clear from Figure 3, showing dielectric data presented as $\tan\delta = \varepsilon''/\varepsilon'$ *vs.* temperature for PBT-PO4 copolymers. In accordance with the Fox-Flory expression [10], the temperature at which the β relaxation takes place shifts toward higher values with the rise in PBT content. At lower temperatures,

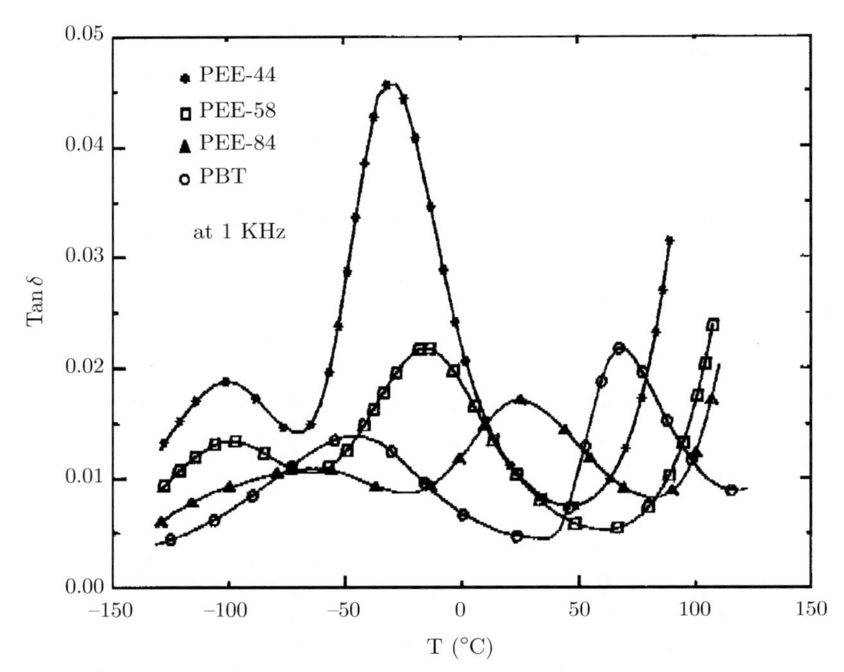

Figure 3. Temperature dependences of $\tan\delta = \varepsilon''/\varepsilon'$ for a series of PBT-PO4 block copolymers at 10^3 Hz. The numbers indicate the hard segment weight fraction [10]

PBT-PO4 block copolymers exhibit two relaxations, clearly detectable in Figure 2 at 0.1 Hz (γ_1 and γ_2 in order of decreasing temperature). These processes correspond to the local molecular motions, which are due to the –COO ester group and to the ether group present in the PO4 [8,9]. Due to the higher rigidity of the ester group, as compared to that of the ether group, the local relaxation of the ester is expected to take place at lower frequencies, *i.e.*, at higher temperatures [9], giving rise to γ_1. At lower temperatures, γ_2 can be attributed to the ether group present in the soft segment. As frequency increases, the maximum loss peaks for both relaxations shift toward higher temperatures and merge together above 10 Hz. The large increase of ε'' at higher temperatures for low frequencies is characteristic of semicrystalline polymers and is due to direct current (DC) conductivity [4–6].

4. Dielectric spectroscopy of multiblock thermoplastic elastomers

Based on PBT, a great variety of thermoplastic elastomers including multiblock terpolymers can be generated. The modification of the nature of the soft and hard segments has a tremendous effect on the relaxation behavior [1,2]. Figure 1b shows a general formula representing a poly(ether ester carbonate) terpolymer (PEEC). The dielectric loss values for three PEEC terpolymers with different PBT/PO4/polycarbonate (PC) weight ratios are shown in Figure 4. The dielectric behavior of 60/40/0 reveals the existence of a broad β relaxation at $\approx 0°C$ for 10 kHz. The β process is accompanied by a lower-temperature γ relaxation, which is observed as a shoulder at $\approx -90°C$ and 10^3 Hz. As the frequency is increased, the maximum loss peaks for both relaxations shift toward higher temperatures. The 60/0/40 copolymer also presents two relaxation processes, β and γ (in order of decreasing temperature). In this case, the β and γ relaxations appear at higher temperatures than those for the 60/40/0 copolymer. Both relaxation maxima are clearly separated from each other at lower frequencies (1, 10, and 100 kHz). The effect of changing the amount of PC and PO4 segments on the dielectric relaxation of the terpolymers is illustrated in Figure 4 for the 60/20/20 terpolymer. The maxima in ε'' corresponding to both β and γ processes shift toward higher temperatures as the amount of PC segments increases and that of PO4 segments decreases.

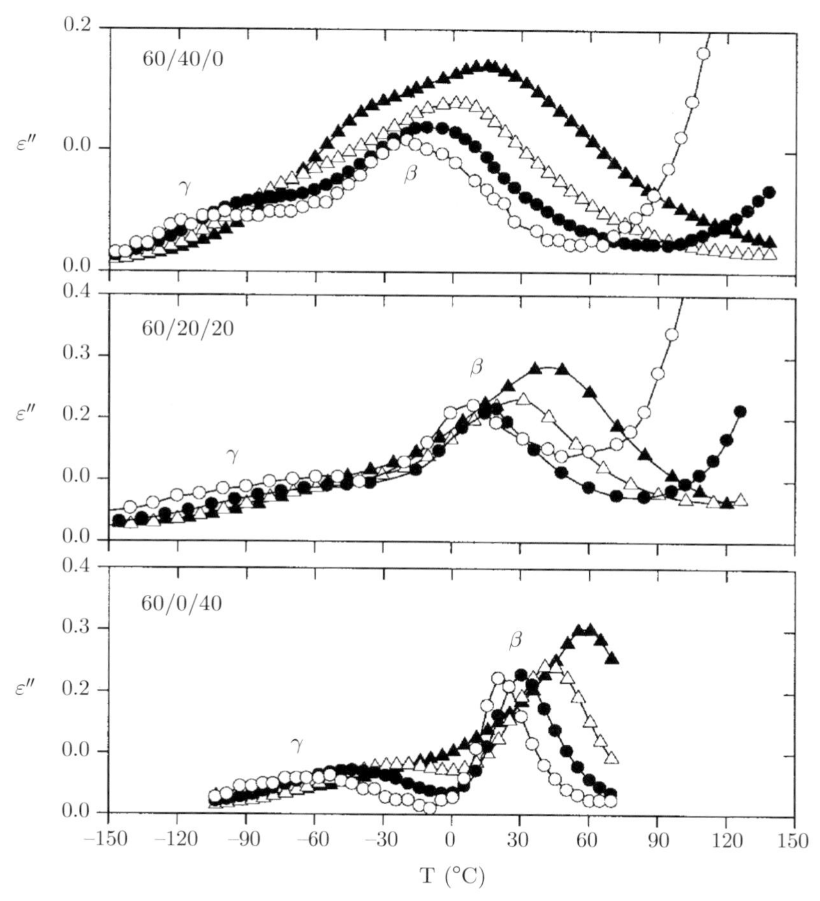

Figure 4. Temperature dependences of ε'' for different PBT/PO4/PC terpolymers at various frequencies: ○ – 10^3 Hz, ● – 10^4 Hz, △ – 10^5 Hz, and ▲ – 10^6 Hz

The appearance of the β process shifts toward higher temperatures with increasing PC content. These observations suggest that the amorphous phase in the terpolymers partially consist of a mixture of PO4 and PC soft segments, and PBT hard segments giving rise to a single glass transition [10–14].

5. Relaxation behavior of poly(ester carbonate) block copolymer across the melting region

Of special interest is to characterize the changes in the relaxation behavior when the elements imparting mechanical stability to the material are modified or eventually destroyed. This can happen, *e.g.*, when the melting temperature of the crystalline microdomains is approached. Block copolymers based on PBT and PC exhibit a larger melting point depression and lower crystallinity than the equivalent PEE block copolymers based on PBT and PO4 [13,15]. This is mainly due to the Flory-Huggins interaction parameter, which is expected to be lower for PBT-PC than for PBT-PO4 on the basis of the stability parameters [16]. The interaction between the hard and soft segments slows down the kinetics of crystallization and decreases the melting temperature when the hard segment concentration decreases. This is illustrated in Figure 5, which shows the differential scanning calorimetry (DSC) heating traces of a series of block copolymers based on PBT and an aliphatic polycarbonate (PC, Duracarb 1.2 from PPG Inc., USA, $\bar{M}_n = 1000$, $\rho = 1.076$ g/cm^3), where the hard segment (PBT) content was varied from 100 to 40 wt%. It is seen that all copolymer

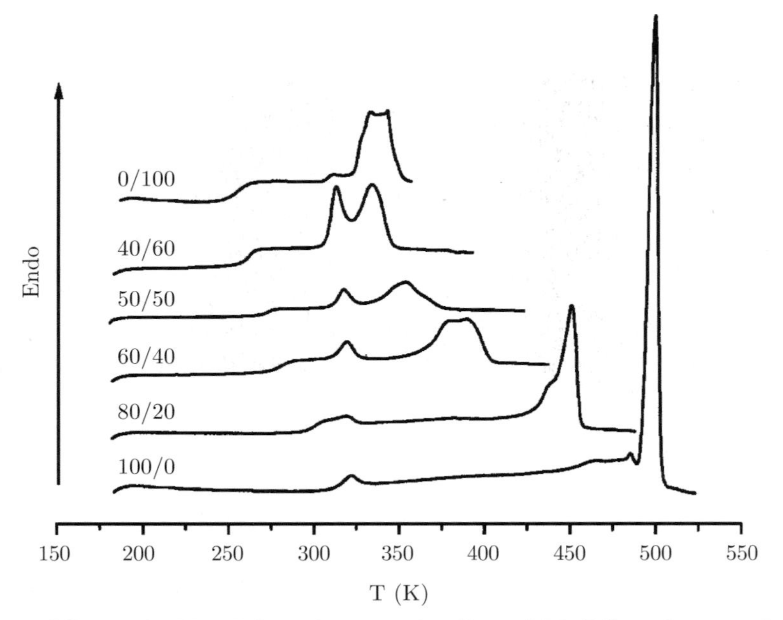

Figure 5. DSC scans for PBT-PC copolymers with different PBT/PC weight ratios; heating rate 20 K min^{-1}

samples show a single glass transition temperature, as revealed by a step-like transition at low temperatures [17]. The value of T_g decreases by decreasing the PBT content. The same holds also for the final melting temperature due to the limited size of the crystallites at lower PBT contents [17].

The ε'' values measured for neat PBT, 40/60, and 60/40 (by wt) PBT/PC copolymers as a function of frequency at three temperatures ($T > T_g$) are shown in Figure 6. The dielectric spectra of all samples reveal the existence of two

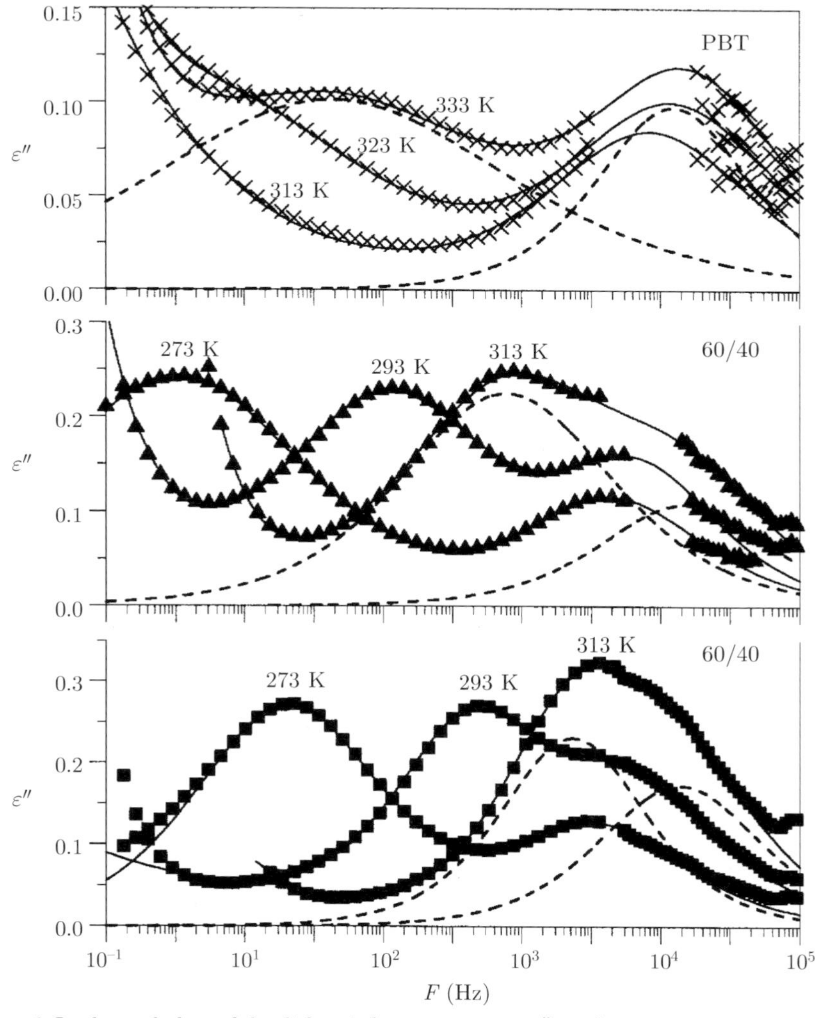

Figure 6. Isothermal plots of the dielectric loss component, ε'', *vs.* frequency, F, in the merging region for neat PBT and two PBT/PC copolymers (weight ratios indicated in the figure). The solid lines represent the results of fitting experimental data to the sum of two Havriliak-Negamy equations (Eq. 1) with a conductivity term. The dashed lines at the highest temperature show the separate contributions of β and γ processes

dielectric processes at lower frequencies: the β relaxation associated with the glass transition and, at higher frequencies, a faster secondary process, which is assumed to be γ relaxation [13,17]. The frequency of maximum loss, F_{max}, for both processes shifts toward higher frequencies with the rise in temperature, and both processes merge at high temperatures in all cases. In addition, at lower frequencies, a strong increment of ε'', associated with an electrical conductivity, is observed. It should be noted that for both relaxations the intensity of ε'' in the neat PBT is lower than that of the copolymers, which is due to its higher degree of crystallinity. In order to determine the specific effect of crystal melting on the relaxation behavior of the 40/60 PBT/PC copolymer, the dielectric measurements for this sample were extended over and above the melting region (Figure 7). In this case, both relaxations remain within the experimental frequency window even in the molten state. The phenomenological Havriliak-Negamy (HN) equation was used to describe the frequency dependence of the complex dielectric constant in the vicinity of dipolar absorption [4,17]

$$\varepsilon^* = \varepsilon_\infty + \frac{\varepsilon_0 - \varepsilon_\infty}{\left[1 + \left(i\omega\tau_{HN}\right)^b\right]^c} \quad (1)$$

where $\omega = 2\pi F$ is the angular frequency, ε_0 and ε_∞ are the relaxed ($\omega = 0$) and unrelaxed ($\omega = \infty$) values of the dielectric constant, τ_{HN} is the central relaxation time of the time distribution function, and b and c $[0 < (b,c) < 1]$ are shape parameters describing the symmetric and asymmetric broadening of the relaxation time distribution function.

An additional contribution of the conductivity is taken into account by adding the term $-i\sigma/(\varepsilon_{vac}\omega^s)$ to Eq. (1). Here, σ is related to the direct current (DC) electrical conductivity, ε_{vac} is the dielectric constant in vacuum and the value of the coefficient $0 < s < 1$ depends on the conduction mechanism [4]. Above T_g, a sum of two HN equations was used to describe both β and γ relaxations, and a symmetric shape was obtained from the fitting procedure ($c = 1$). The dashed lines in Figures 6 and 7 exemplify the separate contributions of both processes at the higher temperatures.

In Figure 8, the relaxation time obtained from the Havriliak-Negami fitting, τ_{HN}, for both β and γ processes is represented as a function of the reciprocal temperature for the different samples. The temperature dependence of the relaxation times for the γ process follows an Arrhenius-type behavior, which is typical of relaxations corresponding to local motions [4]. No change in the apparent activation energy value ($E_a \approx 45$ kJ mol^{-1}) with PBT content in the copolymer is observed. The β process exhibits a characteristic non-Arrhenius temperature dependence similar to that observed for the relaxation associated with the cooperative segmental motions appearing above the glass transition temperature [4]. In this case, it is known that the temperature dependence of the relaxation time, τ_{HN}, follows a Vogel-Fulcher-Tamman (VFT) behavior [4,17].

The dielectric strengths, $\Delta\varepsilon = \varepsilon_0 - \varepsilon_\infty$, for both relaxations corresponding to the different copolymers are depicted in Figure 9 as a function of a

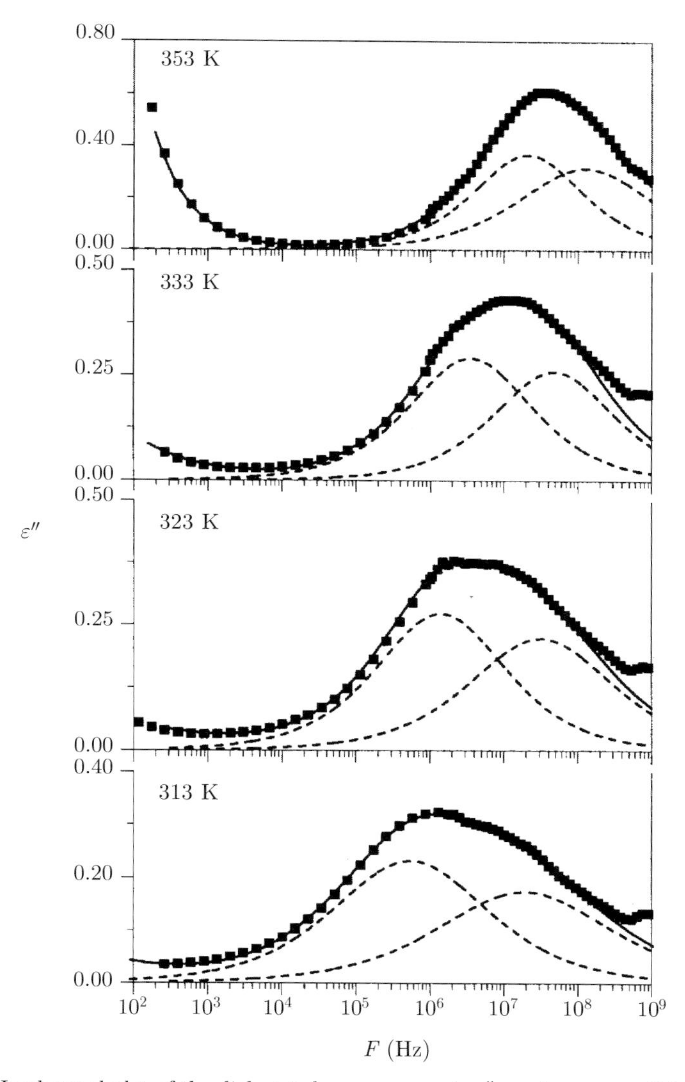

Figure 7. Isothermal plot of the dielectric loss component, ε'', *vs.* frequency, F, in the melting region for a PBT-PC copolymer (40/60 by wt, T_m = 334.3 K). The solid lines represent the results of fitting experimental data to the sum of two HN equations with a conductivity term. The dashed lines show the separate contributions of the β and γ processes

normalized temperature, T_g/T. The dielectric strength is a measure of the amount of relaxing phase [4]. Here, it can be observed that $\Delta\varepsilon_\beta$ for PBT increases with temperature, as frequently observed for semicrystalline polymers [19]. This fact has been explained by a temperature-assisted mobilization of the amorphous phase, which in semicrystalline polymers is dynamically restricted by the presence of crystals. An opposite behavior is observed for

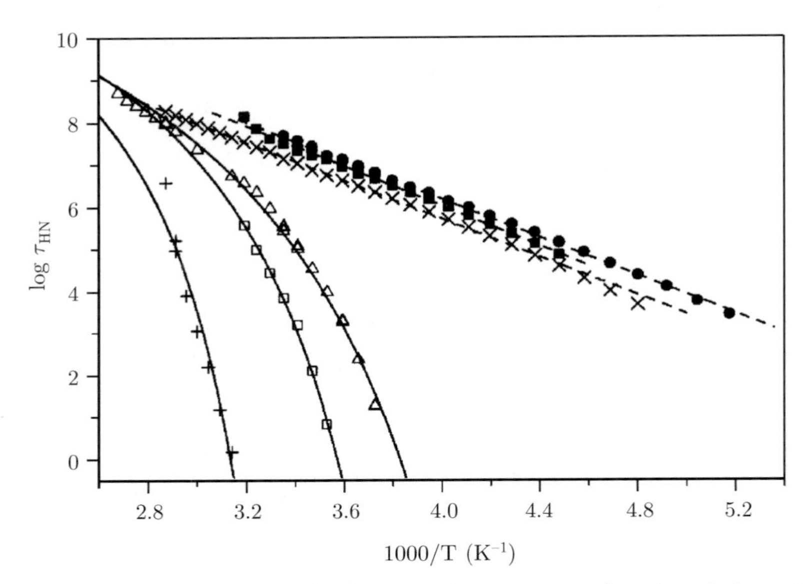

Figure 8. Havriliak-Negami central relaxation time, τ_{HN}, as a function of the reciprocal temperature, $1/T$, for β and γ processes. Experimental data: (▲, △) 60/40 and (■, □) 40/60 (by wt) PBT-PC copolymers; the solid and open symbols represent the β and γ processes, respectively. The symbols $\times$ and $+$ represent β and γ processes for PBT. The solid lines are fittings to the VFT equation

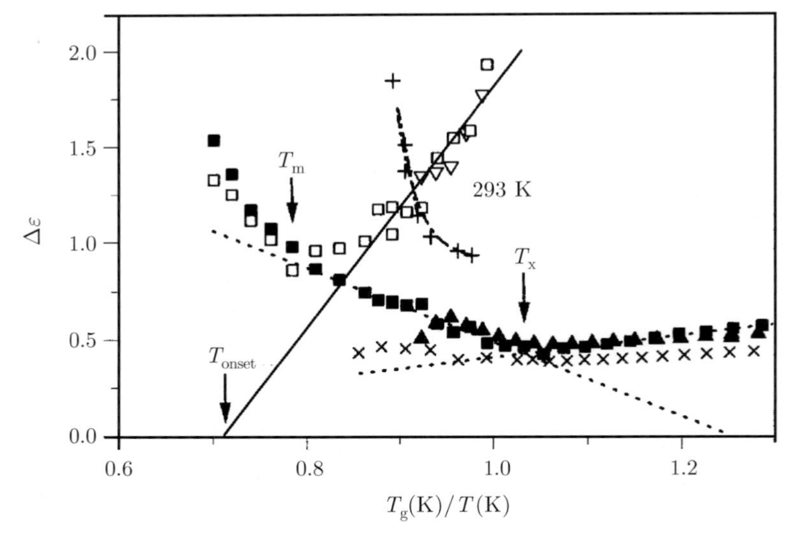

Figure 9. Dielectric strength $\Delta\varepsilon$ obtained from fitting to the HN equation for the β and γ processes as a function of the normalized temperature $T_{\mathrm{g}}(\mathrm{K})/T(\mathrm{K})$. The symbols are the same as in Figure 8. The solid lines are guides for the eye. The T_{g} values are 279.8 K, 261.5 K, and 310.7 K for the 60/40 and 40/60 (by wt) copolymers, and the neat PBT, respectively

the 60/40 and 40/60 (by wt) copolymers. $\Delta\varepsilon_\beta$ of these samples, while semicrystalline, seems to follow qualitatively the Fuoss-Kirkwood equation proposed for amorphous systems [4–6,20]. This can be understood by recalling the fact that in this type of thermoplastic elastomers, the amorphous phase is not as effectively restricted, as it is in semicrystalline homopolymers [21,22]. Above T_g, the dielectric strength of the β process, $\Delta\varepsilon_\beta$, of these copolymers tends to decrease concurrently with an increase in $\Delta\varepsilon_\gamma$. This has been observed in several amorphous polymers [23,24], but not in block copolymers. Moreover, the extrapolation of the $\Delta\varepsilon_\beta$ shown in Figure 9 would indicate a temperature T_{onset} ($\Delta\varepsilon_\beta = 0$) of about 1.40 T_g for the block copolymers. This temperature corresponds to the point where the β and γ relaxations should merge completely and only a common relaxation would be observed [24,25]. However, in these block copolymers, it can be observed that the $\Delta\varepsilon_\beta$ values show a clear departure from the decreasing trend at temperatures above 1.15 T_g.

For the 40/60 (by wt) copolymer, the dielectric measurements in the melting range allow one to establish the influence of crystal melting on the dynamic behavior. The dielectric spectrum for each temperature reveals the presence of both β and γ relaxations above T_m in the measured frequency range. The impact of the crystal melting on the relaxation is reflected by a dramatic change in the trend of $\Delta\varepsilon_\beta$ with temperature. For $T > T_m$ (1.27 T_g), $\Delta\varepsilon_\beta$ undergoes a dramatic increment as a function of temperature. Concurrently, the increase of $\Delta\varepsilon_\gamma$ with temperature becomes stronger. This could be due to an increase in the number of relaxing dipoles. These results, together with the DSC data, suggest a progressive reduction of the crystalline fraction. Therefore, the respective increments of the amorphous phase are expected to be responsible for the observed increment of the dielectric strength for both β and γ relaxations in the melting region.

6. Conclusion

Dielectric spectroscopy was shown to be a powerful technique when dealing with molecular dynamics in thermoplastic elastomers. The combination of dielectric measurements over broad frequency and temperature ranges with a precise structural characterization opens up new possibilities of studying the structure-dynamic relationships in block copolymers (see also Chapter 14).

Acknowledgement

The author cordially thanks the Ministry of Science and Technology (grant FPA2001-2139), Spain, for the generous support of this investigation.

References

1. Legge N R, Holden G and Schroeder H E (Eds.) (1987) *Thermoplastic Elastomers, A Comprehensive Review*, Hanser Publishers, Munich.

2. Baltá-Calleja F J and Roslaniec Z (Eds.) (2000) *Block Copolymers*, Marcel Dekker, New York.
3. Ward I M and Hadley D. W. (1993) *An Introduction to the Mechanical Properties of Solid Polymers*, Wiley, New York.
4. Kremer F and Schönhals A (Eds.) (2002) *Broad Band Dielectric Spectroscopy*, Springer, Berlin.
5. Hedvig P (1977) *Dielectric Spectroscopy of Polymers*, Adam Hilger Ltd., Bristol.
6. Blythe A R (1979) *Electrical Properties of Polymers*, Cambridge University Press, Cambridge.
7. Ezquerra T A, Kremer F and Wegner G (1992) AC Electrical properties of insulator-conductor composites, in *Progress in Electromagnetic Research: Dielectric Properties of Heterogeneous Materials* (Ed. Priou A) Elsevier, Amsterdam, Ch. 7, pp. 273–301.
8. Castles J L, Vallance M A, McKenna J M and Cooper S L (1985) Thermal and mechanical properties of short-segment block copolyesters and copolyether-esters, *J Polym Sci Polym Phys Ed* **23**:2119–2147.
9. North A M, Pethrick R A and Wilson A D (1978) Dielectric properties of phase separated polymer solids: 2. Butanediol terephthalate-poly(tetramethylene oxide terephthalate) copolymers, *Polymer* **19**:923–930.
10. Runt J, Du L, Martynowicz L M, Brezny D M and Mayo M (1989) Dielectric properties and cocrystallization of mixtures of poly(butylene terephthalate) and poly(ester ether) segmented block copolymers, *Macromolecules* **22**:3908–3913.
11. Gallagher K P, Zhang X, Runt J P, Huynh-ba G and Lin J S (1993) Miscibility and cocrystallization on homopolymer-segmented block copolymer blends, *Macromolecules* **26**:588–596.
12. Szymczyk A, Ezquerra T A and Roslaniec Z J (2001) Poly(ether-block-sulfonated ester) copolymers. 2. Mechanical and dielectrical relaxation, *J Macromol Sci–Phys* **B40**:685–708.
13. Roslaniec Z, Ezquerra T A and Baltá-Calleja F J (1995) Dielectric relaxation of poly(ester-ether-carbonate) multiblock terpolymers, *Colloid Polym Sci* **273**:58–65.
14. Fakirov S, Apostolov A A, Boeseke P and Zachmann H G (1990) Structure of segmented poly(ether ester)s as revealed by synchrotron radiation, *J Macromol Sci–Phys* **B29**:379–395.
15. Roslaniec Z (1992) Dynamic and microcalorimetric studies of the phase structure of multiblock elastomers, *Polimery* **37**:328–335 (in Polish).
16. Roslaniec Z (1993) Dynamic mechanical and microcalorimetric studies on the phase structure in blends of two multiblock copolymers: 1. Poly(ether-ester) and poly((ether-carbonate)-urethane) blends, *Polymer* **34**:1249–1255.
17. Alvarez C, Capitán M J, Alizadeh A, Roslaniec Z and Ezquerra T A (2002) Relaxation behavior of poly(ester-carbonate) block copolymer across the melting region, *Macromol Chem Phys* **203**:556–564.
18. Baltá-Calleja F J, Fakirov S, Roslaniec Z, Krumova M, Ezquerra T A and Rueda D R (1998) Microhardness of condensation polymers and copolymers. 1. Coreactive blends of poly(ester ether carbonate) thermoplastic terpolymers, *J Macromol Sci–Phys* **B37**:219–237.
19. Nogales A, Ezquerra T A, Batallán F J, Frick B, López-Cabarcos E and Baltá-Calleja F J (1999) Restricted dynamics in poly(ether-ether-ketone) as revealed by incoherent quasielastic neutron scattering and broad band dielectric spectroscopy, *Macromolecules* **32**:2301–2308.
20. McCrum N G, Read B E and Williams G (1991) *Anelastic and Dielectric Effects in Polymeric Solids*, Dover, New York.

21. Ezquerra T A, Roslaniec Z, López-Cabarcos E and Baltá-Calleja F J (1995) Phase separation and crystallization phenomena in a poly(ester-carbonate) block copolymer: A real-time dielectric spectroscopy and X-ray scattering study, *Macromolecules* **28**:4516–4524.
22. Seymour R W, Overton J R and Corley L S (1975) Morphological characterization of polyester-based elastoplastics, *Macromolecules* **8**:331–335.
23. Bergman R, Alvarez F. Alegría A and Colmenero J (1998) The merging of the dielectric α- and β-relaxations in poly(methyl methacrylate), *J Chem Phys* **109**:7546–7555.
24. Garwe F, Schönhals A, Lockwenz H, Beiner M, Schröter K and Donth E (1996) Influence of cooperative alpha dynamics on local beta relaxation during the development of the dynamic glass transition in poly(*n*-alkyl methacrylate)s, *Macromolecules* **29**:247–253.
25. Casalini R, Fioretto D, Livi A, Lucchesi M and Rolla P A (1997) *Phys Rev B* **56**:3016–3021.

PART III

POLYAMIDE-BASED THERMOPLASTIC ELASTOMERS

Chapter 9

Thermoplastic Poly(Ether-*b*-Amide) Elastomers: Synthesis

F. L. G. Malet

1. Introduction

Thermoplastic polyamide elastomers (TPE-A) are segmented block copolymers with hard blocks consisting of polyamide segments, while the soft blocks usually consist of flexible segments having a low glass transition temperature. Below the melting point of the polyamide segment, T_{mPA}, the elastomeric properties are imparted by the soft, flexible phase, chemically bonded to the semicrystalline polyamide phase, and hydrogen bonds between the amide groups result in a physical crosslinked network. Above T_{mPA}, the disruption of this network leads to the transformation of the material into a regular thermoplastic polymer; hence the "thermoplastic elastomer" denomination. A thermoplastic elastomer is represented schematically in Figure 1.

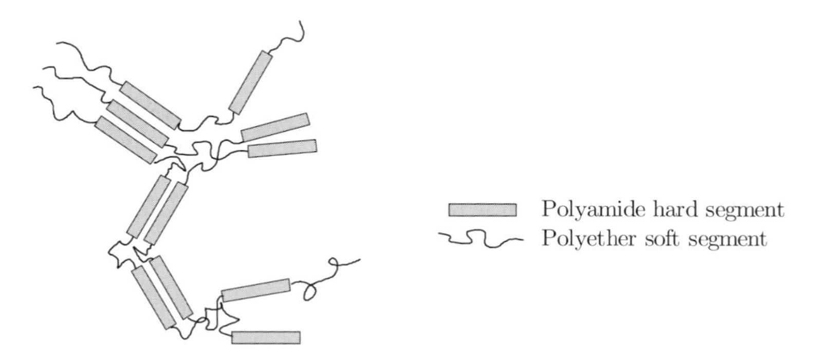

Figure 1. Schematic representation of a thermoplastic elastomer

2. Chemical structure of TPE

Most of the time, the flexible segments used for the preparation of TPE are based on alkylene oxide oligomers, such as poly(tetramethylene oxide) (PTMO, also called PTMG or PTHF), poly(ethylene oxide) (PEO, also called PEG), and poly(propylene oxide) (PPO, also called PPG) (see Figure 2). The nature of the polyether influences the mechanical and physical properties of the final product. For instance, the use of PEO enhances the hydrophilicity and imparts breathability and antistatic properties to the polymer. Slightly better antistatic properties can be achieved by the use of bisphenol A ethoxylates (see Figure 2). In fact, these polyether segments are amorphous, while standard PEO with the same molecular weight is semicrystalline. The conductivity, which is characteristic of the amorphous phase, is thus favored in the case of bisphenol A ethoxylate-based polymers.

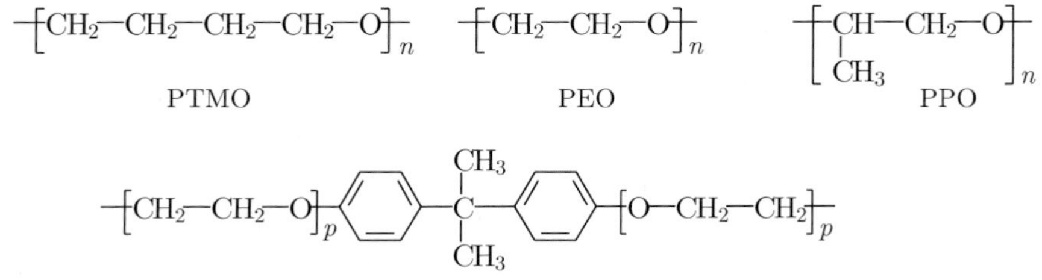

$$-\!\!\left[CH_2-CH_2-CH_2-CH_2-O\right]_n \qquad -\!\!\left[CH_2-CH_2-O\right]_n \qquad -\!\!\left[CH(CH_3)-CH_2-O\right]_n$$

PTMO PEO PPO

$$-\!\!\left[CH_2-CH_2-O\right]_p-C_6H_4-C(CH_3)_2-C_6H_4-\left[O-CH_2-CH_2\right]_p-$$

Bisphenol A ethoxylate (BEO)

Figure 2. Structure of the polyethers most commonly used in TPE synthesis

As for the polyamide (PA) segments, they are usually aliphatic and alicyclic, mainly based on polyamide 6 or 12 (Table 1). TPE based on aliphatic alicyclic PA6.6 [1,2], 6.10 [3,4], 6.12 [4], 10.10 [5], 11.6 [6], or 4.6 [7–9] have also been prepared, as well as TPE containing semicyclic [10], aromatic [11–13], or semi-aromatic (2.T [14], 4.T [15,16], 6.T [17], 11.T [18], 12.T [18]) polyamide blocks. More recently, the growing interest in biodegradable and biocompatible polymers for medical and pharmaceutical applications has led to the preparation of

Table 1. Suppliers, trade names and structures of commercially available TPE-A [24]

Supplier	Trade name	Hard segment	Soft segment
Atofina	PEBAX®	PA12	PTMO
		PA6	PEO
	PLATAMID®	Copolyamide	PEO
Degussa	VESTAMID® E	PA12	PTMO
EMS	GRILAMID® ELY	PA12	PTMO
Sanyo	PELESTAT®	PA12	BEO
		PA6	
Ube	PAE	Copolyamide	PPO

TPE-A containing α-amino acid co-monomers, *e.g.*, glycine and phenylalanine [19]; though in most cases, only AB or ABA block copolymers are prepared.

One can also use a mixture of monomers to prepare the copolyamide segment [20,21]. For instance, the Japanese company Ube patented the synthesis of TPE with polyamide blocks based on lactame 12 and the salt formed by *m*-xylene diamine and a fatty acid dimer [22,23].

The nature of the link between the hard and soft segments depends on the nature of their functional end-groups and can be either an amide, an ester, a urethane, or a urea bond. This will have an impact, *e.g.*, on the thermal and hydrolysis resistance of the final polymer.

TPE-A with polyester [25,26], polycarbonate [27], polybutadiene [28,29], poly(dimethylsiloxane) segments [30], or mixtures thereof [31] can also be prepared. Since these are rarely marketed, this chapter will only be focused on polyether-based TPE-A (see also Chapters 2, 3 and 4).

3. Synthetic methods

Several methods can be employed for the synthesis of polyether-*b*-amide copolymers. Though thermal polymerization is probably the most important one, the synthesis can also be carried out in solution or *via* an interfacial technique. Sometimes a combination of two methods may be required. The polymerization can also be carried out in one or two steps, and in one or two pots. Each method has its advantages and drawbacks, and the choice of the method will usually depend on the nature of the starting material.

3.1. *Polymerization in solution*

The application of a solvent may be necessary when the melting point of the monomers and/or the polymer is too high for bulk conditions. It may also help limit phase separation when the polyamide and the polyether blocks are not compatible enough.

For instance, the synthesis of segmented poly(ether ester amide) copolymers with an aromatic polyamide block was attempted in N,N-dimethylacetamide solution, using a dihydroxy-terminated poly(oxyethylene) oligomer, *m*-phenylenediamine, and isophthaloyl chloride [30]. Unfortunately, only low viscosity materials were obtained, probably due to side reactions involving the amide solvent (transamidation and acylation).

PA4.6 has a high melting temperature, close to 290°C; therefore, the synthesis of PA4.6-based TPE-A under bulk conditions can be quite uneasy, taking into account that the polyether block starts degrading at about 240–270°C, depending on the nature of the polyether [32]. Gaymans *et al.* [7] attempted the synthesis of a PA4.6-PTMO segmented copolymer (50/50 by wt) by a two-step method, the first step consisting in reacting the monomers (the PA4.6 salt and the amine-terminated polyether balanced with adipic acid) at 200–210°C in pyrrolidone for about 1 h. The post-condensation was then carried out for several hours at 255°C under nitrogen. This second step was rather slow since inherent viscosities higher than 1.2 dL/g were only achieved after 16 h. Moreover, ethanol extraction tests revealed a high amount of PTMO-adipic acid oligomers/polymers, suggesting a poor reactivity between the monomers. Pyrrolidone does not seem to be efficient enough to prevent phase separation. Using a better solvent or lowering the molecular weight of the polyether should enhance the reactivity of the system. Gaymans and his co-workers [9] then turned to an amine-terminated PPO instead of PTMO. This time, *m*-cresol was used as the solvent. However, only low molecular weight species were obtained together with a high amount of extractable matter, which proved to be of high polyether content. These results can be explained by an imbalance regarding the stoichiometry of the reactants. In fact, there exists an uncertainty in the functionality and molar mass of the polyether, as well as a potential evaporation of diaminobutane, which has actually been found to take place in the polycondensation of PA4.6 [33].

Another polymerization technique in solution, called the *diisocyanate method*, consists in reacting an α,ω-polyoxyethylene dicarboxylic acid with a dicarboxylic acid and a diisocyanate, either in a one-step or in a two-step process [34]. The carboxylic acid groups can react with the isocyanate groups, leading to the formation of an unstable N-carboxyanhydride intermediate that quickly rearranges to form an amide group [35] (Figure 3).

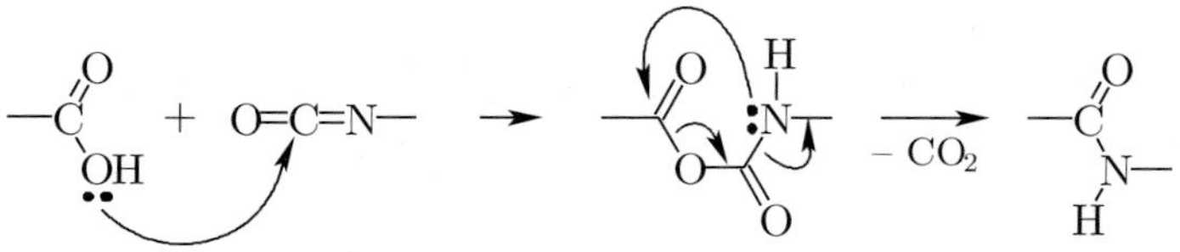

$$-C(=O)OH + O{=}C{=}N- \rightarrow [\text{N-carboxyanhydride}] \xrightarrow{-CO_2} -C(=O)N(H)-$$

Figure 3. Mechanism of amide synthesis *via* N-carboxyanhydride formation [35]

The reaction is usually carried out at high temperatures (of about 200 °C) in a polar solvent, such as tetramethylene sulfone, and the polyamide formation can be accelerated by the addition of 1-phenyl-3-methyl-2-phospholene 1-oxide as catalyst. However, in the case of a two-step process, the reaction time of the first step must be carefully controlled, since the catalyst can also play a role in the formation of carbodiimides from two terminal isocyanate groups [36]. These carbodiimides can then further react and lead to crosslinking [36]. In most cases [34–39], the polymers are prepared with 4,4′-methylene bis(phenylisocyanate) (MDI), using adipic acid, isophtalic acid, azelaic acid, or a mixture of two of them (in order to accelerate the solubilization of the polyamide phase in the solvent) and a polyether based on tetramethylene oxide, ethylene oxide, or a mixture of propylene oxide and ethylene oxide.

Copolymers with urethane links can also be obtained *via* the diisocyanate method, by simply reacting an α,ω-dihydroxy polyether with a dicarboxylic acid and a diisocyanate [40], while copolymers with urea links can be prepared by reacting an α,ω-diisocyanate polyether with an α,ω-diamine polyamide [41,42]. In these last examples, polyethylene oxides were reacted with polyamide prepolymers based on 6-hexanelactam or mixtures of 12-dodecanelactam, 8-octanelactam, and the respective N-methylated lactams.

3.2. *Interfacial polymerization*

The polyamide and the polyether segments are usually incompatible, phase separation often occurs and the reaction between the reactive chain-ends can only take place at the interface. This reaction can be accelerated by using very reactive functional groups, such as acid halides. The synthesis of polyamides and polyesters *via* interfacial polymerization has been extensively reviewed by P. W. Morgan [43] in the mid-sixties. A few years later, Castaldo *et al.* [44] successfully synthesized a poly(ether ester amide) based on PA6.6 and PEO. The α,ω-dihydroxy polyether was first reacted with a diacid chloride for several hours, either in the bulk or in chloroform, and at a rather low temperature (60–90°C). The mixture was then poured into a vigorously stirred aqueous solution of diamine and sodium hydroxide. Later, de Candia *et al.* [45] reproduced this technique to study the physical and mechanical properties of the copolymer. The same polymerization technique was also used to prepare copolymers based on PPO as the polyether segment and PA6.10 as the polyamide block [3,46,47].

3.3. *Direct polycondensation using condensing agents*

This technique [35] originates from the observation of biological mechanisms, and in particular of the formation of peptide bonds through energy-rich acylphosphate intermediates. For the synthesis of polyamides, the reaction is carried out in solution, using a phosphorous compound as activating agent and a base as proton acceptor. Though the early attempts by Sekiguchi [48] on wholly aliphatic polyamides yielded only low molecular weight polymers, the technique was later improved for aromatic or semiaromatic polyamides, especially by Yamazaki [49], who pointed out the importance of pyridine. The introduction of metal salts was also found to be beneficial, partly due to their role as breakers of hydrogen bonds.

To the best of our knowledge, Imai *et al.* [50] were the first to investigate this technique for the synthesis of polyether-amides containing aromatic diamines. The polycondensation was successfully carried out in N-methyl-2-pyrrolidone (NMP) at 100°C with triphenyl phosphite and pyridine in the presence of LiCl and $CaCl_2$. Yamashita *et al.* [51] attempted the synthesis of PA6.6-PEO segmented copolymers, but the yields were rather low.

Later the same polymerization technique proved to be successful in the preparation of multiblock copolymers from poly(oxyethylene)dicarboxylic acid and a polyamide oligomer based on bis(4-aminophenyl)ether and isophthalic acid. The synthesis could be carried out either by a one-pot two-step method [52] or using a two-pot technique by first isolating the polyamide oligomer [11]. The two approaches lead to copolymers with similar thermal and mechanical properties, though this was not the case when a one-pot, one-step method was used [52], probably because of the formation of a more irregular structure.

3.4. *Anionic polymerization*

Anionic polymerization is usually carried out to obtain (co)polymers with controlled architecture and narrow molecular weight distributions. Though polyamides are usually synthesized *via* a polycondensation reaction, it is possible to prepare them *via* the anionic ring-opening polymerization of lactams. In fact, the anionic polymerization of several lactams [53] *e.g.,* 2-pyrrolidone, caprolactam, or laurylllactam [54,55] has been widely reported in the literature. The mechanism is believed to involve the reaction of an activated monomer anion with an N-acyl lactam chain-end, followed by a very fast proton transfer from the monomer to the polymer anion, thus giving rise to a new activated monomer anion and an N-acyl lactam chain increased by one repeat unit. Instead of N-acyl lactam, acyl halides or isocyanates can also be used as activating agents, since they can react rapidly with the monomer or monomer anion to give N-acyl lactam. Thus, the synthesis of ABA triblock copolymers, using functionalized polyethers as activating agents, has been reported. For instance, Yamashita *et al.* [56] have polymerized 2-pyrrolidone and caprolactam in the bulk at 30 °C and 80 °C, respectively, using the bischloroformates of hydroxy-terminated poly(tetramethylene glycol). More recently [57], the polymerization of caprolactam has also been carried out in a twin-screw extruder, using an isocyanate-terminated telechelic PTMO.

3.5. *Thermal polymerization*

This is probably the most important technique, at least on the industrial scale, because it does not involve any potentially toxic organic solvent, nor requires the expensive preparation of raw materials with highly reactive, and even sometimes harmful, functional end-groups.

The thermal polymerization of polyamide and polyether segments with the formation of an *amide link* has been repeatedly described in the literature and numerous patents have been filed by companies, such as Toray Industries, Teijin Ltd., and Asahi Kasei. For instance, Kimura *et al.* [58] successfully prepared a poly(ether-amide) copolymer by polymerizing the "nylon 6.6 salt" together with the isolated salt of a poly(oxyethylene) diglycolic acid and 1,6-diaminohexane, at 270 °C for 5 h in vacuum.

Rasmussen and Smith [59] have reported the preparation in the bulk of several copolymers based on poly(piperazine sebacamide) and poly(oxypropylene) diamine. The reaction was carried out at high temperature (220–245 °C), partly under nitrogen flow and partly in vacuum. Other copolymers were successfully prepared, using a copolyether based on a mixture of ethylene oxide and propylene oxide, as well as the addition of a fatty dimer diamine.

Fradet and co-workers [60] have presented an original preparation of polyether-*b*-amide copolymers using a chain-coupling reaction between 4,4'-disubstituted bisoxazolones and a mixture of amine-terminated polyethers and polyamide oligomers (Figure 4). The reaction could be carried out in a one- or

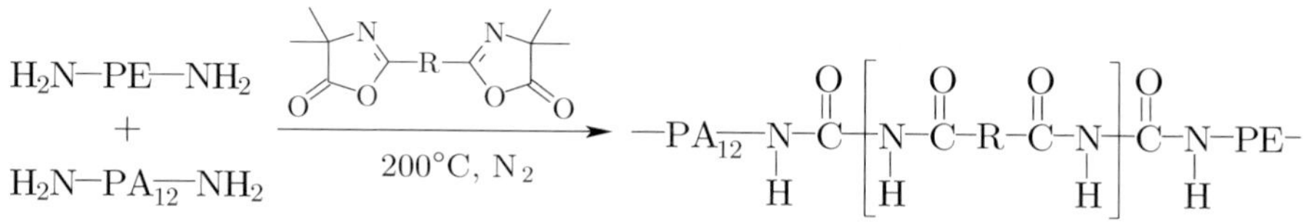

$$H_2N{-}PE{-}NH_2 + H_2N{-}PA_{12}{-}NH_2 \xrightarrow[200^\circ C,\ N_2]{\text{bisoxazolone}} {-}PA_{12}{-}\underset{H}{N}{-}\overset{O}{\overset{\|}{C}}\left[{-}\underset{H}{N}{-}\overset{O}{\overset{\|}{C}}{-}R{-}\overset{O}{\overset{\|}{C}}{-}\underset{H}{N}{-}\right]\overset{O}{\overset{\|}{C}}{-}\underset{H}{N}{-}PE{-}$$

where R = 0 or $-(CH_2)_2-$

Figure 4. Reaction of amine-terminated polyether (PE) and polyamide 12 (PA_{12}) with a bisoxazolone-type coupling reagent, in the bulk at 200°C under nitrogen [60]

two-step procedure and both seem to yield copolymers with comparable viscosities. The resulting copolymers had a melt strength high enough to allow fibers to be drawn from the melt. Compared to conventional thermal polycondensation, this method tends to allow the synthesis of (co)polymers within rather short reaction times at lower temperatures, without the application of vacuum, since there is no by-product formation. Moreover, the stoichiometry between the PA and PE segments should not be adjusted very precisely, since only the stoichiometry between the chain coupling reagent and the amino end-groups is relevant.

In order to improve the mechanical properties and the heat resistance of TPE-A, other links have been investigated. For instance, a poly(ether-*b*-amide) with an *imide link* was prepared by reacting polyamide-forming monomers and a diamino polyether together with trimellitic anhydride [61]. Naphthalene-1,2,5-tricarboxylic acid can also be used for this purpose.

The reaction between alcohol and carboxylic acid groups is much more difficult because of the equilibrium with water and the desired ester. Up to the end of the 1970s, very few references could be found in the literature concerning the synthesis of a poly(ether-*b*-amide) with *ester links*. BASF [62] patented the synthesis of a polyamide obtained by reacting caprolactam and adipic acid with polyethylene glycol at 250°C. *p*-Toluenesulfonic acid was used as catalyst. Unilever [63] attempted the reaction of an ω,ω'-dihydroxy poly(oxyethylene) with oligoamides prepared with a fatty acid dimer and a diamine, at 250°C in vacuum, without the addition of a catalyst. In both cases, only low molecular weight species seemed to have been achieved [64]. The first successful synthesis was carried out by Deleens *et al.* [64], who used a transition metal-based catalyst. This discovery led to the first commercialization of a polyether-*b*-amide by ATO Chimie (now Atofina) in the early eighties.

4. Thermal polymerization of TPE-A with ester links

The synthesis of TPE-A with ester links is rather important on the industrial scale, since it concerns the major suppliers of TPE-A. The influence of the polymerization conditions has been thoroughly studied and will be reviewed in the following sections of this chapter.

4.1. *Polymerization processes*

The polymerization of TPE-A with ester links can be carried out according to various synthetic approaches.

With its range of PEBAX®, Atofina is the leader in the production of TPE-A which are polymerized in two steps [65]. First, an α,ω-dicarboxylic acid-terminated polyamide block is prepared by reacting lactam(s), aminoacid(s) and/or diacid and diamine, usually under pressure at high temperature, in the presence of a chain-terminating diacid (Figure 5). The amount of the latter relative to the amount of the polyamide-forming monomers determines the average molecular weight of the polyamide segment.

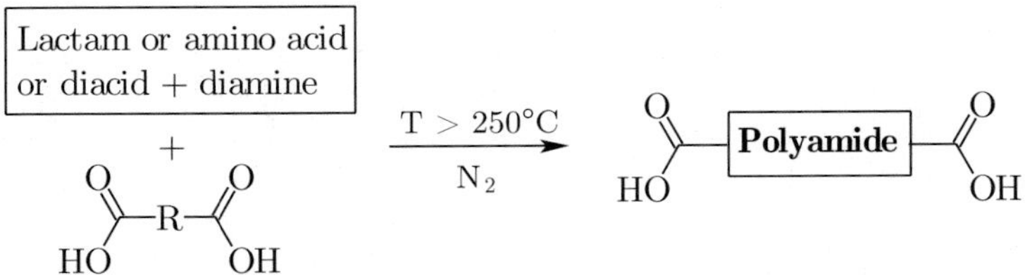

Figure 5. Schematic representation of the synthesis of an α,ω-dicarboxylic acid-terminated polyamide block

This oligomer is then reacted with a commercially available α,ω-dihydroxy-terminated polyether at a lower temperature (in order to limit the degradation of the polyether), in vacuum (in order to drive the esterification equilibrium toward the formation of the ester), and in the presence of a suitable catalyst to enhance the reaction kinetics (Figure 6). After the completion of the reaction, the molten polymer is extruded through a die either by a gear pump or by nitrogen pressure. From the die, strands or ribbons can be solidified and cooled in a water bath, pelletized in a dicer and transferred for drying. Pellets can also be obtained by cooling and slicing the melt right after the die.

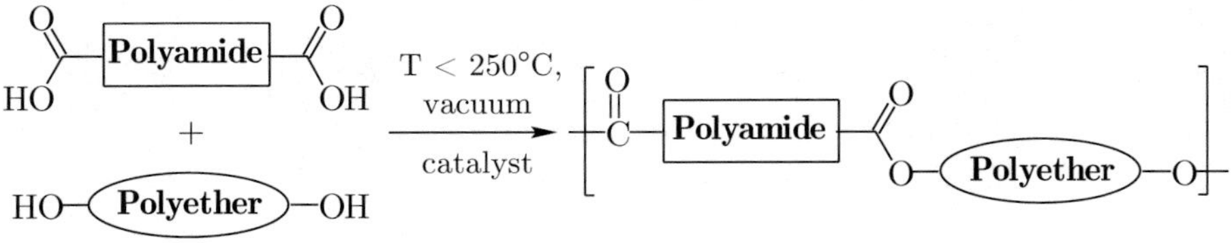

Figure 6. Schematic representation of the esterification step

This technique differs from the standard polyesterification reaction mainly in that the reacting species do not have low molecular weights and therefore the initial viscosity of the reacting media is rather high. Therefore, even if the inherent reactivity of the functional end-groups is the same, the rather low chain mobility makes the reaction more difficult. Another difficulty arises from the low miscibility of the polyamide block and the polyether block, and vigorous stirring is required to prevent phase segregation.

Maréchal and co-workers [66] have carried out a kinetic study on models of the non-catalyzed reaction between α,ω-dicarboxypolyamide-11 and α,ω-dihydroxypolyoxyalkylenes. Their experimental results indicate that the reaction seems to be of second order in acid medium and of first order in alcohol, which was found to be comparable to previous results for esterifications and polyesterificiations in the bulk. The enthalpy and entropy of activation were calculated from experimental data and the rate constant was found to be rather low, suggesting the use of a catalyst in order to achieve polymerization times compatible with industrial concerns.

Another method is to prepare the TPE-A in one step, mixing all the monomers together from the very beginning. Degussa (another important supplier of TPE-A) prepares its range of VESTAMID® in this way [67]. A better compatibility between the reactive monomers is usually achieved, thus enhancing the polymerization kinetics. This leads also to a more statistical distribution of the monomers in the chain. The main drawback of this method is the potential degradation of the polyether segments due to the rather severe process conditions, especially in the case when lactams are used, since they usually require high temperature and pressure for the opening of the cycle.

Because the esterification reaction is usually difficult, it can be advantageous to react the polyamide and the polyether segments *via* transesterification. In fact, in the case of certain less reactive diols, the condensation with the diester will proceed more readily than with the respective acid. Moreover, carboxylic acid esters usually have a lower melting point and the reaction mixture can be more homogeneous, since carboxylic acid esters tend to have a better miscibility with the other reagents [68]. For instance, a PA6.6-PTMO segmented copolymer can be synthesized by first preparing the polyamide segment having hydroxybutylate [2] or butanoate [69] end-groups, that can then react with the polyether diol *via* ester interchange, releasing 1,4-butanediol or *n*-butanol, respectively.

The reaction of polyamides containing carboxylate end-groups with polyethers containing hydroxyl end-groups has been the object of various other patents [70–72]. In particular, a continuous process has also been patented [73], where the monomers are first mixed and heated in a reactor under vacuum. After a reaction time of 1 to 4 h, low molecular weight polymers are obtained and transferred inside a film-type evaporator in order to increase the degree of polymerization. Depending on the required physical property of the product, this second step can take 3 to 30 min. The final product is then discharged, rapidly cooled down and cut into pellets.

4.2. *Nature of the raw materials*

Several parameters have to be taken into account in order to explain the differences in reactivity observed when using various polyamide and polyether blocks. Considering the nature of the polyether block, the nature of the functional end-group is one of the major factors influencing the polyesterification kinetics.

For instance, α,ω-dihydroxy-terminated PEO and PTMO are usually much more reactive than α,ω-dihydroxy-terminated PPO, because it contains secondary alcohols that are less reactive than primary alcohols. This result was confirmed by Maréchal [74] with PA12 as the polyamide segment. However, at a relatively low temperature (around 200°C instead of 220–240°C), the water content in the polyether seems to have a significant impact, since α,ω-dihydroxy-terminated PEO proved to be less reactive than α,ω-dihydroxy-terminated PPO [74], probably because at this temperature water is not eliminated as efficiently as at higher temperatures. In the case of PEO/PPO block copolyethers (Pluronic® or Synperonic®), which contain a mixture of di- and triblock copolymers and, in some other cases, even a PPO homopolymer [75], there is a mixture of primary and secondary alcohol end-groups and the reactivity with a PA12 segment takes an intermediate position compared to those of pure PEG and PPO segments [74]. However, the reactivity of α,ω-dihydroxy-terminated PPO can be increased by the addition of a small amount of a more reactive polyether [76] or by increasing the catalyst content (Table 2).

Table 2. Influence of the more reactive polyether content and of the catalyst content on the reactivity of α,ω-dihydroxy-terminated PPO (600 g/mole) with α,ω-dicarboxylic acid-terminated PA 6 at 240°C in vacuum [76]

Reference	PPO content (wt%)	More reactive polyether Nature	More reactive polyether Content (wt%)	Catalyst content (wt%)	Time* (min)
1	100	–	–	0.8	110
2	100	–	–	1	90
3	100	–	–	2	40
4	90	$PTMO_{2000}$	10	0.8	80
5	90	$Jeffamine_{2000}$	10	0.8	60
6	80	$Jeffamine_{2000}$	20	0.8	40

* From the moment when an abrupt increase in the torque can be observed

In addition to the nature of the functional end-groups, the molecular weight of the polyether has a significant influence on the reactivity. Deleens [64] showed that at a molecular weight higher than 2500 g/mole, the incompatibility with the polyamide segment becomes really hard to overcome. The molecular weight of the polyamide block should also not be higher than 5000 g/mole, because of melt viscosity concerns and reduced mobility of the COOH end-group. Chain extenders linking two (or more) low molecular weight polyethers [77,78] or polyamides [79] can help the synthesis of copolymers with a higher polyether or polyamide content without using high molecular weight oligomers. The chain extender can also be added in a melt-kneading step. In fact, the melt index of a PA12-PTMO copolymer was decreased from 10 g/min (at 200°C, 2.16 kg load) to 6.7 g/min by adding 0.1 part of hexamethylene diisocyanate to 100 parts of the copolymer [80].

As shown by Gaymans and co-workers [78], it is also possible to increase the melting temperature of the polyamide phase by forming longer crystalline lamellae *via* a chain extender, rather than by increasing the molecular weight of the polyamide segment. Depending on the TPE-A targeted, the difference in the solubility parameters between the polyamide block and the polyether will be more or less important, and will partially account for the difference in reactivity. The solubility parameters [in $(cal/cm^3)^{1/2}$] of poly(tetramethylene ether) and PA12 are 8.3 and 10.8, respectively [81] and, therefore, the solubility parameters for the corresponding oligomers should be quite close and phase separation should not be too severe. However, PA6 has a much higher solubility parameter of about 12.4 to 14.3, and in the case of PA6-PTMO copolymers, phase separation cannot be ignored [81]. Reacting all the monomers together (one step process) was not found to be more successful, because the reaction of caprolactam with the diacid rapidly led to the *in situ* formation of the polyamide oligomer that does not react efficiently with the polyether. For this reason, the caprolactam and the polyether were reacted before the addition of the required molar amount of chain terminator [81] and polymers with intrinsic viscosity as high as 1.3 dL/g could be obtained. A similar polymerization technique has also been proposed for the synthesis of PA6.6-PTMO copolymers [82]. Another method consists in first reacting the polyether and a two-fold excess of chain terminator (adipic acid) to obtain the respective dicarboxylic acid-terminated ester. An excess of a PA6.6 salt is then added to the ester in order to form the respective dicarboxylic acid polymer, which is finally reacted with the remaining amount of PTMO to give the desired copolymer [83].

Branched copolymers can also be prepared by introducing multifunctional monomers. For instance, branched PA11-PTMO copolymers were obtained by the addition of 0.25 wt% of 1,3,5-benzenetricarboxylic acid (trimesic acid) to a mixture of PA11 oligomer and polytetramethylene glycol [84].

4.3. *Influence of the catalyst*

Since the early work by Deleens [64], who synthesized TPE-A with $Ti(OBu)_4$ as transition metal catalyst, several other catalytic systems have been proposed, involving other transition metals, such as : Al [85], Ge [85], Hf [86] Sn [85–88], Sb [80,85], or mixtures thereof. In particular, a catalyst based on the Sn/Sb mixture [89] can often be found in the literature.

A drawback of the alkoxide derivatives of transition metals is their sensitivity to hydrolysis; in the presence of water, they can form aggregates (also called condensates, see Figure 7), which possess a catalytic activity only on their surface, thus reducing their overall efficiency by the reduced amount of reactive sites.

$$M(OR)_4 + H_2O \rightarrow (RO)_3M\text{–}OH + ROH$$

$$(RO)_3M\text{–}OH + M(OR)_4 \rightarrow (RO)_3M\text{–}O\text{–}M(OR)_3 + ROH$$

Figure 7. Hydrolysis of transition metal tetra-alkoxides

Even if the reaction medium is thoroughly dried, the water formed during polycondensation can have an impact on the reaction kinetics. This is one of the reasons why working in vacuum is beneficial to the polymerization, since the equilibrium is shifted toward the polyester formation. The increase in catalyst content in order to compensate for the loss due to hydrolysis may further induce undesired thermal or photochemical degradation, which can be catalyzed by the organometallic residues. Another possibility would be to use a catalyst that is not (or much less) sensitive to hydrolysis, such as $(CH_3COO)_2Sn$.

The catalyst content was found to have an impact on the esterification kinetics. Fradet and co-workers [90] confirmed Deleens' observation [64] that there exists an optimal catalyst concentration. Actually, a low catalyst content would not result in a process faster than that carried out without catalyst. This phenomenon has been attributed to the formation of catalyst condensates. However, once a threshold concentration is reached, the polymerization rate will increase with the catalyst concentration, although a too high amount of catalyst can lead to side reactions, which may involve the esterification of the diacid with the catalyst itself.

4.4. *Influence of the molar ratio*

Another important parameter to take into account is the molar ratio of end-groups. In fact, the esterification step requires the reaction of one mole of OH groups with one mole of COOH groups. The excess or shortage of one of the end-groups would result in a limitation of the molecular weight of the final copolymer. If this imbalance is considerable, the synthesis of a material with good mechanical properties would be impossible. Figure 8 illustrates this

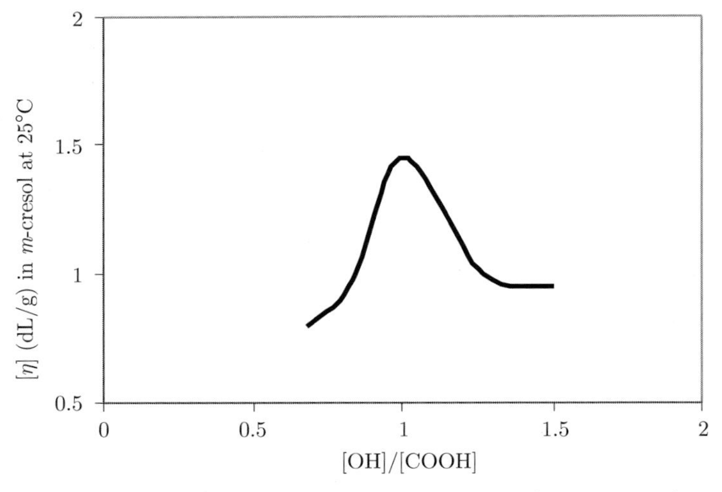

Figure 8. Polycondensation of an ω,ω'-diacid PA11 oligomer (M_n = 2090 g/mole) with an ω,ω'-dihydroxy PEO oligomer (M_n = 2250 g/mole), using 0.3 wt% of $Ti(OBu)_4$ as catalyst, at 260°C under 0.13 mbar. Intrinsic viscosity *vs.* molar ratio dependence [64]

situation in the case of the reaction of a diacid PA11 oligomer with a dihydroxy polyoxyethylene oligomer. The influence of the molar ratio on the molecular weight of the final copolymer was followed by measuring the intrinsic viscosity.

4.5. *Influence of the temperature*

The temperature is another parameter that has to be adjusted in order to achieve the optimum reaction conditions.

As already mentioned for the PEO-based systems, the water content in the polyether may be sometimes too high to be effectively eliminated at a temperature of *ca.* 200°C or lower. This is the reason why the PPO-based systems sometimes react faster despite their secondary alcohol end-groups. Nevertheless, earlier results by Deleens [64] (Figure 9) confirmed that, as could be expected, the increase of the polymerization reaction temperature does have a significant impact on the conversion. However, this temperature should not be too high, not only because of the potential degradation of the polyether [32], but also because of the significant increase in the coloration of the final product, which is rather detrimental for numerous applications. In order to prevent significant degradation, stabilizers can be added to the reaction media. Antioxidants, such as Irganox 1010 [91] or Irganox 1098 (from CIBA Specialty Chemicals), are most often reported in the literature. To further ensure the stability of the copolymer against thermal, thermooxidative or photooxidative aging, other additives can be introduced *via* an extrusion step after polymerization [92,93–95].

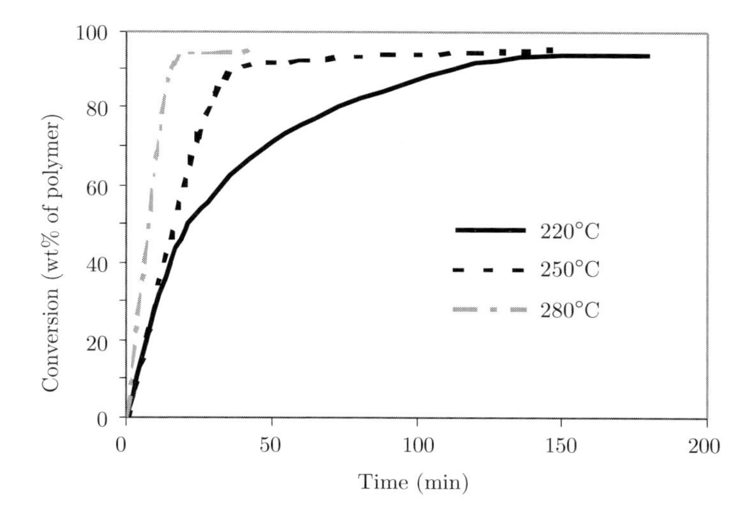

Figure 9. Polycondensation of an ω,ω'-diacid PA 11 oligomer (M_n=2090 g/mole) with an ω,ω'-dihydroxy PEO oligomer (M_n = 975 g/mole), using 0.3 wt% of $Ti(OBu)_4$ as catalyst under 0.13 mbar. Influence of the temperature on the conversion plotted *vs.* the reaction time [64]

4.6. *Influence of the stirring*

Because the polyamide and polyether oligomers are incompatible, the reaction medium always consists of two phases and the esterification reaction can only occur at the interface. For this reason, the reaction kinetics is rather slow at the very beginning of the process. However, as soon as a small quantity of copolymer is formed, it will play the role of compatiblizer of the two phases and lead to faster kinetics. This can be easily observed [96] by the change of the reaction medium from a heterogeneous, opaque system into a transparent, "compatibilized" one.

Nevertheless, a vigorous mixing is rather necessary to increase the dispersion of the two phases and their contact area. Stirring also has an impact on the viscosity of the reaction medium due to a sheer-thinning effect, increasing the diffusion of end-groups. However, the stirring rate should not be too high in order to prevent the copolymer from climbing up along the rotating anchor and sticking to it (Weisenberg effect). In the case of PA11 and polyoxyethylene glycol, Deleens [64] has found that an optimal stirring rate had to be chosen for the polyesterification to be as efficient as possible (Figure 10).

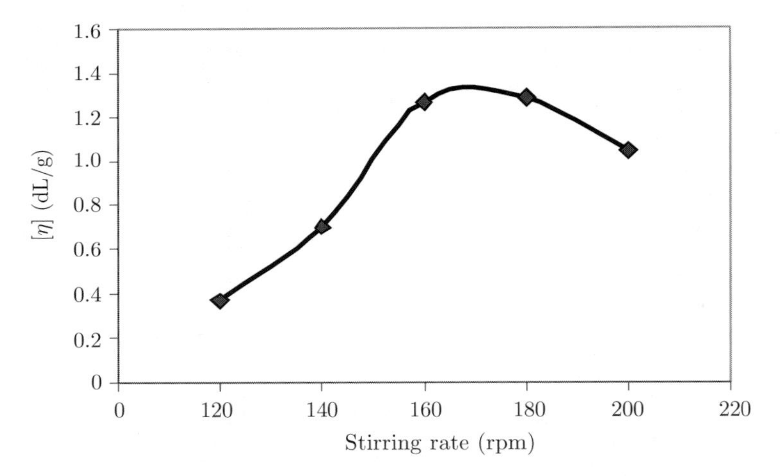

Figure 10. Polycondensation of an ω,ω'-diacid PA 11 oligomer ($M_n = 2090$ g/mole) with an ω,ω'-dihydroxy PEO oligomer ($M_n = 975$ g/mole), using 0.3 wt% of $Ti(OBu)_4$ as catalyst under 1.3 mbar. Influence of the stirring rate on the intrinsic viscosity (in *m*-cresol at 25°C) after 45 min at 270°C [64]

4.7. *Influence of the vacuum level*

Working under reduced pressure is compulsory in order to achieve high molecular weight copolymers, because in this way the equilibrium of the reaction is shifted toward the formation of ester compounds, but also because the con-

Table 3. Polycondensation of an ω,ω'-diacid PA 11 oligomer (M_n = 2090 g/mole) with an ω,ω'-dihydroxy PEO oligomer (M_n = 975 g/mole), using 0.3 wt% of $Ti(OBu)_4$ as catalyst under 1 Torr. Influence of the reduced pressure on the intrinsic viscosity after 45 min at 270°C [64]

Pressure (mbar)	$[\eta]$ (dL/g) in *m*-cresol at 25°C
0.7	1.45
0.9	1.37
1.3	1.27
6.7	0.75
13.3	0.40

ventional catalysts are sensitive to moisture. Table 3 clearly shows the influence of the vacuum level on the final viscosity of the material.

5. Conclusion

Because of their high technical properties, polyamide-based thermoplastic elastomers have attracted a lot of interest and their synthesis has been attempted *via* various polymerization techniques. On the industrial scale, the major companies are producing TPE-A *via* one- and two-step thermal polymerization processes. Many parameters have to be adjusted to ensure optimal reaction efficiency, including catalyst nature and content, temperature, vacuum level and stirring rate. Obviously, all these parameters are also dependent on the nature of the raw materials used, since some polyamide/polyether pairs, depending on their structure and/or the nature of their end-groups, are easier to prepare than others.

References

1. Uenosono T and Imanishi T (1987) Preparation of polyether-polyester-polyamide elastomers with good heat and chemical resistance, JP Patent 63 183,929, to Asahi Glass Co., *Chem Abstr* **110**:77256.
2. Matsuo T (1986) Production of polyether ester amide elastomer, JP Patent 63 035,625, to Asahi Chemical Ind., *Chem Abstr* **109**:151199.
3. Yui N, Tanaka J, Sanui K and Ogata N (1984) Polyether-segmented polyamides as a new designed antithrombogenic material: Microstructure of poly(propylene oxide)-segmented nylon 610, *Makromol Chem* **185**:2259–2267.
4. Chiaki T, Shinobu N and Makoto K (1983) Polyamide elastomers, JP Patent 59 207,930, to Toray Ind., *Chem Abstr* **102**:168169.
5. Liu F X, Zou Y F, Luo X L, Huang Y H and Zhou G F (1992) Study on the condensed state structure of polyamide 1010-poly(tetramethylene oxide) multiblock copolymers, *Polym Mater Sci Eng* **8**:50–55.
6. Chiaki T, Shinobu N and Makoto K (1983) Aliphatic polyamide elastomers, JP Patent 59 213,724, to Toray Ind., *Chem Abstr* **102**:150746.

7. Gaymans R J, Roerdink E, Schwering P J F and Walch E (1988) Manufacture of segmented block copolymer thermoplastic elastomers, EU Patent Appl 360,311 A, to Stamicarbon, *Chem Abstr* **113**:61093.
8. Gaymans R J, Schwering P and de Haan J L (1989) Nylon 46-polytetramethylene oxide segmented block copolymers, *Polymer* **30**:974–977.
9. van Hutten P F, Walch E, Veeken A H M and Gaymans R J (1990) Segmented block copolymers based on polyamide-4,6 and poly(propylene oxide), *Polymer* **31:**524–527.
10. Keiji K and Chiaki T (1982) Polyamide-polyester-polyether rubber, JP Patent 58 206,627, to Toray Ind., *Chem Abstr* **100**:140622.
11. Hirt V P and Herlinger H (1974) Neue Elastomerfasersysteme, *Angew Makromol Chem* **40/41**:71–88.
12. Imai Y, Kajiyama M, Ogata S and Kakimoto M (1984) Synthesis and properties of multi-block copolymers based on poly(oxyethylene)s and aromatic polyamides, *Polymer J* **16(3)**:267–272.
13. Niesten M C E J, Feijen J and Gaymans R J (2000) Synthesis and properties of segmented copolymers having aramid units of uniform length, *Polymer* **41**:8487–8500.
14. Bouma K, Wester G A and Gaymans R J (2001) Polyether-amide segmented copolymers based on ethylene terephthalamide units of uniform length, *J Appl Polym Sci* **80**:1173–1180.
15. Gaymans R J and de Haan J L (1993) Segmented copolymers with poly(ester amide) units of uniform length: synthesis, *Polymer* **34**:4360–4364.
16. Guang L and Gaymans R J (1997) Polyesteramides with mixtures of poly(tetramethylene oxide) and 1,5-pentanediol, *Polymer* **38**:4891–4896.
17. Gaymans R J, Krijgsman J and Husken D (2002) Segmented copolymer containing amide segments, composition for moldings and fibers, WO Patent 2003-070807, to University of Twente, *Chem Abstr* **139**:214917.
18. Chiaki T, Shinobu N and Makoto K (1983) Poly(ether ester amide) elastomers, JP Patent 59 217,726, to Toray Ind., *Chem Abstr* **102**:186464.
19. D'Angelo S, Galleti P, Maglio G, Malinconico M, Morelli P, Palumbo R and Vignola M C (2001) Segmented poly(ether-ester-amide)s based on poly(L,L-lactide) macromers, *Polymer* **42**:3383–3392.
20. Youko F and Chiaki T (1983) Hydrophilic polyamide elastomers, JP Patent 60 112,826, to Toray Ind, *Chem Abstr* **103**:161661.
21. Youko F and Chiaki T (1983) Hydrophilic polyether ester amide, JP Patent 60 063,225, to Toray Ind, *Chem Abstr* **103**:38513.
22. Okamoto H and Okushita Y (1983) Preparation of polyether-amide, JP Patent 59 131,628, to Ube Kosan, *Chem Abstr* **101**:231203
23. Okamoto H and Okushita Y (1983) Production of block polyester-amide, JP Patent 59 193,923, to Ube Kosan, *Chem Abstr* **102**:133328
24. Elvers B, Hawkins S and Russey W (Eds.) (1995) Thermoplastic elastomers, in *Ullmann's Encyclopedia of Industrial Chemistry*, Verlagsgesellschaft, Weinheim, Vol. A26, pp. 633–664.
25. Huet J M and Maréchal E (1974) Synthesis and study of the polycondensate sequences comprising a sequence of poly(ε-caprolactone) II. Synthesis and study of copolycondensate sequences comprising a poly(ε-caprolactone) sequence and polyamide or polycarbonate sequences. An attempt to synthesize of copolycondensate sequences of poly(ethanediol terephthalate) and poly(ε-caprolactone), *Eur Polym J* **10**:771–782 (in French).
26. Okitsu K and Murabayashi K (1985) Polycaprolactone-polyamide elastomer impact absorbers, JP Patent 61,171,731, to Daicel Chemical Ind., *Chem Abstr* **105**:192699.

27. Okushita H and Muramutsu T (2001) Heat-resistant flexible polyamide-polycarbonate block elastomers, EP Patent 1253165, to Ube Ind., *Chem Abstr* **137**:326423.
28. Miyamoto M, Nakanishi H and Tanaka T (1988) Thermally-stable and flexible polyolefin-polyamides, JP Patent 2,113,026, to Mitsubishi Kasei Corp., *Chem Abstr* **113**:172915.
29. Ozgun H B, Kubanek V, Kralicek J and Veruovic B (1986) Synthesis of block copolymers based on polycaprolactam-polybutadiene, *Eur Polym J* **22**:1009–1014.
30. Zdrahala R J, Firer E M and Fellers J F (1977), Block copolymers of poly(*m*-phenylene isophtalamide) and poly(ethylene oxide) or polydimethylsiloxane: Synthesis and general characteristics, *J Polym Sci Polym Chem Ed* **15**:689–705.
31. Okamoto H and Okushita Y (1984) Poly(ether amides), JP Patent 60 158,221, to Ube Ind., *Chem Abstr* **104**:89746.
32. Boulares A, Tessier M and Maréchal E (2000) Synthesis and characterization of poly(copolyethers-block-polyamides) II. Characterization and properties of the multiblock copolymers, *Polymer* **41**:3561–3580.
33. Gaymans R J, Van Utteren T E C, Van den Berg J W A and Schuyer J (1977) Preparation and some properties of Nylon 46, *J Polym Sci Polym Chem Ed* **15**:537–545.
34. Chen A T, Farrissey W J and Nelb R G (1978) Polyester amides suitable for injection molding, US Patent 4,129,715, to The UpJohn Co., *Chem Abstr* **90**:138859.
35. Sekiguchi H and Coutin B (1992) Polyamides, in *Handbook of Polymer Synthesis*, Part A (Ed. Kricheldorf H R) Marcel Dekker Inc., New York, pp. 807–939.
36. Alberino L M, Farrissey W J and Sayigh A A R (1977) Preparation and properties of polycarbodiimides, *J Appl Polym Sci* **21**:1999–2008.
37. Chen A T, Nelb R G and Onder K (1986) New high-temperature thermoplastic elastomers, *Rubber Chem Technol* **59**:615–622.
38. Otsuki T, Kakimoto M-A and Imai Y (1987) Synthesis of multi-block copolymers based on poly(oxyethylene)dicarboxylic acids and polyamides by the diisocyanate method, *Makromol Chem Rapid Commun* **8**:637–640.
39. Otsuki T, Kakimoto M-A and Imai Y (1990) Synthesis and properties of multiblock copolymers based on polyoxyethylene and polyamides by diisocyanate method, *J Appl Polym Sci* **40**:1433–1443.
40. Otsuki T, Kakimoto M-A and Imai Y (1990) Synthesis and properties of new multiblock copolymers based on poly(oxytetramethylene) and polyamides by diisocyanate method, *Polymer* **31**:2214–2219.
41. Raab M, Masař B, Kolařik J and Čefelin P (1979) Mechanical properties of multiblock polyamide-polyoxirane copolymers, *Int J Polym Mater* **7**:219–231.
42. MasaY B, Čefelin P and Šebenda J (1979) Block copolymers of Copolyamides with polyoxirane, *J Polym Sci Polym Chem Ed* **17**:2317–2335.
43. Morgan P W (1965) Condensation polymers: by interfacial and solution methods, in: *Polymer Reviews* (Eds. Mark H F and Immergut E H) Interscience Publishers, New York, London, Sydney, Vol. 10, pp. 1–561.
44. Castaldo L, Maglio G and Palumbo R (1978) Synthesis of polyamide-polyether block copolymers, *J Polym Sci Polym Lett Ed* **16**:643–645.
45. de Candia F, Petrocelli V, Russo R, Maglio G and Palumbo R (1986) Synthesis and physical behaviour of poly(amidoether) block copolymers, *Polymer* **27**:797–802.
46. Yui N, Tanaka J, Sanui K, Ogata N, Kataoka K, Okano T and Sakurai Y (1984) Characterization of the microstructure of poly(propylene oxide)-segmented polyamide and its suppression of platelet adhesion, *Polymer J* **16**:119–128.
47. Ogata N and Yui N (1984) Synthesis of block copolymers by end-reactive oligomers, *J Macromol Sci–Chem* **A21**:1097–1116.

48. Sekiguchi H (1960) Preparation of polybutanamide from α-pyrrolidone, *Bull Soc Chim France* 1827–1830.
49. Yamazaki N, Higashi F and Kawabata J (1974) Studies on Reactions of the N-phosphonium salt of pyridines. XI. Preparation of polypeptides and polyamides by means of triaryl phosphites in pyridine, *J Polym Sci Polym Chem Ed* **12**:2149–2154.
50. Imai Y, Ogata S and Kakimoto M (1984) Synthesis of polyether-amides by direct polycondensation of poly(oxyethylene)dicarboxylic acids with aromatic diamines in the presence of triphenyl phosphite and pyridine, *Makromol Chem Rapid Commun* **5**:47–51.
51. Chujo Y, Hiraiwa A and Yamashita Y (1984) Synthesis of segmented copolyamides by using telechelic prepolymers, *Makromol Chem* **185**:2077–2087.
52. Imai Y, Kajiyama M, Ogata S and Kakimoto M (1985) Improved synthesis of polyether-aramid multi-block copolymers by direct polycondensation, *Polymer J* **17**:1173–1178.
53. Sebenda J (1989) Anionic ring-opening polymerization: lactams, in *Comprehensive Polymer Science – The Synthesis, Characterization, Reactions & Applications of Polymers* (Eds. Eastmond G C, Ledwith A, Russo S and Sigwalt P) Pergamon Press, Oxford, New York, Beijing, Frankfurt, Sao Paulo, Sydney, Tokyo, Toronto, pp. 511–530.
54. Schmid E, Laudonia I, Ernst H and Kaegi W (2001) Preparation of polyamides by anionic polymerization and their recycling, EP Patent 1249465, to EMS-CHEM, *Chem Abstr* **137**:295385.
55. Luisier A, Bourban P E and Manson J A E (2002) Initiation mechanisms of anionic ring-opening polymerization of lactam-12, *J Polym Sci Polym Chem Ed* **40**:3406–3415.
56. Yamashita Y, Matsui H and Ito K (1972) Block copolymerization V. Block anionic polymerization of lactams, *J Polym Sci Polym Chem Ed* **10**:3577–3587.
57. Lee B H and White J L (2002) Formation of a polyetheramide triblock copolymer by reactive extrusion; process and properties, *Polym Eng Sci* **42**:1710–1723.
58. Kimura Y, Sugihara N and Taniguchi I (1983) Novel polycondensations *via* poly(oxyethylene) diglycolic acid diamine salts, *Macromolecules* **16**:1023–1024.
59. Rasmussen J K and Smith H K (1983) Polyamide-polyether copolymers: a new family of impact-resistant thermoplastics, *J Appl Polym Sci* **28**:2473-2482.
60. Acevedo M and Fradet A (1993) Study of bulk chain coupling reactions. II. Reaction between bisoxazolones and amine-terminated polyether: synthesis of polyether-block-polyamides, *J Polym Sci Polym Chem Ed* **31**:1579–1588.
61. Okamoto H and Okushita Y (1984) Poly(ether imide amides), JP Patent 60 158,222, to Ube Ind., *Chem Abstr* **104**:130483.
62. Lautenschlager W and Daumiller G (1968) Polyamide polymers having an antistatic finish, GB Patent 1,110,394 A, to BASF AG.
63. Boylan J, Li T, Schlossman I (1973) Poly(ether-ester-amides) used as antistatic agents for textile materials, Fr Patent 2,178,205 to Unilever-Emery N.V., *Chem Abstr* **81**:14675 (in French).
64. Deleens G, Foy P and Maréchal E (1977) Synthesis and characterization of copolycondensate sequences poly(amide-*seq*-ether) – II – Polycondensation of polyamide-11, ω,ω'-diacid or diester oligomers with polyether ω,ω'-dihydroxy oligomers, *Eur Polym J* **13**:343–360 (in French).
65. Foy P, Jungblut C and Deleens G (1974) Polyether-ester-amides as a product for molding or extrusion, FR Patent 2,273,021, to ATO Chimie, *Chem Abstr* **84**:136532 (in French).
66. Laporte P, Fradet A and Maréchal E (1987) Kinetic study on models of the non-catalysed reaction between α,ω-dicarboxypolyamide-11 and α,ω-dihydroxypolyoxyalkylenes, *J Macromol Sci–Chem* **A24**:1269–1287.

67. Mumcu S, Burzin K, Feldmann R and Feinauer R (1978) Copolyetheramide from laurinlactam, 1,10-decanedicarboxylic acid and α,ω-dihydroxy-(polytetrahydrofuran), *Angew Makromol Chem* **74**:49–60 (in German).
68. Goodman I, Rhys J A (1965) Polyester synthesis and manufacture, in *Polyesters – Volume I – Saturated Polymers* (Eds. Ogorkiewicz R M and Ritchie P D) Iliffe books, London, American Elsevier Publishing Company Inc., New York, pp. 13–37.
69. Matsuo T (1986) Manufacture of polyether-polester-polyamide elastomers, JP 63 097,631 A, to Asahi Chemical Industry Co., *Chem Abstr* **109**:171975.
70. Hashimoto H, Wakumoto H, Nagai K and Todo A (1986) Manufacture of polyamide-polyester-polyether block copolymers, JP Patent 62 246,930, to Mitsui Petrochemical Ind., *Chem Abstr* **109**:55457.
71. Hashimoto H, Wakumoto H, Nagai K and Todo A (1986) Modified polyamide elastomers, JP Patent 63 037,125, to Mitsui Petrochemical Ind., *Chem Abstr* **109**:39241.
72. Hashimoto H, Wakumoto H, Nagai K and Todo A (1986) Modified polyamide elastomers, JP Patent 62 285,920, to Mitsui Petrochemical Ind., *Chem Abstr* **108**:188268.
73. Tsai J H, Chiou J S and Twu Y K (1998) Manufacturing process of polyether-ester amide elastomer and elastic fiber, US Patent 5,917,000, *Chem Abstr* **131**:45914.
74. Boulares A, Tessier M and Maréchal E (2000) Synthesis and characterization of poly(copolyethers-*block*-polyamides) II. Characterization and properties of the multiblock copolymers, *Polymer* **41**:3561–3580.
75. Boulares A, Tessier M and Maréchal E (1998) Synthesis and characterization of poly(copolyethers-*block*-polyamides) I. Structural study of polyether precursors, *J Macromol Sci – Pure Appl Chem* **A35**:933–953.
76. Judas D, Sage J-M (1996) Polymers comprising both polyamide-diacid/polyetherdiol blocks and polyamide-diacid/polyetherdiamine blocks, and their preparation, US Patent 5,574,128A, to Elf Atochem, *Chem Abstr* **123**:84350
77. Maj P and Forichon N (1992) Block polyetheramides, US Patent 5,166,309, to Elf Atochem S.A., *Chem Abstr* **118**:102972
78. Niesten M C E J, Bosch H and Gaymans R J (2001) Influence of extenders on thermal and elastic properties of segmented copolyetheresteramides, *J Appl Polym Sci* **81**:1605–1613.
79. Maj P and Cuzin D (1992) Block copolyetheramides, US Patent 5,213,891, to Elf Atochem S.A., *Chem Abstr* **119**:10281
80. Tanaka C, Kondo M and Yamamoto Y (1984) Polyether ester amides, JP Patent 60 170,624, to Toray Ind., *Chem Abstr* **104**:19959
81. Chung L Z, Kou D L, Hu A T and Tsai H B (1992) Block copolyetheramides. II. Synthesis and morphology of Nylon-6 based block copolyetheramides, *J Polym Sci Polym Chem Ed* **30**:951–953.
82. Toshifumi U and Taichi I (1987) Preparation of polyether-polyester-polyamide elastomers with good heat and chemical resistance, JP Patent 63 183,929, to Asahi Chemical Ind., *Chem Abstr* **110**:77256.
83. Imanishi T and Matsumoto T (1986) Manufacture of block polyether-ester-amide elastomers by melt polymerization, JP Patent 63 035,622, to Asahi Chemical Ind., *Chem Abstr* **109**:151198.
84. Hashimoto H, Todo A, Wakumoto H and Nagai K (1986) Modified polyamide elastomers, JP Patent 62 285,920, to Mitsui Petrochemical Ind., *Chem Abstr* **108**:188268.
85. Hashimoto H, Wakumoto H, Nagai K and Todo A (1986) Manufacture of polyamide-polyester-polyether block copolymers, JP Patent 62 246,930, to Mitsui Petrochemicl Ind., *Chem Abstr* **109**:55457.

86. Ichihara E and Senda H (1995) Antistatic agent imparting no coloration to thermoplastic resins, JP Patent 7,145,368 to Sanyo Chem Ind Ltd, *Chem Abstr* **123**:288408.
87. Liedlof H J (1983) Block poly(ether ester amides), DE Patent 3,428,404, to Inventa A.G., *Chem Abstr* **102**:204499.
88. Tanaka C, Kondo M and Yamamoto Y (1984) Polyether ester amides, JP Patent 60 170,624, to Toray Ind., *Chem Abstr* **104**:19959.
89. Tanaka C, Kondo M and Yamamoto Y (1984) Poly(ether ester amide), JP Patent 60 228,533, to Toray Ind., *Chem Abstr* **104**:187036.
90. Laporte P, Fradet A and Maréchal E (1987) Kinetic study on models of the tetrabutoxyzirconium-catalyzed reaction between α,ω-dicarboxypolyamide-11 and α,ω-dihydroxypolyethylene, *J Macromol Sci–Chem* **A24**:1289–1302.
91. Okitsu K and Go H (1983) Poly(ether ester amide), US Patent 4,536,563, to Daicel Chemical Ind., *Chem Abstr* **103**:124124.
92. Yoko F and Chiaki T (1983) Stabilizers for polyester-polyether-polyamide rubbers, JP Patent 60 053,558, to Toray Ind., *Chem Abstr* **103**:143217.
93. Yoko F and Chiaki T (1983) Stabilizers for polyester-polyether-polyamide rubbers, JP Patent 60 053,557, to Toray Ind., *Chem Abstr* **103**:55273.
94. Yoko F and Chiaki T (1983) Stabilizers for polyamide-polyester-polyether rubbers, JP Patent 60 015,456, to Toray Ind., *Chem Abstr* **103**:7616.
95. Yoko F and Chiaki T (1983) Stabilizers for polyamide-polyester-polyether rubbers, JP Patent 60 015,455, to Toray Ind., *Chem Abstr* **102**:205262.
96. Deleens G, Foy P and Maréchal E (1977) Synthesis and characterization of copolycondensate sequences poly(amide-*seq*-ether) - III. Study of the polycondensation reaction of ω,ω'-dicarboxy polyamide-11 and ω,ω'-dihydroxy polyoxyethylene. Determination of the rate constants and the activation energy, *Eur Polym J* **13**:353–360 (in French).

Chapter 10

Poly(Ether-*b*-Amide) Thermoplastic Elastomers: Structure, Properties, and Applications

R.-P. Eustache

1. Structure and characterization

Poly(ether-*b*-amide) thermoplastic elastomers (PEBA) are segmented block copolymers where the hard blocks consist of polyamide segments, acting as physical crosslinks, and the soft blocks consist of flexible polyether segments having a glass transition temperature well below room temperature and acting as entropic springs.

Consequently, the PEBA resins are first characterized by the chemical composition of the polyamide block (polyamide 6 (PA6), polyamide 66 (PA66), polyamide 11 (PA11), polyamide 12 (PA12), co-polyamide, *etc.*) and polyether block (poly(tetramethylene glycol) (PTMG), poly(ethylene glycol) (PEG), poly(propylene glycol) (PPG), *etc.*), which governs basic physico-chemical properties, such as melting range, density, or the hydrophilic/hydrophobic balance (see Chapter 9 for more information about the chemical composition and nomenclature of the polyamide and polyether structural units).

The characterization based upon the mass ratio of the polyamide and the polyether structural units (PA/PE ratio) is useful because it concerns morphological aspects and the related mechanical behavior: (i) Products where the polyether matrix is the continuous phase (PA/PE < 1) show good elastic properties in a broad deformation range (100–300%) and a low modulus (10–30 MPa). These materials are often transparent. (ii) Products where the polyamide matrix is the continuous phase (PA/PE > 1) show a higher modulus

(150–700 MPa), but a lower elastic deformation (~20%). (iii) Products with a PA/PE ratio close or equal to unit show an intermediate behavior associated to some extent with a co-continuous phase.

1.1. *Crystallinity and morphology*

The morphology of poly(ether-*b*-amide) thermoplastic elastomers, which are multiphase systems, plays an important role in determining the final properties of the material and depends on many factors including not only the chemical composition but also the thermo-mechanical history of the sample. Its characterization is challenging because these multiphase systems are complicated by a crystallization phenomenon possibly inside the polyamide and polyether phases, together with some interphase mixing.

1.1.1. *Thermal transitions*

Numerous techniques have been used to characterize the PEBA supermolecular structure including light microscopy, transmission electron microscopy (TEM), atomic force microscopy (AFM) [1–4], dynamical mechanical thermal analysis (DMTA) [4–7] solid-state nuclear magnetic resonance (solid-state NMR) [7], small-angle X-ray scattering (SAXS) [3,4], wide-angle X-ray scattering (WAXS) [4], small-angle light scattering (SALS) [4], and differential scanning calorimetry (DSC) [4,5,8–10]. The latter is a common technique, often used to give practical information about the maximal service temperature of the materials, as well as additional information relative to the product composition. For instance, Figures 1 and 2 show the effect of the PA12 and PTMG average block length on the thermal response. In Figure 1, one observes the melting endotherm of the polyether crystalline phase below room temperature for PEBA Samples A, B, C, and D with an average PTMG block length equal to 2000 g/mole (see Table 1). This endotherm is not observed in samples E, G, H, and I where the average PTMG block lengths are 1000 and 650 g/mole, respectively (Figure 2).

Table 1. PEBA samples with various PA12 and PTMG block lengths (molecular weight, M_n) and PA/PE ratios

Sample	M_n PA12	M_n PTMG	PA/PE (wt%)
A	600	2000	23/77
B	850	2000	30/70
C	950	2000	38/62
D	2000	2000	50/50
E	1000	1000	50/50
F	1500	1000	60/40
G	2000	1000	67/33
H	4000	1000	80/20
I	5000	650	88/12

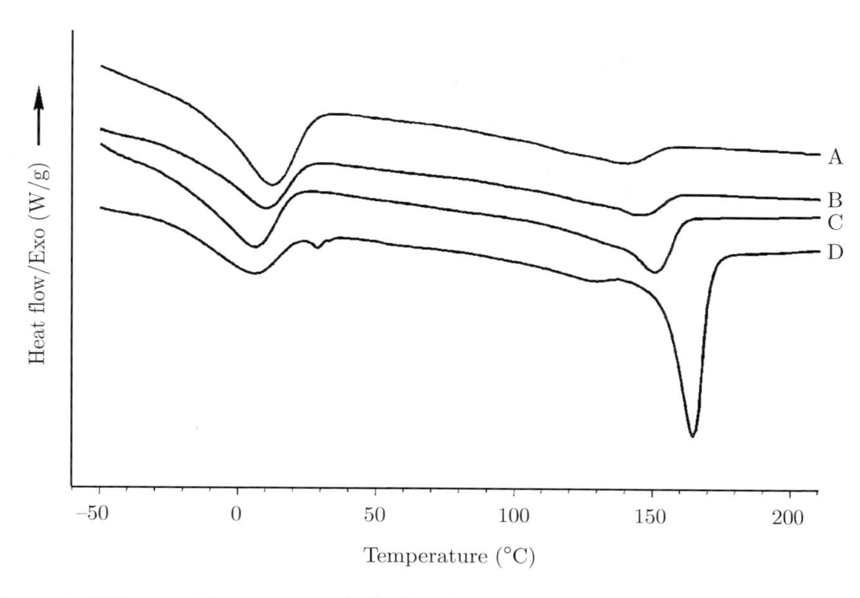

Figure 1. DSC second heating scan (20°C/min) of PEBA Samples A, B, C, D (see Table 1) with a PTMG block molecular weight of 2000 g/mole

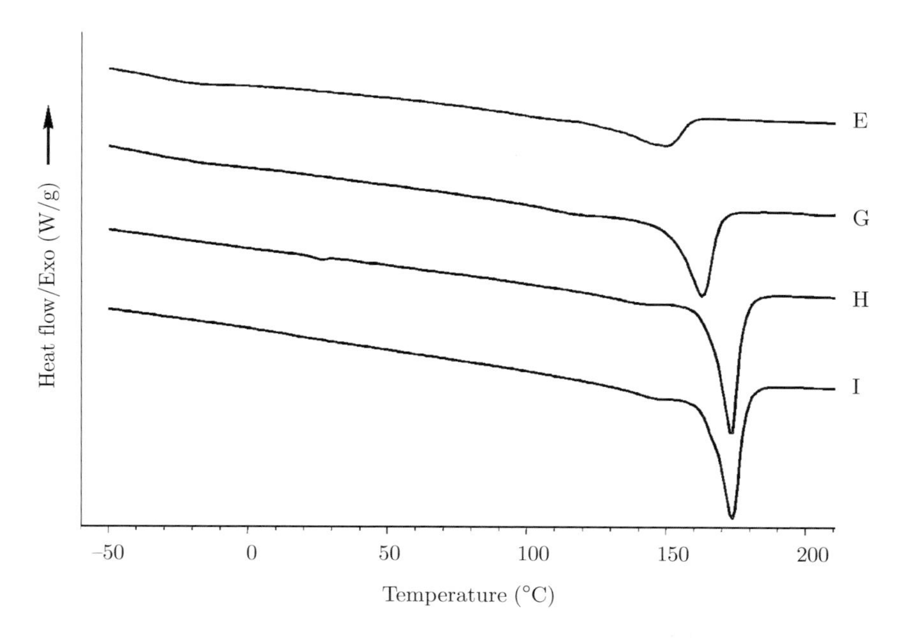

Figure 2. DSC second heating scan (20°C/min) of PEBA Samples E, G, H, I (see Table 1) with an average PTMG block molecular weight below 2000 g/mole

For all samples from A to I, the PA12 melting temperature increases from 130 to 180°C when the PA12 block length varies from 600 up to 5000 g/mole. A relationship can be established for a defined polyamide chemical composition, here PA12 with adipic acid as chain terminator. Thus, knowing the thermal response and the chemical composition, it is possible to estimate the average block length of the polyamide structural unit and *vice versa*. (About the quite similar thermal behavior of poly(ether ester) TPE, see Chapter 6.)

The effect of stretching on the thermal response and on the mechanical behavior of PEBA resins has been reported in the literature [8,9]. Figure 3 shows this effect in Sample A (see Table 1) strained from 0% up to 620%. A new sharp melting endotherm appears above room temperature, indicating strain-induced crystallization of the PTMG segments. This crystallization during elongation gives some stress hardening in the tensile curves at high strain and is responsible for a residual deformation after stress release. Due to the low melting point of these oriented crystals, annealing above 60°C can destroy them and allow the recovery of the initial shape to a great extent. This phenomenon is known as "thermo-shrinkability" and is used in some specific applications (see also Chapter 18).

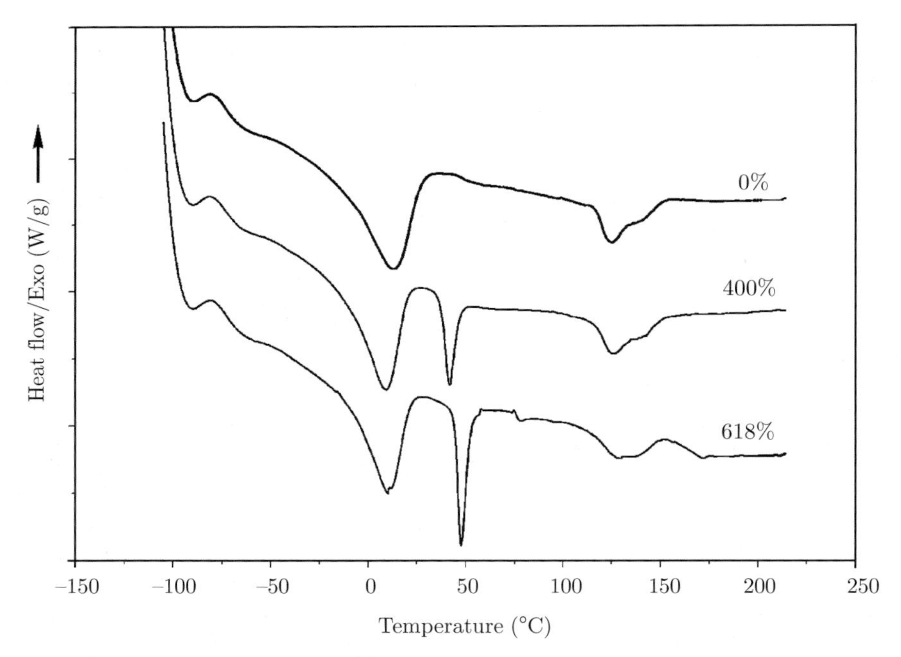

Figure 3. Thermal response (first heating, 20°C/min) of Sample A (see Table 1) stretched at various strains. Measurements performed on prestrained samples

Storage can also affect the product thermal response and modify to some extent the final properties. Figure 4 shows a first heating scan of Sample A

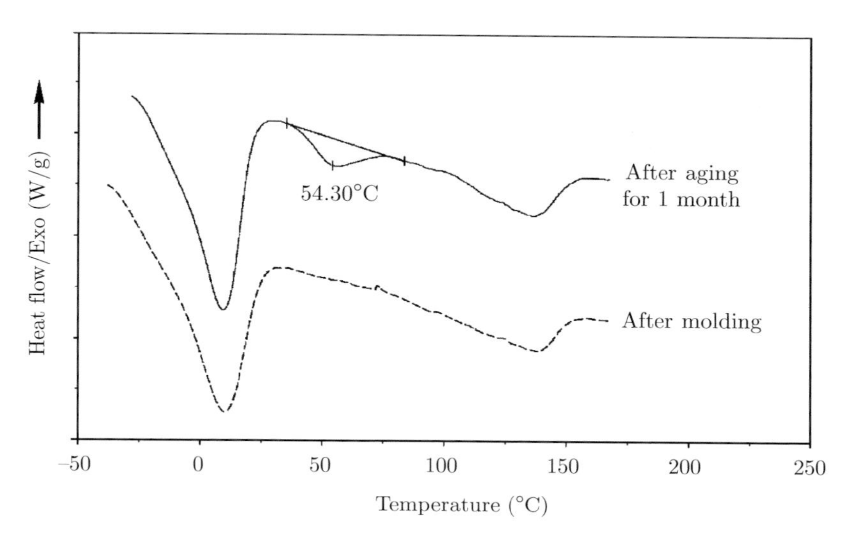

Figure 4. Effect of storage on the thermal response of Sample A (see Table 1)

(see Table 1) just after molding and after storage for one month at room temperature. A new non-negligible endotherm near 50°C is observed after storage. According to studies by Marand *et al.* [11] on other semicrystalline polymers, this "physical aging" observed at a temperature above the polyamide and polyether glass transitions, could be attributed to a secondary crystallization phenomenon inside the polyamide-rich phase during the storage period. This new melting endotherm is related to an increase in the product elastic modulus during the storage period.

1.1.2. *Microscopy and lamellar type morphology*

The crystalline morphology of poly(ether-*b*-amides) has been investigated by light microscopy, as well as by TEM and AFM [1–4]. The polyamide segments crystallize in lamellar structures whatever the PA/PE ratio. Figure 5 shows these PA12 lamellae forming a physical network. The weight percent of these lamellar structures and their interconnectivity in the product is directly related to some mechanical properties, such as elastic modulus and Shore D hardness.

These lamellar structures are able to self-organize in larger spherulitic superstructures easily observed in the case of the rigid grades (PA/PE > 1). The size of the spherulitic superstructures is mainly responsible for the lack of transparency of the rigid PEBA grades, compared to the soft grades. Their size distribution inside a sample depends not only on the PEBA composition, but also on the processing conditions. In Figure 6, polarized light microscopy shows the spherulites in a 2 mm thick injection molded plate. From SALS measurements (Figure 7), the size distribution of the spherulites as a function of the depth from the surface can be deduced (Figure 8).

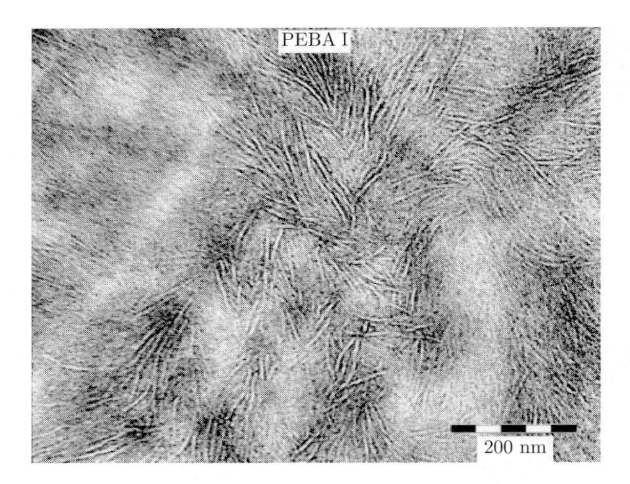

Figure 5. Trasmission electon micrograph of PEBA Sample I (see Table 1).

AFM observations of some soft PEBA grades (PA/PE < 1) under tensile stretching experiments also reveal how the lamellar structures first reorient in the tensile direction before breaking at larger strain and then lead to plastic deformation [2].

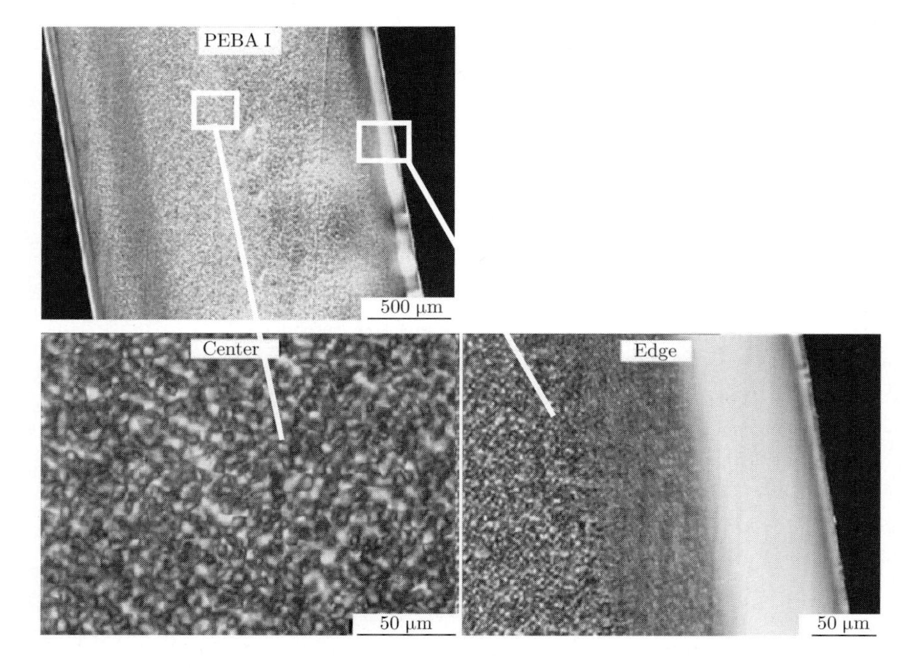

Figure 6. Optical micrograph (in polarized light) of an injection molded plate, 2 mm thick, from Sample I (see Table 1)

Figure 7. SALS patterns of Sample I (see Table 1) at different depths from the plate surface (injection molded), as indicated to the patterns

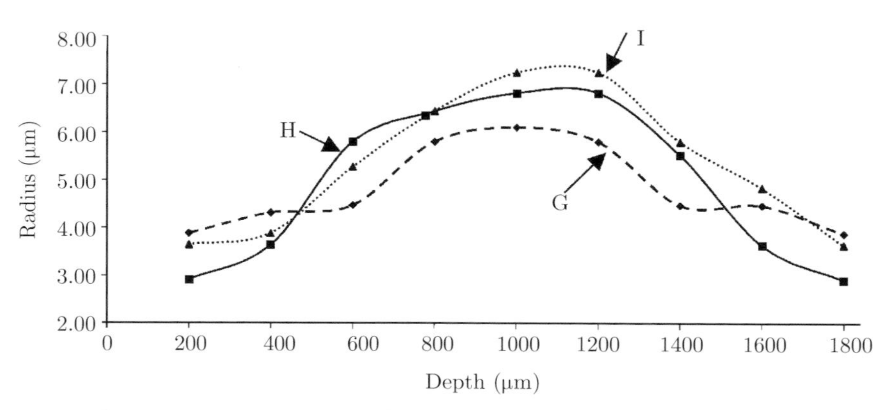

Figure 8. Spherulite size distribution (radius *vs.* depth) in 2 mm thick injection molded plates of PEBA Samples G, H, and I (see Table 1)

1.2. *Microphase separation vs. hard and soft segment block length*

The multiple relaxations in PEBA copolymers have been studied by dynamic mechanical measurements, dielectric measurements, and thermally stimulated current methods [4–7,12,13]. These relaxation methods, in accordance with other spectroscopic techniques, such as solid-state NMR, show that a well defined phase separation occurs in PEBA resins and depends on the nature and the molecular weight of the polyether and polyamide structural units. (These methods are considered also in Chapter 14.)

1.2.1. *Dynamic mechanical thermal analysis*

DMTA is widely used in the literature to characterize microphase separation. The position and origin of the various DMTA relaxation regions in poly(ether-*b*-amide) have been discussed by many research groups [4–7]. Figure 9 shows the evolution of the storage modulus, E', and $\tan\delta$ signals with temperature for various PA/PE ratios. A drop in the storage modulus E' and a peak in the $\tan\delta$ curve characterize the relaxations. Following the standard nomenclature, the multiple relaxations observed in the $\tan\delta$ curves are labelled α, β, and γ, from higher to lower temperatures. The γ peak appearing at about –113°C is

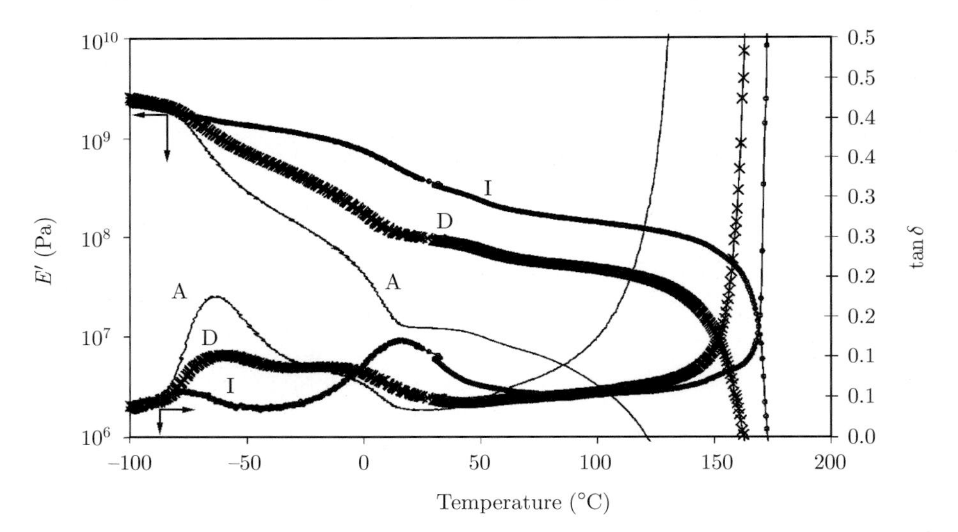

Figure 9. DMTA of PEBA Samples A, D, and I (see Table 1) with various PA/PE ratios: evolution of storage modulus, E', and tan δ with temperature

attributed to local motions of CH_2 groups inside the polyamide and polyether phases and is not shown in Figure 9. The β region between –80°C and –50°C is assumed to be a result of two relaxations with respective magnitude depending on the PA/PE ratio: the main relaxation is the glass transition of the PTMG-rich phase and is clearly predominant for soft PEBA grades (PA/PE < 1). This relaxation can overlap with a secondary relaxation caused by local motions of the free and water-bonded amide groups, well known in the polyamides [14], and observed more easily in the rigid PEBA grades (PA/PE $\gg 1$, see Sample I in Figure 9). The α region, between –20°C and +40°C, is assumed to be the glass transition of the polyamide-rich phase. This relaxation can overlap with another relaxation coming from the melting of PTMG segments when their block molecular weight is equal to or above 2000 g/mole.

The attribution of two well separated glass transitions to the respective amorphous polyamide- and polyether-rich regions [4–7] is in agreement with a well defined microphase separation inside the amorphous regions. It is well documented also for TPEs of the poly(ether ester) type (see Chapter 6).

As indicated in the literature [4,6], although the β peak shows only little variation with the PA/PE ratio, the α peak decreases in temperature for lower PA/PE ratios and a shorter polyamide block length as observed in Figure 10. Such a lowering of the α peak is attributed to an internal plasticization coming from oligoether miscible components inside the polyamide-rich regions. Another possible interpretation could be that, since the immiscible polyamide and polyether components are linked by chemical bonds, the structure and mobility of each block affect the others [15].

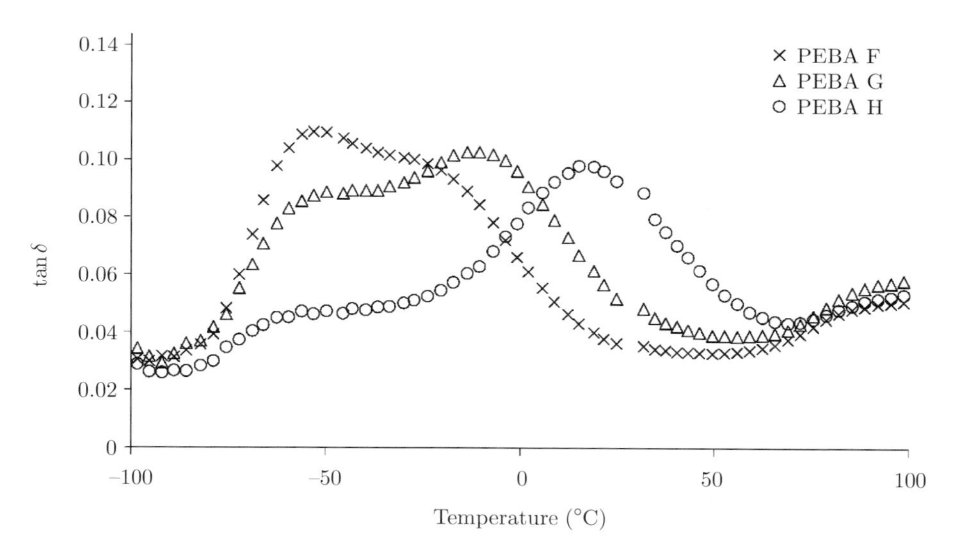

Figure 10. DMTA (tan δ *vs.* temperature) of PEBA Samples F, G, and H with a constant PTMG block length and PA12 block lengths ranging from 1500 to 4000 g/mole

1.2.2. *Spectroscopic investigations of phase organization*

Spectroscopic studies, such as solution NMR and infrared spectroscopy, are often used in order to confirm the PEBA chemical composition [4,7] and to quantify the PA/PE ratio. Few authors have used spectroscopic studies to characterize the PEBA solid-state organization [7]. Figure 11 illustrates an NMR solid-state ^{1}H-Wideline Separation (WISE) experiment with Sample H at room temperature, where the first dimension corresponds to the ^{13}C chemical shifts and provides information about the chemical structure. The second dimension corresponds to the ^{1}H broadband lines and gives information about local fields associated with ^{1}H dipolar couplings somewhat modulated with molecular motions [16]. This experiment reveals a distinct mobility for specific chemical groups of the macromolecular chains associated with polyamide and polyether blocks, respectively. It is an additional confirmation of the good phase separation, especially for PEBA samples where the interpretation of the DMTA spectra is rendered difficult by multiple overlapping relaxations. In Figure 11, PA12 is found to be rigid and PTMG highly mobile, thus confirming the good phase separation between PTMG-rich and PA12-rich regions in Sample H.

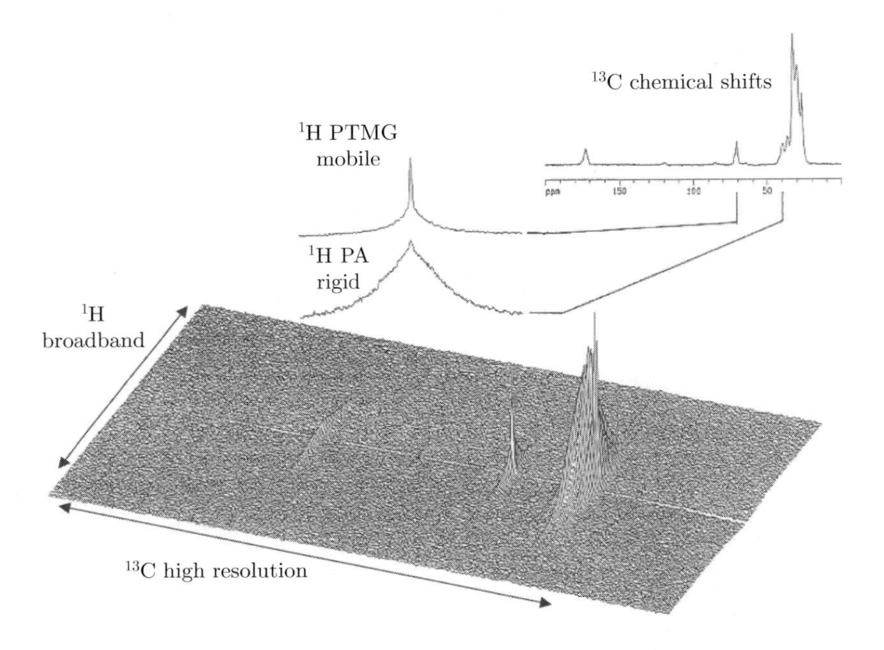

Figure 11. Solid-state NMR WISE experiment with Sample H (see Table 1)

The PTMG crystallization at room temperature of stretched PEBA samples can be also observed in solid-state NMR ^{13}C experiments with cross polarization and magic angle spinning (^{13}C CP-MAS). Figure 12 reveals two ^{13}C chemical shifts at 71.2 and 72.8 ppm for the CH_2–O groups associated with the amorphous and crystalline PTMG regions, respectively. The peak at 72.8 ppm is observed in Figure 12 only for the sample stretched to 400%.

Infrared experiments provide information about the hydrogen-bonding network inside the amorphous polyamide regions [17], which for PEBA depends not only on the temperature, but also on the PA/PE ratio.

2. Properties

Poly(ether-*b*-amide) resins can be hydrophobic or hydrophilic, depending on their chemical composition: PA11, PA12, and PTMG blocks increase hydro-

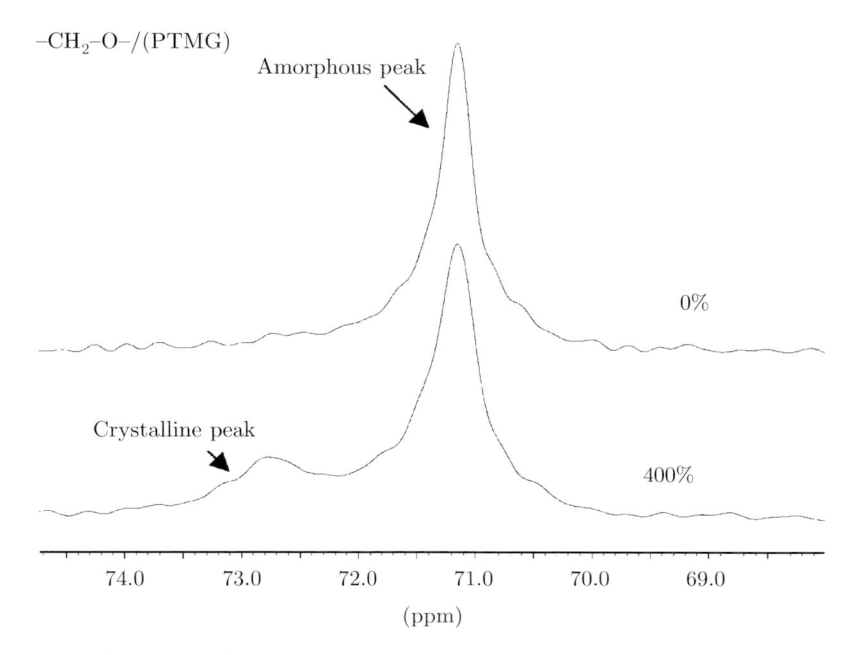

Figure 12. Solid-state CP-MAS spectra at room temperature of Sample A (see Table 1) stretched at 0% and 400%

phobicity while PA6, PA66, and PEG blocks increase hydrophilicity. Depending on the properties required for a given application, like antistatism, breathability, solvent resistance, density, or mechanical behavior, the appropriate chemical composition of the segmental units can be selected.

Due to the considerable morphological changes observed by modification of the PA/PE ratio, the PEBA resins also reveal a great evolution of their mechanical properties from soft (PA/PE < 1) to rigid (PA/PE > 1) grades and the PEBA family is known to bridge the gap between elastomers and thermoplastic products.

2.1. *Mechanical properties*

Figure 13 shows tensile stress-strain curves at room temperature of PEBA with various PA/PE ratios. The rigid grades (Samples G, H, and I with PA/PE > 1, see Table 1) show a small (several %) and linear elastic domain, followed by a well defined yield point, typical of the thermoplastic behavior. As shown in Table 2, the higher the polyamide content, the higher are the tensile strength, yield stress, tensile and flexural modulus values. In the case of the soft grades (Samples A and B with PA/PE < 1, see Table 1), the curves show a more pronounced elastomeric behavior and a yield point is not observed. Sample E with PA/PE $= 1$ reveals an intermediate behavior between rigid and soft grades.

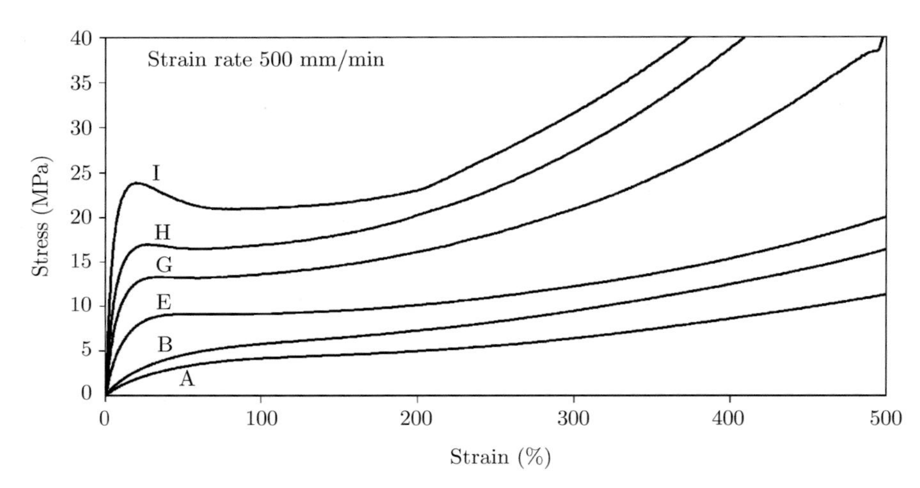

Figure 13. Stress-strain curves at room temperature of PEBA Samples A to I (see Table 1) with various PA/PE ratios

Although the PA/PE ratio and the molecular weights of the polyamide and polyether segments are of prime importance, the thermo-mechanical history of the product can also affect, though to a lower extent, the mechanical properties. As shown in Figure 14, an appropriate annealing of a soft PEBA grade (Sample A, Table 1) not only reduces the tensile modulus and affects the general shape of the stress-strain curve, but also improves its elastic behavior. The plastic deformation and hysteresis for a given strain were found to be reduced. This type of treatment can be of interest for applications where very good elastic properties are required.

As shown in Figure 15, the PEBA grades exhibit low temperature and frequency dependences of their storage modulus, E', in dynamic tensile experiments. They keep a good flexibility when tested well below room temperature. Figure 16 compares the tensile curves of PEBA Sample G and H (Table 1) with an aromatic thermoplastic polyurethane elastomer (TPU) of lower modulus at room temperature. It is seen in Figure 17 that at –20°C this TPU is more rigid than Sample G. The low level of rigidity of the PEBA resins, well below room temperature, is related to the good phase separation observed inside the materials and to the very low glass transition temperature of the polyether-rich phase.

Among other interesting properties of PEBA related to the low glass transition temperature of the polyether-rich phase, one should point out the high impact resistance even at temperatures as low as –40°C (Table 2). Also shown in Table 2 are the very interesting dynamic properties of PEBA resins: a low hysteresis under a given deformation, high resilience, and good resistance to alternating flexure even at low temperatures. Tear and abrasion resistance are also kept on a good level (see also Chapter 17).

Table 2 Mechanical properties of hydrophobic PEBA grades with various PA/PE ratios (see Table 1)

Sample	A	B	E	G	H	I
Modulus of elasticity in flexure (MPa) *ISO 178*	15	20	85	170	290	390
Tensile modulus (MPa) *ISO 527*	10	18	81	161	285	385
Shore D hardness (15 s)	22	26	41	50	58	61
Yield point *ASTM D 638*						
Stress at yield (MPa)	–	–	–	12	17	24
Elongation at yield (%)	–	–	–	25	20	18
Tensile strength at break (MPa)	32	39	42	52	53	54
Elongation at break (%) *ASTM D 638*, 50 mm/min	>800	>650	>550	>450	>350	>350
Impact strength (IZOD) at –40°C *ASTM D 256*, NB = no break						
Notched (–40°C) J/M	NB	NB	NB	NB	109	49
Unnotched (–40°C) J/M	NB	NB	NB	NB	NB	NB
Resistance to cracking by repeated flexure (mm). De Mattia test at –20°C/500000 flexures *ISO 133*	2.5	2.5	-	9	14	-
Tear resistance (KN/m) Unnotched *ISO 34-1*	66	78	115	135	147	177
Abrasion resistance (mm^3) *ISO 4649*	107	96	38	46	34	–
Rebond resilience (%) *ISO 4662*	70	70	65	60	55	55
Hysteresis (Atofina test)						
5%	9.0%	12.0%	–	19.5%	22.0%	–
10%	9.2%	12.1%	–	22.2%	24.2%	–
20%	9.4%	13.9%	–	29.8%	33.0%	–

2.2. *Physico-chemical properties*

Table 3 shows the main physical properties of poly(ether-*b*-amide)s with different PA/PE ratios and chemical structural units. Hydrophilic PEBA Samples K and L (see Table 3) have higher density and water absorption and lower surface resistivity than hydrophobic grades A, D, G, I from various PA/PE ratios. The hydrophilic K and L grades have inherent static dissipative properties and are used in specific antistatic applications.

The chemical resistance depends on the chemical nature of the polyamide and polyether blocks and on the PA/PE ratio. The rigid hydrophobic grades, with polyamide as the continuous phase, have a good resistance to the majority

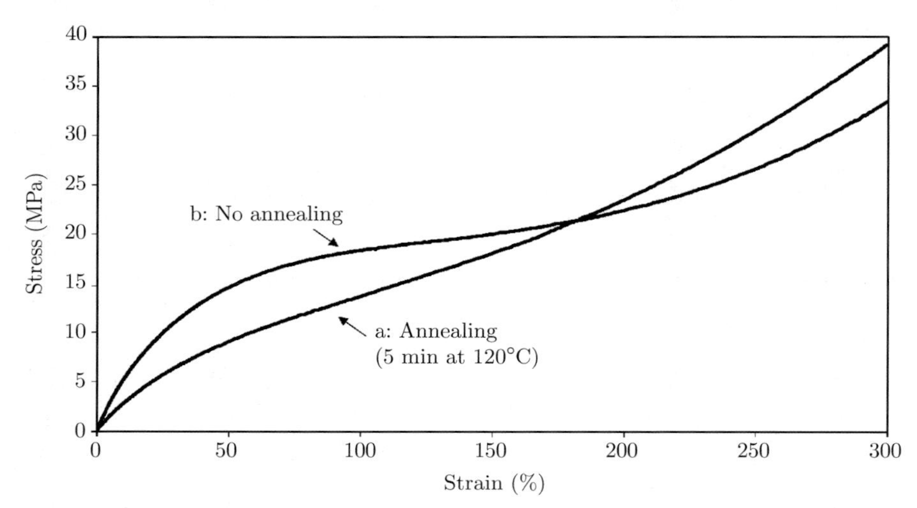

Figure 14. Sample A (see Table 1) with (curve a) and without (curve b) annealing (5 min at 120°C).

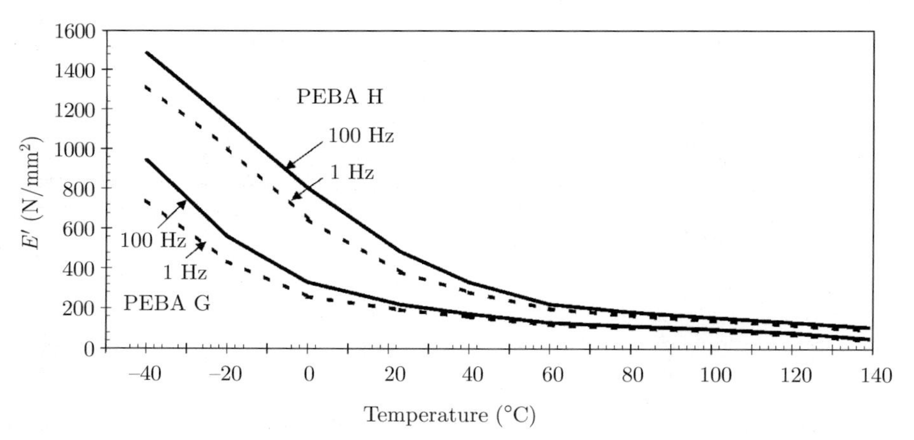

Figure 15. Evolution of the storage modulus, E', of PEBA Samples G and H (see Table 1) with temperature at two frequencies of 1 and 100 Hz

of chemical agents; soft hydrophobic grades can more easily swell in different solvents, but without significant chemical degradation (see also Chapter 11).

2.3. *Processing*

Poly (ether-*b*-amide) products are known to have an excellent processability in injection molding, extrusion (cast films, blown films, sheets, tubes, *etc.*), as well as in assembly processes like overmolding and co-extrusion. This behavior allows the design of thin and sophisticated-shaped materials associating lightweight and good mechanical properties. In order to achieve the best processing

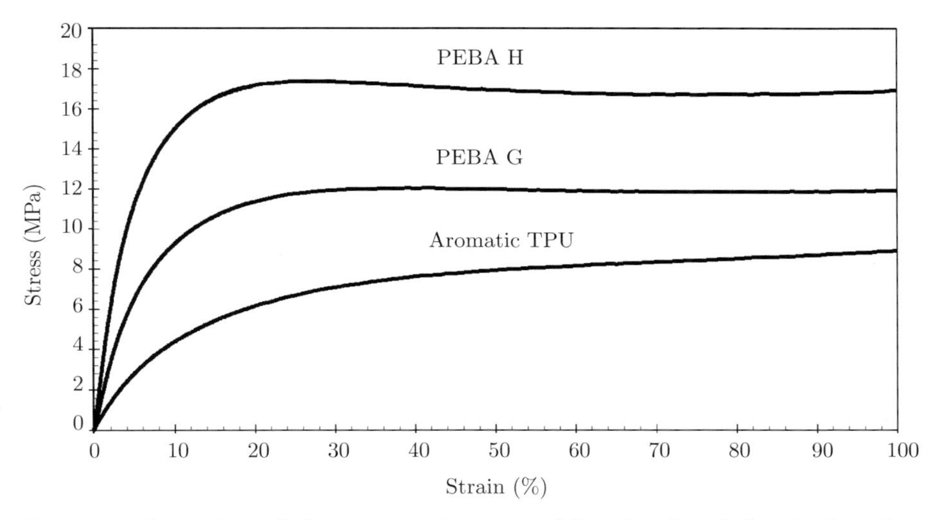

Figure 16. Comparison of the stress-strain curves of Samples G and H with that of an aromatic TPU at room temperature

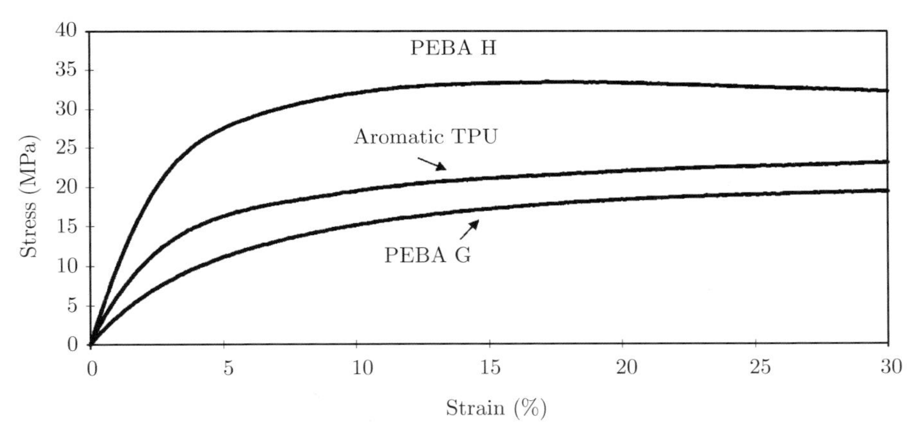

Figure 17. Comparison of the stress-strain curves of Samples G and H at –20°C with that of the same aromatic TPU as in Figure 16

conditions, the semicrystalline character of the PEBA grades, as well as their rheological properties and moisture uptake should be taken into account. The melting point and heat of fusion (see Table 3) help to select the correct processing temperature (about PEBA processability see also Chapter 17).

3. Applications

Among the PEBA family, the PEBAX® commercial resins from Atofina cover a large range of hardness, from 25 Shore D up to 72 Shore D, without any

Table 3. Physical properties. See Table 1 for Samples A, D, G, I. Samples K and I: hydrophilic PEBA grades with PEG as polyether block, and PA12 and PA6 as polyamide block, respectively

Sample	A	D	G	I	K	L
Density (g/cm^3) *ISO 1183*	1.00	1.01	1.01	1.01	1.07	1.14
Melting point (°C) *ISO 11357*	134	160	159	172	158	204
Water absorption at equilibrium (20°C, 50% RH) *ISO 62*	0.4%	0.5%	0.6%	0.7%	1.4%	4.5%
Melt index (g/10 mm) (235°C/1 kg/die 2 mm) *ASTM D 1238*	10	5	7	6	–	–
Linear coefficient of thermal expansion from –40°C to 140°C (1/(10^5 K)) *ISO 11359*	20	19.5	17	16	–	–
Surface resistivity at 20°C and 65% RH (Ohm)*IEC 60093*	310^{12}	510^{12}	310^{12}	410^{13}	310^{8}	1.510^{9}

addition of plasticizers or reinforcing fillers [18,19]*. This is due to the large variation of PA/PE ratios achievable on the industrial scale. The commercial PEBAX® grades find a great variety of applications, depending on the chemical nature of the polyamide and polyether structural units used.

3.1. *Sporting goods*

The sporting goods industry is using the dynamic properties, high resistance to cracking during alternating flexure, and elastic recovery properties of PEBAX® resins, combined with their easy processing, overmolding ability and low density, to produce sole plates, shanks and various components, such as torsion systems, ankle support or straps used in athletic footwear. Many manufacturers of soles with cleats also use these properties in their high-end models to make, *e.g.*, outsoles for football, soccer, rugby, or baseball. The tennis and golf industries also use PEBAX® resins and take advantage of the dynamic properties in their design.

Ski blades, mountaineering, Nordic and Alpine ski boots use the additional high flexibility and high impact characteristics of the PEBAX® in cold conditions, as well as its lightweight.

Among other properties of PEBAX® resins used in sport applications is their good decoration ability, for example with silk-screen, printing, or sublimation processes, as well as their good UV properties and good adhesion on different substrates. The design of top-layers for skis and snowboards makes use of such properties.

* As mentioned earlier, PEBA is an abbreviation used for any polyether-*b*-amide thermoplastic elastomer. In 1981, Atochem launched for the first time on the market a new polyether-*b*-amide thermoplastic elastomer product family under the trade name PEBAX®.

3.2. *Medicine*

PEBAX® resins are often used to make small-diameter medical tubing, known as catheters for demanding medical procedures, such as angioplasty, occlusion, or drug delivery systems. These catheter devices require a material with high kink resistance, high flexibility associated with maximum tensile strength and good tear resistance. Some specific catheter applications to measure cardiac contractibility take also advantage of the low damping coefficient of the PEBAX® resins.

3.3. *Industry*

With their very good mechanical properties, easy processing, and nice surface aspects, PEBAX® resins find a lot of diverse applications, *e.g.*, in the automotive, transportation, or telecommunication industries as seals, automotive clip gaskets, air bags covers, pneumatic hoses, *etc.* The very easy processing of the PEBAX® resins allows the fabrication of very thin and complicated shapes, which find various applications, such as small parts of toys. The soft touch of the PEBAX® resins with a low PA/PE ratio (< 1) is used to give a soft grip, *e.g.*, to cameras. The easy coloring and the non-allergenic properties of the resins are used to make watch wrist bands.

3.4. *Breathable structures*

Hydrophilic PEBAX® resins with polyethylene glycol (PEG) ether structural units allow water vapor to go through compact films at a high rate while being water-proof. These films are permeable to other gases, such as CO_2, O_2, N_2, and C_2H_4.

PEBAX® breathable resins are used as pure films or applied onto textiles, and find applications in a wide range of markets including surgical garments, gloves, films for textile lamination for sporting goods, breathable roofing membranes, and food packaging, among others.

3.5. *Fragrance carrier*

Due to the very good phase separation in the solid state between the polyamide and polyether segments, soft PEBAX® resins (PA/PE < 1) can selectively absorb in their polyether phase a very large amount of many volatile molecules used in the perfume industry while keeping a good structural consistency and mechanical behavior preserved by the polyamide network, which is little affected. The low glass transition temperature of the polyether segments and the large amount of easily incorporated molecules allow their diffusion in a way similar to a liquid state. Thus, soft PEBAX® resins find many applications as fragrance carriers with a controlled release and very good restoration properties of the fragrance composition.

3.6. *Polymer additives and polymer components*

PEBAX® resins with PEG structural units allow the dissipation of static charges and are used at a level varying from 5 to 15 wt% in numerous host matrices (ABS, PS, PC, PE, PP, polyamides, and others) in order to form a conductive network within the host matrix and to impart antistatic properties, even under very dry conditions. End-user markets are computer housings, storage containers, and packaging for electronic industry, smart cards, electronic equipment components, and antistatic fibers.

PEBAX® resins are introduced in various TPU resins, in both dry blends and compounded blends in order to improve their processing. This results in lower torque with accompanying increase in productivity.

Due to the good dynamic and impact properties at temperatures as low as –40°C, PEBAX® resins impart flexibility and toughness to polymer alloys. (Applications of PEBAX® are listed in Chapter 17.)

4. Conclusions and outlook

Poly(ether-*b*-amide) thermoplastic elastomers cover a wide range of flexibility, which allows them to bridge the gap between thermoplastic and elastomeric products. These physically or virtually crosslinked elatomers have an excellent processability in injection molding or extrusion and in assembly processes, such as overmolding. Their specific microphase separation and morphology ensure excellent mechanical and dynamic properties even at temperature as low as –40°C which, associated with product lightweight, chemical resistance and easy processing, make PEBA resins the materials of choice in many fields of application and position them in the upper range of quality among the other thermoplastic elastomers.

PEBA industrial synthesis allows the design of various chemical compositions associated with a broad spectrum of properties. New research efforts are made in order to put on the market new resins with improved performance and additional favorable properties.

References

1. Sauer B B, McLean R S and Thomas R R (2000) Nanometer resolution of crystalline morphology using scanning probe microscopy, *Polym Int* **49**:449–452.
2. McLean R S and Sauer B B (1999) Nano-deformation of crystalline domains during tensile stretching studies by atomic force microscopy, *J Polym Sci Polym Phys Ed* **37**:859–866.
3. Sauer B B, McLean R S, Brill D J and Londono D J (2001) Morphology and orientation during deformation of segmented elastomers studied by SAXS and atomic force microscopy, *Polym Mater Sci Eng* **85**:268–269.
4. Sheth J P, Xu J and Wilkes G L (2003) Solid state structure-property behavior of semicrystalline poly(ether-*block*-amide) PEBAX thermoplastic elastomers, *Polymer* **44**:743–756.

5. Xie M and Camberlin Y (1986) Morphological study of poly(ether-*block*-amide)s, *Makromol Chem* **187**:383–400 (in French).
6. Chosh S, Khastgir D and Bhowmick A K (1998) Effect of block molecular wight on the mechanical and dynamical properties of segmented polyamide, *Polymer* **39**:3967–3975.
7. Boulares A, Tessier M and Maréchal E (2000) Synthesis and characterization of poly(copolyethers-block-polyamides) II. Characterization and properties of the multiblock copolymers, *Polymer* **41**:3561–3580.
8. Okoroafor E and Raul J (1991) Cryodilatation of thermoplastic PEBA elastomers, *J Polym Sci Polym Phys Ed* **29**:1427–1436.
9. Konyukhova E V, Buzin A I and Godovsky Yu K (2002) Melting of polyether block amide (Pebax): the effect of stretching, *Thermochim Acta* **391**:271–277.
10. Di Lorenzo M L, Pyda M and Wunderlich B (2001) Reversible melting in nanophase-separated poly(oligoamide-alt-oligother)s and its dependence on sequence length, crystal perfection and molecular mobility, *J Polym Sci Polym Phys Ed* **39**:2969–2981.
11. Marand H, Alizadeh A, Farmer R, Desai R and Velikov V (2000), Influence of structural and topological constraints on the crystallization and melting behavior of polymers. 2. Poly(arylene ether ether ketone), *Macromolecules* **33**:3392–3403.
12. Chosh S, Khastgir D and Bhowmick A K (1988) Influence of block molecular weight on the dielectric properties of segmented polyamides, *Polym Polym Compos* **6**:323–330.
13. Faruque H S and Lacabanne C (1986) Study of multiple relaxations in PEBAX, polyether block amide (PA12 2135 block PTMG 2032) copolymer using the thermally stimulated current method, *Polymer* **27**:527–531.
14. McCrum N G, Read B E and Williams G (1967) *Anelastic and Dielectric Effects in Polymeric Solids,* John Wiley & Sons, New York.
15. Di Lorenzo M L, Pyda M and Wunderlich B (2001) Calorimetry of nanophase-separated poly(oligoamide-*alt*-oligoether)s, *J Polym Sci Polym Phys Ed* **39**:1594–1604.
16. Schmidt-Rohr K, Clauss J and Spiess H W (1992) Correlation of structure, mobility, and morphology by 2D WISE NMR, *Macromolecules* **25**:3273–3277.
17. Coleman M M, Skrovanek J and Painter P C (1986) Hydrogen bonding in polymers: further infrared temperature studies of polyamides, *Makromol Chem Macromol Symp* **5**:21–33.
18. Legge N R, Holden G and Schroeder H E (1987) *Thermoplastic Elastomers, A Comprehensive Review*, Hanser, Munich.
19. Holden G, Legge N R, Quirk R and Schroeder H E (1996) *Thermoplastic Elastomers,* Hanser, Munich.

Chapter 11

Semicrystalline Segmented Poly(Ether-*b*-Amide) Copolymers: Overview of Solid-State Structure-Property Relationships and Uniaxial Deformation Behavior

J. P. Sheth, G. L. Wilkes

1. Introduction

Block copolymers are unique materials that are designed to advantageously utilize the chemical and physical properties of their constituent blocks. Their backbone architecture is generally of the type A-B or A-B-A and often they are synthesized *via* living anionic polymerization, which allows the synthesis of copolymers with a narrow block molecular weight distribution of 1.1 or less. The degree of polymerization of a constituent block is often about 100. Styrene-isoprene-styrene or styrene-butadiene-styrene based triblock copolymers marketed under the trade name Kraton® are examples of commercial A-B-A triblock copolymers. *Segmented copolymers* are generally considered to be a subset of the block copolymer family. They differ from the A-B or A-B-A block copolymers in that the degree of polymerization of a given segment is often an order of magnitude lower. Typically, segmented copolymers consist of linear macromolecules, which are generally synthesized by a step-growth polymerization technique and possess an $(A\text{-}B)_n$ backbone architecture. Another feature that often distinguishes segmented copolymers from block copolymers is their broader polydispersity of *ca.* 2, which is typical of polymers produced by step-growth polymerization. Lycra® Spandex and Hytrel® are examples of the well known commercial segmented polyurethanes and poly(ether ester)s, respectively. Poly(ether-*b*-amide)s, the subject of this chapter, also belong to this class of copolymers.

In the absence of strong and extensive secondary interactions, such as hydrogen bonding (*e.g.,* Spandex®), segmented copolymers are thermoplastic in nature. A relatively large number of variables, such as backbone chemistry, block molecular weight, the overall molecular weight of the copolymer, *etc.,* can be independently controlled to design materials with targeted properties. The constituent blocks or segments are generally chemically incompatible in their desired "service temperature" range and this is promoted by dissimilar solubility parameters of the respective blocks or segments. In fact, the incompatibility between the blocks in the case of diblock copolymers is generally determined by χN, where χ is the Flory-Huggins interaction parameter and N is the overall degree of polymerization [1]. Such an incompatibility would generally give rise to macroscopic phase separation if the blocks were to be simply blended together. However, in block copolymers, covalent links between the constituent blocks prevent macroscopic phase separation and a microphase-separated morphology results instead. Typically, in the normal service temperature range, one block acts as the hard phase and the other acts as the soft phase. The service temperature falls within a "service window". Its lower limit is determined by the soft block glass transition temperature, T_g, or melting temperature, T_m, in case the soft phase can crystallize. On the other hand, the upper limit of the service window is determined by the T_g or T_m of the hard block. The thermal degradation of the copolymer may become a concern well below the highest transition temperature of the hard block and, consequently, may limit the upper processing time and temperature. In copolymers with a highly microphase-separated morphology, it is desirable that the storage modulus, E', exhibit no (or little) dependence on temperature within the service window. Moreover, within this window, when the soft block dominates the copolymer composition, an elastomeric material results wherein the glassy or the crystalline hard microphases act as physical crosslink sites for the rubbery soft phase.

The concept of *physical or virtual crosslinks* was first introduced by Scollenberger *et al.* [2,3] when they demonstrated that linear segmented ester-based thermoplastic polyurethanes exhibited high extensibility and elasticity in a non-vulcanized form. These materials behaved as *thermoplastic* elastomers (TPE), as opposed to chemically crosslinked elastomers, such as vulcanized natural rubber, which cannot be thermally reprocessed.

In fact, thermoplastic polyurethanes (TPU), a subcategory of TPEs, were among the first to be made commercially available. In the United States, TPUs were marketed for the first time in the 1960s by Goodrich, Mobay, and Upjohn under the trade names Estane®, Texin®, and Pellethane®, respectively. In 1972, DuPont commercialized thermoplastic poly(ether ester)s derived from terephthalic acid, tetramethylene glycol, and polytetramethylene oxide (PTMO) under the trade name Hytrel®. In the late 1960s and 70s, various research groups also explored the possibility of synthesizing poly(ether-*b*-amide)s by covalently linking ether blocks to amide blocks *via* amide, urethane, or urea linkages. However, it was not until the discovery of the tetraalkoxide catalyst family by Atochem (now Atofina) that the synthesis of sufficiently high molecular weight poly(ether-*b*-amide)s with ester linkages was made possible on a commercial scale [4–6]. In the early 1980s, Atochem introduced thermoplastic poly(ether-*b*-amide)s, commercially known as PEBAX® [7]. These TPEs consist of linear chains of hard polyamide (PA) blocks covalently linked to soft polyether (PE) blocks *via* ester groups. The molecular weight of the PE and PA blocks varies from *ca.* 400 to 3000 g/mole and 500 to 5000 g/mole, respectively [8,9]. The general structural formula of the poly(ether-*b*-amide)s which contain *ester linkages* can be represented by Scheme 1. This structure is hereafter referred to as PEBA in this chapter. Due to the fact that these TPEs contain polyether as well as polyamide segments which are interconnected by ester linkages, they have been often included in treatises on polyethers, polyamides, and even polyesters [10–12].

$$\mathrm{HO}\left[\underset{\substack{\|\\ \mathrm{O}}}{\mathrm{C}}-\mathrm{PA}-\underset{\substack{\|\\ \mathrm{O}}}{\mathrm{C}}-\mathrm{O}-\mathrm{PE}-\mathrm{O}\right]_n\mathrm{H}$$

Scheme 1

Detailed information on the synthesis of PEBA is presented in Chapters 2 and 10 and also in [13–16]. However, it is worth mentioning here that commercially available poly(ether-*b*-amide) PEBAX® copolymers are synthesized by melt polycondensation of carboxylic acid-terminated amide blocks with poly(oxyalkylene glycol)s. The polymerization is catalyzed by a metal alkoxide $Ti(OR)_4$ and is carried out at elevated temperatures of *ca.* 230 °C under vacuum. These copolymers are marketed for use in many areas, a few of which include sports equipment, automotive components, applications that require polymers with antistatic properties, and also in biomedical applications, such as tubing and catheter balloons [17]. There is also a growing body of literature on the use of PEBAX® in the manufacture of membranes for gas separations. Vestamid® E-Series elastomers marketed by Degussa-Hüls are another commercial example of PTMO-PA12 based PEBA copolymers.

Since the introduction of PEBAX® by Atochem, many research groups have devoted much effort in the study of the morphology and properties of these copolymers, in addition to other PEBA systems consisting of various polyamides, such as nylon 6, nylon 11, nylon 12, *etc.*, and polyesters, such as PTMO, poly(ethylene oxide) (PEO), and poly(propylene oxide) (PPO). This chapter aims to present an overview of these materials; its scope is limited to the solid-state structure-property relationships and uniaxial deformation behavior of the PEBA copolymers. It is divided into two main sections. PTMO-PA12 systems are first presented and, in the second section, systems based on PEO or PTMO as the soft segment and aliphatic PA, such as nylon 6, or aromatic PA as the hard segment are discussed. Thus it is hoped that this chapter will complement Chapters 2 and 10.

2. PTMO-PA12 based copolymers

One of the most widely studied PTMO-PA12 based PEBA copolymers is the XX33 series of the commercially available PEBAX® materials; the XX designation in XX33 represents the Shore D hardness of the copolymer which ranges from 25 to 72 within this series (see Table 1). In addition to these commercial materials, various research groups have also synthesized PTMO-ester-PA12 based PEBA copolymers [14–16]. The structure-property relationships and the uniaxial deformation behavior of the XX33 series and also other

Table 1. Composition of the PTMO-PA12 based PEBAX® XX33 series

Study	P2533	P3533	P4033	P6333	P7033
	Polyether/polyamide (wt%)				
Bondar *et al.* [23], elemental analysis	80/20	—	53/47	—	—
Sauer *et al.* [20]	—	75/25	52/48	—	—
Sheth *et al.* [22], ^{1}H NMR	85/15	84/16	70/30	37/63	25/75
Others [18,19]	78/22	—	—	—	11/89

PTMO-PA12 based PEBA systems will be addressed in this section. As anticipated, the copolymer composition plays a critical role in determining the morphology and properties of the material. Thus, in studies on the PEBAX® XX33 series, investigators have reported the compositions directly [20,21] or have attempted to determine them in their laboratories by means of ^{1}H NMR [22] or elemental analysis [23]; the calculation of composition in the latter method is based on the fact that nitrogen is only present in the PA repeat unit. From the PEBAX® XX33 compositions listed in Table 1, it is evident that there is a wide variation in the reported composition of a given XX33 sample, even though all studies exhibit the expected trend of increasing Shore D hardness with increasing PA content. In light of the above fact, it is worth noting that based on the synthesis procedure, at a given soft segment (*i.e.,* PE) molecular weight, an increase in the PA content can only be achieved by increasing the PA segment length. To achieve a higher PA content beyond a certain weight percent and still synthesize a *segmented* copolymer, reduction in the starting soft segment molecular weight becomes necessary. From the ensuing discussion, we will see that the issue of segment length is pivotal in explaining the morphology and the thermo-mechanical properties of PEBA copolymers.

2.1. *Mechanical properties*

2.1.1. *Dynamic mechanical thermal analysis*

The dynamic mechanical thermal analysis (DMTA) response of the XX33 series, more specifically P2533, P3533, P4033, P6333, and P7033, reported in a previous publication from our laboratory [22], is shown in Figure 1a,b which provides the data for E' and $\tan\delta$, respectively. The films utilized for this study were compression-molded from the melt. After the application of pressure, the hot mold with a sample film sandwiched between the metal platens was removed from the device and allowed to cool to ambient temperature over a period of *ca.* 15 min. The DMTA scans were conducted at 2 °C/min and 1 Hz under a dry nitrogen atmosphere. It is seen in Figure 1a that the storage modulus behavior tends to display four particular regions although there are some additional subtleties. In brief, beginning at about −100 °C, the systems with the highest PE content undergo an initial softening, which arises from the glass transition behavior of the PE segment. As expected, this transition is not particularly distinct for the two materials with the highest PA contents (P6333 and

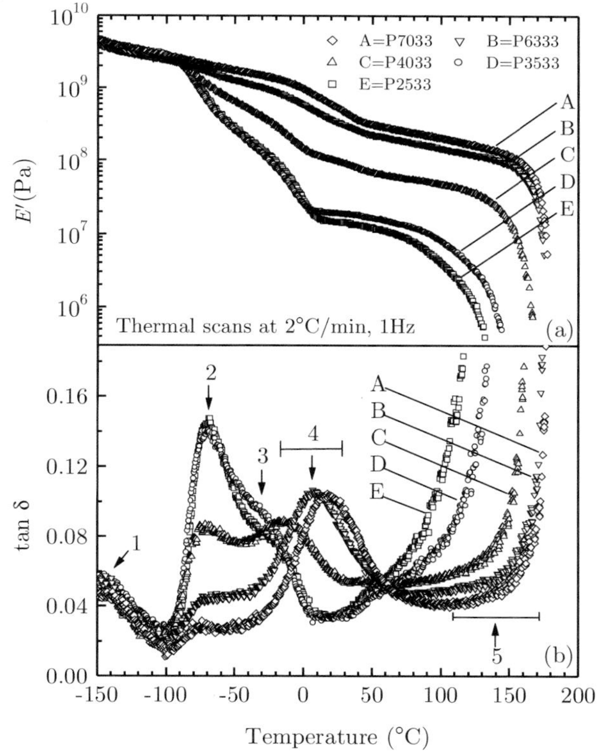

Figure 1. DMTA scans of slowly cooled, melt compression-molded PEBAX® films: (a) storage modulus, (b) tan δ. Transitions are numbered 1 to 5, transitions 4 and 5 being indicated by a general range of temperatures [22]

P7033). Following this transition, there is a "rubbery" plateau behavior in the modulus from about 0 °C to 150 °C, with extensions to higher temperatures for the PA-richer materials. Obviously, this general plateau region is typically the "service window" for many of the applications of these materials, since there is a relatively low dependence of stiffness on temperature. Above the rubbery plateau region, a decline in E' occurs that depends on the melting characteristics of the PA segment; its onset occurs systematically at lower temperatures as the T_m decreases (lower PA content). Further support of this latter observation will also be provided in the discussion of the DSC analysis.

Possibly more revealing with respect to the thermal transition response as noted by DMTA are the corresponding $\tan\delta$ data given in Figure 1b. It can be noted that the transitions are labeled from 1 to 5, both transitions 4 and 5 being indicated by a general temperature range. Beginning with the low temperature region, transition 1 is really a convolution of separate transitions that occur in the respective amorphous PA and PE phases, since both homopolymeric nylon 12 and PTMO display small transitions in this same region arising from the local motion of the methylene sequences in each polymer [24]. Transition 2 distinctly arises from the PTMO glass transition, as confirmed by numerous reports in the literature of such a behavior in segmented polymers containing PTMO as a soft segment [24–26]. It should be mentioned that, as would be expected, transition 2 is at about −70 °C, and is consistent with the soft segment molecular weight range and the relative PE content of the respective samples. Particularly noteworthy are the nearly equal magnitudes of transition 2 in P2533 and P3533, respectively, which is consistent with their almost equal PTMO contents (Table 1). The dependence of the magnitude of transition 2 on the PE content of the sample provides support for the microphase separation behavior of this component. Furthermore, the T_g of the PE phase is relatively independent of the PE content of the sample and occurs at a higher temperature than the T_g of neat PTMO (reported to be at −75 °C by DMTA [27]). The higher T_g of the soft segments no doubt arises from the restrictions placed at the soft segment chain ends by the covalently linked hard segments. On the other hand, the PE content-independent T_g of the soft segments suggests the presence of a PE amorphous phase consisting largely of PTMO segments mixed, at best, with only a small amount of nylon 12 segments. What stands out is the fact that the upper side of transition 2 influences the magnitude of transition 3, which arises from the partial cold crystallization and melting behavior of the PTMO phase where melting is completed at about 10 °C. Further strong support for this interpretation is provided in Section 2.2. However, in the range of transition 3 and extending upward into the region designated as transition 4, the damping increases again because of the glass transition characteristics of the amorphous fraction of the PA component. As is noted for samples P4033, P6333, and P7033, the peak designated as transition 4 is particularly distinct and shifts to higher temperatures as the overall PA content of the sample increases. The other two PA containing segmented materials do not display a significant response of the PA glass transition temperature. Such a behavior is due to the fact that the PA segments are shorter and possibly further mixed with some of the PE material, giving rise to a broadening and a downward shift in its T_g. Thus, transition 4 becomes convoluted into the region where the PE crystalline soft segments melt, which is associated with transition 3. The general characteristics of the PA glass transition temperature can be discerned again by inspecting the three PA-richer compositions when considering the respective E' data in Figure 1a. Finally, with regard to transition 5, which has been addressed when discussing Figure 1a, the rise in $\tan\delta$ is simply due to the onset of melting of the PA crystal phase. Again, further support for this interpretation will be provided from the DSC behavior reviewed in Section 2.2.

To summarize the DMTA response, one notes that the PTMO-PA12 based PEBA systems exhibit a microphase-separated morphology with up to four co-existing phases (similarly to the case of poly(ether ester)s, see Chapter 6), depending upon the copolymer composition and temperature. In systems with a high PTMO content, such as P2533 and P3533, the soft segment length is sufficient to allow cold crystallization of this phase. It completes its melting transition around 10 °C and above this temperature the microphase-separated morphology consists of some isolated short PA segments dissolved in an amorphous matrix of the PE phase. At the other compositional extreme of high PA content (samples P6033 and P7033), the soft segment length is insufficient to allow any significant crystallization of the PE phase, but the PA segments are long enough to form a crystalline phase consisting of folded chain lamellae. Since a molecular weight distribution of the PA segments exists, the shorter PA species within the sample may not be long enough to fold into lamellae and therefore may possibly adopt a more fringed micelle-like semicrystalline texture. Thus, a total of three microphase-separated phases coexist in these samples: (i) a PE amorphous phase with a small, relatively composition-independent amount of isolated PA segments dissolved therein, (ii) a PA amorphous phase in which the concentration of the mixed PE segments depends on the PE segment molecular weight and its content, and (iii) a PA crystalline phase, which melts at about 170 °C. In the copolymer samples with a relatively balanced composition, such as P4033, the segment lengths are sufficient to allow crystallization of both types of segments. Thus, below *ca.* 10 °C a total of four phases can coexist; PE and PA crystalline phases, a partially mixed PE amorphous phase, and a partially mixed PA amorphous phase. In a given partially mixed amorphous phase, one expects the isolated segments of the other component, especially those species associated with the lower end of the molecular weight distribution of this component, to be preferentially dissolved near the boundary or interface formed by the microphase-separated phases. As mentioned earlier, the PE crystalline phase would undergo its melting transition well below ambient temperature. Other reports on DMTA [25,28,29], torsional vibration analysis [11], and thermally stimulated current method [30–34] of PTMO-PA12 based PEBA copolymers provide further support for the above discussion of their general morphological behavior.

2.1.2. *Tensile deformation behavior*

The response of the PEBA systems to tensile deformation can also be used to gain further understanding of their morphology, *i.e.*, the stress-strain tests can complement the morphological model developed from DMTA. The engineering stress, σ_0, *vs.* percent elongation plots of PEBAX® samples P2533, P3533, P4033, P6333, and P7033 reported by Sheth *et al.* [22] are presented in Figure 2. The ambient temperature tests were conducted at a crosshead speed of 15 mm/min on slowly cooled compression-molded dog-bone shaped films having a grip separation distance and gauge length of 10 and 6 mm, respectively. In order to prepare samples for the stress-strain measurements, as received PEBAX® resins were subjected to exactly the same thermal treatment as that used to prepare samples for the DMTA study discussed above. The stress-strain response exhibited by the PEBAX® series is consistent with their relative segment compositions. The DMTA response of P2533 and P3533 indicated that at ambient temperature their morphology consisted of microphase-separated PA domains dispersed in an amorphous matrix, which is compositionally dominated by the PE segments. The long-range connectivity of the hard domains in these two samples is limited due to their relatively low PA hard segment content. Thus, the dispersed crystalline hard domains preferentially

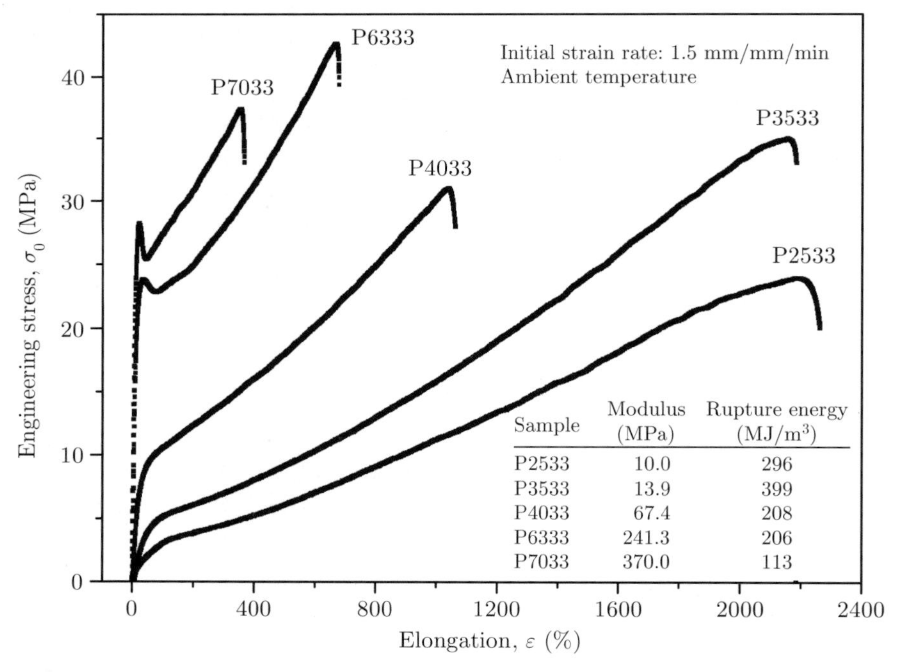

Sample	Modulus (MPa)	Rupture energy (MJ/m³)
P2533	10.0	296
P3533	13.9	399
P4033	67.4	208
P6333	241.3	206
P7033	370.0	113

Figure 2. Stress *vs.* percent elongation behavior of slowly cooled, melt compression-molded PEBAX® films. Inset: Young's moduli and energies of rupture normalized per unit volume of the initial sample [22]

serve as physically crosslinked regions for the amorphous soft segment matrix and enable P2533 and P3533 to behave as elastomers; they exhibit a low Young's modulus and a remarkably high elongation at break. The lack of a distinct yield point in these samples provides further support to their morphological description. Furthermore, it is generally observed that good elastomers exhibit good recoverability, low permanent set, and low mechanical hysteresis. These characteristics of PEBAX®, as dependent of the PA content of the sample, will be addressed in some detail later in this chapter. It may be noted that in the case of P2533 and P3533, the strain at break exceeds 1000%, even when the calculation of the strain values is based on the grip separation distance (10 mm) instead of the gauge length (6 mm) and any slippage of the samples from the grips at high strains is taken into account. On the other hand, a sharp yield point can be observed in both P6333 and P7033. They also exhibit a higher Young's modulus than the other PEBAX® samples reviewed in this chapter and their elongation at break is considerably lower than that of the other members of the series, as might be expected. Such a behavior suggests an increase in the hard phase connectivity/percolation of the PEBAX® materials with increasing PA content. Thus, P6333 and P7033 act more like hard thermoplastics than elastomers. Furthermore, as a result of the increase in the interconnectivity of the hard phase, there is a systematic increase in the Young's modulus (see inset in Figure 2) of the PEBAX® series with PA content. The Young's moduli of the series are in agreement with those reported by Deleens [8]. The toughness of the copolymers, as measured by the energy of rupture normalized per unit volume of the initial sample, is also tabulated in the inset of Figure 2. In general, it decreases with decreasing PE content, although P3533 exhibits greater toughness than P2533 due to a consistently higher sustained load and a comparable percent elongation at break.

2.2. *Thermal analysis*

Differential scanning calorimetry (DSC) can further aid in investigating the effect of segment length and copolymer composition on the crystalline and amorphous phases of PEBA systems. The cooling and the second heating thermal scans conducted at 20 °C/min under a dry nitrogen atmosphere and reported in [22] are presented in Figure 3a and b, respectively. In the cooling scans, the PEBAX® samples with lower PA content show two crystallization peaks, each with a corresponding heat of crystallization, ΔH_c, one for the PA segments at the higher crystallization temperature and the other due to the crystallization of the PE segments at the lower crystallization temperature. The amount of relative crystallinity due to the PE phase, as reflected by ΔH_c, is the highest in P2533 and decreases with decreasing PE content until it cannot be observed in P6533. As expected, the amount of crystallinity due to the PA phase increases with increasing PA content. At the same time, the onset of crystallization shifts to higher temperatures and hence to earlier times in the cooling cycle with increasing PA content, there being one exception to this trend, namely P7033. However, there is no significant change in the crystalliza-

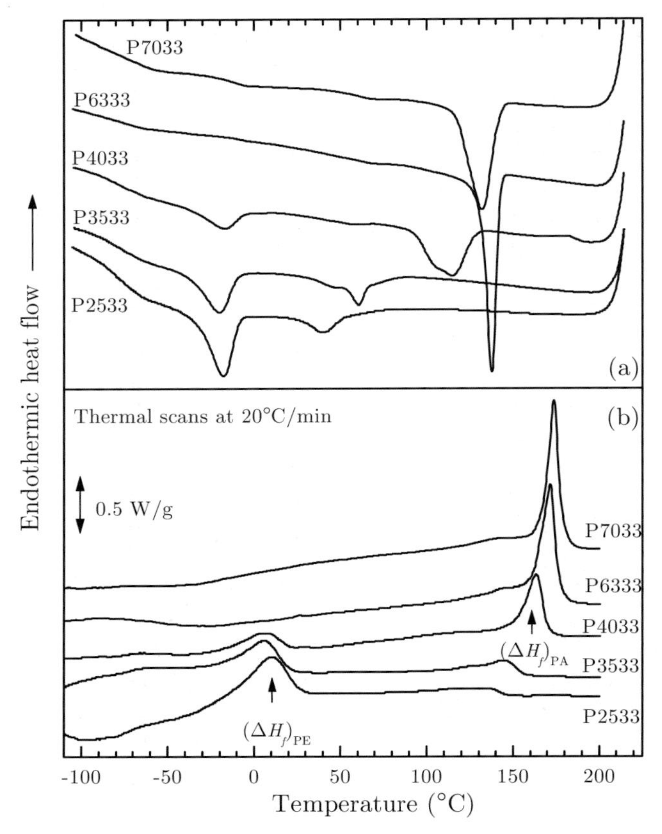

Figure 3. DSC scans of slowly cooled, melt compression-molded PEBAX® films: (a) cooling scans, (b) second heating scans. Traces are displaced vertically for clarity [22]

tion temperature of the PE blocks for the three samples, namely P2533, P3533, and P4033, which show some PE crystallization.

The second heating scans show the melting of these crystalline phases; the PE melting peak in the range of 7–10 °C decreases in intensity until it cannot be observed in P6333. Most of the PE crystals have completely melted in the PE-rich materials, such as P2533 and P3533, by the time the sample reaches room temperature. It might be noted that all the peak T_m values of the PE segments for these undeformed samples were distinctly lower than those reported for high molecular weight PTMO, *i.e.,* 35–43 °C [24,27]. This lower peak T_m behavior of the PE segments undoubtedly arises from their inability to develop crystals of thicker lamellae that can be promoted in the respective homopolymer — in fact, for the systems with low PE content, folded chain PE lamellae would seem to be difficult to form in view of a high concentration of many short segment lengths. Microphase mixing with PA segments could promote crystalline imperfections in the PE crystals and thus may also promote a further lowering of T_m of the PE phase. In contrast to the nearly constant peak melting temperature of the PE phase at *ca.* 7 °C, the PA melting temperature systematically increases with the PA content, as does the associated magnitude of the heat of fusion, ΔH_f, of this transition. The shift of the PA melting point to higher temperatures is very distinct and could be due to an increase in the perfection of the PA crystalline phase and/or increased lamellar thickness (both of which may occur due to a systematic increase in the PA block length with PA content). In related DSC studies, Bornschlegl *et al.* [11] and Di Lorenzo *et al.* [35,36] showed that, irrespective of the PE segment molecular weight, the reciprocal T_m (degrees K) of the PA crystal phase exhibits a linear dependence (with a negative slope) on the reciprocal average degree of polymerization of the PA segment. Based on this behavior, they concluded that the PA segments crystallize and melt as if they were isolated oligomers not linked by the soft segment spacer. Their results are directly in line with earlier observations made by Harrell [37] based on his investigation of the melting behavior of monodisperse hard segments in piperazine-based segmented polyurethanes. The ester hard segments in PTMO-based segmented poly(ether ester) copolymers also exhibit a similar melting behavior [38]. Di Lorenzo *et al.* [35] also showed that in PEBA samples with an average PA segment molecular weight greater than 3700 g/mole (equivalent to a PA content of 80 wt%), the percent crystallinity of the PA phase, calculated from ΔH_f of the PA crystal phase, was in fact slightly higher than that of the neat nylon 12 sample. The facilitation of the ordering of PA segments into a crystal lattice by the more flexible PE segments was offered as an explanation for the increased crystallinity.

The glass transition temperature of the PE phase is difficult to clearly discern from Figure 3 due to the relatively small magnitude of this transition, as compared to the melting endotherms of the PE and PA phases, respectively. In addition, for segmented copolymers, DMTA is a more sensitive technique, as compared to DSC, to examine the copolymer glass transition(s) behavior. It may be recalled that a relatively composition-independent T_g of the PE phase at *ca.* −70 °C was observed by DMTA. The T_g of homopolymeric high molecular weight PTMO as determined by DSC has been reported to be −82 °C [15].

2.3. *Structure determination by scattering and microscopy studies*

In this section, the investigations of the PA crystalline phase in PTMO-PA12 based PEBA copolymers and its ability to pack into lamellae or into a spherulitic superstructure by the techniques of wide-angle X-ray scattering (WAXS), atomic force microscopy (AFM), small-angle light scattering (SALS), and transmission electron microscopy (TEM) are reviewed. In particular, the influence of the PA segment length and crystallization conditions, such as samples cast from a solvent or prepared by compression-molding from the melt, on the development of long-range order in this phase is presented in some detail.

2.3.1. *Wide-angle X-ray scattering*

WAXS patterns at ambient conditions of the unstretched, slowly cooled compression-molded PEBAX® samples P2533, P3533, P4033, P6333, and P7033 reported by Sheth *et al.* [22] are presented in Figure 4. The sample preparation procedure for WAXS was identical to that used to make samples for the DMTA, DSC, and stress-strain measurements. From the WAXS patterns, one notes that the amorphous halo, which is principally due to the PE phase and is distinctly seen in the unstretched P2533 sample, gradually fades as the PA content increases. On the other hand, the Bragg reflections due to the presence of a crystalline PA phase become sharper with increasing PA content of the samples. In the undeformed state, the unit cell of the nylon 12 crystals has been reported to be monoclinic [39]. Since the PEBAX® samples used for the WAXS study were not cooled below room temperature, no reflections that can be attributed to the PE phase were expected to be seen in the unstretched samples. It may be recalled that the DSC second heating scans (Figure 3b) showed that the crystalline PE phase completed its melting transition by about 10°C. WAXS was also utilized by the authors of this same study to investigate the behavior of PEBA copolymers under uniaxial deformation. The results of this study will be addressed in Section 2.4.

2.3.2. *Surface morphology by atomic force microscopy*

Tapping mode AFM has the ability to distinguish the relative stiffness or hardness of different regions on the free surface of a polymer film and present such

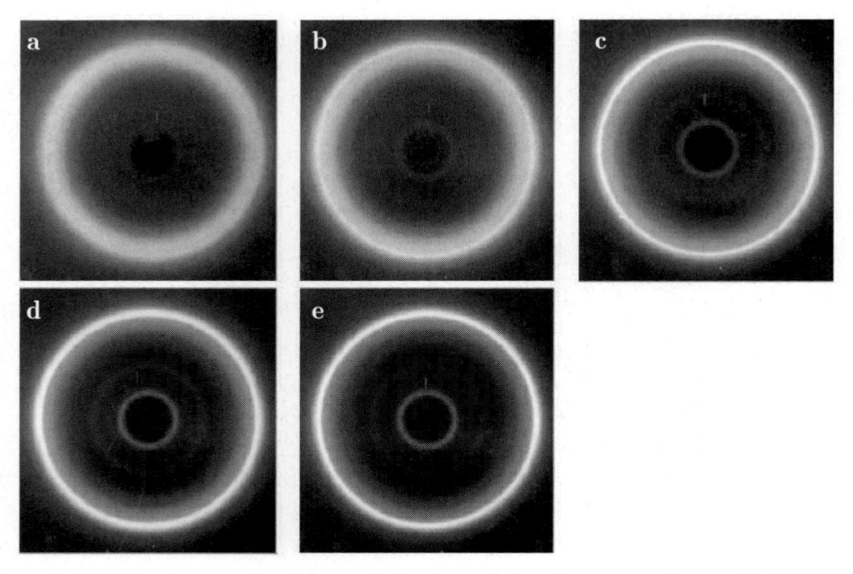

Figure 4. WAXS patterns of undeformed, slowly cooled melt compression-molded PEBAX® films: (a) P2533, (b) P3533, (c) P4033, (d) P6333, and (e) P7033 [22]

a stiffness "map" as a visual image of the morphology of the sample near the surface, up to a depth of a few nanometers. In the particular case of PEBA, at ambient temperature, the PE phase is expected to behave as a soft rubbery phase and thus correspond to regions of low phase angles in the AFM phase images. The interpretation of the high phase angle regions associated with "harder" regions, however, is more complicated as will be discussed below.

The ambient temperature tapping mode AFM phase images of the free surfaces of PEBAX® films prepared from the melt or solution-cast were reported earlier [22] and are reproduced in Figures 5 and 6, respectively. At ambient temperature, the amorphous PA domains behave as a glassy solid (see Figure 3). Thus, it was noted that when the AFM imaging of these samples is undertaken at ambient conditions, the stiffer regions sensed by the AFM tip could possibly correspond to either the amorphous (glassy) or the crystalline regions of the PA blocks. Due to this fact, one cannot easily assign the stiffer regions seen in the tapping mode AFM phase images (of P6333 and P7033 taken at ambient

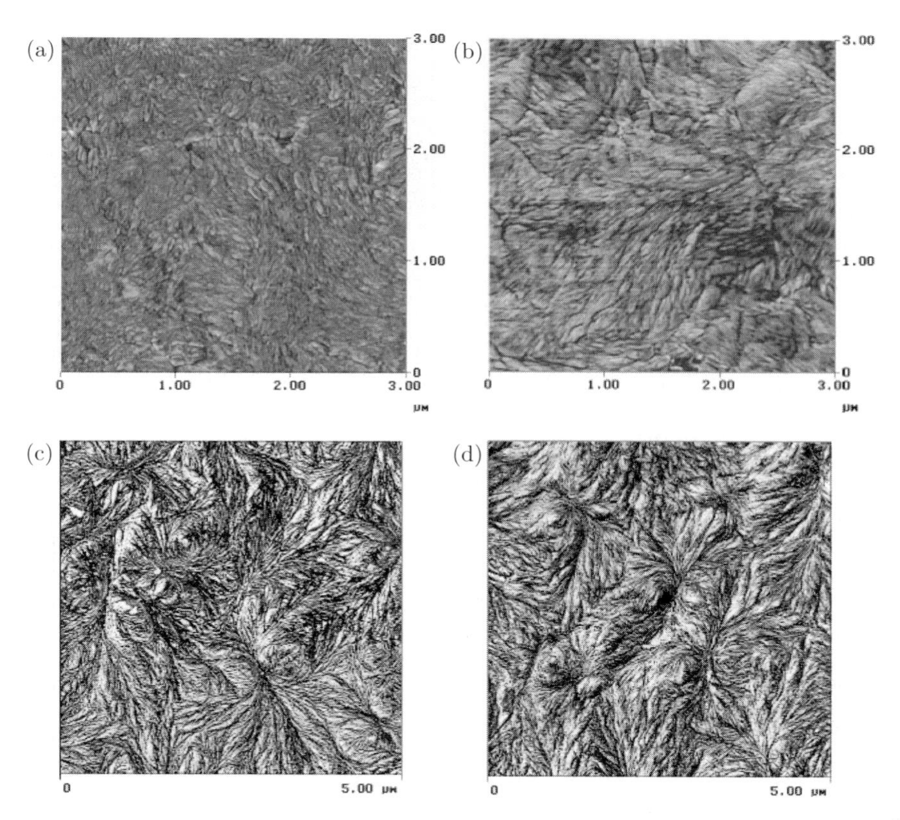

Figure 5. Tapping mode AFM phase images of the "free surface" of melt-pressed PEBAX® films: 3 μm × 3 μm images of (a) P3533, (b) P4033, and 5 μm × 5 μm images of (c) P6333, (d) P7033 [22]

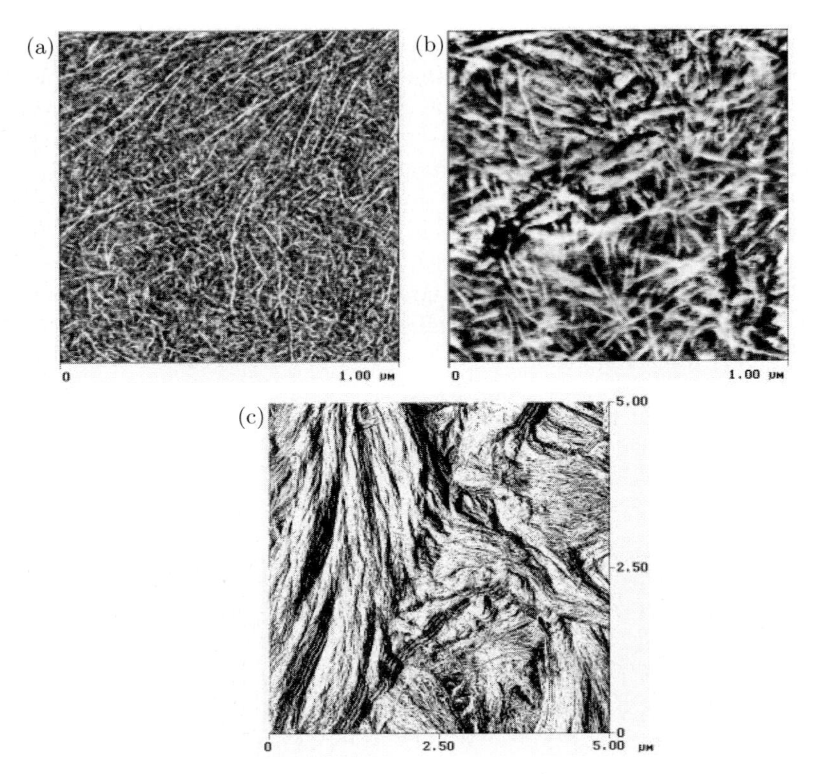

Figure 6. Tapping mode AFM phase images of PEBAX® film cast from 5 wt% dimethyl acetamide solution: (a) 1 µm × 1 µm image of P3533, (b) 1 µm × 1 µm image of P4033, (c) 5 µm × 5 µm image of P7033 [22]

temperature) exclusively to the crystalline PA regions. However, as seen in Figure 5, the presence of a spherulitic morphology on the free surface of the melt-pressed P6333 and P7033 films is clearly evident. Furthermore, for this crystallization condition, a larger spherulite is indicated in P7033 than in P6333. As it will be discussed later, the results obtained using SALS are in agreement with the AFM results. It may be noted that in the case of poly(ether ester)-based segmented TPEs, various researchers [40–43] have shown the ability of the polyester hard blocks to crystallize and also give rise to a spherulitic superstructure. Melt compression-molded samples of P2533 (not shown), P3533, and P4033 did not clearly indicate the presence of a spherulitic superstructure and the SALS data (presented later) provided further support to the AFM results. However, it may be recalled from the DSC results (Figure 3) that P2533, P3533, and P4033 still possess a PA crystalline phase.

When cast from warm solution (5 wt% in dimethyl acetamide), the presence of long and slender stiff PA regions in the ambient temperature phase images of samples P3533 and P4033 (Figure 6a,b) can be clearly seen. In our previous discussion, it was noted that since the amorphous PA phase behaved as a glassy solid in the vicinity of the ambient temperature, it was difficult to separate the high phase angle regions seen in the AFM phase images into crystalline PA and amorphous PA phases, respectively. The same problem is encountered in the present case; however, the presence of long and slender stiff PA regions can be clearly seen. From such an observation, the authors inferred that when cast from moderately dilute solutions, the PA crystalline phase in P3533 and P4033 formed lamellar crystals that were rather randomly ordered. Upon comparison of Figures 6a and b, it appears that P4033 has thicker lamellae than P3533. The authors assumed that such a visual observation could result when the lamellae are not oriented perfectly normal to the free surface and thus did not necessarily suggest that the lamella thickness in solution-cast PEBAX® samples increases with the PA content. On the other hand, P7033, which has the highest PA content in the PEBAX® series investigated (Table 1), was shown to give rise to a highly interconnected PA hard phase. The presence of such a high interconnectivity within the PA hard phase can be clearly seen in the AFM phase image of the free surface of the solution-cast P7033 film (Figure 6c). However, there is no indication of the formation of a spherulitic morphology, as was seen on the free surface of the melt compression-molded P7033 film.

2.3.3. *Bulk morphology by small angle light scattering*

An optically anisotropic spherulitic superstructure in a polymer film can generally be detected by SALS [44] if sufficient light (preferably $> 70\%$) is transmitted through the sample. AFM is a surface technique, whereas SALS can be used to investigate the bulk morphology if scattering due to the surface roughness can be minimized by immersing the sample in a non-interacting liquid (*e.g.*, silicone oils) with refractive index matching that of the sample. SALS has been utilized [22] to determine whether the spherulitic structure observed in P6333 and P7033 by tapping mode AFM was also present throughout the bulk of a given PEBAX® sample. H_v SALS images of the slowly cooled unstretched samples were obtained and are shown in Figure 7. It may be noted that an H_v SALS pattern is obtained when the plane of the incident polarized light and the plane of the analyzer are perpendicular to each other, the latter being horizontal. Samples P6333 and P7033 clearly displayed the presence of optically anisotropic spherulites. A symmetric four-leaf clover pattern was, however, not seen in P2533, P3533, and P4033 suggesting the absence of a sufficiently well defined optically anisotropic superstructure in these samples. In fact, no significant H_v light scattering intensity was noted for these three samples. Thus, the SALS results were in good agreement with earlier AFM results of melt compression-molded PEBAX® films.

From the SALS patterns, the average diameter of the spherulites [44] in P6333 and P7033 was calculated to be 7.3 and 18.1 μm, respectively. On comparison of the average size of the spherulites in P6333 and P7033 as determined by SALS and AFM (Figure 5), it was noted that the AFM images indicated a smaller spherulite size (higher nucleation density). This is believed to be due to the fact that melt compression-molded P6333 and P7033 films

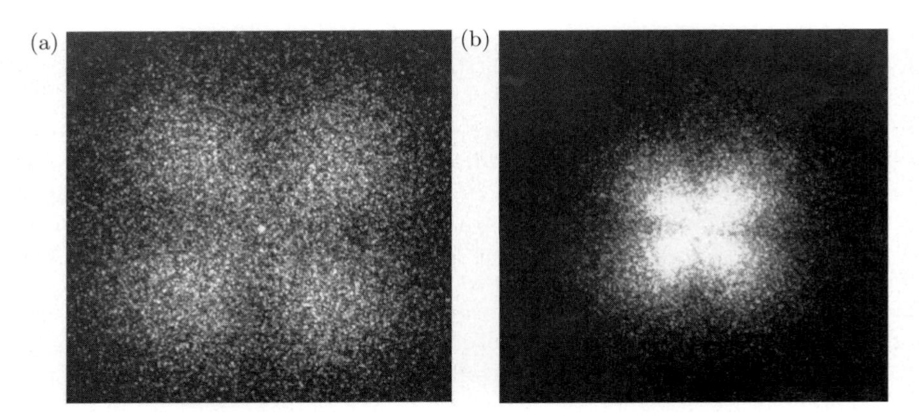

Figure 7. H_v SALS patterns of unstretched, slowly cooled melt compression-molded PEBAX® films: (a) P6333, (b) P7033. Sample-to-film distance is 340 mm [22]

were cooled at a higher rate in the hot stage as compared to the cooling rate achieved when the samples were freely cooled in air outside the hot stage. It was also observed in the same study that melt compression-molded and ice water-quenched films of P6333 and P7033 showed no indication of a symmetric four-leaf clover pattern, suggesting the absence of a defined optically anisotropic spherulitic superstructure in the quenched samples.

2.3.4. *Bulk morphology by transmission electron microscopy*

TEM was used by Goldbach *et al.* [12] to study the bulk morphology of PTMO-PA12 based PEBA copolymers synthesized by Mumcu *et al.* [14]. Thin films were prepared from the melt and stained with a mixture of allylamine and osmium tetraoxide to enhance the contrast. The TEM images of nylon 12 and PEBA samples with 75, 50, and 40 wt% PA12 content are reproduced in Figure 8a-d. The presence of a spherulitic superstructure in the TEM image of nylon 12 (Figure 8a) can be distinctly seen. The diameter of the spherulite, as seen from the image, is *ca.* 2.5 μm. At a higher magnification (image not shown) 6–8 nm thick lamellae were noted, which Goldbach *et al.* calculated to essentially be equivalent to four nylon 12 repeat units, based on the assumption that the chains are oriented perpendicular to the surface of the lamellae. The TEM images of the three PEBA samples (Figure 8b-d) also exhibit the presence of a spherulitic superstructure. However, the lamellae appear to be less tightly packed within the superstructure of the respective PEBA samples than in the neat nylon 12 spherulite. In particular, they are most loosely packed in the PEBA sample with the lowest PA12 content within the series, namely 40 wt% (Figure 8d). The diameter of the spherulitic superstructure in this sample, however, is *ca.* 2.5 μm, (similar to that in neat nylon 12), suggesting that the nucleation density in nylon 12 and in 40 wt% PA12 containing PEBA samples is similar under these preparation conditions. However, the amount of PA 12 phase crystallinity in

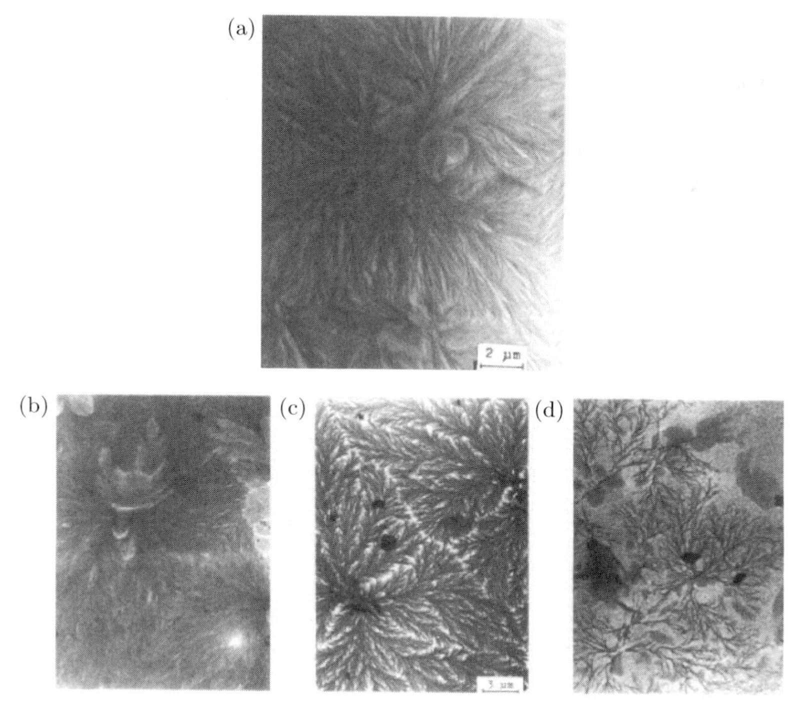

Figure 8. TEM micrographs of melt-crystallized films of: (a) nylon 12 (contrast enhanced by Pt/C shadowing), (b) PEBA with 75 wt% PA12, (c) 50 wt% PA12, and (d) 40 wt% PA12 [12]

the PEBA samples decreases in general with decreasing PA content, as was noted earlier from the ΔH_f values in the second heating DSC scans shown in Figure 3b. Thus, given a similar nucleation density (and hence a similar superstructure size), but a lower PA12 phase crystallinity, it is not surprising to note superstructures in the PEBA samples more loosely packed with lamellae than those in neat nylon12. The measurement of the superstructure in the other two PEBA materials could not be made from the images shown. However, Goldbach *et al.* noted that at higher magnification, the TEM image (not shown) of the 50 wt% PA12 containing PEBA sample exhibited 6–8 nm thick lamellae, which is similar to the finding in the neat nylon 12 sample.

2.4. *Uniaxial deformation behavior*

In the preceding sections, a morphological picture was presented of a range of unstretched PEBA materials created by thermo-mechanical, scattering, and microscopy techniques. The effect of the constituent segment lengths and thus copolymer composition, and the crystallization conditions on the morphology was discussed. It was also shown from the stress-strain response of PEBA that when the PE component dominates the copolymer composition, elastomeric materials result, exhibiting an elongation at break in excess of 1000%.

Conversely, PA phase-rich compositions result in hard thermoplastic materials with high Young's moduli. However, the deformational features of the morphology cannot be easily or fully explained by the stress-strain response alone. Therefore, in this section, the influence of uniaxial deformation on the morphology of PEBA is further reviewed. Depending upon the PA content of the sample, the techniques of SALS [22] or AFM in conjunction with 2D-SAXS [20,45] have been used by investigators to study the morphological changes in PEBA under relatively small uniaxial deformation. At larger deformations (above *ca.* 500%), WAXS and DSC have been used [20,22,46] to investigate both the orientation of the hard PA segments and the strain-induced crystallization of the soft PTMO phase in particular.

2.4.1. *Morphological changes induced by small uniaxial deformation*

PTMO-PA12 based PEBA with a high PA content. The composition of PEBAX® samples P6333 and P7033 consists of *ca.* 63 and 75 wt%, respectively, of PA12 segments (Table 1). At such a high PA content, the PA12 segments are long enough to form a crystalline phase consisting of folded chain lamellae, which organize into a spherulitic superstructure when the PEBAX® films are prepared by slow cooling from the melt. The occurrence of a spherulitic superstructure was confirmed earlier by AFM (Figure 5) and SALS (Figure 7). The optically anisotropic nature of the PEBAX® spherulites was advantageously utilized by Sheth *et al.* [22] to study the effect of uniaxial deformation, which is of both practical and fundamental interest, on the superstructural features of P6333 and P7033.

The SALS images of compression-molded, slowly cooled unstretched, stretched, relaxed, and annealed samples reported in [22] are reproduced in Figure 9. The P6333 and P7033 films were uniaxially stretched up to 150% and SALS images were taken at 15 and 150% strain, respectively. At 15% strain, P7033 is within the linear region of the stress-strain curve (see Figure 2). Even at such a low deformation, as can be seen from Figure 9b, elongational and shear stresses propagate through the PA crystal phase, resulting in deformation of the spherulites. At 150% strain, P7033 has passed through the yield point (at *ca.* 25% strain) and is well within the non-linear region of the stress-strain curve. Significant deformation of the spherulites occurred at such high strains (Figure 9c). The stretched P7033 sample was allowed to *freely relax* at ambient temperature and was then annealed at 90°C for 90 min and slowly cooled in air. In the same study, it was reported that P7033 freely relaxed to a strain of 102% and, after annealing, exhibited a permanent set of 70%. The SALS image of the annealed sample is shown in Figure 9d. On the other hand, sample P6333 was reported to exhibit a permanent set of 50% after annealing. As expected, this indicates that the permanent set is a function of the PA content of the sample. For the sake of comparison, an undeformed P7033 film was stretched to 70% to match the value of the permanent set of the annealed sample. This sample was allowed to freely relax and was then subjected to annealing at 90°C

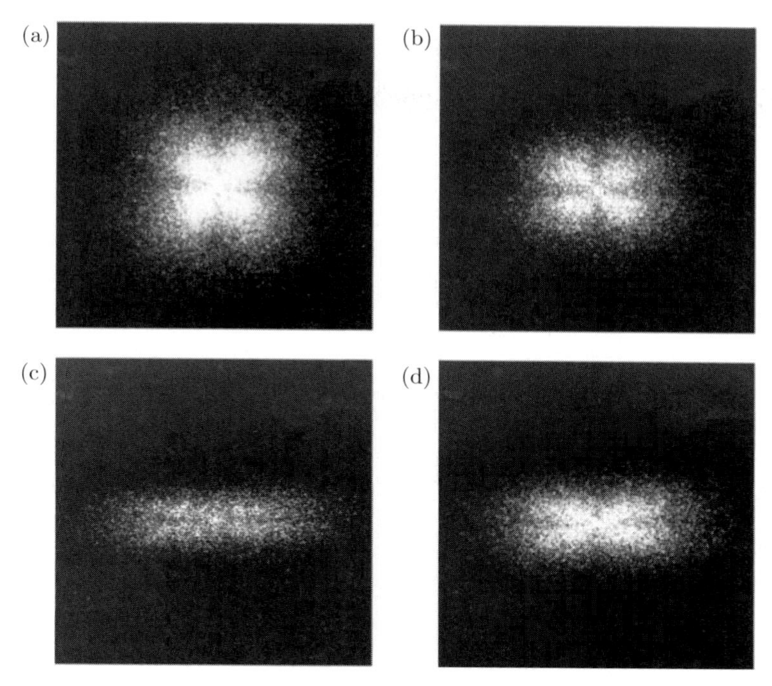

Figure 9. H_v SALS patterns of slowly cooled, melt compression-molded PEBAX® films: (a) unstretched P7033, (b) P7033 stretched to 15%, (c) P7033 stretched to 150%, and (d) stretched P7033 allowed to relax and annealed at 90°C for 90 min. After annealing the sample relaxed to 70% strain. Sample-to-film distance 340 mm and the stretch direction is vertical [22]

for 90 min, followed by slow cooling in air. After annealing, a permanent set of 31% was exhibited by P7033.

PTMO-PA12 based PEBA with a high PE content. Tapping-mode AFM in conjunction with 2D-SAXS was utilized by Sauer *et al.* [20,45] to investigate the effect of uniaxial deformation on the morphology of solution-cast PEBAX® sample P3533. Films of P3533, which has a high PE content of *ca.* 80 wt%, were cast from a 7 wt% solution in dimethyl acetamide. They were uniaxially stretched to a draw ratio of 3.2×, and AFM images and 2D-SAXS patterns were taken at various elongations, as shown in Figure 10. At this point, it must be reemphasized that AFM is a surface imaging technique, whereas SAXS gives information about the entire bulk of the sample. Thus, in the following discussion, tapping-mode AFM phase images were only used to "indirectly" aid the interpretation of the 2D-SAXS patterns.

In the AFM phase image of the undeformed P3533 film in Figure 10, the presence of randomly oriented, long ribbon-like lamellae with a high aspect ratio were observed (also see Figure 6a). The azimuthally independent reflection in the corresponding SAXS pattern confirmed this random orientation of the PA phase lamellae; the long spacing, L, in the sample being 24 nm. At a 1.5×

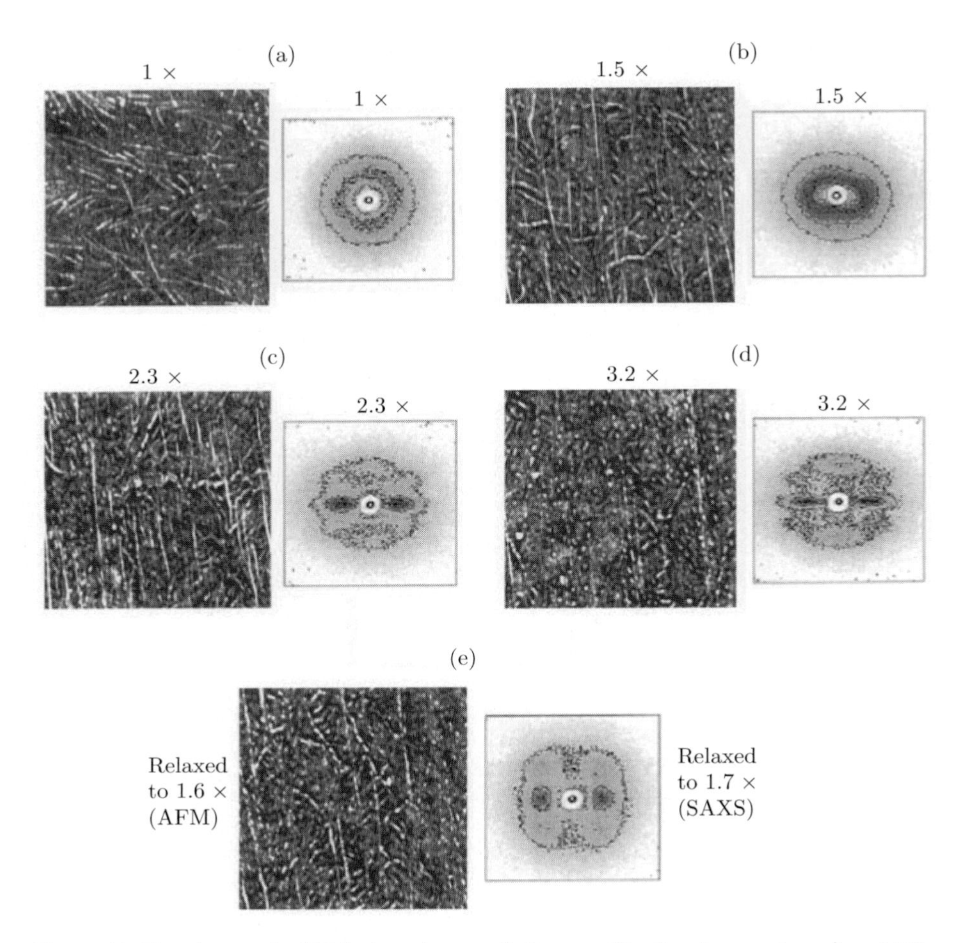

Figure 10. Tapping mode AFM phase images (left, magnification 1 mm × 1 mm) and 2D-SAXS patterns (right) of PEBAX® 3533 films cast from a dimethyl acetamide solution and stretched: (a) 1×, (b) 1.5×, (c) 2.3×, (d) 3.2×, and (e) relaxed. Stretch direction is vertical in the 2D-SAXS patterns and close to vertical in the AFM images [20,45]

draw ratio, orientation of some lamellae in the stretch direction occurred. Also seen by AFM was the concurrent large-scale break-up of the ribbon-like lamellae into smaller hard domains of lower aspect ratio. The circular SAXS pattern of the unstretched sample became elliptical at 1.5×. The long spacing, L, in the meridional direction was reported to increase to 38 nm and that in the transverse direction remained essentially the same at *ca.* 20 nm. Thus, AFM and SAXS data suggest that elongational and shear forces set-up in the film upon drawing, causing breakup of the ribbon-like lamellae that were initially oriented in the stretch direction. The resulting smaller domains are then pulled further apart and begin to orient in the stretch direction. Sauer *et al.* [20] suggested that these small hard domains, interconnected by the soft segments, formed nanofibrillar structures. They estimated the thickness of the hard domains to be *ca.* 10 nm. Thus, with $L = 38$ nm in the meridional direction, the estimated length of the slightly extended soft segments between the hard domains in the nanofibrillar structures would be 20 nm. From the 1.5 × SAXS pattern, Sauer *et al.* noted an increase in the scattered intensity at low scattering angles, possibly arising because of the formation of the aforementioned nanofibrillar structure and also due to hard domains formed by the deformation-induced association of the dissolved hard segments. Elongation to 2.3 × resulted in the breakdown of most of the lamellae, which were oriented in the stretch direction, as seen from the AFM image. However, the corresponding SAXS pattern was more revealing. It showed a strong reflection along the equator that arose from the longer lamellae and smaller hard domains that were highly oriented in the stretch direction. L along the equator remained unchanged at *ca.* 20 nm, but that in the meridional direction, not surprisingly, was reported to increase marginally to *ca.* 44 nm. Sauer *et al.* also noted weak reflections around $\pm 45°$ azimuthal angles, which arise from the small hard domains and nanofibrils that are not perfectly oriented along the stretch direction. At 3.2 ×, all ribbon-like lamellae, irrespective of their orientation, were broken up into smaller domains. The SAXS pattern revealed first and second order reflections along the meridian (one reflection being very close to the beam stop), which undoubtedly arose due to the improved long-range order of the hard domains. The reflection along the equator sharpened due to an increase in the hard domain orientation in the stretch direction (see also Chapters 6 and 7).

Upon removal of the external load, the solution-cast P3533 film was unable to revert back to its original length and displayed a residual draw ratio of *ca.* 1.6 × or a permanent set of 60%. The ability of some of the hard domains to re-associate into ribbon-like lamellae, which are predominantly oriented in the stretch direction can be seen from the AFM image of the freely relaxed film taken only a few minutes after the removal of the external load. However, a vast majority of the smaller hard domains were unable to re-associate. From this behavior, Sauer *et al.* concluded that even after the external load was removed, some residual stresses remain locked in. These stresses can be relieved if the sample is annealed above the T_g of the hard phase [47]. The SAXS pattern of the freely relaxed sample, aged at room temperature for 20 h, revealed an equatorial reflection arising from lamellae and other smaller hard domains oriented in the stretch direction. The hard domains tilted away from the stretch direction gave rise to more diffuse reflections around the $\pm 45°$ azimuthal angles. After annealing of this sample above the T_g of the hard phase, one would expect it to recover close to its original dimensions and the 2D-SAXS pattern to closely resemble that of the unstretched sample.

2.4.2. *Morphological changes induced by large uniaxial deformation*

In the preceding section, it was shown that, at relatively small deformations, the hard PA domains become activated and act as load-bearing structures.

Upon elongation, they begin to rotate in the stretch direction and gradually break up into smaller domains under increased deformation. On the other hand, the soft PTMO segments do not appear to exhibit any large-scale activation. However, Warner [46] demonstrated that strain-induced crystallization of the PTMO chains occurred when the sample was deformed above *ca.* 500% strain at ambient temperature. Upon strain-induced crystallization, the T_m of the PTMO crystalline phase increased from below room temperature to *ca.* 45°C. Thus, the sample P2533 hardened due to the presence of a PE crystalline phase at ambient temperature. The change in the melting temperature of the PE phase as a function of elongation, reported by Warner, is presented in Figure 11.

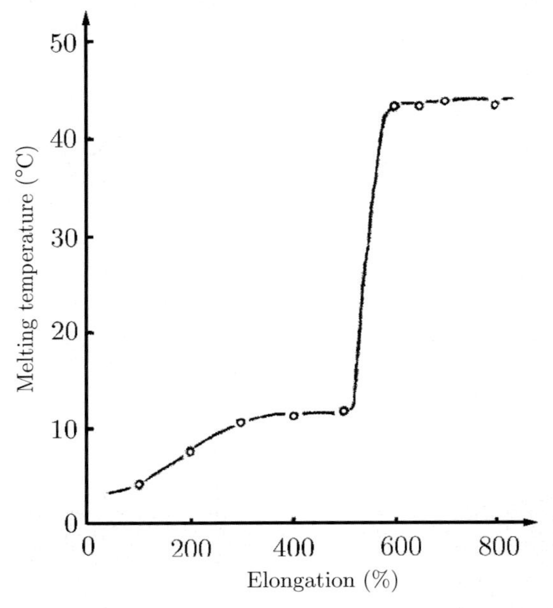

Figure 11. Effect of elongation on the soft segment melting temperature of PEBAX® P2533 [46]

WAXS patterns of slowly cooled compression-molded P2533 and P3533 films, stretched to a draw ratio of 8×, have been reported [22]. They are reproduced in Figure 12. The expected PE crystal phase equatorial reflections [48] can be clearly seen in the WAXS patterns shown in Figure 12a. When the stretched films were allowed to *freely relax* and later subjected to a heating scan in the DSC beginning from ambient temperature, the melting of the PE crystal phase was distinctly observed by DSC (Figure 12b). Furthermore, and as expected, the amount of crystallinity (determined by ΔH_f) due to the strain-induced crystalline PE phase was higher in P2533 than in P3533 when the two samples were deformed under identical conditions. In order to determine the level of permanent set in these two samples, the respective inelastic samples (due to strain-induced crystallization upon stretching to 700%) were allowed to freely

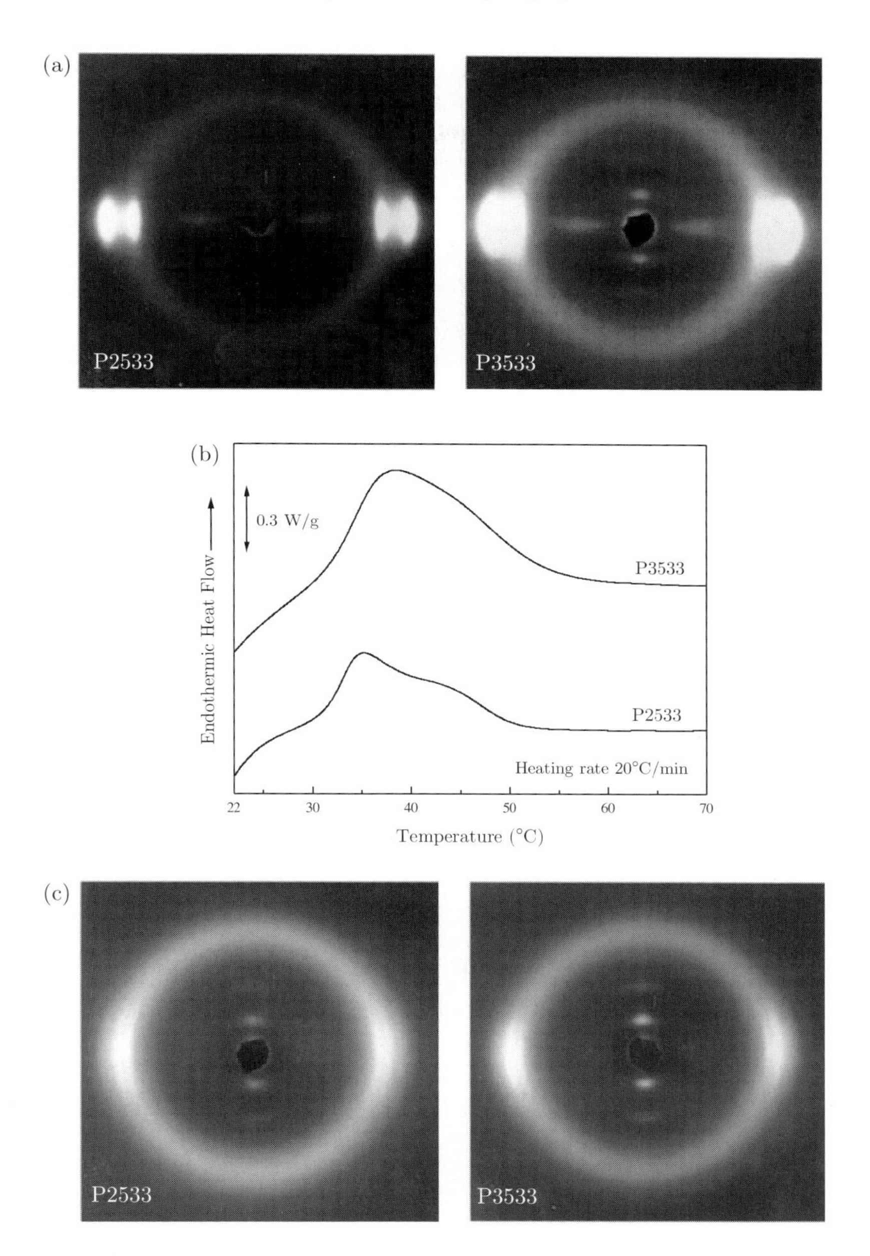

Figure 12. Strain-induced crystallization of slowly cooled, melt compression-molded P2533 and P3533: (a) WAXS images of samples stretched to a draw ratio of 8×, (b) DSC thermal scans of freely relaxed samples (traces displaced vertically for clarity), and (c) WAXS images of freely relaxed samples annealed at 70°C for 90 min. Annealed samples P2533 and P3533 exhibited a permanent set of 63% and 97%, respectively, when measurements were made immediately after cooling of the annealed samples to room temperature. Stretch direction in WAXS patterns is vertical [22]

relax and then annealed at 70 °C for 90 min, and slowly cooled to ambient temperature. The annealed samples exhibited a permanent set of 63 and 97%, respectively, when measurements were made immediately after the annealed samples cooled down to room temperature. The melting of the PE crystals in these two samples was also observed by WAXS; patterns are reproduced in Figure 12c. Annealing of the inelastic samples at 70°C resulted in a loss of intensity of the equatorial reflections due to the PE crystal phase and a concomitant darkening of the amorphous halo as compared to the stretched and unannealed samples. Furthermore, the equatorial reflections due to the residual oriented PA crystal phase were also noted. The equatorial and meridional PA crystalline reflections were darker in P3533 than in P2533.

Of note is the similarity in the strain-induced crystallization and general melting range behavior of the PE segments in samples P2533 and P3533 to that of the well known strain-induced crystallization characteristics of very lightly crosslinked natural rubber [49]. Specifically, this latter material, when stretched sufficiently, can be promoted to develop a high enough crystalline content to induce inelastic behavior (high permanent set) which arises from a row-nucleated crystalline phase [50]. Upon heating such an inelastic material, the crystals melt at *ca.* 30 °C and elasticity is again restored, as it also was when the strain-induced crystals of the PE phase melted as well. While the PE strain-induced crystalline amount was undoubtedly lower than that promoted in highly stretched (racked) [51] natural rubber, these PE crystals, in conjunction with those already present in the crystalline PA phase, provide sufficient rigidity so that inelastic behavior (and high permanent set) occurred. Again, when the PE crystals melted, elasticity was restored. Due to this similarity in strain-induced crystallization and melting behavior, the samples P2533 and P3533 were considered to be heat shrink materials, as has been the case of racked natural rubber [46,52–55].

2.4.3. *Stress relaxation*

Polymers typically exhibit a time-dependent relaxation behavior when subjected to a constant deformation. Stress relaxation tests can be used to follow this phenomenon. The stress relaxation behavior of the PEBAX® systems within different regions of their respective stress-strain curves reported in [22] is presented in Figure 13. Measurements were made at 7.5% strain (within the linear region of the stress-strain curve) and at 150% strain (well beyond the yield point of all the PEBAX® samples investigated). The samples were deformed at an initial strain rate of 40 mm/mm/min and the tests were conducted at room temperature. Stress relaxation of a sample was quantified as the ratio of the absolute decrease in the stress at $t = 10\,000$ s to the stress immediately recorded after the sample was first stretched.

At 7.5% strain (Figure 13a), in samples with relatively high PE content (P2533, P3533, P4033), the relaxation occurred primarily in the activated PE phase, which has a higher degree of mobility at ambient temperature, and these

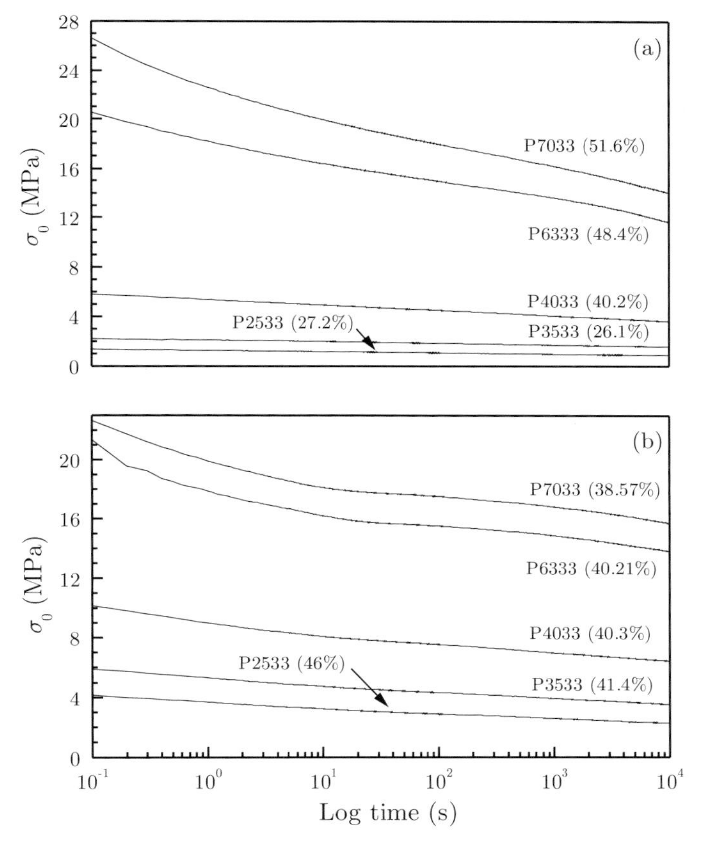

Figure 13. Stress relaxation curves of slowly cooled, melt compression-molded PEBAX® films: (a) at 7.5% strain, (b) at 150% strain. Numbers in brackets indicate the ratio of the absolute decrease in samples stress at $t = 10\,000$ s to the stress immediately recorded after the sample was first stretched [22]

samples exhibited an almost linear stress relaxation behavior with log time. However, in samples P6333 and P7033 having a lower PE content, the stress relaxation behavior deviated markedly from linearity, suggesting that as the PE content of the sample decreases, even the PA phase experiences the applied deformation. Therefore, the local PE phase deformation is strongly dependent on the PA content of the sample. The relaxation of the activated PE phase coupled with a significant relaxation of the deformed PA phase thus resulted in a non-linear stress relaxation behavior in P6333 and P7033. Furthermore, the difference in the stress relaxation between P2533 and P3533 was statistically insignificant and it was concluded that the degree of stress relaxation exhibited by a sample at small deformation increases with PA content.

At 150% strain (Figure 13b), the authors noted that the percent stress relaxation values of the PEBAX® samples were more or less independent of the

PA content of the sample. This somewhat surprising behavior was attributed to the fact that at a sufficiently high strain, the PA hard phase experiences a significant degree of deformation and results in the breakup of this partially interconnected rigid phase. The reader may recall the deformation of melt compression-molded P6333 and P7033 films studied by SALS (Figure 7) where a strain of 150% resulted in a significant deformation of the four-leaf clover pattern indicating the considerable deformation experienced by the PA hard phase. Thus, it was concluded that the local PE phase deformation is almost independent of the PE content of the sample and is solely a function of the applied deformation. Consequently, it was noted that the overall "normalized" stress relaxation of a sample was almost independent of the PA content of the sample.

2.4.4. *Mechanical hysteresis*

In addition to stress relaxation, knowledge of the hysteresis behavior is also of practical importance since the latter is a direct measure of the energy dissipated per cycle as heat. The mechanical hysteresis (MH) behavior of P2533, P4033, and P7033 is presented in Figure 14. Each of the three materials was subjected to a strain level of either 15 or 150% for three uniaxial deformation cycles each. A given deformation cycle was started immediately after the completion of the previous one. The tests were conducted on slowly cooled compression-molded samples (see Section 2.1.1) at ambient temperature and a constant crosshead speed of 15 mm/min. Under these test conditions, one notes a systematic increase in the MH value and also the permanent set of the corresponding cycles, both at 15 and 150% strain levels, as the PA content of the sample (Table 1) increases. As expected, throughout the series, for a given strain cycle at 150%, higher % MH and permanent set are exhibited than for the corresponding 15% strain cycle.

From the morphological model of the PEBAX® samples presented earlier in this chapter, we have seen that as the PA content of the sample increases, the long-range connectivity of this hard semicrystalline phase also increases. In P2533, due to its low PA content, the long-range connectivity of this phase is very limited. Thus, when P2533 is uniaxially stretched, one expects the rubbery PE amorphous matrix to be deformed more than the dispersed PA domains. Nonetheless, there would still be some rearrangement of the original microstructure and also loss of energy as heat, which leads to the small % MH and permanent set values exhibited by P2533 (Figure 14a) when the load is removed. However, the relative changes in the microstructure during subsequent cycles appear to be smaller than that during the first deformation cycle, as indicated by their lower % MH and permanent set values. Such behavior is believed to be due to the fact that once the original microstructure becomes disrupted during the first cycle, it does not have enough time to completely "heal" before the next cycle is initiated. The disruption of the microstructure would undoubtedly be more severe when P2533 is cyclically deformed to a strain

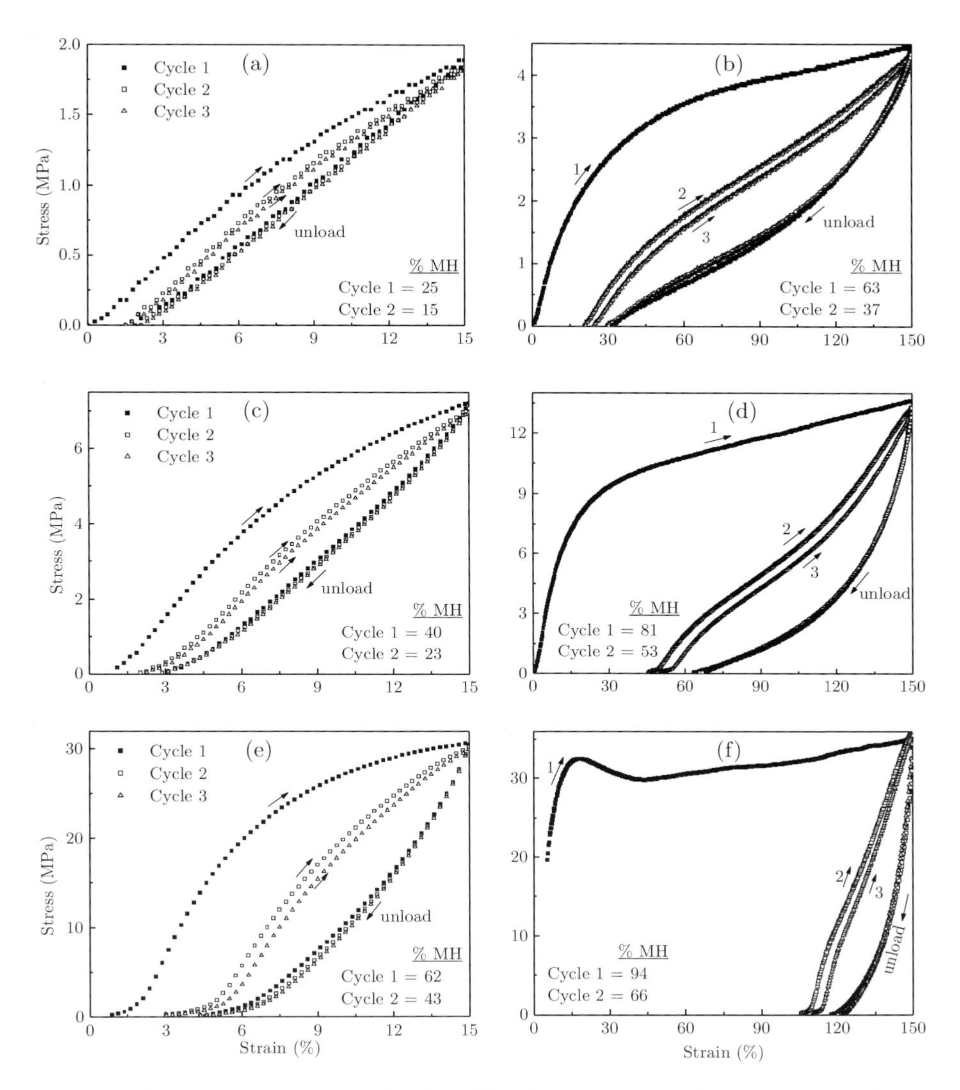

Figure 14. Mechanical hysteresis of PEBAX® P2533 (a) and (b), P4033 (c) and (d), and P7033 (e) and (f) at 15 and 150% strain, respectively. Tests were conducted at ambient temperature, and a crosshead speed of 15 mm/min

level of 150% as a result of which one notes higher % MH and permanent set values in the corresponding cycle than at the 15% strain level. Thus, the good recoverability characteristics of P2533 along with its low Young's modulus and very high elongation at break (Figure 2) allow P2533 to behave as a good elastomer under these test conditions.

At the other compositional extreme (P7033), the PA content is high enough to allow the formation of a continuous spherulitic PA superstructure when the sample is crystallized from the melt. Upon deformation, the disruption of this PA superstructural texture and the overall morphological rearrangements would therefore be much more severe for P7033 than for P2533. This fact is reflected in the significantly higher % MH and permanent set values exhibited by P7033 as compared to P2533 at both 15 and 150% strain levels. It may be noted that during the second and other consequent deformation cycles, the disrupted PA microphases can no longer effectively serve as load bearing structures. Thus, after the first deformation cycle, P7033 behaves more like P2533, as indicated by the similar MH values exhibited by P7033 during Cycle 2 (66%) and by P2533 during Cycle 1 (63%) at a uniaxial deformation of 150% strain level. In general, the thermoplastic behavior of P7033 is evident from its relatively poor recoverability, high Young's modulus, and lower extension at break (Figure 2). As expected, the mechanical hysteresis behavior of P4033 is intermediate between the elastomeric P2533 and the thermoplastic P7033.

3. Other poly(ether-*b*-amide) copolymers

3.1. *Poly(ethylene oxide)-based PEBA*

In the preceding discussion, PEBA copolymers with PTMO as the soft segment were reviewed. PEO is more hydrophilic than PTMO and investigators [23,56] have explored the possibility of using PEO-based PEBA copolymers for membranes in gas separation applications. Hatfield *et al.* [57,58] investigated the morphology and the polymorphism in the PA crystalline phase of PEO-PA6 and PEO-PA12 based PEBA commercial materials (acquired from Atofina); each contained *ca.* 58 wt% PEO. Using DSC and solid-state ^{13}C NMR, they demonstrated that the two samples possessed a microphase-separated morphology. In general, the thermal transitions noted in the DSC traces of the two samples were similar to that in PTMO-based P4033 (see Figure 3b), although their positions and profiles were specific to the nature of the constituent segments. In particular, Hatfield *et al.* noted two endothermic peaks, attributable to the PEO and the PA crystalline phases. As expected, a depression in T_m of the crystalline phases in the two copolymers, as compared to their respective homopolymers, was evident. The melting of the PE phase showed the larger depression of the two, from 64 °C to *ca.* 12 °C. Thus, at room temperature, it was noted that the microphase-separated morphology of the two copolymers consisted of a crystalline phase of the PA segments and partially mixed amorphous phases of PEO and PA segments. The reader may recall the discussion in Section 2.1.1 on the possible nature of mixing within these amorphous phases that are dominated by the PA and PEO segments, respectively. In addition, a broadening of the endothermic melting peaks and a decrease in the ΔH_f values of the transitions also occurred. Hatfield *et al.* suggested that this behavior indicated a decrease in both the percent crystallinity and the perfection of the crystalline phases as compared to neat PEO, nylon 6, and nylon 12.

Particularly interesting was their investigation of the polymorphism in the PA phase of the two samples. The crystalline phases of nylon 6 and nylon 12 can either adopt the α or the γ form. However, at ambient temperature, the thermodynamically stable forms in nylon 6 and nylon 12 are α and γ, respectively. The α form in nylon 6 consists of extended chains with hydrogen bonds between antiparallel chains, thus producing an arrangement of parallel sheets [59]. It gives rise to a set of two reflections with d spacings of *ca.* 3.75 and 4.4 Å in its WAXS pattern. In nylon 12, the thermodynamically stable γ form of the crystals consists of hydrogen bonded parallel chains. Thus an arrangement of alternate parallel and antiparallel sheets is produced wherein the amide plane in nylon 12 is twisted out of the methylene plane by 62° [59]. The γ crystalline form in nylon 12 is characterized by a single reflection at *ca.* 4.15 Å in its WAXS pattern. Not surprisingly, the crystalline forms of the PA phase in PEO-PA6 and PEO-PA12 were consistent with those of the respective homopolymeric nylons. As noted from the discussion on DSC, the PE crystal phase completes its melting transition well below room temperature. Thus, one does not expect the characteristic PEO reflections at the d spacings of 3.81 and 4.63 Å to be seen in the WAXS patterns taken at ambient temperature. Drawing and annealing, or chemical treatment was employed by Hatfield *et al.* [58] and Apostolov *et al.* [60] to induce the transformation of a particular unit cell form of the PA crystal phase in PEO-PA6 and PEO-PA12 based PEBA.

3.2. *PEBA based on aromatic PA segments of uniform length*

To this point, the structure-property relationships and uniaxial deformation behavior of aliphatic PEBA copolymers synthesized by polycondensation of dihydroxy-terminated PE oligomers, such as PTMO or PEO, with dicarboxylic acid-terminated amide monomers have been reviewed. *Via* this synthetic route, at a fixed soft segment length, the hard segment length increases in direct proportion to the overall PA content of the sample and, in addition, a hard segment molecular weight distribution (of *ca.* 2) also occurs. Gaymans *et al.* attempted to overcome the latter feature by synthesizing the hard segments first. It was hoped that a narrower molecular weight distribution of the hard segments would translate into an increase in both the order and percent crystallinity of the hard phase, which in turn could improve the extent of microphase separation in the copolymer. Furthermore, the latter would result in materials with higher E', greater elongation at break, and a decreased temperature dependence of E' within the service window. In a controlled manner, they synthesized aromatic hard segments first and later fractionated them by recrystallization to isolate segments of uniform length. Transesterification and polycondensation of these aromatic hard segments of uniform length with dihydroxy-terminated PTMO or PEO macromonomers resulted in PEBA copolymers. Although a melt polymerization route was explored, PEBA with intrinsic viscosity greater than 1 dL/g (which is indicative of a high molecular weight) resulted only when the synthesis of hard segments was carried out in solution and subsequent polycondensation in the melt [61]. The structure of the repeat unit of the PTMO-based aromatic PEBA copolymer is given in Scheme 2 and it is hereafter represented as PTMO-TΦT, where the TΦT designation represents the aromatic hard segment.

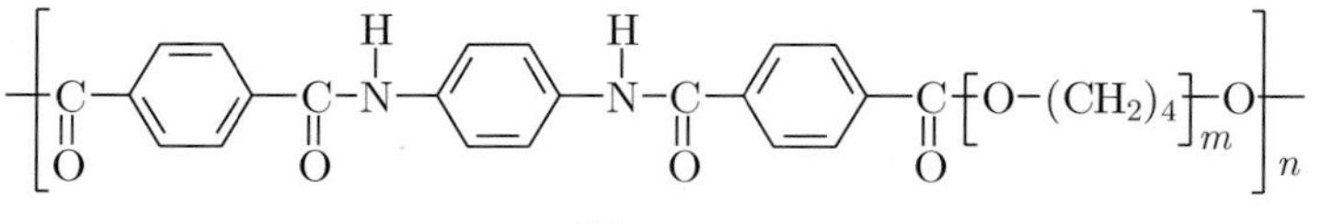

$$\left[-\underset{\substack{\|\\O}}{C}-C_6H_4-\underset{\substack{\|\\O}}{C}-\overset{H}{N}-C_6H_4-\overset{H}{N}-\underset{\substack{\|\\O}}{C}-C_6H_4-\underset{\substack{\|\\O}}{C}-\left[O-(CH_2)_4\right]_m-O-\right]_n$$

Scheme 2

The defining characteristic of the synthesis leading to the above segmented copolymer is that the overall PA content of the sample can only be increased by decreasing the soft segment molecular weight. In a series of studies [61–76], aromatic PEBA with various modifications of the backbone chemistry were investigated. In the ensuing discussion, the solid-state structure-property relationships and deformation behavior of aromatic PEBA will be briefly reviewed. The discussion will be mostly limited to PTMO-TΦT based systems.

3.2.1. *Dynamic mechanical thermal analysis*

The temperature dependence of the storage modulus of PTMO-TΦT based PEBA copolymers was reported by Gaymans *et al.* [61]. The soft segment molecular weight within the series ranged from 650 to 2900 g/mole, as presented in Figure 15. The DMTA data of injection molded samples was collected by a

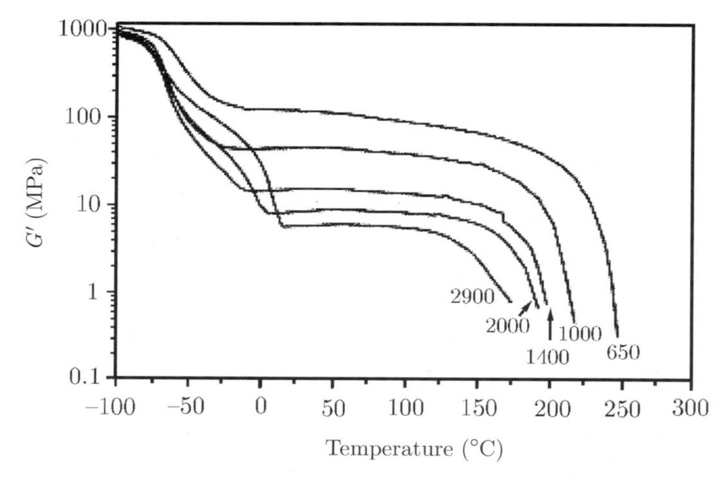

Figure 15. Temperature dependent shear storage modulus, G', of injection molded PTMO-TΦT based PEBA copolymers. The DMTA scans were conducted at a heating rate of 1°C/min and 1 Hz. The number next to each curve indicates the PTMO molecular weight [61]

torsion pendulum at a frequency of 1 Hz and a heating rate of 1°C/min. The general features of the DMTA response are similar to the PTMO-PA12 based systems discussed in Section 2.1.1 (see Figure 1a). Starting from *ca.* –80 °C, one notes a relatively sharp and soft segment molecular weight-independent glass transition at ca. –70 °C. The T_g of homopolymeric PTMO, as mentioned earlier, is *ca.* –75 °C. Samples with PTMO molecular weight greater than 1000 g/mole displayed a shoulder in the –50 to 20 °C range, arising from the crystallization and melting of the PTMO crystal phase. A temperature-insensitive rubbery plateau followed thereafter. As expected, the average magnitude of the storage modulus within this service window increased with increasing hard segment content (or decreasing soft segment molecular weight) due to an increase in the number of physical crosslink sites and also an improved connectivity of the hard phase.

Following the rubbery plateau, a decrease in the storage modulus above 175°C occurred due to the softening of the sample, induced by the melting of the TΦT crystalline phase. DSC analysis of these PEBA copolymers revealed an endothermic peak above 200 °C in samples with a TΦT content greater than 20 wt%, attributable to the melting of the TΦT crystalline phase. Moreover, T_m increased with increasing TΦT content. Similarly, the onset of the softening of these PEBA copolymers, as seen from their DMTA response in Figure 15, is also dependent upon the hard segment content in the sample. The hard segment content-dependent T_m (as noted from DSC) and the copolymer softening temperature (as determined from DMTA) is indeed surprising, considering the very narrow distribution of the hard segment lengths. It may be recalled that in aliphatic PTMO-PA12 based systems, the melting temperature of the hard segments increased with their length, irrespective of the soft segment molecular weight. Such a behavior was suggested to be due to an increase in the PA lamellar thickness and a general improvement in the crystal perfection with increasing hard segment length. In the case of aromatic PTMO-TΦT PEBA systems, Gaymans *et al.* explained the particular melting behavior of the hard segments on the basis of a "solvent effect" produced by the soft segments. This concept was first formulated for a random copolymer by Flory [77] to explain the effect of the non-crystallizable units covalently linked to the segment ends of the crystallizable component on the melting temperature at which the last crystal of the crystallizable component melts, as expressed by Eq. (1)

$$\frac{1}{T_m} - \frac{1}{T_m^0} = -\left(\frac{R}{\Delta H_f}\right) ln\, p \tag{1}$$

where T_m is the temperature at which the last crystal of the crystallizable block/segment melts, T_m^0 is the melting temperature of the neat homopolymer, ΔH_f is the latent heat of fusion, and R is the universal gas constant. The quantity p is the sequence probability propagation of a crystallizable unit, or in other words, it is the probability of a crystallizable group being succeeded by another crystallizable group.

As predicted by Eq. (1), for the PTMO-TΦT based PEBA copolymers addressed in Figure 15, the plot of the T_m^{-1} *vs.* –ln p was linear with a positive slope. A similar linear behavior was exhibited by aromatic PEBA copolymers in which the central phenyl ring of the hard segment of uniform length was replaced by either $(CH_2)_2$ or $(CH_2)_4$, thereby again giving some support for the use of Eq. (1). To facilitate the calculation, the quantity p was replaced by X_A, the molar fraction of the crystallizable units, based on the assumptions that all the hard segments crystallized and the molar volumes of both the crystallizable and non-crystallizable fractions were equal. The former assumption was made on the basis of the observation that in the DMTA response of the aromatic PEBA, the PTMO T_g appeared to be independent of the hard segment content and no separate hard segment T_g was detected. It may be qualified that in segmented copolymers wherein there is strong and extensive hydrogen bonding, it is generally difficult to clearly distinguish a hard segment T_g, even though an amorphous phase of the hard segments exists [78]. Furthermore, in the particular case of segmented copolymers, p does not scale with X_A as it does in random copolymers and, for this reason, replacing p with X_A may not be entirely justifiable for segmented copolymers. However, it may be recalled from Section 2.2 that a series of PEBAX®, poly(ether ester)s, and segmented polyurethanes with increasing hard segment length were also shown to exhibit a good fit to the plot of T_m^{-1} *vs.* ln n, where n is the crystallizable segment length.

To summarize the DMTA response, PTMO-TΦT based PEBA copolymers exhibit a microphase-separated morphology. In general, at ambient temperature, a PTMO amorphous phase, possibly with a few dissolved isolated TΦT segments, coexists with a TΦT crystalline phase. As the TΦT content of the sample increases, the long-range connectivity of the hard phase improves, thereby resulting in an increase in the average rubbery plateau modulus. As compared to their aliphatic PTMO-PA12 counterparts, the upper temperature limit of the service window is higher in PTMO-TΦT based PEBA due to the higher temperature stability of the aromatic TΦT hard phase.

3.2.2. *Uniaxial deformation behavior*

To successfully utilize materials in engineering applications, knowledge of both the morphology and the response of such a morphology to deformation is required. In this section, the effect of deformation on PTMO-TΦT based PEBA will be reviewed. Gaymans *et al.* [65] reported the stress-strain response of PTMO-TΦT based PEBA with PTMO molecular weight of 1000, 2000, and 2900 g/mole, corresponding to a hard segment content of 23, 13, and 10 wt%, respectively. Their results are presented in Figure 16a. The three samples displayed in general a low Young's modulus and a high elongation at break, which is typical of elastomeric materials. Elastomers generally do not display a yield point in their stress-strain response, although the $PTMO_{1000}$ based PEBA sample does exhibit a yield point, suggesting the presence of long-range connectivity of the hard phase, generated by the higher hard segment content than in the other samples

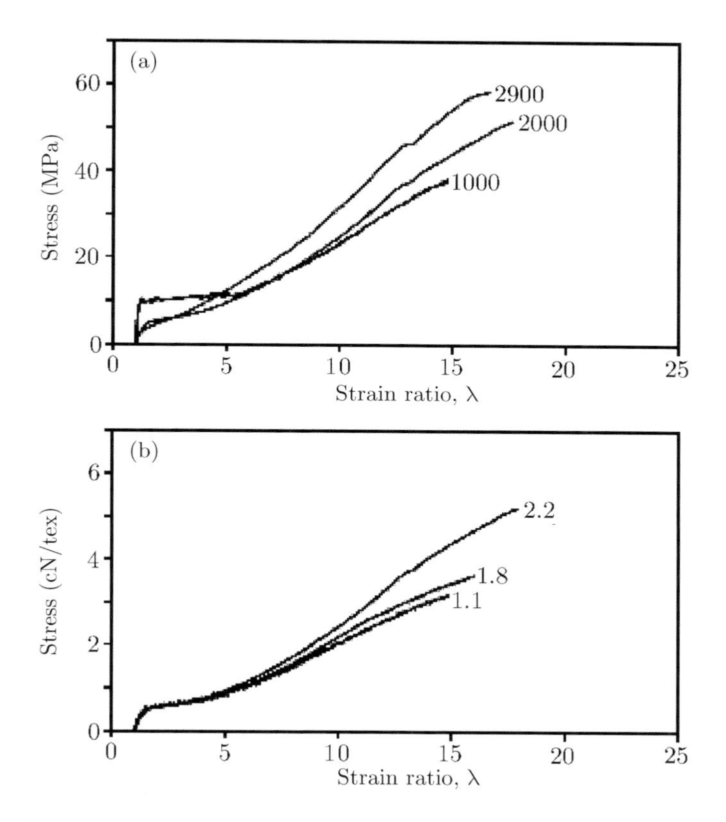

Figure 16. Stress *vs.* strain behavior of melt-extruded PTMO-TΦT based PEBA measured at an initial strain rate of 250 mm/min with an initial clamp distance of 25 mm. The number next to each curve indicates in (a) the PTMO molecular weight (g/mole) and in (b) the inherent viscosity (dL/g) of the $PTMO_{2000}$-TΦT copolymer [65]

within the series. The elongation at break of the three samples is comparable to aliphatic PTMO-PA12 based PEBA of comparable hard segment content (see Figure 2) and appears to increase with increasing PTMO molecular weight. Such a behavior may arise from differences in the overall molecular weight of the copolymer. Indeed, as seen in Figure 16b for aromatic PEBA with $PTMO_{2000}$, the strain at break increased with increasing inherent viscosity of the copolymer. Inherent viscosity can be related to the molecular weight of the copolymer, although the Mark-Houwink-Sakurada equation cannot be easily applied to calculate the molecular weight due to the presence of strong secondary interactions in these copolymers. Nonetheless, the higher strain at break with increasing inherent viscosity (and thus, molecular weight) is no doubt due to an improved entanglement density, which allows greater chain orientation without chain slippage. The toughness of the copolymers, as measured by the energy of rupture normalized per unit volume of the initial sample, also increased with increasing inherent viscosity of the copolymers. The strain hardening at high strains, as seen in Figure 16a,b, is due to the strain-induced crystallization of the PTMO segments. Furthermore, the stress at break increased with increasing PTMO molecular weight (Figure 16a) due to the fact that the fraction of PTMO chains able to crystallize under strain would be higher in $PTMO_{2900}$ than in $PTMO_{2000}$ or $PTMO_{1000}$. Thus, there would be a larger PTMO crystal fraction in $PTMO_{2900}$ based PTMO-TΦT PEBA to resist the applied load without a large deformation.

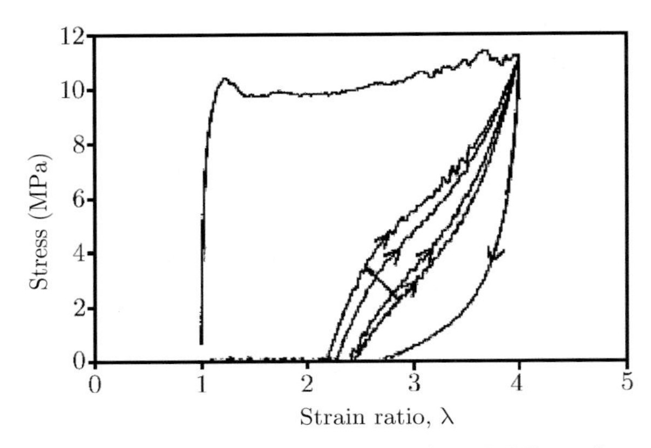

Figure 17. Cyclic tensile test for melt-extruded $PTMO_{1000}$-TΦT copolymer. Initial strain rate 200 mm/min, grip separation distance 50 mm. Relaxation time (no load applied) between each cycle increases in the direction of the arrow: 0 s, 30 s, 30 min, and 16 h [65]

The results of a cyclic loading test, also reported by the research group of Gaymans *et al.* [65], are reproduced in Figure 17. A $PTMO_{1000}$-TΦT based PEBA sample (hard segment content of *ca.* 23 wt%) was subjected to uniaxial strain of 300% to study its ambient temperature hysteresis and recovery behavior. The second cycle was initiated instantaneously at the end of the first and the consecutive cycles after an increasing relaxation time interval during which no load was applied to the sample. Time intervals of 0 s, 30 s, 30 min, and 16 h were employed between consecutive cycles. As seen in Figure 17, at the end of the first loading-unloading cycle the aromatic PEBA sample exhibits a permanent set of *ca.* 170% strain. The Young's modulus systematically increases with each consecutive cycle, as more time is allowed for the sample to heal. Such a behavior suggests that as the sample relaxes, the crystalline lamellar domains of the hard TΦT that are disrupted during the loading cycle are able to reassociate (to form larger domains) with time. Hence, the sample is able to partially recover more of its original dimensions, as suggested by the lower strain at which each consecutive cycle begins, and also its load-bearing capacity. Obviously, such a recovery would be aided at elevated temperatures.

The stress relaxation (SR) of the PTMO-TΦT based PEBA copolymers, stretched to a strain of 100% within 5 s was found to increase with increasing TΦT content. SR in this particular study was defined as

$$SR_{x\%} = \frac{|\Delta\sigma_{x\%}|}{\Delta \log t} \tag{2}$$

where $\sigma_{x\%}$ is the stress at x% strain and t is time in seconds.

The $PTMO_{2900}$-TΦT sample with a TΦT content of *ca.* 10 wt% had a SR of 0.37. SR values of 0.48 and 0.82 were exhibited by $PTMO_{2000}$-TΦT (13 wt% TΦT) and $PTMO_{1000}$-TΦT (23 wt% TΦT), respectively.

To better understand the specifics of hard *vs.* soft segment orientation during deformation and the mechanical response of the copolymer, various investigators have used the technique of infrared dichroism. For instance, numerous studies have focused on the orientation behavior of the hard and soft segments in segmented polyurethanes during uniaxial deformation [44,78,79]. In the case of PTMO-TΦT based PEBA, infrared dichroism was utilized by the research group of Gaymans *et al.* [68] to simultaneously examine the orientation of both the hard TΦT and the soft PTMO segments. They monitored the amide carbonyl stretching at 1645 cm^{-1} to measure the orientation function of the hard TΦT segments. The orientation function of the soft PTMO segments was measured by following the IR absorbance of the aromatic ester carbonyl stretching at 1715 cm^{-1} or the ether stretching at 1098 cm^{-1}. In general, the hard and soft segment orientation behavior of the aromatic PEBA was found to be similar to conventional segmented polyurethane copolymers [80,81]. Initially, at low strain, the TΦT hard phase lamellae oriented perpendicular to the stretch direction due to their high aspect ratio. Disruption of these lamellae due to elongational and shear forces occurred at higher strain and the lamellae of reduced aspect ratios gradually rotated to preferentially orient the hard segments along the stretch direction. The orientation of the PTMO segments in the stretch direction was negligible until *ca.* 450% strain, above which the onset of strain-induced crystallization of these segments was observed. As expected, the PTMO segments in the phase that underwent strain-induced crystallization became highly oriented along the deformation axis.

The ability of the PTMO segments in aliphatic PEBA copolymers to undergo strain-induced crystallization was reviewed earlier in Section 2.4.2. Gaymans *et al.* [68] investigated the effect of the molecular weight of the PTMO segments on the strain-induced crystallization behavior of aromatic TΦT based PEBA copolymers. Samples with a PTMO molecular weight of 1000 and 2000 g/mol, respectively, were stretched to 1000% strain, allowed to relax and thereafter subjected to a DSC heating scan. As can be seen in Figure 18, the PTMO melting peak in the $PTMO_{2000}$-TΦT based PEBA sharpened, as compared to the unoriented sample, and the T_m increased from below room temperature to *ca.* 45 °C. On the other hand, the T_m of the crystalline PTMO phase (that underwent strain-induced crystallization) of the $PTMO_{1000}$-TΦT sample increased only to about room temperature. Therefore, when the oriented sample was allowed to relax and subjected to a DSC heating scan, no distinct difference in the PTMO melting behavior between the oriented and the unoriented samples was observed. It may be recalled from Section 2.4.2 that the PTMO-PA12 based

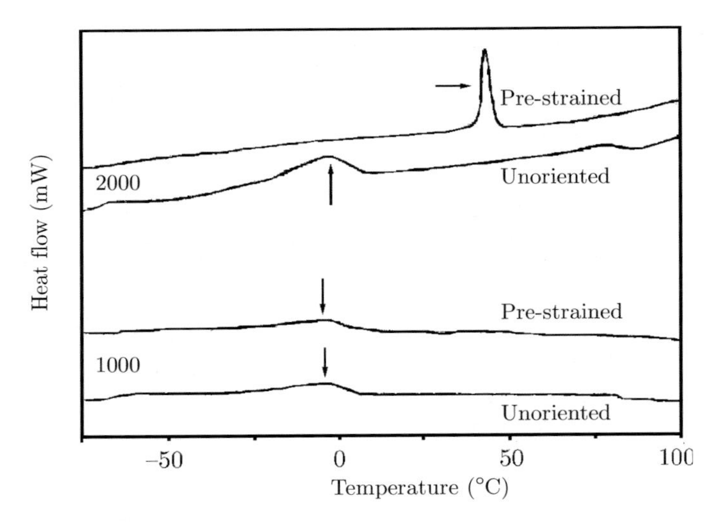

Figure 18. PTMO melting peaks of oriented and 1 000% pre-strained $PTMO_{1000}$-TΦT and $PTMO_{2000}$-TΦT copolymers. DSC heating scans were conducted at 20 °C/min [68]

P2533 material underwent strain-induced crystallization above 500% strain and the T_m of the resultant PE crystalline phase was also *ca.* 45 °C.

4. Conclusion

The structure-property behavior and the uniaxial deformation behavior of poly(ether-*b*-amide) copolymers have been reviewed in this chapter. Copolymers based on either aliphatic or aromatic (of uniform length) hard segments and PTMO or PEO as the soft segments were partially explored. These PEBA copolymers possess a complex morphology consisting of microphase-separated hard and soft phases. The extent of microphase separation in these materials is a function of the relative composition of the hard and soft phases and is also dependent on the segment molecular weight.

As expected, the mechanical properties of the PEBA copolymers depend on the relative composition of the hard and soft components, as well as upon their respective levels of microphase separation. For the PTMO based PEBA copolymers, when the PA volume fraction is less than 30 wt%, soft elastomeric materials generally result, showing a remarkably high extension at break of over 1 000%, low permanent set, and low hysteresis. On the other hand, when the PA phase is the major component, hard thermoplastic materials result. As expected, the PA-rich copolymers display a higher permanent set and also a higher mechanical hysteresis, as compared to the PEBA copolymers with lower PA content. In the PTMO based PEBA samples with aromatic hard segments of uniform length, the upper temperature limit of the service window increases towards 200 °C because of the improved thermal stability of the aromatic hard segments. The overall molecular weight of the copolymer also affects the mechanical properties of the material. For instance, the extension at break in PTMO based aromatic PEBA systematically increases with increasing intrinsic viscosity (which serves as an indirect measure of the molecular weight) of the copolymer.

In PTMO based PEBA copolymers where the PE phase dominates the composition and the molecular weight of the PE segments is higher than *ca.* 2000 g/mole, strain-induced crystallization of the PE phase occurs at ambient temperature when the sample is uniaxially deformed above *ca.* 500%. Upon strain-induced crystallization, the T_m of the PE phase increases from below room temperature to *ca.* 45 °C and the initially soft elastomeric sample becomes inelastic until the PE phase undergoes melting.

Thus, poly(ether-*b*-amide) copolymers allow great flexibility in designing materials with targeted mechanical properties. A wide range of application requirements can easily be accommodated by these materials simply by adjusting their relative segment composition. The constituent polyether or polyamide segments can also be modified to meet specific demands. Hence, together with poly(ether ester)s and segmented polyurethanes, poly(ether-*b*-amide)s constitute an important class of polymers that can be used to overcome some application specific limitations of conventional homopolymers or polymer blends.

References

1. Bates F S and Fredrickson G H (1999) Block copolymers-designer soft materials, *Physics Today* **52**:32–38.
2. Scollenberger C S (1955) Simulated vulcanizates of polyurethane elastomers, US Patent 2,871,218, to B F Goodrich.
3. Scollenberger C S, Scott H and Moore G R (1962) Polyurethane VC, virtually crosslinked elastomer, *Rubber Chem Tech* **35**:742–752.
4. Deleens G, Foy P and Maréchal E (1977) Synthesis and characterization of block polyamideether copolycondensates. I. Synthesis and study of difunctional ω,ω′-poly(amide 11) oligomers, *Eur Polym J* **13**:337–342.
5. Deleens G, Foy P and Maréchal E (1977) Synthesis and characterization of block polyamideether copolycondensates. II. Polycondensation of ω,ω′-diacid- and ω,ω′-diester-poly(amide 11) oligomers with ω,ω′-dihydroxy-polyether oligomers, *Eur Polym J* **13**:343–351.
6. Deleens G, Foy P and Marechal E (1977) Synthesis and characterization of block polyamideether copolycondensates. III. Study of the polycondensation of ω,ω′-dicarboxylic-poly(amide 11) and ω,ω′-dihydroxy-polyoxyethylene. Determination of the rate constants and activation energies, *Eur Polym J* **13**:353–360.
7. Deleens G (1981) ANTEC-81:419-420.
8. Deleens G (1987) Polyether block amide thermoplastic elastomers, in *Thermoplastic Elastomers: A Comprehensive Review* (Eds. Legge N R, Holden G and Schroeder H E) Hanser Publishers, New York, Ch. 9B, pp. 217–230.
9. Flesher Jr J R (1986) PEBAX polyether block amides – a new family of engineering thermoplastic elastomers, in *High Performance Polymers: Their Origin and Development* (Eds. Seymour R B and Krishenbaum G S), Elsevier, New York, pp. 401–408.

10. Kohan M I (1995) *Nylon Plastic Handbook*, Hanser Publishers, New York.
11. Bornschlegl E, Goldbach G and Meyer K (1985) Structure and properties of segmented polyamides, *Prog Colloid Polym Sci* **71**:119–124.
12. Goldbach G, Kita M, Meyer K and Richter K P (1986) Structure and properties of polyamide 12 alloys, *Prog Colloid Polym Sci* **72**:83–96.
13. Nelb R G and Chen A T (1996) Thermoplastic elastomers based on polyamides, in *Thermoplastic Elastomers: A Comprehensive Review* (Eds. Legge N R, Holden G, Quirk R and Schroeder H E) Hanser Gardner Publishers, New York, Ch. 9, pp. 228–255.
14. Mumcu S, Burzin K, Feldmann R and Feinauer R (1978) Copolyetheramides from laurolactam, decane-1,10-dicarboxylic acid and α,ω-dihydroxy(polytetrahydrofuran), *Angew Makromol Chem* **74**:49–60.
15. Boulares A, Tessier M and Maréchal E (2000) Synthesis and characterization of poly-(copolyether-block-polyamides) II. Characterization and properties of multiblock copolymers, *Polymer* **41**:3561–3580.
16. Castaldo L, Maglio G and Palumbo R (1978) Synthesis of polyamide-polyether block copolymers, *J Polym Sci Polym Lett Ed* **16**:643–645.
17. Koch R B (1982) PEBAX (polyether block amide), *Adv Polym Tech* **2**:160–162.
18. Konyukhova E V, Buzin A I and Godovsky Y K (2002) Melting of polyether block amide (Pebax): the effect of stretching, *Thermochim Acta* **391**:271–277.
19. Clayden N J and Pendlebury R (2001) NMR study of the effect of electron beam processing on a poly(ether-*block*-amide), *Polymer* **42**:8373–8377.
20. Sauer B B, McLean R S, Brill D J and Londono D J (2002) Morphology and orientation during the deformation of segmented elastomers studied *via* small-angle X-ray scattering and atomic force microscopy, *J Polym Sci Polym Phys Ed* **40**:1727–1740.
21. Private communication from Atofina *via* Sauer B B.
22. Sheth J P, Xu J and Wilkes G L (2003) Solid state structure-property behavior of semicrystalline poly(ether-block-amide) PEBAX® thermoplastic elastomers, *Polymer* **44**:743-756.
23. Bondar V I, Freeman B D and Pinnau I (1999) Gas sorption and characterization of poly(ether-b-amide) segmented block copolymers, *J Polym Sci Polym Phys Ed* **37**:2463–2475.
24. Wunderlich B (1980) *Macromolecular Physics*, Academic Press, New York, Vol. 3, p. 52.
25. Boubil H, Okorofor E, Belhoucine and Rault J (1989) Morphology of polyamide and polyether block amide blends, *Polym Eng Sci* **29**:679–684.
26. Legge N R, Holden G and Schroeder H E (Eds.) (1987) *Thermoplastic Elastomers: A Comprehensive Review*, Hanser Publishers, New York, pp. 427, 472.
27. McCrum N G, Read B E and Williams G (1967) *Anelastic and Dielectric Effects in Polymeric Solids*, Dover Publications, New York, p. 561.
28. Alberola N (1988) Micromechanical properties of polyether block amide copolymers, *J Appl Polym Sci* **36**:787–804.
29. Alberola N, Vassel A and Helluin C (1989) Mechanical relaxation processes in polyether block amide copolymer (PEBA), *Makromol Chem Macromol Symp* **23**:219–224.
30. Faruque H S and Lacabanne C (1986) Study of multiple relaxations in PEBAX, polyether block amide (PA12 2135 block PTMG 2032), copolymer using the thermally stimulated current method, *Polymer* **27**:527–531.
31. Faruque H S and Lacabanne C (1987) Anelastic and dielectric properties of polyether-polyamide copolymer PEBAX studied by a thermally stimulated depolarisation current method, *J Phys D Appl Phys* **20**:939–944.

32. Faruque H S and Lacabanne C (1987) A thermally stimulated current technique for measuring the molecular parameters of Pebax, a polyether-block amide copolymer, *J Mater Sci* **22**:675–678.
33. Faruque H S and Lacabanne C (1990) Thermally stimulated depolarisation of poly(ether block amide) copolymer, *J Mater Sci* **25**:321–324.
34. Demont P, Chatain D, Lacabanne C and Glotin M (1989) Retardation and relaxation behavior of poly(ether-block-amide)s (PEBA) copolymers by thermally stimulated creep (TSCr) and current (TSCu), *Makromol Chem Macromol Symp* **25**:167–186.
35. Di Lorenzo M L, Pyda M and Wunderlich B (2001) Calorimetry of nanophase-separated poly(oligoamide-alt-oligoehter)s, *J Polym Sci Polym Phys Ed* **39**:1594–1604.
36. Di Lorenzo M L, Pyda M and Wunderlich B (2001) Reversible melting in nanophase-separated poly(oligoamide-alt-oligoether)s and its dependence on sequence length, crystal perfection, and molecular mobility, *J Polym Sci Polym Phys Ed* **39**:2969–2981.
37. Harrell Jr L L (1969) Segmented polyurethanes. Properties as a function of segment size and distribution, *Macromolecules* **2**:607–612.
38. Wegner G (1987) Polyether block amide thermoplastic elastomers, in *Thermoplastic Elastomers: A Comprehensive Review* (Eds. Legge N R, Holden G and Schroeder H E) Hanser Publishers, New York, Ch. 9B, p. 418.
39. Cojazzi G, Ficher A, Garbuglio C, Malta V and Zannetti R (1973) The crystal structure of polylauryllactam (Nylon 12), *Makromol Chem* **168**:289–301.
40. Zhu L-L and Wegner G (1981) The morphology of semicrystalline segmented poly(etherester) thermoplastic elastomers, *Makromol Chem* **182**:3625–3638.
41. Cella R J (1973) Morphology of segmented polyester thermoplastic elastomers, *J Polym Sci Polym Symp* **42**:727–740.
42. Lilaonitkul A, West J C and Cooper S L (1976) Properties of poly(tetramethylene oxide)-poly(tetramethylene terephthalate) block polymers, *J Macromol Sci–Phys* **B12**:586–623.
43. Mody P C, Wilkes G L and Wagener K B (1981) Structure-property relationships of a new series of segmented polyether-polyester copolymers, *J Appl Polym Sci* **26**:2853–2878.
44. Ward I M (Ed.) (1977) *Structure and Properties of Oriented Polymers,* Chapman and Hall, New York.
45. McLean R S and Sauer B B (1999) Nano-deformation of crystalline domains during tensile stretching studied by atomic force microscopy, *J Polym Sci Polym Phys Ed* **37**:859–866.
46. Warner S (1990) Strain-induced crystallization and melting behavior of polyether-amide block copolymer, *J Elast Plast* **22**:167–173.
47. Akay M and Ozden S (1995) The influence of residual stresses on the mechanical and thermal properties of injection molded ABS copolymer, *J Mater Sci* **30**:3358–3368.
48. Cesari M, Perego G and Mazzei A (1965) The crystal structure of poly(tetrahydro-furan), *Makromol Chem* **83**:196–206.
49. Toki S and Hsiao B S (2003) Nature of strain-induced structures in natural and synthetic rubbers under stretching, *Macromolecules* **36**:5915–5917.
50. Andrews E H (1964) Crystalline morphology in thin films of natural rubber. II. Crystallization under strain, *Proc Roy Soc* London **A277**:562–570.
51. Oth J F M and Flory P J (1958) Thermodynamics of shrinkage of fibrous (racked) rubber, *J Am Chem Soc* **80**:1297–1304.
52. Desgrand J V (1832) Method of weaving elastic fibers, UK Patent 6,334.
53. Mertens A (1841) Improvements in the manufacture of plaited fabrics, UK Patent 9,186.

54. Baare F and Garelly J G (1859) Manufacturing corrugated fabrics, US Patent 24,691.
55. Sheperd T L (1935) Improvements in and relating to the manufacture of a fabric from elastic threads, US Patent 423,997.
56. Barbl B, Funari S S, Gehrke R, Scharnagl N and Stribeck N (2003) SAXS and gas transport in polyether-block-polyamide copolymer membranes, *Macromolecules* **36**:749–758.
57. Hatfield G R, Guo Y, Killinger W E, Andrejak R A and Roubicek P M (1993) Characterization of structure and morphology in two poly(ether-block-amide) copolymers, *Macromolecules* **26**:6350–6353.
58. Hatfield G R, Bush R W, Killinger W E and Roubicek P M (1994) Phase modification and polymorphism in two poly(ether-block-amide) copolymers, *Polymer* **35**:3943–3947.
59. Malta V, Cojazzi G, Fichera A, Ajo D and Zannetti R (1979) A reexamination of the crystal structure and molecular packing of nylon 6, *Eur Polym J* **15**:765–770.
60. Apostolov A A, Bosvelieva E, Du Chesne A, Goranov K and Fakirov S (1993) Multiblock poly(ether-ester-amide)s based on polyamide-6 and poly(ethylene glycol). 2. Effect of composition and soft-segment length on the structure of poly(ether-ester-amide)s as revealed by X-ray scattering, *Makromol Chem* **194**:2267–2277.
61. Niesten M C E J, Feijen J and Gaymans R J (2000) Synthesis and properties of segmented copolymers having aramid units of uniform length, *Polymer* **41**:8487–8500.
62. van Hutten P F, Mangus R M and Gaymans R J (1993) Segmented copolymers with polyesteramide units of uniform length: structure analysis, *Polymer* **34**:4193–4202.
63. Gaymans R J and de Haan J L (1993) Segmented copolymers with poly(ester amide) units of uniform length: synthesis, *Polymer* **34**:4360–4364.
64. Guang L and Gaymans R L (1997) Polyesteramides with mixtures of poly(tetramethylene oxide) and 1,5-pentanediol, *Polymer* **38**:4891–4897.
65. Niesten M C E J and Gaymans R J (2001) Tensile and elastic properties of segmented copolyetheresteramides with uniform aramid units, *Polymer* **42**:6199–6207.
66. Niesten M C E J and Gaymans R J (2001) Segmented copolyetheresteramids with extended poly(tetramethyleneoxide) segments, *Polymer* **42**:1461–1469.
67. Niesten M C E J and Gaymans R J (2001) Influence of type of uniform aromatic amide units on segmented copolyetheresteramides, *Polymer* **42**:931–939.
68. Niesten M C E J, Harkema S, van der Heide E and Gaymans R J (2001) Structural changes of segmented copolyetheresteramides with uniform aramid units induced by melting and deformation, *Polymer* **42**:1131–1142.
69. Niesten M C E J and Gaymans R J (2001) Influence of extenders on thermal and elastic properties of segmented copolyetheresteamides, *J Appl Polym Sci* **81**:1605–1613.
70. Niesten M C E J and Gaymans R J (2001) Comparison of properties of segmented copolyetheresteramides containing uniform aramid segments with commercial segmented copolymers, *J Appl Polym Sci* **81**:1372–1381.
71. Bouma K, Wester G A and Gaymans R J (2001) Polyether-amide segmented copolymers based on ethylene terephthalamide units of uniform length, *J Appl Polym Sci* **80**:1173–1180.
72. Niesten M C E J, Krijgsman J, Harkema S and Gaymans R J (2001) Melt spinnable Spandex fibers from segmented copolyetheresteramides, *J Appl Polym Sci* **82**:2194–2203.
73. Sorta E and della Fortuna G (1980) Poly(ester amide)-polyether block copolymers: preparation and some physicochemical properties, *Polymer* **21**:728–732.
74. Perego G, Cesari M and della Fortuna G (1984) Structural investigations of segmented block copolymers. I. The morphology of poly(ether esteramide)s based on poly(tetramethylene oxide) soft segments, *J Appl Polym Sci* **29**:1141–1155.

75. Perego G, Cesari M and della Fortuna G (1984) Structural investigations of segmented block copolymers. II. The morphology of poly(ether ester)s based on poly(tetramethylene oxide) soft segments: comparison between poly(ether ester)s and poly(ether esteramide)s, *J Appl Polym Sci* **29**:1157–1169.
76. Cernia E and D'Ilario L (1985) Thermoplastic elastomers: structural and morphological aspects of ether-esteramide copolymers, *J Polym Sci Polym Phys Ed* **23**:49–57.
77. Flory P J (1955) Theory of crystallization in copolymers, *Trans Faraday Soc* **51**:848–857.
78. Cooper S L, Estes G M and Seymour R W (1971) Infrared studies of segmented polyurethane elastomers. II. Infrared dichroism, *Macromolecules* **4**:452–457.
79. Siesler H W (1988) Rheo-optical Fourier-transform infrared spectroscopy of polymers. 14. Segmental orientation and strain-induced crystallization of a poly(ether-urethane-urea) elastomer, *Phys Chem Chem Phys* **92**:641–645.
80. Moreland J C and Wilkes G L (1991) Segmental orientation behavior of flexible water-blown polyurethane foams, *J Appl Polym Sci* **43**:801–815.
81. Yeh F, Hsiao B S, Sauer B B, Michel S and Siesler H W (2003) *In situ* studies of structure development during deformation of a segmented poly(urethane-urea) elastomer, *Macromolecules* **36**:1940–1954.

PART IV

POLYURETHANE-BASED THERMOPLASTIC ELASTOMERS

Chapter 12

Thermoplastic Polyurethane Elastomers in Interpenetrating Polymer Networks

O. Grigoryeva, A. Fainleib, L. Sergeeva

1. Introduction

Numerous kinds of *interpenetrating polymer networks* (IPNs) are available as commercial products [1–7], but they are often not recognized as such. Sperling [1] defined IPNs as combinations of two (or more) polymers in the shape of networks, at least one of which is synthesized and/or crosslinked in the presence of the other(s). Ideally, unlike graft or block copolymers, there are no covalent bonds between the components in IPNs. Now it is known [3–6] that most IPNs are two-phase microheterogeneous systems with some interpenetration between the constituent phases and with molecular level of mixing in each phase, because of the impossibility of separation of these phases due to the conditions of IPN preparation. Depending on the production method used or on the characteristics of the structure formed, interpenetrating polymer networks are classified as sequential, simultaneous, latex, gradient, or grafted IPNs [1,2,6]. Compositions consisting of linear (or branched) and crosslinked polymers are termed semi-IPNs.

Although this definition still fits most materials called IPNs, now some hybrid polymer blends, such as *thermoplastic apparent* IPNs, call for a broader concept [3,7,8]. In contrast to the chemically crosslinked IPNs, physical bonds are characteristic of the crosslinking in thermoplastic apparent IPNs. These physical bonds are glassy domains of block copolymers, ionic clusters in ionomers, or crystalline domains in semicrystalline polymers. The components of thermoplastic IPNs are capable to form physical networks and are characterized by interpenetration of phases. Thermoplastic apparent IPNs take an intermediate position between blends of linear polymers and real IPNs, since they behave like chemically crosslinked polymers at relatively low temperatures and as thermoplastics at elevated temperatures [8]. The blends based on combinations of physically crosslinked polymer and linear polymer, or physically crosslinked polymer and chemically crosslinked (thermoset) polymer, where the physically crosslinked polymer network constitutes the continuous phase and the other component disperses into domains, also exhibit the properties of thermoplastic compositions [8].

Segmented thermoplastic polyurethanes (TPUs) are an important class of thermoplastic elastomers [9], often used as one of the components in IPNs. Their versatile properties are generally attributed to their microphase-separated morphology, consisting of microdomains rich in hard segments (HS microdomains) and a microphase rich in soft segments (SS microphase), and arising from the thermodynamic immiscibility of HS and SS [9]. Because of their advantageous mechanical properties, the IPNs based on TPUs and other polymers are widely studied and used as commercial materials [1–8]. It is important to note that when TPUs form a continuous phase, the final IPNs behave as thermoplastic elastomers.

In this chapter, structure-property relationships in new thermoplastic apparent IPNs (AIPNs) and in semi-IPNs containing a polyurethane thermoplastic elastomer are discussed.

2. Polyurethane elastomer-based thermoplastic apparent interpenetrating polymer networks

2.1. *Semicrystalline polyurethane/(styrene-acrylic acid) copolymer-based thermoplastic AIPNs produced in the melt*

Thermoplastic AIPNs of several compositions prepared by mechanical blending in a roll mill of crystallizable polyurethane (CPU) and styrene-acrylic acid random copolymer (S-*co*-AA) have been investigated using different techniques by Vatalis *et al.* [10,11]. The CPU was based on toluene diisocyanate (TDI, mixture of 2,4- and 2,6-isomers, molar ratio 65/35) and oligomeric butylene adipate glycol (BAG, molar mass 2000) in a molar ratio of 1.01/1.00. The S-*co*-AA was obtained by bulk radical copolymerization of S and AA, and the comonomer molar ratio in S-*co*-AA was *ca.* 72/28. Both neat CPU and S-*co*-AA are physically crosslinked polymers. CPU is crosslinked by means of hydrogen bonds and microcrystallites of BAG acting as effective crosslinking sites with a degree of crystallinity of *ca.* 50%. Won *et al.* [12] have shown that in S-*co*-AA, physical crosslinking is due essentially to strong dimer hydrogen bonding of carboxyl groups.

The wide- and small-angle X-ray scattering curves (WAXS and SAXS, respectively) [10] (Figure 1a,b) reflect some interactions between the components, affecting their microphase structure in the thermoplastic AIPNs studied. The experimental curves of the CPU/S-*co*-AA compositions differ greatly from

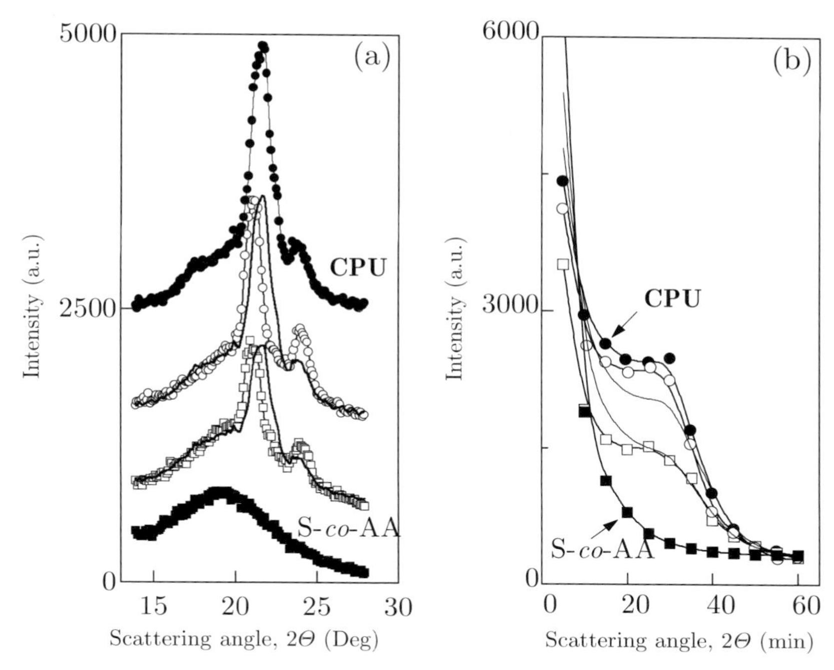

Figure 1. Typical experimental (points) and additive (solid lines) WAXS (a) and SAXS (b) curves for neat CPU (●) and S-*co*-AA (■), as well as for thermoplastic AIPNs with (○) 80 and (□) 50 wt% CPU. Except for the two lowest WAXS curves, each next curve is shifted upwards by 500 arbitrary units

the additive curves (solid lines in Figure 1a,b), calculated from the scattering intensities of neat CPU and S-*co*-AA, taking into account their mass fraction in the blends (for instance, the additive intensity of the 80/20 blend $I_{add} = 0.8\, I_{\mathrm{CPU}} + 0.2\, I_{\mathrm{S}\text{-}co\text{-}\mathrm{AA}}$, under the assumption of additivity of the CPU and S-*co*-AA contributions to the scattering ability of the CPU/S-*co*-AA blends). The observed deviations from additivity are caused by the changes in the microstructure of the CPU/S-*co*-AA compositions.

It can be seen in Figure 2a that the experimental degree of crystallinity, X_{exp}, of the CPU component (calculated from WAXS data by the Mattews equation [13]) exceeds the additive one. This means that the introduction of the amorphous S-*co*-AA in the crystallizable CPU matrix promotes the aggregation of the flexible CPU blocks into crystallites, resulting in a gradual increase of the crystallite size, D (calculated from WAXS data by the Scherer equation [13]) at higher S-*co*-AA contents (Figure 2b). This was explained [11] by an integration of S-*co*-AA macrochain fragments inside the crystallites of flexible CPU blocks during the CPU crystallization.

SAXS investigations have shown [10] that the introduction and gradual increase of the S-*co*-AA content in the CPU matrix results in the increase of the packing degree of the flexible blocks in the CPU crystallites and in a gradual

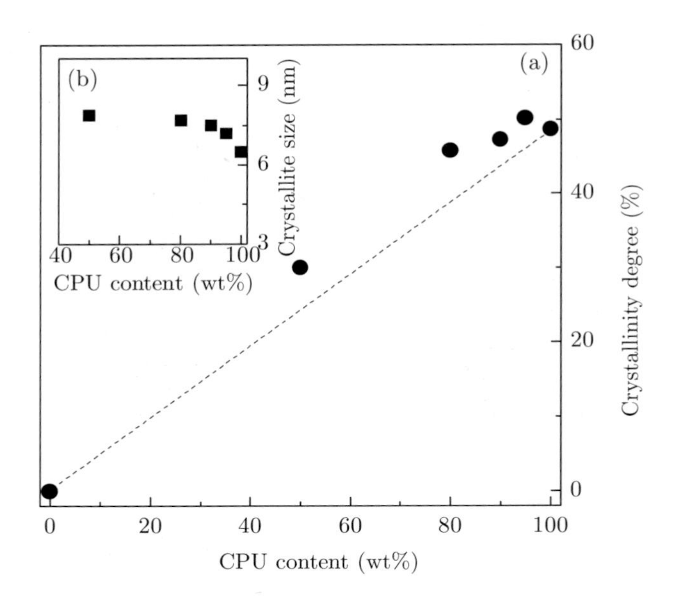

Figure 2. Concentration dependence of (a) crystallinity degree and (b) crystallite size of the CPU component in thermoplastic AIPNs. The straight line is additive

worsening of the crystallite quality in the volume unit of thermoplastic AIPNs. It was found that the average distance between the crystallite centers, L (*i.e.*, the periodicity calculated by the Bragg equation from the SAXS data [13]), increased from 17.5 nm for neat CPU to 19.5 nm for CPU in thermoplastic AIPNs and the crystal lattice spacing, d (*i.e.*, the distance between reflecting planes, calculated by the Bragg equation from the WAXS data [13]), increased from *ca.* 0.40 nm to 0.42 nm.

Table 1 lists differential scanning calorimetry (DSC) results published in [11]. It is seen that melting of BAG crystallites is already observed at 5 wt% of CPU, suggesting that the CPU component partly retains its specific characteristics in thermoplastic AIPNs in the entire composition range. The compositions with 10 to 95 wt% of S-*co*-AA exhibit lower T_m values compared to neat CPU, which is in good agreement with the above WAXS and SAXS data and results from the formation of less perfect crystallites due to integration of S-*co*-AA macrochain fragments. The glass transition temperature of CPU, T_{g1}, is practically independent of composition in the range studied (Table 1). On the contrary, T_{g2} of S-*co*-AA increases in general with the rise in CPU content, particularly at low CPU contents. This is explained in terms of physical interactions of the COOH groups of AA in S-*co*-AA with the ester groups of the flexible CPU blocks, which results in an improved microphase separation of AA and S in the S-*co*-AA component. It was concluded [10,11] that hydrogen bonding between AA and CPU promotes microphase separation of both components, CPU and S-*co*-AA.

Table 1. Transition temperatures (DSC results) for the CPU and S-*co*-AA components in thermoplastic AIPNs [11]

Composition (wt% CPU)	CPU		S-*co*-AA
	T_m (°C)	T_{g1} (°C)	T_{g2} (°C)
0	—	—	114
5	51	—	113
10	48	—	117
20	53	—	121
35	52	—	128
50	50	—	128
80	50	–33	129
90	54	–34	—
95	55	–33	—
97	56	–33	—
100	55	–34	—

The mechanical properties of the thermoplastic AIPNs, such as density, ρ, flow limit, σ_f, elasticity modulus, E, and tensile strength, σ, change non-additively with composition, showing extreme values at low contents (5 to 10 wt%) of CPU or S-*co*-AA (Table 2). This behavior confirms that each component in the thermoplastic AIPNs affects the microphase structure and properties

Table 2. Physical-mechanical properties of thermoplastic AIPNs [10]

Composition (wt% CPU)	Flow limit, σ_f (MPa)	Tensile strength, σ (MPa)		Elasticity modulus, E (MPa)		Density, ρ (g/cm^3)	
		Exp.	Add.	Exp.	Add.	Exp.	Add.
0	*	20.0	20.0	2060	2060	1.113	1.113
5	*	39.8	19.6	2610	1977	1.113	1.117
20	*	28.9	18.4	1890	1727	1.128	1.129
50	9.0	11.7	16.1	1350	1227	1.134	1.152
80	15.1	15.5	13.7	645	727	1.187	1.175
90	15.6	15.6	12.9	401	561	1.190	1.183
95	16.1	16.1	12.5	328	477	1.195	1.187
100	12.1	12.1	12.1	394	394	1.191	1.191

* Sample showing brittle fracture

of the other component. This is only possible by the formation of the interpenetrating structure of thermoplastic AIPNs [14] due to strong physical interactions between the functional groups of CPU and S-*co*-AA. It is known [12] that the acid groups of AA in S-*co*-AA can be considered to be anionic groups (COO^-H^+) with low degree of ionization, which are capable of taking part in intermolecular physical (electrostatic) interactions with polar groups of the flexible and rigid CPU blocks. However, since the content of rigid blocks in CPU is only about 8 wt%, the ester groups of the flexible CPU blocks play the decisive role in the intermolecular physical interactions with COOH groups of AA in S-*co*-AA. This should be the reason for the partial dilution of the S-*co*-AA macromolecules in the flexible phase of the CPU matrix. The thermoplastic AIPNs can be divided into two groups: thermoplastic AIPNs with CPU content ≥ 50 wt%, showing essentially the behavior of neat CPU, and thermoplastic AIPNs with CPU content $\leq 35\%$, showing essentially the behavior of neat S-*co*-AA.

Some secondary relaxations of the components in thermoplastic AIPNs have been investigated by thermally stimulated depolarization current (TSDC) techniques and thermally stimulated conductivity (TSC) measurements [10,11]. It was found that upon addition of S-*co*-AA to CPU, the secondary γ and β CPU peaks (at *ca.* –140 °C and *ca.* –100 °C, respectively) shift slightly to lower temperatures, *i.e.*, the corresponding relaxations become faster, these shifts being more pronounced at low S-*co*-AA contents. The shifts can be related to physical interactions between the IPN components and to their partial miscibility. Rizos *et al.* [15] have shown that as a result of such interactions, changes in the local free volume may occur, affecting the secondary relaxation times. The same changes in the β relaxation of PU have been found in polyurethane/polystyrene IPNs by Pandit and Nadkarni [16].

2.2. *Semicrystalline polyurethane/(styrene-acrylic acid) copolymer-based thermoplastic AIPNs produced in solution*

Thermoplastic AIPNs based on the same CPU and S-*co*-AA (Section 2.1), but prepared by mechanical blending of the components in a common solvent were investigated by dynamic mechanical thermal analysis (DMTA), DSC, WAXS, SAXS, TSDC, DRS, and others techniques by Sergeeva, Kyritsis *et al.* [17–20]. Mechanical blending of two (or more) polymers in the melt mentioned in the previous section is among the basic methods of producing thermoplastic IPNs [8]. This method is characterized by a higher technological effectiveness than, *e.g.*, polymer blending in solution. However, in order to solve some technological problems related to the production of, *e.g.*, thin films, membranes, *etc.*, polymer blending in solution could be irreplaceable. Therefore, it seemed quite interesting to study the structure-property relationships in thermoplastic AIPNs produced by the solution technique.

WAXS measurements [18,19] have shown the existence of regions with the structure of the neat components in all thermoplastic AIPNs. It follows that both the crystallizability of BAG in CPU and the mean size of BAG microcrystallites are rather little affected by S-*co*-AA (Figure 3). SAXS data [19] have shown that the microheterogeneity of CPU is preserved in thermoplastic AIPNs and the average distance between the centers of nearest crystallites, L, does not practically change with composition (L is *ca.* 21.0 nm for both neat CPU and CPU in AIPNs). However, an increase of SAXS intensity observed at 2Θ of about 20 min indicates the increase of structural heterogeneity in these

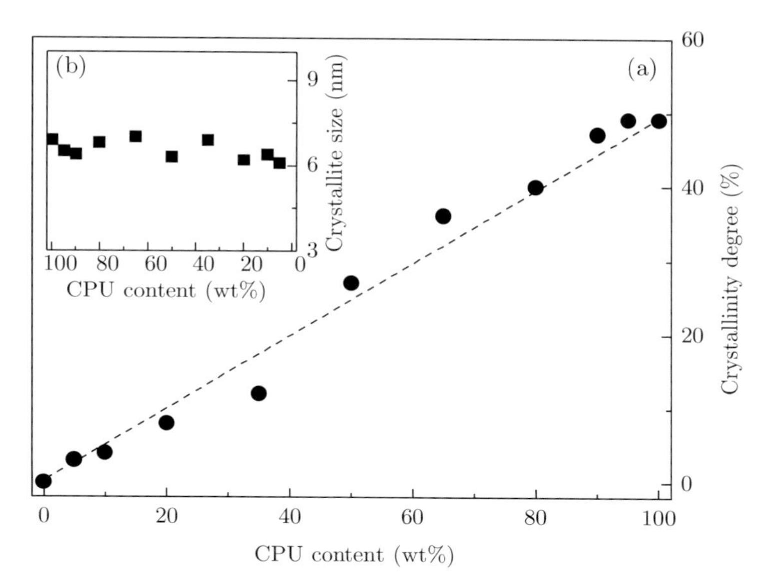

Figure 3. Concentration dependence of (a) crystallinity degree and (b) crystallite size of the CPU component in thermoplastic AIPNs prepared in solution. The straight line is additive

systems. On the other hand, the observed deviations between the additive straight lines and the experimental WAXS and SAXS data indicate some partial miscibility of CPU and S-*co*-AA in thermoplastic AIPNs, in agreement with DSC results [17]; it was found that a mixed microphase of fixed composition (CPU/S-*co*-AA = 90/10) coexists with essentially pure microphases of CPU and S-*co*-AA.

Comparison of the experimental degree of crystallinity, X_{exp}, and the theoretical (additive) value, X_{add}, in Figure 3 shows that X_{exp} is lower than X_{add} when the amorphous S-*co*-AA component is prevailing (S-*co*-AA > 50 wt%). For compositions with S-*co*-AA content < 50 wt%, X_{exp} is higher than X_{add}. The same tendency of the concentration dependence of X_{exp} has been obtained from DSC data [17]. This means that the formation of CPU crystallites in thermoplastic AIPNs takes place easier when CPU is the continuous phase.

DMTA measurements (Figure 4) [18,19] have shown that thermoplastic AIPNs can be considered as multiphase systems having at least one crystalline and two amorphous phases, and regions of mixed compositions. Their mechanical properties are determined by the heterogeneity of the neat polymers, as well as by the heterogeneity caused by the thermodynamic immiscibility of the components. The degree of incompatibility is determined to a great extent by the ratio of intra- and intermolecular hydrogen bonds between the functional groups of CPU and S-*co*-AA. For thermoplastic AIPNs with CPU content up to 10 wt%, CPU–S-*co*-AA interactions are mainly taking place, whereas at higher CPU contents CPU–CPU and S-*co*-AA–S-*co*-AA interactions are prevailing.

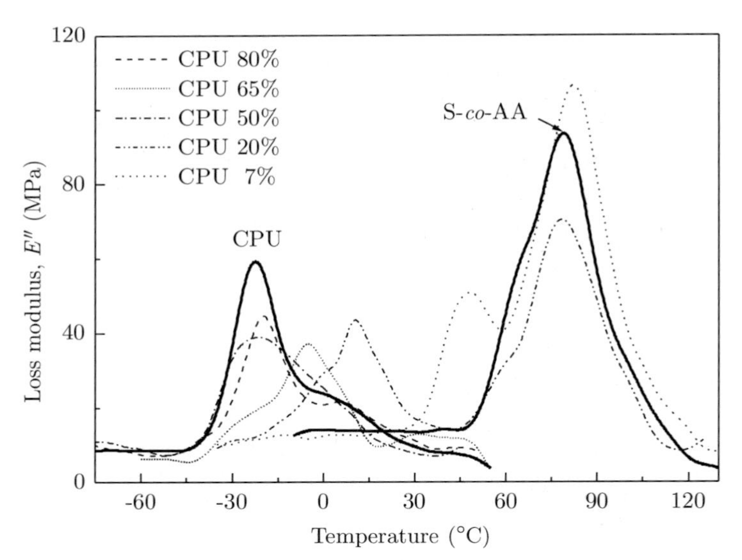

Figure 4. Typical temperature dependences of the loss modulus, E'', for neat CPU and S-*co*-AA (solid lines), as well as for thermoplastic AIPNs with various CPU contents

It is important to note that the physical crosslinking, *i.e.*, crystallites of CPU, inter-, and intramolecular hydrogen bonds of CPU and S-*co*-AA, is destroyed at elevated temperatures [8].

The elasticity modulus and tensile strength of thermoplastic AIPNs change non-additively with composition (Figure 5) [19]. Their extrema are observed

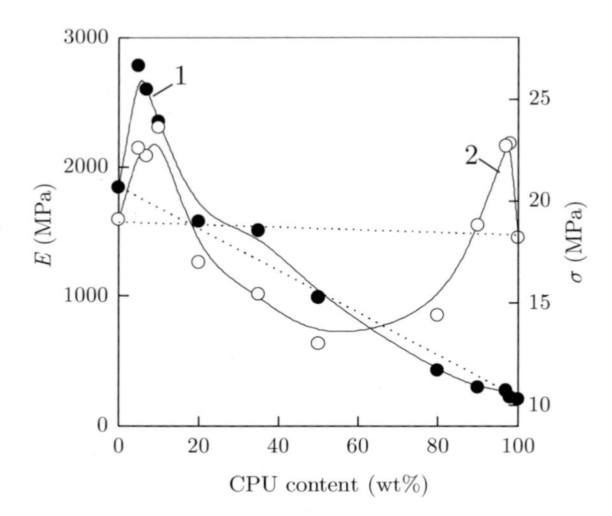

Figure 5. Experimental (points) and additive (dotted lines) concentration dependences of the elasticity modulus, E (1), and tensile strength, σ (2), of the thermoplastic AIPNs prepared in solution

at low contents (5 to 15 wt%) of CPU or S-*co*-AA, *i.e.,* at the compositions characterized by phase mixing effects observed in X-ray scattering and DMTA measurements. For instance, the tensile strength of thermoplastic AIPNs with 5 to 10 wt% S-*co*-AA is higher by about 20% than that of neat CPU. From the technological point of view, this observation suggests a principle possibility of obtaining composites with desired properties by changing the composition and the method of preparation of thermoplastic AIPNs.

The results obtained are explained [19] by the formation of networks of inter- and intramolecular hydrogen bonds, resulting in the creation of a double-phase continuity. The intermolecular hydrogen bonds between the functional groups of the components promote the improvement of the compatibility of the latter in thermoplastic AIPNs. On the other hand, the degree of segregation of the CPU and S-*co*-AA microphases increases as AA and BAG leave the S-*co*-AA and CPU phases, respectively, because their functional groups take part in intermolecular hydrogen bonding. Intramolecular hydrogen bonding promotes the microphase separation of components in thermoplastic AIPNs.

The thermodynamic state of thermoplastic AIPNs produced in solution has been investigated by Sergeeva *et al.* [20] in order to get a deeper insight in their structure as a function of composition. The thermodynamic state was estimated by calculation of the free energy of mixing of the constituent polymers, applying the vapor sorption method and using a vacuum installation and a McBain balance. To this purpose, the chloroform vapor sorption by neat CPU and S-*co*-AA, and by thermoplastic AIPNs with CPU content of 5, 10, 50, 90, and 95 wt% was studied. The thermodynamic parameters were calculated from the experimental data on chloroform vapor sorption according to the method reported in [21,22]. Briefly, from the concentration dependence of the free energy of mixing of chloroform and the system components, using the thermodynamic cycles [21], we have calculated the changes in free energy of mixing, Δg^x, of CPU and S-*co*-AA in thermoplastic AIPNs within the composition range chosen. It was found that the Δg^x values are slightly negative ($\Delta g^x = -0.3 \div -3.6$ Jg^{-1}) at low contents (up to 10 wt%) of each component, while the free energy of mixing is positive ($\Delta g^x = +0.5$ Jg^{-1}) for thermoplastic AIPNs containing 50 wt% CPU. Therefore, in thermoplastic AIPNs containing CPU or S-*co*-AA up to 10 wt%, the components are thermodynamically miscible, and thus thermoplastic AIPNs are thermodynamically stable. The microphase separation of the components takes place in thermoplastic AIPNs of medium compositions and, probably, in other compositions, where the systems are thermodynamically unstable. It can be assumed that in these cases the complete phase separation of the components in thermoplastic AIPNs does not occur, this process being prevented by the intermolecular physical network formation. It should be noted that the extreme changes in all properties of thermoplastic AIPNs were observed in the transition region from thermodynamically stable to thermodynamically unstable (metastable) state.

2.3. *Semicrystalline polyurethane/(styrene-acrylic acid) ion-containing block copolymer thermoplastic AIPNs*

Thermoplastic AIPNs based on the same CPU (Section 2.1) and (styrene-acrylic acid) ion-containing (K^+) block copolymer (S-*b*-AA(K)) prepared by blending of the components in a common solvent have been studied using DMTA and DSC techniques by Bartolotta *et al.* [23]. S-*b*-AA(K) was obtained by emulsion block copolymerization of styrene and acrylic acid, followed by its partial neutralization using KOH solution (0.1 N). The S/AA molar ratio was *ca.* 90/10 and the degree of neutralization was *ca.* 0.3. As already mentioned, CPU is physically crosslinked by means of strong hydrogen bonds and microcrystallites acting as effective crosslinking sites. S-*b*-AA(K) is physically crosslinked by both ionic and residual hydrogen bonds.

The local and cooperative molecular mobility of neat CPU and S-*b*-AA(K), and a thermoplastic AIPN of composition CPU/S-*b*-AA(K) = 80/20 wt% was investigated. The experimental results obtained with the thermoplastic AIPN were compared to those obtained with the neat components. The presence of distinct calorimetric and mechanical transitions, which are unambiguously attributed to the two components is a clear indication of a multiple-phase heterogeneous structure. These observations reflect the thermodynamic incompatibility of the components, even though a limited miscibility is inferred by small but noticeable variations of the magnitudes of local (γ-relaxation) and cooperative (α_a-relaxation and melting events) transitions, which deviate significantly from a simple dilution effect. The shift of T_g of the CPU component to higher temperatures and the largest Δc_p jump observed in the DSC thermogram of the thermoplastic AIPN, as compared to the thermal characteristics of neat CPU, as well as the decreasing crystallinity of the CPU component result from the limited miscibility in the thermoplastic AIPN. The growth affinity between the CPU and S-*b*-AA(K) components is believed to be the result of interactions (hydrogen and ionic bonds) between their functional groups. Finally, the differences revealed in the behaviors of the calorimetric and mechanical glass transition temperatures are regarded as experimental evidences of the existence of locally heterogeneous relaxation motions, which are probed by the different length scales associated with the calorimetric and mechanical techniques.

2.4. *Polyaminourethane/polyurethane ionomer-containing thermoplastic AIPNs*

Thermoplastic AIPNs based on anion-containing polyurethane (PU_1) and polyaminourethane (PU_2) were prepared and investigated using different techniques by Stepanenko *et al.* [24] and Tsonos *et al.* [25,26]. The PU_1 was based on oligo-oxytetramethylene glycol (OTMG, molar mass 1000), TDI (mixture of 2,4- and 2,6-isomers, molar ratio 65:35), and magnesium salt of *p*-hydroxybenzoic acid (Mg-HBA), and the molar ratio of the reagents was 1/2/2. The PU_2 was based on the same OTMG and TDI, and N-methyl-diethanolamine (NMDA), at the same molar ratio. Polymer blends containing 5, 10, 30, and 50 wt% of PU_1

were obtained; the details on the preparation conditions are described in [24,25]. The molecular mobility, microphase morphology and their dependence on the composition of the blends were investigated by TSDC, DSC, Fourier transform infrared (FTIR), SAXS, and others techniques [24-26].

Dielectric TSDC measurements [25] of the blends revealed four relaxation mechanisms. The subglass secondary γ-relaxation (at ~120 K) is associated with local motions of parts of the molecular chain. The β-relaxation (at ~160 K) is attributed to the motions of the polar carbonyl groups of the polymer chain. A systematic change of the magnitude (maximum of the current) and position (temperature of the maximum current) of these two relaxations depending on the composition of the ionomer (PU_1 content) was not observed.

The α-relaxation is related to the glass transition, T_g, of the amorphous soft phase of the PU_1 and PU_2 (Table 3). The Maxwell-Wagner-Sillars (MWS) relaxation is related to interfacial polarization caused by the motion of ions released during the glass transition. The temperature at which α- and MWS-relaxations (T_g and T_{MWS}, respectively) appear in the TSDC plots and the magnitude of these relaxations are strongly affected by the degree of microphase separation of the PU_1 and PU_2 in the blends (Table 3). Based on TSDC measurements, the parameter m_{TSDC}, a criterion expressing the relative degree of phase mixing, was introduced [25]. This parameter takes into account all the factors affecting phase mixing expressed by the partial parameters. According to the definition of m_{TSDC}, higher values of m_{TSDC} correspond to higher degrees of phase mixing of the components in the blends. The values of m_{TSDC} for all the PU_1/PU_2 compositions are presented in Table 3. According to this parameter, the blends obey the following order of decreasing phase mixing (PU_1/PU_2 in wt%): 50/50 > 0/100 > 5/95 > 10/90 > 100/0 > 30/70. It was concluded that, as a result of the complexity of the systems, their structures and properties are not simple functions of their composition.

Table 3. Transition temperatures and general relative degree of phase mixing, m_{TSDC}, (TSDC data), as well as some mechanical properties of the PU_1/PU_2 blends [24,25]

Composition (PU_1/PU_2 wt%)	T_α (T_g) (K)	T_{MWS} (K)	m_{TSDC}	Tensile strength, σ (MPa)	Elongation, ε (%)	Elasticity modulus, E (MPa)
0/100	241	264	0.866	18.0	800	2.0
5/95	223	257	0.254	25.2	540	3.9
10/90	230	259	0.211	27.8	490	2.9
30/70	223	254	0.158	20.4	350	6.0
50/50	232	250	1	13.9	110	12.6
100/0	221	266	0.175	21.0	300	7.0

It was concluded that neat PU_2 and the PU_1/PU_2 blend (50/50) exhibit the best phase mixing, in accordance with the analysis based on TSDC results. SAXS measurements showed that the neat PU_1 ionomer has a high degree of microphase separation. In this polyurethane, the high SAXS intensity may be a result of an influence of magnesium ions and of the formation of strong ionic and ion-molecular bonds by the ionic groups of PU_1 (*e.g.*, $Mg^{2+} \cdots O{=}C$, $Mg^{2+} \cdots OH$), in addition to the physical network of intermolecular hydrogen bonds. These additional bonds increase the association of the hard blocks, resulting in the increased degree of microphase separation in PU_1. Neat PU_2 has a low degree of microphase separation and an imperfect macrolattice of the hard domains in its volume. The low packing density of the hard blocks in PU_2 domains is related to the presence of the bulky CH_3 groups in their structure. These groups, in combination with the low molecular weight of PU_2, prevent the formation of strong hard domains. It was shown that the PU_1/PU_2 blends had a higher degree of phase mixing than neat PU_1. The addition of PU_2 to PU_1 causes defects in the hard domains of the latter and leads to the mixing of soft and hard microphases. In the PU_1/PU_2 blends studied, dissolution of hard segments into the continuous soft phase and the appearance of a broad size distribution of the hard microdomains were observed.

3. Polyurethane-containing semi-IPNs

In papers reviewed in this section, the thermoplastic polyurethane elastomer was used [27–52] as a linear component of semi-IPNs in which, as already mentioned in Section 1, the second component should be a chemically crosslinked polymer. As a second component the polyurethane [27], polyepoxyisocyanurate [28–33] and polycyanurate [34–52] networks were used. Once again, it is important to note that if the thermoplastic PU forms a continuous phase, the final semi-IPNs exhibit the properties and behavior of thermoplastic elastomers.

The final morphology of semi-IPNs is usually characterized by a microheterogeneous structure caused by phase separation [6]. The latter proceeds in the course of the reaction due to the appearance of thermodynamic immiscibility of the matrix-forming homopolymers. A significant change in the compatibility of the components can be achieved by introducing fillers into such systems [53]. The direction and degree of the microphase separation process greatly depend on whether the filler was introduced into a blend of miscible or immiscible polymers [54,55]. By introducing the filler into the semi-IPNs of miscible polymers, microphase separation can take place. For compositions consisting of immiscible polymers, either an increase or a decrease in the degree of phase separation may be observed. The driving force for the morphology change results from adsorption and/or chemisorption of the components onto the filler surface. As a rule, the selective adsorption of one of the components onto the filler surface leads to intensification of the phase separation. In this way, the introduction of fillers into polymeric blends or IPNs may generate dramatic changes in their phase morphology and, consequently, it may greatly affect their physical-mechanical properties. The effects of fillers on the compatibility of the components in IPNs have been reported in few works [56–58].

3.1. *Polyurethane/polyurethane semi-IPNs with miscible components*

The viscoelastic properties of semi-IPNs based on a thermoplastic polyurethane (TPU) and a chemically crosslinked polyurethane (PU network) formed in the TPU solution, as well as of the same semi-IPNs filled with γ-Fe_2O_3 powder have been studied by Brovko *et al.* [27]. The TPU (molar mass 40000 g/mole) was synthesized from poly(1,4-butylene)glycol adipate (molar mass 1000 g/mole), 4,4′-methylenebis(phenyl isocyanate) and 1,4-butane diol as a chain extender. The PU network was based on a vinyl chloride/vinyl acetate/vinyl alcohol terpolymer with an adduct of 1,1,1-trimethylolpropane and 2,4-toluene diisocyanate; the appropriate for crosslinking NCO/OH ratio was chosen to be 0.55. The non-filled semi-IPNs exhibited only one T_g situated between those of TPU and PU network. Therefore, the components of the semi-IPNs may be assumed to be compatible. It was found that the semi-IPNs filled with a small amount of γ-Fe_2O_3 (up to 1.4 vol%) exhibited one T_g shifted toward higher temperatures. When the concentration of γ-Fe_2O_3 in the semi-IPNs was increased to 6.5 and 10.5 vol%, two distinct peaks of tan δ were observed, both shifted to higher temperatures, as compared to those of neat TPU and PU network. This observation was explained by the following processes taking place during the formation of semi-IPNs filled with γ-Fe_2O_3: (i) adsorption of TPU and of the PU network constituents on the filler surface, (ii) chemisorption of the isocyanate component of PU network on the filler due to interaction of NCO groups with the filler surface OH groups, and (iii) a random covalent chemical bonding (grafting) of the TPU to the PU network chains.

The semi-IPNs filled with 6.5 to 10.5 vol% of γ-Fe_2O_3 are characterized by dual phase continuity. A very important characteristic of the phase separation in the IPNs is the degree of segregation, α, giving the fraction of material, which undergoes phase separation. When $\alpha = 1$, the system is fully phase-separated; at $\alpha = 0$, it is fully miscible. The value of α can be estimated from the characteristic maxima of the mechanical loss curves according to the equation proposed by Rosovitsky and Lipatov in [59]:

$$\alpha = \frac{h_1 + h_2 - (h_1 l_1 + h_2 l_2)/L}{h_1^0 + h_2^0} \tag{1}$$

where h_1^0 and h^0 are the intensities of the maxima of tanδ of the neat polymers, h_1 and h_2 are the intensities of the maxima of tanδ of the respective polymer components in the semi-IPNs, l_1 and l_2 correspond to the shifts of maxima of tanδ on the temperature scale, L is the interval between the maxima of tanδ of the neat polymers. The value of α increased from 0.2 to 0.3 with the rise of the γ-Fe_2O_3 content from 6.5 to 10.5 vol%, evidencing the increase of microphase separation of the components in filled semi-IPNs.

3.2. *Polyurethane-containing semi-IPNs with immiscible components*

3.2.1. *Polyurethane/polyepoxyisocyanurate grafted semi-IPNs*

Grafted semi-IPNs based on the same TPU as described in Section 3.1 and a curable system of thermally treated 4,4'-methylenebis(phenyl isocyanate) (TMPI) and epoxy oligomer (EO) based on bisphenol A have been synthesized and investigated [28-33]. A detailed study [32] of the chemical reactions occurring during semi-IPNs synthesis, as well as of the reactions occurring in the binary systems TPU/TMPI, EO/TMPI, and TPU/EO, has shown some grafting of TPU to the growing polyepoxyisocyanurate network. These reactions were cyclotrimerization and cyclodimerization of isocyanate, including the formation of mixed cyclodimers with carbodiimide groups arising at thermal pre-treatment of TMPI, reaction of isocyanate and epoxy components yielding oxazolidinone cycles, interaction of TPU and isocyanate with formation of allophanate structures, and others. It was noted that homopolymerization of epoxy oligomers could also take place. The reactions occurring in binary blends were observed in the ternary system as well; the content of different structures in the final material depends on the reaction rates and on the interference of the participating components.

3.2.2. *Polyurethane/polyepoxyisocyanurate grafted semi-IPNs filled with γ-Fe_2O_3*

The influence of finely dispersed γ-Fe_2O_3 on the phase structure of grafted semi-IPNs [28-31], as well as on the kinetics and interrelation of the reactions occurring [32] has been studied. Semenovich, Fainleib *et al.* [32] have concluded that γ-Fe_2O_3 inhibits cyclotrimerization and, at the same time, catalyzes cyclodimerization of isocyanate groups and their reaction with carbodiimide groups forming mixed cyclodimers. It was also found that the filler catalyzed the reaction of isocyanate groups with epoxy and hydroxy groups of the epoxy resin, leading to the formation of oxazolidinone and urethane fragmens, respectively. The balance between the reactions depends on the filler content in the system.

It was found that the introduction of the filler into the semi-IPNs influences to some degree the microphase separation of the components. The temperature dependences of $\tan\delta$ of a TPU, a TMPI/EO network, and TPU-TMPI/EO semi-IPNs containing 0 to 17.0 vol% of γ-Fe_2O_3 are shown in Figure 6 [28]. In contrast to non-filled semi-IPNs, the filled materials exhibit two distinct $\tan\delta$ peaks even at a very low γ-Fe_2O_3 content. It can be seen that for all compositions studied the temperature positions of the two $\tan\delta$ peaks are between the T_gs of the neat components, evidencing incomplete phase separation of the components in the filled semi-IPNs. In order to estimate the degree of microphase separation, the degree of segregation, α, was calculated using Eq. (1). It was found that in the different compositions the values of α varied non-monotonically within the limits from 0.10 to 0.25 [28,29]. Taking into account the neg-

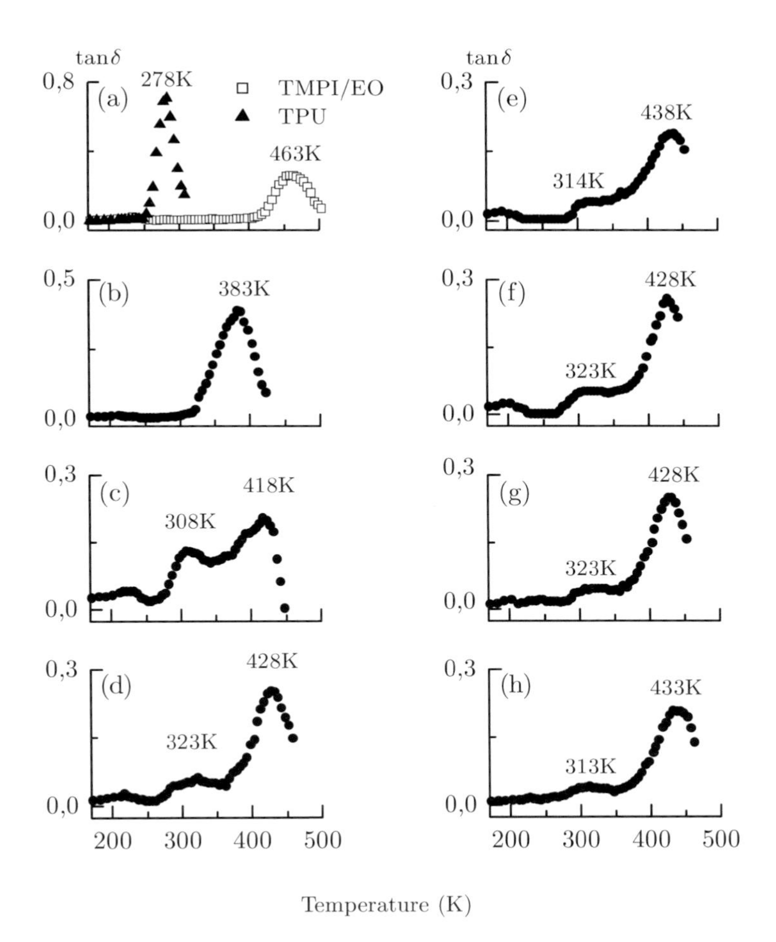

Figure 6. Dynamic mechanical measurements of tan δ (loss factor) *vs.* temperature for neat TPU and TMPI/EO (a), and grafted semi-IPNs of constant TPU content (33 wt%) filled with different amounts of γ-Fe_2O_3 (vol%): 0 (b), 0.5 (c), 1.3 (d), 5 (e), 9.5 (f), 13.4 (g), 17 (h)

ligible breadth of the α variations, it was concluded that the degree of microphase separation of the components was practically independent of the γ-Fe_2O_3 content in the filled semi-IPNs studied. Sergeeva *et al.* [29] studied additionally the effect of the surface activity of the filler on the viscoelastic properties and phase structure of the material. In some experiments, the γ-Fe_2O_3 powder was pre-treated with a surfactant (potassium salt of dialkylpoly(ethylene glycol) orthophosphate) [29].

The polyurethane/polyepoxyisocyanurate grafted semi-IPNs filled with γ-Fe_2O_3 can be used as a working layer for magnetic recording media [33]. Special tests have shown the favorable mechanical characteristics of the polyurethane/polyepoxyisocyanurate/γ-Fe_2O_3 working layer of magnetic tapes even after aging at 200°C for 4 h. Under the same aging conditions, a standard working layer sample was completely destroyed [33].

3.2.3. *Polyurethane/polycyanurate grafted semi-IPNs*

In 1992-1994, the first reports [34–38] on the synthesis, kinetic peculiarities, and characterization of structure-property relationships in polyurethane/polycyanurate grafted semi-IPNs have been published. Efforts have been directed to the creation of a new material from the widely used thermoplastic polyurethane elastomer (as a linear component) and the intensively developing high performance cyanate ester resins (CERs, as thermosets). Linear segmented TPU exhibits rubbery characteristics and thermoplasticity, which are directly related to its structure. CER formulations offer a variety of excellent thermal and good mechanical properties, which commend them for use in advanced technology (*e.g.,* high-speed electronic circuitry and aerospace composite matrices). For the electronics market, the attractive features of polycyanurates (PCNs) synthesized by polycyclotrimerization of CERs (Figure 7) are their low dielectric loss characteristics, dimensional stability at molten solder temperatures (220–270°C), high purity, inherent flame-retarding properties (providing the potential to eliminate brominated flame retardants), and excellent adhesion to conductor metals at temperatures up to 250°C [39,60–62]. What are their disadvantages? The system must be cured at high temperatures in order to achieve complete conversion, *i.e.,* the respective production costs are high. However, the main drawback, which hinders the more extensive application of the cured materials, is their low toughness. Modification of the materials by chemical and physical approaches is expected to solve these problems.

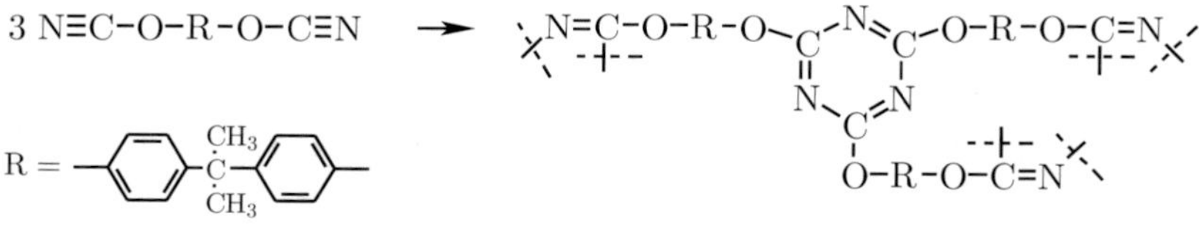

3 N≡C–O–R–O–C≡N →

$R = -C_6H_4-C(CH_3)_2-C_6H_4-$

Figure 7. Cyclotrimerization of the dicyanate of bisphenol A

Synthesis, chemical interaction between components, reactive grafting, and compatibilization. An attempt was made to combine the best properties of both materials (polyurethane and polycyanurate) in their semi-IPNs. The possibility of chemical interaction between the components in polyurethane/polycyanurate semi-IPNs leading to the formation of chemically grafted semi-IPNs was first reported by Fainleib and co-authors [34], and then confirmed in later works [35–38].

Polyurethane/polycyanurate grafted semi-IPNs were prepared by polycyclotrimerization of dicyanate ester of bisphenol A (DCEBA) in the immediate presence of TPU, as described in Section 3.1. During semi-IPN synthesis, at least two competitive chemical processes were established: (i) formation of the polycyanurate network by polycyclotrimerization (Figure 7) of DCEBA in the presence of TPU, resulting in microphase separation at a certain reaction time due to incompatibility of the components and (ii) chemical interaction between the forming network and the polyurethane, preventing the microphase separation of the components. By that time, no references concerning the cyanate–urethane chemical reactions were found. The chemical interaction between cyanate and urethane was investigated initially with model monofunctional low molecular mass organic compounds [63]. 4-*Tert*-butylphenyl cyanate (BPC) and phenyldodecanol urethane (PDU) were used as the model compounds.

High performance liquid chromatograms (HPLC) of the products obtained at the molar ratio BPC/PDU = 5:1 after heating at 150°C are plotted in Figure 8 [63]. It was found that the heights of the elution time peaks of BPC and PDU (9.2 min and 25.4 min, respectively) decrease with increasing reaction time. Three main products were eluted at $t_e = 5.3$ min, $t_e = 17.3$ min, and $t_e = 26.9$ min, and two additional products were eluted at $t_e = 30.0$ min and $t_e = 31.3$ min. The products with elution times 5.3 min and 26.9 min were identified as 4-*tert*-butylphenol (BP) and 4-*tert*-butylphenoxytriazine (BPT), respectively. As far as the separate heating of BPC and PDU under the same conditions has not led to the formation of the product eluted at $t_e = 17.3$ min, it was reasonable to assume that it resulted from a direct interaction between BPC and PDU.

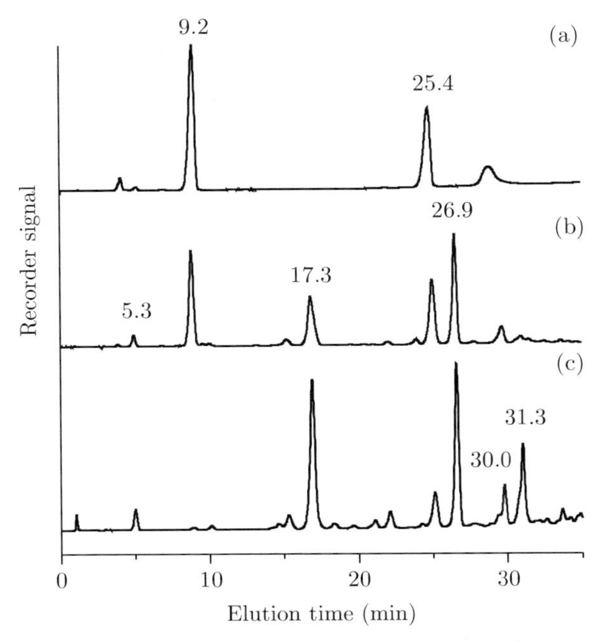

Figure 8. HPL chromatograms of BPC/PDU (molar ratio of 5/1) after heating at 150°C for: (a) 0 min, (b) 60 min, (c) 180 min

The chemistry of the processes occurred in the above system was studied by FTIR and matrix-assisted laser desorption ionization technique with time of flight detection (MALDI-TOF). On the basis of the results obtained, a scheme of the simultaneous and sequential reactions with formation of BP, BPT, and some intermediate and additional products, including possible substituted cyanurate, isocyanurate, and mixed triazine cycles was proposed [63].

The existence of hybridization effects *via* cyanate–urethane chemical interaction was evidenced in polyurethane/polycyanurate semi-IPNs by multiple attenuated total reflection (ATR) infrared spectroscopy [40]. It was calculated that at least 10–30% of the cyanate groups participate in the reaction with TPU at a 20 to 60% TPU content in the composition. It is noteworthy that the above estimates are in good agreement with FTIR/HPLC experiments performed on a low-molecular model cyanate/urethane blend [63]; it was shown that, in addition to polycyclotrimerization, *ca.* 20–30% of the cyanate groups formed co-products with the urethane.

Kinetic peculiarities. The study of the kinetics of synthesis of polyurethane/polycyanurate grafted semi-IPNs [40] has shown that the process of polycyclotrimerization of DCEBA is characterized by an induction period, which strongly decreases by the introduction of TPU in the reaction system and disappears at TPU contents in the composition higher than 20%. The higher the content of TPU in the system, the higher is the rate of conversion of the cyanate groups.

Vilensky, Fainleib *et al.* [41,42] have performed DSC studies of the kinetic peculiarities of DCEBA polycyclotrimerization in the presence of two different segmented thermoplastic polyurethanes: TPU described in Section 3 and TPU-2 (molecular mass 40000 g/mole) based on oligooxytetramethylene glycol (molecular mass 1000 g/mole). The initial DCEBA/TPU blend was immiscible, whereas the other initial blend, DCEBA/TPU-2, proved to be miscible. The analysis of the two processes of DCEBA melting and further DCEBA polycyclotrimerization in the presence of TPUs showed significant differences depending on the miscibility or immiscibility of DCEBA and TPU. The catalytic effect of TPUs on DCEBA polycyclotrimerization was quite strong in the immiscible DCEBA/TPU blend, suggesting that the heterogeneous catalysis taking place during the curing of DCEBA/TPU is more advantageous than homogeneous catalysis related to the curing of the DCEBA/TPU-2 blend.

Phase structure and relaxation behavior. In previous publications [38,40,43–49], several methods were used for characterization of the microphase structure of the semi-IPNs studied. WAXS, SAXS, DSC [40], dielectric relaxation spectroscopy (DRS), and TSDC [48] measurements showed that pure PCN is characterized by a typical homogeneous structure, but in segmented TPUs microphase separation was observed on the level of the thermodynamically immiscible hard and soft domains. As for semi-IPNs, the destruction of the microphase-separated morphology of TPU was observed and microphase separation between the PCN and TPU phases, expected from the difference in their solubility parameters, was not found. Grigoryeva *et al.* [43] noted that the values of the solubility parameters of TPU and PCN are not very close; one can conclude that PCN has to be thermodynamically immiscible with TPU and the TPU/PCN compositions should have a microheterogeneous structure.

In addition to their close solubility parameters, miscible polyblends are characterized by a single glass transition temperature, which depends on the relative weight fractions of the components and their respective T_g values. Such a "large-scale" homogeneity was confirmed by the existence of a single T_g in TPU/PCN semi-IPNs, estimated by DMTA [44–46], DSC [40], DRS and TSDC [48] techniques, the experimental data being presented in Figure 9 and showing the same trend. The theoretical compositional dependence of T_g of the TPU/PCN semi-IPNs can be obtained for DMTA data [38] according to the Fox equation

$$1/T_g = x_1/T_{g1} + x_2/T_{g2} \tag{2}$$

and for DSC data [40] according to the Couchman-Karasz equation

$$T_g = \frac{x_1 \Delta C_{p1} T_{g1} + x_2 \Delta C_{p2} T_{g2}}{x_1 \Delta C_{p1} + x_2 \Delta C_{p2}} \tag{3}$$

where x_1 is the weight fraction of LPU in the semi-IPNs, T_{g1} is the glass transition temperature of TPU, ΔC_{p1} is the specific heat increment at the glass transition of TPU, and x_2, T_{g2}, and ΔC_{p2} are the respective parameters of PCN.

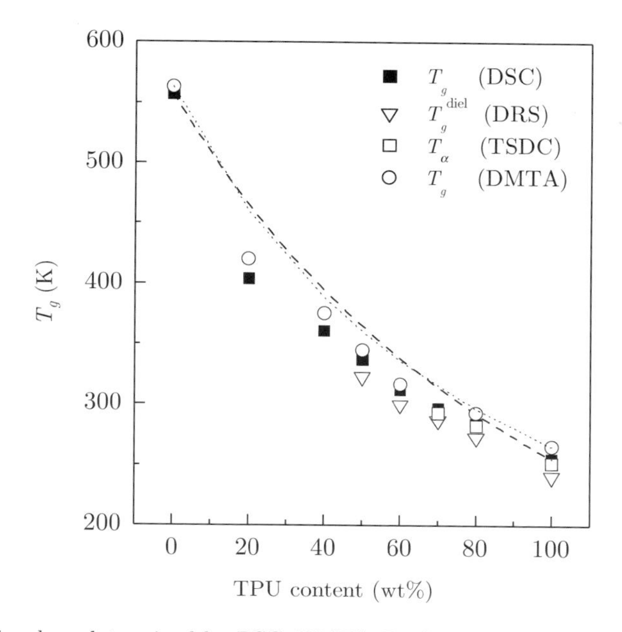

Figure 9. T_g values determined by DSC, DMTA, DRS, and TSDC *vs.* TPU content. The dotted line is a fit of Fox equation (Eq. 2) to the DMTA data and the dashed line is a fit of the Couchman-Karasz equation (Eq. 3) to the DSC data

Figure 9 shows the concentration dependences of T_g obtained from the predictions based on Fox and Couchman-Karasz equations.

A slightly negative deviation from both equations is observed. Such a deviation indicates that there is some interaction between the components in the system, which is in good agreement with the results of model research and FTIR measurements of TPU/PCN systems (Section 3.2.3) concerning the urethane-cyanate (cyanurate) chemical interaction (grafting). We assume that this grafting is responsible for the large-scale heterogeneity observed with the TPU/PCN semi-IPNs.

At the same time, combined laser-interferometric creep rate spectroscopy/DSC analysis [40] indicated a pronounced nano-scale ($\leq$2 nm) dynamic heterogeneity within or below the extremely broad glass transition in these single-phase materials.

Influence of carbon fiber fillers on the formation and phase structure. The effect of two carbon fiber fillers: a basic filler (CF) and CF with a surface modification by orthophosphoric acid residues (PCF) on the kinetics of epoxypolycyanurate (EPCN)/TPU grafted semi-IPN formation and phase structure was studied [50,51]. The network component in TPU/EPCN semi-IPNs was synthesized from the reactive blend of EO and DCEBA (Section 3.2.1). Acceleration of the occurring reactions was observed by the introduction the both carbon fiber fillers into the semi-IPNs studied, as seen in Figure 10 from the kinetic curves for a semi-IPN with 20 wt% of TPU. This acceleration is assumed to be associated with the additional chemical reactions between the components, with the catalytic effect of TPU on cyanate cyclotrimerization [41,42], and with interactions of polymers with the filler surface discussed in [50].

The influence of the fillers on the phase structure of the semi-IPNs was studied by DMTA. In the temperature dependence of E'' for the 50/50 semi-IPN, a broad single peak was observed and explained by the forced compatibility of the components [51]. This apparent compatibility results from a competition between the rates of the chemical interactions and phase separation (diffusion). The effect of the introduction of a filler, as well as of its surface modification on the degree of microphase separation in semi-IPNs was also reported in [51].

In Figure 11, the mechanical spectra of unfilled 50/50 semi-IPN and of those containing from 4 to 24 vol% of CF are shown. It is seen that the introduction of small amounts (4 vol%) of CF into the semi-IPN leads to the appearance of a clearly observable relaxation maximum in the mechanical spectrum relating to TPU and of a downward shoulder in the temperature range from 325 to 425 K relating to relaxation processes conditioned by molecular mobility at the interface. Thus, the introduction of the CF filler generates microphase separation in the semi-IPNs studied. The further increase of filler content enhances the microphase separation. As it is seen in Figure 11, the intensity of the peak in the range of 325 to 425 K increases and two peaks with maxima close to the T_g values of TPU and EPCN are observed in the mechanical

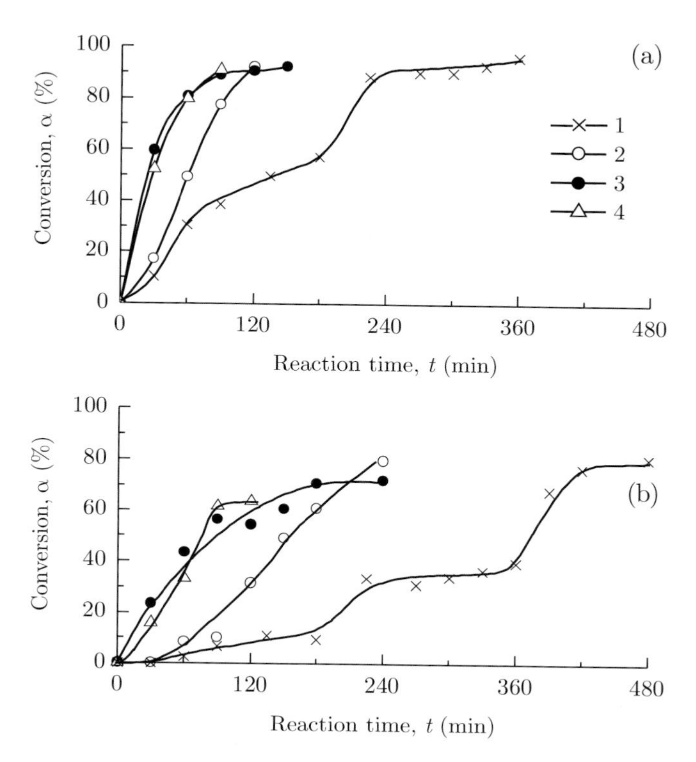

Figure 10. Kinetic curves for the conversion of (a) cyanate and (b) epoxy groups in: (1) EPCN, (2) EPCN/TPU = 80/20 non-filled semi-IPN, (3) filled with 1.5 vol% of CF, and (4) with 1.5 vol% of PCF

spectrum of semi-IPN filled with 16 vol% of CF. At CF contents of 20 to 24 vol%, the spectrum looks rather like a curve with one broad peak in the temperature range between the T_gs of TPU and EPCN. Analysis of the mechanical spectra shown in Figure 11 [51] suggested adsorption of TPU onto the CF surface and subsequent formation of two phases: a polymer phase rich in EPCN and a phase rich in TPU adsorbed onto the filler particles at low filler concentrations. At higher filler contents, it is supposed that the two components adsorb onto the filler surface and the degree of the phase separation in the system decreases. Comparative analysis of the influence of the filler surface activity on the phase structure of the semi-IPNs filled with CF and PCF has shown that the PCF-filled system is characterized by a lower level of heterogeneity. As far as the filler was introduced into the system at a stage of network formation, the physical adsorption of TPU onto the PCF surface is probably compensated by a competitive chemisorption of EPCN due to chemical reactions of epoxy and cyanate groups of the epoxycyanate oligomers with the PCF surface functional groups, hampering the phase separation process [50].

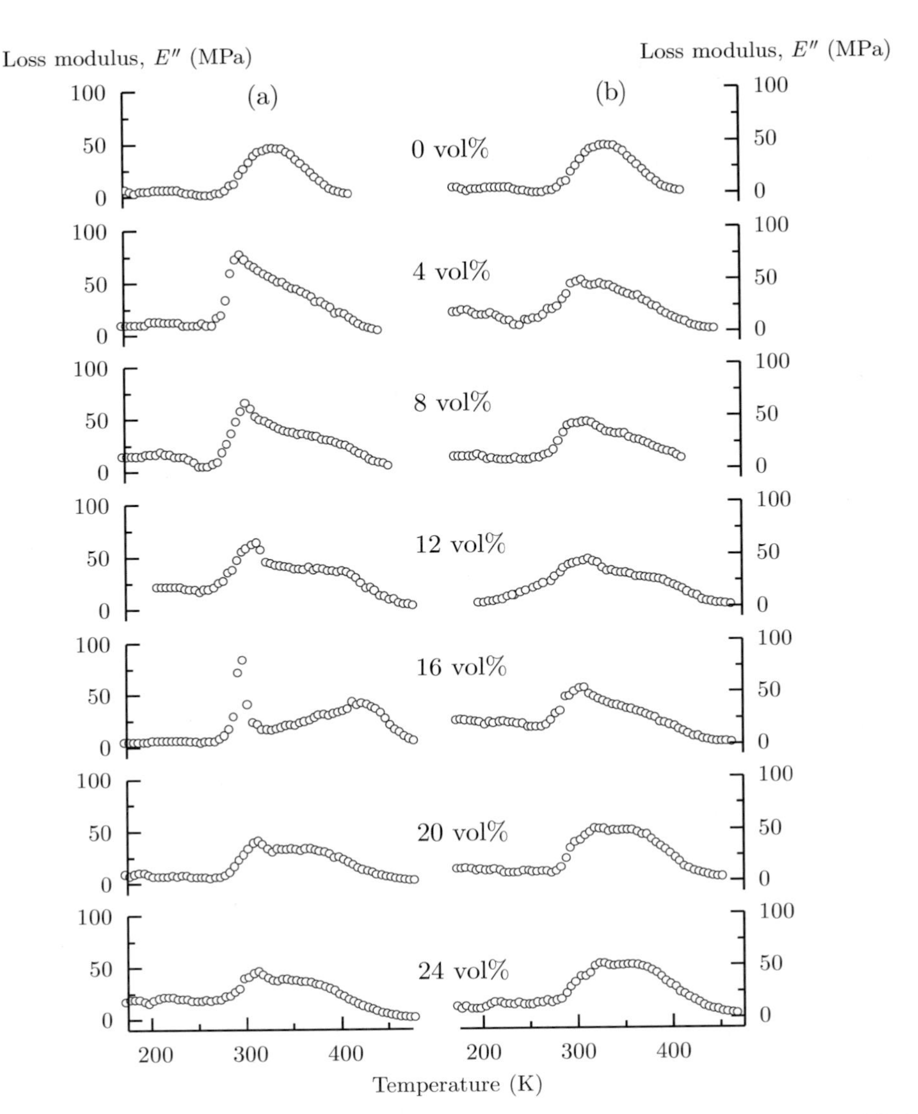

Figure 11. Temperature dependences of the loss modulus, E'', of a 50/50 semi-IPN filled with CF (a) and PCF (b)

A substantial increase in specific electrical resistance, reduction of combustibility, and existence of a piezoelectric effect have been also observed [52] in polyurethane-polycyanurate grafted semi-IPNs filled with PCF.

Adhesion to metals. The adhesion characteristics of TPU-modified (on the principle of semi-IPNs) polycyanurate have been studied by Grigoryeva *et al.* [43]. Aluminum and titanium plates were used as substrates to prepare the joints bonded by DCEBA cured in the presence of TPU (Figures 12 and 13).

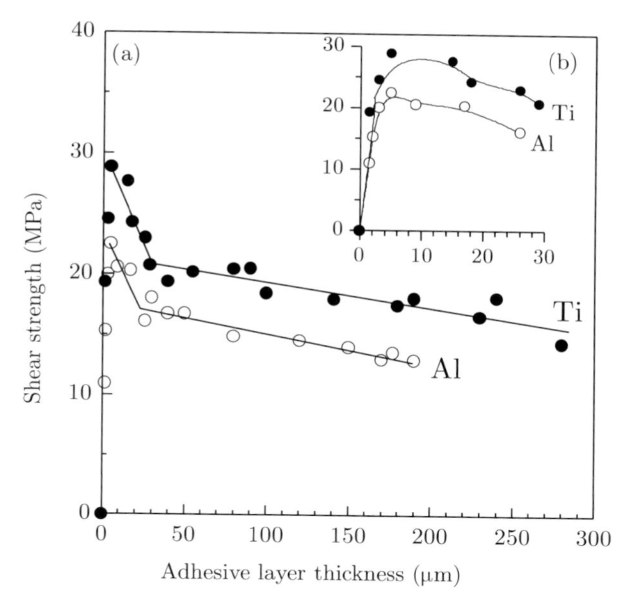

Figure 12. Adhesive strength *vs.* adhesive layer thickness for the pure PCN network (Al and Ti metal substrates). An extended scale is used in the inset

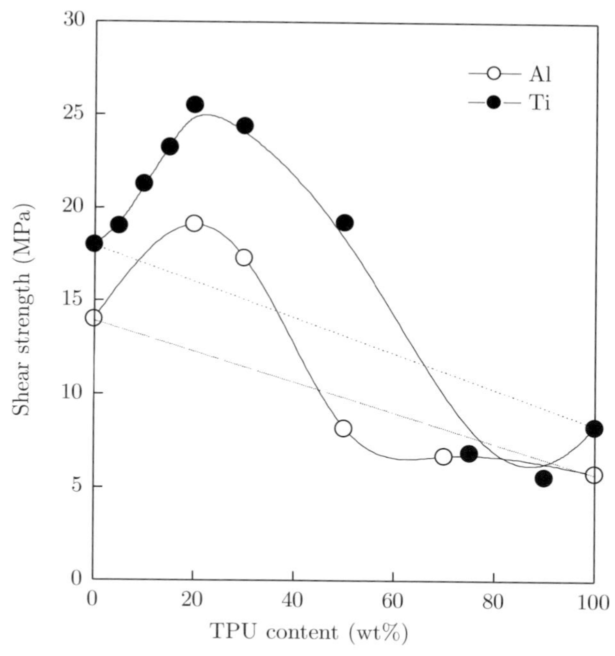

Figure 13. Composition dependence of the adhesive strength for aluminum and titanium (layer thickness of 150 μm)

A dependence of the adhesive strength on the adhesive layer thickness has been observed with all compositions studied. Generally, the thicker the adhesive layer, the higher is the adhesive strength. The introduction of TPU first increases the shear strength of PCN to aluminum and titanium. The maximal values of adhesive (shear) strength were achieved at TPU contents of 20 to 25 wt% corresponding to the formation of the hybrid PCN/TPU network only, owing to the chemical incorporation of TPU into PCN. At higher TPU contents, in addition to the semi-IPNs based on the hybrid network, non-incorporated TPU is observed in the adhesive layer. It has been concluded [43] that the presence of non-incorporated TPU in the adhesive layer leads to a reduction of the adhesive strength. It has been noted [43] that polycyanurates modified with thermoplastic polyurethane elastomers could be used as high-temperature heating-melt adhesives, coatings, and sealants, as well as matrices for high-performance composites.

4. Conclusions

Thermoplastic polyurethane elastomers, characterized by thermoplasticity and causing internal phase separation of hard and soft blocks in polymer chains, when used in the preparation of thermoplastic and interpenetrating polymer networks, have a strong influence on the phase structure of the other components and on the final structure of the IPNs obtained. The molecular structure of TPU itself can also undergo essential changes under the influence of the IPN components. In the cases discussed in this chapter, they are caused by interpenetration of the chains of thermoplastic PU and other polymer chains, as well as by physical and chemical interactions occurring between the reactive groups of different polymer molecules. PU can impart their thermoplasticity to the matrix system and contribute to the homogeneity of the resulting IPN structure. The structure and properties of IPNs based on thermoplastic polyurethane elastomers can be directly controlled by changes of the polyurethane chemical structure, the nature and structure of the second component, the IPN composition, and the production method. These materials, characterized by a broad range of desireable properties, can be used as coatings, membranes, adhesives, and matrices for different types of composites applicable in the automotive, construction, agricultural, electronic, aircraft, and other industries.

Acknowledgements

All the systems described in the chapter were synthesized in the Department of Interpenetrating Polymer Networks and Systems (Prof. L. M. Sergeeva, Head) of the Institute of Macromolecular Chemistry at the National Academy of Sciences of Ukraine (IMC, NASU) and jointly studied with partners from leading research groups at institutes and universities in Russia, Italy, Greece, Germany, and France.

References

1. Sperling L H (1981) *Interpenetrating Polymer Networks and Related Materials*, Plenum, New York.
2. Lipatov Y S and Sergeeva L M (1979) *Interpenetrating Polymer Networks*, Naukova Dumka, Kyiv (in Russian).
3. Klempner D and Frisch K C (Eds.) *Advances in Interpenetrating Polymer Networks*, Technomic, Lancaster, PA, Vol. I (1989), Vol. II (1990), Vol. III (1994).
4. Klempner D, Sperling L H and Utracki L A, (Eds.) (1994) *Interpenetrating Polymer Networks*, Adv Chem Ser No. 239, ACS Books, Washington, DC.
5. Sperling L H and Mishra V (1996) The current status of interpenetrating polymer networks, *Polym Adv Techn* **7**:197–208.
6. Lipatov Y S (2001) *Phase-Separated Interpenetrating Polymer Networks,* Ukrainian State Chemical-Technology University, Dnepropetrovsk.
7. Sperling L H (1986) *Multi-Component Polymer Materials*, Adv Chem Ser No. 211, ACS Books, Washington, DC.
8. Ali S A M and Hourston D J (1994) Thermoplastic interpenetrating polymer networks, in *Advances in Interpenetrating Polymer Networks* (Eds. Klempner D and Frisch K C) Technomic, Lancaster, PA, Vol. IV, pp. 17–43.
9. Schrader S, Pohl G, Carius H-E and Goering H (1990), Structure and relaxation behaviour of physical networks based on segmented polyurethane, in *Physical Networks. Polymers and Gels* (Eds. Burchard W and Ross-Murphy S B) Elsevier Applied Science, London, New York.
10. Vatalis A, Delides C, Grigoryeva O, Sergeeva L, Brovko A, Zimich O, Shtompel V, Georgoussis G and Pissis P (2000) Thermoplastic apparent interpenetrating polymer networks of polyurethane and styrene/acrylic acid copolymer obtained by melt mixing. Structure-property relationships, *Polym Eng Sci* **40**:2072–2085.
11. Vatalis A, Delides C, Georgoussis G, Kyritsis A, Grigoryeva O, Sergeeva L, Brovko A, Zimich O, Shtompel V, Neagu E and Pissis P (2001) Characterization of thermoplastic interpenetrating polymer networks by various thermal analysis techniques, *Thermochimica Acta* **371**:87–93.
12. Won H J, Yong K K and Ick H K (1991) Phase behavior of ternary polymer blends of poly(styrene-co-acrylic acid), poly(ethylene oxide), and poly(methyl methacrylate), *Macromolecules* **24**:4708–4712.
13. Campbell D and White J R (1989) *Polymer Characterization*, Chapman and Hall, London.
14. Hermant I, Damyanidu M and Meyer G C (1983) Transition behaviour of polyurethane-poly(methyl methacrylate) interpenetrating polymer networks, *Polymer* **24**:1419–1424.
15. Rizos A K, Fytas G, Ma R J, Wang C H, Abetz V and Meyer G C (1993) Local molecular motion in polyurethane-poly(methyl methacrylate) interpenetrating polymer networks, *Macromolecules* **26**:1869–1875.
16. Pandit S B and Nadkarni V M (1994) Sequential interconnected interpenetrating polymer networks of polyurethane and polystyrene. 1. Synthesis and chemical structure elucidation, *Macromolecules* **27**:4583–4594.
17. Sergeeva L, Grigoryeva O, Zimich O, Privalko E, Shtompel V, Privalko V, Pissis P and Kyritsis A (1997) Structure-property relationships in thermoplastic pseudo-IPNs. I. Phase morphology, *J Adhesion* **64**:161–171.
18. Sergeeva L, Grigoryeva O, Brovko A, Zimich O, Nedashkovskaya N, Slinchenko E and Shtompel V (1997) Pseudo-thermoplastic interpenetrating polymer networks based on polyurethane and styrene-co-acrylic acid, *J Prikladn Khim* **70**:2038–2045 (in Russian).

19. Kyritsis A, Pissis P, Grigoryeva O, Sergeeva L, Brovko A, Zimich O, Privalko E, Shtompel V and Privalko V (1999) Structure-property relationships in thermoplastic apparent interpenetrating polymer networks based on crystallizable polyurethane and styrene-acrylic acid copolymer, *J Appl Polym Sci* **73**:385–397.
20. Sergeeva L, Karabanova L, Grigoryeva O, Zimich O and Gorbach L (1997) Thermodynamics of the pseudo-thermoplastic interpenetrating polymer networks based on polyurethane and styrene-co-acrylic acid, *Ukr Khim J* **63**:65–69 (in Russian).
21. Tager A A (1972) Thermodynamic stability of polymer–solvent and polymer–polymer systems, *Vyssokomol Soedin* **19**:2690–2706 (in Russian).
22. Karabanova L V, Boiteux G, Gain O, Seytre G, Sergeeva L M and Lutsyk E D (2001) Semi-IPNs based on polyurethane and polyvinyl-pyrrolidone. I. Thermodynamic state and dynamic mechanical analysis, *J Appl Polym Sci* **80**:852–862.
23. Bartolotta A, Carini G, D'Angelo G, Di Marco G, Farsaci F, Grigoryeva O, Sergeeva L, Slisenko O, Starostenko O and Tripodo G (2004) Local and cooperative molecular mobility in thermoplastic polymers, *Phil Mag* **84**:1591–1598.
24. Stepanenko L, Novikova T, Sergeeva L, Shtompel V and Chernobay A (2002) Investigation of the structure and properties of blends based on polyurethane ionomer and polyaminourethane, *J Appl Chem* **75**:1341–1345 (in Russian).
25. Tsonos C, Apekis L, Viras K, Stepanenko L, Karabanova L and Sergeeva L (2000) Investigation of the microphase separation in blends of polyurethane-based ionomers, *J Macromol Sci–Phys* **B39**:155–174.
26. Tsonos C, Apekis L, Viras K, Stepanenko L, Karabanova L and Sergeeva L (2001) Electrical and dielectric behavior in blends of polyurethane-based ionomers, *Solid State Ionics* **143**:229–249.
27. Brovko O O, Sergeeva L M, Kuznetsova V P and Lemeshko V N (1998) Dynamic mechanical studies of polyurethane-polyurethane semi-interpenetrating polymer networks filled with γ-Fe_2O_3, *Eur Polym J* **35**:2045–2050.
28. Brovko A A, Sergeeva L M, Slinchenko E A and Fainleib A M (1996) Dynamic mechanical study of filled semi-interpenetrating polymer networks: Influence of γ-Fe_2O_3 on microphase structure, *Polym Int* **40**:299–305.
29. Sergeeva L M, Slinchenko E A, Brovko A A, Fainleib A M and Nedashkovskaya N S (1996) Effect of γ-Fe_2O_3 on microphase separation in semi-interpenetrating polymer networks on base of polyurethane and polyepoxyisocyanurate, *Polym Sci* **B38**:225–230.
30. Brovko A A, Slinchenko E A, Sergeeva L M and Fainleib A M (1996) Semi-interpenetrating polymer networks based on polyurethane and polyepoxyisocyanurate filled of γ-Fe_2O_3, *Ukr Chem J* **62**:133–139.
31. Brovko A A, Slinchenko E A, Fainleib A M, Sergeeva L M and Nedashkovskaya N S (1996) Influence of filler on the compatibility of the components in semi-interpenetrating polymer networks, *Compos Polym Mater* **57**:28–35 (in Russian).
32. Semenovich H M, Fainleib A M, Sergeeva L M, Nedashkovskaya N S and Slinchenko E A (1997) Effect of γ-iron oxide on the cure of a polyurethane-diisocyanate-epoxidian oligomer system, *Polym Sci* **A39**:1168–1172.
33. Fainleib A M, Sergeeva L M, Danchuk L V and Gukalov S P (1992) Composition for magnetic recording medium, USSR Patent 1,826,082.
34. Fainleib A M, Novikova T I, Shantalii T A and Sergeeva L M (1992) Kinetic of formation semi-interpenetrating polymer networks based on crosslinked polycyanurate and linear polyurethane, *Vyssokomol Soedin* **B33**:60–67 (in Russian).
35. Fainleib A M, Novikova T I, Shantalii T A and Sergeeva L M (1992) Synthesis, structure and some properties of the polycyanurate-polyurethane semi-IPNs, *Polym Mater Sci Eng* **66**:131–132.

36. Lipatov S Yu, Fainleib A M, Shantalii T A and Sergeeva L M (1992) Semi-interpenetrating networks based on the oligomers of dicyanates and linear polyurethanes, *Polym Sci* **34**:407–410.
37. Fainleib A M, Shantalii T A and Sergeeva L M (1993) Polycyanurate-polyurethane semi-IPNs, *Compos Polym Mater* **54**:14–16 (in Russian).
38. Brovko A A, Fainleib A M, Shantalii T A, Sergeeva L M and Davidenko V V (1994) Structure and viscoelastic properties of polycyanurate-polyurethane semi-interpenetrating polymer networks, *Polym Sci* **36**:934–938.
39. Fainleib A M, Sergeeva L M and Shantalii T A (1991) Triazinecontaining interpenetrating polymer networks, *Compos Polym Mater* **50:**63–72 (in Russian).
40. Bershtein V A, Egorova L M, Ryzhov V P, Yakushev P N, Fainleib A M, Shantalii T A and Pissis P (2001) Structure and segmental dynamics heterogeneity in hybrid polycyanurate-polyurethane networks, *J Macromol Sci-Phys* **B40**:105–131
41. Vilensky V A, Fainleib A M, Goncharenko L A and Danilenko I Yu (2002) Influence of "inclusion" polymer and components thermodynamic affinity on bisphenol A dicyanate ester polycyclotrimerization, *Reports Natl Acad Sci Ukraine* N1:142–148 (in Russian).
42. Fainleib A, Vilensky V, Goncharenko L, Grytsenko V and Brostow W (2003) Thermochemical study of process of cyanate ester resin polycyclotrimerization in immediate presence of polymeric modifiers, *Thermochim Acta*, in press.
43. Grigoryeva O P, Fainleib A M, Pissis P and Boiteux G (2002) Effect of hybrid network formation on adhesion properties of polycyanurate/polyurethane semi-interpenetrating polymer networks *Polym Eng Sci* **42**:2440–2448.
44. Bartolotta A, Di Marco G, Lanza M, Carini G, D'Angelo G, Tripodo G, Fainleib A M, Slinchenko E A and Privalko V P (1997) Molecular mobility in semi-IPNs of linear polyurethane and heterocyclic polymer networks, *J Adhesion* **64**:269–286.
45. Bartolotta A, Di Marco G, Carini G, D'Angelo G, Tripodo G, Fainleib A, and Privalko V (1998) Relaxation in semi-interpenetrating polymer networks of linear polyurethane and heterocyclic polymer networks, *J Non-Cryst Solids* **235–237**:600–604.
46. Bartolotta A, Di Marco G, Lanza M, Carini G, D'Angelo G, Tripodo G, Fainleib A M, Slinchenko E A, Shtompel V I and Privalko V P (1999) Synthesis and physical characterization of semi-IPNs of linear polyurethane and heterocyclic polymer network, *Polym Eng Sci* **39**:549–558.
47. Baltá Calleja F J, Privalko E G, Fainleib A M, Shantalii T A, and Privalko V P (2000) Structure-microhardness relationships for semi-interpenetrating polymer networks, *J Macromol Sci–Phys* **B39**:131–141.
48. Georgoussis G, Kyritsis A, Bershtein V A, Fainleib A M and Pissis P (2000) Dielectric studies of chain dynamics in homogeneous semi-interpenetrating polymer networks, *J Polym Sci Polym Phys Ed* **38**:3070–3087.
49. Pissis P, Georgoussis G, Bershtein V A, Neagu E and Fainleib A M (2002) Dielectric studies in homogeneous and heterogeneous polyurethane/polycyanurate interpenetrating polymer networks, *J Non-Cryst Solids* **305**:150–158.
50. Seminovych G M, Fainleib A M, Slinchenko E A, Brovko A A, Sergeeva L M and Dubkova V I (1999) Influence of carbon fibre on formation kinetics of cross-linked copolymer from bisphenol A dicyanate and epoxy oligomer, *React Funct Polym* **40**:281–288.
51. Brovko O O, Fainleib A M, Slinchenko E A, Dubkova V I and Sergeeva L M (2001) Filled semi-interpenetrating polymer networks: formation kinetics and properties, *Compos Polym Mater* **23**:85–91 (in Russian).

52. Sergeeva L M, Dubkova V I, Fainleib A M, Alekseenko V I, Brovko O O and Maevskaya O I (2000) The role of active carbon-fibrous filler in decrease of combustibility of semi-interpenetrating polymer networks, *Intern J Polym Mater* **47**:31–41.
53. Lipatov Y S (1991) *Physical-Chemical Basis of Filling Polymers*, Chemistry, Moscow.
54. Lipatov Y, Nesterov A, Ignatova T, Gudima N and Gristsenko O (1986) Influence of solid surface on equilibria in polymer mixtures, *Eur Polym J* **22**:83–87.
55. Shifrin V, Lipatov Y and Vasilenko O (1987) Influence of different nature fillers on changing thermodynamics of interactions in binary polymer blends, *Reports Natl Acad Sci Ukraine* **B**(1):56–59 (in Russian).
56. Lipatov Y, Sergeeva L, Karabanova L, Rosovitskij V, Skiba S and Babkina N (1986) On change of miscibility of components of interpenetrating polymer networks under an influence of fillers, *Reports Acad Sci USSR* **291**:635–638 (in Russian).
57. Lipatov Y, Sergeeva L, Karabanova L, Rosovitskij V, Skiba S and Babkina N (1988) Influence of fillers on viscoelastic properties and miscibility of components of interpenetrating polymer networks based on polyurethane and polyesteracrylate, *Vyssokomol Soedin* **A30**:649–655 (in Russian).
58. Lipatov Y, Karabanova L and Sergeeva L (1994) Thermodynamic state of reinforced interpenetrating polymer networks, *Polym Int* **34**:7–13.
59. Rosovitsky V F and Lipatov Y S (1985) To determination of segregation degree in two phase polymer systems on parameters of relaxation maxima, *Reports Acad Sci USSR* **283**:910–913 (in Russian).
60. Fainleib A M, Shantalii T A and Pankratov V A (1991) Copolymers of cyanate esters and plastics based on them, *Compos Polym Mater* **49:**39–53 (in Russian).
61. Hamerton I (Ed.) (1994) *Chemistry and Technology of Cyanate Ester Resins*, Chapman & Hall, Glasgow.
62. Nair C P R, Mathew D and Ninan K N (2000) Cyanate ester resins, recent developments, *Adv Polym Sci* **155**:1–99.
63. Fainleib A M, Kay M, Buffel K, Bauer J and Bauer M (2000) Chemical transformations in blends of monofunctional organic cyanate and urethane, *Reports Natl Acad Sci Ukraine* N12:170–174 (in Russian).

Chapter 13

Polyurethane Thermoplastic Elastomers Comprising Hydrazine Derivatives: Chemical Aspects

Yu. Savelyev

1. Introduction

The broad range of various approaches to the preparation of thermoplastic polyurethane elastomers is quite obvious and the possibilities of obtaining a specific structure of the macromolecular chain are numerous. Therefore, polyurethanes should be considered not only as a class of polymers; they could be also very efficient in the deliberate creation of macrochains with desired structures [1].

Polyurethanes (PU) containing hydrazide fragments in their macrochains are of special interest for the development of new polymers. This is because: (i) hydrazine and its derivatives are bifunctional reagents, readily forming polymers with various classes of organic multifunctional compounds and (ii) polymers based on hydrazine and its acylated derivatives are characterized by their valuable operational properties [2].

Chain extenders, such as macroheterocyclic compound (MC) derivatives, *e.g.*, macroheterocyclic dihydrazides (derivatives of the respective crown ethers), are of significant interest for the creation of a new generation of segmented PU. The synthesis in a rather short term of MC having the remarkable feature to combine in complexes with high-selectivity ions of varying nature strongly influenced the organic chemistry itself and the development of new technologies [3]. To the best of our knowledge, there is no information so far in the literature about the application of crown ethers in the production of segmented PU.

This chapter offers an overview of the research efforts in the field of the creation of thermoplastic elastomers of the polycondensation type on the basis of hydrazine and its derivatives, and of the results obtained during the last two decades.

2. Synthesis and properties of polyurethane thermoplastic elastomers comprising hydrazine derivatives

Hydrazine and its derivatives form high molecular compounds with various bifunctional electrophilic reagents. Polymers with valuable properties, which found practical application were thus obtained [2].

2.1. *Polyurethane-semicarbazides*

Polyurethane-semicarbazides (in earlier reports, they are also called polyurethane-acylsemicarbazides) are linear polymers containing urethane and (acyl)semicarbazide groups in their macromolecules. Sulfur or phosphorus can also be included in the macrochains [2].

The polyurethane-semicarbazides V are prepared in two steps [2]. As the first step, one obtains the prepolymer III through the interaction of an oligoether or oligoester I and a diisocyanate II. In the second step, the prepolymer III is extended by a stoichiometric amount of a dicarboxylic acid dihydrazide (IV, a difunctional derivative of hydrazine):

$$\underset{\text{I}}{HO{-}R{-}OH} + 2\,\underset{\text{II}}{OCN{-}R'{-}NCO} \longrightarrow \underset{\text{III}}{OCN{-}R'{-}NH\underset{\overset{\|}{O}}{C}O{-}R{-}O\underset{\overset{\|}{O}}{C}NH{-}R'{-}NCO}$$

$$n\,(\text{III}) + n\,\underset{\text{IV}}{H_2NNH\underset{\overset{\|}{O}}{C}{-}R''{-}\underset{\overset{\|}{O}}{C}NHNH_2} \longrightarrow$$

$$\underset{\text{V}}{\left[\underset{\overset{\|}{O}}{C}NH{-}R'{-}NH\underset{\overset{\|}{O}}{C}O{-}R{-}O\underset{\overset{\|}{O}}{C}NH{-}R'{-}NH\underset{\overset{\|}{O}}{C}NHNH\underset{\overset{\|}{O}}{C}{-}R''{-}\underset{\overset{\|}{O}}{C}NHNH\right]_{n-1}}$$

where R, R', R'' = Ar or Alk

The solution-cast polyurethane-semicarbazide films are elastic and transparent, and are characterized by high mechanical properties. The stability of these polymers at low temperatures grows with the increase of the macrochain flexibility, because of the weakening of the intermolecular interactions at lower concentrations of urethane and semicarbazide groups.

2.2. *Sulfur-containing polyurethane-semicarbazides*

The sulfur-containing polyurethane-semicarbazides are usually called polyurethane-sulfosemicarbazides. The chemical and thermal properties of these polymers depend on the valent state of the sulfur atom included in the macrochain, as well as on the chemical structure of the groups to which it is bonded. The repeat unit of polyurethane-semicarbazides can be represented as VI:

$$\left[\underset{\underset{O}{\|}}{C}NH{-}R'{-}NH\underset{\underset{O}{\|}}{C}O{-}(RO)_{\overline{n}}\underset{\underset{O}{\|}}{C}NH{-}R'{-}NH\underset{\underset{O}{\|}}{C}NHNH\overset{\overset{O}{\|}}{\underset{\underset{O}{\|}}{S}}{-}R''{-}\overset{\overset{O}{\|}}{\underset{\underset{O}{\|}}{S}}NHNH\right]_{x\text{-}1}$$

VI

where R = $CH(CH_3)CH_2$, $(CH_2)_4$; R' = 2,4-$CH_3C_6H_3$, 4,4'-$C_6H_4CH_3C_6H_4$;
R'' = $(CH_2)_4$, C_6H_4-C_6H_4, C_6H_4-A-C_6H_4; A = CH_2, O, S, SO_2

The sulfur-containing polymers are synthesized by a method similar to that described above for the polyurethane-semicarbazides V in donor solvents (DMF, DMSO, HMF, dioxane, *etc.*). Tertiary amines, organotin compounds, and transition metal β-diketonates are active catalysts in this synthesis, leading to the obtaining of polymers with a maximal molecular mass of 50 000–60 000 [4].

The high flexibility of the macrochain and less intensive intermolecular interactions, as compared to those in classical polyurethane-semicarbazides, lead to a higher solubility in amide solvents (DMF, DMA, HMF, DMSO), *m*-cresol and dioxane, and better frost-resistance of polyurethane-sulfosemicarbazides. One should also point out the better solubility of the initial reagents (dihydrazides of disulfoacids), which facilitates considerably the polymer synthesis. Compared to classical polyurethane-semicarbazides, the polyurethane-sulfosemicarbazides are less stable to water and dilute aqueous solutions of acids and alkalis, their alkaline hydrolysis being faster, especially at elevated temperatures. The incorporation of the sulfohydrazide group in the macromolecule accelerates the polymers degradation by UV irradiation.

2.3. *Phosphorus-containing polyurethane-semicarbazides*

The flammability of polymer materials can be reduced by the chemical incorporation of phosphorus atoms in their macromolecules, using either phosphoric acid dihydrazides as chain extenders, or phosphorus-containing oligoethers IX–XII in the prepolymer synthesis [5].

The phosphorus-containing polyurethane-semicarbazides VIII are obtained in two steps: (i) preparation of the phosphorus-containing prepolymer VII and (ii) chain extension by isophthalic or O-substituted phenylthiophosphoric acid dihydrazide:

$$HO{-}(RO)_{\overline{n}}{-}H + 2\,OCN{-}R'{-}NCO \longrightarrow OCN{-}R'{-}NH\underset{\underset{O}{\|}}{C}O{-}(RO)_{\overline{n}}\underset{\underset{O}{\|}}{C}NH{-}R'{-}NCO \quad \text{VII}$$

$$x\,(\text{VII}) + x\,H_2NNH\underset{\underset{O}{\|}}{C}{-}C_6H_4{-}\underset{\underset{O}{\|}}{C}NHNH_2 \longrightarrow$$

$$\left[\underset{\underset{O}{\|}}{C}NH{-}R'{-}NH\underset{\underset{O}{\|}}{C}O{-}(RO)_{\overline{n}}\underset{\underset{O}{\|}}{C}NH{-}R'{-}NH\underset{\underset{O}{\|}}{C}NHNH\underset{\underset{O}{\|}}{C}{-}C_6H_4{-}\underset{\underset{O}{\|}}{C}NHNH\right]_{x\text{-}1}$$

VIII

where

$$R = (CH_2CH_2O)_2-\underset{CH_3}{\overset{O}{\overset{\|}{P}}}-(CH_2CH_2)_2,\ (CH_2CH_2O)_2-\underset{C_6H_5}{\overset{O}{\overset{\|}{P}}}-(CH_2CH_2)_2,\ (CH_2CH_2O)_2-\underset{OC_6H_5}{\overset{O}{\overset{\|}{P}}}-(CH_2CH_2)_2,$$

IX, X, XI

$$-CH_2CH_2-\underset{CH_3CH_2O-\underset{O}{\underset{\|}{P}}-OCH_2CH_3}{N}-CH_2CH_2-;$$

XII

$R' = 1,3\text{-}CH_3C_6H_4,\ C_6H_4CH_2C_6H_4,\ C_6H_5OP(S){=}$

The study of the ignition time, the time of self-extinction, and the weight losses at burning has shown that the synthesized phosphorus-containing polyurethane-semicarbazides are among the conventional fire-proof or self-extinguishing materials.

2.4. *Ionomeric polyurethane-semicarbazides*

The possibility of using the polyurethane ionomeric resins as water dispersions, free of organic solvents and surface-active agents, simplifies the development of environmentally friendly production facilities [6–8]. Hydrazine-containing polyurethanes having ionic groups in their macrochains combine the properties of ionomeric resins with those of polyurethanes based on hydrazine derivatives [6,9].

2.4.1. *Cation-active polyurethane-semicarbazides*

Cation-active polyurethane-semicarbazides are obtained either as DMF solutions, or as water dispersions. Both methods comprise two stages. A prepolymer containing tertiary nitrogen atoms that can be protonized by a quaternizing reagent [9] is first formed by interaction of the oligoether (or oligoester) XIII, the diisocyanate XIV, and a compound bearing the tertiary nitrogen atom XV. Then the prepolymer XVI extension is combined with disperson in water and salt formation, the chain extension by hydrazine and/or carboxylic acid dihydrazides XVII being the fastest process.

$$n\,\underset{XIII}{HO-(RO)_x-H} + m\,\underset{XIV}{OCN-R'-NCO} + \underset{XV}{HO-(CH_2)_2-\underset{CH_3}{N}-(CH_2)_2-OH} \longrightarrow$$

$$OCN\left[R'-NH\underset{O}{\underset{\|}{C}}O-(RO)_x-\underset{O}{\underset{\|}{C}}NH-R'-NH\underset{O}{\underset{\|}{C}}O-(CH_2)_2-\underset{CH_3}{N}-(CH_2)_2-O\underset{O}{\underset{\|}{C}}NH\right]_n //$$

$$\left[R'-NH\underset{O}{\underset{\|}{C}}O-(CH_2)_2-\underset{CH_3}{N}-(CH_2)_2-O\underset{O}{\underset{\|}{C}}NH\right]_{m-n}R'-NCO \quad \text{XVI}$$

where $R = (CH_2)_4,\ CH(CH_3)_2$; $R' = (CH_3)C_6H_3,\ C_6H_4CH_2)C_6H_4$; $n = 1–4$, $m = 1–6$

The second stage of the obtaining of cation-active polyurethane semicarbazides XVIII can be represented schematically as:

$$p\,\mathrm{XVI} + p\,\mathrm{H_2NNH\underset{\underset{O}{\|}}{C}{-}R''{-}\underset{\underset{O}{\|}}{C}NHNH_2}\ \ (\mathrm{XVII}) \xrightarrow{HA}$$

$$\Big\{\underset{\underset{O}{\|}}{C}\mathrm{NH}\Big[\mathrm{R'{-}NH\underset{\underset{O}{\|}}{C}O{-}(RO)_{\mathit{x}}{-}\underset{\underset{O}{\|}}{C}NH{-}R'{-}NH\underset{\underset{O}{\|}}{C}O{-}(CH_2)_2{-}\overset{H\ A^-}{\underset{CH_3}{\overset{|+}{N}}}{-}(CH_2)_2{-}O\underset{\underset{O}{\|}}{C}NH}\Big]_n //$$

$$\Big[\mathrm{R'{-}NH\underset{\underset{O}{\|}}{C}O{-}(CH_2)_2{-}\overset{A^-\ H}{\underset{CH_3}{\overset{|+}{N}}}{-}(CH_2)_2{-}O\underset{\underset{O}{\|}}{C}NH}\Big]_{m-n}\mathrm{R'{-}NH\underset{\underset{O}{\|}}{C}{-}NHNH\underset{\underset{O}{\|}}{C}{-}R''{-}\underset{\underset{O}{\|}}{C}NHNH}\Big\}_{p-1}$$

XVIII

where R" = $(CH_2)_{0\text{-}8}$, 1,3-C_6H_4; A = Cl^-, CH_3COO^-, $ClCH_2COO^-$

The spontaneous dispersion of the cation-active polyurethane-semicarbazide in water takes place at the ionic groups (Table 1) either by precipitation or by phase inversion in polar, water-miscible solvents, such as acetone or dioxane. The salt formation is effected with hydrochloric, acetic, or chloroacetic acid.

Table 1. Characteristics of the aqueous dispersions of cation-active polyurethane-semicarbazides

R	R′*	R″	Salt former	N^+ (meq/g polymer)	Particle size (μm)
$(CH_2)_4$	TDI	NH_2NH_2	HCl	0.81	0.06
$(CH_2)_4$	TDI	$CH_2(CONHNH_2)_2$	HCl	0.80	0.15
$(CH_2)_4$	TDI	$(CH_2)_8(CONHNH_2)_2$	HCl	0.77	0.15
$(CH_2)_4$	TDI	1,3-$C_6H_4(CONHNH_2)_2$	HCl	0.75	0.13
$(CH_2)_4$	TDI	$(CH_2)_4(CONHNH_2)_2$	$ClCH_2COOH$	0.78	0.10
$(CH_2)_4$	TDI	1,3-$C_6H_4(CONHNH_2)_2$	$ClCH_2COOH$	0.75	0.07
$CH(CH_3)CH_2$	TDI	1,3-$C_6H_4(CONHNH_2)_2$	HCl	0.50	0.15
$CH(CH_3)CH_2$	MDI	$(CH_2)_4(CONHNH_2)_2$	CH_3COOH	0.48	0.27
$CH(CH_3)CH_2$	TDI	1,3-$C_6H_4(CONHNH_2)_2$	HCl	0.75	0.15
$(CH_2)_4$	TDI	1,3-$C_6H_4(CONHNH_2)_2$	HCl	0.58	0.12
$(CH_2)_4$	TDI	1,3-$C_6H_4(CONHNH_2)_2$	CH_3COOH	0.75	0.13

*MDI – 4,4′-methylene *bis*(*p*-phenyl isocyanate), KDI – xylelenediisocyanate (mixture of 2,4- and 2,6-isomers, 70:30, used in Table 2), TDI – toluenediisocyanate (mixture of 2,4- and 2,6-isomers, 80:20)

The factors determining the colloid-chemical properties of the cation-active polyurethane-semicarbazides are the nature and concentration of the ionic centers, and their position in the repeat unit of the macromolecule. The aqueous dispersions of the cation-active polyurethane-semicarbazides obtained have long shelf lives and form hydrophilic films.

2.4.2. *Anion-active polyurethane-semicarbazides*

Anion-active polyurethane-semicarbazides XIX are obtained using bifunctional anion-containing chain extenders, such as dihydroxy- or diaminocarboxylic acids, dihydroxy- or diaminosulfoacids [9]:

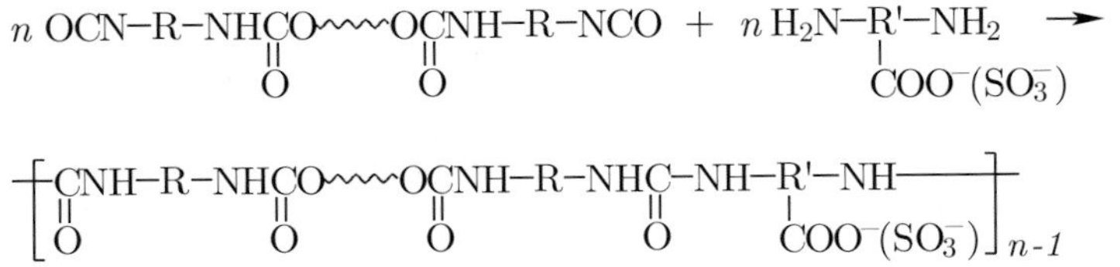

$$n\,\mathrm{OCN{-}R{-}NH\underset{\underset{O}{\|}}{C}O}\!\sim\!\sim\!\sim\mathrm{O\underset{\underset{O}{\|}}{C}NH{-}R{-}NCO} + n\,\mathrm{H_2N{-}\underset{\underset{COO^-(SO_3^-)}{|}}{R'}{-}NH_2} \longrightarrow$$

$$\left[\mathrm{\underset{\underset{O}{\|}}{C}NH{-}R{-}NH\underset{\underset{O}{\|}}{C}O}\!\sim\!\sim\!\sim\mathrm{O\underset{\underset{O}{\|}}{C}NH{-}R{-}NH\underset{\underset{O}{\|}}{C}{-}NH{-}\underset{\underset{COO^-(SO_3^-)}{|}}{R'}{-}NH}\right]_{n\text{-}1}$$

XIX

Stable (shelf life 5–18 months) dispersions are obtained, with particle size of 30–35 nm. From these dispersions, stable elastic films with high mechanical properties (tensile strength, σ, of 8.9–40.0 MPa and elongation at break, ε, in the range of 400–1500%) can be formed.

From the prepolymers based on aromatic diisocyanates containing sulfite groups, aqueous dispersions of anion-active polyurethane-semicarbazides are obtained by phase inversion [10]. The respective repeat units XX can be represented as:

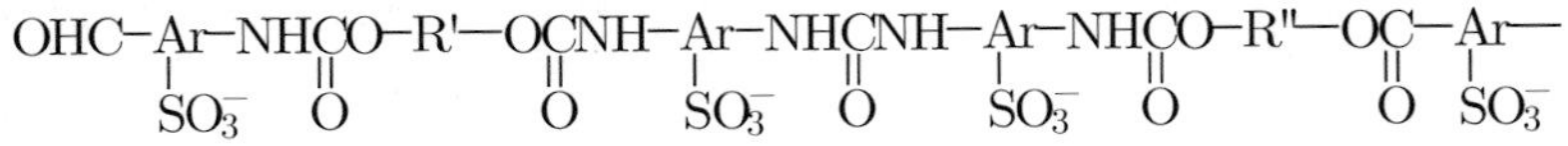

$$\mathrm{OHC{-}\underset{\underset{SO_3^-}{|}}{Ar}{-}NH\underset{\underset{O}{\|}}{C}O{-}R'{-}O\underset{\underset{O}{\|}}{C}NH{-}\underset{\underset{SO_3^-}{|}}{Ar}{-}NH\underset{\underset{O}{\|}}{C}NH{-}\underset{\underset{SO_3^-}{|}}{Ar}{-}NH\underset{\underset{O}{\|}}{C}O{-}R''{-}O\underset{\underset{O}{\|}}{C}{-}\underset{\underset{SO_3^-}{|}}{Ar}{-}}$$

XX

where Ar is an aromatic diisocyanate radical; R' is a hydrocarbon radical of poly(propylene oxide) or poly(tetramethylene oxide); R'' is a hydrocarbon radical of poly(propylene oxide), poly(tetramethylene oxide), or diethylene glycol

Water, hydrazine, β-hydrazinoethanol, adipic or isophthalic acid dihydrazide are used as chain extenders. The dispersions (particle size of 650–705 nm) form stable elastic films with $\sigma = 10.2$–28.8 MPa and $\varepsilon = 450$–700%.

2.5. *Polyurethane thermoplastic elastomers with macroheterocyclic fragments in the main chain*

2.5.1. *Polyurethane-semicarbazides based on crown ether sulfonyl derivatives*

The incorporation of macroheterocyclic compounds and their derivatives in polymer structures leads to the obtaining of polymers with specific behavor upon complex formation [3,11,12]. Polyurethane-semicarbazides comprising crown ether sulfonyl derivatives in the main chain were obtained by a conventional two-stage method [13–16]. The structure of the repeat unit of the resulting crown ether-containing thermoplastic elastomers XXI can be represented as:

$$-\!\!\left[OCNH-R^1-CONH-R^2-NHCOO-R^3-OCONH-R^2-NHCO\right]_n\!-R^4$$

XXI

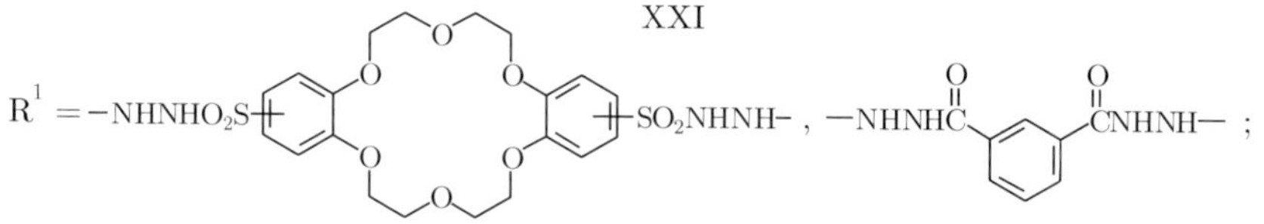

$R^1 = -NHNHO_2S-$(dibenzo-18-crown-6)$-SO_2NHNH-$, $-NHNHC(O)-C_6H_4-C(O)NHNH-$;

DHSDB18C6 DHIA

$R^2 = -C_6H_4-CH_2-C_6H_4-$; $R^3 = -[(CH_2)_4-O]_n-$; $R^4 =$ (benzo-18-crown-6)$-SO_2NHNH$

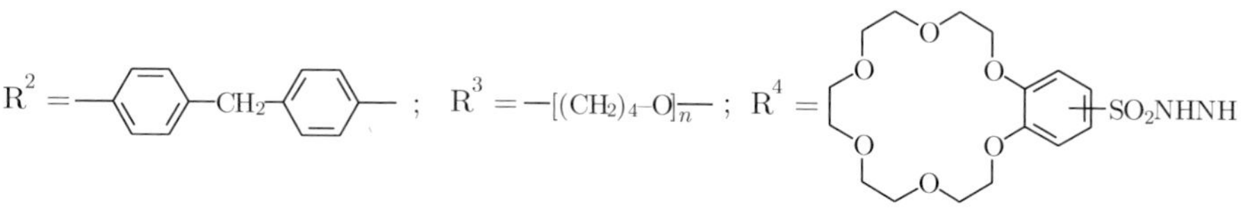

The dihydrazide of disulfonyldibenzo-18-crown-6 (DHSDB18C6) was used as chain extender, while the end-groups were formed by the hydrazide of sulfonylbenzo-18-crown-6 (HSB18C6). DHSDB18C6 and HSB18C6 were synthesized using a technique similar to that reported in [17]. The compositions and some characteristics of the crown ether-containing thermoplastic elastomers XXI are given in Table 2.

The crown ether-containing thermoplastic elastomers obtained are soluble in amide solvents (DMF, DMA, DMSO). Their intrinsic viscosity, $[\eta]$, is within the range of 0.22 to 0.41 dL/g (in DMF at 25°C).

Table 2. Composition and mechanical properties of the thermoplastic elastomers with dibenzo-18-crown-6 fragments in the main chain

Sample	Oligo-ether*	Diiso-cyanate**	Chain extender (end-group)***	σ (MPa)	ε (%)
PU-1	P-1050	MDI	DHSDB18C6	12.5	260
PU-2	L-1050	MDI	DHSDB18C6	10.8	320
PU-3	L-1050	KDI	DHSDB18C6	12.0	250
PU-4	PEO-400	TDI	DHSDB18C6	24.8	185
PU-5	P-1050	MDI	DHSDB18C6:(HSB18C6)=4:(1)	10.0	280
PU-6	P-1050	MDI	DHSDB18C6:DHIA = 1:1	18.4	295
PU-7	P-1050	MDI	DHSDB18C6:DHIA: (HSB18C6) = 1:1:(0.5)	13.8	365
PU-8	P-1050	MDI	DHSDB18C6:DHIA:(HSB18C6) = 1:1:(1)	10.9	325
PU-9	P-1050	MDI	DHSDB18C6:DHIA:(HSB15C5) = 1:1:(1)	20.5	350
PU-10	P-1050	MDI	DADB18C6	20.3	450
PU-16	P-1050	MDI	DHIA	31.7	325
PU-19	P-1050	MDI	DHDPhMSA	29.5	375

* P-1050 – poly(tetramethylene oxide), 1050 g/mole, L-1050 – poly(propylene oxide), 1050 g/mole, or poly(ethylene oxide) (PEO), 1050 g/mole; ** Same as in Table 1; *** DADB18C6 – diaminodibenzo-18-crown-6, DHIA – isophthalic acid dihydrazide, HSB15C5 – hydrazide of sulfonylbenzo-15-crown-5, DHDPhMSA – diphenylmethane disulfonic acid dihydrazide

2.6. *Other hydrazine-containing polyurethane thermoplastic elastomers*

2.6.1. *Polymers based on asymmetric dimethylhydrazine*

The asymmetric dimethylhydrazine (ADMH) has been used in the chemistry of macromolecular compounds mostly as polymer modifier [18]. The peculiarities of the reactivity of the tertiary nitrogen atom in ADMH result in the formation of hydrazinium compounds. Only at steric hindrance by quaternization, the nitrogen atom of the amine group can become reactive.

At a non-stoichiometric ratio of the reagents, one can prepare a crosslinked polymer by two-step synthesis. A different result is obtained in the reaction of the polyurethane prepolymer (PP, dissolved in DMF) with ADMH [19,20]. At a 2-, 5-, and even 10-fold increase of the PP molar fraction, DMF-soluble macromolecular compounds are formed. The respective solution-cast films have mechanical properties similar to those of classical polyurethane elastomers (Table 3).

Table 3. Composition and some properties of polyurethane elastomers based on asymmetric dimethylhydrazine

Sample	PP:ADMH (molar ratio)	$[\eta]$ (dL/g)	ρ (g/cm^3)	σ (MPa)	ε (%)
PU(H) 1/1	PP:ADMH (1:1)	0.34	–	–	–
PU(H) 2/1	PP:ADMH (2:1)	0.72	1.1235	24.5	6630
PU(H) 5/1	PP:ADMH (5:1)	0.57	1.1126	23.3	7725
PU(H) 10/1	PP:ADMH (10:1)	0.51	1.1111	22.2	8800
PU(H) 20/1	PP:ADMH (20:1)	0.66	1.1017	26.1	9960

The number average molecular weights (MW) of polymers are determined by the analysis of their terminal groups. The number of amide and associated hydrazide groups is determined by potentiometric non-aqueous titration. The number of nitrogen atoms in polymer samples is evaluated according to Kjeldahl and the MW of some elastomers is determined ebullioscopically. Such a complex approach has made it possible to evaluate the number average MW of the elastomers obtained, and to draw conclusions on the possible structure of their macromolecules (Figure 1). As for the case of model reactions [21], formation of two types of structures is found. Only in the cases of PP:ADMH = 1:(6–1) and 2:1 (catalytic reaction), one obtains oligomers or low molecular elastomers with a linear structure (Figure 1, XXII), while in all other cases, elastomers with probably branched (comb-like) macrochains (Figure 1, XXIII and XXIV) are obtained by the formation of biuret structures. The biuret formation is confirmed by the fact that the introduction of phosphoric acid in the reaction system prevents the formation of biuret bonds and results in a decrease of the polymer MW, as was observed in the present case.

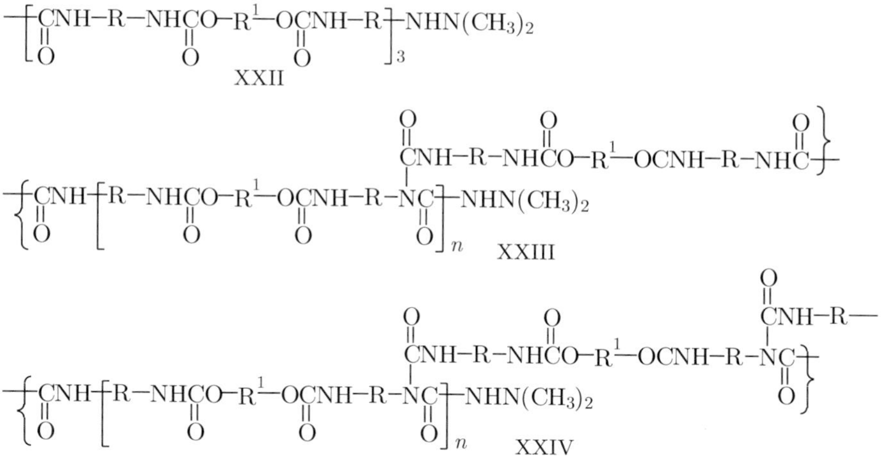

$$-\!\left[\underset{\|}{\overset{}{C}}NH-R-NH\underset{\|}{C}O-R^1-O\underset{\|}{C}NH-R\right]_3\!\!-NHN(CH_3)_2 \quad (C=O)$$

XXII

$$\left\{\underset{O}{\underset{\|}{C}}NH-\left[R-NH\underset{O}{\underset{\|}{C}}O-R^1-O\underset{O}{\underset{\|}{C}}NH-R-\underset{\underset{\overset{\|}{C}NH-R-NH\overset{\|}{C}O-R^1-OCNH-R-NH\overset{\|}{C}\}}{|}}{N}\underset{O}{\underset{\|}{C}}\right]_n\!-NHN(CH_3)_2\right.$$

XXIII

$$\left\{\underset{O}{\underset{\|}{C}}NH-\left[R-NH\underset{O}{\underset{\|}{C}}O-R^1-O\underset{O}{\underset{\|}{C}}NH-R-\underset{\underset{\overset{\|}{C}NH-R-NH\overset{\|}{C}O-R^1-OCNH-R-\underset{\underset{\overset{\|}{C}NH-R-}{|}}{N}\underset{O}{\underset{\|}{C}}\}}{|}}{N}\underset{O}{\underset{\|}{C}}\right]_n\!-NHN(CH_3)_2\right.$$

XXIV

Figure 1. Possible macrochain structures of the polyurethanes based on ADMH: XXII – linear, XXIII – low branched, XXIV – branched

The intrinsic viscosity, $[\eta]$, varies in different ways with the increase of the PP content (Table 3). Usually, the decrease of the viscosity of polymers (of the same molecular weight) is an evidence of macromolecules of the branched type, except for the case of comparable lengths of the branches and of the main chain, when the molecule has the shape of a star [1].

The structural analysis of the samples PP:ADMH = 2:1 obtained at 60 and 80°C in the absence of catalysts shows the formation of a branched structure at temperatures lower than those required for the formation of allophanate (120–140°C) and biuret (100°C) structures [1]. The ability of the ADMH to form hydrazinium cations leads to the assumption that the elastomer synthesis with the participation of ADMH involves formation of reactive centers of ionic nature, similarly to the Ritter reaction, where the synthesis of the N-substituted amides of carboxylic acids passes through a stage of carbonium ion formation [22]. This assumption is confirmed by the following observations: (i) the highest activity in catalytic processes is shown by tin acetylacetonate dichloride, the latter providing the possibility of formation of hydrazinium chloride and (ii) the process becomes much faster by the addition of hydrochloric acid to the reaction system.

2.6.2. *Ionomeric polyurethane thermoplastic elastomers*

Conditions are established for the preparation of hydrophilic and water-soluble polyurethane elastomers based on ADMH and other hydrazine derivatives, 1,6-hexamethylene diisocyanate and poly(tetramethylene oxide) (MW 1000 g/mole) [23]. The hydrophilicity of the final products can be increased by one order of magnitude and depends on the concentration of ionic groups and on the functionality of the quaternizing agents. The polyurethane elastomers obtained are in the form of organic or aqueous-organic solutions and, in some cases, they are thermoreversible gels.

Amphoteric polyurethane elastomers containing tertiary nitrogen atoms (capable of quaternization) in some macrochain fragments and carboxyl groups (capable of salt formation) in the others, are also synthesized [24]. Depending on the ratio of these groups in the macrochain, one obtains stable ionomeric aqueous dispersions with particle size from 185 to 520 nm (pH of the medium is in the range of 3.15–3.95). The films obtained from these dispersions are characterized by good mechanical properties ($\sigma = 20$ MPa, ε up to 800%). Their hydrophilicity varies from 20 to 150%.

2.7. *Polyurethane-semicarbazides and polyurethane-sulfosemicarbazides modified by transition metal β-diketonates*

Polyurethanes with a desired set of physico-chemical properties can be prepared by varying their chemical structure, as well as by the proper choice of catalysts for urethane formation and modifiers; in this respect, the ideal choice would be a single substance combining these two functions. By the addition of various transition metal β-diketonates, hydrazine-containing polyurethanes with specific properties were prepared [25–27].

Polyurethane-semicarbazides and polyurethane-sulfosemicarbazides were obtained by the conventional two-stage synthesis [2] in the presence of a transition metal acetylacetonate; the polymers obtained have the following structure of the repeat unit, XXV:

$$\left[\underset{\underset{O}{\|}}{C}NH{-}R{-}NH\underset{\underset{O}{\|}}{C}O{-}(R'O)_m{-}\underset{\underset{O}{\|}}{C}NH{-}R{-}NH\underset{\underset{O}{\|}}{C}NHNH{-}R''{-}NHNH\right]_n$$

XXV

R = 4,4′-$C_6H_4CH_2C_6H_4$; $CH_3C_6H_3$ (mixture 2,4- and 2,6-isomers);
R′ = $(CH_2)_4$, $CH_2CH(CH_3)CH_2$, $OC(CH_2)_4COO(CH_2)_4$;
R″ = $O_2SC_6H_4$-A-$C_6H_4SO_2$, OCC_6H_4CO; A = CH_2, O, S, SO_2.

Transition metal acetylacetonates ($M(AcAc)_n$, β-diketonates) were added to the reaction mixture. The nature of the metal atom and the structure of the chain extender determine to a great extent the character of the process of complex formation (see Section 3.2.1), which, in turn, strongly affects the thermal stability of the final products. The catalytic additives $Zn(AcAc)_2$, $Co(AcAc)_2$, or $Fe(AcAc)_3$ increase the temperature of 5% weight loss (τ_5) of the polyurethane-semicarbazides by 40–50°C (Table 4). Polyurethane sulfosemicarbazides with substantially increased thermal stability are obtained by the addition of the modifier in an amount of $\geq 2\%$, which provides the optimal condition for complex formation, *i.e.*, $NHNHSO_2$: $M(AcAc)_n$ = 1:1. The inhibition of the thermal degradation can be explained by the formation of a denser chain packing as a result of the coordination crosslinking caused by $M(AcAc)_n$. This denser packing hampers the diffusion of oxygen into the polymer volume; up to 300°C, the thermal degradation rate is proportional to the concentration of diffusing

Table 4. Composition and some properties of polyurethane-semicarbazides and polyurethane-sulfosemicarbazides (XXV)* with $M(AcAc)_n$ additives

R′	R″	M^{n+}	M^{n+} content (%)	ε (%)	δ (MPa)	τ_5 (°C)
$[(CH_2)_4]_{14}$	$1,3\text{-}C_6H_4(CO_2)_2$	-	-	325	62,3	267
$[(CH_2)_4]_{14}$	$1,3\text{-}C_6H_4(CO_2)_2$	Zn^{2+}	0.0006	375	61.0	312
$[(CH_2)_4]_{14}$	$1,3\text{-}C_6H_4(CO_2)_2$	Co^{3+}	0.0006	385	59.8	310
$[(CH_2)_4]_{14}$	$1,3\text{-}C_6H_4(CO_2)_2$	Fe^{3+}	0.0006	380	60.1	313
$[(CH_2)_4]_{14}$	$1,3\text{-}C_6H_4(CO_2)_2$	Cr^{3+}	0.0006	345	61.3	281
$[(CH_2)_4]_{14}$	$(C_6H_4SO_2)_2O$	-	-	375	41.7	196
$[(CH_2)_4]_{14}$	$(C_6H_4SO_2)_2O$	Fe^{3+}	2.0	290	147.1	251
$[(CH_2)_4]_{14}$	$(C_6H_4SO_2)_2CH_2$	-	-	440	37.5	193
$[(CH_2)_4]_{14}$	$(C_6H_4SO_2)_2\ CH_2$	Zn^{2+}	2.5	300	34.5	232
$[(CH_2)_4]_{14}$	$(C_6H_4SO_2)_2$	-	-	100	6.9	241
$[(CH_2)_4]_{14}$	$(C_6H_4SO_2)_2$	Ni^{2+}	0.6	140	9.0	243
$[(CH_2)_4]_{28}$	$(C_6H_4SO_2)_2\ CH_2$	-	-	550	46.6	200
$[(CH_2)_4]_{28}$	$(C_6H_4SO_2)_2\ CH_2$	Ni^{2+}	0.6	500	54.3	222
$[CH_2CH(CH_3)CH_2]_{14}$	$(C_6H_4SO_2)_2\ CH_2$	-	-	550	47.9	196
$[CH_2CH(CH_3)CH_2]_{14}$	$(C_6H_4SO_2)_2\ CH_2$	Ni^{2+}	0.6	330	37.9	212
$[CH_2CH(CH_3)CH_2]_{28}$	$(C_6H_4SO_2)_2\ CH_2$	-	-	1150	4.4	218
$[CH_2CH(CH_3)CH_2]_{28}$	$(C_6H_4SO_2)_2\ CH_2$	Ni^{2+}	0.6	570	7.8	220

* Prepolymer based on 4,4′-methylene-*bis*(*p*-phenyl isocyanate) ($R = 4,4'\text{-}C_6H_4\text{-}CH_2\text{-}C_6H_4$)

oxygen. An additional effect of $M(AcAc)_n$ on the degradation processes is that these additives enhance the destruction of the hydroperoxide, which is a branching agent.

The $M(AcAc)_n$ additives noticeably increase the stability of hydrazine-containing elastomers to UV-irradiation [27]. The extent of inhibition of the photolytic processes is determined mainly by the nature of the metal atom. The largest increase of photostability upon exposure to UV-light for 200 h is observed with polyurethane-sulfosemicarbazides containing $Ni(AcAc)_2$. Probably, the role of nickel(II) acetylacetonate in the inhibition of the photodegradation processes is mainly to terminate the hydroperoxide homolytic decomposition. Studies of the influence of UV irradiation on polymers in the absence of oxygen (in inert atmosphere) show the inhibiting effect of $Ni(AcAc)_2$ on photodegradation. Nickel(II) acetylacetonate acts through the mechanism of suppression of excited chromophores and as a UV absorber. This is in agreement with the conclusions drawn in [28] about the thermal and thermooxidative destruction of conventional polyurethane elastomers extended by diols. Obviously, the role of $Fe(AcAc)_3$ in the inhibition of the polyurethane-sulfosemicarbazides degradation is to prevent the process of photolysis.

The addition of $M(AcAc)_n$ affects the radiolysis taking place in polyurethane-sulfosemicarbazides upon irradiation with accelerated electrons [29]. The stability of polyuretnane elastomers to radiolysis is evaluated by means of coefficients characterizing the preservation of the polymer sample tensile strength (K_σ) and elongation to break (K_ε):

$$K_\sigma = (\sigma_{exp}/\sigma_{init})\cdot 100\%, \; K_\varepsilon = (\varepsilon_{exp}/\varepsilon_{init})\cdot 100\%,$$

where σ_{exp} and ε_{exp} are experimental values, and σ_{init} and ε_{init} are initial values.

After irradiation by a dose of 100 Mrad, the value of K_σ increases up to 200% for polyurethane-sulfosemicarbazide added with $Co(AcAc)_3$ and K_ε takes its minimal value of 78%. Processes of gel formation evidence that $Fe(AcAc)_3$ and $Cr(AcAc)_3$ have little effect on this process. Upon irradiation of the polyurethane-sulfosemicarbazides modified by $Zn(AcAc)_2$ with a doze of 50 Mrad, the coefficients K_σ and K_ε are 49.2 and 48.5%, respectively. The inhibitory activity of $M(AcAc)_n$ in the process of radiolysis of the elastomers studies obeys the order: Co(III) > Cr(III) > Fe(III) ≫ Zn(II). The stabilizing effect of $M(AcAc)_n$ upon radiolysis of polyurethane-sulfosemicarbazides is possibly related to redox processes and to a decrease in the oxidation number of the metal upon the destruction of hydroperoxide.

3. Structure and performance of hydrazine-containing polyurethane thermoplastic elastomers

The structure and properties of polyurethane thermoplastic elastomers that are block copolymers of the $[AB]_n$ type are mainly determined by the chemical structure of the macromolecule and by the intermolecular bonds in the system where the soft and hard blocks exist as thermodynamically incompatible microphases. The effect of the macromolecular structure on the properties of the polymers obtained is discussed in this section.

3.1. *Macrochain structure and supramolecular organization of hydrazine-containing polyurethane elastomers*

In this section, two non-trivial and rarely discussed polymer types are considered: polyurethanes based on asymmetric dimethylhydrazine and polyurethanes based on the hydrazides of macroheterocylic compounds.

3.1.1. *IR spectroscopy*

Polyurethane elastomers based on asymmetric dimethylhydrazine. The analysis of the IR data of ADMH-based polyurethanes (for their synthesis see Section 2.6.1) suggested an analogy between the model reaction of urethane formation [21] and the polyurethane formation concerning the intensities of the bands of carbonyl IR absorption depending on the ratio of the starting reagents. Similarly to the case of the model reaction, one may conclude that in the process of polyurethane formation, the extension by ADMH of the macrochain comprising NCO end-groups took place with formation of biuret-type fragments.

The IR spectra reveal all the characteristic adsorption bands of polyurethanes: CH_2 groups (2865 and 2950 cm^{-1}), COC groups (1115 cm^{-1}), CO groups (associated (1650 cm^{-1}) and free of hydrogen bonds (1710 cm^{-1})), NH groups (3320 cm^{-1}). The intensity of the band due to NH groups is much higher than that for 4,4′-methylene *bis*(*p*-phenyl isocyanate). The band at 2270 cm^{-1} determined by vibrations of NCO groups is not observed, *i.e.*, all these groups participate in the reaction of polyurethane formation. It should be noted that the appearance of a band at 1670 cm^{-1} can be attributed to the biuret group. The spectra of all the polymers studied are identical.

Polyurethanes including crown ether fragments in the main chain. According to the classical concepts [30], IR spectroscopy provides information on the chemical and steric structures of polyurethanes and their structure-property relationships. Important data on the molecular orientation of polymers can also be collected by examining the orientation of specific chain segments in complex multiphase polymer systems, *e.g.*, segmented block copolymers including polyurethanes based on crown ether derivatives and some heterocyclic compounds [13,31–34].

The IR spectroscopic data make it possible to establish the structure of the synthesized polyurethanes. The IR spectra of samples PU-1, PU-3, PU-4, PU-5, PU-7, and PU-8 (Table 2) exhibit absorption bands characteristic of the stretching (3280–3320 cm^{-1}) and bending (1538–1555 cm^{-1}) vibrations of NH groups. In the 1230–1248 cm^{-1} range, the stretching vibrations of COC groups are observed; the bands at 1720–1740 cm^{-1} are due to the stretching vibrations of free CO groups [35]. The IR spectra of the polymers based on DHSDB18C6 (PU-1) and DADB18C6 (PU-10) differ substantially in the region below 1500 cm^{-1}. The bands at 1170 and 1340 cm^{-1} in the spectrum of PU-1 are attributed to vibrations of SO_2 groups. In the region of the bending vibrations of CH_2 groups (1400–1500 cm^{-1}), the spectra of both samples exhibit bands at 1420 and 1480 cm^{-1}; however, the intensity distribution in the spectra of these samples differs drastically. The bending vibrations of CH_2 groups due to both polyester fragments and crown ether cycles appear in this region. This finding is in accordance with literature data [36]. Possibly, the differences in the intensity of these bands can be assigned to a change in the conformation of the macrocycle. For both samples, the characters of vibrations due to polyether fragments and urethane groups are identical. The differences in the bands at 1000 and 1300 cm^{-1} can be assigned to a change in the nature of the substituents at the benzene ring.

Interestingly, the absorption in the region of bonded CO groups (1650 cm^{-1}) noticeably increases in the multiple attenuated total reflectance (ATR) spectra taken from the film surface. Comparison of the intensities of absorption bands for the PU-1, PU-2, PU-6, PU-9, and PU-10 samples suggests that they increase from PU-1 to PU-6 (PU-6 shows the maximum intensity), and then decrease. In other words, PU-6 is characterized by the highest concentration of bonded CO groups on the surface, while in the case of PU-10, their concentration is the lowest. This implies that the samples studied have different supramolecular structures. It should be noted that the absorption band typical of symmetric vibrations of SO_2 groups (1160 cm^{-1}), which is present in the transmission spectrum, is also observed in the multiple ATR spectra, *i.e.*, crown ether fragments are present both in the polymer bulk and on its surface. The band intensities suggest that the distribution of these fragments is uniform.

The results of polarization experiments carried out under deformation on PU-9 films are shown in Figure 2, where the IR absorption spectra of PU-9 films with a variable degree of uniaxial stretching are presented. The spectrum of the undeformed film (Figure 2a) displays absorption bands at 1730, 1710, 1690, 1680, 1665, and 1650 cm^{-1} corresponding to carbonyl groups and absorption bands at 1555, 1540, 1525, and 1510 cm^{-1} due to the bending vibrations of NH groups. In the spectrum of the 50% stretched film (Figure 2b), the bands at 1555, 1540, and

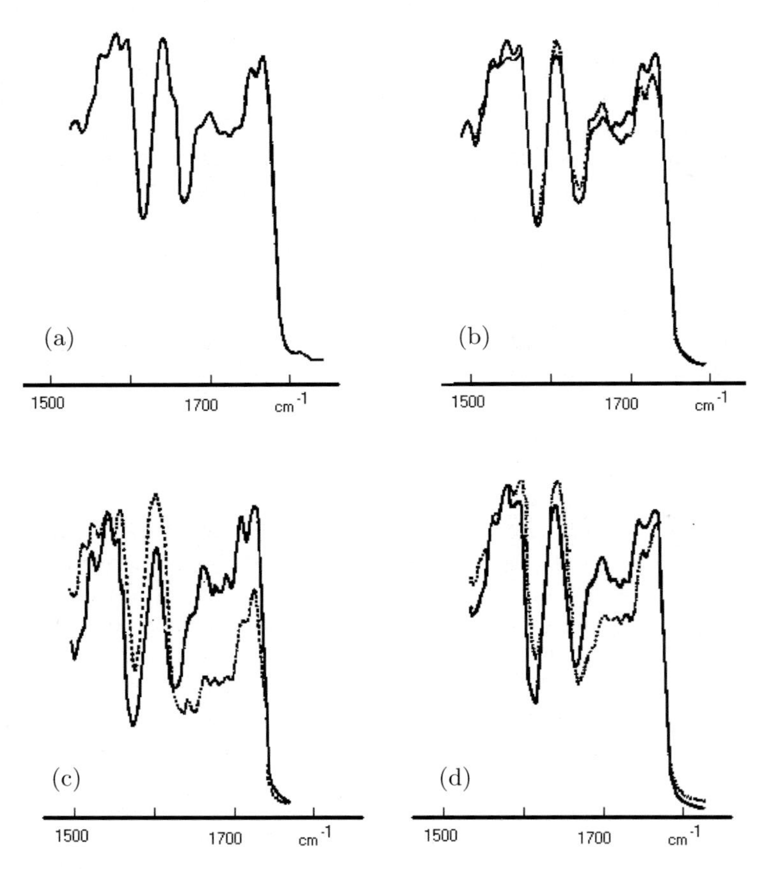

Figure 2. IR-spectra of the PU-9 sample: (a) unstrained, (b) uniaxially stretched to 50%, (c) stretched to 350%, (d) "relaxed" for 5 days; residual elongation of 40%. Perpendicular polarization of the IR absorption with respect to the stretch direction (solid lines); collateral polarization (dotted lines)

1525 cm^{-1} show perpendicular dichroism, the band at 1600 cm^{-1} (the benzene ring) is equidistant, the carbonyl bands at 1730, 1710, 1690, and 1680 cm^{-1} are perpendicular, and the bands at 1665 and 1650 cm^{-1} are equidistant. A quite different pattern is observed after 350% elongation (Figure 2c). The bands corresponding to vibrations of NH groups become equidistant, the band at 1600 cm^{-1} is also equidistant, while all the bands due to carbonyl groups are perpendicular. After relaxation for 5 days, the film shows a residual elongation of about 40% and the absorption bands (Figure 2d) are characterized by the same dichroism, but on a lower level.

The IR dichroism inversion is detected only for some absorption bands of the same chemical origin (NH or CO groups). This effect was extensively studied in [35] for a great number of block copolymers. Our data prove an analogous pattern of the IR dichroism inversion.

All the absorption bands in Figure 2 refer to vibrations in hard segments. Upon sample deformation, domains of hard blocks are initially oriented as a whole and the spectrum exhibits the orientation of the transition moments of certain vibrations, *i.e.*, the orientation of various groups relative to the stretching direction. Depending on the conformation of the hard block, these groups (urethane, semicarbazide, benzene rings) reveal their typical arrangement upon longitudinal orientation of this block. When a specific external loading is achieved, the glassy domain undergoes deformation as a result of a change in the conformation of hard blocks rather than to a change in the distance between the blocks (this would lead to its destruction). In this case, the reorientation of some groups is observed in the IR spectrum, while the other groups are characterized by longitudinal orientation.

3.1.2. *Small-angle X-ray scattering*

Polyurethane elastomers based on ADMH. All samples of the polyurethane elastomers studied (for synthesis see Section 2.6.1) are amorphous, since the poly(tetramethylene oxide) used in their synthesis loses its recrystallization ability within the urethane-containing polymers. For this reason, the areas of heterogeneity existing in the bulk of polyurethane elastomers are domains of hard blocks. As it follows from the small-angle X-ray scattering (SAXS) data (Table 5), sample PU(H) 1/1, which is characterized by structural heterogeneity,

Table 5. Parameters of the microheterogeneous structure of polyurethane elastomers synthesized at various PP:ADMH ratios

Sample	PP:ADMH molar ratio	$2\theta_{max}$ (min)	L (nm)	α_{segr}
PU(H) 1/1	1:1	25	21	0.23
PU(H) 2/1	2:1	30	17.5	0.17
PU(H) 5/1	5:1	32	16	0.29
PU(H) 10/1	10:1	42	12.5	0.54

shows only a weak ordering of the hard blocks (long period, $L \sim 21$ nm). The twofold decrease in the ADMH quantity (PU(H) 2/1) causes some decrease in the segregation level, α_{segr}, [37] of the soft and hard blocks. An improvement of the spatial arrangement of domains takes place, the latter forming a spatial macrolattice with a lower value of $L \sim 17.5$ nm. As it follows from the character of the changes in the X-ray scattering, the further decrease of the ADMH content (5- and 10-fold) results in the respective increase of α_{segr} of the soft and hard blocks in samples PU(H) 5/1 and PU(H) 10/1, and in the decrease of the long period of the spatial macrolattice of hard domains (Table 5).

One can draw conclusions on the processes of structure formation in the samples from the gradual increase of the scattering intensity (the area of the interference peak) and from the increasing scattering angle values ($2\theta_{max} \sim 25'$, $30'$, $35'$, and $42'$). Thus, the decrease in the ADMH amount results in a gradual increase of the segregation level. As a whole, the polyurethane elastomers based on ADMH are microheterogeneous amorphous systems, their degree of microphase separation being determined by the compatibility of soft and hard fragments of the microchain.

Polyurethanes containing crown ether fragments in the main chain. Information on the microphase morphology of polyurethane elastomers based on crown ethers (for synthesis see Section 2.5.1) obtained by SAXS measurements can be found in [13,16,33,34] (Table 6, Figure 3). The diffuse interference maxima

Table 6. Calorimetric and SAXS data for polyurethanes containing crown ether fragments in the main chain*

Sample	W	T_{g1} (K)	ΔC_{p1} (J/g deg)	T_{s2} (K)	L (nm)
PU-1	0.60	230	0.40	385	10.0
PU-6	0.49	220	0.20	380	19.3
PU-7	0.55	211	0.29	385	14.0
PU-8	0.54	—	—	—	14.0
PU-16	0.53	216	0.24	375	14.0

* W – soft segment weight fraction, T_{g1} – glass transition temperature of the soft segment, ΔC_{p1} – specific heat change at T_{g1}, T_{s2} – softening temperature of the hard segment phase, L – long period

in the SAXS curves obtained with polyurethane elastomers suggest the existence of spatial periodicity in the hard segment (HS) domains dispersed in a continuous matrix, rich in soft segments (SS). The relatively narrow size distribution of the HS domains suggests a statistical distribution of 4,4′-methylene *bis*(*p*-phenyl isocyanate), isophthalic acid dihydrazide, and disulfonyldibenzo-18-crown-6-dihydrazide fragments in the hard segments studied. This follows from the assumprion of equal reactivity of the NH_2 groups in the different dihydrazides with respect to the NCO groups of the prepolymer.

As it follows from the SAXS data for polyurethane elastomers (Figure 3), the intensity of the interference maxima in their scattering curves, the charac-

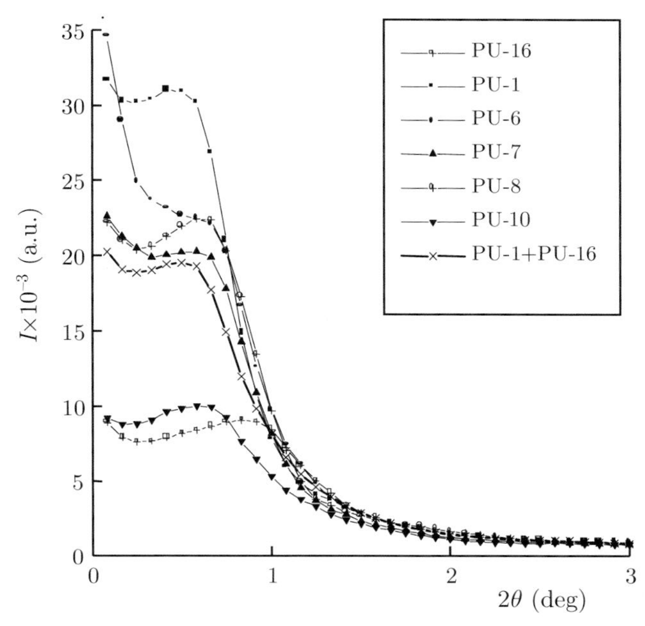

Figure 3. SAXS profiles for samples of some elastomeric polyurethanes

teristic level of local order of the HS in the microdomains, and hence the segregation level, α_{segr}, decrease in the order: PU-1 > PU-6 ≈ PU-7 ≈ PU-8 ≫ PU-10 (DADB18C6) ≈ PU-16 (DHIA) [38]. These results are similar to the DSC data for the polyurethane elastomers reported in [16], their interpretation leading to the following conclusions. Competition between the ether moieties in SS and carbonyl moieties in HS in the formation of intermolecular hydrogen bonds with the proton-donor parts of the HS polar groups is responsible for the disordering of the HS portion in the matrix of macroheterocyclic compounds of sample PU-16; this results in a relatevely low value of α_{segr} for this elastomer. It should be pointed out that the increase of the dimension of the macrolattice of hard domains in PU-6 does not suggest an additivity of the contributions of hard blocks with different structure, as verified by results obtained with the PU-1+PU-16 mixture. This last observation proves the validity of the assumption of "electrostatic equilibrium" of hard domains with increased electrostatic potential present in the macrolattice.

3.2. *Behavior of hydrazine-containing polyurethane thermoplastic elastomers in the process of complex formation*

3.2.1. *Complex formation in the systems polyurethane-sulfosemicarbazides–metal β-diketonates*

Complex formation studies in model systems and in polymer systems [26] by IR spectroscopy showed that intraspheric coordination of the carbonyl group takes place in the case of the polyurethane-semicarbazides (PU-16), and of the oxygen atom of the SO_2 group in the hydrazide fragment of the polyurethane-sulfosemicarbazides (PU-19) with respect to the metal atom, as well as extra-spheric coordination of the H-bond of the NH group to the atom of the first coordination sphere $M(AcAc)_n$. A complex dependence of the catalytic activity of the $M(AcAc)_n$ studied on the number of the 3d-electrons of the central atoms is found; the catalytic activity increases with the rise in the number of electrons for $M(AcAc)_n$ of bivalent metals and decreases in the case of trivalent metals.

For the system PU-19·$Fe(AcAc)_3$, identical variations of the frequency bands for the stretching vibrations of the SO_2 group are observed, with a simultaneous frequency shift of the M–O band and the acetylacetonate cycle in $Fe(AcAc)_3$, and with frequency variations of the band characterizing the stretching vibrations of the NH group. By the addition of $Zn(AcAc)_2$ to PU-19, the band of stretching vibrations of the NH group is shifted by 35 cm^{-1} to longer wavelengths, and absorption emerges in the range of 1550–1650 cm^{-1}; the intensity of the bands related to symmetrical and asymmetrical vibrations of the SO_2 group decreases (1160–1380 cm^{-1}).

In this respect, the ATR spectra taken from the surface of films provide even more reliable information. Upon consideration of the processes of complex formation, one should take into account that the oligoether fragments include electron-donor groups and complex formation by $M(AcAc)_n$ can also take place with the glycol component of the macrochain. UV spectroscopic data for the system PU-19·$Fe(AcAc)_3$ confirm this assumption, since variations in the intensity of electron absorption spectra are observed by the increase of the oligoether concentration, *i.e.*, complex formation takes place. Thus, the nature of the metal atom and the structure of the chain extender determine to a great extent the character of complex formation in polyurethane systems.

3.2.2. *Complex formation of crown ether-containing polyurethanes with metal cations*

The analysis of the IR spectra of crown ether-containing polyurethanes [39] (PU-1, PU-10, for synthesis see Section 2.5.1), after addition of KNO_3, suggests that the urethane groups do not participate in complex formation with K^+. Variations are registered in the region of bending vibrations (σ_{CH_2} and ν_{CH_2}) of the crown ether cycle; this observation is in agreement with the conclusion drawn in [40] that the K^+ cation changes its configuration when incorporated into the crown ether cavity. The nature of complex formation of the polyurethane samples with potassium perchlorate is studied by conductivity measurements. The conductivity of the potassium perchlorate solution decreases by the addition of polyurethanes containing dibenzo-18-crown-6. As pointed out in [3], this is caused by formation of slow-moving complex cations of larger size. According to [41], in the case of an alkali metal perchlorate, which is completely dissociated in DMF, the observed equivalent conductivity, Λ, is to be proportional to the dibenzo-18-crown-6/potassium perchlorate stoichiometric ratio.

The dependence of the observed equivalent conductivity on the crown ether fragment-to-alkali metal cation concentration ratio ($[CE]/[K^+]$) for the samples studied (Figure 4) shows that the stoichiometric ratio $[CE]/[K^+]$ equals unit [38].

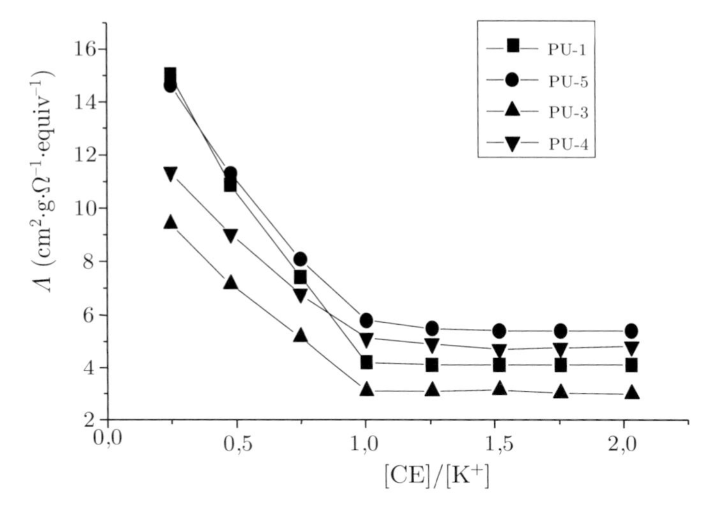

Figure 4. Dependence of the equivalent conductivity, Λ, of samples of crown ether-containing polyurethane elastomers on the dibenzo-18-crown-6/potassium perchlorate concentration ratio, DMF, 25°C

Table 7 shows the experimental data on the extraction ability of the polyurethane elastomers with DHSDB18C6 and HSB18C6 fragments in the main chain. Crown ether-containing polyurethane elastomers show a high extraction ability with respect to the alkali (Na^+ and K^+) and alkali earth (Mg^{2+} and Ca^{2+}) metal cations. The addition of sulfonylbenzo-15(18)-crown-5(6) hydrazides as terminal groups leads to an increase in this extraction by 13–17% [39]. The extraction ability of this type of elastomers with respect to metal cations is determined by complying with the principle of structural correspondence and by the "macroeffect" [3,12].

Table 7. Extraction ability of crown ether-containing polyurethane elastomers*

Sample	K^+	Na^+	Ca^{2+}	Mg^{2+}
PU-1	68.7	76.9	80.4	57.4
PU-3	87.5	97.8	93.7	96.7
PU-4	61.1	55.5	62.9	22.9
PU-6	55.7	84.6	67.1	69.0
PU-7	82.6	91	90.8	92.8
PU-8	74.1	97.1	97.6	94.1

*Picric acid 7×10^{-5} mole/L, metal hydroxide 1×10^{-2} mole/L, chain extender 3.5×10^{-4} mole/L, 25°C

In addition to the described cases of manifestation of the "macroeffect" in the extraction processes, at which the polyether fragment "wraps" the metal ion, with formation of a complex of the O,O-type [3,42], formation of complexes of the O,O- and O,N-type also occurs in the systems studied, with the participation of hydrazide fragments of the macromolecule [27], *i.e.*, intermolecular complexes "macromolecule–metal ion" (schematic model XXVI) and "macromolecule–metal ion–macromolecule" (schematic model XXVII) are formed.

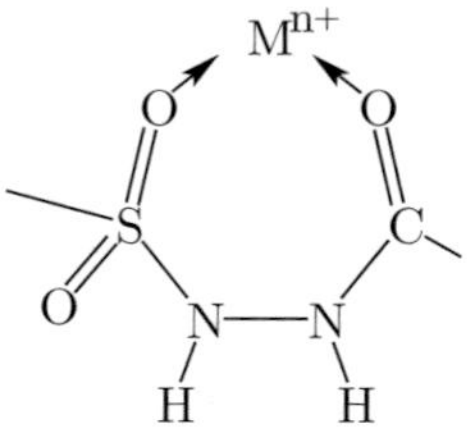

XXVI 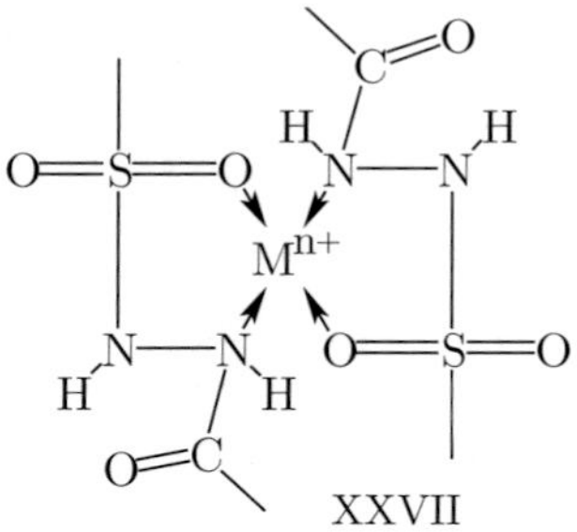 XXVII

One can evaluate the contribution of the "macroeffect" (schematic models XXVI, XXVII) by the extraction ability of sample PU-19 (polyurethane elastomer based on diphenylmethane disulfoacid dihydrazide). The value of the "macroeffect" is estimated to be 14.3% with respect to K^+. One should mention that the addition of 18-crown-6 as modifier to this polyurethane elastomer in amounts of up to 10 moles per 1 mole of polymer leads to an increase in the extraction ability to only 43.6%. In other words, the extraction ability does not obey the additivity principle, due to blocking of the crown ether cavity by the polymer matrix.

The incorporation in the macrochain of open-chain analogs of the crown ethers, such as diethylene glycols, triethylene glycols and poly(ethylene oxide)s makes it possible to estimate the contribution of the macroeffect to the complex formation, because of participation of the oxygen-containing fragments of the macrochain, with formation of the O,O-type complexes [3]. The maximal extraction ability value of such elastomers (free of crown ether cycles), with respect to Li^+, Sr^{2+}, and Ce^{3+}, is about 30%.

3.3. *Biological activity of hydrazine-containing polyurethane thermoplastic elastomers*

During the last decades, the destruction of various (polymer) materials by microorganisms became a rather important problem. The annual losses resulting from biodegradation worldwide amount to billions of dollars. More particularly, the polyurethane thermoplastic elastomers are subject to degradation induced by microorganisms because of the large number of simple and complex ether bonds in the macrochain that are unstable to their action. The preparation of thermoplastic elastomers stable to biodegradation is of great importance not only from the viewpoint of technology and economy, but also for ecological considerations.

The construction of a macromolecule including properly chosen elements and fragments [43] with synergistic effect makes it possible to impart a biological activity to polymer materials. One of the routes to the creation of such materials is the synthesis of polyurethane-sulfosemicarbazides modified by transition metal β-diketonates (see Section 2.7.1), as well as polyurethane elastomers based on macroheterocyclic compounds.

3.3.1. *Fungicidal activity of polyurethane-sulfosemicarbazides modified by transition metal β-diketonates and polyurethane elastomers based on crown ethers*

The fungicidal (fungistatic) activity of polymers is estimated according to [44] by application to indicator types of mold fungi. Polyurethane-sulfosemicarbazides modified by transition metal β-diketonates (Table 4) show high stability to the action of various mold fungi [45]. The highest fungicidal activity is observed with the elastomers containing zinc acetylacetonate.

Some of the samples of thermoplastic elastomers based on crown ethers show fungicidal activity. The polymers practically completely preserve their mechanical characteristics after the biological tests (Figure 5) [46]. The presence

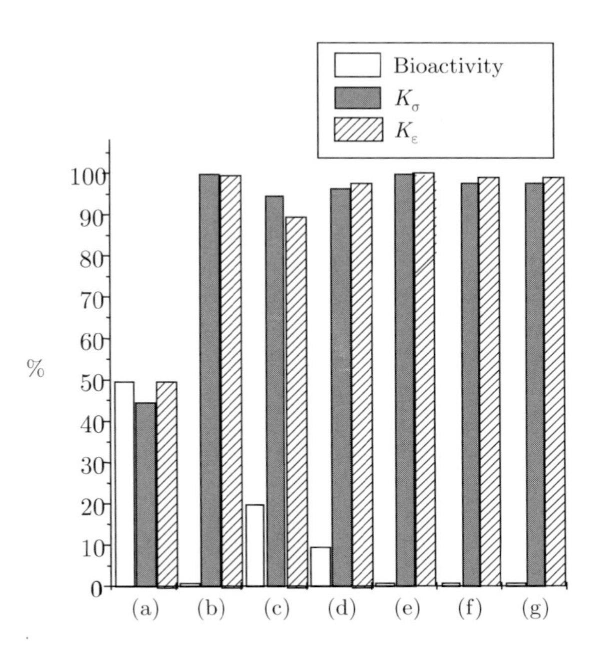

Figure 5. Bioactivity and mechanical stability (K_σ, K_ε) of polyurethane elastomers: (a) PU-16, (b) PU-1, (c) PU-3, (d) PU-8, (e) PU-6, (f) PU-7, (g) PU-5

in the PU structure of crown ether fragments is a necessary, but insufficient factor determining the biological activity. Thus, PU on the basis of SBDH18N6 (PU-1) has fungicidal properties, while the sample based on PU with DAB18C6 (PU-10) is characterized by a total absence of biological activity. This can be explained by the different accessibility of the biophores to the receptor caused mainly by differences in the supramolecular organization of the polymers.

According to the data of multiple ATR spectroscopy taken from the film surfaces, the concentration of the associated CO groups (connected with crown ethers fragments) on the surface is maximal for PU-1 and minimal for PU-10, *i.e.*, the accessibility of biophores and their activity in the polyurethane elastomers under consideration is different.

3.3.2. *Bactericidal activity*

The bactericidal activity of polyurethane elastomers containing crown ether hydrazides (samples PU-1, PU-7, and PU-8) was tested against pathogenic bacteria, *i.e., Staphylococcus aureus, Pseudomonas aeruginosa,* and *Bacillus Subtilis* (Figure 5). The antimicrobial effect of the elastomers was evaluated by the presence or absence of growth of microorganisms under sample discs and around them [47]. The evaluation of the bactericidal activity of the samples was carried out immediately after their preparation and after four years of storage in normal (*i.e.*, non-sterile) conditions [46]. All the polymers tested revealed antimicrobial properties [45]. Their bactericidal activity is preserved in time because of the presence of covalent bonds between the biologically active monomers and the macrochains.

3.3.3. *Toxicity*

The application of polyurethanes in any area of human activity should be preceded by an obligatory evaluation of their effect on humans.

Histotoxicity. The estimations of the histotoxicity [47] of the samples having bactericidal activity (see Section 3.3.2) have shown that their histotoxicity index is within the limits 0.78 ± 0.01–0.80 ± 0.01, *i.e.*, they can be considered non-toxic.

Allergenic effect. Tests of PU elastomer extractions on rabbits [38] have not revealed any pathological deviations. Skin irritation effects on rats (hyperemia, hypostasis, dry skin, peeling, formation of ulcers) were not found.

Hemolytic action. PU elastomers in concentrations of 0.9–1.3% do not cause hemolysis, *i.e.,* they do not include hemolytically acting agents and are not toxic [38].

3.3.4. *Physiological activity*

The possibility of controlling some structural parameters of block copolymers, such as free volume fraction, distance between the macrochains, chain packing and mobility, determining their gas-barrier properties creates the basis for the achievement of some beneficial effects on the physiological properties of plants.

Aqueous dispersions of ionic PU based on ADMH and isophthalic or terephthalic acid hydrazide (see Section 2.6.2) can form thin, water vapor- and carbon dioxide-permeable films on the mesophyle surface of plants. At the same time, these films reduce the evaporation of water from the leaves of agricultural crops (cereals, winter wheat in particular) under arid conditions; the transpiration intensity is decreased to a greater extent than photosynthesis intensity, *i.e.*, the efficiency of water use by winter wheat plants is increased (Table 8) [48].

Table 8. Influence of aqueous dispersions of ionomeric polyurethanes on the productivity and protein content in grains of winter wheat under arid conditions

Experimental conditions	Mass (g)					Protein content per grain (%)
	Leaves	Stem	Ear	Grains in Wagner's pot	1000 grains	
Optimum moisture content in soil (70%)	14.4±0.5	26.5±0.6	29.4±0.9	25.6±0.8	43.4±0.7	15.6±0.3
Aridity (30% moisture)	6.6±0.2	14.6±0.5	20.5±0.6	14.2±0.3	30.2±0.6	10.2±0.5
Aridity + PU	11.1±0.3	22.2±0.5	27.6±0.4	22.5±0.5	39.2±0.4	12.6±0.4

Under arid conditions, a single treatment of the winter wheat with an aqueous dispersion of ionomeric PU resulted in increased stability to drought of the plants upon earing up (the most sensitive to moisture losses vegetation stage), thus reducing the crop losses and enhancing the protein content per grain.

4. Conclusions

The development of polymeric materials with high functionality is among the most promising fields in the chemistry of macromolecular compounds. The low molecular macrochain extenders offer ample opportunities of variation of the macromolecular chemical structure, thus influencing the properties of the final polymeric materials. Hydrazine and its derivatives rank high in this respect, as it is confirmed in the present chapter. The combination of various valuable properties of hydrazine-based thermoplastic elastomers brings them in line with the best commercial polymers.

The further development of the chemistry and technology of polyurethanes will be undoubtedly defined mainly by the efforts in the creation of new types of hydrazine-containing polyurethanes: latexes revealing the specific properties of polyurethanes and ionic compounds, biologically active polyurethanes and composite materials for medical purposes on their basis, multifunctional polyurethanes.

Acknowledgement

The author gratefully acknowledges the assistance of Dr. Vitaly Ya. Veselov and his valuable comments on the original manuscript.

References

1. Saunders J H and Frish K C (1962) *Polyurethanes. Chemistry and Technology. Part I. Chemistry*, Interscience Publishers, John Wiley & Sons, New York-London.
2. Grekov A P, and Sukhorukova S A (1976) *Polymers Based on Hydrazine,* Naukova dumka, Kiev (in Rusian).
3. Hiraoka M (1982) *Crown Compounds. Their Characteristics and Applications*, Elsevier, Amsterdam-Oxford-New York.
4. Savelyev Yu V, Veselov V Ya, Grekov A P, Perekhod'ko G D, Koval E Z and Sidorenko L P (1993) Method for the obtaining of elastic polyurethanes, USSR Patent 1824409, to the Institute of Macromolecular Chemistry, National Academy of Sciences of Ukraine, Bul. 24 (in Russian).
5. Goncharova L B, Grekov A P, Batjaeva S I and Troev K D (1986) Phosphorus-containing cation-active polyurethane semicarbazides, *Ukr Khim Zh* **52**:1100–1104 (in Russian).
6. Urban D and Takamura K (Eds) (2002) *Polymer Dispersions and Their Industrial Applications,* Wiley-VCH Verlag GmbH, Weinheim.
7. Bechara I, Chang B H and Ilmenev P (2002) Cationic polyurethane dispersion and composition containing same, US Patent 6,339,125, to Crompton Corp.
8. Zhao C L, Wistuba E, Roser J, Fitzgerald P and Spitzer J (2001) Use of polymer dispersions as binding agents for sealing compounds and coating compounds, US Patent 6,242,515, to BASF AG.
9. Grekov A P and Yakovenko A G (1987) *Polyurethane Latexes, Physical Chemistry and Modification of Polymers*, Naukova dumka, Kiev, pp. 100–115 (in Russian).
10. Sukhorukova S A, Chumak L A, Grekov À P and Travinkaya T V (1988) Synthesis and examination of anion-active polyurethane latexes, *Vyssokomol Soedin Ser A* **30:**1206–1210 (in Russian).
11. Veselov V Ya, Savelyev Yu V and Grekov A P (1993) Polymeric sorbents of metal ions based on crown ethers and their linear analogs, *Compos Polym Mater* **55:**3–31 (in Russian).
12. Vögtle F and Weber E (Eds.) (1985) *Host-Guest Complex Chemistry of Macromolecules. Synthesis, Structures, Applications,* Springer Verlag, Berlin-Heidelberg-New York-Tokyo.
13. Savelyev Yu V, Akhranovitch E R, Grekov A P, Privalko E G, Korskanov V V, Shtompel V I, Privalko V P P, Pissis P and Kanapitsas A (1998) Influence of chain extenders and chain end-groups on the properties of segmented polyurethanes. 1. Phase morphology, *Polymer* **39**:3425–3429.
14. Pissis P, Kanapitsas A, Savelyev Yu V, Akhranovitch E R, Privalko E G and Privalko V P (1998) Influence of chain extenders and chain end-groups on the properties of segmented polyurethanes. 2. Dielectric study, *Polymer* **39**:3431-3435.
15. Georgoussis G, Kyritsis A, Pissis P, Saveleyev Yu V, Akhranovitch E R, Privalko E G and Privalko V P (1999) Dielectric studies of molecular mobility and microphase separation in segmented polyurethanes, *Eur Polym J* **35**:2007–2017.
16. Privalko V P, Shapoval R L, Privalko E G, Akhranovitch E R, Savelyev Yu V, Pissis P and Geogoussis G (1997) Influence of chain extenders and chain end-groups on

the properties of segmented polyurethanes. Steric immolization effect, *Reports Acad Sci Ukraine* N10:153–156.

17. Savelyev Yu V, Grekov À P, Akhranovitch E R and Veselov V Ya (1997) Polyurethanes containing macroheterocyclic fragments in the backbone as fungicidal materials and polymeric sorbents and method for their preparation, Ukrainian Patent 15147A, to the Institute of Macromolecular Chemistry, National Academy of Sciences of Ukraine (in Ukrainian).
18. Lopurev V A, Dolgushin G V and Voronkov M G (1998) Applied chemistry of the 1,1-dimethylhydrazine and its derivatives, *Zh Prikl Khim* **71:**1233–1248 (in Russian).
19. Savelyev Yu V, Kharitonova V K, Veselov V Ya and Perekhrest A I (2003) Polyurethanes based on asymmetrical dimethylhydrazine, *Reports Acad Sci Ukraine* N8:138–143 (in Russian).
20. Savelyev Yu V, Veselov V Ya and Grekov A P (2003) Method of obtaining of elastic polyurethanes, Ukrainian Patent 54533, to the Institute of Macromolecular Chemistry, National Academy of Sciences of Ukraine (in Ukrainian).
21. Savelyev Yu V, Khranovsky V A, Veselov V Ya, Grekov A P and Savelyeva O A (2003) Features of the interaction of 1,1-dimethylhydrazine with phenyl isocyanate, *Zh Org Khim* **39**:105–108 (in Russian).
22. Lee T C L and Pearce M E (1971) New polyamides via Ritter reaction, *J Polym Sci A-1* **9**:557–561.
23. Savelyev Yu V (2003) Private communication.
24. Savelyev Yu V, Levchenko N I, Grekov A P and Veselov V Ya (2003) Method for the obtaining of ionomeric water-soluble polyurethanes, Ukrainian Patent 53729, to the Institute of Macromolecular Chemistry, National Academy of Sciences of Ukraine (in Ukrainian).
25. Grekov A P, Savelyev Yu V and Veselov V Ya (1987) Effect of some transition metal β-diketonates on the mechanical properties and thermooxidative destruction of polyurethanes with hydrazine fragments, *Vyssokomol Soed Ser A* **29**:1904–1909 (in Russian).
26. Savelyev Yu V, Fedorenko O M, Grekov A P, Veselov V Ya and Khranovskij V A (1990) Processes of complex formation in polyurethanes containing hydrazine fragments with some transition metal β-diketonates, *Ukr Khim Zh* **56**:306–310 (in Russian).
27. Grekov A P, Savelyev Yu V, Veselov V Ya and Fedorenko O M (1990) Role of some transition metal β-diketonates for increasing of the photostability of linear polyurethanes, *Vyssokomol Soedin Ser B* **32**:499–503 (in Russian).
28. Timm Th (1984) Modern concepts of physical and chemical processes during thermal and thermooxidative destruction of polyurethane elastomers. Part 2. Chemical processes. III. Photooxidation, *Kautsch Gummi Kunstst* **37**:1021–1040 (in German).
29. Savelyev Yu V, Veselov V Ya and Grekov A P (1992) Influence of the metal β-diketonates on the radiolysis of polyurethanes, *Ukr Khim Zh* **58**:84-89 (in Russian).
30. Lipatov Yu S, Kercha Yu Yu and Sergeeva L S (1970) *Structure and Properties of Polyurethanes,* Naukova dumka, Kiev (in Russian).
31. Savelyev Yu V, Grekov A P, Akhranovitch E R and Veselov V Ya (1995) Investigations of the structure of novel crown ether-containing polyurethanes, *Reports Acad Sci Ukraine* N8:122–125 (in Russian).
32. Saveleyev Yu V, Khranovskij V A and Akhranovich E R (2002) Polyurethanes based on crown ether derivatives: study of spectral effects, *Polym Sci Ser A* **44**:2236–2240.
33. Savelyev Yu V, Grekov A P and Akhranovitch E R (1995) Crown ether-containing polyurethanes: structure, properties, *J Polym Sci B Polym Phys Ed* **37**:544–548.
34. Savelyev Yu V, Grekov A P, Akhranovitch E R, Veselov V Ya and Shtompel V I (1999) Features of the structure and some properties of polyurethanes with macroheterocyclic fragments, *Vyssokomol Soedin Ser A* **41**:534–538 (in Russian).

35. Khranovskij V A (1987) Conformational state and molecular organization of block copolyurethanes, DSc Thesis (Chemistry), Institute of Macromolecular Chemistry, Academy of Sciences of Ukraine, Kiev (in Russian).
36. Yakshin V V, Tsarenko N A, Zhukova N G and Laskorin B N (1992) Synthesis of polybenzocrown ethers on a styrene-divinylbenzene matrix, *Reports Acad Sci Russia* **325**:77–79 (in Russian).
37. Bonart R and Müller E H (1974) Phase separation in urethane elastomers as judged by low-angle X-ray scattering. I. Fundamentals, *J Macromol Sci* **10:**177–189.
38. Savelyev Yu V (2000). Polymers based on macroheterocyclic compounds and hydrazine derivatives, DSc Thesis, Institute of Macromolecular Chemistry, National Academy of Sciences of Ukraine, Kiev (in Ukrainian).
39. Savelyev Yu, Akhranovich E R, Kranovskij V A and Veselov V Ya (2003) Behaviour of polyurethanes with macroheterocyclic fragments in the processes of complex formation, *Vyssokomol Soedin Ser A* **45**:1–7 (in Russian).
40. Tsivadze A Yu, Varnek A A and Khutorskoj V E (1991) *Coordinational Compounds of Metals with Crown-Ligands,* Mir, Moscow (in Russian).
41. Takeda Y (1983) A conductance study of 18-crown-6 and dibenzo-18-crown-6 complexes with K^+ in various solvents, *Bull Chem Soc Japan* **56**:866–869.
42. Pomogajlo A V and Uflyand I E (1991), *Chemistry of Macromolecular Ligands,* Khimiya, Moscow (in Russian).
43. Barenbojm G M and Malenkov A G (1986) *Biologically Active Materials*, Nauka, Moscow (in Russian).
44. State Standard of the USSR 9.049 9.053-89 (1990) Materials, polymeric. Methods of laboratory experiments on their stability against mold fungi, State Committee on Standards of the USSR, Moscow (in Russian).
45. Savelyev Yu V, Grekov A P, Koval E Z, Sidorenko L P and Veselov V Ya (1997) Biostability of linear polyurethanes, *Ukr Khim Zh* **63**:70–73 (in Russian).
46. Savelyev Yu V (1997) Polyurethanes having biological activity, *Reports Acad Sci Ukraine* N11:147–151 (in Russian).
47. Galatenko N A, Konstantinov Yu B, Lujk A I, Maximenko V B and Oshkad'orov S P (1998) Methodological materials on toxicological-hygienic examinations of polymeric materials and products on their basis for medical applications, Official edition of the Ministry of Public Health of Ukraine, Kiev (in Ukrainian).
48. Grugoryk I P, Saveleyev Yu V, Mashkovskaya S P, Levchenko N I and Nikolajchuk V I (2003) Efficiency of the activity of ionomeric polyurethanes on the photosynthetic apparatus and productivity of winter wheat under arid conditions, *Letters Uzhgorod National University* N12:73–81 (in Ukrainian).

Chapter 14

Molecular Dynamics and Ionic Conductivity Studies in Polyurethane Thermoplastic Elastomers

P. Pissis, G. Polizos

1. Introduction

Segmented polyurethanes (PUs) are typical representatives of linear block copolymers of the type $(A\text{-}B)_n$ and an important class of thermoplastic elastomers. They are composed of short alternating blocks of soft (SS) and hard segments (HS). The SS impart elastomeric character to the copolymer, whereas the HS form a solid phase (HS microdomains) through intermolecular association and impart dimensional stability to the array of macromolecules. Low molecular mass polyethers and polyesters are typically used as SS, while HS normally consist of an aromatic diisocyanate that has been chain extended with a low molecular weight diol or diamine to form an oligomeric aromatic urethane or urethane urea segment [1–3].

Polyurethane thermoplastic elastomers (PTE) are characterized by microphase separation into an SS microphase and HS microdomains. In fact, the versatile physical and thermomechanical properties of TPE, which make them attractive for several technological applications, are based on their microphase-separated structure. The glass transition temperature, T_g, of the amorphous SS microphase is typically below 0°C, so that at room temperature the material behaves as an elastomer. At high temperatures (typically above 100°C), dissociation of the physical bonds occurs, the HS microdomains are destroyed and the material flows as a linear polymer. In addition to microphase separation, intermolecular hydrogen bonding and partial crystallization of the SS microphase often contribute to the thermoplastic elastomeric behavior of PTE.

The microphase separation in PTE results from the thermodynamic incompatibility of HS and SS. However, the microphase separation is not complete, even in the case of fully incompatible segments, due to the existence of covalent bonds between them [4]. The factors affecting the degree of microphase separation (DMS) include the segment polarity and length, the crystallizability of either segment, the overall composition and the flexibility of HS [5,6]. According to Koberstein *et al.* [7], a critical HS length is required for microphase seraration, whereas DMS generally improves with the increase in HS content. Chu *et al.* [8] pointed to the significance of the kinetic factor (in addition to the thermodynamic one) for the phase structure formation.

The investigation of structure-property relationships, *i.e.*, of the relationships between composition, processing, structure/morphology, molecular dynamics and final properties, is a fundamental issue in materials science. The profound understanding of these relationships is a prerequisite for the optimization of the composition and the processing, resulting in materials with predicted properties tailored to specific applications. Much effort has been devoted to the investigation of these relationships in PTE and a variety of experimental techniques have been employed to this aim. Dielectric techniques are widely used in such investigations to provide information on molecular dynamics [9–11]. The main advantage of dielectric techniques, as compared to other techniques for measuring the molecular mobility, is the extremely broad frequency range, typically 10–12 decades, which can be covered relatively easily [12]. In addition, ionic conductivity measurements, typically combined with dielectric measurements, provide information on molecular dynamics, since the motion of ions is often governed by segmental dynamics [10]. Moreover, ionic conductivity studies can be utilized for morphological characterization, as the moving ions probe the local morphology [13].

Over the last years we employed dielectric techniques and ionic conductivity measurements to investigate molecular dynamics in several PU systems, in close collaboration with Prof. V. P. Privalko, Dr. Yu. V. Savelyev, Prof. V. V. Shilov, Prof. E. V. Lebedev, Dr. Ye. P. Mamunya and their coworkers at the Institute of Macromolecular Chemistry at the National Academy of Sciences of Ukraine in Kiev, typically in the framework of the investigation of structure-property relationships combining several experimental techniques. The systems investigated include neat PTE with a large variety of materials used as SS, HS and chain extenders, PUs blended with a second polymer, composites and nanocomposites with a PU matrix, interpenetrating polymer networks based on PUs (full, semi, thermoplastic, gradient interpenetrating polymer networks) and PU ionomers. The present chapter summarizes our results with selected systems, with particular emphasis on PU ionomers, where ionic conductivity studies may be of direct relevance for practical applications. In addition to a systematic presentation and discussion of the results obtained with various PTE, we focus on selected examples to illustrate the potential of dielectric and ionic conductivity methods for molecular dynamics studies in complex multicomponent systems on the basis of methodologies developed to that aim.

The organization of the present chapter is as follows. Dielectric techniques for molecular dynamics studies, in particular broadband dielectric spectroscopy (DS) and thermally stimulated depolarization currents (TSDC) techniques are shortly presented in the next section. Section 3, devoted to ionic conductivity measurements and analysis, focuses mostly on analysis, as the measuring techniques and equipment are often similar to those used for DS. The micro-phase separation and morphology of segmented PUs is discussed in the following Section 4, which completes the first introductory part of the chapter. Results obtained with selected PTE are presented in Section 5, followed by a larger Section 6 devoted to PU ionomers with ionic moieties in either of the HS and SS. PU ionomers of the latter type are often based on poly(ethylene oxide) (PEO) as the SS component and, for this reason, Section 6 includes a discussion of telechelics based on PEO, which may serve as model systems for PU ionomers. In Section 7, we discuss recent results obtained with nanocomposites based on PTE, a topic attracting much current interest, before we conclude with Section 8.

2. Dielectric techniques for molecular dynamics studies

2.1. *Broadband dielectric spectroscopy*

Dielectric techniques are a powerful tool for studying molecular dynamics in various materials. Their main advantage over other techniques of measuring molecular dynamics is the extremely broad frequency range covered, which extends from about 10^{-5} to about 10^{11} Hz [14–16]. Obviously, this broad frequency range cannot be covered by a single technique.

In most cases, the measurements are carried out isothermally in the frequency domain and the terms dielectric spectroscopy (DS) and dielectric relaxation spectroscopy (DRS) are then used. Other terms frequently used for DRS are impedance spectroscopy and admittance spectroscopy. Impedance spectroscopy is usually used in connection with electrolytes and electrochemical studies, whereas admittance spectroscopy often refers to semiconductors and devices. Isothermal measurements in the time domain are often used, either as a convenient tool for extending the range of measurements to low frequencies (slow time-domain spectroscopy, dc transient current method, isothermal charging-discharging current measurements) or for fast measurements corresponding to the frequency range of about 10 MHz – 10 GHz (time-domain spectroscopy or time-domain reflectometry). Finally, TSDC is a special dielectric technique in the temperature domain, which will be discussed in Section 2.2.

For measurements in the frequency domain, capacitance bridges, impedance analyzers, frequency response analyzers, radio-frequency reflectometers and network analyzers are typically employed. Figure 1 shows schematically the frequency range of dielectric measurements covered by different techniques and equipments [17]. The principle of these measurements is as follows. The sample

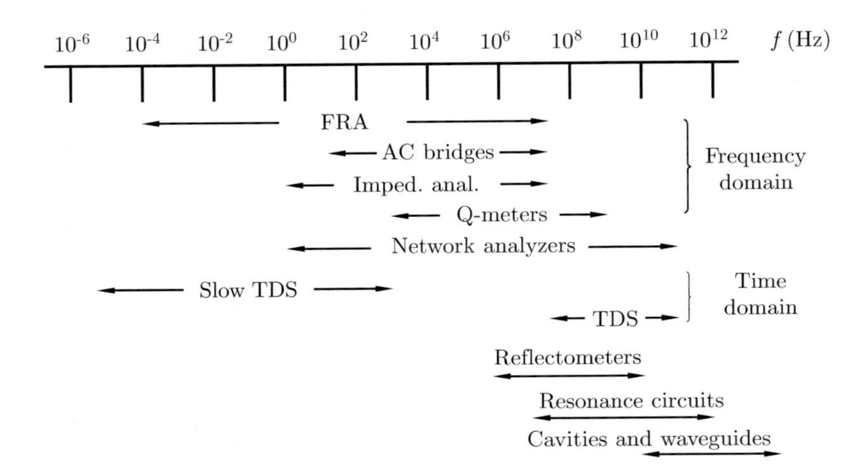

Figure 1. Techniques and equipment for dielectric measurements. FRA means frequency response analyzer, TDS is time domain spectroscopy

under investigation is placed in a capacitor with empty capacitance C_0, which becomes a part of an electric circuit. A sinusoidal voltage with angular frequency ω is applied to the circuit and the complex impedance, $Z(\omega)$, of the sample is measured. The complex dielectric permittivity $\varepsilon\ (\omega) = \varepsilon'(\omega) - \mathrm{i}\varepsilon^2(\omega)$, defined by

$$\varepsilon(\omega) = \frac{C}{C_0} \tag{1}$$

where C is the capacitance of the filled capacitor, is then obtained from

$$\varepsilon(\omega) = \frac{1}{i\omega Z(\omega) C_0} \tag{2}$$

In slow time domain spectroscopy, a voltage step V_p is applied to the sample and the polarization or depolarization current $I(t)$ is measured as a function of time. The time-dependent dielectric permittivity $\varepsilon(t)$ is then given by

$$\varepsilon(t) = \frac{C(t)}{C_0} \quad \text{and} \quad \frac{d\varepsilon}{dt} = \frac{I(t)}{C_0 V_p} \tag{3}$$

Usually, the depolarization current is measured to avoid the dc conductivity contribution. The dielectric relaxation spectrum is then obtained by Fourier transform or approximate formulas, *e.g.*, the Hamon approximation [14]. By carefully controlling the sample temperature and accurately measuring the depolarization current, precision measurements of the dielectric permittivity down to 10^{-6} Hz are possible [18]. In fast time domain spectroscopy or reflectometry, a step-like pulse propagates through a coaxial line and is reflected from the sample section placed at the end of the line. The difference between the reflected and the incident pulses recorded in the time domain contains the information on the dielectric properties of the sample [19,20].

In addition to dielectric permittivity, ε, which is the most physically meaningful dielectric function to describe the material response, under the condition that in the experiments the electric field is the independent variable and the charge is the dependent one (*i.e.*, ε is a compliance), the electric modulus M is often employed to analyze relaxation phenomena in ionic conductors [21].

$$M = \frac{1}{\varepsilon} = M' + iM'' = \frac{\varepsilon'}{\varepsilon'^2 + \varepsilon''^2} + i\frac{\varepsilon''}{\varepsilon'^2 + \varepsilon''^2} \tag{4}$$

The formalisms of $\varepsilon(\omega)$ and $M(\omega)$ are equivalent. Transformation from one to another may emphasize and, therefore, help resolve particular aspects of the relaxation process (as demonstrated later in this chapter), but no new information can be extracted.

Independently of the specific dielectric technique used, the results of dielectric measurements are usually analyzed in the form of complex dielectric permittivity $\varepsilon(\omega) = \varepsilon'(\omega) - ie^2(\omega)$ at constant temperature by fitting empirical relaxation functions to $\varepsilon(\omega)$. In the examples to be given later in this chapter, the two-shape-parameter Havriliak-Negami (HN) expression [22]

$$\varepsilon(\omega) - \varepsilon_\infty = \frac{\Delta\varepsilon}{[1 + (i\omega\tau)^{1-\alpha}]^\beta} \tag{5}$$

is fitted to the experimental data for a relaxation mechanism. In this equation, $\Delta\varepsilon$ is the dielectric strength, $\Delta\varepsilon = \varepsilon_s - \varepsilon_\infty$, where ε_s and ε_∞ are the low- and high-frequency limits of ε', respectively, τ is the relaxation time, $\tau = 1/2\pi f_{HN}$, where f_{HN} is a characteristic frequency closely related to the loss peak frequency f_{max}, and α, β are the shape parameters describing the shape of the $\varepsilon''(\omega)$ curve below and above the frequency of the peak, respectively, $0<\alpha\leq 1$ and $0<\beta\leq 1$. This expression becomes the single Debye form for $\alpha = 0$, $\beta = 1$, the symmetric Cole-Cole form for $\alpha \neq 0$, $\beta = 1$, and the asymmetric Cole-Davidson form for $\alpha = 0$, $\beta \neq 1$ [22,23]. A proper sum of HN expressions is fitted to $\varepsilon(\omega)$ in the case of more than one overlapping mechanisms plus a term for the contribution of conductivity, if the latter has a contribution at the temperature of measurements [10,13]. For each relaxation mechanism, there are then three sources of information: the time scale of the response (τ or f_{max}), the dielectric strength ($\Delta\varepsilon$), and the shape of the response (α, β). By measuring $\varepsilon(\omega)$ at several temperatures, the time scale of the response is analyzed in terms of the Arrhenius equation for secondary relaxations and the Vogel-Tammann-Fulcher (VTF) equation for the primary α relaxation, and valuable information on the activation parameters is obtained [10,13]. Examples will be given later in this chapter.

The dielectric permittivity $\varepsilon(\omega)$ describes the material response to the application of an alternating electric field $E(\omega)$. For small electric field strengths, a linear relationship holds between E and the polarization P

$$P(\omega) = (\varepsilon(\omega) - 1)\varepsilon_0 \, E(\omega) \tag{6}$$

where ε_0 is the permittivity in vacuum [14]; $\varepsilon(\omega)$ is related by the theory of dielectric relaxation to the correlation function $\Phi(t)$ of the polarization fluctuations [12,17],

$$\frac{\varepsilon(\omega) - \varepsilon_\infty}{\varepsilon_s - \varepsilon_\infty} = \int_0^\infty \left[-\frac{d\Phi(t)}{dt} \right] \exp(-i\omega t) dt \tag{7}$$

and

$$\Phi(t) = \frac{< \Delta P(t) \Delta P(0) >}{< \Delta P(0)^2 >} \tag{8}$$

where ΔP denotes a fluctuation of the polarization around its equilibrium value and the brackets denote the averaging over an ensemble or time t.

Next to DS, several other experimental techniques, such as mechanical spectroscopy, nuclear magnetic resonance, neutron scattering, dynamic light scattering and Raman spectroscopy, are often being employed to investigate molecular dynamics in polymeric systems. When comparing the results obtained with the same material and different techniques, one should take into account the difference in time scale and the difference in local (spatial) scale. The first is obvious: *e.g.*, the main chain relaxation (α relaxation) related to the glass transition is measured at higher temperatures in high-frequency DRS than in low-frequency dynamic mechanical analysis. The latter deserves a closer examination and is the object of intense investigation: different spectroscopies look at different species and thus probe the mobility at different spatial scales. At first glance, discrepancies among the results of different spectroscopies on the same material turn out to be very informative in identifying the molecular units with probed mobility [24].

2.2. *Thermally stimulated depolarization currents techniques*

The TSDC method is a dielectric method in the temperature domain, which allows for a fast characterization of the dielectric response of the material under investigation. The method consists of measuring the thermally activated release of stored dielectric polarization. It corresponds to measuring dielectric losses against temperature at constant low frequencies of 10^{-2}–10^{-4} Hz [25,26]. The low equivalent frequency is a characteristic feature of the TSDC method, which is often used to extend the range of dielectric measurements down to low frequencies. In this method, the sample is inserted between the plates of a capacitor and polarized by the application of an electric field E_p at temperature T_p for time t_p, which is long, as compared to the relaxation time at T_p of the dielectric dispersion under investigation. With the electric field still applied, the sample is cooled to temperature T_0 (which is sufficiently low to prevent depolarization by thermal excitation) and then is short-circuited and reheated at a constant rate b. A discharge current is generated as a function of temperature, which is measured with a sensitive electrometer. The resultant

TSDC spectrum often consists of several peaks, their shapes, magnitudes and locations providing information on the time scale and on the dielectric strength of the various relaxation mechanisms present in the sample [25]. In contrast to DS (isothermal measurements in the frequency domain), the stages of polarization and depolarization (stimulus and response) are separated in the TSDC method. This is beneficial with respect to the conductivity effects in ionically conducting polymers, where dipolar processes (typically the α relaxation associated with the glass transition) are often overlapped by the ionic conductivity in DS measurements, but not in TSDC measurements [27]. The method is characterized by a high sensitivity and, owing to its low equivalent frequency [25], by a high resolving power. In addition, it provides special variants of the experimental analysis of complex relaxation mechanisms into approximately single responses [25–27]. Examples will be given later in this chapter.

3. Ionic conductivity measurements and analysis

For ionic conductivity measurements, the same equipment may be used as for the dielectric measurements described in Section 2. The term impedance spectroscopy is preferably used in this case instead of DS, the latter being reserved mostly for dipolar dielectric materials. In addition to characterizing the bulk electrical properties of the materials, impedance spectroscopy includes also the investigation of effects at the interfaces between the sample and the electronically conducting electrodes [16]. In recent years, the frequency range of the ionic conductivity measurements has been significantly extended to higher frequencies, spanning more than 17 decades and ranging from less than 10^{-3} Hz to more than 10^{14} Hz. For this broad frequency range, conductivity spectroscopy is a better term [28]. It includes, in addition to the traditional impedance regime, which is below a few MHz, the radio regime, which ranges from a few MHz to a few GHz, the microwave regime, which ranges from a few GHz to about 150 GHz, and the far-infrared regime, which is above 150 GHz. Electrodes are used only in the first regime, whereas guided and unguided electromagnetic waves are employed in the second, the third, and the fourth regime, respectively [28]. Results to be presented and discussed later in this chapter refer exclusively to the impedance regime.

In the basic impedance spectroscopy experiment, an electrical stimulus (a known voltage or current) is applied to the electrodes and the response (the resulting current or voltage) is measured [16]. Similar to dielectric measurements (Section 2), in most cases the measurements are performed in the frequency domain by applying a single-frequency voltage to the sample under investigation and measuring the phase shift and amplitude, or real and imaginary parts, of the resulting current at that frequency. The information obtained from impedance measurements refers either to the bulk material (conductivity, mobilities of charges, generation-recombination rates, *etc.*) or to the material-electrode interface (capacitance of the interfacial region, adsorption-reaction rate constants, *etc.*). In this chapter, our interest is focused on the investigation of the bulk material properties. Special methodologies of the measurements and data analysis allow to separate bulk from interfacial effects [16].

The results of impedance spectroscopy measurements are often presented in the (complex) impedance plot. Referring again to the basic experiment of applying a sinusoidal voltage $V(t) = V_0 \sin(\omega t)$ to the sample (Section 2), we record now the resulting steady state current $I(t) = I_0 \sin(\omega t + \varphi)$ and define the impedance $Z(\omega) \equiv V(t)/I(t)$ with magnitude $|Z(\omega)| = V_0/I_0(\omega)$ and phase angle $\varphi(\omega)$. The impedance is plotted in the complex plane (Argand diagram) as a planar vector using rectangular and polar coordinates (Figure 2)

$$\begin{gathered} Z(\omega) = Z' + iZ'' = |Z|\exp(i\varphi) \\ Z' = \mathrm{Re}(Z) = |Z|\cos\varphi \\ Z'' = \mathrm{Im}(Z) = |Z|\sin\varphi \\ \varphi = \tan^{-1}(Z''/Z') \\ |Z| = [(Z')^2 + (Z'')^2]^{1/2}. \end{gathered} \tag{9}$$

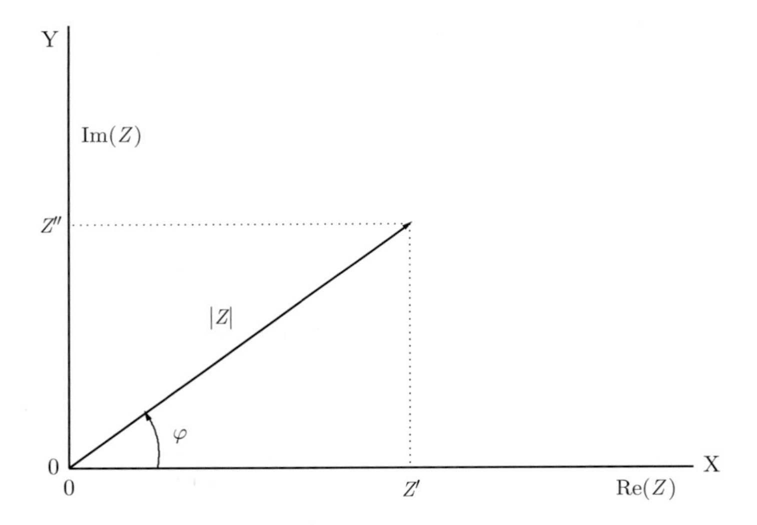

Figure 2. Presentation of the impedance, Z, using rectangular and polar coordinates

In conventional impedance spectroscopy, Z is measured (nowadays mostly automatically) as a function of frequency over a wide frequency range in the linear regime (small voltage signal for linear response). Plots of $Z(\omega)$ against ω provide then information on the electrical properties of the material under investigation. Very popular and comprehensive are the plots of $-\mathrm{Im}(Z)$ (or $\mathrm{Im}(Z)$, if we write $Z(\omega) = Z' - iZ''$ instead of Eq. (9)) against $\mathrm{Re}(Z)$ with the frequency ω as a parameter (complex impedance plots), which provide (often by extrapolation) directly the value of the resistance and, by knowing the geometrical characteristics of the sample, the value of the dc conductivity, σ_{dc}, at the temperature of measurements. Examples will be given later in this chapter.

When the emphasis is put on the bulk electrical properties of the material under investigation in comparison with the predictions of theoretical models, ac conductivity plots, *i.e.*, plots of ac conductivity, σ_{ac}, against frequency, ω, at constant temperature, are well suited for presenting and discussing the results of the ionic conductivity measurements [16,28]. The data are recorded isothermally with variation of the frequency, ω, and $\sigma_{ac}(\omega)$ is calculated from these data (in fact, the real part, σ'_{ac}, of the complex conductivity), *e.g.*, in the admittance presentation (Eq. (2))

$$\sigma_{ac}(\omega) = \omega\ \varepsilon_0\varepsilon''(\omega) \tag{10}$$

Jonscher suggested the following power law dependence (universal dynamic or dielectric response) [23]

$$\sigma_{ac}(\omega) = \sigma_{dc} + A\omega^s, \qquad 0.5 < s < 0.7 \tag{11}$$

where A and s are temperature dependent parameters. Equation (11) and its modifications are frequently used for fitting to the experimental data. Examples will be given later in this chapter.

4. Microphase separation and morphology of segmented polyurethanes

The solid state properties of PUs are determined to a great extent by their microphase-separated structure. Much effort has been devoted and a variety of experimental techniques, including mainly small-angle X-ray scattering (SAXS), differential scanning calorimetry (DSC), infra-red spectroscopy, DS, and dynamic mechanical thermal analysis, have been employed to investigate the microphase separation and morphology in PUs.

The polyaddition reaction bonding HS to SS by means of chain extenders is a stochastic process. This aspect, as well as the rather wide molecular weight distributions of component segments, is the natural explanation for both a broad dispersion of the size of HS microdomains separated by a continuous SS phase, and a fairly low DMS [29,30]. Runt and coworkers pointed out that poly(urethane urea) multiblock copolymers have relatively low overall DMS (20–40%), in contrast to the common notion that these copolymers are well phase-separated materials [31].

In a series of papers, Koberstein *et al.* employed SAXS and DSC to investigate the microphase separation and morphology in segmented PUs based on poly(oxypropylene) end-capped with poly(oxyethylene), 4,4′-diphenylmethane diisocyanate (MDI), and 1,4-butanediol (BD). DMS was found to decrease with decreasing HS content and/or decreasing HS length; when the latter decreased below a critical value, corresponding to a chain with 3–4 MDI residues, the HS were assumed to be dissolved within the continuous SS phase [7,32]. The intrinsic flexibility of HS was found to have also significant influence on the morphology of PUs: the fairly high DMS in segmented PU, prepared from polycaprolactone, hexamethylene diisocyanate (HDI) and hydrogenated bisphenol A, was reduced to nearly zero by simply replacing the flexible HDI by the stiffer MDI [33]. Also in model segmented PUs from either poly(propylene glycol) (PPG) or poly(butylene adipate) (PBAD) and three different diisocyanates (all-*trans*-4,4′-dicyclohexylmethane diisocyanate, 100% *t,t*-HMDI; HMDI with 20% of trans isomers, 20% *t,t*-HMDI; MDI) DMS, investigated by SAXS, DSC, DS and TSDC, was found to increase systematically in the order MDI, 20% *t,t*-HMDI, 100% *t,t*-HMDI [6,34].

The typical morphologies observed for the HS microdomains in segmented PUs are ellipsoids and lamellae [7,33]. In polyolefin-based PUs [9,35], the non-polar SS force the polar HS to be well organized in HS microdomains due to inter-urethane hydrogen bonding. The presence of a mixed interfacial region was detected in these PUs, which increases with increasing HS content [9,35] and decreases with increasing emulsifier content [35]. Nuclear magnetic resonance was employed to examine the existence of the interfacial region between the crystalline and rubbery regions in PU elastomers prepared from *p*-phenylene diisocyanate, poly(tetramethylene oxide) (PTMO) and BD [36]. Block copolymers were used as emulsifiers to control and minimize the DMS during synthesis in the bulk [35]. Different preparation routes were found to induce various characteristics and tensile properties of the corresponding PUs attributed mainly to premature microphase separation occurring during the polymerization and the competition between the phase separation rate and the polymerization rate. Dielectric spectroscopy was employed to follow the process of microphase separation in these PUs [35].

Several researchers investigated the influence of the type and molecular weight (MW) of the polyether or the polyester used on the final morphology of PUs. DMS was found to increase with increasing MW, *i.e.*, with increasing chain length, of the polyether used in the cases of PPG [29] and poly(tetramethylene glycol) (PTMG) [37]. DSC and SAXS measurements showed that DMS in polyesterurethanes increases with increasing block length, the shortest block lengths being almost single-phase [38]. Several researchers stressed the significant influence of the processing conditions, in addition to that of the chemical structure, on the multiphase structure of PUs. Runt and coworkers employed SAXS to investigate the dependence of the microphase separation in poly(urethane urea) block copolymers on the preparation conditions, which were varied by adjusting the temperature and vacuum during solution casting [39]. In PUs based on MDI, BD and several polyethers and polyesters, DMS first increased with increasing annealing temperature from room temperature, reached a maximum at *ca.* 107°C, and then decreased [40]. In these PUs, the interdomain spacing was found to increase with increasing HS content and with increasing annealing temperature [40]. Different HS structures were observed in melt-crystallized and compression-molded specimens of crystallizable PUs by simultaneous synchrotron SAXS-DSC experiments [41]. Thermal treatment was found to induce a coarsening of HS microdomains and an increase in the interfacial region, leading, to some extent, to a weakening of the mechanical properties, as observed above the glass transition temperature of the SS matrix [42]. The polymorphism of the HS crystallites of a commercial PU was investigated as dependent on the melt processing conditions by means of wide-angle X-ray scattering (WAXS) and DSC, the results allowing to establish a processing-structure-property relation [43].

Ryan and coworkers investigated the effect of chain extenders on the morphology of flexible PU foams by using SAXS, Fourier transform infrared spectroscopy and rheometry [44]. By the addition of chain extenders, the onset of microphase separation was delayed and the interdomain spacing was increased. The effect was found to strongly depend on the chemical structure of the chain extenders and the compatibility between chain extenders and urea segments [44]. Runt and coworkers investigated the phase morphology in poly(urethane urea)s, in which diamines were used as the chain extenders rather than polyols [45]. The overall phase separation was significantly reduced when ethylenediamine (chain extender) was replaced by a mixture of ethylenediamine and 1,4-diaminocyclohexane. The microphase structure and the microphase separation kinetics were investigated by means of synchrotron SAXS in PUs with MDI and BD as the HS and PTMO and PPO-PEO as the SS [8]. A more complete microphase separation was observed in the PTMO-based sample, which could be explained as partially due to a kinetic factor. The microphase separation of a segmented PU, which was quenched from the homogeneous melt state to lower temperatures, was found to behave like a relaxation process, with a rate depending strongly on the annealing temperature [46]. Kang and Stoffer showed that the kinetic factor (HS mobility and the viscosity of the system) can control the morphology of some PU systems [47].

The significance of microphase separation and morphology for controlling the final properties of PUs has been a strong motivation for developing methods to calculate the microphase composition and to characterize the DMS; the early developments were based solely on DSC [48]. Koberstein and Leung [32] proposed a method based on a combined analysis of DSC and SAXS experiments and the use of a modified Fox equation for the glass transition behavior of mixed microphases. The results of these calculations were found to show trends, which are consistent with SAXS and DSC data in PUs synthesized from MDI, BD and PPO-PEO [32]. The same method was successfully applied in the case of PU ionomers, with the ions being incorporated in the SS microphase [49]. Chen *et al.* [4] proposed a simple method to calculate the microphase composition and DMS, provided that the phase-separated morphology does not affect the microdomain glass transition behavior. However, as these authors stated, there is experimental evidence against this assumption [4].

5. Neat polyurethane thermoplastic elastomers

DRS and TSDC were systematically employed in a series of recent papers to investigate the molecular dynamics and ionic conductivity in neat PTE. In many cases, the results were combined with those of morphological characterization techniques, in particular SAXS and DSC, and of water sorption/diffusion measurements, aiming to contribute to a better understanding of the structure-property relationships [5,6,10,13,34,50]. Several parameters of PU preparation were systematically varied in these investigations, including the type of SS and HS [6,34] and of chain extender and chain end-groups [5,10,50].

TSDC studies proved very powerful for characterizing the various mechanisms of molecular mobility in PUs. In order to assist the discussion of the TSDC data in this and the following sections, Figure 3 shows a TSDC thermogram typical of PUs. It has been recorded with a PU based on PTMG (MW 1000 g/mole), MDI and a mixture of dihydrazide of isophthalic acid and 1,4-di-N-oxy-2,3-bis-(oxymethyl)-quinoxaline as chain extender [5]. Four dispersions (TSDC peaks) are well discerned in Figure 3 due to unfreezing of

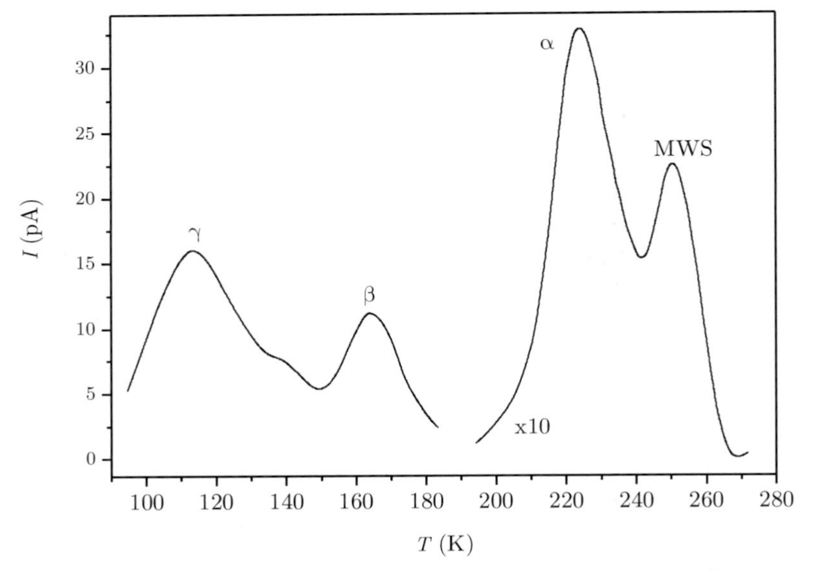

Figure 3. Typical TSDC thermogram for a polyurethane sample

molecular motions during heating of the previously polarized sample. The first two peaks, γ and β in the order of increasing temperature, are due to secondary relaxations, associated with local motions of chain segments. In agreement with previous work [51,52], γ is associated with crankshaft motions of the $(CH_2)_n$ sequences with attached dipolar groups, whereas β is attributed to the motion of the polar carbonyl groups with participation of attached water molecules. The third peak (α peak) is dipolar and located close to the calorimetric glass transition temperature, T_g. In fact, the peak temperature T_α of the TSDC α peak in polymers is a good measure of T_g [24]. For PUs, the α peak arises from the reorientation of the polar soft chain segments during the glass transition of the SS-rich microphase [10]. The fourth peak above T_g is due to Maxwell-Wagner-Sillars (MWS) interfacial polarization between the SS-rich microphase and the HS-rich microdomains, *i.e.*, it is related to morphology changes at the glass transition [10,50]. This interpretation is consistent with the results of measurements with different polarizing fields or with blocking electrodes (thin insulating foils between the sample and the capacitor plates), which indicate that all four peaks are due to bulk (volume) polarization [13].

Figure 4 shows representative DRS results obtained with a PTE based on PTMG, MDI and asymmetrical dimethylhydrazine (DMH) as chain extender over a wide range of frequencies and temperatures [53]. The glass transition temperature of this sample is $T_g = 217$ K, as revealed by DSC measurements. At low temperatures, a broad secondary relaxation is observed as a step in $\varepsilon'(f)$ and as a peak in $\varepsilon''(f)$ (centered at about 10^4 Hz at 173 K), attributed to

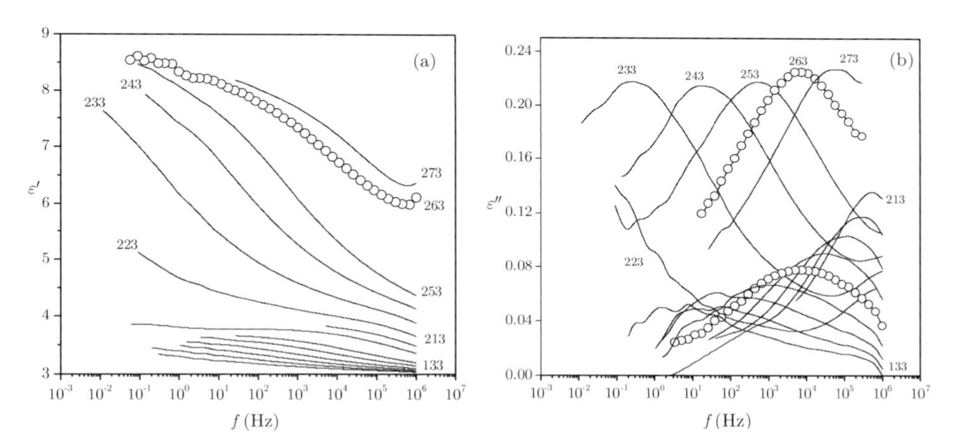

Figure 4. Real part, $\varepsilon'(\alpha)$, and imaginary part, $\varepsilon''(\beta)$, of the complex dielectric permittivity against frequency f for a PTE at several temperatures in steps of 10 K. For the sake of clarity, experimental points are shown only for $T = 263$ K in (a) and $T = 173$ and 263 K in (b), whereas data at other temperatures are given as continuous lines [53]

the γ relaxation. The broadening of the response at the low-frequency side of the spectrum is due to the presence of the slower β relaxation. Compared to the TSDC (Figure 3), the β relaxation is weaker in DRS measurements. This is due to the fact that the sample is more dry in the DRS apparatus (Figure 4), where dry nitrogen gas has been used for the temperature control [53]. TSDC and DRS measurements on PTE samples hydrated at various levels of relative humidity supported this hypothesis [54]. The strong loss peak at $T > T_g$, shifting significantly with temperature, is due to the α relaxation associated with the glass transition, in agreement with the DSC and the TSDC data and the results obtained by a detailed analysis of the DRS data [54].

The DRS counterpart of a TSDC thermogram of the type of Figure 3 is shown in Figure 5 for the same PTE of Figure 4: ε'' against temperature at a constant frequency of 100 Hz. The data have been recorded isothermally (Figure 4)

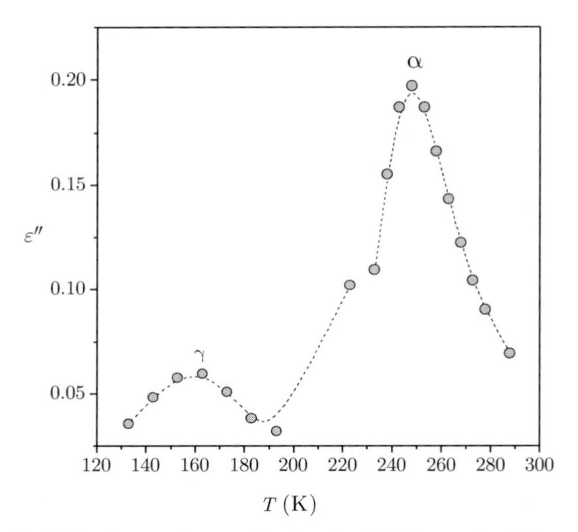

Figure 5. Isochronal $\varepsilon''(T)$ plot at $f = 100$ Hz for the PTE sample of Figure 4. The symbols are experimental points and the lines are guides for the eye [53]

and replotted here. The frequency of 100 Hz is higher than the equivalent frequency of the TSDC measurements, which is in the range of 10^{-2}–10^{-4} Hz [25], so the peaks are shifted to higher temperatures with respect to TSDC. Interestingly, no peak corresponding to the TSDC MWS peak is observed in Figure 5, in agreement with the fact that no DRS counterpart of this peak has been recorded in Figure 4.

Three sources of information are associated with each DRS loss peak (and TSDC peak), as already mentioned in Section 2.1: the frequency (temperature) position of the peak, which provides information on the time scale of the response, the magnitude of the peak, which provides information on the concentration of dipolar units contributing to the loss peak and on dipole moments, and the shape of the peak, which provides information about interactions on the molecular level [12,17,22,23]. In the case of DRS loss peaks, this information is rationalized in terms of the Havriliak-Negami (HN) expression (Eq. (5)) and other similar expressions discussed in Section 2.1. The values of the fitting parameters can then be discussed in terms of molecular dynamics in relation to structure and morphology. An interesting question is whether the evaluation of the α loss peak provides any information on the microphase separation and any measure of DMS in PTE [6,10,13]. In the case of TSDC measurements, the same question refers to the combination of the α and the MWS peak, as the latter clearly reflects properties of the phase-separated morphology [54].

In addition to DRS and TSDC, a variety of experimental techniques (DSC, WAXS, SAXS and water sorption/diffusion measurements) were employed to investigate the microphase separation and molecular dynamics in model PTE based on either PPG or PBAD and three different diisocyanates (100% *t,t*-HMDI, 20% *t,t*-HMDI and MDI, see Section 4) [6,34]. The samples investigated

Table 1. Composition of PTE used to investigate effects of the type of HS and SS on the microphase separation and molecular dynamics

Sample	Isocyanate	Polyol	Hard block (wt%)
1	20% *t,t*-HMDI	2000 MW PPG	30
2	100% *t,t*-HMDI	″	30
3	MDI	″	30
4	20% *t,t*-HMDI	2000 MW PPG	50
5	100% *t,t*-HMDI	″	50
6	MDI	″	50
7	20% *t,t*-HMDI	2000 MW PBAD	30
8	100% *t,t*-HMDI	″	30
9	MDI	″	30
10	20% *t,t*-HMDI	2000 MW PBAD	50
11	100% *t,t*-HMDI	″	50
12	MDI	″	50

PPG = poly(propylene glycol); PBAD = poly(butylene adipate)

are listed in Table 1. Figure 6 compares T_g determined by DSC (midpoint) and the peak temperature T_α of the TSDC α peak for these samples. As a general remark, it should be noted that T_α follows the changes of T_g with the sample,

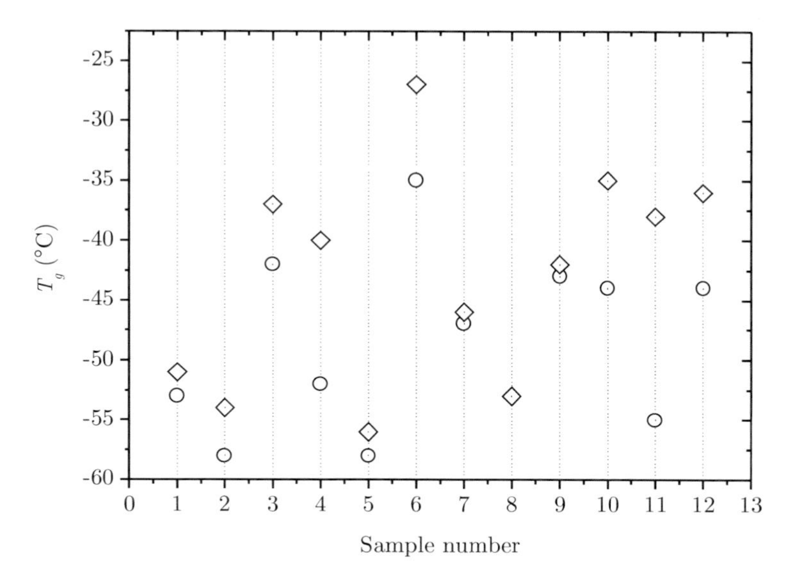

Figure 6. Calorimetric glass transition temperatures, T_g (○), and TSDC peak temperatures, T_α (◇), for the samples listed in Table 1

suggesting that T_α is a good measure of T_g. The characteristics of the TSDC α peak and the DRS α loss peak [6] were used to establish changes in morphology and molecular dynamics induced by replacing the polyether PPG by the polyester PBAD, by increasing the amount of HS from 30 to 50 wt% and by varying the composition of the diisocyanates (Table 1). No effects of the type and of the amount of the SS on DMS were observed. It is interesting to note, however, that the DRS α loss peak shifts to lower temperatures (isochronal plots) and its strength $\Delta\varepsilon$ (Eq. (5)) decreases by replacing PPG by PBAD. Also, the peak is shifted to higher temperatures, in agreement with the TSDC results, and is broader for the samples with 50 wt% HS, in agreement with the DSC results [6,34]. These results suggest that by increasing the HS content, the mobility of the SS is restricted, shifting the α peak to higher temperatures and the increased molecular interactions result in a broader distribution of relaxation times. The DMS was found to systematically change with the composition of the diisocyanates used and to increase in the order MDI, 20% *t,t*-HMDI, 100% *t,t*-HMDI, as evidenced by the systematic shift of T_g (DSC), T_α (TSDC) and the peak temperature T_m (DRS) to lower temperatures in the order given above and by the narrowing of the DSC signal. Additional support for the samples with 100% *t,t*-HMDI showing the highest DMS is provided by the systematically larger values of the magnitude I_n of the TSDC MWS peak [10,13,50] and by the observation of cold crystallization for these samples if the soft phase is PBAD. Surprisingly, the samples with 100% *t,t*-HMDI show the lowest values of water uptake, h, and water content bonded to primary hydration sites, h_m [6], whereas the samples with MDI show the highest values of specific heat capacity increment at the glass transition, Δc_p, and of dielectric strength of the α relaxation (both in TSDC and in DRS) [6,34].

The effects of the type of chain extender on the microphase separation and molecular dynamics in PTE were investigated by means of DSC, SAXS, DRS, and TSDC [5,10,50]. The PTE were based on PTMG (MW = 1000 g/mole), MDI and various fragments, including crown ethers, as chain extenders and/or chain end-groups [5]. Results obtained by the various techniques suggested that DMS increases with increasing concentration of crown ethers as chain extenders and/or chain ends. With increasing DMS, the normalized (to the same experimental conditions) magnitude of the TSDC α peak and of the MWS interfacial peak were found to increase, whereas the MWS peak shifted to lower temperatures towards the glass transition temperature, T_g. The results were less clear with respect to the characteristics of the DRS α loss peak. Finally, the dc conductivity, σ_{dc}, obtained by analysing the DRS data, was found to increase with increasing DMS [50]. Thus, from the methodological point of view, these results suggest that the normalized magnitude of the TSDC α peak and of the TSDC MWS peak, the peak temperature of the TSDC MWS peak, and σ_{dc} are sensitive to variations of the microphase morphology and can be used to measure DMS. At the same time, however, attention was drawn to the need that these and similar criteria should be further quantified and tested on selected systems [50].

In two recent publications [53,55], DRS and TSDC were employed to investigate the molecular dynamics in novel PTE in order to obtain information on the microphase separation and to compare it with DSC and SAXS studies. PTE in [53] were prepared in a two-step process: formation of a prepolymer (PP) with reactive NCO groups by the reaction of an oligoether or oligoester (OE) with a diisocyanate (DIC) at the first stage, which is extended with a diol or diamine (chain extender, CE) at the second stage. PTMG (MW = 1000 g/mole), MDI or 2,6-toluenediisocyanate (TDI) and asymmetrical DMH (I) or a DMH derivative (II) of the chemical structure given in Scheme 1, were used as OE,

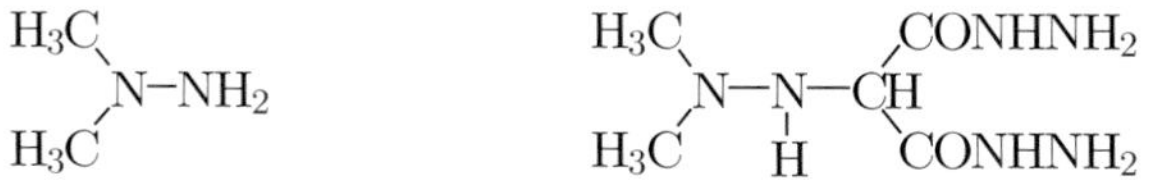

$(H_3C)_2N{-}NH_2$ (I)

$(H_3C)_2N{-}NH{-}CH(CONHNH_2)_2$ (II)

Scheme 1. Structure of asymmetrical dimethylhydrazine (DMH) (I) and of a DMH derivative (II) used as chain extenders in [53,55]

DIC and CE, respectively. The materials prepared are listed in Table 2. Special attention was paid to the investigation of the α relaxation by DRS and TSDC and to the interfacial MWS relaxation by TSDC. Representative results of the DRS measurements (raw data in the frequency and temperature domain) were

Table 2. Materials used, composition, and density of the PUs studied in [53]

PU	DIC	CE	PP:CE	Composition (wt%)			ρ (g/cm^3)
				DIC	OE	CE	
PU1	MDI	I	1:1	31.9	63.8	4.3	1.0866
PU2	MDI	I	10:1	33.2	66.4	0.4	1.1046
PU3	TDI	I	10:1	25.7	73.9	0.4	1.0853
PU4	MDI	II	1:1	28.3	56.5	15.2	1.4877

already shown in Figures 4 and 5. Figure 7 shows the DRS α loss peak on sample PU1 of Table 2 measured at several temperatures. The HN expression (Eq. (5)) was fitted to the data in order to obtain information on the time scale, τ, the magnitude, $\Delta\varepsilon$, and the shape, α, β, of the response. The latter was preferably quantified in terms of the Kohlrausch-Williams-Watts parameter, β_{KWW}, obtained by the approximate equation [53]

$$\beta_{KWW} = \frac{1.14}{w} \tag{12}$$

where w is the full width at half maximum (in decades) and 1.14 that of a single Debye peak. $\Delta\varepsilon$ and β_{KWW} change with temperature. In Table 3, the

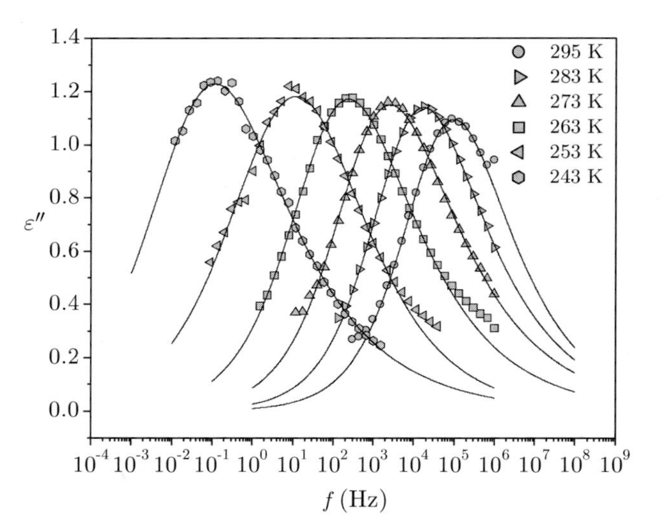

Figure 7. $\varepsilon''(f)$ plots for the α relaxation in PU1 (Table 2) at several indicated temperatures. The lines are fits of the HN expression, Eq. (5), to the experimental data (symbols) [53]

corresponding values are listed at a temperature $T_g + 37$ K, so that the results obtained for different samples can be compared.

The information obtained from each TSDC peak is quantified in terms of the peak temperature, corresponding to the time scale of the response in the DRS experiments, the full width at half maximum (FHW) as a measure of the shape of the response, and the dielectric strength, $\Delta\varepsilon$, calculated from the measured depolarization charge [25–27]. These values are listed in Table 4 for the α and the MWS peaks. The difference $\Delta T = T_{MWS} - T_\alpha$ is also included. Note that the magnitude and the shape of the response change with temperature and this should be considered in the comparison. Results not shown here indicate that the DRS α peak becomes broader and $\Delta\varepsilon$ increases slightly with decreasing temperature [53].

The time scale of the DRS α response was further studied by means of Arrhenius plots (activation diagrams). As an example, Figure 8 shows the Arrhenius plot (log f_{max} *vs.* reciprocal temperature, where f_{max} is the frequency of maximum dielectric loss) for two samples, PU1 and PU4 of Table 2. The data were fitted to the Vogel-Tammann-Fulcher (VTF) equation [56]

$$f_{max} = \mathrm{A}\exp\left[-\frac{\mathrm{B}}{T - \mathrm{T}_0}\right] \tag{13}$$

which is characteristic of cooperative α transitions. A, B and T_0 (Vogel temperature) are empirical temperature-independent parameters. The quality of the fits is good and reasonable values (not shown here) were obtained for A, B and T_0 for the four samples under investigation. Included in the plot are also

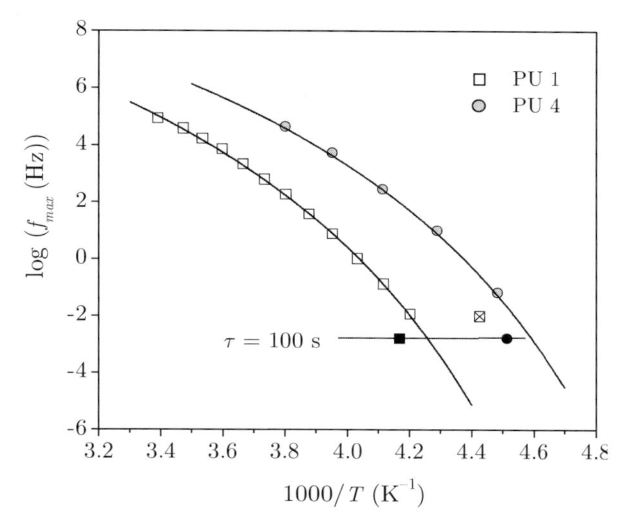

Figure 8. Arrhenius plot for the α relaxation in PU1 and PU4 (Table 2). The open symbols are for the DRS data, full symbols for the TSDC data and the crossed symbol for DSC. The lines are fits of Eq. (13) (VTF) to the DRS data [53]

the TSDC T_α temperatures at the equivalent frequency [25–27] of 1.6 mHz (corresponding to τ = 100 s) and for sample PU1 T_g = 226 K, which was obtained from DSC measurements, is in reasonable agreement with the DRS data. The VTF plots were extended to lower frequencies and the dielectric glass transition temperature $T_{g,diel}$, defined by the convention $\tau(T_{g,diel}) = 100$ s [13,50], has been determined and listed in Table 3.

To facilitate the discussion of the dielectric data in terms of microphase separation, we have included in Table 3 the segregation level, α_{seg}, which is

Table 3. DRS data for the α relaxation and the segregation level, α_{seg}, from SAXS for the PUs of Table 2 [53]

PU	SAXS data	DRS data		
	α_{seg}	$\Delta\varepsilon_\alpha$	β_{KWW}	$T_{g,diel}$ (K)
PU1	0.23	7.5	0.46	235
PU2	0.36	1.8	0.32	224
PU3	0.44	2.5	0.30	226
PU4	0.67	2.9	0.43	217

the overall DMS determined by SAXS; $\alpha_{seg} = 1$ for an ideally separated material and becomes zero for complete mixing [53]. Based on the SAXS data, the DMS increases by reducing the molar ratio of CE, by replacing MDI by TDI, and

Table 4. TSDC data for the α and MWS peaks of the PUs of Table 2 [53]

PU	TSDC data						
	T_α (K)	$(FHW)_\alpha$ (K)	$\Delta\varepsilon_\alpha$	T_{MWS} (K)	$(FHW)_{MWS}$ (K)	$\Delta\varepsilon_{MWS}$	ΔT (K)
PU1	233	15	8.3	255	11	4.6	22
PU2	222	23	2.9	263	32	3.1	40
PU3	219	27	2.5	261	21	8.6	41
PU4	216	28	2.9	297	32	2.7	82

by replacing CE I by CE II, the latter effect being more significant. As a general remark, the DRS and the TSDC data in Tables 3 and 4 are in reasonable agreement with each other and/or show the same trends. The most direct information, with respect to microphase separation, from the DRS/TSDC data for the α relaxation comes from the measures they provide for T_g. They show similar trends in the series of PUs (Table 2) taking into account that the range of changes is rather limited. Both dielectric techniques give, in particular, the highest values of T_g for PU1 and the lowest values for PU4, in very good agreement with the SAXS data, which show that PU1 has the lowest and PU4 the highest DMS. It is interesting to note that the T_g values determined by DSC for PU1 (226 K) and PU2 (217 K) show a similar trend. Thus, the dielectric techniques provide measures for T_g, which reflect variations of DMS in good agreement with the SAXS and DSC studies. With respect to the TSDC MWS peak, the results in Tables 3 and 4 show that there is good agreement between T_{MWS} and $\Delta T = T_{MWS} - T_\alpha$ on the one hand and α_{seg} on the other, and suggest that T_{MWS} and ΔT reflect the state of microphase separation [53]. It is interesting to note that in PTE with various fragments, including crown ethers, as chain extenders and/or chain end-groups, $\Delta T = T_{MWS} - T_\alpha$ was found to decrease with increasing DMS, as reported above. Note, however, that this was due to the concomitant increase in the dc conductivity, which shifts the MWS peak to lower temperatures [50].

Table 5 lists the composition, PP:CE molar ratio, density, and segregation level determined by SAXS and the T_g determined by DSC of a PTE series where the ratio PP:CE was systematically varied over extreme ranges, and DRS and TSDC were employed to investigate the molecular dynamics and to draw conclusions on the microphase separation. Figure 9 shows representative $\varepsilon''(f)$ plots in the region of the α relaxation and Table 6 displays results obtained by analyzing the DRS data for some of the samples listed in Table 5. The kinetic free volume fraction at T_g, f_g, was calculated by [57]

$$f_g = \frac{T_g - \mathrm{T}_0}{\mathrm{B}} \tag{14}$$

using the values of T_0 and B obtained from the VTF fits.

$\Delta\varepsilon$ and the shape parameters α and β in Eq. (5) are temperature dependent and should not be compared for different samples at the same absolute

Table 5. Prepolymer-to-chain extender molar ratio (PP:CE), composition, density, ρ, segregation level, α_{seg}, and glass transition temperature, T_g, of the PUs studied in [55]

PU	PP:CE	Composition (wt%)			ρ (g/cm^3)	α_{seg}	T_g (K)
		MDI	OE	CE			
PU1	1:1	31.9	63.8	4.3	1.0866	0.23	226
PU2	2:1	32.5	65.3	2.2	1.0955	0.17	222
PU5	5:1	33.0	66.1	0.9	1.1235	0.29	215
PU6	6:1	33.1	66.2	0.7	1.1050	0.27	226
PU7.5	7.5:1	33.1	66.3	0.6	1.0951	0.15	235
PU10	10:1	33.2	66.4	0.4	1.1046	0.36	217

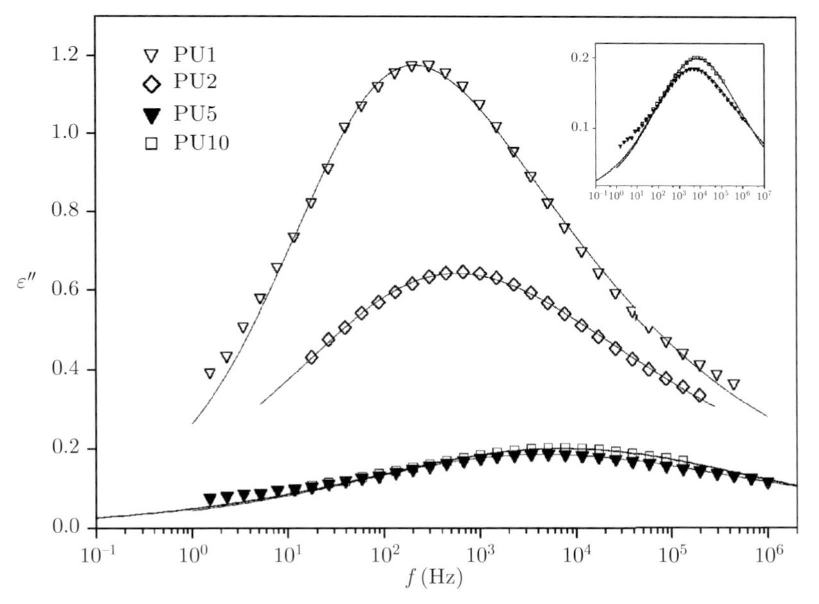

Figure 9. Comparative $\varepsilon''(f)$ plots at 263 K for four samples indicated on the plot (Table 5). The lines are fits of the HN expression (Eq. (5)) to the experimental data (points). The inset shows details for PU5 and PU10 [55]

Table 6. Results of DRS measurements for the α relaxation: relaxation strength, $\Delta\varepsilon_\alpha$, shape parameter, β_{KWW} (both at 263K), VTF parameters B and T_0, dielectric glass transition temperature, $T_{g,diel}$, and kinetic free volume fraction at T_g, f_g, for selected samples of Table 5 [55]

PU	$\Delta\varepsilon_\alpha$	β_{KWW}	B (K)	T_0 (K)	$T_{g,diel}$ (K)	f_g
PU1	7.1	0.46	1988	178	235	0.024
PU2	5.1	0.40	1357	182	232	0.030
PU5	1.7	0.36	1226	187	226	0.023
PU10	1.7	0.32	2090	169	224	0.023

temperature, but at the same distance from the T_g. However, the temperature dependence of $\Delta\varepsilon$, α, and β (not shown here) is rather weak and T_g is not very different for the various samples (Table 5), so the values listed in Table 6 for $\Delta\varepsilon$ have been taken at the same temperature for all samples, namely at 263 K. It is seen in Figure 9 and in Table 6 that $\Delta\varepsilon_\alpha$ decreases significantly at lower CE contents (first three samples) and the peak becomes broader. The f_g values are in the range typically determined for polymers [57] and do not show any systematic variation with the PP:CE ratio.

Figure 10 shows the results of TSDC measurements for several samples of Table 5 in the region of the α and the MWS peaks [55]. Table 7 lists results of the analysis of the TSDC data with respect to the time scale, the magnitude, and the shape of the response. The time scale of the TSDC α peak can be compared with that of the DRS α peak on the basis of the T_α (Table 7) and

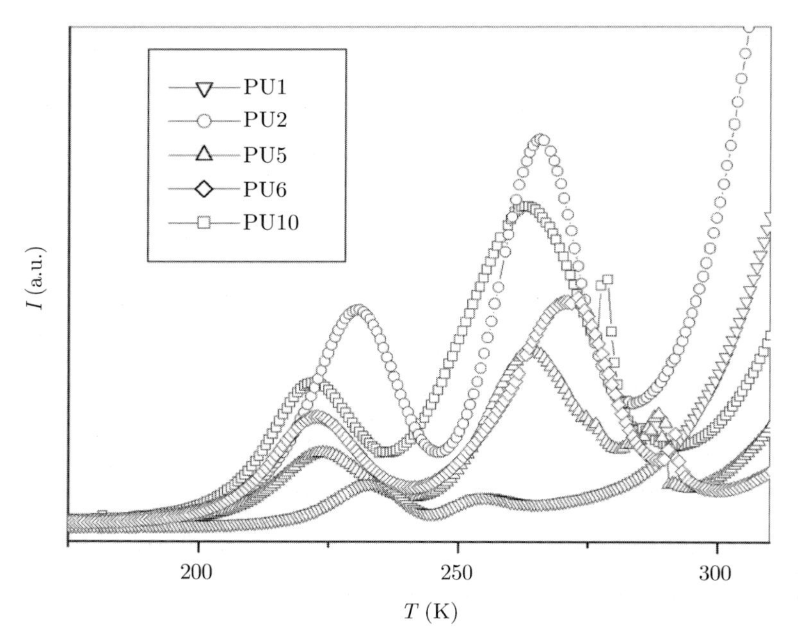

Figure 10. TSDC thermograms in the region of the α and MWS peaks for several samples of Table 5 indicated on the plot [55]

the $T_{g,diel}$ values (Table 6). They show the same trend in the series of the PTE of Table 5, both decreasing at lower amounts of CE in the PTE mixture, as well as close absolute values. Also, with respect to the shape of the response, both β_{KWW} in Table 6 and W_α in Table 7 show the same trend, namely a broadening of the response at lower CE amounts. Finally, the $\Delta\varepsilon_\alpha$ values determined by TSDC (Table 7) are close to those determined by DRS (Table 6) and show the same trend of decreasing with decreasing amount of CE [55].

Table 7. Results of TSDC measurements for the α relaxation and the MWS peak: peak temperature, full width at half maximum, W, relaxation strength, $\Delta\varepsilon$, and $\Delta T = T_{MWS} - T_{\alpha}$ for selected samples of Table 5 [55]

PU	T_{α} (K)	W_{α} (K)	$\Delta\varepsilon_{\alpha}$	T_{MWS} (K)	W_{MWS} (K)	$\Delta\varepsilon_{MWS}$	ΔT (K)
PU1	233	15	5.6	255	11	4.6	22
PU2	231	21	3.4	266	21	7.3	35
PU5	224	23	1.2	264	16	3.7	40
PU6	223	22	1.8	270	24	5.6	47
PU10	222	23	0.9	263	32	3.1	41

With respect to the microphase separation, both the SAXS and the DSC results in Table 5 suggest that DMS increases in general, at lower CE contents, but shows a significant scattering. The TSDC data for the width, W_{MWS}, and the relaxation strength, $\Delta\varepsilon_{MWS}$, of the interfacial MWS relaxation in Table 7 do not show any systematic variation with the PP:CE ratio. However, $\Delta T = T_{MWS} - T_{\alpha}$ clearly increases at the higher PP:CE ratios, except for PU10 with the extreme ratio 10:1, indicating that DMS also increases [3,24], thus confirming the interpretation of the SAXS and the DSC data. The DRS and the TSDC data for the α relaxation in Tables 6 and 7, respectively, show, in excellent agreement with each other, that with increasing PP:CE ratio the response becomes faster, smaller, and broader. We interpret the systematic decrease of T_{α} and $T_{g,diel}$ with the rise in the PP:CE ratio as a clear indication that DMS increases at higher PP:CE ratios. The results for the magnitude and the width of the relaxation can be understood, if one assumes that, by increasing the PP:CE ratio, a branched structure is developed. In fact, preliminary measurements of the solubility and viscosity of solutions of the PTE of Table 5 in organic solvents indicate increased branching at higher PP:CE ratios [55].

6. Ionomers based on polyurethane thermoplastic elastomers

6.1. *Ionomers with ionic moieties in the hard segments*

Ionomers are macromolecules containing a small number of ionic groups covalently bonded to a non-polar chain [58–60]. They are classified into anionomers, cationomers and zwitterionomers depending on the nature of the ionic groups. Ionomers must be distinguished from polyelectrolytes, which are often defined as having one ionic group per repeat unit, whereas ionomers contain a maximum of 0.15–0.20 ionic groups per repeat unit. The structure and properties of polymers are modified by the introduction of ionic groups. It is generally accepted that the aggregation of ionic groups into microdomains (multiplets and clusters), forming physical crosslinks, gives rise to many of the unique properties of ionomers [58–60].

PU ionomers combine the favorable properties and abilities of PUs with those of ionomers and offer vast possibilities for the preparation of new materials tailored to meet specific end use requirements. Conventional PU ionomers contain randomly distributed ionic groups in the HS. The block copolymer precursors are typically prepared by incorporation of pendant acid groups in the hard segment units. The results of detailed studies suggest that, in general, ionic interactions are the primary driving force for improved microphase separation [61,62].

The final properties of conventional PU ionomers, which are controlled to a large extent by microphase separation, can be further varied by incorporation of appropriately chosen additives. In a recent study, the effect of a liquid crystalline oligomer (LCO) on the DMS and on the molecular dynamics of a PU ionomer was investigated by DRS and TSDC [63]. The final chemical structure of the PU ionomer with a concentration of ionic centers of 8% is shown in Scheme 2. Composites with the LCO cholesteric ester of caprylic acid (PU/LCO composites) were obtained by casting from a common solvent (dimethyl formamide); LCO mass fractions w = 0, 0.02, 0.10, and 0.20.

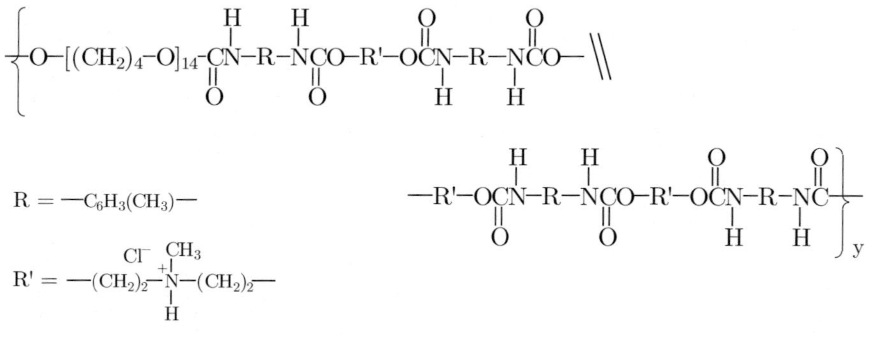

$$\Big\{ -O-[(CH_2)_4-O]_{14}-\underset{O}{\overset{}{C}}\overset{H}{N}-R-\overset{H}{N}\underset{O}{C}O-R'-O\overset{O}{C}\underset{H}{N}-R-\underset{H}{N}\overset{O}{C}O- $$

$$ -R'-O\underset{O}{C}\overset{H}{N}-R-\overset{H}{N}\underset{O}{C}O-R'-O\overset{O}{C}\underset{H}{N}-R-\underset{H}{N}\overset{O}{C}- \Big\}_y $$

$R = -C_6H_3(CH_3)-$

$R' = -(CH_2)_2-\overset{+}{N}(CH_3)(H)\,Cl^{-}-(CH_2)_2-$

Scheme 2. Chemical structure of the PU ionomer with ionic groups in the HS [63]

DSC, WAXS and SAXS measurements with the PU/LCO composites indicate a mutual influence of the components on their states of aggregation (most significantly, promoting smearing-out of the interfaces between HS and SS of the PU ionomer into broad interfacial regions of intermediate composition). The composite with w = 0.10 showed the highest degree of crystallization of the LCO component. Finally, the DMS of the PU ionomer was found to decrease and the size distribution of HS to become broader with the rise in LCO content [63].

Table 8 lists TSDC results for the α and the MWS relaxations of the PU/LCO composites, which are expected to be closely related to morphology. The peak temperatures of the two relaxations, T_α and T_{MWS} respectively, do not practically change with composition. The corresponding normalized (to the same experimental conditions) current density maxima, J_α and J_{MWS}, representative

Table 8. TSDC results for the α and the MWS relaxations in the PU ionomers. T_α and T_{MWS} are the peak temperatures, J_α and J_{MWS} the normalized peak current densities, FHW the full half-width of the MWS peak and W the activation energy of the MWS peak [63]

W	T_α (K)	T_{MWS} (K)	J_α (A/m^2)	J_{MWS} (A/m^2)	FHW (K)	W (eV)
0.00	208	256.0	2.9x10^{-8}	3.3x10^{-7}	14.1	1.0
0.02	207	259.5	2.3x10^{-8}	3.7x10^{-7}	11.0	1.2
0.10	208	255.5	1.2x10^{-8}	0.7x10^{-7}	12.3	1.1

for the relaxation strength $\Delta\varepsilon$ of each peak, *i.e.*, for the number of relaxing units contributing to the peak [25–27], do not behave additively and are significantly reduced for the w = 0.10 composite. The decrease in the magnitude of the MWS peak for this composite may indicate a decrease of DMS, in agreement with the results obtained by DSC and SAXS [63]. The MWS peak becomes narrower in the composites, whereas its activation energy, calculated by the initial rise method [25–27], does not practically change with composition.

A comparative plot of the frequency dependence of the dielectric permittivity, $\varepsilon'(f)$, of the PU/LCO composites at 298 K is shown in Figure 11. An overall decrease of molecular mobility is observed in the composites, confirmed also by plots of $\varepsilon''(f)$ and $\sigma_{ac}(f)$ at the same temperature (not shown here). The decrease of ε' in the composites is much stronger than that corresponding to

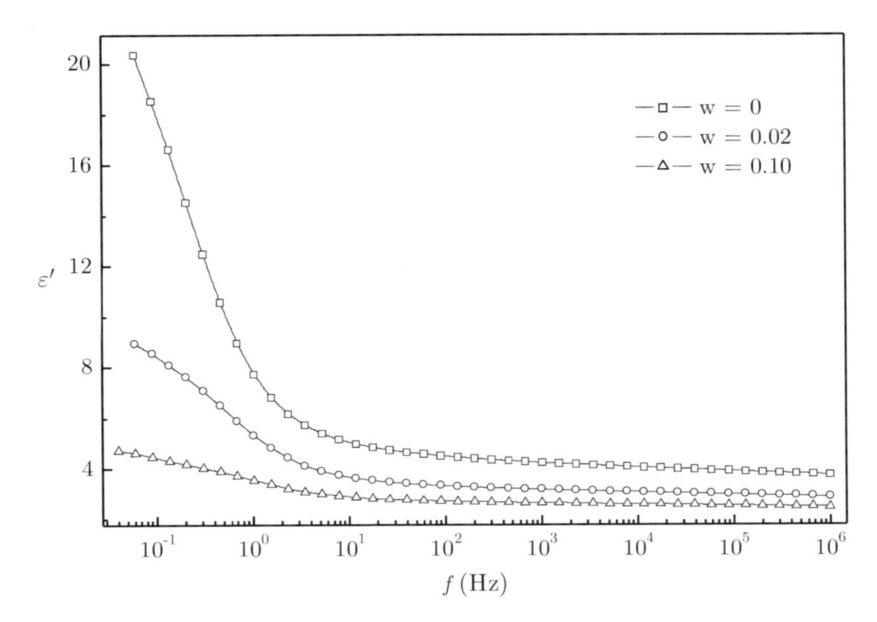

Figure 11. Frequency dependence of dielectric permittivity, $\varepsilon'(f)$, at 298 K for the PU ionomers indicated on the plot [63]

additivity. The high values of ε' at low frequencies are related to conductivity effects. The overall decrease of molecular mobility in the composites is not confined to room temperature, but it shows up in the entire temperature range of DRS measurements, 150–350 K, in particular also at $T < T_g$. Figure 12 shows $\varepsilon''(T)$ at a fixed frequency of 100 Hz. The data have been recorded isothermally in the frequency range 10^{-2}–10^6 Hz with temperature steps of 5 or 10 K and replotted in Figure 12. In addition to the overall, non-additive decrease of the molecular mobility in the composites, indicated by $\varepsilon'(f)$ in Figure 11 and by

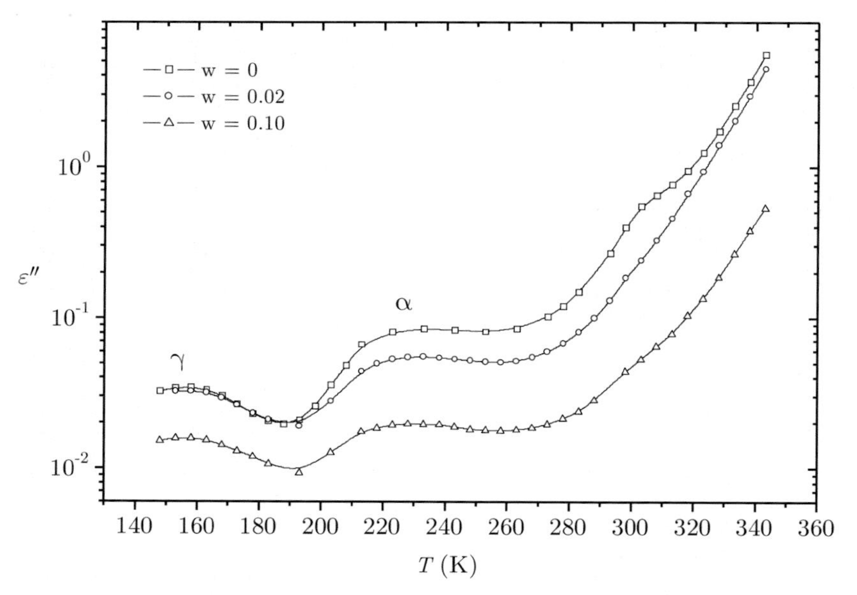

Figure 12. Temperature dependence of dielectric loss, $\varepsilon''(T)$, at 100 Hz for the PU ionomers indicated on the plot [63]

$\varepsilon'(T)$ (not shown here) [63], two relaxations (loss peaks) are observed in Figure 12, in agreement with the results of TSDC measurements [63]: the γ relaxation at about 150 K and the α relaxation at about 220 K. The temperature/frequency position of the two relaxations does not change with composition, in agreement with the TSDC results. For the α relaxation, this result is also in agreement with the DSC data for the glass transition temperature, which is composition-independent [63]. The shape of the response does not change with the composition for both relaxations, as indicated by normalized plots, $\varepsilon''/\varepsilon''_{max}$ *vs.* f/f_{max} (not shown here). The main result in Figure 12 is the significant reduction of the magnitude of the two relaxations in the composites, in particular for the sample with w = 0.10. For this sample, ε''_{max} decreases by a factor of about 2 for the γ relaxation and by a factor of about 4 (as compared to a factor of about 2.5 in TSDC measurements, see Table 8) for the α relaxation, with respect to the initial polymer. This much stronger (with respect to additivity) reduction of the intensity of the γ and α relaxations suggests some interactions between LCO and SPU and a modification of the composite morphology. Probably, smearing-out of interfaces between HS and SS, as indicated by SAXS and DSC [63], is responsible for these effects. It is interesting to note also that the reduction of the intensity of the cooperative α relaxation is larger than that of the local γ relaxation.

The large increase of ε'' at high temperatures (Figure 12) is related to conductivity effects. These effects were further studied within the conductivity formalism (Eq. (10)) and the modulus formalism (Eq. (4)), since the motion of ions, giving rise to these effects, may be used as a probe of the local morphology. Figure 13 shows comparative plots of the frequency dependence of the ac conductivity, $\sigma_{ac}(f)$, and of the imaginary part of electric modulus, $M''(f)$, at 343 K for the three samples studied. At low frequencies, $\sigma_{ac}(f)$ in Figure 13 becomes frequency-independent and the plateau value gives the dc conductivity, σ_{dc}. This was determined at several temperatures from plots similar to those shown in Figure 13. Figure 14 shows the Arrhenius plot of

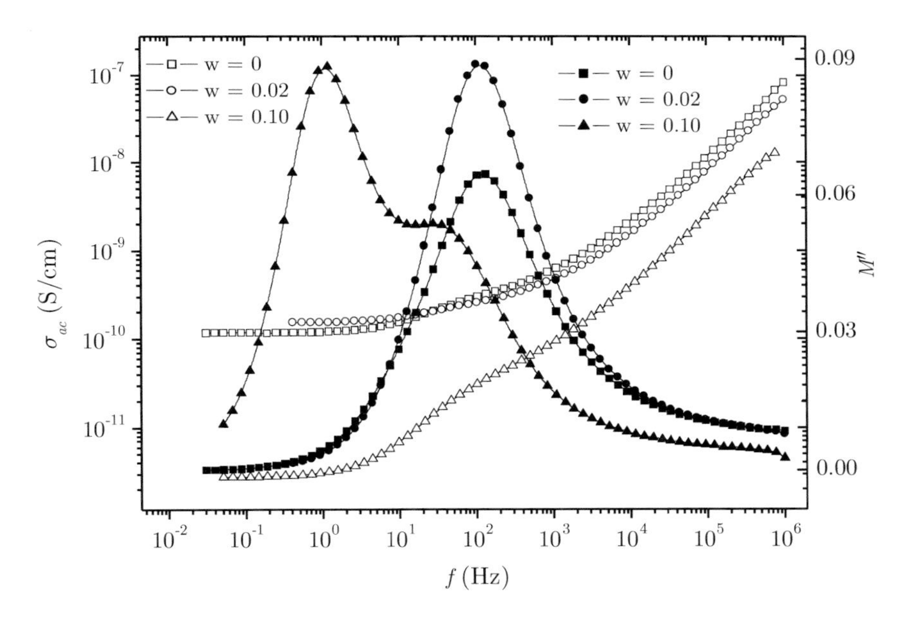

Figure 13. Ac conductivity, σ_{ac} (open symbols), and imaginary part of electric modulus, M'', (filled symbols) against frequency, f, at 343 K for the PU ionomers indicated on the plot [63]

σ_{dc}. It is seen that at each temperature, σ_{dc} is slightly higher in the composite with w = 0.02 and significantly reduced in the composite with w = 0.10, as compared to neat PU. The data in Figure 14 can be described by straight lines, indicating that the temperature dependence of σ_{dc} obeys the Arrhenius equation

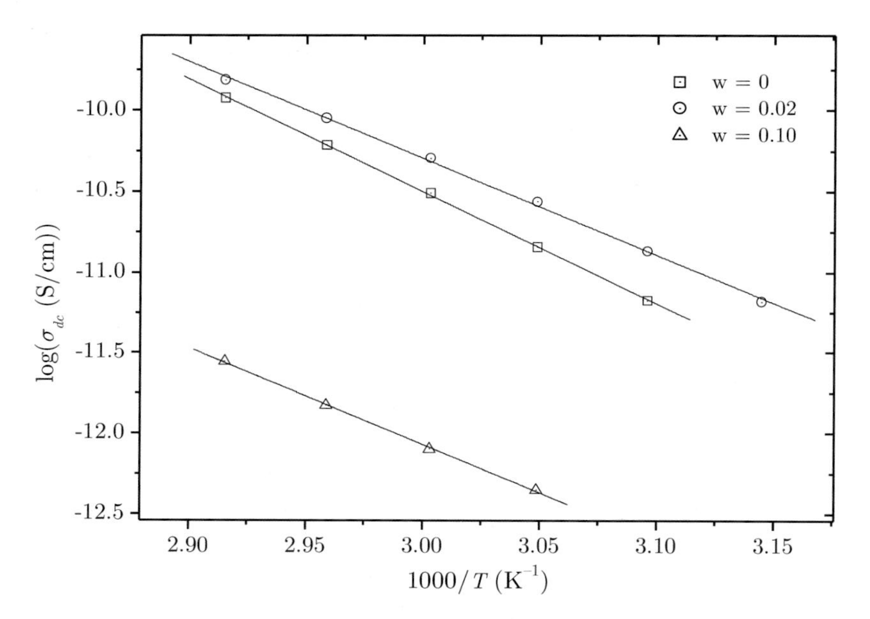

Figure 14. Arrhenius plot of the dc conductivity σ_{dc} for the PU ionomers indicated on the plot [63]

$$\sigma_{dc}(T) = \sigma_0 \exp\left(-\frac{W}{kT}\right) \tag{15}$$

where σ_0 is a constant, k is Boltzmann's constant and W is the activation energy of conductivity, 1.3 eV for the PU ionomer and 1.2 eV for the two composites (±0.1 eV). Two comments should be made here. First, the validity of the Arrhenius equation, instead of the VTF equation, suggests a decoupling of the conductivity from the segmental motion [63]. Second, the values of W are similar to those of the TSDC MWS peak in Table 8 and support the interpretation given for this peak [13,64].

For the composite with w = 0.10, a peak and a shoulder are observed in $M''(f)$ (Figure 13) corresponding to the conductivity relaxation (the change from σ_{dc} to σ_{ac} [28]) and the conductivity current relaxation (CCR) [64], respectively. Detailed studies show that CCR corresponds to the TSDC MWS peak in Table 8 and that in the samples with w = 0 and 0.02, the peak in Figure 13 corresponds to CCR, whereas the conductivity relaxation, a shoulder on the low-frequency side of the $M''(f)$ peak, is not resolved at the temperature of measurements in Figure 13 [63]. Finally, the Kohlrausch-Williams-Watts stretched exponential shape parameter, β_{KWW}, of CCR in Figure 13, calculated by Eq. (13), increases in the composites, in agreement with the decrease of the half-width of the TSDC MWS peak in Table 8, suggesting a homogenization of the conducting paths of the MWS relaxation in the composites [63].

6.2. *Ionomers with ionic moieties in the soft segments*

Recently, the preparation of two series of PTE of a new type with acid or ionic groups in the polyether SS and the investigation of their stucture-property relationships have been reported [49,65–69]. The first series was based on PTMG (M series) and the second on PEG (E series). The main interest was focused on understanding the influence of structure and morphology on the mechanisms of molecular mobility and ionic conductivity, as well as the relation between segmental mobility and ionic conductivity. For structural and morphological studies, DSC, WAXS, SAXS and Fourier transform infrared photoacoustic spectroscopy (FTIR PAS) were employed. The molecular dynamics and ionic conductivity were studied by DRS and TSDC.

Ionomers of the M series were synthesized on the basis of PTMG (MW 1040), MDI, BD, and pyromellitic anhydride (PMA) as starting materials and dimethyl formamide (DMF) as solvent [65,67]. A carboxyl-containing oligoether (I) (SS precursor), terminated by OH-groups and denoted 2M1H, was prepared using PTMG and PMA in the ratio 2:1. A series of isocyanate compounds (II) (HS precursors), terminated by NCO-groups, was synthesized using MDI and BD in the ratios 1:0, 2:1, 3:2, 4:3, and 6:5 and denoted B0, B1, B2, B3, and B5, respectively. The PUs containing B0, B1, B2, B3, and B5 hard chain segments with acid groups in SS were prepared by reaction of (I) with (II) with NCO/OH = 1 and denoted M1B0H, M1B1H, M1B2H, M1B3H and M1B5H, respectively. Li-, Na-, and K-salts of the PUs were prepared by neutralization in DMF of the carboxylic groups with lithium hydroxide, sodium hydroxide, and potassium acetate, respectively.

The general molecular structure of the PTMG-based ionomers is shown in Scheme 3. The length of the PU HS, which is proportional to the weight fraction of HS, w_2, increases in each series of samples, *e.g.*, in the Na series, from M1B0Na to M1B5Na.

Table 9 lists DSC data for the glass transition and melting of crystallites and the parameters of the microphase-separated morphology of the precursor 2M1H, the acid PUs M1B0H–M1B5H, and the salt ionomers M1B0Na and M1B1Na. The microphase composition was calculated following the formalism

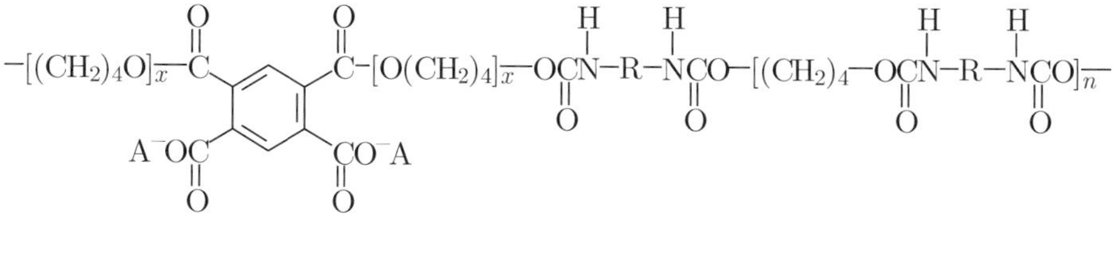

$$-[(CH_2)_4O]_{\overline{x}}-\overset{O}{\overset{\|}{C}}-C_6H_2(COO^-A^+)_2-\overset{O}{\overset{\|}{C}}-[O(CH_2)_4]_{\overline{x}}-O\underset{O}{\underset{\|}{C}}\overset{H}{\overset{|}{N}}-R-\overset{H}{\overset{|}{N}}\underset{O}{\underset{\|}{C}}O-[(CH_2)_4-O\underset{O}{\underset{\|}{C}}\overset{H}{\overset{|}{N}}-R-\overset{H}{\overset{|}{N}}\underset{O}{\underset{\|}{C}}O]_{\overline{n}}-$$

Scheme 3. General molecular structure of the PTMG-based PU ionomers. R is 4,4′-$C_6H_4CH_2C_6H_4$, $x = 14$, $n = 0, 1, 2, 3, 5$ (B0–B5 series, respectively), A is H^+, Li^+, Na^+, K^+ [65]

developed by Koberstein *et al.* [7,32], details of the calculations being given in [65]. Crystallinity is observed only for the precursor 2M1H and the acid M1B0H. It is interesting that in the salt counterpart of the latter, M1B0Na, the crystallinity is suppressed. The (midpoint) glass transition temperature, T_g, of the SS-rich phase is in the range from –74 to –50°C. The HS content dependence of T_g shows a maximum of –50°C for sample M1B2H. Comparison of the data for acid and salt ionomers with the same HS content suggests that the incorporation of ionic groups into the molecular structure leads to a considerable decrease of T_g and of ΔC_p.

The parameters of the microphase-separated morphology in Table 9 show that in the acid PUs the overall weight fraction of the SS-rich phase W_1 decreases monotonically with increasing HS weight fraction w_2 and by far exceeds the SS weight fraction $1-w_2$. Already for the sample M1B1H, the W_1

Table 9. DSC data and the parameters of the microphase-separated structure of the polyurethane ionomers: w_2 is the HS weight fraction, $T_{m,s}$ the melting temperature of the SS-rich phase, $T_{g,s}$ the glass transition temperature of the SS-rich phase, $\Delta T_{g,s}$ the width of the glass transition temperature of the SS-rich phase, T_g^0 the calculated glass transition temperature of the sample using Fox's equation, $\Delta C_{p,s}$ the specific heat capacity jump at T_g, W_1 the overall weight fraction of the SS-rich phase, $\omega_{1,s}$ the weight fraction of the SS component in the SS-rich phase, and $\omega_{1,h}$ the weight fraction of the HS component in the SS-rich phase [65]

Sample	w_2	$T_{m,s}$ (°C)	$T_{g,s}$ (°C)	$\Delta T_{g,s}$ (°C)	T_g^0 (°C)	$\Delta C_{p,s}$ (J/(gK))	W_1	$\omega_{1,s}$	$\omega_{1,h}$
2M1H	0	+17	–74	19	–74	0.522	–	–	–
M1B0H	0.113	+10	–61	12	–63	0.458	0.985	0.872	0.128
M1B1H	0.218	–	–52	23	–51	0.360	0.871	0.792	0.208
M1B2H	0.301	–	–50	20	–40.5	0.275	0.680	0.775	0.225
M1B3H	0.370	–	–60	24	–31	0.209	0.464	0.863	0.137
M1B5H	0.470	–	–60	30	–18	0.115	0.255	0.863	0.137
M1B0Na	0.113	–	–67	15	–63	0.383	0.790	0.923	0.077
M1B1Na	0.218	–	–59	19	–51	0.325	0.722	0.854	0.146

value indicates the appearance of SS-rich microdomains, which make up about 87 wt% of the material. This means that the extension of the hard chain segment of the acid PUs is favorable to the onset of the microphase-separated morphology that is common to many segmented PUs [6,50]. The concentration of HS in the SS-rich microphase varies from about 13 to about 23%, with a maximum for the sample M1B2H. The neutralization of the acid groups leads to an abrupt decrease in W_1. This clearly demonstrates the occurrence of some ion aggregation regions within the SS-rich matrix.

SAXS measurements in the series of the acid PUs (Table 9) show maxima (except for the 2M1H and the M1B0H samples), which become more intensive upon HS extension, typical of most microphase-separated PUs [5]. The values of the Lorentz-corrected interdomain spacing calculated from the maximum position of both the intensity curve, D_1, and the one-dimensional correlation function, d_{1D}, are listed in Table 10 [65,67] together with the calculated values of the thickness of the HS lamellar-like microdomains, T_h, and the values of the HS length, L_h. T_h increases upon HS extension, however less sharply than L_h, suggesting folding of the chains in the HS microdomains. SAXS measurements of the salt ionomers show that the microphase separation into HS microdomains and SS microphase of the acid PUs is preserved, even if less pronounced [49]. The

Table 10. SAXS-based parameters of the microphase-separated structure of the acid polyurethanes of Table 9. D_1 and d_{1d} are values of the interdomain spacing, T_h the calculated thickness of the HS lamellar-like microdomains, and L_h the HS length [65]

Sample	D_1 (nm)	d_{1d} (nm)	T_h (nm)	L_h (nm)
M1B1H	14.4	12.0	2.1	3.15
M1B2H	14.4	12.0	2.9	5.00
M1B3H	14.8	12.5	3.8	6.85
M1B5H	15.7	14.0	5.6	10.55

ionomer peak, due to aggregation of the ionic groups into multiplets [58–60], observed in model PU ionomers with ionic groups in the HS [70], was observed in all the ionomers studied here. It follows that the PU ionomers with ionic groups in the SS are characterized by the coexistence of the segment microphase separation and the ionic group segregation morphologies. For the M1B1Na sample, the PU microdomain spacing is 17 nm and the ionic aggregate size is about 6 nm. Thus, the ionic aggregates are located within the interdomain SS layers, which are roughly twice thicker than the size of the aggregates. The PU ionomers with ionic groups in the SS are characterized by more pronounced ionic aggregate morphology and lower ionic group concentration in the aggregates, as compared to model PU ionomers with chemically similar structure [49].

Figure 15 shows TSDC thermograms obtained with M1B1H and M1B1Na, representative of the PU ionomers with ionic groups in the SS based on PTMG. The γ, β, α, and MWS relaxations are observed in the order of increasing temperature. Due to overlapping of the α peak with the more intense MWS peak, it is difficult to judge, on the basis of the TSDC data, whether there is any systematic variation of T_g with the HS content and the type of ion. The larger magnitude of the MWS peak in M1B1H than in M1B1Na in Figure 15 indicates a higher DMS in the acid than the salt form, in close correlation with and in support of the results of structural characterization reported above. The MWS peak is double in M1B1Na, suggesting a more complex morphology, very probably related to the coexistence of the segment microphase separation and the ionic group segregation morphologies [49,66].

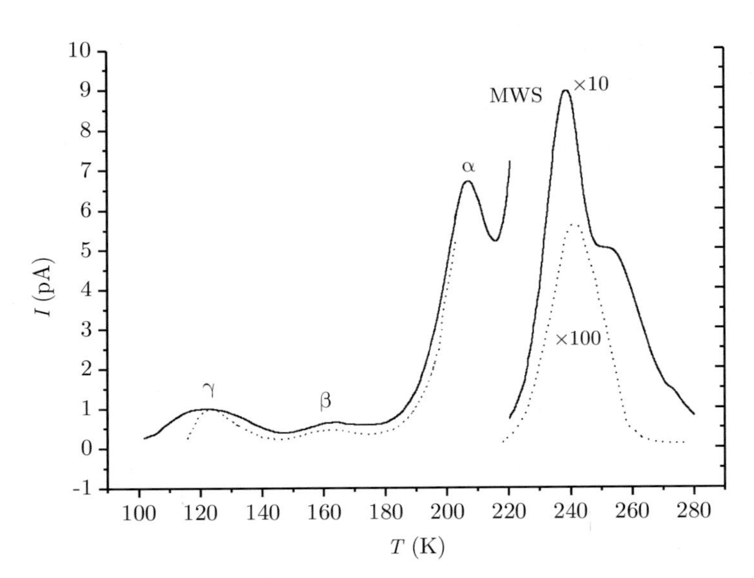

Figure 15. TSDC thermograms of the acid (...) and of the Na (—) PTMG-based polyurethane ionomers, M1B1H and M1B1Na

Figure 16 shows $\varepsilon''(f)$ spectra obtained with M1B1Na in wide frequency and temperature ranges. At low frequencies and high temperatures, one observes high ε'' values related to conductivity effects. The broad and weak loss peak shifting rapidly from Hz to MHz with increasing temperature is due to the α

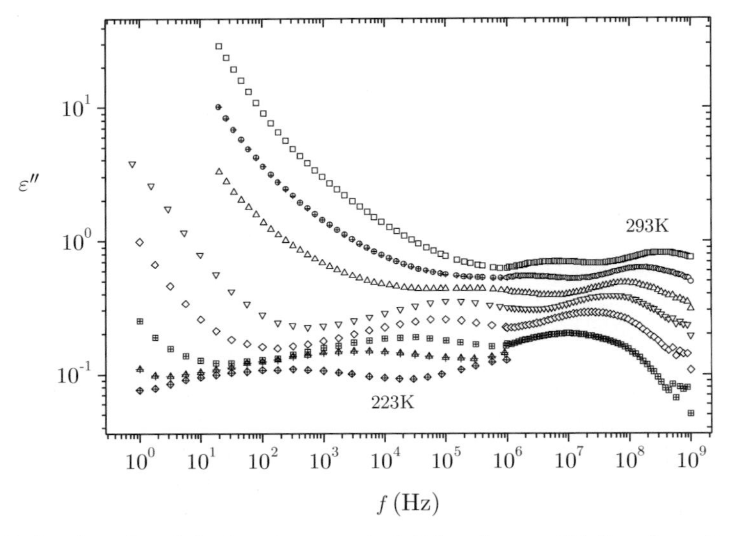

Figure 16. Log-log plot of the imaginary part of dielectric permittivity, ε'', *vs.* frequency, *f*, for the M1B1Na polyurethane ionomer at several temperatures in steps of 10 K [66]

relaxation, whereas the peak at higher frequencies is attributed to the γ relaxation, in agreement with results obtained with other PTE (Figure 4). Similar results were obtained also with the other PU ionomers of Table 9. The Arrhenius plot (activation diagram) corresponding to the data in Figure 16 is shown in Figure 17. The VTF Eq. (13) was satisfactorily fitted to the data for the α relaxation with reasonable values of the fitting parameters. The Arrhenius equation

$$f_{max} = f_0 \exp\left(-\frac{W}{kT}\right) \tag{16}$$

where W is the activation energy, k is Boltzmann's factor and f_0 is a frequency factor, was fitted to the data for the γ relaxation and W determined to be 0.40 eV, which is within the range of W values obtained for the γ relaxation in PUs [3,6,10,49].

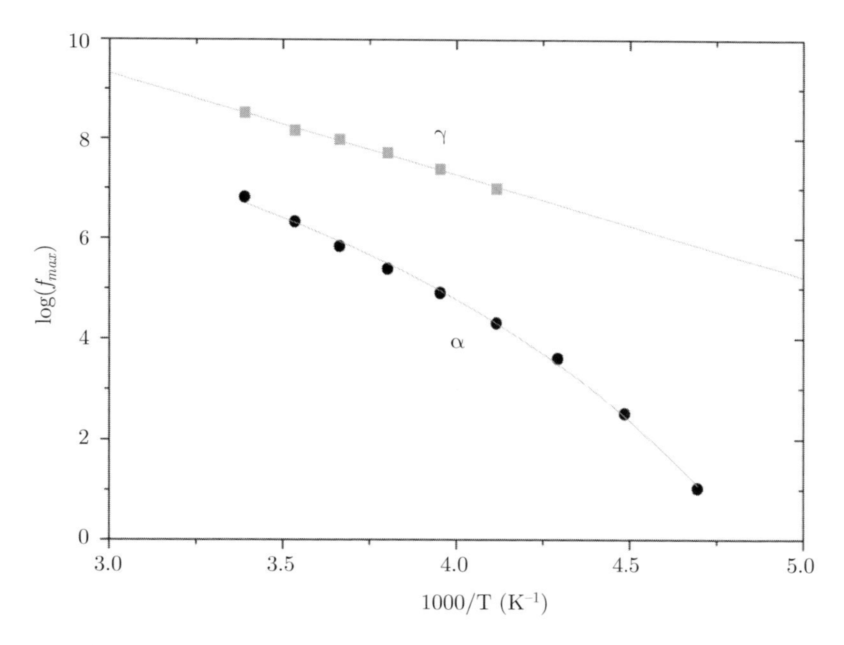

Figure 17. Arrhenius plot for the α and γ relaxations in M1B1Na polyurethane ionomer

The ionic conductivity in the PU ionomers was studied in detail by following the measuring and analytical procedures briefly described in Section 3. Figure 18 shows ac conductivity plots, $\sigma_{ac}(f)$, of the protonic conductors M1B1H–M1B5H at 25°C. At low frequencies, σ_{ac} becomes frequency-independent and the plateau value gives the dc conductivity, σ_{dc}. σ_{dc} in Figure 18 decreases with increasing n in the polymer structure (Scheme 3), *i.e.*, with increasing HS content. These results indicate, in correlation with those of the structural and morphological

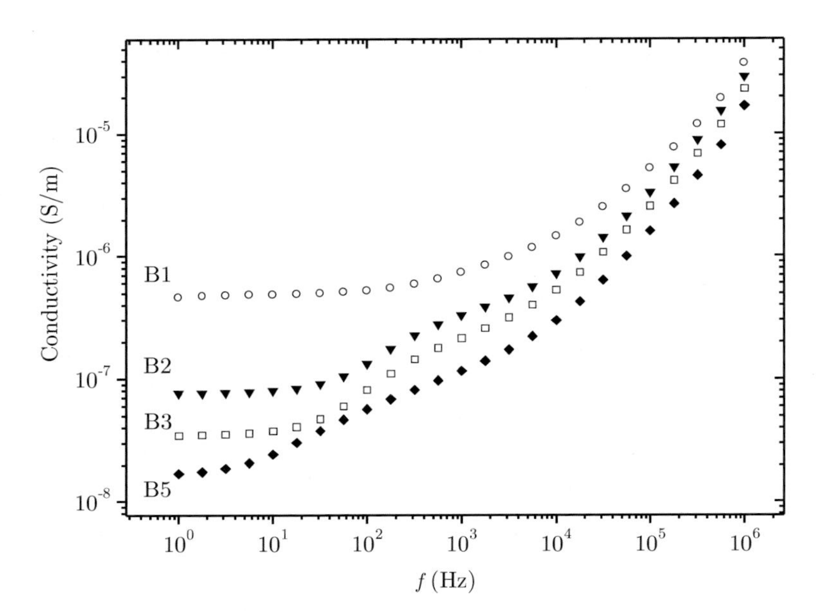

Figure 18. Log-log plot of the ac conductivity, σ_{ac}, *vs.* frequency, *f*, in M1B1H-M1B5H polyurethane ionomers at 25°C [67]

characterization (Table 9), that the electrical conductivity occurs through the SS-rich matrix. They can be rationalized in terms of the effective medium theory [52,71,72], the two phases being the SS-rich matrix and the HS-rich microdomains, however the exact composition of the phases (Table 9) and details of the internal sample topology should be taken into account [49,65]. A feature is observed in the kHz frequency region in Figure 18, which may be indicative of MWS interfacial polarization. Results for the dependence of σ_{ac} on the type of ion in the M1B-based ionomers (not shown here) indicate that σ_{dc} of the salt ionomers is, in general, by about two orders of magnitude lower as compared to that of the acid ionomers [49]. This finding correlates well with those of the structural and morphological characterization with respect to ionic segregation [49].

The frequency dependence of σ_{ac} in the PU ionomers studied at several temperatures is similar to that observed in many organic and inorganic ionic conductors [28,49,65–69]. With increasing temperature, σ_{dc} increases and the frequency range of σ_{dc} extends to higher frequencies. The "knee" in $\sigma_{ac}(f)$, *i.e.*, the frequency region where $\sigma_{ac}(f)$ begins to depart significantly from the plateau value, corresponds to the conductivity relaxation [28], discussed previously with respect to Figure 13. The Arrhenius plots of σ_{dc} for M1B1H and M1B1Na are shown in Figure 19. At each temperature, the absolute value of σ_{dc} is by 1–2 orders of magnitude lower in the ionic than in the acid ionomer. The lines are fits to a VTF-type equation [67]

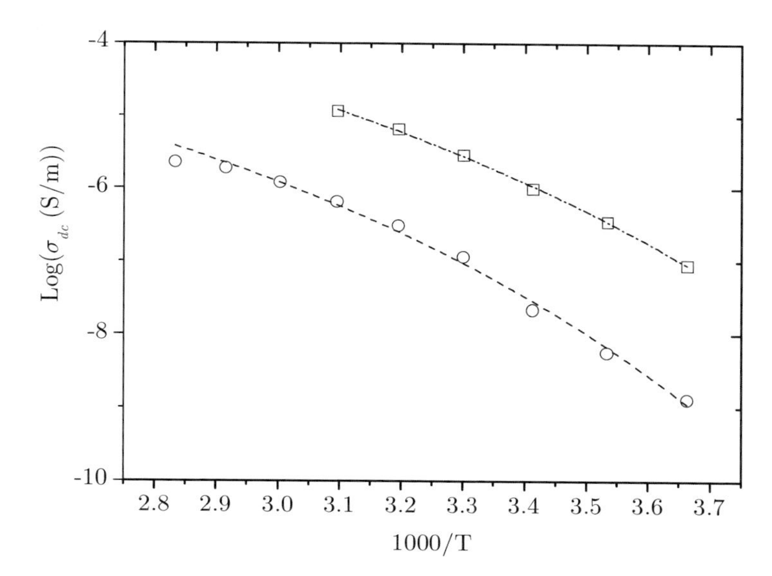

Figure 19. Arrhenius plot of σ_{dc} in M1B1H (□) and M1B1Na (○). The lines are fits of Eq. (17) to the data

$$\sigma_{dc} = \sigma_0 \exp\left(-\frac{B}{T - T_0}\right) \tag{17}$$

with reasonable values for the empirical constants σ_0, B, and T_0 [49,66]. The results suggest that in both samples the charge carrier transport is governed by the motion of the polymeric chains.

Figure 20 shows the Arrhenius plot of the α relaxation associated with the glass transition of the SS-rich microphase in the acid M1B1H ionomer, in the ε formalism, of the conductivity relaxation in the M formalism and of σ_{dc}. The lines are fits of the VTF Eqs. (13) to the ε and the M data and (17) to the σ_{dc} data. Included in the plot are peak temperatures of 210 and 241 K of two TSDC peaks assigned to dipolar α relaxation and to interfacial MWS relaxation, respectively (Figure 15) at the equivalent frequency of 1.6 mHz, corresponding to α relaxation time τ = 100 s [13]. They are in general agreement with the DRS data, suggesting, in particular, a close relation between the conductivity relaxation and the MWS TSDC peak. Table 11 lists the fitting parameters of the fits in Figure 20 and of that of the α relaxation in the M formalism (not shown in Figure 20). Of particular importance are the values of B, related to the apparent activation energy. Those for the α relaxation are similar to each other within the two formalisms. Those for the conductivity relaxation and for σ_{dc}, similar to each other, are by a factor of about 1.6 larger than those for the α relaxation. These results can be understood in terms of significant decoupling of the protonic conductivity from the structural relaxation [67].

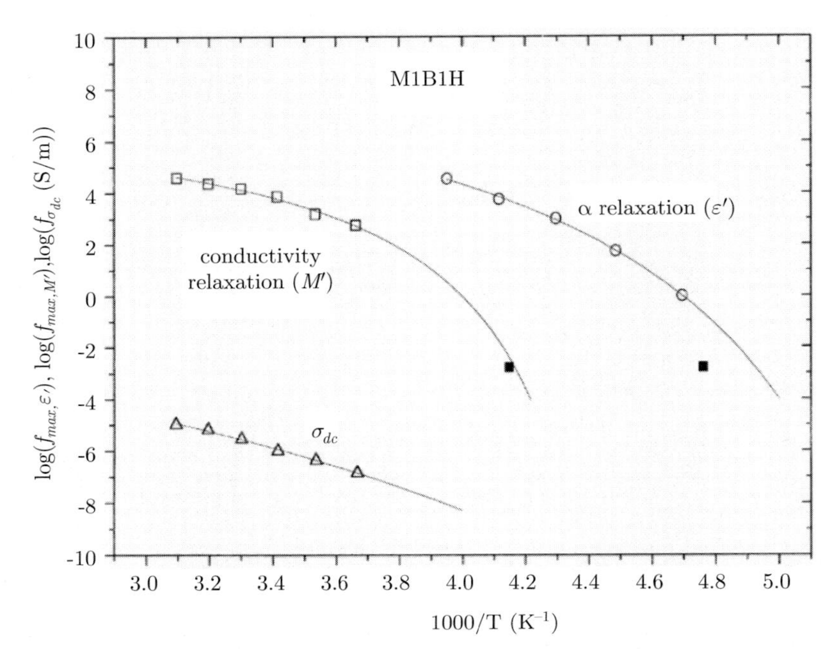

Figure 20. Arrhenius plot for the α relaxation in the permittivity formalism (○), the conductivity relaxation in the modulus formalism (□), and the dc conductivity, σ_{dc} (△), for the acid ionomer M1B1H. Included are TSDC peak temperatures (■) at the equivalent frequency of 1.6 mHz. The lines are fitting of the VTF Eqs. (13) and (17) to the ε and M data and to the σ_{dc} data, respectively [67]

Table 11. VTF parameters (Eqs. (13) and (17)) for the acid M1B1H ionomer [67]

α relaxation									Conductivity relaxation		
ε''			M''			σ_{dc}			M''		
A (Hz×10^{-9})	B (K)	T_0 (K)	A (Hz×10^{-8})	B (K)	T_0 (K)	A (S/m×10^{2})	B (K)	T_0 (K)	A (Hz×10^{-9})	B (K)	T_0 (K)
1.4	842	173	0.20	758	167	8.70	1348	173	0.34	1336	173

The electrical conductivity of the PTMG-based ionomers investigated in this section is rather low, probably because of poor solvating properties of PTMG. PEG is expected to behave better in that respect, so that ionomers based on PEG may exhibit higher σ_{dc} values, at least at temperatures higher than those of crystallite melting. A second series of PU ionomers (E series) was synthesized by following procedures similar to those described above for the M series and replacing PTMG by PEG (MW 1500). These ionomers have been designated in a similar way as those of the M series, with M1 being replaced by E1.5, *e.g.* E1.5B1Na. Their structure is shown in Scheme 4.

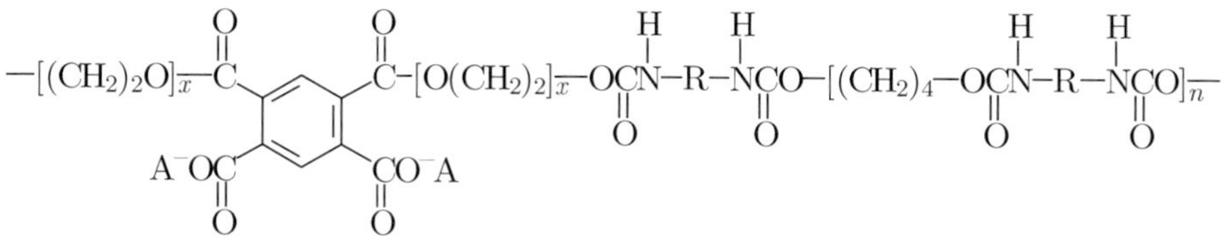

Scheme 4. General molecular structure of the PEG-based ionomers. R is 4,4′-$C_6H_4CH_2C_6H_4$, $x = 35$, $n = 0, 1, 2$, A is H^+, Li^+, Na^+, K^+ [68]

DSC thermograms of the PEG-based ionomers show, in addition to the glass transition of the SS-rich microphase, crystallization and melting events of PEG. Measurements in the acid and the ionic form (Li^+, Na^+, K^+) of the samples based on E1.5B0, E1.5B1, and E1.5B2 show that crystallization (also in the subsequent second heating scan) is observed in all samples of the first two series. In the third series, E1.5B2, a small melting endotherm is observed in the thermogram of the acid sample, E1.5B2H (however, not in the second heating scan, indicating a slow rate of recrystallization for this sample), whereas the samples become amorphous upon neutralization. For the acid ionomers, T_g was found to systematically increase with increasing HS content, from 229 K for E1.5B0H to 241 K for E1.5B2H, whereas no systematic variation of T_g with the ion type was observed [68].

For the acid ionomers, SAXS curves show the intensity maximum, characteristic of the microphase-separated PUs [7], only for E1.5B2H. For the salt ionomers, a second intensity maximum is observed at higher q values, characteristic of ionic aggregation [58–60]. The results suggest that for the acid ionomers the DMS increases upon HS extension [7]. For the salt ionomers the microphase separation is preserved, but is less pronounced [49]. The values of the Lorentz-corrected interdomain spacing are about 16 nm for the microphase-separated morphology and about 6 nm for the ionic aggregation, *i.e.*, similar to those in the M series.

Photoacoustic infrared spectroscopy (FTIR PAS) was employed to characterize the different hydrogen bonds formed and to determine binding modes and chain orientation [69]. As an example, results of fitting analysis for the band in the C=O stretching region are shown in Table 12: a sum of two bands, corresponding to free and hydrogen-bonded carbonyls, respectively, was fitted

Table 12. Data of the FTIR PAS fitting for the bands in the C=O stretching region

Sample	ν_b^a (cm^{-1})	ν_f^a (cm^{-1})	A_b/A_f [b]
E1.5B1Na	1706	1730	0.93:1
E1.5B2Na	1703	1730	1.19:1
M1B1Na	1705	1733	1.93:1

[a] ν_f and ν_b are the frequencies of free and hydrogen-bonded C=O, respectively,
[b] A_f and A_b are the band areas corresponding to ν_f and ν_b

to the measured complex band. It is seen that the relative amount of hydrogen-bonded C=O increases with increasing HS extension and is larger for PTMG-based ionomers than for PEG-based ones [69].

Figure 21 shows results for the frequency dependence of the ac conductivity, σ_{ac}, for E1.5B1H at several temperatures. As expected, σ_{ac} and σ_{dc}, obtained by extrapolation to zero frequency, increase with increasing temperature. However, a discontinuity is observed in the spectra with a large step between 303 and

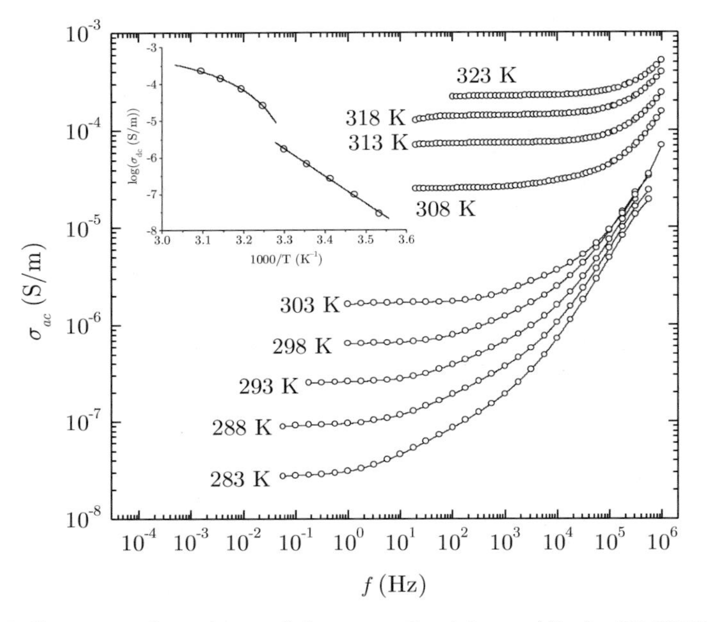

Figure 21. Frequency dependence of the ac conductivity, $\sigma_{ac}(f)$, for E1.5B1H at several temperatures indicated on the plot. The inset shows the corresponding Arrhenius plot of the dc conductivity, σ_{dc}. The lines in the inset are fits of the Arrhenius Eq. (15) at low temperatures and the VTF Eq. (17) at high temperatures to the experimental data

308 K, as well as values very close to each other for 313 to 323 K. These results correlate very well with those of DSC studies showing melting of crystallites at about 309 K [68]. The corresponding Arrhenius plot, shown in the inset to Figure 21, indicates a change of the conductivity mechanism at the melting temperature, from Arrhenius-type (Eq. (15)) at $T < 303$ K with $W = 1.50$ eV to VTF-type (Eq. (17)) at $T > 303$ K. In the $\sigma_{ac}(f)$ plot for the salt ionomer E1.5B2Na (not shown here) no gap is observed, in agreement with the results of DSC measurements, showing that this ionomer is amorphous [68]. In consistency with that, the corresponding Arrhenius plot for σ_{dc} does not present any discontinuity and the VTF Eq. (17) gives a satisfactory fit to the data with reasonable values of the fitting parameters, suggesting that the conductivity mechanism is governed by the motion of the polymer chains [68].

The results reported above for the PU ionomers based on PTMG (M series) and on PEG (E series) show rather high values of protonic conductivity of their acid form, which might be of interest for various technological applications [65–69]. Conductivity occurs through the SS-rich microphase and is governed by the motion of the polymeric chains in the case of amorphous SS-rich microphase. Conductivity increases by 1–2 orders of magnitude by replacing PTMG by PEG (at temperatures higher than the melting temperature of crystallites of the PEG-rich microphase) and decreases upon neutralization and upon HS extension. Table 13 lists σ_{dc} values of several PU ionomers at 25 and 50°C [68].

Table 13. Values of the dc conductivity, σ_{dc}, of several PU ionomers at 25 and 50°C

Sample	σ_{dc} (S/m)		Sample	σ_{dc} (S/m)	
	$T = 25$°C	$T = 50$°C		$T = 25$°C	$T = 50$°C
E1.5B1H	6.6x10^{-7}	2.2x10^{-4}	M1B5H	2.0x10^{-8}	–
E1.5B2H	7.6x10^{-7}	4.1x10^{-5}	E1.5B1Na	4.4x10^{-8}	1.2x10^{-4}
M1B1H	1.5x10^{-6}	1.0x10^{-5}	E1.5B2Na	2.7x10^{-7}	2.9x10^{-5}
M1B2H	8.0x10^{-8}	–	M1B1Na	4.0x10^{-8}	6.5x10^{-7}
M1B3H	3.5x10^{-8}	–	M1B2Na	2.5x10^{-9}	–

6.3. *Telechelics based on poly(ethylene oxide)*

In order to further investigate the structure-property relationships in PU ionomers and to prepare materials with improved electrical conductivity, telechelics based on PEO capped with hydroxyl groups at both ends and with charge groups in the flexible chains were prepared and studied [73,74]. By ionization of the end-groups of such telechelics, model ionomers, the so-called halato-telechelic polymers, can be prepared [58]. PEO was chosen as the basic component because of its excellent solvation properties. To avoid crystallization, PEO of low molecular weight (300) was used. To achieve the single-ion transport mechanism required for a number of applications, one type of ion (acid groups in the flexible chains, protonic conductivity) was used [73].

Three telechelics were prepared and investigated (Scheme 5). They are designated as 2E0.3/2H/2OH, 3E0.3/4H/2OH, and 4E0.3/6H/2OH, respectively, in the order of increasing number of carboxyl groups. 2E0.3, 3E0.3, and 4E0.3 stand for the number of PEO-300 groups, 2H, 4H and 6H for the number of the carboxyl groups and 2OH for the hydroxyl groups at both ends.

Figure 22 shows $\varepsilon''(f)$ plots in wide frequency and temperature ranges for 3E0.3/4H/2OH, representative also for the other two telechelics. The lines are fittings of a sum of an HN expression (Eq. (5)) for the α relaxation and a conductivity term accounting for the large increase of ε'' at low frequencies and high temperatures [73]. Equation (10) was used to calculate the frequency

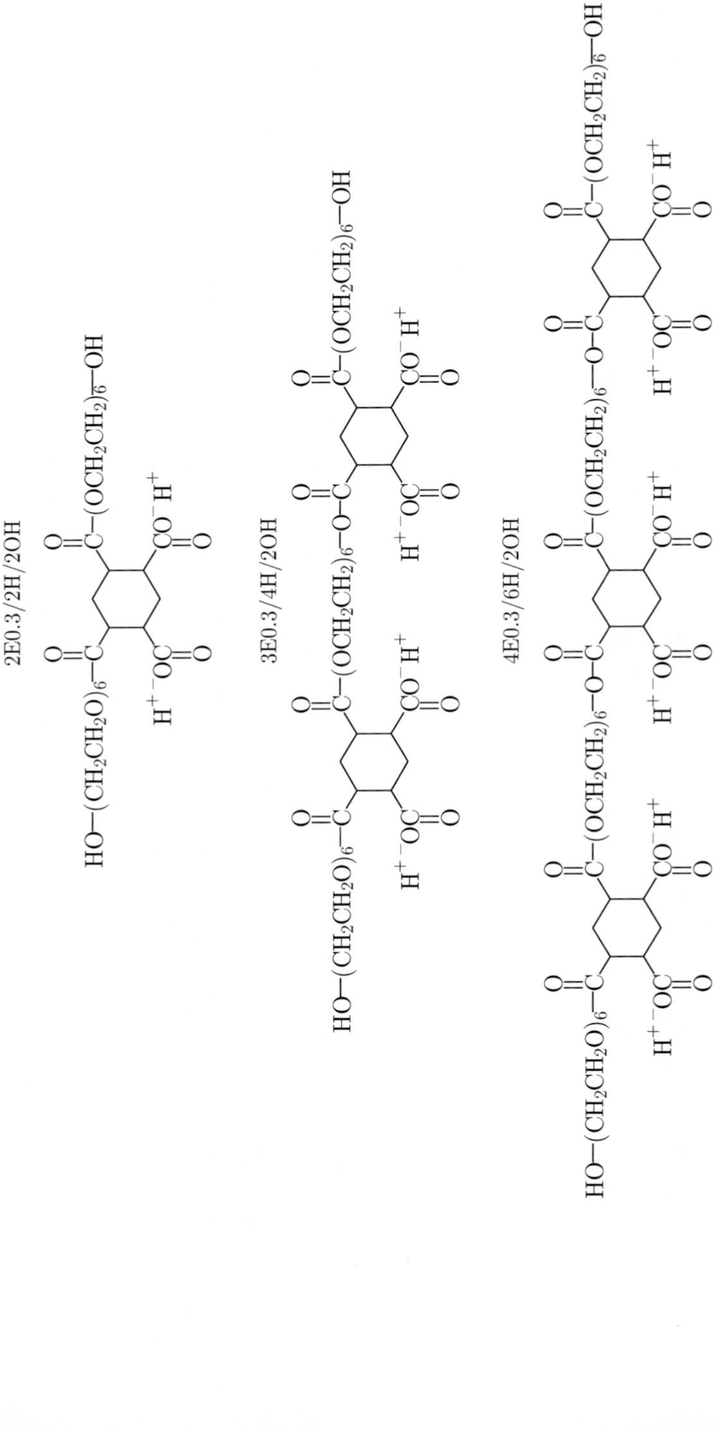

2E0.3/2H/2OH

$HO-(CH_2CH_2O)_6-C(=O)-C_6H_{10}(COO^-H^+)_2-C(=O)-(OCH_2CH_2)_6-OH$

3E0.3/4H/2OH

$HO-(CH_2CH_2O)_6-C(=O)-C_6H_{10}(COO^-H^+)_2-C(=O)-(OCH_2CH_2)_6-O-C(=O)-C_6H_{10}(COO^-H^+)_2-C(=O)-(OCH_2CH_2)_6-OH$

4E0.3/6H/2OH

$HO-(CH_2CH_2O)_6-C(=O)-C_6H_{10}(COO^-H^+)_2-C(=O)-(OCH_2CH_2)_6-O-C(=O)-C_6H_{10}(COO^-H^+)_2-C(=O)-(OCH_2CH_2)_6-O-C(=O)-C_6H_{10}(COO^-H^+)_2-C(=O)-(OCH_2CH_2)_6-OH$

Scheme 5. Molecular structure of the telechelics based on PEO-300 [73]

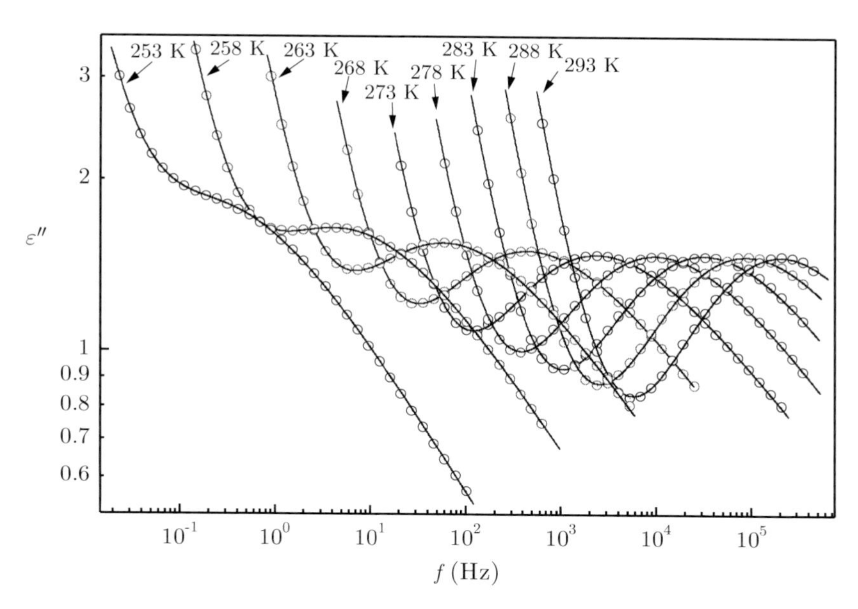

Figure 22. Imaginary part, ε'', of the dielectric permittivity for 3E0.3/4H/2OH *vs.* frequency at several temperatures indicated on the plot [73]

dependence of the ac conductivity, $\sigma_{ac}(f)$, at each temperature. The corresponding plot [73] (not shown here) exhibits features typical of ionic conductors [28,49,65–69]: increase of σ_{dc} and extension of the frequency region of σ_{dc} behavior to higher frequencies with increasing temperature.

The results presented in Figure 22 and similar results obtained for the other two telechelics were used, after analysis, to construct the Arrhenius plot shown in Figure 23. Included in the plot are also DSC and TSDC data at the equivalent frequency of 1.6 mHz [73]. The lines are fits of the VTF Eq. (13) to the experimental data with reasonable values of the fitting parameters. The fits allow to calculate $T_{g,diel}$, defined by $\tau(T_{g,diel}) = 100$ s, where $\tau = 1/2\text{p}f_{max}$ [73]. This condition gives $T_{g,diel}$ values very close to those of T_g determined by DSC and TSDC (Table 14). The inset to Figure 23 shows the normalized plot with respect to T_g determined by DSC [75], which suggests similar fragilities of the

Table 14. Values of the dc conductivity, σ_{dc}, of telechelics at 298 K, 333 K and, for comparison, at $T/T_{g,DSC} = 1.2$, and glass transition temperatures determined by DSC, TSDC, and DRS [73]

Sample	σ_{dc} (S/cm)		σ_{dc} (S/cm)	$T_{g,DSC}$	$T_{g,TSDC}$	$T_{g,diel}$
	298 K	333 K	$T/T_{g,DSC} = 1.2$	(K)	(K)	(K)
2E0.3/2H/2OH	$8.8\text{x}10^{-8}$	$1.4\text{x}10^{-6}$	$3.0\text{x}10^{-8}$	241 K	237 K	235 K
3E0.3/4H/2OH	$1.6\text{x}10^{-9}$	$5.1\text{x}10^{-8}$	$2.3\text{x}10^{-9}$	251 K	251 K	246 K
4E0.3/6H/2OH	$3.8\text{x}10^{-10}$	$2.4\text{x}10^{-8}$	$2.7\text{x}10^{-9}$	260 K	-	256 K

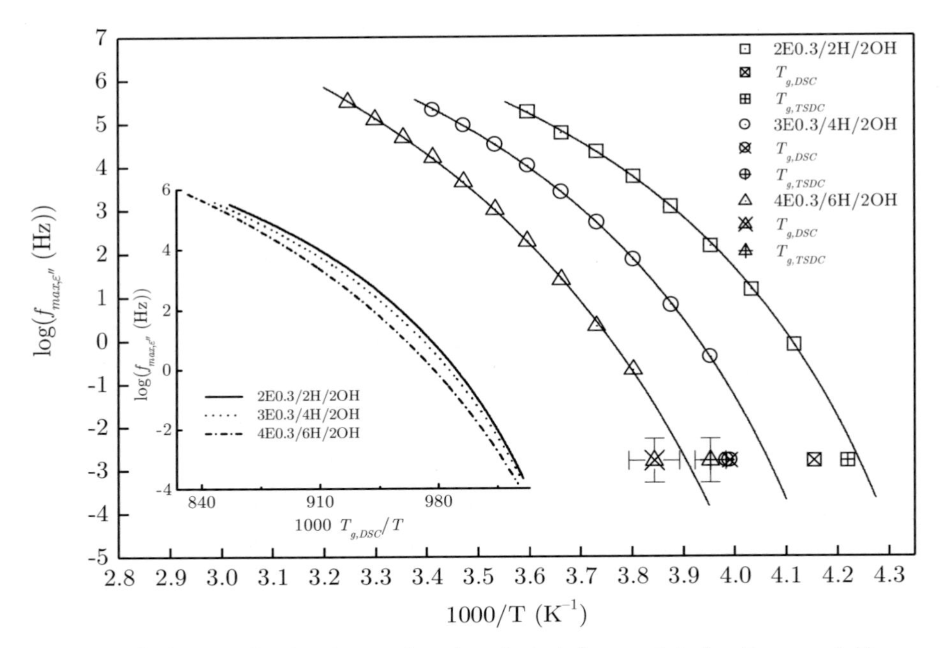

Figure 23. Arrhenius plot for the α relaxation. Included are points for $T_{g,DSC}$ and $T_{g,TSDC}$ at the equivalent frequency of 1.6 mHz with estimated error bars. The inset shows the normalized plot with respect to T_g [73]

three telechelics [73]. The temperature dependence of the dc conductivity, σ_{dc}, in the three telechelics investigated is described by the VTF Eq. (17), indicating that their protonic conductivity is governed by the motion of the polymeric chains. The absolute values of σ_{dc} (listed in Table 14), derived from the σ_{ac} plots, are higher for the 2E0.3/2H/2OH sample, as compared to the other two samples. The question arises whether this is due to the lower glass transition temperature of that ionomer, since the conductivities in Table 14 were compared at the same temperature. Results for σ_{dc} at $T/T_{g,DSC} = 1.2$ included in Table 14 suggest that the higher σ_{dc} value for the 2E0.3/2H/2OH sample is not only due to its lower glass transition temperature, in agreement with a more thorough analysis by means of T_g-normalized Arrhenius plots [73].

In several amorphous ionically conducting systems, including polymers, a separation between α relaxation and the dc conductivity (in general, between rotational and translational modes) has been observed with the temperature decreasing towards T_g [76]. This effect, known as decoupling, was further studied with the three telechelics of Scheme 5 [74]. Figure 24 shows comparative Arrhenius plots of the α relaxation and of dc conductivity for the sample 3E0.3/4H/2OH. The two plots approach each other with the temperature decreasing towards T_g. To compare the time scales of α relaxation and σ_{dc} in absolute values, the inset to Figure 24 shows the Arrhenius plot of f_{max} for the

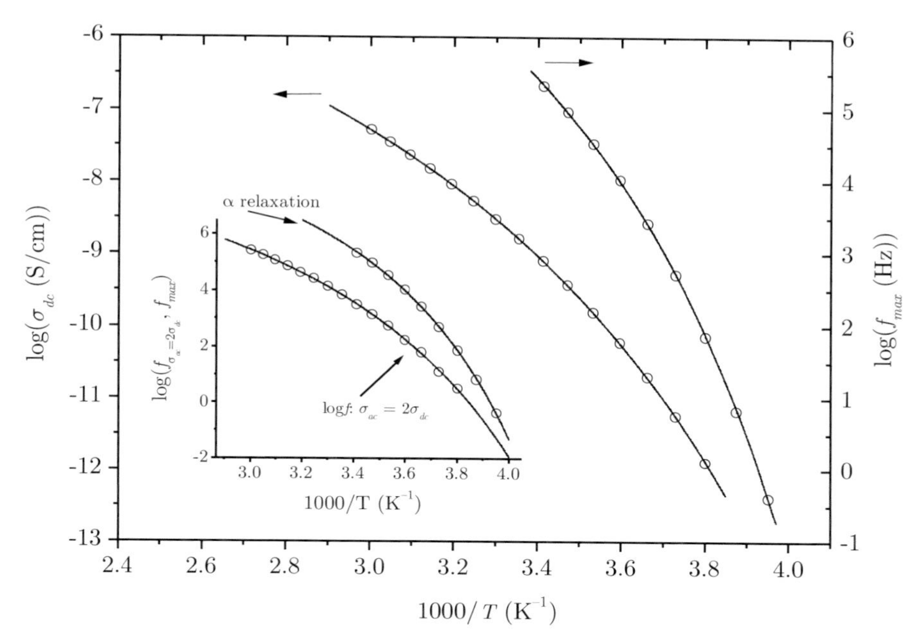

Figure 24. Arrhenius plot of the maximum frequency, f_{max}, of the α relaxation and of the dc conductivity σ_{dc} for the sample 3E0.3/4H/2OH. The inset shows the Arrhenius plot of a relaxation and of conductivity relaxation for the same sample [74]

α relaxation and of the frequency where σ_{ac} becomes $2\sigma_{dc}$, $f_{\sigma_{ac}=2\sigma_{dc}}$. This frequency has been widely considered as characteristic for the conductivity relaxation [77]. The α relaxation is shown to remain faster than the conductivity relaxation, even at temperatures close to T_g, where the time scales of the two relaxations become similar [74]. For several inorganic ionic conductors, the decoupling index R, defined by the ratio of the relaxation times of the structural relaxation (α) and the conductivity relaxation, both at T_g [78], is higher by orders of magnitude than 1 at T_g and decreases towards unit at high temperatures, by extending the definition of R to variable temperatures [79]. To compare with this behavior, we calculated the relaxation times τ for both processes and the decoupling index R for 4E0.3/6H/2OH. The T_g-normalized plot of R for this sample [74] shows qualitatively the same temperature dependence as in inorganic conductors [79], an issue which should be further followed in future work.

The conductivity in the telechelics was investigated also at various pressures at fixed temperature (298 K). Figure 25 shows complex impedance plots for 3E0.3/4H/2OH at three different pressures. The lines are semicircular fits, indicating a Debye conductivity mechanism. $R_{dc}(=1/G_{dc})$ is obtained by extrapolation to low frequencies and σ_{dc} by $\sigma_{dc} = G_{dc}d/S$, where d is the thickness and S is the surface of the sample. As expected, σ_{dc} decreases with increasing pressure. The plot in the left inset to Figure 25 shows the variation of σ_{dc} with pressure for this sample. The linear fit of the equation [80]

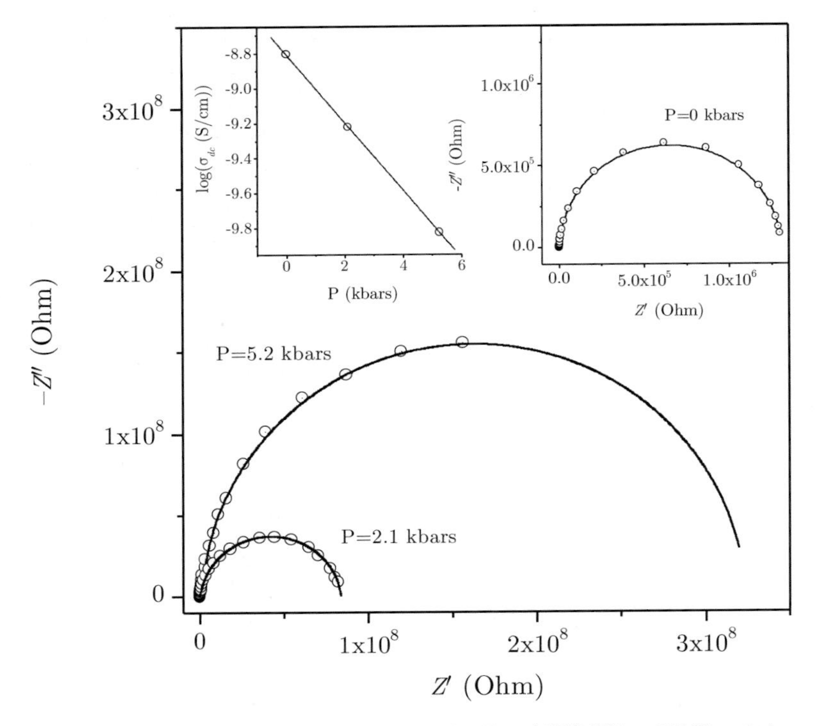

Figure 25. Complex impedance plot for the sample 3E0.3/4H/2OH at 298 K and the pressure indicated on the plot. The inset shows the variation of σ_{dc} with pressure [74]

$$\left(\frac{\partial \ln \sigma}{\partial P}\right)_T = -\frac{\Delta V^*}{\mathrm{R}T} \tag{18}$$

where R is the universal gas constant, gives $\Delta V^* = 4.7\,\mathrm{cm}^3\,\mathrm{mol}^{-1}$ for the apparent activation volume.

7. Nanocomposites based on polyurethane thermoplastic elastomers

Polymer nanocomposites have attracted much interest in recent years [81–84]. The mechanical and the physical properties of the nanocomposites are significantly improved, as compared to those of the polymer matrix, at much smaller filler contents than would be required for conventional macroscale or microscale composites. They also exhibit distinctive properties related to the small particle size and correspondingly small mean interparticle spacing (typically also in the nanometer range). So far, there is no satisfactory theoretical explanation of the origin of property improvements in polymer nanocomposites, but it is generally accepted that the large surface-to-volume ratio of the nanoscale inclusions plays a significant role. The presence of an interfacial layer between the bulk polymer and the filler surface, with altered structure and chain mobility, has been established by various techniques [85,86]. The increase in free volume due to loosened molecular packing of the chains confined between the nanoparticles may also contribute to a modification of the polymer structure and dynamics [87]. DRS and TSDC techniques, widely employed in the framework of the investigation of structure-property relationships in polymeric systems, are well suited for studying modifications of polymer dynamics and interfacial effects in polymer nanocomposites [88–90].

PTE are often being used as matrices in polymer nanocomposites. Much effort has been devoted to optimizing the composition and the preparation and processing conditions for the obtaining of polyurethane/layered silicate nanocomposites of exfoliated structure with improved barrier properties [91,92]. An interesting question both from the fundamental and the technological point of view, with respect to microphase separation, is whether the latter is modified by the presence of nanofiller. DRS and TSDC measurements in epoxy resin/layered silicate nanocomposites indicated that the large-scale heterogeneity of the structure of the epoxy resin matrix becomes less pronounced on addition of the nanoparticles [88].

Figure 26 shows DRS results for the α relaxation associated with the glass transition of the polymeric matrix obtained with polyurethane/montmorillonite nanocomposites [93]. The PU matrix was synthesized in two stages from PTMG (MW 1040), MDI, and asymmetrical DMH as chain extender (sample PU2 in Table 5). The samples were designated as HD/X, where X gives the content of montmorillonite, X = 0–15 wt%. The results in Figure 26 indicate that the time

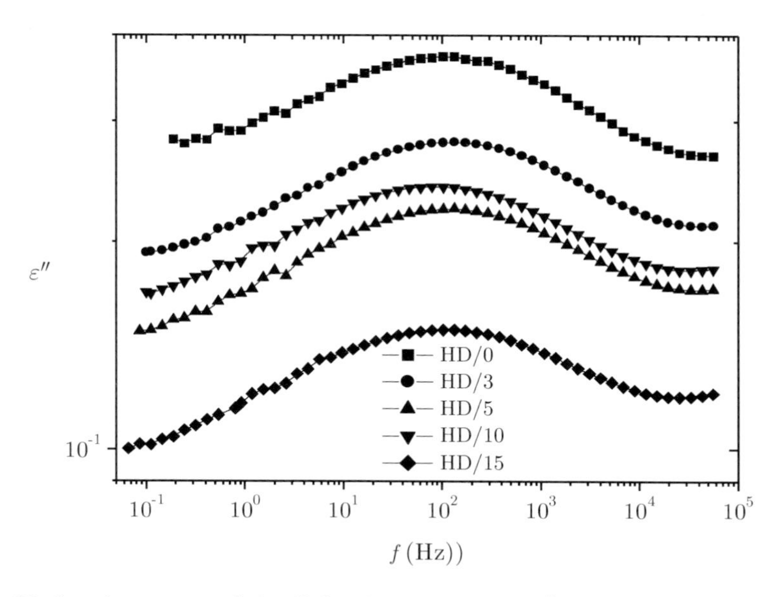

Figure 26. Imaginary part of the dielectric permittivity, ε'', *vs.* frequency, f, for the α relaxation for the nanocomposites indicated on the plot at $T = 223$ K

scale and the shape of the relaxation, quantified in terms of the HN expression (Eq. (5)), are not influenced by the presence of the nanoparticles, whereas the magnitude of the relaxation decreases on addition of montmorillonite more than additivity would require. These results, confirmed also by TSDC measurements [93], suggest that a part of the polymer chain is immobilized on the montmorillonite surfaces, whereas the rest behaves like the neat matrix, *i.e.*, without any change in the microphase-separated morphology. Results for the frequency dependence of the ac conductivity, σ_{ac}, calculated by Eq. (10), indicate a reduction of σ_{ac} and σ_{dc} in the nanocomposites, as compared to the neat matrix, in agreement with the improvement of barrier properties, established by water sorption/diffusion measurements [93].

In a series of recent papers, the preparation and the investigation of structure-property relationships in organic-inorganic nanocomposites (OICs) based on linear and crosslinked PUs and sodium silicate were reported [94–96]. The OICs were formed by joint polymerization of the urethane oligomer and the aqueous solution of sodium silicate. The urethane oligomer was synthesized from oligooxypropylene glycol (OPG) of various MW and functionality $f = 2$ or 3 and 2,4-toluene diisocyanate (TDI). The OPGs were marked by 1052, 2102, 3603 and 5003 according to MW and f (*e.g.*, OPG-1052 means MW $=$ 1050 and $f = 2$). The TDI/OPG molar ratio was 2/1 for the bifunctional and 3/1 for the trifunctional OPGs. Thus, urethane oligomers with various MW and linear or branched structure were obtained. The OICs were marked by OIC-1 to OIC-5 according to the OPG used. The corresponding initial PUs were obtained by polymerization of the urethane oligomers with the air moisture and were marked by PU-1 to PU-5 [96].

WAXS, SAXS, DSC, dynamic mechanical thermal analysis, stress-strain measurements, water sorption/diffusion, DRS, and TSDC were employed to investigate structure-property relationships in the OICs. The WAXS data in Figure 27 show one diffraction peak of the diffusion type for the initial PUs, which shifts slightly to lower q values for the branched, as compared to the linear PUs. These results indicate that the initial PUs are amorphous and that the branched samples are characterized by a slightly larger distance between the atomic layers than the linear ones. For the OICs, the data in Figure 27 show additional diffraction maxima superimposed to the diffusive diffraction maximum of PU. These maxima are caused by the crystalline structure of the inorganic component (IC). The estimation of the size of the IC crystallites by Sherrer's method gives values around 7.5 nm [96]. WAXS data obtained with sample OIC-3, which has been exposed to water (immersion) and then dried prior to the WAXS measurements (curve $4'$ in Figure 27), show a decrease in the intensity of the additional maxima and suggest, in agreement with SAXS data not shown here, partial degradation of the crystal phase of the IC in the OICs. The SAXS data (not shown here) suggest that the initial PUs are characterized by a weak microphase separation with Bragg's distance between the microranges of heterogeneity of about 5 nm in the linear and 7.5 nm in the

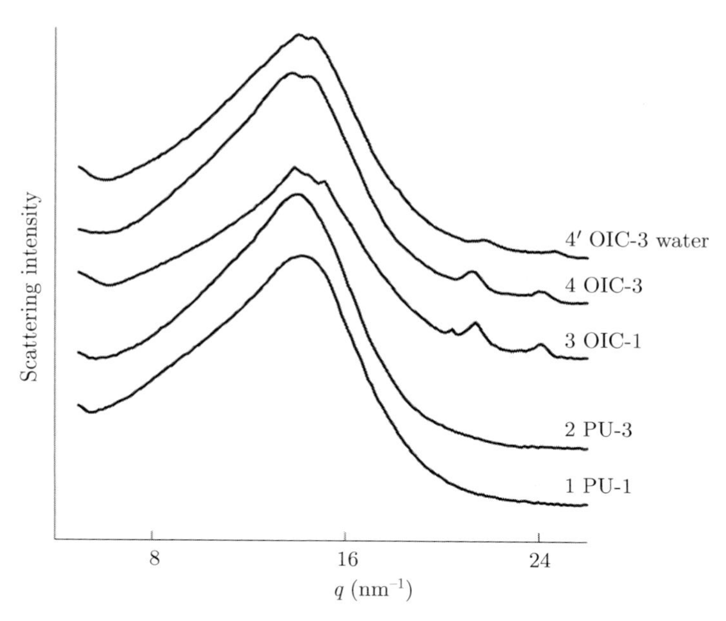

Figure 27. WAXS profile of initial polyurethanes (PUs) and composites (OICs) with 20 wt% inorganic component indicated on the plot

branched samples [96]. SAXS data obtained with the OICs suggest a decrease in the degree of heterogeneity with increasing length of the flexible chain component, *i.e.*, with the transition from OIC-1 to OIC-2 and from OIC-3 to OIC-5, obviously due to the corresponding decrease of the HS fraction. It is interesting to note the similar size of the two scales of heterogeneity in the OICs, *i.e.*, of the microphase separation of the PU matrix and of the IC crystallites.

Figure 28 shows results of DRS measurements: real part of the dielectric permittivity, ε', against frequency, f, at 25°C in PU-1 and the corresponding nanocomposites with 10, 20, and 30 wt% IC [94]. ε' is a measure of the molecular mobility, as it reflects the ability of dipolar units to follow the changes of the applied electric field. ε' and, thus, the molecular mobility increases in the nanocomposites, as compared to the neat PU matrix, the changes being more significant for OIC-1-10 wt%. The overall increase in molecular mobility observed in Figure 28 is possibly caused by an increase of free volume due to loosened packing of polymer chains in the nanocomposites, in agreement with results obtained with other polymeric nanocomposites [87]. The significant increase of ε' with decreasing frequency in the nanocomposites with 20 and 30 wt% IC in Figure 28 obviously does not reflect a dipolar behavior and is related to conductivity effects and space charge polarization. The overall increase of the molecular mobility, indicated by DRS measurements, is supported also by the results of TSDC measurements [95]. The corresponding TSDC thermograms exhibit

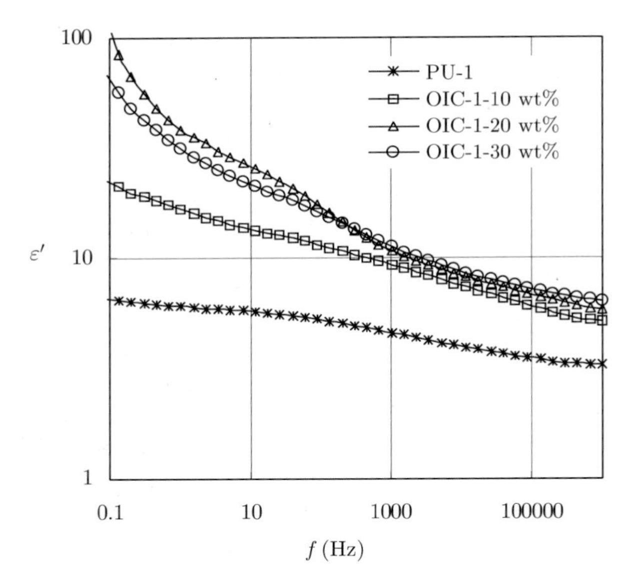

Figure 28. Real part of dielectric permittivity, ε', against frequency in PU-1 and various OICs-1 at 25°C

secondary relaxations at lower temperatures and the α relaxation asociated with the glass transition at about −20°C, which increase in magnitude on addition of the IC. The TSDC α peak shifts slightly to higher temperatures on addition of the IC, probably due to a contribution of interfacial polarization to the peak, as DSC measurements on the same samples indicate no shift of T_g, in agreement with the coexistence of the microphase-separated morphology [96].

8. Conclusions

We presented results on the chain dynamics and on the ionic conductivity in polyurethane thermoplastic elastomers and complex systems on their basis obtained in the framework of the investigation of structure-property relationships in these materials. The results were obtained by dielectric techniques, including broadband dielectric spectroscopy (DS) and thermally stimulated depolarization currents (TSDC), and were discussed in relation to those obtained by other experimental methods. Dielectric techniques proved very powerful in these investigations, their main characteristic features/advantages over other techniques being the extremely broad frequency range covered by DS and the high sensitivity and high peak resolving power of TSDC.

Polyurethane thermoplastic elastomers, like other segmented polyurethanes, are characterized by microphase separation into hard-segment microdomains and a soft-segment microphase, which is the cause for their versatile physical and thermomechanical properties. Considerable work has been devoted to introduce quantitative measures of the degree of microphase separation in segmented polyurethanes by various experimental techniques. The suitability of dielectric techniques to that aim was discussed in the present chapter.

The microphase-separated morphology of polyurethane thermoplastic elastomers is preserved in complex multicomponent systems on their basis; the degree of microphase separation may be, however, modified by the presence of additional components. Two examples of such systems are discussed in some detail in the present chapter. In the first example, polyurethane ionomers with ionic moieties in the soft segments, the chain microphase separation morphology coexists with the ionic group segregation morphology. In the specific example of the M1B1Na ionomer of Section 6.2., the polyurethane microdomain spacing is 17 nm and the size of the ionic aggregates, located within the interdomain soft-segment layers, is about 6 nm. In the second example of organic-inorganic polymer nanocomposites, consisting of a polyurethane organic phase and a mineral inorganic phase, the second morphology coexisting with the chain microphase separation morphology is that of the nanoparticles of the inorganic phase, the two morphologies having similar characteristic lengths (Section 7).

Acknowledgement

The authors wish to thank Prof. V. P. Privalko, Dr. Yu. V. Savelyev, Prof. V. V. Shilov, Prof. E. V. Lebedev and Dr. Ye. P. Mamunya from the Institute of Macromolecular Chemistry of the National Academy of Sciences of Ukraine, Kiev, Ukraine; Prof. A. Vassilikou-Dova from the University of Athens; Dr. A. Kyritsis, Dr. A. Kanapitsas and Ms. S. Kripotou from the National Technical University of Athens, for providing data and/or for valuable discussions.

References

1. Noshay A and McGrath J E (1997) *Block Copolymers*, Academic Press, New York.
2. Legge N R, Holden G and Schroeder H E (Eds.) (1987) *Thermoplastic Elastomers – a Comprehensive Review*, Hanser Publishers, Munich.
3. Spathis G, Kontou E, Kefalas V, Apekis L, Christodoulides C, Pissis P, Ollivon M and Quinquenet S (1990) Relaxation phenomena and morphology of polyurethane block copolymers, *J Macromol Sci-Phys* **B29**:31–48.
4. Chen K T, Chui Y J and Shieh S T (1997) Glass transition behaviors of a polyurethane hard segment based on 4,4′-diisocyanatodiphenylmethane and 1,4-butanediol and the calculation of microdomain composition, *Macromolecules* **30**:5068–5074.
5. Savelyev Yu V, Akhranovich E R, Grekov A P, Privalko E G, Korskanov V V, Shtompel V I, Privalko V P, Pissis P and Kanapitsas A (1998) Influence of chain extenders and chain end groups on properties of segmented polyurethanes. I. Phase morphology, *Polymer* **39**:3425–3429.
6. Kanapitsas A, Pissis P, Gomez Ribelles J L, Monleon Pradas M, Privalko E G and Privalko V P (1999) Molecular mobility and hydration properties of segmented polyurethanes with varying structure of soft and hard chain segments, *J Appl Polym Sci* **71**:1209–1221.
7. Koberstein J T, Galambos A F and Leung L M (1992) Compression-molded polyurethane block copolymers. 1. Microdomain morthology and thermomechanical properties, *Macromolecules* **25**:6195–6204.

8. Chu B, Gao T, Li Y, Wang J, Desper C R and Byrhe C A (1992) Microphase separation kinetics in segmented polyurethanes: effects of soft segment length and structure, *Macromolecules* **25**:5724–5729.
9. Cuve L, Pascault J P, Boiteux G and Seytre G (1991) Synthesis and properties of polyurethanes based on pdyolefine: 1. Rigid polyurethanes and amorphous segmented polyurethanes prepared in polar solvents under homogeneous conditions, *Polymer* **32**:343–352.
10. Pissis P, Kanapitsas A, Savelyev Yu V, Akhranovich E R, Privalko E G and Privalko V P (1998) Influence of chain extenders and chain end-groups on properties of segmented polyurethanes. II. Dielectric study, *Polymer* **39**:3431–3435.
11. Fruebing P, Krueger H, Goering H and Gerhard-Multhaupt R (2002) Relaxation behavior of thermoplastic polyurethanes with covalently attached nitroaniline dipoles, *Polymer* **43**:2787–2794.
12. Kremer F and Schoenhals A (Eds) (2003) *Broadband Dielectric Spectroscopy*, Springer, Berlin.
13. Georgoussis G, Kanapitsas A, Pissis P, Savelyev Yu V, Veselov V Ya and Privalko E G (2000) Structure-property relationships in segmented polyurethanes with metal chelates in the main chain, *Eur Polym J* **36**:1113–1126.
14. Daniel V N (1967) *Dielectric Relaxation*, Academic Press, London.
15. Hill N E, Vaughan W, Price A H and Davies M (1969) *Dielectric Rroperties and Molecular Behaviour*, Van Nostrand, London.
16. Macdonald J R (Ed.) (1987) *Impedance Spectroscopy*, Wiley, New York.
17. Runt J P and Fitzgerald J J (Eds.) (1997) *Dielectric Spectroscopy of Polymeric Materials*, American Chemical Society, Washington, DC.
18. Takeishi S and Mashimo S (1992) Dielectric relaxation measurements in the ultralow frequency region, *Rev Sci Instrum* **53**:1155–1159.
19. Nozaki S and Bose T K (1990) Broadband complex permittivity measurements by time-domain spectroscopy, *IEEE Trans Instrum Meas* **39**:945–951.
20. Feldman Y, Andrianov A, Polygalov E, Ermolina I, Romanychev G, Zuev Y and Milgotin B (1996) Time domain dielectric spectroscopy: an advanced measuring system, *Rev Sci Instrum* **67**:3208–3216.
21. Moynihan C T, Boesch L P and Laberge N L (1973) Decay function for the electric field relaxation in vitreous ionic conductors, *Phys Chem Glasses* **14**:122–125.
22. Havriliak S Jr and Havriliak S J (1997) *Dielectric and Mechanical Relaxation in Materials*, Hanser, Munich.
23. Jonscher A K (1983) *Dielectric Relaxation in Solids*, Chelsea Dielectrics, London.
24. Vatalis A S, Kanapitsas A, Delides C G and Pissis P (2001) Relaxation phenomena and morphology in polymer blends based on polyurethanes investigated by various thermal analysis techniques, *Thermochim Acta* **372**:33–38.
25. Van Turnhout J (1980) Thermally stimulated discharge of electrets, in *Electrets Topics in Applied Physics* (Ed. Sessler G M) Springer, Berlin, Vol. 33, pp. 81–215.
26. Pissis P, Anagnostopoulou-Konsta A, Apekis L, Daoukaki-Diamanti D and Christodoulides C (1991) Dielectric effects in water-containing systems, *J Non-Cryst Solids* **131-133**:1174–1181.
27. Kyritsis A, Pissis P, Gomez Ribelles J L and Monleon Pradas M (1994) Depolarization thermocurrent studies in poly(hydroxyethyl acrylate)/water hydrogels, *J Polym Sci Part B Polym Phys Ed* **32**:1001–1008.
28. Funke K and Cramer C (1997) Conductivity spectroscopy, *Current Opinion in Solid State & Materials Science* **2**:483–490.

29. Privalko V P, Khaenko E S, Grekov A P and Savelyev Yu V (1994) Structure-property relationships for a series of crown ether-containing polyurethane ureas, *Polymer* **35**:1730–1738.
30. Apekis L, Pissis P, Christodoulides C, Spathis G, Niaounakis M, Kontou E, Schlosser E, Schoenhals A and Goering H (1992) Physical and chemical network effects in polyurethane elastomers, *Colloid Polym Sci* **90**:144–150.
31. Garret J T, Runt J and Lin J S (2000) Microphase separation of segmented poly(urethane urea) block copolymers, *Macromolecules* **33**:6353–6359.
32. Koberstein J T and Leung L M (1992) Compression-molded polyurethane block copolymers. 2. Evaluation of microphase compositions, *Macromolecules* **25**:6205–6213.
33. Li Y, Kang W, Stoffer J O and Chu B (1994) Effect of hard-segment flexibility on phase separation of segmented polyurethanes, *Macromolecules* **27**:612–614.
34. Privalko V P, Privalko E G, Shtompel V I, Pissis P, Kanapitsas A, Monleon Pradas M and Gomez Ribelles J L (1999) Influence of the structure of soft and stiff chain fragments on properties of segmented polyurethanes. I. Phase morphology, *Polym Eng Sci* **39**:1534–1539.
35. Boiteux G, Cuve L and Pascault J-P (1994) Synthesis and properties of polyurethanes based on polyolefin. 3. Monitoring of phase separation by dielectric relaxation spectroscopy of segmented semicrystalline polyurethane prepared in bulk by the use of emulsifiers, *Polymer* **35**:173–178.
36. Ishida M and Yoshinaga K (1996) Solid-state ^{13}C NMR analyses of the microphase-separated structure of polyurethane elastomer, *Macromolecules* **29**:8824–8829.
37. Chen W-P and Schlick S (1990) Study of phase separation in polyurethanes using paramagnetic labels: effect of soft-segment molecular weight and temperature, *Polymer* **31**:308–314.
38. Velankar S and Cooper S L (1998) Microphase separation and rheological properties of polyurethane melts. 1. Effect of block length, *Macromolecules* **31**:9181–9192.
39. Garrett J T, Lin J S and Runt J (2002) Influence of preparation conditions on microdomain formation in poly(urethane urea) block copolymers, *Macromolecules* **35**:161–168.
40. Li Y, Liu J, Yang H, Ma D and Chu B (1993) Multiphase structure of segmented polyurethanes: its relation with spherulite structure, *J Polym Sci B Polym Phys Ed* **31**:853–867.
41. Koberstein J T and Galambos A F (1992) Multiple melting in segmented polyurethane block copolymers, *Macromolecules* **25**:5618–5624.
42. Etienne S, Vigier G, Cuve L and Pascault J-P (1994) Microstructure of segmented amorphous polyurethanes: small-angle X-ray scattering and mechanical spectroscopy studies, *Polymer* **35**:2737–2743.
43. Pompe G, Pohlers A, Poetschke P and Pionteck J (1998) Influence of processing conditions on the multiphase structure of segmented polyurethane, *Polymer* **39**:5147–5153.
44. Li W, Ryan A J and Meier I K (2002) Effect of chain extenders on the morphology development in flexible polyurethane foam, *Macromolecules* **35**:6306–6312.
45. Garret J T, Runt J and Lin J S (2000) Microphase separation of segmented poly(urethane urea) block copolymers, *Macromolecules* **33**:6353–6359.
46. Li Y, Gao T and Chu B (1992) Synchrotron SAXS studies of the phase-separation kinetics in a segmented polyurethane, *Macromolecules* **25**:1737–1742.
47. Kang W and Stoffer J O (2000) Effect of hard segments on morphology of polyurethanes, *Polym Prepr* **41**:1132–1133.

48. Leung L M and Koberstein J T (1986) DSC annealing study of microphase separation and multiple endothermic behavior in polyether-based polyurethane block copolymers, *Macromolecules* **19**:706–713.
49. Shilov V V, Shevchenko V V, Pissis P, Kyritsis A, Georgoussis G, Gomza Yu P, Nesin S D and Klimenko N S (2000) Morphology, dielectric relaxation and conductivity of the novel polyurethanes with acid and ionic groups in the polyether segments, *J Non-Cryst Solids* **275**:116–136.
50. Georgoussis G, Kyritsis A, Pissis P, Savelyev Yu V, Akhranovich E R, Privalko E G and Privalko V P (1999) Dielectric studies of molecular mobility and microphase separation in segmented polyurethanes, *Eur Polym J* **35**:2007–2017.
51. McCrum G N, Read E B and Williams G (1967) *Anelastic and Dielectric Effects in Polymeric Solids*, Wiley, New York.
52. Hedvig P (1977) *Dielectric Spectroscopy in Polymers*, Adam Hilger, Bristol.
53. Roussos M, Konstantopoulou A, Kalogeras I M, Kanapitsas A, Pissis P, Savelyev Yu and Vassilikou-Dova A (2004) Comparative dielectric studies of segmental mobility in novel polyurethanes, *e-Polymers 2004* no. 042 (http://www.e-polymers.org, ISSN 1618-7229).
54. Pissis P, Apekis L, Christodoulides C, Niaounakis M, Kyritsis A and Nedbal J (1996) Water effects in polyurethane block copolymers, *J Polym Sci Part B Polym Phys* **34**:1529-1536.
55. Raftopoulos K, Zegkinoglou I, Kanapitsas A, Kripotou S, Christakis I, Vassilikou-Dova A, Pissis P and Savelyev Yu (2004) Dielectric and hydration properties of segmental polyurethanes, *e-Polymers 2004* no. 043 (http://www.e-polymers.org, ISSN 1618-7229).
56. Donth E (2001) *The Glass Transition*, Springer, Berlin.
57. Privalko V P (1999) Glass transition in polymers: dependence of the Kohlrausch stretching exponent of kinetic free volume fraction, *J Non-Cryst Solids* **255**:259–263.
58. Tant M R, Mauritz K A and Wilkes G L (Eds.) (1997) *Ionomers*, Chapman and Hall, London.
59. Pineri M and Eisenberg A (Eds.) (1987) *Structure and Properties of Ionomers*, Reidel Publishing Co, Dordrecht.
60. Eisenberg A and King M (1977) *Ion-Containing Polymers, Physical Properties and Structure*, Academic Press, New York.
61. Goddard R G and Cooper S L (1995) Polyurethane cationomers with pendant trimethylammonium groups. 1. Fourier transform infrared temperature studies, *Macromolecules* **28**:1390–1400.
62. Goddard R G and Cooper S L (1995) Polyurethane cationomers with pendant trimethylammonium groups. 2. Investigation of the microphase separation transition, *Macromolecules* **28**:1401–1406.
63. Charnetskaya A G, Polizos G, Shtompel V I, Privalko E G, Kercha Yu Yu and Pissis P (2003) Phase morphology and molecular dynamics of a polyurethane ionomer reinforced with a liquid crystalline filler, *Eur Polym J* **39**:2167–2174.
64. Yamamoto K and Namikawa H (1988) Conduction current relaxation of inhomogeneous conductor I, *Jpn J Appl Phys* **27**:1845–1851.
65. Shilov V V, Shevchenko V V, Pissis P, Kyritsis A, Gomza Yu P, Nesin S D and Klimenko N S (1999) Morphology and protonic conductivity of the polyurethanes with acid groups in the flexible segment, *Solid State Ionics* **120**:43–50.
66. Pissis P, Kyritsis A and Shilov V V (1999) Molecular mobility and protonic conductivity in polymers: hydrogels and ionomers, *Solid State Ionics* **125**:203–212.

67. Pissis P, Kyritsis A, Georgoussis G, Shilov V V and Shevchenko V V (2000) Structure-property relationships in proton conductors based on polyurethanes, *Solid State Ionics* **136-137**:255–260.
68. Polizos G, Kyritsis A, Pissis P, Shilov V V and Shevchenko V V (2000) Structure and molecular mobility studies in novel polyurethane ionomers based on poly(ethylene oxide), *Solid State Ionics* **136-137**:1139–1146.
69. Polizos G, Georgoussis G, Kyritsis A, Shilov V V, Schevchenko V V, Gomza Yu P, Nesin S D, Klimenko N S, Wartewig S and Pissis P (2000) Structure and electrical conductivity in novel polyurethane ionomers, *Polym Int* **49**:987–992.
70. Visser S A and Cooper S L (1991) Analysis of small-angle X-ray scattering data for model polyurethane ionomers: Evaluation of hard-sphere models, *Macromolecules* **24**:2584–2593.
71. Fokin A G (1996) Macroscopic conductivity of random inhomogeneous media. Calculation methods, *Physics-Uspekhi* **39**:1009–1032.
72. Pelster R and Simon U (1999) Nanodispersions of conducting particles: preparation, microstructure and dielectric properties, *Colloid Polym Sci* **277**:2–14.
73. Polizos G, Shilov V V and Pissis P (2001) Molecular mobility and protonic conductivity studies in telechelics based on poly(ethylene oxide) capped with hydroxyl groups at both ends, *Solid State Ionics* **145**:93–100.
74. Polizos G, Shilov V V and Pissis P (2002) Temperature and pressure effects on molecular mobility and ionic conductivity in telechelics based on poly(ethylene oxide) capped with hydroxyl groups at both ends, *J Non-Cryst Solids* **305**:212–217.
75. Angell C A (1991) Relaxation in liquids, polymers and plastic crystals – strong/fragile patterns and problems, *J Non-Cryst Solids* **131-133**:13–31.
76. Fan J, Marzke R F, Sanchez E and Angell C A (1994) Conductivity and nuclear spin relaxation in superionic glasses, polymer electrolytes and the new polymer-in-salt electrolyte, *J Non-Cryst Solids* **172-174**:1178–1189.
77. Ngai K L (1996) A review of critical experimental facts in electrical relaxation and ionic diffusion in ionically conducting glasses and melts, *J Non-Cryst Solids* **203**:232–245.
78. Angell C A (1990) Dynamic processes in ionic glasses, *Chem Rev* **90**:523–541.
79. Moynihan C T (1998) Contributions of dynamic heterogeneity to non-exponential electrical relaxation in ionically conducting glasses and melts, *J Non-Cryst Solids* **235-237**:781–788.
80. Duclot M, Alloin F, Brylev O, Sanchez J Y and Souquet J L (2000) New alkali ionomers: transport mechanism from temperature and pressure conductivity measurements, *Solid State Ionics* **136-137**:1153–1160.
81. Giannelis E P, Krishnamoorti R and Manias E (1999) Polymer-silicate nanocomposites: Model systems for confined polymers and polymer brushes, *Adv Polym Sci* **138**:107–147.
82. Alexandre M and Dubois P (2000) Polymer-layered silicate nanocomposites: preparation, properties and uses of a new class of materials, *Mater Sci Eng* **28**:1–63.
83. Sanchez C, Lebeau B, Chaput F and Boilot J-P (2003) Optical properties of functional hybrid organic-inorganic nanocomposites, *Adv Mater* **15**:1969–1994.
84. Kickelbick G (2003) Concepts for the incorporation of inorganic building blocks into organic polymers on a nanoscale, *Progr Polym Sci* **28**:83–114.
85. Tsagaropoulos G and Eisenberg A (1995) Dynamic mechanical study of the factors affecting the two glass transition behavior of filled polymers. Similarities and differences with random ionomers, *Macromolecules* **28**:6067–6077.

86. Litvinov V M, Barthel H and Weis J (2002) Structure of a PDMS layer grafted onto a silica surface studied by means of DSC and solid-state NMR, *Macromolecules* **35**:4356–4364.
87. Bershtein V A, Egorova L M, Yakushev P N, Pissis P, Sysel P and Brozova L (2002) Molecular dynamics in nanostructured polyimide-silica hybrid materials and their thermal stability, *J Polym Sci Part B Polym Phys Ed* **40**:1056–1069.
88. Kanapitsas A, Pissis P, and Kotsilkova R (2002) Dielectric studies of molecular mobility and phase morphology in polymer-layered silicate nanocomposites, *J Non-Cryst Solids* **305**:204–211.
89. Poetschke P, Dudkin S M and Alig I (2003) Dielectric spectroscopy on melt processed polycarbonate-multiwalled carbon nanotube composites, *Polymer* **44**:5023–5030.
90. Kramarenko V Y, Shantalii T A, Karpova I L, Dragan K S, Privalko E G, Privalko V P, Fragiadakis D and Pissis P (2004) Polyimides reinforced with the sol-gel derived organosilicon nanophase as low dielectric permittivity materials, *Polym Adv Technol* **15**:144–148.
91. Xu R, Manias E, Snyder A J and Runt J (2001) New biomedical poly(urethane urea)-layered silicate nanocomposites, *Macromolecules* **34**:337–339.
92. Song L, Hu Y, Li B, Wang S, Fan W and Chen Z (2003) A study on the synthesis and properties of polyurethane/clay nanocomposites, *Int J Polym Anal Charact* **8**:317–326.
93. Kripotou S, unpublished results.
94. Mamunya Ye P, Kanapitsas A, Pissis P, Boiteux G and Lebedev E (2003) Water sorption and electrical/dielectric properties of organic-inorganic polymer blends, *Macromol Symp* **198**:449–459.
95. Zois H, Kanapitsas A, Pissis P, Apekis L, Lebedev E and Mamunya Ye P (2004) Dielectric properties and molecular mobility of organic/inorganic polymer composites, *Macromol Symp* **205**:263–272.
96. Mamunya Ye P, Shtompel V I, Lebedev E V, Pissis P, Kanapitsas A and Boiteux G (2004) Structure and water sorption of polyurethane nanocomposites based on organic and inorganic components, *Eur Polym J* **40**:2323–2331.

PART V

BLENDS, COMPOSITES, APPLICATIONS, AND RECYCLING OF THERMOPLASTIC ELASTOMERS

Chapter 15

Polymer Blends Containing Thermoplastic Elastomers of the Condensation and Addition Types

J. Karger-Kocsis, S. Fakirov

1. Introduction

1.1. *General remarks*

Thermoplastic elastomers are usually characterized by a high ultimate elongation (>100%) and low set properties (<40%). The compression and tension sets reflect the elastic recovery of elastomeric materials after the corresponding loading. For an ideal rubber the set value is 0%, whereas for an ideal thermoplastic it is 100%. Thermoplastic elastomers (TPEs) are either block copolymers [1,2] or blends [3,4]. TPEs produced by polycondensation and polyaddition reactions belong to the group of the block (segmented) copolymers. The immiscibility of the two blocks leads to a microphase-separated morphology. "Hard" domains dispersed in the "soft", rubbery phase form the thermally reversible "physical crosslinks" (Figure 1). The soft phase has a glass transition, T_g, below the service temperature. The hard domains are either amorphous or crystalline. At elevated temperatures, the hard domains soften (amorphous) or melt (crystalline) and the material can be processed similar to conventional thermoplastic resins. After cooling the hard domains are restored *via* phase segregation (amorphous) and crystallization (crystalline) processes.

The TPEs treated in this chapter are linear, segmented copolymers, typically of the $(AB)_n$ build-up, where A and B represent the hard and soft blocks, respectively. The physical crosslinks in the TPEs usually originate from crystalline

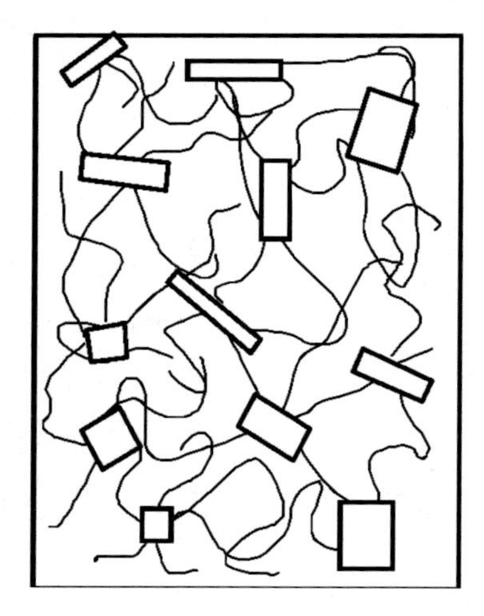

Figure 1. Schematic representation of the physical network structure of TPEs of the segmented copolymer type. "Knots" in block copolymers: (i) amorphous domains (phase segregation); (ii) crystalline domains (crystallization) – lamellae (folded chain block) and micelles (fringed micelle); (iii) ionic aggregates, clusters (ionic bonding); the rectangular elements are the "knots"

domains. They are mostly poly(ether amide)s (PEBA, where B is staying for "block"; another abbreviation for this kind of TPE is COPA) and poly(ether ester)s (PEBES; another usual abbreviation is COPE, or the most frequently used PEE) from the "condensation-type" family, whereas poly(ether urethane)s (PEBU) and poly(ester urethane)s (PESBU) considered here are produced by polyaddition. The latter are also abbreviated by TPU (thermoplastic polyurethanes) when no distinction is made between polyether- and polyester-based ones. The chemical build-up, synthesis, and properties of these TPEs are already discussed in detail in other chapters of this book (see Chapters 2 and 3).

The properties of TPEs can be tailored upon request in a broad range *via* the respective synthesis route. Figure 2 depicts the temperature dependence of the complex modulus, E^*, for TPUs of Shore A and Shore D hardness. Note that the most important characteristic property of rubbers of both thermoplastic and thermoset types is their hardness. Shore A 80 and Shore D 65 grades set by the chemical build-up (*i.e.*, hard-to-soft block ratio) do not represent the lower and upper hardness thresholds of the related TPEs (Figure 2). Nevertheless, the large difference in the stiffness between the TPUs is obvious in Figure 3, displaying the hardness of various polymers and rubbers. This figure already suggests the potential market of TPEs: replacement of traditional thermoset rubbers in various applications.

Since the properties of TPEs can be adjusted by the synthesis, their modification *via* blending is less explored. For this reason, TPEs are predominantly used as modifiers (*i.e.*, minor phase components) in blends and

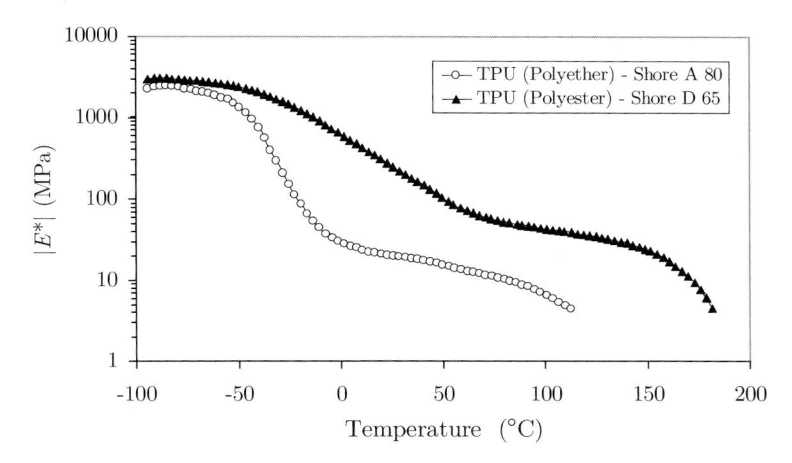

Figure 2. Complex modulus, E^*, as a function of temperature for TPUs of various build-up and Shore hardnes

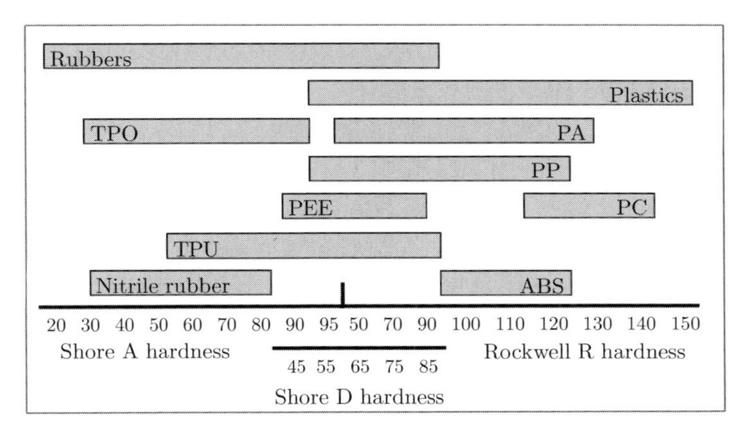

Figure 3. Hardness range of selected polymers and TPEs. TPO – thermoplastic polyolefin elastomer; all other abbreviations are disclosed in the text

scarcely as matrix (*i.e.*, major or continuous phase) materials. It should be emphasized that blends with PEBA and PEBES are commercially less relevant than those with TPUs. On the other hand, keen scientific interest was devoted to blends containing other TPEs than polyurethanes. This was due to compatibility and cocrystallization issues. As it will be shown later, some blends with PEE are fully miscible, which is rather seldom for polymer blends. Further, the rules, if any, for the cocrystallization are not yet fully understood. Blends composed of PEBA and PEBES with polyamide (PA) and polyester, respectively, which contain the same crystallizable segments are "ideal" materials to study the cocrystallization phenomenon.

1.2. *Miscibility vs. compatibility*

The performance of polymer blends is strongly affected by the compatibilization phenomenon [5]. For this reason, it seems appropriate to list some basic considerations. Over four decades, scientists and engineers used the terms *miscibility* and *compatibility* loosely and/or interchangeably. For the sake of absolute clarity and understanding, we define them here following Ref. [6]. *Thermodynamic miscibility* is attributed to polymer blends that are completely miscible and homogeneous down to the molecular level and do not show any phase separation. *Practical compatibility* describes polymer blends with useful properties in commercial practice. It should be emphasized that most of the commercially useful blends have practical compatibility, even though they do not have thermodynamic miscibility; in fact, they usually form multiphase morphologies that produce a synergistic advantage in the balance of properties, an effect that is not available from any single polymer [6]. Considering compatibility, one has to note further that a polymer blend, when thermodynamically immiscible, separates into two or more phases, thus resulting in poor practical properties. In such cases, it is generally assumed that either the sizes of the dispersed domains are not optimal or the immiscibility of the two phases produces weak interfaces, leading in turn to easy failure of the blend under stress. The most popular way of trying to solve this problem is the addition of a third ingredient, a *compatibilizer*. In rare cases, the compatibilizer actually produces thermodynamic miscibility. In most cases, however, it simply acts as a surfactant to reduce the domain size, or as an adhesive to strengthen the interface between the two immiscible polymer phases or components. Success is judged by the improvements in various practical properties as a result of practical compatibility [6–8].

1.3. *Chemical interactions in blends of condensation polymers*

These introductory notes on compatibility are important for the better understanding of one of the main advantages resulting from an inherent property of the blends prepared from condensation or functionalized polymers. At appropriate thermal treatment conditions (sufficiently high temperature and treatment duration, vacuum, catalyst, *etc.*), chemical reactions (*additional condensation* and *transreactions*) between condensation polymers in the melt [9], as well as in the solid state [10] can take place at the interfaces, as shown schematically below:

Additional condensation

$$\text{---COOH} + \text{H}_2\text{N---} \rightleftharpoons \text{---CONH---}$$

Transreaction

$$\begin{array}{ccc} \text{----CO} & & \text{O----} \\ | & + & | \\ \text{----NH} & & \text{OC----} \end{array} \rightleftharpoons \begin{array}{c} \text{----COO---NHCO----} \\ \\ \text{----NHCO---COO----} \end{array}$$

homopolymers copolymers

These reactions result in the formation of a copolymeric layer at the interface, playing the role of a compatibilizer, *i.e.*, one deals with a *self-compatibilization effect* in that there is no need to introduce an extra synthesized copolymer of the blend components, as in the common approach [5]. Compatibilization can be effective only in the initial stages of chemical interaction. During prolonged thermal treatment, the reaction goes to completion and finally involves the entire amount of the molten blend components in the formation of block copolymers, thus transforming the homopolymeric (molten) blend into copolymers.

$$(A)_n + (B)_m \longrightarrow \text{-----}(A)_x(B)_y(A)_z\text{-----}$$

In the case when $(A)_n$ and $(B)_m$ are crystallizable homopolymers, the block copolymers also crystallize. With the progress of the chemical interactions, the block copolymers may, however, convert into random copolymers.

-----AABBBBBBBBBBAAAAAAB----- ⟶ -----ABABBABABAA-----

An important result of the randomization is the loss of the ability of the matrix to crystallize, which can influence the overall behavior of the blend [7,8]. The occurrence of the described chemical changes and of physical transformations in the blend components is repeatedly proved by various techniques. For instance, in the wide angle X-ray scattering (WAXS) patterns of blends based on poly(ethylene terephthalate) (PET) and polyamide 6 (PA6) after prolonged treatment (25 h) at a relatively high temperature (240°C) at which PA6 is in the molten state, one can see that PA6 does not crystallize upon cooling, in contrast to the case of shorter annealing (4–5 h). The same conclusion can be drawn from the differential scanning calorimetry (DSC) measurements carried out on the same PET/PA6 system [11]: the melting peak of PA6 does not appear after prolonged thermal treatment. The dynamic mechanical thermal analysis (DMTA) curves, taken after the various steps of thermal and mechanical treatment of the PET/PA6 blend [12], indicate that after the final treatment (25 h at 240°C) PA6 disappears as an amorphous phase, too, since its T_g is no longer observed. One might assume that PA6 has been totally removed, *e.g.*, by selective extraction from the system, but this is not the case; instead, it has reacted with PET to form a random copolymer. In such a case, PA6 can neither crystallize nor form amorphous domains large enough to display their own T_g. That is, one of the blend component (PA6 in the present case), is non-crystallizable and, what is more, it is chemically bonded to PET. This conclusion is supported also by IR measurements after selective extraction, selective dyeing, and solubility tests [13]. All these methods indicated an involvement of PA6 in a copolymer with PET. Further details about the melting-, crystallization-, and miscibility-induced sequential reordering in condensation copolymers can be found in [14].

In order to improve the compatibility during melt blending, the same approach was extended to different polycondensation blends; a catalyst was added in some cases [15,16]. Blends of PET/PA6, polyamide 66 (PA66)/PA6 and PET/PA6/PA66 in weight ratios of 40/60, 40/60, and 20/60/20, respectively, were used for these studies. The described peculiarities of polymer blends can be expected and realized when the blend partners are condensation polymers or, at least, functionalized polyolefins.

2. Thermoplastic blends with polycondensation elastomers

The best review on thermoplastic blends with various TPEs of polycondensation and polyadditon types was given by Kalfoglou *et al.* [17].

2.1. *Thermoplastic blends with poly(ether amide)s*

Note that PEBA consists of PA blocks covalently bonded to soft polyether blocks *via* ester groups. The PA block may be of PA6, PA11 or PA12, whereas the polyether is composed of poly(ethylene oxide) (PEO), poly(propylene oxide) (PPO) or poly(tetramethylene oxide) (PTMO) [18].

As the stiffness of both the traditional rubbers and the thermoplastic elastomers is low, fillers and reinforcements are usually incorporated in order to improve their stiffness and strength. In the past, numerous attempts were made to produce polymer/polymer composites in which the reinforcement is caused also by a polymer. The favored polymers for *in situ* reinforcement were thermotropic (thermoplastic) liquid crystalline polymers (polyesters; TLCP). A TLCP has rigid chain architecture and possesses exceptional properties in fiber form. Its tensile strength and modulus are higher than 1 GPa and lower than 100 GPa, respectively [19]. From blends containing TLCP, *in situ* composites can be produced if the TLCP component is deformed into a fibrillar type morphology *via* suitable techniques [19]. Champagne *et al.* [20] found that the stiffness of PEBA with 30 wt% TLCP can reach 1 GPa when the draw ratio is higher than 12. This means a 50-fold increase in the modulus compared to neat PEBA. The authors also investigated how the mean TLCP fibril diameter depended on the draw ratio. An interesting finding was that only large TLCP drops were transformed into microfibers, whilst small ones were not. The cited authors also showed that for the in-plane stiffness of the TLCP microfiber-reinforced sheets, the Halpin-Tsai equation can be used. Note that this equation is widely used to predict the stiffness of composites with unidirectional fiber reinforcements.

2.2. *Thermoplastic blends with poly(ether ester) elastomers*

Similar to PEBA-containing blends (not reported above due to missing reports in the open literature), the compatibility between PEE and other polymers may arise from electron donor-electron acceptor interactions (between the carbonyl and ether groups of PEE, and H^+-X^- dipoles of the other blend component, where X is typically halogen), π-electron interactions between aromatic rings of the blend components, interchange reactions (*e.g.*, between ester and amide groups) and transesterification reactions (between ester groups in various chemical environments, as shown above) [17].

Among PEE blends, those with poly(vinyl chloride) (PVC) have been most extensively studied [17]. Using various test methods, it was shown that PEE/PVC blends exhibited a single T_g when the PEE content was between 25 and 50 wt%. At higher PEE contents (60–65 wt%), a shoulder in the T_g peak could be resolved, whilst at 80 wt% PEE, two T_g peaks appeared indicating immiscibility. The T_g of the PEE/PVC blends decreased linearly with added PEE [17,21]. Annealing of these blends at $T > 100°C$ yielded improved toughness owing to some phase separation [17]. This was also corroborated by other researchers [22], who demonstrated enhanced segregation between the hard and soft segments in PEE and a "compaction" accompanied by a T_g increase for the PVC. The peak of the mechanical loss factor (tan δ) was very broad and prominent, and for this reason the respective blends were foreseen for acoustic damping applications. In order to improve the damping properties, which means a further broadening of the tan δ as a function of temperature, a third component (poly(methyl acrylate)) was introduced in the blend, which was miscible neither with PVC, nor with PEE. The outcome of this strategy confirmed the expectations, since the tan δ peak was further "stretched" along the temperature axis [23]. Ternary blends were also produced by adding poly(ether urethane)s, PEBU, in PVC/PEE. The compatibility of PVC/PEE/PEBU depended on the composition. Blends with 1/1/1 and 1/2/1 weight ratios were compatible, whereas that composed of PVC/PEE/PEBU = 2/1/1 was miscible [24]. PEE/phenoxy (*i.e.*, polyhydroxy ether of bisphenol A) blends were also found to be miscible in a certain range [25,26]. The miscibility of this blend was attributed to interchange reactions (alcoholysis) occurring between the hydroxyl groups of the phenoxy and the proton acceptor groups of the PEE (carbonyl or ether).

Blending of PEE with poly(butylene terephthalate) (PBT) as the hard segment has been extensively studied ([17,25] and references therein). As already pointed out, the interest in this blend combination was to elucidate some open questions related to cocrystallization. The commercial interest in PBT/PEE blends is more trivial, *i.e.*, cost reduction *via* dilution with PBT. If the hard segment content is > 80 wt%, PEE is completely miscible with PBT. However, when the hard segment content is < 60 wt%, the blends show incomplete miscibility. Commercial PBT/PEE blends (PEE content > 50 wt%) are offered for the automotive sector (bumper, fascia applications). Such blends have good surface finish and good paint adhesion without any primer. PBT/PEE blends with < 10 wt% PEE are available as tough injection molding polyester grades [27].

2.2.1. *Cocrystallization phenomena in blends of PBT and PEE as revealed by X-ray analysis*

Blending of a crystallizable homopolymer and a copolymer containing blocks of the same homopolymer could result in *cocrystallization* of the common component and provide a large amount of tie molecules linking the two blend components. To explore this opportunity, a blend of PBT and a poly(ether ester) thermoplastic elastomer, the latter being a copolymer of PBT (hard segments) and poly(ethylene glycol) (PEG) (soft segments), was used [28]. This system is attractive not only because the two polymers have the same crystallizable component (PBT) but also because the copolymer, being an elastomer, would strongly affect the mechanical properties of the blend. In addition to the expected cocrystallization effect, the system is distinguished by the potential possibility of chemical interactions (mostly *via* transreactions) between the two components [8].

The PEE was prepared on a semicommercial scale [29] from PBT and PEG of molecular weight 1000 in a weight ratio of 1:1. Neat PBT and PEE were cooled in liquid nitrogen, then finely ground and blended (1:1 by wt). Films of the blend and of the respective neat components were prepared by means of a capillary rheometer equipped with metal rolls for immediate pressing the extruded bristle. The films obtained were 3–4 mm wide and 100–150 μm thick depending on the extrusion rate and the force applied to the rolls; after cold drawing ($\lambda = 4$–8) they were annealed with fixed ends for 6 h at 170°C (neat PBT and PEE) or at 170, 180, 190, and 200°C (PBT/PEE blend) in a vacuum oven [28].

Small-angle X-ray scattering (SAXS) curves of the blend and its components annealed at various temperatures are shown in Figure 4 [30]. Curves

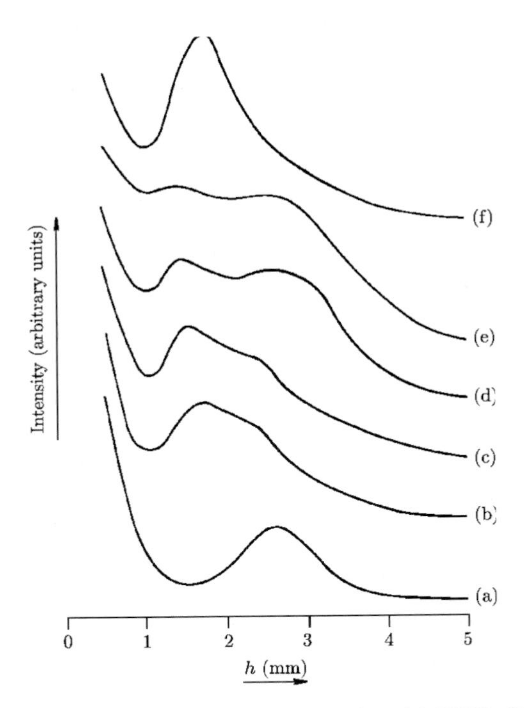

Figure 4. SAXS curves of drawn and annealed samples: (a) PBT, $T_a = 170$°C; (b)–(e) PBT/PEE blend, $T_a = 170$, 180, 190 and 200°C, respectively; (f) PEE, $T_a = 170$°C. The scattering angle 2θ (rad) is related to h (mm) as $2\theta = h/200$ [30]

(a) and (f), corresponding to the neat PBT and PEE, respectively, reveal a well expressed long spacing that is characteristic of oriented and annealed samples. A maximum comprising two overlapping peaks positioned at different angles is present in the curves of the blend (Figure 4, curves (b)–(e)). The observed tendency toward the appearance of two peaks with increasing annealing temperature can be considered as an additional indication of the coexistence of two types of PBT crystallites differing in their density and spatial arrangement. In order to check this assumption, SAXS curves of the blend pre-annealed at 200°C were recorded at different temperatures. The results are displayed in Figure 5.

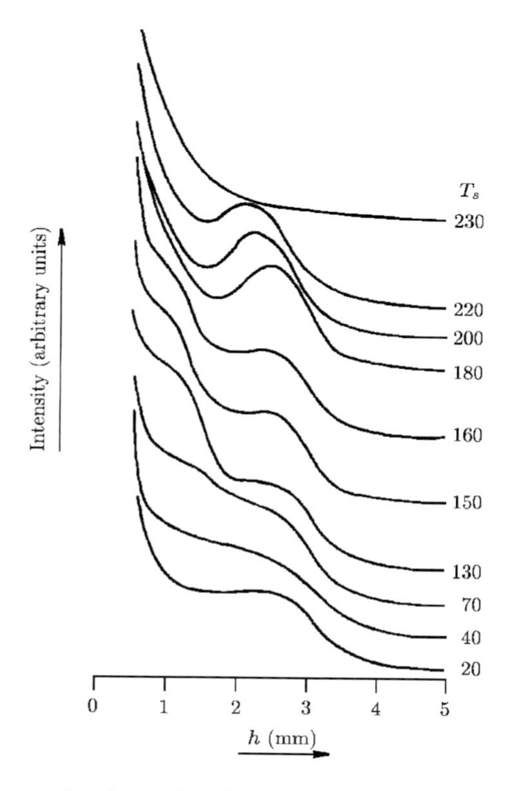

Figure 5. SAXS curves of a drawn PBT/PEE blend annealed at T_a = 200°C, taken at different temperatures, T_s (°C), as indicated; h as in Figure 4 [30]

At low temperatures (below 100°C), only one weak maximum is observed. With the rise of the sample temperature during the SAXS measurement, T_s, a new maximum appears at a smaller angle. The intensity of the two maxima increases up to T_s = 180°C; at this temperature the maximum at the smaller angle disappears, as does the other one at T_s = 230°C. Bearing in mind the melting temperatures of the two components [28], it is easy to assign the maximum at the smaller angle to the PBT in PEE and the other one to the neat PBT. This conclusion is also supported by the observation that the copolymers are distinguished by a much larger long spacing (at the expense of amorphous regions) than the homopolymer, in accordance with previous findings [31] (about the long spacing in copolymers, see also Chapter 6). Figure 6 shows WAXS patterns of films of the blend annealed at 170, 200, and 240°C.

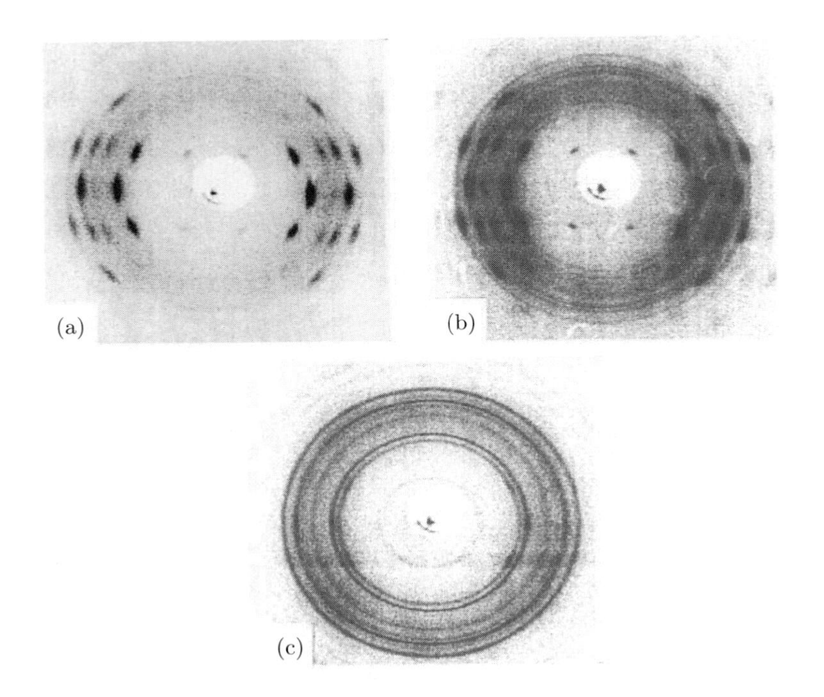

Figure 6. WAXS patterns taken at room temperature of a drawn PBT/PEE blend annealed at: (a) 170°C, (b) 200°C, and (c) 240°C [30]

As seen in Figure 6a, the PBT crystallites are highly oriented, while concentric circles due to isotropic PBT are observed in Figure 6b in addition to the arcs originating from highly oriented PBT crystallites. The PEE weight fraction in the blend is 1/2. The fraction of PBT in the PEE is also 1/2, and approximately only half of the PBT blocks are crystalline. It follows that the weight fraction of randomly oriented crystallites in the blend amounts to only 1/8 of the sample mass, and for this reason the circles corresponding to this isotropic fraction (Figure 6b) have a much lower intensity.

Annealing at 170°C (Figure 6a) results just in perfection of the oriented crystallites originating from PEE and neat PBT (as proved by DSC [28] and SAXS measurements, Figure 5), while annealing at 200°C leads to melting of crystallites from the PEE (with $T_m = 181$°C). The molten phase crystallizes again during the subsequent cooling, but no longer in the oriented state (Figure 6b). Actually, the coexistence of PBT reflections in the shape of arcs (from the neat PBT) and isointensity circles (from the PEE, Figure 6b) indicates again the presence of two types of PBT crystallites. By annealing at a sufficiently high temperature (240°C) they pass into the isotropic state and crystallize again during subsequent cooling (Figure 6c). Here, we have the quite uncommon situation when two types of crystallites (highly oriented and randomly distributed) of the same chemical composition (PBT in the present case) coexist in the same blend sample (Figure 6b).

Another interesting peculiarity of the system is that before annealing the blend above T_m of the lower-melting PEE, the two components (neat PBT and neat PEE) are in a highly oriented state (Figure 6a), as can be concluded from the WAXS pattern, obtained from the overlapping of the scattering reflexes arising from the crystallites of the two components. Only after annealing at 200°C, *i.e.*, above the melting of PEE, the latter is losing its orientation and during the subsequent cooling it crystallizes again, this time in an isotropic state (Figure 6b). In this way, the reflexes belonging to the homo-PBT crystallites can be easily distinguished because they characterize a polymer in a highly oriented state (as in Figure 6a) and those produced by the crystallites of PEE are typical of a isotropically crystallized polymer.

As a matter of fact, the thermal treatment above the melting of PEE transforms the originally drawn PBT/PEE blend into a polymer-polymer composite. In the latter, the matrix (PEE) is reinforced by microfibrils of the highly oriented component (homo-PBT).

It is noteworthy that this new type of polymer-polymer composites, the *microfibrillar reinforced composites* (MFC), was developed during the last decade [32–35]. In contrast to the classical composites, *i.e.*, those reinforced by discontinuous or continuous fibers, MFCs cannot be manufactured by conventional melt blending of the starting components, the matrix and the reinforcing materials.

The manufacturing of MFCs involves blending of immiscible polymeric partners, differing in their melting temperatures, T_m. The essential steps of MFC preparation are: (i) blending and extrusion, (ii) drawing (with good orientation of all components), and (iii) treatment at a temperature above the T_m of the lower-melting component but below the T_m of the higher-melting one. During the drawing step, the blend components are oriented, and nanostructured fibrils are created (*fibrillization step*). In the subsequent thermal treatment, when melting of the lower-melting component occurs (*isotropization step*), the oriented fibrillar structure of the higher-melting component should be preserved. It is important to note here that MFCs are based on polymer blends, but they should not be considered as "drawn blends" since the izotropization step results in the formation of an isotropic matrix reinforced by fibrils of the higher-melting component (as in the present case, Figure 6b) *i.e.*, one finally deals with a typical composite material [36].

Concerning the behavior of the non-crystalline portion, one can conclude from the results of the dynamic mechanical measurements [28] that two well defined amorphous phases are present in the thermally untreated PBT/PEE blend. The first one originates from the soft segments (PEG) of the copolymer (with $T_g^{PEG} = -50°C$) and the second one represents amorphous PBT belonging to neat PBT and to the PBT hard segments of the copolymer (with $T_g^{PBT} = 40°C$). With respect to whether the amorphous PBT phases originating from neat PBT and PEE are spatially separated, as crystallites do, the DMTA measurements are irrelevant since amorphous PBT has almost the same T_g, regardless of its origin (about T_g of these polymers see also Chapter 6).

The results of thermal analysis [28] and X-ray scattering experiments (Figures 4–6) strongly suggest that PBT and PEE are incapable of cocrystallization, at least when blended in a weight ratio of 1:1. This conclusion is in agreement with the results of a thorough study on the miscibility and cocrystallization of neat PBT and PEE containing tetramethylene oxide as the soft segment [37]. In this study, Gallagher *et al.* concluded that the amorphous phase is transformed from completely immiscible into miscible when the PBT content in PEE was raised from 75 to 91 wt%. Cocrystallization was observed in the miscible blends under all crystallization conditions applied [37].

Special attention should be given to the observation that the blend annealed ($T_a = 200°C$) above the melting ($T_m = 181°C$) of the copolymer for 6 h showed a rather high melting point of the crystallites arising from PEE (higher by 30°C than that of the blend annealed at $T_a = 170°C$ or of the neat copolymer), while T_m of neat PBT changes slightly (by about 5°C) [28]. Such behavior is also reported by Gallagher *et al.* [37], but only for PEE containing 65 wt% PBT, or more. In order to understand the melting behavior of PEE in the blend, one has to take into account two facts. First, previous investigations on PEE with different PBT/PEG ratios led to the conclusion that the thermal treatment of PEE results in dephasing of the two types of segments, *i.e.*, annealing at 200°C causes the formation of a chemically more uniform PBT melt in the molten PEE ($T_m = 181°C$). Second, at the same temperature ($T_a = 200°C$), the crystallites of the neat PBT in the blend are not molten ($T_m = 225°C$) and play the role of nucleation agents for the subsequent crystallization of the PBT hard segments during cooling. Actually, one deals with some kind of "epitaxial" growth, as noted by Gallagher *et al.* [37]. One can explain the unusually high melting temperature of PEE in the blend only by the effect of the pre-existing crystallites of neat PBT.

This "epitaxial" growth of PBT crystallites from PEE on neat PBT crystallites can be considered as *partial cocrystallization*, *i.e.*, formation of continuous crystals consisting of two spatially unseparated, crystallographically identical populations of crystallites, differing in their size, perfection, origin, and time of appearance, as represented schematically in Figure 7.

Obviously, the epitaxial effect of the neat PBT crystallites contributes to the perfection of the PBT crystallites arising from the copolymer, as concluded from

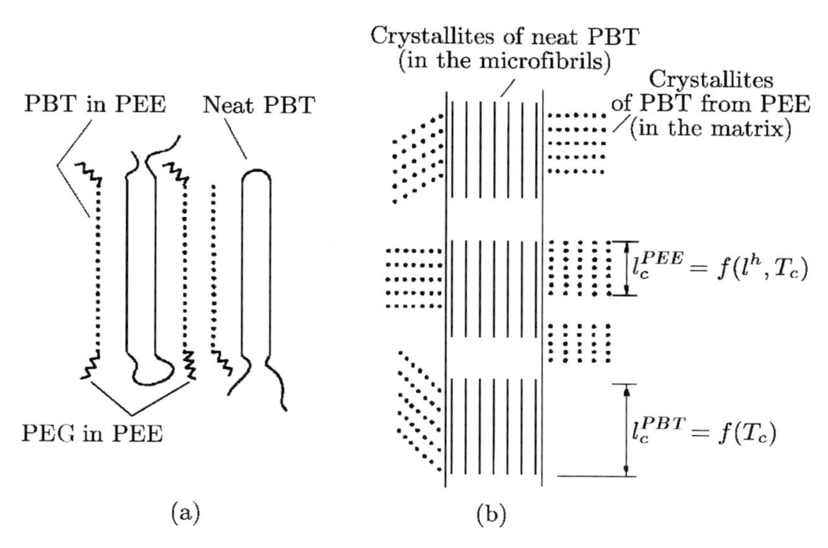

Figure 7. Schematic representation of the two types of cocrystallization of neat PBT and PEE with different PBT contents in microfibrillar reinforced polymer composites: (a) complete cocrystallization (PBT in PEE is 75–91 wt% [37]); (b) partial cocrystallization (PBT in PEE is below 75 wt%; in the present case it is 50 wt% [28]). l_c = lamella thickness, T_c = crystallization temperature, l^h = length of the hard segments (PBT) in PEE [30]

their T_m, but this effect is insufficient to rule out the structural peculiarity of the copolymer (short PBT segments diluted by PEG) in order to produce crystallites of the same perfection as those of neat PBT (Figure 7b). This peculiarity of PEE explains the finding of Gallagher *et al.* [37] that *complete cocrystallization*, *i.e.*, formation of uniform crystallites with simultaneous participation of PBT from both homopolymer and copolymer, is only observed with PEEs containing 75–91 wt% PBT (Figure 7a). Such PEEs are distinguished by sufficiently long hard segments (longer than 170 Å [31]). PEEs that do not crystallize (typical ones contain 50 or 65 wt% PBT) [37] have an average hard-segment length of 50 Å, or slightly longer, and are incapable of producing the thermodynamically required lamellar thickness, l_c, as shown in our previous study [28] and also by Gallagher *et al.* [37] (Figure 7).

Although in the present case one does not observe complete cocrystallization, the partial cocrystallization contributes substantially to the miscibility of neat PBT and PEE and, in this way, improves the adhesion between the matrix (PEE) and the reinforcing microfibrils (PBT). Actually, the isotropic crystallization of PEE after annealing at 200°C occurs on the surface of highly oriented crystalline microfibrils (Figure 7b). The improved adhesion explains the improved mechanical integrity of the MFC based on PBT and PEE after annealing at 200°C. This conclusion is supported also by the observed tendency to equalization of the measured and calculated Young's moduli with the rise of the annealing temperature [28].

The formation of new chemical linkages between neat PBT (fibrils) and the PEE copolymer (matrix) could hardly be expected since the temperature of thermal treatment (T_a = 180–200°C) is insuffcient for the occurrence of trans-reactions [9,10]. Even the thermal treatment at a higher temperature (250°C) during sample preparation does not result in any significant transesterification reactions. Using the DSC technique, Huo *et al.* [38] studied the possibility of transreaction occurring at 250°C in the miscible PBT/polyarylate blend and, for times shorter than 10 min, they did not detect any significant transesterification. This result is consistent with the earlier report by Kimura *et al.* [39] indicating that transesterification is considerable only after holding the same blend at 250°C for more than 100 min. The above findings explain the much smaller probability of occurrence of transreactions in the present case: (i) the annealing time is too short (5 min) and (ii) the PBT/PEE blend is immiscible and the annealing at 250°C is "static", *i.e.*, stirring is not applied.

In conclusion, cocrystallization, even partial, seems to be an attractive approach to overcome the immiscibility in polymer blends, and it can be applied to both condensation polymers and polyolefins having a common crystallizable component.

2.2.2. *Polymorphic transitions in PBT/PEE blends*

Polymorphic transitions in semicrystalline polymers and the microhardness technique. When polymers are crystallized from the melt or from solution, their crystalline region may exhibit various types of polymorphic modifications, depending on the cooling rate, evaporation rate of the solvent, temperature, and other factors. These modifications differ in their molecular and crystal structures, as well as in their physical properties. Many types of crystalline modifications have been reported [40].

Some polymorphic modifications can be converted into one another by a change in temperature. Phase transitions can also be induced by an external stress field. Phase transitions under tensile stress can be observed in natural rubber when it orients and crystallizes under tension and reverts to its original amorphous state by relaxation [41]. Stress-induced transitions are also observed in some crystalline polymers, such as PBT [42,43] and its block copolymers with poly(tetramethylene oxide) (PTMO) [44], poly(ethylene oxide) (PEO) [45,46], polyoxycyclobutane [47], nylon 6 [48], poly(vinylidene fluoride) [49,50], polypivalolactone [51], keratin [52,53], and others. These stress-induced phase transitions are either reversible, *i.e.*, the crystal structure reverts to the original structure on relaxation, or irreversible, *i.e.*, the newly formed structure does not revert after relaxation. Examples of substances with the former include PBT, PEO, and keratin.

PBT (also referred to as poly(tetramethylene terephthalate)) became the subject of keen interest in the mid-1970s when three groups of researchers [42,43,54] independently reported that it could crystallize in two distinct polymorphic forms. The α-form was found in a relaxed sample, whereas the

β-form could only be observed when the sample was held under strain. There have been a number of attempts to determine the unit cell parameters for the two crystalline forms [54–57], but some degree of controversy still exists. However, the general consensus is that in the α-form, the molecular chain is not fully extended, probably with the glycol residue in a *gauche-trans-gauche* conformation, whereas in the β-form, the chain is fully extended with the glycol residue in the all-*trans* conformation:

$$\text{—O—CH}_2\text{—CH}_2\text{—CH}_2\text{—CH}_2\text{—O—}$$

$$\alpha\text{: GGT}\bar{\text{G}}\bar{\text{G}}$$

$$\beta\text{: TTTTT}$$

There has been considerable interest in the mechanisms of the $\alpha \rightarrow \beta$ transition, and this has been modeled for static and dynamic measurements [58-60]. X-ray diffraction patterns and infrared (IR) and Raman spectra show specific changes through this $\alpha \rightarrow \beta$ transition [42–44,61–63]. The stress and strain dependence of the molar fraction of the β-form, X_β, increases drastically above the critical stress f^*; the fraction X_β is almost linearly proportional to the strain; the transition is reversible; the stress-strain curve has a plateau of the critical stress f^*, *i.e.*, the curve is divided into three regions: the elastic deformation of the α-phase (0–4% strain), the $\alpha \rightarrow \beta$ transition (plateau region; 4–12% strain), and the elastic deformation of the β-phase ($> 12\%$ strain). These experimental results indicate that this stress-induced phase transition is a thermodynamic first-order transition.

It has also been pointed out that, in block copolymers of PBT and PTMO, the transition is smeared over a wide range of stresses because of the effect of the PTMO soft segments [44].

Microindentation with a point indenter, causing deformations on a very small scale, is one of the simplest ways to measure the mechanical properties of a material [64]. The method uses a diamond pyramid that penetrates the surface of a specimen upon application of a given load at constant rate for a given time (Figure 8). Because of its simplicity, microindentation has become a common technique to measure the micromechanical behavior of polymers and its correlation with microstructure [65,66]. *Microhardness*, *H*, is obtained

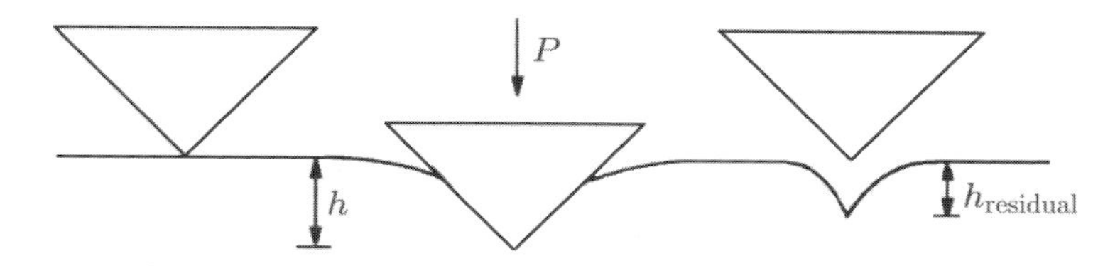

Figure 8. Contact geometry for a pyramid indenter at zero load P (left), maximum load (center), and complete unload (right). The residual penetration depth after load removal is given by h_{residual} [69]

by dividing the peak contact load by the projected area of impression, which is measured under a light microscope (imaging method). Hardness defined in this way is an indication of the irreversible plastic deformation process. The strain boundaries for plastic deformation below the indenter have been shown to depend on the polymer morphology (crystal size and perfection, degree of crystallinity, *etc.*) [67]. Typical loads of 10^2 mN, when applied to the surface of a conventional polymer, such as PET, produce penetration depths of about 2–3 μm. From a macroscopic point of view, H is directly correlated to the yield stress of the material, σ_y, through Tabor's relation $H \approx k\sigma_y$ where $k \sim 2$–3. Tabor's relation is well substantiated for unoriented polymers when creep is minimized [67,68].

Strain-induced polymorphic transition in neat PBT and neat PEE as revealed by microhardness measurements in real time. Usually, the polymorphic modifications were characterized by microhardness, WAXS, or other techniques *after the completion* of the polymorphic transition, irrespective of the factor inducing the latter. In the present subsection, an attempt is made to characterize the respective polymorphic phases during the transition itself, *i.e.*, the microhardness measurements have to be performed during stress-induced polymorphic transitions, rather than *post mortem* as, for example, in many other cases [68].

Bristles of commercial PBT (diameter of about 1 mm) were drawn at room temperature *via* neck formation (final diameter of about 0.5 mm, draw ratio of 3–4) and thereafter annealed with fixed ends at 200°C for 6 h in a vacuum oven.

The microhardness was measured again at room temperature using a Leitz tester supplied with a square-pyramidal diamond indentor and a stretching device. The microhardness measurement was carried out at a deformation step of $\varepsilon = 5\%$. Loads of 147 and 245 mN were employed (loading cycle of 0.1 min) to eliminate the instant elastic contribution. Ten indentations were made for each point [69]. H measurements were performed under strain up to 20% relative deformation, ε, defined as $\varepsilon = (l - l_0)/l_0$ where l_0 and l are the starting and a given length of the sample, respectively.

The effect of strain on the H value is illustrated in Figure 9. One can see a very well defined decrease in the H values (from 150 to 120 MPa) in a rather narrow deformation range ($\varepsilon = 5$–8%). Before this drop in H ($\varepsilon = 0$–5%), the microhardness has a relatively constant value of about 150 MPa, typical of semicrystalline PBT [70]. After reaching its lowest value ($H = 118$ MPa), around $\varepsilon = 10\%$, H starts to increase almost linearly up to 150 MPa, with the further rise of deformation up to $\varepsilon = 20\%$ when the sample breaks.

Although the stress-induced $\alpha \rightarrow \beta$ polymorphic transition in PBT is well documented, comparative WAXS measurements of the samples have been also carried out in the same deformation range where the H measurements were performed. In addition, the size, D_{hkl}, of the coherently diffracting domains (the crystal size in the (100), (010) and ($\bar{1}04$) (c-axis) directions) during

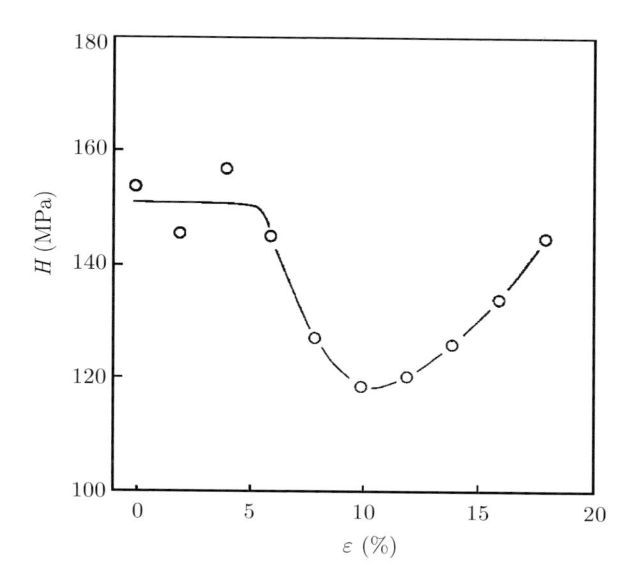

Figure 9. Effect of the relative tensile deformation, ε, on the microhardness, H, of neat PBT [69]

stretching was calculated from the integral breadth of the equatorial reflections according to $\delta\beta = 1/D_{hkl}$ [69,71].

The fact that the abrupt drop in the H value (Figure 9) coincides with the well documented $\alpha \rightarrow \beta$ transition just in the deformation range of 4 to 10–12% [43] shows that the observed changes in H are related exclusively to the stress-induced polymorphic transition. Thus, the starting value of $H = 155$ MPa should be typical of PBT containing the crystalline α-phase, and the lower value of $H = 118$ MPa can be assigned to PBT comprising in addition the crystalline β-phase. The observation that $H^{\alpha} > H^{\beta}$ is obviously related to the fact that the α-modification is distinguished by a denser chain packing, as can be concluded from the ideal crystal densities.

According to the comparison [55] of the published crystalline polymorphic structures, the volume of the unit cell for the α-phase varies between 260.0 and 262.8 Å^3 and that of β-phase between 268.9 and 285.0 Å^3. As mentioned above, it has been demonstrated [72] that in cases when w_c and l_c do not change, the crystal hardness, H_c, depends exclusively on the chain packing within the crystalline phase. In this way, one can explain why $H^{\alpha} > H^{\beta}$, although other factors acting in the same direction could also exist.

The observed strong increase in H, after the completion of the stress-induced polymorphic transition (from 118 up to 150 MPa), can be explained by the additional chain orientation in the amorphous regions which results in an additional densification of the total structure. In fact, this densification effect should take place during the earlier stages of deformation when the indentation anisotropy also increases. However, in this case it is compensated for by the effect on H of the stress-induced polymorphic transition. Furthermore, it is hard to accept that the observed increase in H values after the polymorphic transition (Figure 9) is due to an increase in crystallinity during stretching since such a change cannot be found in the WAXS diffractograms.

In conclusion, the presented results show that microindentation hardness allows one to follow the strain-induced polymorphic transition in PBT in real time [73]. The observed rather abrupt change in the H values (within 2–4% external relative deformation) makes the method competitive, with respect to sensitivity, with other commonly used techniques, such as WAXS, IR, or Raman spectroscopy, and others [42,43,54–58,61]. What is more, the same sensitivity of the technique has been observed when applied to strain-induced polymorphic transitions in multiblock copolymers of PBT.

An additional reason for performing this study on PEE is the fact that the copolymers of PBT with PEG have not been studied so far with reference to polymorphic transitions. The starting material is a polyblock poly(ether ester) [74] based on PBT and PEG (with molecular weight 1000 and polydispersity 1.3 according to size exclusion chromatography analysis [75]) in a weight ratio of 57/43. The synthesis was carried out on a semicommercial scale, as described elsewhere [76].

The sample used in the present study was shaped as a bristle, drawn at room temperature to $5\times$ its initial length, and then annealed with fixed ends for 6 h in vacuum at 170°C. The starting material was highly oriented, the α-modification of PBT being predominant [74].

Microhardness measurements were carried out as in the above case of neat PBT. Deformation measurements up to 80% overall relative deformation, ε, after which the sample broke, were performed by means of a stretching device [74]. The dependence of the microhardness on the deformation for drawn and annealed bristles of PEE (PBT/PEG = 57/43 by wt) is plotted in Figure 10. It is seen that H behaves quite differently depending on the stress applied. At the lowest deformations (ε up to 25%), H is almost constant (~33 MPa) and thereafter, in a very narrow (2–3%) deformation range, the H value suddenly drops by 30%, reaching the value of 24 MPa, which is maintained in the ε range between 30% and 40%. With the further rise in ε (between 40% and 60%), H increases to nearly its initial value, followed again by a decrease to 25 MPa for ε of about 60% to 80%.

The most striking change in H is the sharp decrease at $\varepsilon = 25$–27%. Taking into account the similar behavior of neat PBT (Figure 9) where the same change was observed (a drop in H by 20%) due to the well documented, stress-induced polymorphic transition in PBT, one can assume that, in the present case, the same transition takes place. The fact that the $\alpha \rightarrow \beta$ polymorphic transition in the PEE under investigation occurs at a much higher deformation (around $\varepsilon = 27\%$ *vs.* 5–10% for neat PBT (Figure 9)) can be explained by the peculiar behavior of thermoplastic elastomers.

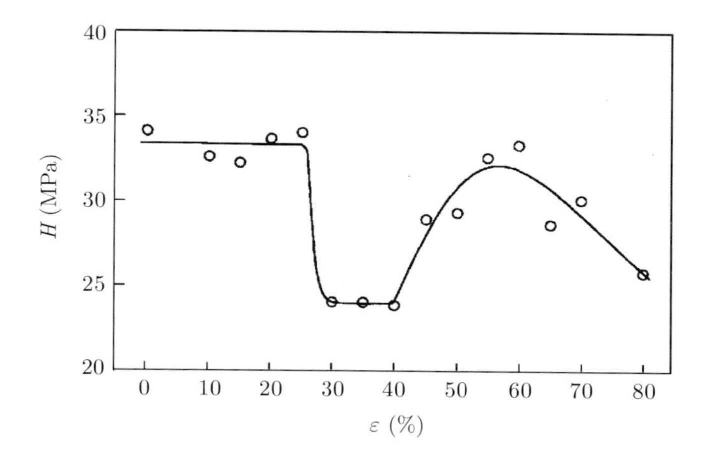

Figure 10. Microhardness, H, *vs.* overall relative tensile deformation, ε, of drawn and annealed PEE (PBT/PEG = 57/43 by wt) bristles [69]

In an earlier study [75], it was shown that the deformation of PEE (with varying PBT/PEG weight ratio, deformed in the 25–50% range) is related to conformational changes exclusively in the amorphous regions, the crystallites remaining unaffected. The stretching and relaxation of these amorphous segments cause an affine and reversible change of the measured X-ray long spacing [28,75,77–79]. With the further increase of deformation (for PEE with the same composition as in the present case, starting at ε = 25–30%), the external load is transferred to the crystallites, contributing to the observed polymorphic transition (Figure 10). For more details on the deformation mechanism of PEE, see Chapters 6 and 7, and [80].

The drop in the microhardness is due to the fact that the β-phase is less dense than the α-phase [55], as discussed above. An additional support for the assumption concerning the stress-induced polymorphic transition around ε = 27% has been found in our WAXS study on the same samples, in the same deformation range [69].

An interesting question arises about the reason for the observed increase in H (Figure 10) at higher deformations (above ε = 40 %). It hardly can be a result of regeneration of the α-phase, distinguished by a higher microhardness, H^{α}, because the sample is under stress. In order to explain the observed H increase, one has to recall a peculiarity of the system: PEE is a physical network in which the crystallites are the crosslinks. In addition, the crystallites are embedded in a soft amorphous matrix, *i.e.*, they are "floating" in a relatively low-viscosity matrix. During the deformation, the network will stretch further depending on the relative "softness" of the various components of the system. The stretching of the amorphous segments restricts the mobility of the crystallites and makes the system harder. In other words, the decrease in H caused by the replacement of crystallites of higher H by others of lower H is compensated by this hardening effect, caused by the stretching of the network [74]. In addition to the assumed hardening of the non-crystalline phases, the crystallization of the PEG soft segments in the same deformation range acts simultaneously in the same direction, as proved by WAXS measurements on the same sample [69].

The subsequent decrease of H in Figure 10, again down to 25 MPa, at the highest deformation ($\varepsilon = 80\%$) can be explained by a partial destruction of crystallites due to pullout of hard segments. This mechanism has been demonstrated by SAXS [75,79] and WAXS [81] measurements. The introduction of defects in the crystallites by this pullout may cause a decrease in their H value.

Strain-induced two-stage polymorphic transition in blends of PBT and PEE as revealed by microhardness measurements in real time. The results of the polymorphic transition obtained in real time on a blend of PBT and a PEE thermoplastic elastomer, the latter being a copolymer of PBT and PEG, are considered here. It should be mentioned that these blends are well characterized by DSC, SAXS, DMTA, and static mechanical measurements [28,78–81].

For the preparation of the blend, both the homopolymer PBT and PEE were cooled in liquid nitrogen and finely ground. PBT was blended with PEE in a weight ratio of PBT/PEE = 51/49. Bristles of the blend were prepared by means of a capillary rheometer loaded with the dried powdered material, flushed with argon and heated to 250°C. The melt was kept in the rheometer for 5 min and then extruded through the capillary (diameter of 1 mm). These bristles were annealed for 6 h at 170°C in vacuum with fixed ends [82].

Microhardness was measured at room temperature up to 30% overall relative deformation, ε (when fracture occurs). As in the aforementioned measurements of microhardness under strain, in the present case also a deformation step of $\varepsilon = 5\%$ was adopted.

Figure 11 shows the dependence of the microhardness on the external relative deformation for the PBT/PEE blend. H drops abruptly from 50 to 40 MPa at the very beginning of stretching (around $\varepsilon = 2$–3%), keeping this lower value up to $\varepsilon = 25\%$. In accordance with the previous cases (Figures 9 and 10), this abrupt drop in H can be assigned to the $\alpha \rightarrow \beta$ polymorphic transition in PBT crystallites. A further increase in deformation from $\varepsilon = 5\%$ to $\varepsilon = 25\%$ does not cause any pronounced changes in H. Such a constance of H suggests that no polymorphic transitions take place in this deformation interval. The next transition starts at $\varepsilon = 25\%$ and seems to be completed in a rather narrow deformation range between $\varepsilon = 25$ and 30% (Figure 11). The occurrence of two distinct transitions, starting at quite different deformation values ($\varepsilon = 5$ and 25%), suggests that the PBT crystallites differ significantly in their response to the external mechanical load. DSC and X-ray results show the existence of the following two types of PBT crystallites: those of the neat PBT and those of the PBT segments from the multiblock PEE, as discussed in the previous section [28,78]. One can assume that the first strain-induced polymorphic transition arises from the neat PBT crystallites and the second one, appearing at a higher deformation range, can be assigned to PBT crystallites belonging

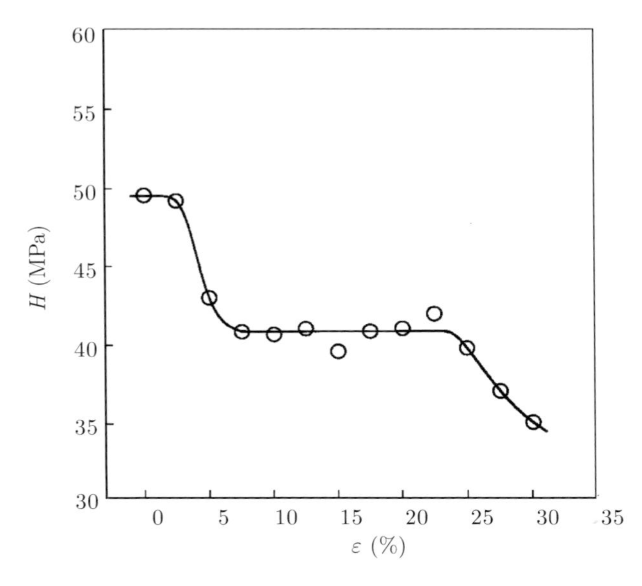

Figure 11. Variation of microhardness, H, with increasing relative tensile deformation, ε, of the blend PBT/PEE = 51/49 by wt (PEE with PBT/PEG = 49/51 by wt) [69]

to PEE. This assumption is supported by the curves presented in Figure 12, where the results from Figure 11 are replotted together with the previous data for neat PBT (Figure 9) and its multiblock copolymer PEE (Figure 10).

On the one hand, one can see the fairly good agreement between the deformation range for the blend and that for the neat PBT (Figure 12, curves (a) and (b)), and between that for the blend and that for the copolymer, on the other (Figure 12, curves (b) and (c)). In the first case, the strain-induced polymorphic transition is observed at deformations around $\varepsilon = 5\%$ and, in the second case, at ε around 25% (Figure 12, curves (a) and (b), and (b) and (c), respectively). Comparison of the curves presented in Figure 12 for neat PBT (top), the blend (middle), and the PBT-PEG copolymer (bottom) suggests that the first drop in H at around $\varepsilon = 5\%$ (Figure 11) originates from the strain-induced polymorphic transition in the crystallites comprising only neat PBT segments. The second decrease in H at around $\varepsilon = 25$–30% is related to the strain-induced polymorphic transition in the PBT crystallites comprising PBT segments from its multiblock copolymer PEE. Another striking observation in Figure 12 is the fact that the numerical value for the experimentally measured H of the blend (Figure 12, curve (b)) is much lower than that calculated according to the additivity law, using the values for neat PBT and PEE (Figure 12, curves (a) and (c), respectively). One possible explanation could be related to the presence of two types of PBT crystallites differing significantly in their perfection, as demonstrated earlier [28]. The second one could be connected with the strong influence on H of the crystal surface free energy [67,83]. The latter would contribute to the decrease in crystal hardness, particularly because of more imperfect PBT crystallites arising from the PEE copolymer [82].

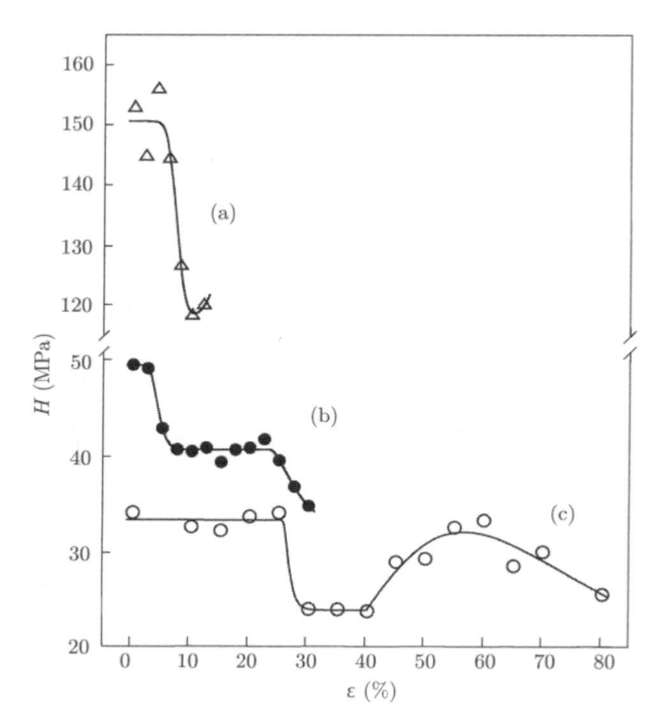

Figure 12. Variation of microhardness, H, with the relative tensile deformation, ε, for: (a) neat PBT; (b) a blend PBT/PEE = 51/49 by wt (PEE with PBT/PEG = 49/51 by wt); (c) multiblock PBT-PEE copolymer (PEE with PBT/PEG = 57/43 by wt) [69]

Cocrystallization in PBT/PEE blends as revealed by microhardness measurements. The study of the strain-induced polymorphic transitions by the microhardness technique offers the opportunity to gain additional information not only on the deformation behavior of more complex polymer systems, such as polymer blends, but also on the possible occurrence of cocrystallization. Since polymer blends are usually multiphase systems, it is interesting to establish the manner of reaction between independent phases under the external load. In addition to the SAXS studies described above (Section 2.2.1), the polymorphic transition will reflect the behavior of the crystalline phases, provided that a strain-induced polymorphic transition is available [82].

From the above results (Figures 11 and 12), one may ask what could be the reason for the different behaviors of the two types of crystallites with respect to their two-step response to the external mechanical field. Before answering this question, let us recall some structural peculiarities of the system under investigation, discussed also in Section 2.2.1. (i) The present blend contains the same crystallizable component in both the homopolymer and the copolymer, as already emphasized above. In other words, the PBT crystallites can arise as a result of *complete cocrystallization*, *i.e.*, formation of uniform crystallites with simultaneous participation of PBT from both the homopolymer and the copolymer, as found for PEE containing 75–91wt% PBT [78] and schematically represented in Figure 7a. (ii) For reasons related to the length of the crystallizable blocks in the PEE copolymers [28], the more frequent case is that of *partial cocrystallization*, *i.e.*, formation of (continuous) crystals consisting of two spatially non-separated, crystallographically identical populations of crystallites, differing in their size, perfection, origin, and time of appearance. Such a type of cocrystallization is observed when the PBT/PEE blends are drawn and thereafter annealed at a temperature between the melting temperatures of the two crystallite species [28]. This case is represented schematically in Figure 7b. (iii) The third possibility of crystallization is when *no cocrystallization* takes place, *i.e.*, formation of two populations of PBT, originating from the neat PBT and the PEE, respectively.

From the microhardness data presented in Figure 11, one could suggest that, in the present case, one deals with the third case for the following reasons. If PBT crystallites resulted from complete cocrystallization, one would expect a single strain-induced polymorphic transition in the entire deformation range. In case of partial cocrystallization, one would observe a more or less continuous polymorphic transition between the deformation ranges typical of neat PBT and of PEE. This interpretation was adopted in [30] and [69], based exclusively on the assumption of a continuous polymorphic transition, which actually is not observed (Figure 11).

The experimental results in Figure 11 show two rather abrupt transitions, separated by a deformation range of 20%. At first glance, this finding supports the assumption that, for the blend under investigation, one deals with two crystallographically identical, but spatially separated species of crystallites differing in the origin of the PBT segments.

In order to answer the question whether the observed behavior of H with the progress of ε (Figure 11) means the existence of only spatially separated crystalline species, one has to consider in more detail the peculiarities in the deformation mechanism of the two blend components, the homo-PBT and the PEE.

The finding that the two species of crystallites respond to the mechanical field separately (Figure 11), first the neat PBT crystallites and later those arising from PEE, means that the neat PBT crystals are probably dispersed within PEE in such a way that they experience the mechanical field from the very beginning of loading. What is more, one can assume that some internal stress and/or strain preexists in the blend, since the strain-induced polymorphic transition starts at even lower deformations (about $\varepsilon = 2$–3%, Figure 11) than in the case of neat PBT ($\varepsilon = 5\%$; Figure 9). In addition to the well pronounced mechanical response of the neat PBT crystallites, one observes conformational changes (stretching) only in the amorphous intercrystalline layers, as concluded from previous SAXS measurements [75,79,81] and described in more detail in Chapter 6, where the deformation behavior of PET, PBT, and PEE in tensile mode is compared. This comparison suggests that the conformational changes occurring only in amorphous layers are reversible and they are observed up to a deformation level $\varepsilon = 8\%$ (for PET), 17% (for PBT) and 50–75% (for PEE). This also means that the segments in the crystallites do not experience the external load before reaching the respective strain level, *i.e.*, the crystallites remain unaffected. For the PBT/PEE blends studied, these processes dominate up to deformations of about $\varepsilon = 25\%$ when the applied stress is also experienced by the crystallites arising from the PEE copolymer (Figure 11).

Namely this peculiarity of PEE to demonstrate relatively high (reversible) deformations only at the expense of the amorphous areas is the probable reason for the observed well separated (by $\varepsilon = 20\%$) two polymorphic transitions processes in the PBT/PEE blend (Figure 11). Nevertheless, if the PBT crystallites arising from PEE grow on the surface of those of homo-PBT or if the two crystallite species are spatially separated, the polymorphic transition in PEE can start only after achieving an appropriate level of strain ($\varepsilon = 25\%$ in the present case). As a matter of fact, this deformation characteristic of PEE thermoplastic elastomers leads to the observed two separate polymorphic transitions in the PBT/PEE blend (Figure 11). In other words, the results displayed in Figure 11 can hardly be considered as an argument against the assumption of the existence of partial cocrystallization (Figure 7b) formulated in the preceeding Section 2.2.1 on the basis of X-ray studies of the same system.

The systematic variation under strain of microhardness performed on (i) neat PBT [73], (ii) its multiblock copolymer PEE [74,84], and (iii) PBT/PEE blends [82] reveals the ability of these systems to undergo a strain-induced polymorphic transition. The possibility to follow this strain-induced polymorphic transition, even in complex systems, such as polymer blends, allows one to draw conclusions concerning also such basic phenomena as cocrystallization. In the case of the PBT/PEE blend, two distinct, well separated (with respect to the deformation range), strain-induced polymorphic transitions arising from the two species of PBT crystallites are observed. From this observation, it is concluded that (i) neat PBT and the PBT segments from the PEE copolymer undergo polymorphic transitions separately, *i.e.*, no complete cocrystallization takes place, and (ii) the two types of crystallites do not respond simultaneously to the external load.

3. Thermoplastic blends with polyaddition thermoplastic elastomers

The interest in blending of thermoplastics with thermoplastic polyurethanes (TPUs) is equivocal: some properties of TPUs, such as erosion and abrasion resistance are excellent [85,86], which already justify the blending efforts, but the urethane linkage (–NH–CO–O–) in TPUs is unstable at high temperatures. Isocyanate and hydroxyl groups form *via* reversible dissociation of the urethane bond, and they are very reactive with the usual functional groups (amine, epoxy, carboxyl, *etc.*) contained in compatibilizing blend components. To exploit these reactions whereby producing new blends with useful properties is the major task of the respective R&D activities.

Blending of PVC with TPU (mostly of the PESBU type) started already in the 1960s. Binary and ternary (acrylonitrile-butadiene-styrene copolymer (ABS), PA, or acrylonitrile-butadiene rubber (NBR) as the third component) blends were developed for extrusion processing (cable and wire insulation, packaging). Such blends proved to be useful antislip shoesole materials, as well [87].

The commercially most relevant TPU application is the impact modification of polyoxymethylene (POM) and up to 30 wt% TPU is used for POM toughening [87]. During injection molding of the related blends, a strong skin-core structure may appear that influences the (fracture) mechanical properties [88]. It was recently reported that the impact strength of POM/TPU blends can be further improved when diphenylmethane diisocyanate (MDI) is used as compatibilizer [89]. MDI probably crosslinks the TPU, thus making it less prone to aggregation in the disperse phase. It was also observed that the impact performance of POM/TPU blends depends not only on the TPU type and content, but also on the crystallinity and spherulite size of the POM matrix [90]. The latter aspect holds generally for blends with semicrystalline polymer matrices [91]. The good performance of POM/TPU blends is usually attributed to hydrogen bond formation between the components. POM/amorphous copolyester/TPU (PEBU or PESBU) ternary blends were produced and improvements in the mechanical properties were achieved [92].

TPU has been used to toughen polycarbonate (PC), and enhance its resistance to *environmental stress corrosion cracking* (ESC) [87]. Note that TPEs are often used to improve the ESC behavior of semicrystalline polymers, as shown recently for PEE-toughened PBT blends [93]. PC/TPU blends, where either PC or TPU is the matrix polymer, are commercially available and used in the production of gears, tubing, and bushes where the excellent abrasion resistance of TPU is exploited. It should be noted, however, that the reason behind the outstanding resistance of TPU to abrasion and erosion is not yet known [86]. Archondouli *et al.* [94] investigated PC/PESBU blends and found a good PESBU dispersion up to 30 wt% polyurethane content. On the other hand, the degree of compatibilization between PESBU and PC was markedly lower than for PESBU/linear polyester (PET, PBT) blends. This was attributed to the moderate reactivity of the aromatic –OH groups in PC towards the isocyanate groups and to the thermal instability of the resulting copolymer (which should act as compatibilizer). An alternative explanation of the increased miscibility is hydrogen bonding between the –NH– groups of the TPU and carbonyl groups of the PC [95]. Ester-amide interchange reactions were assumed to be responsible for the good dispersion of the polyurethane and for the strong adhesion between the phases in PBT/PESBU blends [96]. This resulted in improved mechanical properties, because the copolymer formed by the interchange reaction acted as an efficient compatibilizer in the blend.

ABS/TPU blends, which combine high toughness, frost resistance (ensured by TPU), paintability, and heat resistance (ensured by ABS), were introduced on the market long time ago [87]. PS/PEBU blends were successfully compatibilized by the addition of styrene-*co*-maleic ahydride copolymer (SMA). Further, by peroxide-induced crosslinking of PEBU, the tendency to aggregation of the latter was considerably reduced [97].

Contradictory results were published on the structure-property relationships of PA/TPU systems [17,98]. Note that in these blends the possible H-bonding between the carbonyl group of the urethane linkage and secondary amine groups of the amide should yield some compatibilization effect.

Great efforts were undertaken to produce polyolefin/TPU blends, as well. This development was aimed at combining the beneficial properties of polyolefins (low price, low density, easy processing, chemical and hydrolytic resistance, *etc.*) with those of TPUs (adhesion, paintability, haptics, abrasion resistance) [99–103]. Various functional (ethylene-*co*-acrylic acid, maleic anhydride-grafted polyolefins, SMA) and non-functional polymers (styrene/ethylene-butylene/styrene rubber (SEBS)) were tried as compatibilizers. A major conclusion of the cited authors was that the interfacial reaction, achieved by reactive extrusion, strongly depends on the blend compatibility and interfacial properties. Polymer pairs showing a great difference in the surface energies can hardly be functionalized (and thus no copolymer forms), since their functional groups are kept away from the interface [104]. However, exactly this phenomenon can be exploited to produce, *e.g.*, electrically conductive blends. Segal *et al.* [105] showed that carbon black particles in TPU may form a chain-like structure (percolation network). If TPU experiences phase continuity in the other blend component (PP) which is also present as a continuous phase, a double percolation network (electrical and structural) emerges.

In order to shed light on the reactive compatibilization issue, Lu *et al.* [106] have studied the reactivity of common functional groups with urethanes using model substances. According to their ranking, the reactivity of the functional groups toward urethanes is

$$\text{primary amine} > \text{secondary amine} \gg \text{hydroxyl} \sim \text{acid} \sim \text{anhydride} \gg \text{epoxide.}$$

Based on this information, Lu and Macosko studied the compatibility of blends composed of functionalized PPs and TPU [107]. In this work, maleated PP (PP-g-MA), primary and secondary amine-functionalized PPs (PP-g-NH_2 and PP-g-NHR, respectively) were blended with TPU. The compatibility of the blends decreased as follows

$$\text{PP-g-NHR} = \text{PP-g-NH}_2 \gg \text{PP-g-MA.}$$

This was attributed to the higher reactivity of amines with the urethane linkage compared to the anhydride. However, this ranking contradicts the usual reactivity order, since primary amines are more reactive than secondary ones. The authors explained this finding by a melt amination process affected by some by-products. Blends with amine-functionalized PPs exhibited a finer and more stable morphology and superior mechanical properties than those containing maleated PP, *i.e.*, PP-g-MA.

Information on the relative reactivity of various functional groups with urethanes will definitely give a new impetus to develop novel blends, including polyolefin/TPU formulations.

4. Rubber blends with thermoplastic elastomers

Reports on blending of rubbers with TPEs are scarce. This is likely due to the very conservative attitude of the rubber industry and the high price of the TPE grades.

Based on the miscibility of PVC with PEE, the blending of PEE with polar rubbers, such as NBR, chlorinated polyethylene (CPE), and chlorosulfonated polyethylene (CSM) attracted research interest. Recall that PVC/NBR blends are fully miscible, similar to PVC/PEE ones. Therefore, it was intuitive to check the compatibility between PEE and NBR in uncured state [17,108]. The interest behind CPE/PEE and CSM/PEE blends was due to the supposed dipole interactions between the components yielding good compatibility [17,109].

As already mentioned, TPUs have excellent wear and abrasion resistance in addition to good tear and tensile properties. The latter mechanical properties are jeopardized, however, at service temperatures beyond *ca.* 70°C, mostly owing to superimposed creep. Dassin *et al.* [110] developed a grafting process to prepare "self-crosslinkable" TPU. According to their approach, MDI was first reacted with the –NH– groups of the urethane linkages. Next, the free isocyanate groups were reacted by amine-functionalized trialkoxysilane yielding allophanate junctions between the TPU chains and the aminosilane. The alkoxy groups of the latter can be crosslinked *via* polycondensation reactions in the presence of water. Note that water-assisted crosslinking of silane-grafted polyolefins is an industrial process since the mid-1970s ([110,111] and references therein). The process of Jin and Fritz [112] is somewhat analogous to the above one. However, these authors produced thermoplastic dynamic vulcanizates (TDV, [4]) by compounding TPU with polyolefins. The polyolefin phase was dynamically vulcanized using an organosilane/water carrier crosslinking system. The crosslinked polyolefin was finely dispersed in the TPU matrix. Recall that dynamic vulcanization means the selective vulcanization of one component of the blend during mixing, which finally becomes the disperse phase [4]. It can be predicted that thermoplastic elastomers composed of TPU and polyolefins will appear on the market soon, as this developement is fueled by cost savings.

5. Thermoset resin blends with thermoplastic elastomers

It is well known that the toughness of thermosets creates problems; they can be solved by adding functionalized rubbers, which form a disperse phase during curing. The particle size of the rubber is typically in the range of 1 to 10 μm, and 1–2 μm is preferred. Major toughness improving mechanisms are alleviation of the triaxial stress state *via* rubber cavitation followed by crack bifurcation. In the latter stage, stretching of the rubber particles and shear yielding of the matrix resin are involved. In addition to phase-segregated rubber inclusions, other "obstacles" produced *in situ* or added into the resin prior to its curing in preformed shape can also work as toughness modifiers. The failure scenario depends on numerous factors for which the interested reader is addressed to monographs [113,114].

Among liquid rubbers, functional NBR grades are favored for thermoset toughening. However, Wang *et al.* [115,116] produced hydroxyl-, amine-, and anhydride-terminated polyurethane prepolymers and toughened bifunctional epoxy (EP) resins with them. The particle size was found between 0.5 and 5 μm with a mean size of *ca.* 1.5 μm. The fracture energy was doubled by adding 5 parts of functional prepolymer to 100 parts resin (5 phr). At 15 phr rubber content with respect to the functionality, the fracture energy changed according to the following ranking: hydroxyl (10 times) $>$ amine (6 times) $>$ anhydride (4-fold improvement compared to the neat resin).

Attempts were also made to use both amorphous and semicrystalline polymers for resin toughening [113,114]. In this way, both the T_g and the hot/wet performance of the thermosets are less affected. Various polymers were dispersed either in the form of fine powder (< 50 μm) in the resins or produced *in situ*. In the latter case, the modifier polymer was first dissolved in the resin (eventually using co-solvents) before the resin curing caused phase segregation. Among the polymers tried, PBT and its elastomeric versions proved to be the most efficient [116–120]. This effect was called by numerous authors *phase transformation toughening*. Note that phase transformation toughening is known for some steels and especially for zirconia-containing ceramics. Stress-induced phase transition is associated with volume increase (dilation) due to which the crack tip may become under compressive stresses. This means a special type of crack tip blunting which greatly improves the toughness of the corresponding material. It was speculated that PBT particles undergo $\alpha \rightarrow \beta$ phase transition, as discussed in Section 2.2.2. Recall that this is accompanied by a *ca.* 9% volume increase [118]. Unfortunately, there is no direct proof of the onset of the $\alpha \rightarrow \beta$ transition, since the latter is reversible and the particles on the fracture surface do not indicate this change. Note that phase transition should be the more efficient the higher the crystallinity and the more prefect the related crystallites are. Albeit there are some hints for that [119–121], it is still questionable that phase transition, affecting the release of the hydrostatic stresses at the crack tip, occurred at all. Figure 13 shows the fracture surfaces of an EP resin toughened by various amounts of PBT copolymer particles. Based on the fracture surface, one can claim that crack pinning (generating a "tail" after each particle) and bifurcation associated with shear deformation of the matrix (causing steps and welts in the matrix) are responsible for the toughness improvement. Crack bifurcation is favored just by the stress concentration action of the particles belonging to planes adjacent to that of the final crack. So, for the onset of

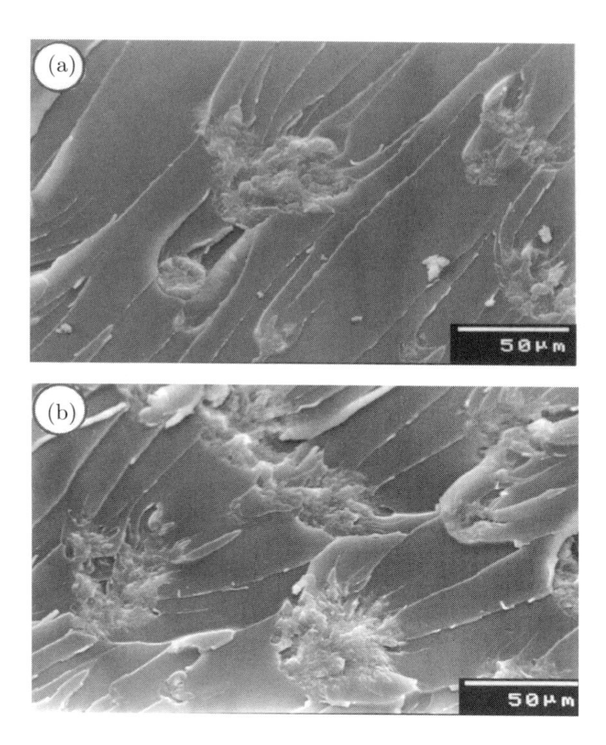

Figure 13. Fracture surfaces of particle-toughened resins containing (a) 5 and (b) 15 wt% PBT copolymer. Fracture surface produced at room temperature under static loading

crack bifurcation, no $\alpha \rightarrow \beta$ transition should be supposed. Further, the surface of the broken PBT particles does not indicate that their fracture was preceded by considerable plastic deformation, which is called particle bridging in this case. On the other hand, particle bridging would be a strong support for the supposed phase transition.

6. Summary and outlook

Blending of thermoplastic elastomers (TPEs) of both polycondensation and polyaddition types with other polymers is justified when the resulting blend exhibits a unique property combination of the blend components. Such value-added engineering polymers can be prepared by alloying polyolefins with suitable TPEs. Polyolefins belong to the family of commodity plastics and are inexpensive. On the other hand, recent achievements in their polymerization technology (metallocene synthesis) may rise the interest to transform them into "specialities". To that purpose, blending of polyolefins with TPEs is a straightforward alternative. It can be predicted that for the production of new TPEs, monomers and oligomers from renewable resources will be more extensively used. The outcome of such research may have an impact on the blending activities, as well. It has to be emphasized again that the versatile synthesis of TPEs along with the related property tailoring possibilities are the factors which are strongly opposed to the blending strategy.

Acknowledgements

This work was supported for J. Karger-Kocsis by grants of the German Science (DFG; Ka 1202/14) and Fonds der Chemischen Industrie (FCI). S. Fakirov expresses his appreciation to the Alexander von Humboldt Foundation, Germany, for the financial support through the Institute Partnership Program (Project V-Fokoop-DEU/100 77 94), making possible his stay at the Institute for Composite Materials Ltd. of the Technical University of Kaiserslautern where this chapter was prepared.

References

1. Walker B M and Rader C P, Eds. (1988) *Handbook of Thermoplastic Elastomers*, Van Nostrand Reinhold, New York.
2. Holden G, Legge N R, Quirk R P and Schroeder H, Eds. (1996) *Thermoplastic Elastomers*, Hanser, Munich.
3. De S K and Bhowmick A K, Eds. (1990) *Thermoplastic Elastomers from Rubber–Plastic Blends*, Ellis Horwood, Chichester.
4. Karger-Kocsis J (1999) Thermoplastic rubbers via dynamic vulcanization, in *Polymer Blends and Alloys* (Eds. Shonaike G O and Simon G P) Marcel Dekker, New York, pp. 125–153.
5. Utracki L A (1989) *Polymer Alloys and Blends*, Hanser, Munich.
6. Deanin R D and Chuang C H (1993) Polyolefin polyblends, in *Handbook of Polyolefins: Synthesis and Properties* (Eds. Vasile C and Seymour R B) Marcel Dekker, Inc., New York, pp. 779–798.
7. Xanthos M and Warth H (1999) Effects of transreactions on the compatibility and miscibility of blends of condensation polymers, in *Transreactions in Condensation Polymers* (Ed. Fakirov S) Wiley-VCH, Weinheim, Ch. 10, pp. 411–427.
8. *Transreactions in Condensation Polymers* (1999) (Ed. Fakirov S) Wiley-VCH, Weinheim.
9. Flory P J (1953) *Principles of Polymer Chemistry*, Cornell University Press, Ithaca, New York.
10. Fakirov S (1990) Solid state reactions in linear polycondensates, in *Solid State Behavior of Linear Polyesters and Polyamides* (Eds. Schultz J M and Fakirov S) Prentice Hall, Englewood Cliffs, NJ, Ch. 1, pp. 1–74.
11. Fakirov S, Evstatiev M and Friedrich K (1995) Interfacial interactions in microfibrillar composites from condensation polymers, in *High Technology Composites in Modern Applications* (Eds. Paipetis S A and Youtsos A G) University of Patras, Patras, Greece, pp. 12–20.
12. Serhatkoulu T, Bahar I, Erman B, Fakirov S, Evstatiev M and Sapundjieva D (1995) Chemical interactions in poly(ethylene terephthalate)/polyamide 6 blends as revealed by dynamic–mechanical studies, *Polymer* **36**:2371–2377.
13. Evstatiev M, Nicolov N and Fakirov S (1996) Morphology of microfibrillar reinforced composites from polymer blends, *Polymer* **37**:4455–4463.

14. Fakirov S and Denchev Z (1999) Sequential reordering in condensation copolymers, in *Transreactions in Condensation Polymers* (Ed. Fakirov S) Wiley-VCH, Weinheim, Ch. 8, pp. 319–389.
15. Evstatiev M, Schultz J M, Petrovich S, Fakirov S, Georgiev G and Friedrich K (1998) In situ polymer-polymer composites from PET/PA6/PA66 blends, *J Appl Polym Sci* **67**:723–737.
16. Evstatiev M, Fakirov S and Friedrich K (1995) Effect of blend composition on the morphology and mechanical properties of microfibrillar composites, *Appl Compos Mater* **2**:93–106.
17. Kalfoglou N K, Kallitsis I K and Koulouri E G (2000) Multiblock copolymer/homopolymer blends in *Block Copolymers* (Eds. Baltá Calleja F J and Roslaniec Z) Marcel Dekker, New York, pp. 357–379.
18. Sheth J P, Xu J and Wilkes G L (2003) Solid state structure–property behavior of semicrystalline poly(ether-block-amide) PEBAX® thermoplastic elastomers, *Polymer* **44**:743–756.
19. Baird D G and McLeod M A (2000) Liquid crystalline polymer blends in *Polymer Blends, Vol.2: Performance* (Eds. Paul D R and Bucknall C B) Wiley, New York, pp. 429–453.
20. Champagne M F, Dumoulin M M, Utracki L A and Szabo J P (1996) Generation of fibrillar morphology in blends of block copolyetheresteramide and liquid crystal polyester, *Polym Eng Sci* **36**:1636–1646.
21. Hourston D J and Hughes I D (1979) Polymeric systems for acoustic damping. I. Poly(vinyl chloride)-segmented polyether ester blends, *J Appl Polym Sci* **21**:3093–3103.
22. Hourston D J and Hughes I D (1979) Annealed segmented polyether ester-poly(vinyl chloride) blends, *Polymer* **20**:823–826.
23. Hourston D J and Hughes I D (1981) Polymeric systems for acoustic damping. III. Blends of poly(vinyl chloride), segmented poly(ether ester), and poly(methyl acrylate), *J Appl Polym Sci* **26**:3487–3491.
24. Hourston D J and Hughes I D (1981) Binary and ternary blends of poly(vinyl chloride), Hytrel, and a polyurethane, *J Appl Polym Sci* **26**:3467–3473.
25. Gaztelumendi M and Nazábal J (1998) Phenoxy/Hytrel blends. I. Miscibility and melt-state reactions, *J Appl Polym Sci* **70**:185–193.
26. Gaztelumendi M and Nazábal J (1999) Phenoxy/Hytrel blends. II. Dynamic and tensile properties of unreacted miscible blends, *J Appl Polym Sci* **72**:85–93.
27. Akkapeddi M K (2003) Commercial polymer blends, in *Polymer Blends Handbook* (Ed. Utracki L A) Kluwer Academic Publishers, Dordrecht, pp. 1099–1100.
28. Apostolov A A, Fakirov S, Sezen B, Bahar I and Kloczkowski A (1994) Structural and mechanical studies of a blend of poly(butylene terephthalate) and poly(ether ester) based on poly(butylene terephthalate) and poly(ethylene glycol), *Polymer* **35**: 5247–5255.
29. Fakirov S and Gogeva T (1990) Poly(ether ester)s based on poly(butylenes terephthalate) and poly(ethylene oxide)glycols. Part 1. Poly(ether ester)s with various polyether/polyester ratios, *Makromol Chem* **191**:603–614.
30. Fakirov S, Evstatiev M and Friedrich K (2002) Nanostructured polymer composites from polymer blends, in *Handbook of Thermoplastic Polyesters* (Ed. Fakirov S) Wiley-VCH, Weinheim, Ch. 23, pp. 1093–1132.
31. Fakirov S, Apostolov A A and Fakirov C (1992) Long spacing in segmented block copoly(ether ester)s – origin and features, *Int J Polym Mater* **18**:51–70.
32. Evstatiev M and Fakirov S (1992) Microfibrillar reinforcement of polymer blends, *Polymer* **33**:877–880.

33. Fakirov S, Evstatiev M and Schultz J M (1993) Microfibrillar reinforced composite from drawn poly(ethylene terephthalate)/nylon 6 blend, *Polymer* **34**:4669–4479.
34. Fakirov S, Evstatiev M and Petrovich S (1993) Microfibrillar reinforced composite from binary and ternary blends of polyesters and nylon 6, *Macromolecules* **26**:5219–5226.
35. Fakirov S, Evstatiev M and Friedrich K (2000) From polymer blends to microfibrillar reinforced composites, in *Polymer Blends, Vol. 2: Performance* (Eds. Paul D R and Bucknall C B) J Wiley & Sons, Inc., New York, Ch. 33 pp. 455–475.
36. Karger-Kocsis J (1996) Composites (structure, properties, and manufacturing), in *Polymeric Materials Encyclopedia* (Ed. Salamone J C) CRC Press, Boca Raton, Vol. 2, pp. 1378–1383.
37. Gallagher K P, Zhang X, Runt J P, Huynh-ba G and Lin J S (1993) Miscibility and cocrystallization in homopolymer-segmented block copolymer blends, *Macromolecules* **26**:588–596.
38. Huo P P, Cebe P and Capel M (1993) Dynamic-mechanical relaxation and X-ray scattering study in poly(butylene terephthalate)/polyarylate blends, *Macromolecules* **26**:4275–4282.
39. Kimura M, Porter R S and Salee G (1983) Blends of poly(butylene terephthalate) and polyarylate before and after transesterification, *J Polym Sci Polym Phys Ed* **21**:367-378.
40. Tashiro K and Tadokoro H (1987) Crystalline Polymers, in *Concise Encyclopedia of Polymer Science and Engineering* (Ed. Kroschwitz J I) J. Wiley, New York, pp. 221-228.
41. Mandelkern L (1964) *Crystallization of Polymers*, McGraw-Hill Inc., New York.
42. Jakeways R, Ward I M, Wilding M A, Hall I H, Desborough I J and Pass M G (1975) Crystal deformation in aromatic polyesters, *J Polym Sci Polym Phys Ed* **13**:799–813.
43. Yokouchi M, Sakakibara Y, Chatani Y, Tadokoro H, Tanaka T and Yoda K (1976) Structures of two crystalline forms of poly(butylene terephthalate) and reversible transition between them by mechanical deformation, *Macromolecules* **9**:266–273.
44. Tashiro K, Hitamatsu M, Kobayashi M and Tadokoro H (1986) Stress-induced crystalline phase transition in block copolymers of poly(butylene terephthalate) and poly(tetramethylene oxide). II. Phase transition and inhomogeneous deformation, *Sen'i Gakkaishi* **42**:659–664.
45. Takahashi Y, Sumita I and Tadokoro H (1973) Structural studies of polyesters. IX. Planar zig–zag modification of poly(ethylene oxide), *J Polym Sci Polym Phys Ed* **11**:2113–2122.
46. Tashiro K and Tadokoro H (1977) Elastic modulus and mechanical deformation in polymer crystals, *Kagaku* (Kyoto) **32**:848–850 (in Japanese).
47. Takahashi Y, Osaki Y and Tadokoro H (1980) Structures of three crystal modifications of poly(3,3-dimethyloxacyclobutane), *J Polym Sci Polym Phys Ed* **18**:1863–1878.
48. Miyasaka K and Ishikawa K (1968) Effects of temperature and water on the $\gamma \rightarrow \alpha$ crystalline transition of nylon 6 caused by stretching in the chain direction, *J Polym Sci Part A-2* **6**:1317–1329.
49. Lando J B, Olf H G and Peterlin A (1966) Nuclear magnetic resonance and X-ray determination of the structure of poly(vinylidene fluoride), *J Polym Sci Part A* **4**:941-951.
50. Hasegawa R, Takahashi Y, Chatani Y and Tadokoro H (1972) Crystal structures of three crystalline forms of poly(vinylidene fluoride), *Polymer J* **3**:600– 610.
51. Prud'homme R E and Marchessault R H (1974) $\alpha \rightarrow \beta$ Transformation in polypivalolactone, *Macromolecules* **7**:541–545.
52. Astbury W T (1933) X-ray interpretation of fiber structure, *J Soc Dyers Colourists* **49**:168–179.

53. Hearle J W S, Chapman B M and Senior G S (1971) Interpretation of the mechanical properties of wool, *Appl Polym Symp* **18**:775–794.
54. Boye C A and Overton J R (1974) Reversible, stress-induced, solid-phased transition in poly(tetramethylene terephthalate), *Bull Am Phys Soc* **19**:352.
55. Desborough I J and Hall I H (1977) A comparison of published crystalline structures of poly(tetramethylene terephthalate), *Polymer* **18**:825–830.
56. Stambaugh B, Koenig J L and Lando J B (1979) The unstressed α phase in poly(tetramethylene terephthalate), *J Polym Sci Polym Phys Ed* **17**:1053–1062.
57. Mencik Z (1975) The crystal structure of poly(tetramethylene terephthalate), *J Polym Sci Polym Phys Ed* **13**:2173–2181.
58. Brereton M G, Davies G R, Jakeways R, Smith T and Ward I M (1978) Hysteresis of the stress-induced crystalline phase transition in poly(butylene terephthalate), *Polymer* **19**:17–26.
59. Davies G R, Smith T and Ward I M (1980) Dynamic mechanical behavior of oriented poly(tetramethylene terephthalate) as a function of static extension, *Polymer* **21**:221-225.
60. Hall I H and Pass M G (1978) Measurement of integrated intensities from polymeric–fiber X-ray diffraction photographs, *J Appl Crystallogr* **8**:60–64.
61. Tashiro K, Nakai Y, Kobayashi M and Tadokoro H (1980) Solid-state transition of poly(butylene terephthalate) induced by mechanical deformation, *Macromolecules* **13**:137–145.
62. Jakeways R, Smith T, Ward I M and Wilding M A (1976) Reversible crystal deformation and conformational changes in poly(tetramethylene terephthalate), *J Polym Sci Polym Lett* **14**:41–46.
63. Ward I M and Wilding M A (1977) Infrared and Raman spectra of poly(methylene terephthalate), *Polymer* **18**:327–335.
64. Tabor D (1951) *The Hardness of Metals*, Oxford C. Press, Glasgow.
65. Baltá Calleja F J (1994) Microhardness studies of polymers and their transitions, *Trends Polym Sci* **2**:419–425.
66. Deslandes Y, Alva Rosa E, Brisse F and Meneghini T (1991) Correlation of microhardness and morphology of poly(ether-ether-ketone) films, *J Mater Sci* **22**:2769–2777.
67. Baltá Calleja F J (1985) Microhardness relating to crystalline polymers, *Adv Polym Sci* **66**:117–148.
68. Baltá Calleja F J and Fakirov S (2000) *Microhardness of Polymers*, Cambridge University Press, Cambridge.
69. Fakirov S and Baltá Calleja F J (2002) Strain-induced polymorphic transition in poly(butylene terephthalate), its copolymers and blends, in *Handbook of Thermoplastic Polyesters* (Ed. Fakirov S) Wiley-VCH, Weinheim, Ch. 21, pp. 927–964.
70. Giri L, Roslaniec Z, Ezquerra T A and Baltá Calleja F J (1997) Microstructure and mechanical properties of PBT–PC block copolymers: influence of composition, structure and physical aging, *J Macromol Sci-Phys* **B36**:335–343.
71. Baltá Calleja F J and Vonk C G (1989) *X-ray Scattering of Synthetic Polymers* (Ed. Jenkins A D) Elsevier, p. 143.
72. Baltá Calleja F J and Kilian H G (1985) A novel concept in describing elastic and plastic properties of semicrystalline polymers: polyethylene, *Colloid Polym Sci* **263**:697–707.
73. Fakirov S, Boneva D, Baltá Calleja F J, Krumova M and Apostolov A A (1998) Microhardness under strain. Part I. Effect of strain–induced polymorphic transition of poly(butylene terephthalate) on microhardness, *J Mater Sci Lett* **17**:453–457.
74. Apostolov A A, Boneva D, Baltá Calleja F J, Krumova M and Fakirov S (1998) Microhardness under strain. 2.Microhardness behavior during stress-induced polymorphic

transition in block copolymers of poly(butylene terephthalate), *J Macromol Sci-Phys* **B37**:543–555.
75. Fakirov S, Fakirov C, Fischer E W and Stamm M (1991) Deformation behavior of poly(ether ester) thermoplastic elastomers as revealed by small–angle X-ray scattering, *Polymer* **32**:1173–1180.
76. Fakirov S and Gogeva T (1990) Poly(ether ester)s based on poly(butylenes terephthalate) and poly(ethylene oxide)glycols. Part 1. Poly(ether ester)s with various polyether/polyester ratios. Part 2. Effect of polyether segment length. Part 3. Effect of thermal treatment and drawing on the structure of the poly(ether esters), *Makromol Chem* **191**:603–614, 615–624, 2341–2354.
77. Stribeck N, Apostolov A A, Zachmann H G, Fakirov C, Stamm S and Fakirov S (1994) Small angle X-ray scattering of segmented block copolyetheresters during stretching, *Int J Polym Mater* **25**:185–200.
78. Gallagher K P, Zhang X, Runt J P, Huynh-ba G and Lin J S (1993) Dynamic mechanical relaxation and X-ray scattering study of poly(butylenes terephthalate)/polyarylate blends, *Macromolecules* **26**:4275–4282.
79. Fakirov S, Fakirov C, Fischer E W and Stamm M (1992) Deformation behaviour of poly(ether ester) thermoplastic elastomers with destroyed and regenerated structure as revealed by small-angle X-ray scattering, *Polymer* **33**:3818–3827.
80. Fakirov S and Stribeck N (2002) Flexible copolyesters involving PBT: strain-induced structural changes in thermoplastic elastomers, in *Handbook of Thermoplastic Polyesters* (Ed. Fakirov S) Wiley-VCH, Weinheim, Ch. 15, pp. 672–716.
81. Stribeck N, Sapundjieva D, Denchev Z, Apostolov A A, Zachmann H G, Stamm M and Fakirov S (1997) Diffraction behavior of a poly(ether ester) copolymer as revealed by small– and wide–angle X-ray radiation from synchrotron, *Macromolecules* **30**:1329–1339.
82. Boneva D, Baltá Calleja F J, Fakirov S, Apostolov A A and Krumova M (1998) Microhardness under strain 3. Microhardness behaviour during stress-induced polymorphic transition in blends of poly(butylene terephthalate) and its block copolymers, *J Appl Polym Sci* **69**:2271–2276.
83. Baltá Calleja F J, Santa Cruz C, Bayer R K and Kilian H G (1990) Microhardness and surface free energy in linear polyethylene: the role of entanglements, *Colloid Polym Sci* **268**:440–446.
84. Baltá Calleja F J, Boneva D, Krumova M and Fakirov S (1998) Microhardness under strain. 4. Reversible microhardness in polyblock thermoplastic elastomers with poly(butylene terephthalate) as hard segments, *Macromol Chem Phys* **199**:2217–2220.
85. Barkoula N-M and Karger-Kocsis J (2002) Processes and influencing parameters of the solid particle erosion of polymers and their composites, *J Mater Sci* **37**:3807–3820.
86. Zhang Z, Barkoula N-M, Karger-Kocsis J and Friedrich K (2003) Artificial neural network predictions on erosive wear of polymers, *Wear* **255**:708–713.
87. Utracki L A (1989) Introduction to polymer blends, in *Polymer Blends Handbook* (Ed. Utracki L A) Kluwer Academic Publishers, Dordrecht, Vol. 1, Ch. 1, pp. 1–122.
88. Pecorini T J, Hertzberg R W and Manson J A (1990) Structure-property relations in an injection-moulded, rubber-toughened, semicrystalline polyoxymethylene, *J Mater Sci* **25**:3385–3395.
89. Mehrabzadeh M and Rezaie D (2002) Impact modification of polyacetal by thermoplastic elastomer polyurethane, *J Appl Polym Sci* **84**:2573–2582.
90. Gao X, Qu C, Zhang Q, Peng Y and Fu Q (2004) Brittle-ductile transition and toughening mechanism in POM/TPE/$CaCO_3$ ternary composites, *Macromol Mater Eng* **289**:41–48.

91. Karger-Kocsis J (1995) Microstructural aspects of fracture in polypropylene and its filled, chopped fiber and fiber mat reinforced composites, in *Polypropylene: Structure, Blends and Composites* (Ed. Karger-Kocsis J) Chapman and Hall, London, Vol. 3, Ch. 4. pp. 142–201.
92. Lam K L, Abu Bakar A, Mohd Ishak Z A and Karger-Kocsis J (2004) Unpublished results.
93. Kuipers N B, Riemslag A C, Lange R F M, Janssen M, Marissen R, Dijkstra K and Bakker A (2003) The environmental stress cracking of a PBT/PBA co-poly(ester ester), in *Fracture of Polymers, Composites and Adhesives II – ESIS Publ. 32* (Eds. Blackman B R K, Pavan A and Williams J G), Elsevier, Oxford, pp. 115–126.
94. Archondouli P S, Kallitsis J K and Kalfoglou N K (2003) Compatibilization and property characterization of polycarbonate/polyurethane polymeric alloys, *J Appl Polym Sci* **88**:612–626.
95. Fambri L, Penati A and Kolarik J (1997) Modification of polycarbonate with miscible polyurethane elastomers, *Polymer* **38**:835–843.
96. Archondouli P S and Kalfoglou N K (2001) Compatibilization and properties of PBT/PU polymeric alloys, *Polymer* **42**:3489–3502.
97. Navarro Cassu S and Felisberti M I (2002) Polystyrene and polyether polyurethane elastomer blends compatibilized by SMA: Morphology and mechanical properties, *J Appl Polym Sci* **83**:830–837.
98. Pesetskii S S, Fedorov V D, Jurkowski B and Polosmak N D (1999) Blends of thermoplastic polyurethanes and polyamide 12: Structure, molecular interactions, relaxation, and mechanical properties, *J Appl Polym Sci* **74**:1054–1070.
99. Wallheinke K, Pötschke P and Stutz H (1997) Influence of compatibilizer addition on particle size and coalescence in TPU/PP blends, *J Appl Polym Sci* **65**:2217–2226.
100. Pötschke P, Wallheinke K, Fritsche H and Stutz H (1997) Morphology and properties of blends with different thermoplastic polyurethanes and polyolefins, *J Appl Polym Sci* **64**:749–762.
101. Pötschke P, Wallheinke K and Stutz H (1999) Blends of thermoplastic polyurethane and maleic-anhydride grafted polyethylene. I: Morphology and mechanical properties, *Polym Eng Sci* **39**:1035–1048
102. Pötschke P, Pionteck J and Stutz H (2002) Surface tension, interfacial tension, and morphology in blends of thermoplastic polyurethanes and polyolefins. Part I. Surface tension of melts of TPU model substances and polyolefins, *Polymer* **43**:6965–6972.
103. Stutz H, Heckmann W, Pötschke P and Wallheinke K (2002) Structural effects of compatibilizer location and effectivity in termoplastic polyurethane–polyolefin blends, *J Appl Polym Sci* **83**:2901–2905.
104. Stutz H, Pötschke P and Mierau U (1996) Chemical reactions during reactive blending of polyurethanes: Fiction or realty? *Macromol Symp* **112**:151–158.
105. Segal E, Tchoudakov R, Narkis M and Siegmann A (2003) Sensing of liquids by electrically conductive immiscible polypropylene/thermoplastic polyurethane blends containing carbon black, *J Polym Sci Part B Phys* **41**:1428–1440.
106. Lu Q-W, Hoye T R and Macosko C W (2002) Reactivity of common functional groups with urethanes: Models for reactive compatibilization of thermoplastic polyurethane blends, *J Polym Sci Part A Chem* **40**:2310–2328.
107. Lu Q-W and Macosko C W (2004) Comparing the compatibility of various functionalized polypropylenes with thermoplastic polyurethane (TPU), *Polymer* **45**:1981–1991.
108. Hourston D J and Hughes I D (1981) Dynamic mechanical behaviour of polyether ester–nitrile rubber blends, *Polymer* **22**:127–129.

109. Hourston D J and Hughes I D (1980) Binary and ternary blends from poly(vinyl chloride), segmented poly(ether ester) and chlorosulphonated polyethylene, *Polymer* **21**:469–471.
110. Dassin S, Dumon M, Mechin F and Pascault J-P (2002) Thermoplastic polyurethanes (TPUs) with grafted organosilane moieties: A new way of improving thermomechanical behavior, *Polym Eng Sci* **42**:1724–1739.
111. Narkis M, Tzur A, Vaxman A and Fritz H G (1985) Some properties of silane-grafted moisture-crosslinked polyethylene, *Polym Eng Sci* **25**:857–862.
112. Jin F and Fritz H G (2001) Reactive compounding of thermoplastic polyurethanes and EO-copolymers, in *Proceedings of PPS Regional Meeting*, Antalya, Turkey, p. 233.
113. Yee A F, Du J and Thouless M D (2000) Toughening of epoxies, in *Polymer Blends, Vol.2: Performance* (Eds. Paul D R and Bucknall C B) Wiley, New York, pp. 225-267.
114. Pascault J-P, Sautereau H, Verdu J and Williams R J J (2002) *Thermosetting Polymers*, Marcel Dekker, New York, Ch. 13, pp. 389–419.
115. Wang H-H and Chen J-C (1995) Toughening of epoxy resin by reacting with functional terminated–polyurethanes, *J Appl Polym Sci* **57**:671–677.
116. Wang H-H and Chen J-C (1996) Chemical modification and crystalline polymer particle filled epoxy resins, *J Adv Mater* July:25–31.
117. Iijima T, Miura S, Fujimaki M, Taguchi T, Fukuda W and Tomoi M (1996) Toughening of aromatic diamine–cured epoxy resins by poly(butylene phthalate)s and the related copolyesters, *J Appl Polym Sci* **61**:163–175.
118. Kim J K and Robertson R E (1992) Toughening of thermoset polymers by rigid crystalline particles, *J Mater Sci* **27**:161–174
119. Nichols M E and Robertson R E (1994) The toughness of epoxy-poly(butylene terephthalate) blends, *J Mater Sci* **29**:5916–5926.
120. Kim S, Jo W H, Kim J, Lim S H and Choe C R (1999) The effect of crystalline morphology of poly(butylene terephthalate) phases on toughening of poly(butylene terephthalate)/epoxy blends, *J Mater Sci* **34**:161–168.
121. Kim S T, Kim J, Lim S, Choe C R and Hong S I (1998) The phase transformation toughening and synergism in poly(butylene terephthalate)/poly(tetramethyleneglycol) copolymer–modified epoxies, *J Mater Sci* **33**:2421–2429.

Chapter 16

"Nanoreinforcement" of Thermoplastic Elastomers

J. Karger-Kocsis

1. Introduction

Polymeric nanocomposites (organic/inorganic hybrid materials) have attracted considerable interest in both academia and industry. This is due to the outstanding properties achieved at low "nanofiller" content (usually less than 5 wt%). The nanocomposites contain inorganic fillers the size of which, at least in one dimension (thickness), is in the nanometer range. In contrast, traditional fillers and reinforcements have a lateral dimension of the order of micrometers (> 5 μm). Albeit the term "*nanocomposites*" was introduced recently, some of their components are natural products, while others find industrial application since more than half a century (carbon black reinforcement of rubbers). Like traditional fillers and reinforcements, the nanoparticles can be grouped according to their shape and aspect ratio (length-to-thickness ratio), and one can distinguish one- (*e.g.*, carbon nanotubes), two- (*e.g.*, layered silicates) and three-dimensional (*e.g.*, nanoparticles, zeolites, mesoporous glasses) versions.

The unexpected beneficial behavior of nanocomposites is due to filler/matrix and filler/filler interactions. Owing to the huge specific surface of well dispersed nanoparticles, it is often quoted that the properties of nanocomposites are controlled by interface/interphase characteristics. The distinction between interface and interphase depends on whether or not this region has a finite thickness. Polymer nanocomposites are already commercialized, *e.g.*, films of improved barrier properties and low-density injection moldable resins with high heat distortion temperature. Although the structure-property relationships in nanocomposites are poorly understood, the reader may get useful information to this topic from some recent monographs [1–7]. It should be noted that few reviews are devoted to nanotube reinforcement, and more works describe the incorporation of spherical nanoparticles. On the other hand, the overwhelming majority of the R&D works is dedicated to the use of modified layered silicates.

Nanoreinforcement strategies have been exhaustively adopted for polymers of both thermoplastic and thermoset types. Interestingly, thermoplastic elastomers (TPE) are scarcely studied. This may be due to several reasons, as pointed out in this chapter. Therefore, the aim of this chapter is to draw attention on possible property improvements *via* nanoreinforcement instead of summarizing the related achievements with TPEs. Most of the nanomodification work has been performed on thermoplastic polyurethanes (TPUs). The author is convinced that the related knowledge may serve as a guide for other TPEs, as well. For this reason, strong emphasis is put on the polyaddition-type TPEs, rather than on polycondensation ones. (TPE-based nanocomposites are discussed also in Chapter 14.)

2. Concepts and realization of nanosclale reinforcements

Composites are defined as materials consisting of two or more distinct *components* with an interface between them. This definition is mostly used for materials containing reinforcements, which are characterized by a high aspect ratio. This is the case of one- and two-dimensional particles. Particles with a low aspect ratio (the aspect ratio of a spherical particle is 1) are termed fillers, as their incorporation in the matrix is usually accompanied by both cost reduction and property degradation (strength, toughness). The distinction between *filler* and *reinforcement* is arbitrary in many cases even for macro- and micro-composites, and hardly practicable for nanoparticles. The stress transfer from the "weak" matrix to the "strong" reinforcement occurs *via* the interface and there are many analogies between macro- and nanocomposites. The challenging question is whether the reinforcing effect can be amplified by using nanoscale particles instead of traditional microscale (considering one lateral dimension) ones. The answer is positive. It was demonstrated in several papers that the property profile achieved by adding 30–40 wt% filler (spherical chalk, platy talk) or 10–20 wt% reinforcement (discontinuous glass fiber) can also be realized by dispersing 4–5 wt% layered silicate ([1,2,6] and references therein). It was also shown that effects usually caused by reinforcements could also be achieved by well dispersed spherical nanoparticles, which are appropriately bonded to the matrix [4]. Since layered silicates serve predominantly as nanoreinforcing materials for polymers [1,2,6,8], it is worth to briefly show their action and potential.

Layered silicates of natural (clays) and synthetic origin possess a layer thickness of *ca.* 1 nm. The lateral dimension of this platy (disc-shaped) reinforcement may reach several micrometers. However, the aspect ratio of clays (bentonite, montmorillonite) is usually less than 300. Isomorphic substitution of higher valence cations in the silicate framework by lower valence ones renders the layers negatively charged. This negative surface change of the layers is counterbalanced by alkali cations located between the clay galleries. They can be replaced by a bulky cationic organic surfactant (termed *intercalant*) by ion-exchange reactions. By this treatment, the initially hydrophilic silicate becomes organophilic. At the same time, the interlayer spacing (basal or d spacing) also increases. To reach an interlayer distance greater than 1.5 nm seems to be necessary in order to facilitate the intercalation with the polymer molecules. *Intercalation* means that the initial interlayer spacing increases, but a still well detectable long-range order remains between the layers. In the case of *exfoliation*, the individual layers are dispersed in the matrix without any structural ordering (Figure 1).

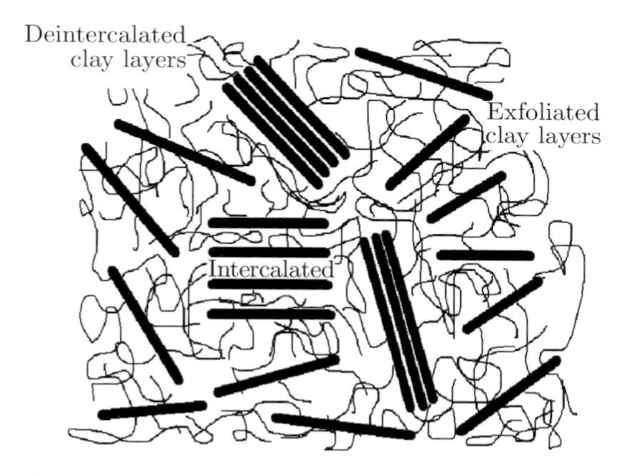

Figure 1. Scheme of the dispersion stage of layered silicates in polymers

What is the driving force of intercalation? Intercalation is governed by thermodynamic and kinetic parameters. This holds for all kinds of intercalation (in the melt, *via* solvents, *etc.*) and in all kinds of hosts (two-dimensional layered stack, three-dimensional framework). It is intuitive that the polymer molecules penetrating in between the silicate layers lose their conformational freedom, which is equal to entropy loss. On the other hand, for a molecular dispersion, the Gibb's free energy term should be negative. Note that the organophilic silicate layer can be regarded as a stiff polymer backbone to which molecules of the cationic tenside are tethered ("brush molecules" composed of inorganic and organic parts [8]). The entropy loss from the side of the bulk polymer molecules enhances the free energy instead of causing its reduction. An overall entropy gain may occur, however, when taking into account the organic intercalants tethered to the clay layers. Their conformational freedom may be strongly enhanced by intermingling and entangling with chains of the bulk polymer. This is the possible scenario from the entropic point of view (*i.e.*, resulting in entropy increase). Further, energetically favored interactions may take place between molecules of the intercalant and the bulk polymer. These interactions

(H-bonding, chemical reactions, acid – base and dipole interactions, *etc.*) may reduce the enthalpy term and thus the overall free energy, as well. All these aspects should be considered when selecting the proper "organophilization" of the layered silicate for a given polymer. It is worth noting that positively charged layered inorganic systems are also available. They can be made organophilic by anion exchange using suitable anionic surfactants (*e.g.*, [9]). However, according to the open literature, such nanoparticles were less used to reinforce polymers and definitely not yet adopted for TPEs.

There are several ways to produce nanocomposites (*e.g.*, [10]). The related strategies are so versatile that they cannot be easily grouped. In this chapter, the production methods will be discussed on the example of TPU nanocomposites containing layered silicates. The related production methods are grouped as follows: (i) *in situ* techniques (polymerization, sol-gel chemistry), (ii) solvent-assisted methods (solution, latex), and (iii) melt compounding.

As emphasized above, the concept of nanoreinforcement is far less practiced for TPEs than for thermoplastics [1,2,4,6,8], thermosets [1,2,8,11] and even for traditional rubbers [3,12]. This remark holds especially for addition and condensation TPEs. On the contrary, considerable amount of work was done on olefinic and styrenic TPEs. What is the reason for paying little attention to the nanoreinforcement of TPEs? This is likely due to the fact that the properties of addition and condensation TPEs can be tailored upon request *via* the related synthesis. Recall that this was also the major argument for their less explored blending with other polymers. A further analogy with blending is that markedly more reports addressed the nanoreinforcement of TPUs compared to poly(ether amide) and poly(ether ester) block copolymers (see also Chapter 15).

3. Nanoreinforcement of thermoplastic polyurethanes

Poly(ether-*b*-urethane)s and poly(ester-*b*-urethanes)s are designated as TPUs. Interestingly, nanoreinforcement is considered as a suitable tool for the property improvement of crosslinked polyurethanes. The interest in this development is due to the traditional application of crosslinked polyurethanes (coatings, varnishes, *etc.*) where the modification with nanoparticles may be very promising (*e.g.*, yielding scratch and abrasion resistance). Therefore, it is appropriate to have a look at the related works and adopt the results for thermoplastic grades. Note that the related teaching can be rather easily transferred from thermoset to thermoplastic polyurethanes.

3.1. *In situ techniques*

In situ techniques involve polymerization and sol-gel reactions. In the former case, polyurethanes are produced from suitable polyols and polyisocyanates in the presence of nanoparticles or their sources (*e.g.*, an organoclay with mean particle size initially in the range of a few micrometers). According to the sol-gel route, nanoparticles are produced *in situ via* suitable chemical reactions (*e.g.*, [10]).

3.1.1. *Polymerization*

Petrović *et al.* [13] produced segmented polyurethanes of various soft segment contents (50 and 70 wt%) and reinforced them by incorporating colloid silica (up to 30 wt%). The mean diameter of the spherical colloid silica was 12 nm. It was found that the nanoparticles are evenly distributed in both hard and soft phases. The nanoparticles strongly influenced the spherulitic morphology of the hard phase. The size of the crystalline lamellae decreased and the formation of the spherulite-forming fibrils was suppressed, too. These findings are very important and probably give the answer to the question why crystalline TPEs are not favored for nanoreinforcement. Recall that in both addition and condensation TPEs the hard (crystalline) domains are the "knots" of the physical network structure. A fundamental change in the built-up, crystalline organization of these knots usually has a negative impact on the mechanical performance. This is in conflict with the expected tendency. The basic challenge with the nanoreinforcement of TPEs is to develop a "second" physical network. In the latter, the knots should be provided by the nanoparticles with strongly adhering, immobilized molecules around them. Accordingly, the nanoparticle-induced network should form exclusively in the amorphous (soft) phase. The expected outcome is a dual physical network: a nanoparticle-induced one in the amorphous phase which does not interfere with the development of the inherent physical network created by the crystalline (hard) domains. Unfortunately, this double network, if any, can hardly be established.

In situ polyaddition of polyisocyanates and polyols together with organophilic layered silicates was used to produce polyurethane nanocomposites. Zilg *et al.* ([14] and references therein) used organophilic fluoromica (intercalated with bis(hydroxyethyl)-methyl-dodecyl ammonium cations) as a nanofiller. It was found that this hydroxy-functionalized silicate exfoliates upon shearing in poly(oxypropylene)glycol if the molecular mass of the latter is high enough. The resulting nanoreinforced crosslinked polyurethanes exhibited improved tensile strength and even elongation at break values. *In situ* techniques are preferred to avoid the nanoparticle aggregation. Chen *et al.* [15] found that aggregation is not easy to circumvent due to the high surface energy of nanosilica and its tendency to form hydrogen bonds. Similar results were reported also by Zhou *et al.* [16]. As pointed out before, several works discussed the organoclay reinforcement of slightly or tightly crosslinked polyurethanes (*e.g.*, [17–21]). According to X-ray diffraction (XRD) results, the authors concluded that the organoclay layers were intercalated by the polyurethane molecules [17–21]. Exfoliation of the clay was only observed when the clay intercalant had, in addition to the alkyl chain, also ligands with hydroxyl groups. Even in the case of intercalation, a strong improvement in tensile stiffness and strength of the nanocomposites was observed. Contrary to the expectations, the ductility (strain at break) was less affected. This may be attributed to the "plasticizing effect" of the gallery onium ions (intercalant). "Plasticizing" is due to the intermingling of the molecules of the intercalant with those of the polymer matrix [17]. Thus, nanoreinforcement was often not associated with a shift in the glass transition temperature, T_g, toward higher temperatures (*e.g.*, [20,22]). Such a shift is expected due to the immobilization of adhered polymer molecules (*e.g.*, [15]). On the contrary, the plasticizing effect should result in a T_g shift toward lower temperatures. The position and intensity of the T_g peak depends on the dispersion stage of the silicate. Exfoliation usually results in an intensity reduction and in a shift toward higher temperatures. In a recent paper, Dai *et al.* [23] reported the structure-property relationships of an *in situ* produced organoclay-reinforced (cetyltrimethylammonium bromide intercalant) segmented polyether-polyurethane. The authors found an abrupt increase in the tensile strength and Young's modulus as a function of the organoclay content. In addition, the elongation at break has doubled in the presence of 2 wt% organoclay. This was attributed to the exfoliation of the organoclay on the basis of XRD and transmission electron microscopic (TEM) results.

3.1.2. *Sol-gel chemistry*

The sol-gel route is an interesting alternative in nanocomposite preparation. As aggregation/agglomeration phenomena can be avoided by this technique, it is usually followed when transparent products are targeted. The sol-gel method starts with molecular precursors and results in a nanoscaled metal oxide framework by hydrolysis and condensation reactions (*e.g.*, [10]). The final product is usually SiO_2, for which various alkoxysilanes are used as precursors. The precursor may be dispersed in some of the polymerizing components, added to a solution containing the polymeric product, by silane functionalization of the related reactants, *etc.* Cho *et al.* [24] produced a nanoparticle-filled polyurethane from a tetraetoxysilane precursor. The mechanical properties (tensile strength, stiffness and elongation at break) were improved by the addition of tetraethoxysilane up to 10 wt%. Beyond this threshold, the stress and elongation at break values dropped while the stiffness remained unaffected. Cuney *et al.* [25] compared the properties of polyurethanes crosslinked either *via* a silica-rich disperse phase (sol-gel route) or by urethane chemistry. Note that the silane compound was introduced in the polymerizing mixture. The aim of these authors was to check whether, for the same soft component, the sol-gel or the polyurethane chemistry is more promising.

3.2. *Solvent-assisted methods*

3.2.1. *Solution technique*

In order to obtain polyurethanes with highly regular chains, the polymerization should be carried out in solution. Dimethylformamide (DMF), dimethylacetamide, dimethylsulfoxide, *etc.*, may serve as solvents. Tien and Wei [26] studied the effect of benzidine-modified montmorillonite (MMT) on the hydrogen bonding in the hard segments of TPU. The latter was produced in random one-shot synthesis in DMF. It was established that the H-bonding in the hard segments was reduced owing to the presence of silicate layers. As a consequence, the morphology of the segmented polyurethane changed considerably. Note that this finding is in line with that of Petrović *et al.* [13]. The extent of the H-bond reduction depended on the amount of the organoclay and its dispersion stage. Chang and An [27] studied the effect of the organoclay intercalant on the TPU properties. The polyurethane was dissolved in dimethylacetamide. Hexadecylamine (primary amine) and quaternary amine compounds (dodecyl-trimethyl and dimethyl-hydrogenated tallow-(2-ethylhexyl) ammonium salts; the latter is the intercalant of a commercial organoclay, *i.e.*, Cloisite® 25A of Southern Clay) were used as intercalants of the organoclay. According to XRD spectra and TEM pictures, intercalated structures were found in the nano-composites containing up to 6 wt% organoclay. The tensile characteristics (strength, modulus, ultimate strain) passed through a maximum depending on the organoclay loading. The maximum was located in the range of 2 to 4 wt%. The best mechanical performance was achieved with a commercial organoclay (Cloisite® 25A of Southern Clay). This is the right place to mention that solution intercalation is often practiced in laboratories. Researchers believe that the best dispersion achievable by this way can be considered as a benchmark for melt intercalation. Han *et al.* [28] investigated the effect of MMT cointercalation in polyurethane nanocomposites. MMT was modified by cetyltrimethylammonium bromide and/or aminoundecanoic acid. The related nanocomposites were produced by solution intercalation of a TPU synthesized by bulk polymerization. The tensile strength, modulus, elongation at break, and tear strength of the plain polyurethane were 10.6 MPa, 5.1 MPa, 1090%, and 2.8 MPa, respectively. In the presence of 5 wt% organoclays of various modifications, these values changed to 13.9–15 MPa, 7–7.1 MPa, 1470–1630%, and 3.6–3.9 MPa, respectively. XRD and TEM results corroborated the onset of an intercalated clay structure. The water absorption was reduced and the thermooxidative resistance improved with the incorporation of the organocaly.

Choi *et al.* [29] showed an elegant way to use the polyurethane prepolymer (termed by the authors "organifier") as clay intercalant. Further, to get a better organoclay dispersion, sonication was used. Owing to the clay intercalation with the organifier, the initial intergallery distance was enhanced from 1.18 (Na-MMT) to 2.29 nm. In the corresponding nanocomposites containing up to 5 wt% organoclay, a further small gallery expansion was noticed (up to *ca.* 2.6 nm). Due to the presence of the organoclay, the oxygen permeability was reduced by *ca.* 30% and the resistance to thermal degradation improved. It is worth noting that the gas permeation depends not only on the clay content and its dispersion, but also on the matrix morphology [30]. A strong enhancement in mechanical properties was reported [29]. The tensile strength and modulus of the nanocomposites scattered between 27.2 and 43.2 MPa and the Young's modulus between 7.2 and 11.4 MPa. This corresponds to a pronounced property upgrade, as the unfilled matrix had a tensile strength of 14.7 MPa and a modulus of 6.5 MPa. The scatter in the above values is due to the organoclay content (1 to 5 wt%) and sonication time of the solution (0 and 60 min, respectively).

Wang *et al.* [31] used the solution intercalation of an organoclay to get peculiar optical properties. In this case, the clay (saponite) was intercalated with a suitable compound bearing chromophore moieties. Ni *et al.* [32] tried to use an amine-functionalized attapulgite as chain extender (instead of a diamine) in the polymerization reaction. The best mechanical performance was achieved by dispersing 2–5 wt% organophilic attapulgite in the polyurethane.

3.2.2. *Latex technique*

Organic solvents cause several problems (toxicity, flammability, disposal, recovery, *etc.*) and for this reason there is a growing tendency to their elimination. A promising way in this direction is to disperse the polymer in an aqueous medium. Polymeric products, including TPUs, can be emulsified and thus brought in a latex form in many cases. There is a further benefit with this strategy: pristine clays and layered silicates, especially those in sodium forms, can be well dispersed in water. Water acts as a swelling agent *via* hydration of the intergallery Na^+ ions. Note that pristine layered silicates (LS) cost markedly less than the organophilic versions. Needless to say that latex intercalation is extensively practiced with rubber latices [12,33,34]. Varghese *et al.* [33,34] produced thermoplastic poly(ester-urethane) nanocomposites by blending the related latex with 10 wt% LS. Similar to solution intercalation, the test films were produced by casting. Figure 2 shows the XRD spectra of the pristine LS along with that of the polyurethane nanocomposite film.

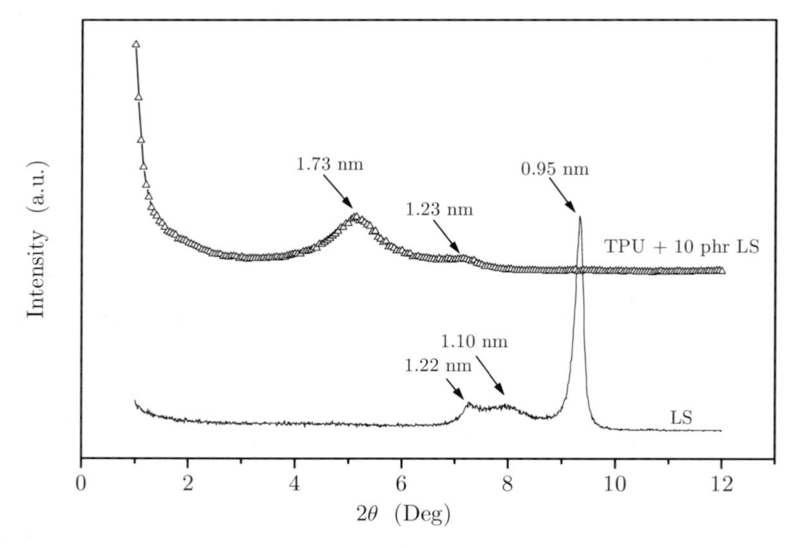

Figure 2. XRD spectra of the pristine layered silicate (LS; Na-fluorohectorite) and cast TPU nanocomposite containing 10 parts per hundred resin (phr) LS

The major peak of the nanocomposite indicates that the intergallery distance expanded up to 1.73 nm from the initial 0.95 nm. This good intercalation was also proved by transmission electron microscopy (TEM) (Figure 3).

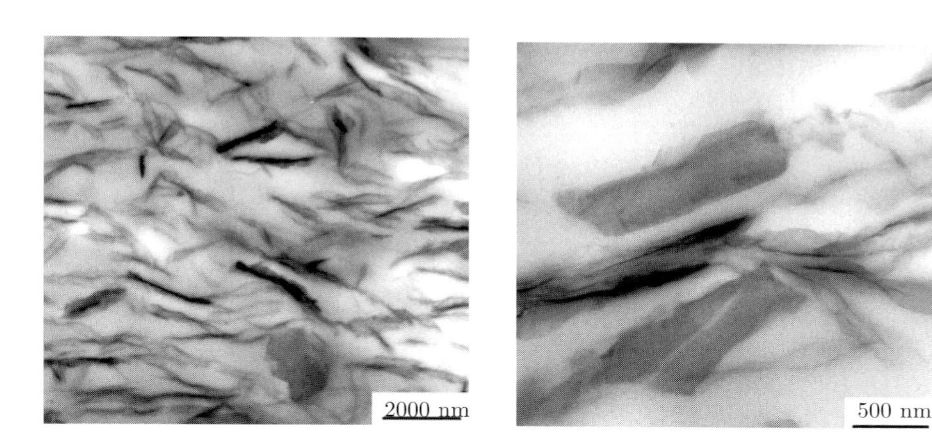

Figure 3. TEM pictures at various magnifications taken from a latex-cast TPU nanocomposite containing 10 phr LS

Figure 3 shows that a part of this synthetic layered silicate (Na-fluorohectorite) is even exfoliated. One can also see in these TEM pictures the very high aspect ratio of this synthetic LS (clay layers displaced from the view plane). The dynamic-mechanical thermal analysis (DMTA) spectra in Figure 4 display how

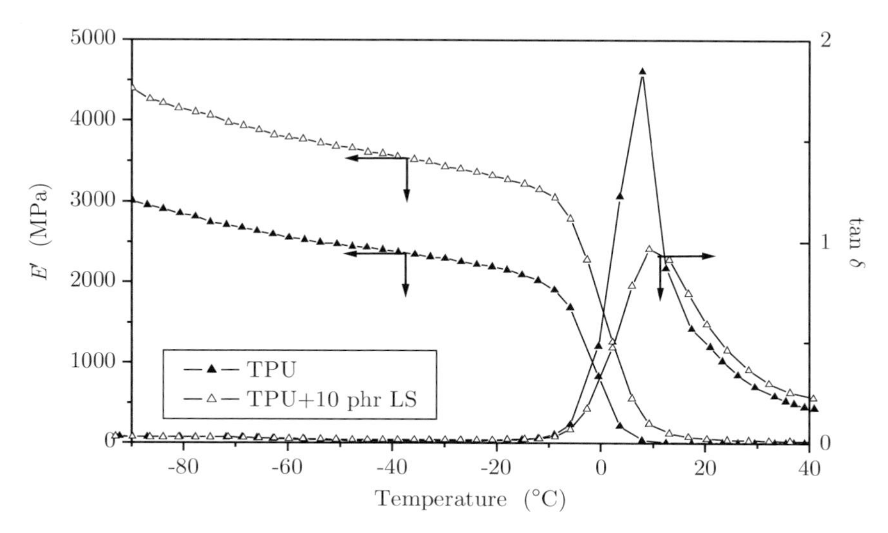

Figure 4. Storage modulus, E', and mechanical loss factor, tan δ, as a function of temperature, T, for a latex-cast TPU with and without 10 phr LS

markedly the stiffness below the T_g of TPU was enhanced by the LS incorporation. This, along with the strong reduction in the intensity of the T_g peak in the $\tan\delta - T$ trace, are clear indications of the nanoreinforcement. The reduction in the T_g relaxation is due to the strong interaction between the polymer molecules and the clay layers. This kind of "bonding" reduces the segment mobility of the molecular chains, resulting in a reduction of the T_g peak intensity. The mechanical properties (before and after thermal aging) of the neat and nanoreinforced thermoplastic polyurethane are summarized in Table 1.

Table 1. Tensile mechanical properties of latex-casted TPU films with and without 10 phr LS (Na-fluorohectorite)

Property	TPU		TPU+LS (10 phr)	
	As cast	Aged (70°C, 7 days)	As cast	Aged (70°C, 7 days)
Tensile strength (MPa)	4.0	10.5	15.9	17.9
Tensile modulus (MPa) at:				
100% elongation	0.8	1.1	5.6	7.6
200% elongation	0.9	1.4	7.8	10.7
300% elongation	1.1	1.8	10.1	13.5
Elongation at break (%)	932	772	543	444

3.3. *Melt compounding*

Melt compounding is a very attractive way to produce commercial nanocomposites. Major arguments for this claim are the available industrial compounding facilities and the environmentally friendly preparation. Albeit the intercalation/exfoliation phenomena during melt compounding obey also the thermodynamic and kinetic rules, there is a further parameter which strongly affects the clay dispersion in nanocomposites. This is linked to the locally emerging shear and elongational flow fields. Extensive shear and elongational stresses peel apart the silicate layers, as shown in several works (*e.g.* [2,6,35]). It should be born in mind, however, that the thermal stability of the usual organoclays is limited. Results obtained by melt compounding did not meet the expectations at all. The author's group studied the effect of commercial organoclays intercalated with primary (octadecylamine modified montmorillonite, Nanomer® I.30P of Nanocor) and quaternary amine compounds (intercalant methyl-tallow-bis(2-hydroxyethyl) ammonium salt, Cloisite® 30B of Southern Clay) on the mechanical performance of TPUs of various hardness values. The poly(ether-*b*-urethane) (PEBU) and poly(ester-*b*-urethane) involved in this extrusion compounding exhibited Shore A80 and Shore D65 hardness values. Albeit the XRD spectra suggested intercalation with Nanomer® I.30P and exfoliation with Cloisite® 30B in PEBU (Figure 5), only slight improvements in the mechanical

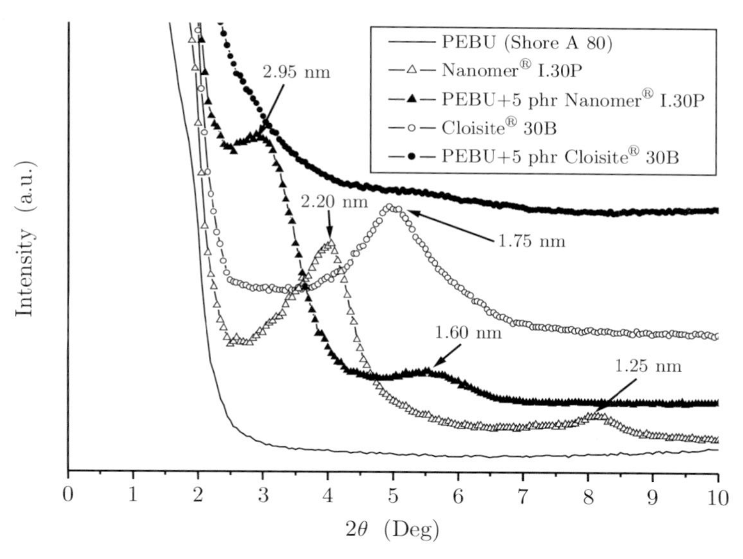

Figure 5. XRD spectra taken form the organoclay powders in reflexion and from the melt-compounded PEBU nanocomposites containing 5 phr organoclay in transmission mode

performance were achieved. The possible exfoliation with Cloisite® 30B was due to its hydroxyl functional intercalant. The effect of the organoclay was, however, far more prominent for the low-hardness polyether-based TPU than for the high-hardness polyester version. For the latter TPU, practically no improvement in the mechanical performance was found. On the contrary, the DMTA traces of the nanocomposites with a PEBU matrix showed some stiffness enhancement, however, mostly above the T_g (Figure 6) [36]. Recall that an efficient nanoreinforcement should yield a considerable stiffness increase also below the T_g of the matrix (as shown in Figure 4).

Reasons for this moderate (if any) property improvement are discussed in the exhaustive work of Finnigan *et al.* [37]. The authors produced TPU/organoclay (Cloisite® 30 B) nanocomposites by melt blending (twin-screw extrusion) and solution intercalation. Note that not only the extrusion compounding, but also the selected commercial organoclay were similar to those in our work [36]. A further analogy with our earlier work was the broad hardness range of the polyether-type TPU selected (Shore 80 A and 55 D). The published results, namely that the tensile strength and elongation values were not improved in contrast to the Young's modulus, also agree with ours [36]. The most interesting result of Finnigan *et al.* [37] was that the mean molecular mass of the TPU significantly decreased upon both solvent (during which sonication was used) and melt intercalation. This was attributed to effects of the sonication (solution method) and thermal/thermooxidative degradation (melt compounding), respectively. DMTA, differential scanning calorimetry (DSC) and wide-angle X-ray scattering (WAXS) measurements revealed a strong alteration in the

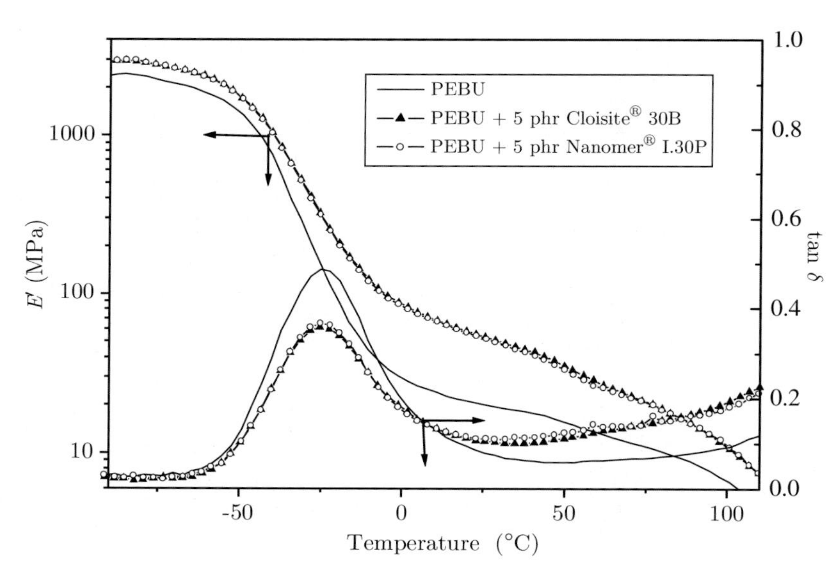

Figure 6. Storage modulus, E', and mechanical loss factor, $\tan\delta$, as a function of temperature, T, for a melt-compounded poly(ether-*b*-urethane) (PEBU, Shore A 80) with and without 5 phr organoclay; the organoclays contained primary amine (Nanomer® I.30P) or hydroxyl functionalized quaternary ammonium intercalants (Cloisite® 30B)

microphase morphology. These findings are consistent with results reported earlier [13]. Ma *et al.* [38] used an organophilic modified rectorite for the melt compounding with a poly(ester-*b*-urethane). For the organophilization of the rectorite, two different quaternary ammonium salts and benzidine were used. According to the authors, the clay was intercalated up to a loading of 5 wt%, while beyond this threshold, clay agglomeration took place. The mechanical properties were improved by clay intercalation and decreased by the onset of agglomeration. Interestingly, the tear strength was less sensitive to the clay dispersion than the tensile properties. Eckel *et al.* [39] emphasized in their work, which covered organoclay-reinforced TPU as well, that the clay dispersion can hardly be characterized adequately solely by XRD. According to the cited authors, the right tool for that is TEM. This claim seems to be of great relevance and is in conflict with the usual belief that XRD traces are good enough to distinguish between intercalated and exfoliated layered silicates (nanocomposites based on TPU are considered also in Chapter 14).

4. Nanoreinforcement of "condensation" thermoplasic elastomers

The author was not able to find reports in the open literature dealing with nanocomposites having poly(ether-*b*-amide) (PEBA) and poly(ether-*b*-ester) (PEE) matrices. This is very surprising owing to several reasons. First, polyamides are those polymers where the nanoreinforcement concept really works, irrespective of whether *in situ* polymerization or melt compounding techniques are adopted. It is worth noting that the pioneering work at Toyota Central Research Labs was performed also on polyamides (PAs), mostly on PA6 ([1,2,6,40] and references therein). Later, numerous studies were published on organoclay-reinforced PA12 [1,2,6,41,42]. Note that PA12 is the usual crystalline segment in PEBA grades. Second, similar to PAs, thermoplastic polyester-based nanocomposites were successfully produced *via* various intercalation techniques [1,2,6]. This holds also for poly(butylene terephthalate) (PBT) (*e.g.*, [43,44]) which is the usual crystalline block in PEE. Third, numerous reports are available dealing with systems composed of organoclays and polymers with oxyethylene (or similar units) chains (*e.g.*, [45,46]). In addition, poly(ethylene oxide) is a favored polymer for nanoreinforcement *via* the sol-gel route [47]. Note that polyalkyleneoxides form usually the soft segments in the thermoplastic elastomers produced by polycondensation reactions. Thus, there are no arguments against the clay intercalation/exfoliation with such segmented block copolymers. The possible reason behind this lack of interest is the very versatile synthesis of these elastomers *via* which their properties can be tailored in a wide range.

5. Outlook and future trends

As predicted above, the potential of the nanoreinforcement (including carbon nanotubes) will be checked in condensation thermoplastic elastomers in the very near future. A possible driving force in this respect may be linked to coloration issues, since the organoclay would be a suitable carrier for several organic colorants (*e.g.*, disazo compounds). In this way, migration and bleed problems could be avoided. Attempts will also be made to use new types of nanofillers, such as acid-activated Ca-bentonite, anion-exchanged hydrotalcites, boehmites, layered graphites, *etc.* Owing to cost reduction, the sol-gel methods will require cheaper precursors than the presently used organosilane and organotitanate compounds. Nevertheless, the strong competition in respect to property modification between synthesis and nanoreinforcement will remain.

Acknowledgements

The author thanks the Fonds der Chemischen Industrie and the German Science Foundation for the support of his research providing the results included in this chapter.

References

1. Pinnavaia T J and Beall G W, Eds. (2000) *Polymer-Clay Nanocomposites*, Wiley, Chichester, UK.
2. Sinha Ray S and Okamoto M (2003) Polymer/layered silicate nanocomposites: a review from preparation to processing, *Progr Polym Sci* **28**:1539–1641.
3. Chazeau L, Gauthier C, Vigier G and Cavaillé JY (2003) Relationships between microstructural aspects and mechanical properties of polymer-based nanocomposites,

in *Handbook of Organic-Inorganic Hybrid Materials and Nanocomposites* (Ed. Nalwa H S), Amer Sci Publ, Los Angeles, Vol.2, Ch. 2, pp. 63–111.

4. Zhang M Q, Rong M Z and Friedrich K (2003) Processing and proprtuies of nonlayered nanoparticle reinforced thermoplastic composites, in *Handbook of Organic-Inorganic Hybrid Materials and Nanocomposites* (Ed. Nalwa H S), Amer Sci Publ, Los Angeles, Vol. 2, Ch. 3, pp. 113–150.
5. Andrews R and Weisenberger M C (2004) Carbon nanotube polymer composites, *Current Opinion in Solid State and Materials Science* **8**:31–37.
6. Karger-Kocsis J and Zhang Z (2004) Structure-property relationships in nanoparticle/ semicrystalline themoplastic composites, in *Mechanical Properties of Polymers Based on Nanostructure and Morphology* (Eds. Baltá Calleja F J and Michler G H), Marcel, New York, in press.
7. Hammel E, Tang X, Trampert M, Schmitt T, Mauthner K, Eder A and Pötschke P (2004) Carbon nanofibers for composite applications, *Carbon* **42**:1153–1158.
8. Giannelis E P, Krishnamoorti R and Manias E (1999) Polymer-silicate nanocomposites: model systems for confined polymers and polymer brushes, in *Advances in Polymer Science Vol. 138*, Springer, Berlin, pp. 107–147.
9. Mülhaupt R, Engelhardt T and Schall N (2001) Nanocomposites – en route to application, *Kunststoffe* **91**(10):178–190.
10. Kickelbick G (2003) Concepts for the incorporation of inorganic building blocks into organic polymers on a nanoscale, *Progr Polym Sci* **28**:83–114.
11. Karger-Kocsis J, Gryshchuk O, Fröhlich J and Mülhaupt R (2003) Interpenetrating vinylester/epoxy resins modified with organophilic layered silicates, *Compos Sci Technol* **63**:2045–2054.
12. Karger-Kocsis J and Wu C-M (2004) Thermoset rubber/layered silicate nanocomposites. Status and future trends, *Polym Eng Sci* **44**:1083–1093.
13. Petrović Z S, Cho Y J, Javni I, Magonov S, Yerina N, Schaefer D W, Ilavskỳ J. and Waddon A (2004) Effects of silica nanoparticles on morphology of segmented polyurethanes, *Polymer* **45**:4285–4295.
14. Zilg C, Dietsche F, Hoffmann B, Dietrich C and Mülhaupt R (2001) Nanofillers based upon organophilic layered silicates, *Macromol Symp* **169**:65–77.
15. Chen X, Wu L, Zhou S and You B (2003) In situ polymerization and characterization of polyester-based polyurethane/nano-silica composites, *Polym Int* **52**:993–998.
16. Zhou S-X, Wu L-M, Sun J and Shen W-D (2003) Effect of nanosilica on the properties of polyester-based polyurethane, *J Appl Polym Sci* **88**:189–193.
17. Wang Z and Pinnavaia T J (1998) Nanolayer reinforcement of elastomeric polyurethane, *Chem Mater* **10**:3769–3771.
18. Zilg C, Thomann R, Mülhaupt R and Finter J (1999) Polyurethane nanocomposites containing laminated anisotropic nanoparticles derived form organophilic layered silicates, *Adv Mater* **11**:49–52.
19. Song M, Hourston D J, Yao K J, Tay J K H and Ansarifar M A (2003) High performance nanocomposites of polyurethane elastomer and organically modified layered silicate, *J Appl Polym Sci* **90**:3239–3243.
20. Zhang X, Xu R, Wu Z and Zhou C (2003) The synthesis and characterization of polyurethane/clay nanocomposites, *Polym Int* **52**:790–794.
21. Moon S-Y, Kim J-K, Nah C and Lee Y-S (2004) Polyurethane/montmorillonite nanocomposites prepared from crystalline polyols, using 1,4-butanediol and organoclay hybrid as chain extender, *Eur Polym J* **40**:1615–1621.
22. Rhoney I, Brown S, Hudson N E and Pethrick R A (2004) Influence of processing method on the exfoliation process for organically modified clay systems. I. Polyurethanes, *J Appl Polym Sci* **91**:1335–1343.

23. Dai X, Xu J, Guo X, Lu Y, Shen D, Zhao N, Luo X and Zhang X (2004) Study on structure and orientation action in polyurethane nanocomposites, *Macromolecules* **37**:5615–5623.
24. Cho J W and Lee S H (2004) Influence of silica on shape memory effect and mechanical properties of polyurethane-silica hybrids, *Eur Polym J* **40**:1343–1348.
25. Cuney S, Gérard J F, Dumon M, Pascault J P, Vigier G and Dušek K (1997) Hydroxyl-terminated oligomers crosslinked by alkoxysilane sol-gel or polyurethane chemistries: A comparison, *J Appl Polym Sci* **65**:2373–2386.
26. Tien Y I and Wei K H (2001) Hydrogen bonding and mechanical properties in segmented montmorillonite/polyurethane nanocomposites of different hard segment ratios, *Polymer* **42**:3213–3221.
27. Chang J-H and An Y U (2002) Nanocomposites of polyurethane with various organoclays: Thermomechanical properties, morphology, and gas permeability, *J Polym Sci Part B Phys* **40**:670–677.
28. Han B, Cheng A, Ji G, Wu S and Shen J (2004) Effect of organophilic montmorillonite on polyurethane/montmorillonite nanocomposites, *J Appl Polym Sci* **91**:2536–2542.
29. Choi W J, Kim S H, Kim Y J and Kim S C (2004) Synthesis of chain-extended organifier and properties of polyurethane/clay nanocomposites, *Polymer* **45**:6045–6057.
30. Osman M A, Mittal V, Morbidelli M and Suter U W (2003) Polyurethane adhesive nanocomposites as gas permeation barrier, *Macromolecules* **36**:9851–9858.
31. Wang W-J, Chin W-K and Wang W-J (2002) Synthesis and structural characterizations of [chromophore]$^+$-saponite/polyurethane nanocomposites, *J Polym Sci Part B Phys* **40**:1690–1703.
32. Ni P, Li J, Suo J and Li S (2004) Study on mechanical properties of polyurethane-attapulgite nanocomposites, *J Mater Sci* **39**:4671–4673.
33. Varghese S, Gatos K G, Apostolov A A and Karger-Kocsis J (2004) Morphology and mechanical properties of layered silicate reinforced natural and polurethane rubber blends produced by latex compounding, *J Appl Polym Sci* **92**:543–551.
34. Varghese S and Karger-Kocsis J (2005) Layered silicate/rubber nanocomposites via latex and solution intercalations, in *Polymer Composites from Nano- to Macroscale* (Eds. Friedrich K, Fakirov S and Zhang Z), Springer, Berlin, in press.
35. Fornes T D, Yoon P J, Keskkula H and Paul D R (2001) Nylon 6 nanocomposites: the effect of matrix molecular weight, *Polymer* **42**:9929–9940.
36. Karger-Kocsis J (2001) unpublished results.
37. Finnigan B, Martin D, Halley P, Truss R and Campbell K (2004) Morphology and properties of thermoplastic polyurethane nanocomposites incorporating hydrophilic layered silicates, *Polymer* **45**:2249–2260.
38. Ma X, Lu H, Liang G and Yan H (2004) Rectorite/thermoplastic polyurethane nanocomposites: Preparation, characterization, and properties, *J Appl Polym Sci* **93**:608–614.
39. Eckel D F, Balogh M P, Fasulo P D and Rodgers W R (2004) Assessing organo-clay dispersion in polymer nanocomposites, *J Appl Polym Sci* **93**:1110–1117.
40. Kawasumi M (2004) The discovery of polymer-clay hybrids, *J Polym Sci Part A Chem* **42**:819–824.
41. Kim G-M, Lee D-H, Hoffmann B, Kressler J and Stöppelmann G (2001) Influence of nanofillers on the deformation process in layered silicate/polyamide-12 nanocomposites, *Polymer* **42**:1095–1100.
42. McNally T, Murphy W R, Lew C Y, Turner R J and Brennan G P (2003) Polyamide-12 layered silicate nanaocomposites by melt blending, *Polymer* **44**:2761–2772.

43. Chang J-H, An Y U, Ryu S C and Giannelis E P (2003) Synthesis of poly(butylene terephthalate) nanocomposite by in-situ interlayer polymerization and characterization of its fiber (I), *Polym Bull* **51**:69–75.
44. Scatteia L, Scarfato P and Acierno D (2004) Rheology of PBT-layered silicate nanocomposites prepared by melt compounding, *Plast Rubb Compos* **33**:85–91.
45. Ruiz-Hitzky E and Aranda P (2000) Electroactive polymers intercalated in clays and related solids, in *Polymer-Clay Nanocomposites* (Eds. Pinnavaia T J and Beall G W), Wiley, Chichester, UK, Ch. 2, pp. 19–46.
46. Öl E, Alemdar A, Güngör N and Hayashi S (2002) Influences of nonionic poly(ethylene glycol) polymer PEG on electrokinetic and rhelogical properties of bentonite suspensions, *J Appl Polym Sci* **86**:341–346.
47. Manaf I and Mascia L (2004) Compatibilization of water born poly(ethylene oxide)-silica hybrids, *Adv Polym Technol* **23**:125–134.

Chapter 17

Commercial Condensation and Addition Thermoplastic Elastomers: Composition, Properties, and Applications

O. Gryshchuk

1. Introduction

Thermoplastic elastomers (TPE) are a unique class of engineering materials combining the look, feel, and elasticity of conventional thermoset rubbers with the processing efficiency of thermoplastics. Because TPE are thermoplastics, their melt processability makes them very suitable for high-volume injection molding and extrusion. They can also be reclaimed and recycled (see also Chapter 19). As elastomers, TPE exhibit true elasticity. The range of TPE grades encompasses rubber-like properties, various hardness values, low compression set, and high elongation. The most widespread commercial thermoplastic elastomers of the condensation type are polyester- (TPE-E) and polyamide-based (TPE-A), whereas among the addition types the polyurethane-based (TPE-U) thermoplastic elastomers are of major importance. These TPE compounds are used in various applications, including personal care, packaging, sporting goods, houseware, hardware and electronics. The synthesis of condensation TPE is considered in Chapters 2, 3, and 9. In this chapter, the properties and applications of some commercial condensation and addition TPE are briefly reviewed. By no means it intends to cover all available commercial condensation TPE, and only typical examples of the various types of condensation and addition TPE are the subject of this chapter. It is also not intended to compare these commercial products.

2. Polyester-based thermoplastic elastomers (TPE-E)

TPE-E are high-performance engineering materials, bridging the gap between the rubbery elastomers and rigid plastics. They can replace successfully thermoset rubbers, such as polychloroprene, chlorosulfonated polyethylene, nitrile and epichlorohydrin. TPE-E copolyesters possess exceptional toughness, impact resistance, load bearing capacity, and low temperature flexibility. Their creep resistance, flexural fatigue resistance and resilience make them suitable for applications in dynamic regimes. Their chemical resistance to aliphatic hydrocarbon solvents, oils and greases is very good, even at elevated temperatures. Dilute acids and bases, as well as polar solvents (water and alcohols) are compatible with copolyesters at room temperature, but not at elevated temperatures. (The deformation behavior of TPE-E is discussed in Chapters 6, 7, and 16.)

2.1. *Hytrel® engineering thermoplastic elastomers*

Hytrel® is a registered trademark [1] of DuPont for its family of engineering thermoplastic elastomers. Hytrel® engineering thermoplastic elastomers are block copolymers, consisting of a hard (crystalline) segment of poly(butylene terephthalate) (PBT) and a soft (amorphous) segment based on long-chain polyether glycols. The properties are determined by the hard-to-soft segment ratio and by the build-up of the segments. Hytrel® thermoplastic polyester elastomers have the flexibility of rubbers, the strength of plastics, and the processability of thermoplastics. They are ideal for parts requiring excellent flexural fatigue resistance and a broad service temperature range. What is more, they are strongly resistant to tearing, flex-cut growth, creep, and abrasion. Hytrel® mechanical properties provide good strength and stiffness, in addition to an outstanding toughness, while the chemical properties make these products highly resistant to hydrocarbons and numerous other fluids. Hytrel® grades are available in the full range of Shore D hardness from 30 to 82. Special grades include heat-stabilized, flame-retardant, and blow-molding ones. Concentrates offered include black pigments, UV protection additives, hydrolysis-resistant additives, heat stabilizers and flame retardants. Hytrel® can be readily formed into high-performance parts by a variety of thermoplastic processing techniques, including injection molding, extrusion, blow molding, rotational molding and melt casting. For injection molding, temperatures between 175 and 260°C are selected depending on the polymer type. All grades have a sharp melting point and very good melt stability.

There are two main groups of products depending on the type of soft segment used. These are *standard grades* offering the best balance of cost and performance and ranging in Shore D hardness from 35 to 82, and *high performance grades* providing an even better performance and service life in applications where properties, such as abrasion resistance and tear strength are critical. They range in Shore D hardness from 40 to 72. Within each of these groups, a range of hardnesses or flexural moduli is obtained by varying the hard-to-soft segment ratio. Available grades of each group are shown in Table 1.

Table 1. Available Hytrel® grades

Hardness	Flexural modulus (MPa)	Hytrel® grades (Standard)	Hytrel® grades (High performance)
35D	35	G3548L	–
40D	55–67	G4074/G4078W	4056/4068
45D	94	–	4556
47D	117	G4774	–
55D	193-207	G5544	5526/5556/5586
63D	300	6358	6356
72D	570	7248	7246
104R*	1207	8238	–

*Rockwell R

The excellent properties of the Hytrel® engineering thermoplastic elastomer make it quite appropriate for a number of demanding applications where mechanical strength and durability are required in a flexible article. Examples include seals, belts, bushings, pump diaphragms, gears, protective boots, hoses and tubing, springs and couplings, hinges, impact and sound absorbing devices. In many of these applications, Hytrel® allows a multipiece rubber or even a metal composite assembly to be replaced by a single part. Some of the industries where Hytrel® is used include automotive, electrical/electronic, appliances and power tools, sporting goods, furniture, and off-road transportation/equipment.

2.2. *Ecdel® engineering thermoplastic elastomers*

Ecdel® elastomers are unique poly(ester ether)s marketed by Eastman Chemical Co. [2]. These materials offer clarity and toughness over an extremely wide range of temperatures and chemical resistance required for a variety of demanding applications, including flexible medical and pharmaceutical packaging. Ecdel® elastomers are well known for their low levels of extractables. They may be sterilized in autoclaves and by gamma irradiation, ethylene oxide, or E beam. Three grades of Ecdel® elastomers, PCCE 9965, PCCE 9966, and PCCE 9967, are commercially available to meet the requirements of the most common polymer processing operations, such as extrusion, injection molding, and extrusion blow molding. Additionally, Ecdel® elastomers can be coextruded with other resins to provide unique blends for the most demanding applications. Their typical physical properties are shown in Table 2.

All three formulas may be injection molded and extruded into film and tubing, but only Ecdel® elastomer 9967 can be extrusion blow molded. The melt temperatures generally range from 205 to 260°C. Optimization of the melt

Table 2. Typical physical properties of Ecdel® elastomers

Properties	ASTM method	Unit	Typical value
Thermal properties			
Inherent viscosity	EMN-A-AC-G-V-1	DL/g	1.05–1.23
Flow rate (230°C/2.16 kg)	D1238	g/10 min	4–20
Crystalline peak melting point, T_m	D3418	°C	205–207
Crystallization temperature on cooling, T_c	DSC	°C	140
Glass transition temperature, T_g	DSC	°C	–3°
Heat of fusion	E793	kJ/kg	27
Thermal conductivity	C 177	W/m·K	0.19
Coefficient of linear thermal expansion	D696	mm/mm·°C	15×10^{-5}
Brittleness temperature	D746	°C	<–75
Vicat softening temperature at 1 kg load	D 1525	°C	170
Mechanical properties			
Density	D792	g/cm^3	1.13
Durometer hardness:			
Shore D scale	D2240	–	55
Shore A scale		–	95
Tensile stress at break	D638	MPa	20–23
Tensile stress at yield	D638	MPa	13–14
Elongation at yield	D638	%	30–38
Elongation at break	D638	%	300–400
Tensile modulus	D638	MPa	170
Flexural modulus	D790	MPa	150
Izod impact strength, notched at –40°C	D256	J/m	40–50
Torsional modulus temperature			
at 240 MPa	D 1043	°C	–28
at 930 MPa	D 1043	°C	<–70
Water absorption, 24 h immersion	D570	%	0.35–0.4

temperature for a given process is desirable because the range varies from process to process. Ecdel® elastomers may be extrusion blow molded directly into bags or extruded into films for the later production of bags. The latter, when made of Ecdel® elastomer, usually require an overwrap to maintain the drug concentration at the desired level for the shelf life of the product. The barrier properties can be improved through coextrusion with other materials, such as polypropylene or poly(propylene-ethylene) copolymers. Certain applications involve exposure to conditions ranging from subfreezing temperatures to autoclaving. Ecdel® elastomers can often provide the heat stability and flexibility needed for use in such environments.

2.3. *RTP® engineering thermoplastic elastomers*

RTP® 1500 Series copolyesters possess exceptional toughness, impact resistance, load bearing capacity and low temperature flexibility. Their creep resistance, flexural fatigue resistance, and resilience make them suitable for applications in dynamic regimes. As a specialty compounder, RTP Company [3] formulates thermoplastic compounds to meet specific customer and application requirements. The summarized product data sheets (see Table 3) show the property profiles of the products. RTP® 1500 Series copolyesters are tough and resilient, creep resistant, flexible at low temperatures, resistant to chemicals, oils, and solvents, available in 30D to 80D durometer hardness, and show an excellent high temperature performance.

Typical markets of RTR® TPE are the automotive, fluid power, and electrical/electronic industries; they are also used in the production of appliances, power tools, sporting goods, footwear, wires and cables, furniture, transportation, and for medical purposes.

Table 3. Properties of RTP® 1500 Series elastomers

Property	ASTM method	Unit	Typical value
Injection pressure	–	MPa	69–103
Injection cylinder temperature	–	°C	193–238
Mold temperature	–	°C	21–66
Miscellaneous			
Density	D792	g/cm^3	1.17–1.25
Molding shrinkage, 3.18 mm section	D955	%	0.5–1.9
Water absorption, 24 h at 23°C	D570	%	0.5–0.6
Mechanical properties			
Tensile strength	D412	MPa	8–40
Tensile elongation	D412	%	200–550
Shore D hardness	D2240	–	40–70
Electrical properties			
Volume resistivity	D257	Ohm·cm	10^3–10^{16}
Thermal properties			
Deflection temperature at 1.82 MPa	D648	°C	49–54

2.4. *Kopel*® *engineering thermoplastic elastomer*

Kopel® is a thermoplastic polyester elastomer of Kolon Industry [4]. It offers benefits in processing, productivity, and economical-performance balance. Kopel® consists of polymerized hard and soft segments. The hard segments are crystalline PBT and the soft segments are amorphous polyether-type compounds. This structure ensures Kopel® rubber-like behavior and provides superior flexibility, durability, performance in a wide temperature range, and mechanical strength for versatile applications. Typical properties of Kopel® elastomers are listed in Table 4, while Table 5 shows some of their applications.

Table 4. Typical properties of Kopel® elastomers

Property	ASTM method	Unit	Typical value
	Miscellaneous		
Density	D792	g/cm³	1.15–1.27
Hardness, durometer	D2240	–	40–72
Melt index, 230°C	D1238	g/10 min	12–20
Water absorption	D570	%	0.3–0.7
Mold shrinkage	D955	%	1–1.6
	Thermal properties		
Melting point	DSC	°C	170–218
Heat deflection temperature (HDT), 0.46 MPa	D648	°C	65–148
Solenoid brittle point	D746	°C	<–70–<–65
Flammability	UL94	–	HB**
	Mechanical properties		
Tensile strength, 23°C	D638	MPa	25–39
Tensile elongation, 23°C	D638	%	400–850
Flexural modulus, 23°C	D790	MPa	49–510
Izod notched impact strength	D256		
23°C		J/m	255–N/B*
–40°C		J/m	54–N/B*
Abrasion resistance CS-17 wheel	D1044	mg/kcycles	3.0–10.0

* N/B: no break; ** HB: lowest flame resistance

2.5. *Arnitel*® *engineering thermoplastic elastomer*

In 1991, DSM took over *Arnitel*® from Akzo Nobel, and subsequently the Arnitel® applications have been considerably extended. Since 1991, the

Table 5. Applications of Kopel® TPE

Grade	Description	Applications
KP3340	High flexibility (D40)	Fire hose, tubing, pad, sheet, film seals, gaskets, jackets
KP3355	High flexibility, resilience (D55)	Tubing, belting, film sheets, hose, boots, roto-molded tire
KP3363	Mechanical strength, resilience (D63)	Tubing, hose, film, insulation coatings, fuel tanks, gears
KP3372	Mechanical strength (D72)	Wire&cable coatings, tubing, film&sheet, molded shapes
KP3755	High heat resistance (D55)	High viscosity, extrusion, insulation coating

activities involving Arnitel® have grown by more than 10% per year, making DSM [5] one of the global players in this market. Arnitel® has an exceptional flexibility and can perform or even outperform functions that normally require conventional rubbers. Available in a wide hardness range, Arnitel® can replace metals, thermoplastics, leather and rubber, often with a reduction in finished part costs. Arnitel® can be used over a wide range of temperatures; it has exceptional fatigue and creep resistance, as well as resistance to oils, greases, and many other chemicals. It is characterized by excellent strength over a wide range of temperatures, excellent dynamic properties, *e.g.*, creep and fatigue, high degree of versatility in processing, easy coloring using master batches, surface quality from high gloss to textured one, excellent heat resistance (long-term 165°C), good electrical insulation properties, low moisture absorption, excellent dimensional stability, and easy flow with fast cooling times.

Arnitel® is extensively used in the automotive industry for applications requiring exceptional fatigue resistance and resistance to oils and greases. It is used, for instance, in joint boots for kinetic wheel couplings, not only due to its durability, light weight and flexibility, but also because it retains oil and keeps out the influence of external factors, such as dust, salt, and mud. Arnitel® is also used in covering airbags. The cover can thus be opened under all types of conditions, providing easy access to the airbag without damaging it. The steering wheel is one of the components of a car that is exposed to sunlight, heat and cold, and is subject to frequent use. With Arnitel®, the effect of these factors is minimized. Leather steering wheel covers also adhere better to steering wheels that contain Arnitel®.

Arnitel® is used in large quantities for the protection of electrical cables and signaling cables for railways, mainly in areas exposed to high temperatures. It can also be found in "railway pads", *i.e.*, rubber mats that enable the use of railway tracks for longer periods without maintenance, and for damping the vibrations in the rail network. DSM offers a broad portfolio for this market; there are different types for use in passenger lines (for instance, the TGV

Madrid-Barcelona-France border) and for freight lines (USA). The use of Arnitel® increases the total life by a factor of 4 in comparison to the traditional systems. In view of its longer life and the low maintenance requirements, Arnitel® contributes to bringing down train delays.

Arnitel® is used to a large extent in medical environments, where hygiene, safety, and reliability are primary considerations. It can be sterilized without any problems and is resistant to bacteria, viruses, and infections. These properties make it also suitable for use in operating-theatre sheets. Dressings and other bandages are more comfortable for the patient thanks to its elasticity and capacity to "breath".

Arnitel® finds applications in the production of sporting goods, *e.g.*, it improves the spring force of golf balls, so that they travel farther. Thanks to Arnitel®, the top layer of snowboards can now be printed with sharper, brighter and more visible images, they show also resistance to scratches. Clothing with Arnitel®, for example sailing jackets, is very light and is well known for its "breathing" characteristics, and also offers excellent protection against cold.

In construction, Arnitel® is used to improve the grip of tools; it creates a pleasant, rubber-like grip and is resistant to greases. Tools with Arnitel® also last longer and the manufacturers of quality tools provide life-long guarantees on their instruments. Foils that are used, among others, in the construction of roofs and walls are stronger, more flexible, more elastic, and also allow the passage of air, thanks to Arnitel®, as a result of which they have a longer life and require less maintenance.

Arnitel® can be used in a large number of applications "in and around the house". For instance, it is used in box spring mattresses, particularly because of its spring force. In washing and drying machines, it is used for rings that last longer than rubber components and are resistant to high temperatures.

2.6. *Keyflex® engineering thermoplastic elastomers*

Keyflex® BT is the trade name of LG Chem [6] for its engineering thermoplastic polyester elastomers. This TPE-E has flexibility and elastic resiliency similar to those of rubber, as well as mechanical strength, heat- and weather-resistance better than those of rubber. All grades of Keyflex® BT are block copolymers consisting of a hard (crystalline) segment of PBT and a soft (amorphous) segment based on long-chain polyether glycols. Similar to the previous cases, its properties are determined by the ratio of hard to soft segments. Keyflex® BT offers a unique combination of mechanical, physical, and chemical properties that qualifies it for a wide range of applications.

The highly desirable characteristics of Keyflex® BT resins include excellent resistance to creep, impact, and flexural fatigue, high impact strength and flexibility at low temperatures, good retention of properties at elevated temperatures, high resistance to chemicals, oils, solvents, and weathering, high tear strength and abrasion resistance, easy and economical processing, good recyclability.

Keyflex® BT 1040D and 1047D are low modulus injection molding grades. These offer excellent low temperature properties, flexural fatigue resistance and creep resistance. Keyflex® BT 1055D is a medium modulus grade. This product maintains good balance between flexibility and mechanical properties. It has excellent mechanical strength, abrasion resistance, and flexibility both at high and low temperatures. Keyflex® BT 1063D and 1072D are high modulus injection molding grades. These offer high service temperature, and have excellent resistance to oils, fuels, and solvents. Due to their extremely low fuel permeability, they are adequate products for fuel tank parts. The applications of Keyflex® BT are listed in Table 6.

Table 6. Applications of Keyflex® BT grades

Grade	Shore D hardness	Applications
Keyflex® BT-1040D	40	Hose jackets, belts for hydraulic hoses and tubes, and sealing materials
Keyflex® BT-1047D	47	Films and parts for sports goods including skins of golf balls
Keyflex® BT-1055D	55	Molding materials, automobile parts, belts, tubes and hoses
Keyflex® BT-1063D	63	Fuel tanks, gears, tubes, and hoses
Keyflex® BT-1072D	72	Coil tubes, wire coatings, and inner liners of power cables

2.7. *Pibiflex®* *engineering thermoplastic elastomer*

Pibiflex® is a thermoplastic product of P-GROUP [7] with marked elastomeric properties and good performance behavior with regard to many chemicals, oils, and solvents. It is well suited for engineering applications where toughness and elasticity, high creep resistance, impact strength and flexural fatigue resistance in a broad temperature range are mainly required. Table 7 lists the basic properties of Pibiflex® elastomer. This product finds applications mainly in the shoe industry, electrical industry, and automotive industry.

2.8. *Riteflex®* *engineering thermoplastic elastomers*

Riteflex® elastomeric products combine toughness, tear and flexural fatigue resistance with the ability to perform over a wide temperature range of –40 to 121°C. Their chemical relationship to the other members of Ticona's polyester family [8] not only provides Riteflex® elastomers with their thermal properties, but also imparts an excellent chemical resistance to common solvents, oils, greases, and dilute acids and bases. Riteflex® polyester elastomers are available as unreinforced polymers in a wide range of Shore D hardnesses.

Table 7. Basic properties of Pibiflex® elastomers

Property	Test method	Unit	Typical value
	Miscellaneous		
Density	ASTM D 792	g/cm^3	1.07–1.29
Water absorption (24 h, 23°C)	ASTM D 570	%	0.16–4.0
Melting point (DSC)	ASTM D 3417	°C	200–215
Melt flow index (220°C, 2.16 kg)	ASTM D 1238	g/10 min	1–175
	Mechanical properties		
Tensile strength at break	ASTM D 638	MPa	11–40
Elongation at break	ASTM D 638	%	400–900
Flexural modulus	ASTM D 790	MPa	30–400
Tear strength (method B)	ASTM D 724B	N/mm	35–190
Fatigue resistance	ASTM D 1052	mm/kcycles	0.1/300–6/50
Abrasion resistance (taber HI8 – 1 kg)	ASTM D 1044	mg/kcycles	50–120
Shore D hardness	ASTM D 2240		26–100
	Thermal properties		
Vicat softening point (1 kg; 120°C)	ASTM D 1525	°C	80–201
HDT – method B (0.4552 MPa)	ASTM D 648	°C	46–128
	Flammability properties		
Oxygen index	ASTM D 2863	–	20–28
	Electrical properties		
Volume resistivity	ASTM D 257		
at + 23°C		Ohm·cm	$>10^{14}$–$>10^{13}$
at + 100°C		Ohm·cm	$>10^{13}$–$>10^{11}$
Dielectric strength (at 1.6 mm) at 2.0 mm, 23°C	ASTM D 149	kV/mm	16–18

In general, the harder versions have enhanced heat and chemical resistance, while the softer materials possess good low temperature mechanical properties. Table 8 lists some basic properties of these elastomers.

The range of properties available from Riteflex® elastomers is reflected in the diversity of applications, in which these versatile materials have been used, *e.g.*, hoses, tubing, seals, gaskets, belts, pump diaphragms, wire coatings, hooks, fasteners, films, sheets, non-wovens and monofilaments, to name some of the more outstanding.

Table 8. Basic properties of Riteflex® elastomers

Property	Test standard	Unit	Typical value
Miscellaneous			
Density	ISO 1183	g/cm^3	1.14–1.29
Molding shrinkage (parallel)	ISO 294-4	%	0.8–2.2
Molding shrinkage (normal)	ISO 294-4	%	1.4–2.2
Mechanical properties			
Tensile modulus	ISO 527-2/1A	MPa	60–690
Stress at 50% strain	ISO 527-2/1A	MPa	8–35
Stress at break (50 mm/min)	ISO 527-2/1A	MPa	20–31
Strain at break (50 mm/min)	ISO 527-2/1A	%	50–350
Flexural modulus (23°C)	ISO 178	MPa	68–750
Flexural strength (23°C)	ISO 178	MPa	2.4–11
Shore hardness D scale, 15 s value	ISO 868	-	35–77
Thermal properties			
Melting temperature (10°C/min)	ISO 11357-1,-2,-3	°C	163–217
Tempature of deflection under load (0.45 MPa)	ISO 75-1/-2	°C	52–114
Linear thermal expansion coefficient (parallel)	ISO 11359-2	10^{-4}/°C	14–22
Electrical properties			
Relative permittivity (1 MHz)	EC 60250	–	3.3–4.7
Dissipation factor (1 MHz)	EC 60250	10^{-4}	200–400
Volume resistivity	EC 60093	Ohm·cm	10^{13}–10^{12}
Electric strength	EC 60243-1	kV/mm	13–16
Comparative tracking index	EC 60112	–	>600

2.9. *Skypel® engineering thermoplastic elastomer*

Skypel® is the registered trademark of a polyester-based thermoplastic engineering elastomer manufactured by SK Chemicals [9]. It can be processed in a wide variety of products from ultra small precision parts to spread sheets by injection, extrusion, and others. Skypel® is a block copolymer consisting of hard crystalline phase and soft amorphous phase, and the relative proportions of the two phases determine the material properties. Skypel®, ranging over a broad hardness spectrum, offers excellent features such as toughness, resilience, resistance to creep, impact and flexural fatigue, flexibility at low temperatures, thermal stability at elevated temperatures, and resistance to many industrial chemicals, oils, and solvents. Table 9 shows some typical properties of Skypel®.

Table 9. Typical properties of Skypel® grades

Property	ASTM method	Unit	Skypel® grade							
			G130D	G140D	G155D	G163D	G168D	G172D	G175D	G182D
Hardness	D2240	Shore D	30	40	55	63	68	72	75	77
Density	D792	g/cm^3	1.07	1.16	1.18	1.21	1.24	1.25	1.27	1.27
Water absorption, 24 h	D570	%	0.8	0.6	0.5	0.3	0.3	0.3	0.3	0.2
Mold shrinkage	D955	%	0.4	0.8	1.2	1.5	1.6	1.7	2.0	2.2
Tensile stress at 5% strain	D638	MPa	0.6	2.4	7.5	11.0	15.0	23.0	26.0	30.0
Tensile stress at 10% strain	D638	MPa	1.8	4.4	12.5	17.5	24.0	30.0	36.0	40.0
Tensile stress at break	D638	MPa	22.0	27.0	39.0	44.0	46.0	47.0	49.0	50.0
Elongation at break	D638	%	900	600	550	500	450	420	380	350
Flexural modulus	D790	MPa	28	68	210	320	470	550	780	900
Tear strength	D1004	N/mm	95	115	165	180	193	205	230	260
Izod impact strength/notched	D256	J/m	N/B	N/B	N/B	N/B	230	120	40	40
Resilience	D2632	%	72	57	55	53	47	–	–	–
Melting point	D3418	°C	174	155	202	212	215	218	220	222
Heat distortion temperature	D648	°C	–	70	105	130	140	150	154	156
Melt flow rate	D1238	g/10 min	18	7	8	10	13	13	10	8
at temperature		°C	220	190	220	230	230	230	230	230

The applications of this product in the automotive industry are in the production of backup rings, ball joint bushes, bumper fascia, door latch strikers, emblems, mounting taps for antennas, stoppers, seat belt parts, slide plates, protector tubes, air-duct hoses, CVJ boots, rack and pinion boots, suspension boots. Antenna jackets, controller buttons, electrical cable connectors, electrical curlers, noiseless gears, cable jacketing, and retractable telephone coiled cord are some examples of its electrical applications. This product can be successfully applied in the manufacture of gears and sprockets, hammer handles, packings, belts (V-shaped, round), conveyor belts, fire hoses, fluid hoses, gas pipe liners, hydraulic hoses, diaphragms, flexible couplings, fasteners and clamps, frames for glasses, hair blow brushes, shoe ornaments and other parts, ski boots (cross country), watch bands, golf ball covers, modifiers, sports ball bladders, *etc.*

3. Polyamide-based thermoplastic elastomers (TPE-A)

Thermoplastic polyamide elastomers are block copolymers also containing soft and hard segments. TPE-A can be processed and recycled in the same way as conventional thermoplastic materials, but possess properties and performance similar to those of vulcanized rubber at the service temperature. In general, polyethers, such as polyoxypropylene or poly(tetramethylene ether) glycol, are often chosen as the soft segments, and the hard segments are polyamides. Both types of segments are usually chemically bonded by ester linkages from a condensation process of prepolymers and diacids, which serve as chain extenders. Among the TPE-A, the nylon 6-based TPE-A has a high potential for industrial usage because of the low cost of caprolactam (the properties of TPE-A are also discussed in Chapters 10 and 11).

3.1. *Pebax®* *engineering thermoplastic elastomers*

One of the more recent contributions to the overall growth of the thermoplastic elastomer family is a polyether-*b*-amide (PEBA) resin, best known by the trademark *Pebax®* [10] (see also Chapter 9). The PEBA structure combines a regular linear chain of rigid polyamide segments interspaced with flexible polyether segments. This two-phase crystalline and amorphous structure creates properties bridging the gap between thermoplastics and rubbers. Pebax® resins from Atofina are polyether-*b*-copolyamide polymers. By varying these block types and ratios, a wide variety of physical and mechanical properties can meet the demands of a versatile number of applications. Pebax® grades have been developed and utilized in high performance areas from industrial to sporting goods and to high tech garments. Table 10 lists some typical properties of the Pebax® grades.

Pebax® grades fall into the following main categories:

Designated as Pebax® 33 Series, these grades are designed to be readily extruded or injection molded. This series is available with a Shore D hardness ranging from 25 to 72. The broad spectrum of mechanical properties gives the

Table 10. Typical properties of Pebax® grades

Property	ASTM method	Unit	Pebax® grade						
			7233	7033	6333	5533	4033	3533	2533
Density	D792	g/cm^3	1.02	1.02	1.01	1.01	1.01	1.01	1.01
Water absorption equilibrium (20°C)	D570	%		0.64		0.5	0.5	0.5	0.5
Immersion for 24 h	D570	%		0.83		1.2	1.2	1.2	1.2
Hardness	D2240		72D	69D	63D	55D	40D	35D	25D
Tensile strength, ultimate	D638	MPa	63	57	56	50	39	38	34
Elongation, ultimate	D638	%	360	400	300	430	390	580	640
Flexural modulus	D790	MPa	738	462	338	200	90	19	15
Izod impact, notched	D256								
20°C		J/m	75	N/B	N/B	N/B	N/B	N/B	N/B
–40°C		J/m	75	51	80	N/B	N/B	N/B	N/B
Abrasion resistance H 18/1000 g	D1044	mg/kcycle	29	57	84	93	94	104	161
Tear resistance, notched	D624C	N/mm	1400	900	850	650	400	260	220
Melting point	D3418	°C	176	174	172	168	168	152	148
Vicat softening point	D1525	°C	164	165	161	144	132	74	60
HDT	D648	°C	106	98	90	66	52	46	42
Compression set (24 h, 70°C)	D395A	–	–	6	6	10	21	54	62

designer a wide selection for such applications, as sporting goods, footwear, mechanical components, tubing and belting.

Some Pebax® grades are used to enhance the characteristics of other thermoplastics, in particular as antistatic agents, since they naturally disperse electrostatic charges, and as such can be added to a large number of thermoplastic matrices (acrylonitrile-butadiene-styrene (ABS), polystyrene, polyoxymethylene (POM), polyvinylchloride (PVC), polymethylmethacrylate (PMMA) and polycarbonate (PC)) to impart permanent antistatic properties, whatever the ambient relative moisture or surface friction they may be subjected to. These antistatic compounds are suitable for countless applications in office and electronic equipment (photocopier components, printer components, *etc.*) as well as in technical packaging.

Pebax® products can be successfully used as impact modifiers and mechanical strength modifiers, since these thermoplastic elastomers have a glass transition temperature below –50°C. They can therefore be used as impact modifiers (Pebax® Series 33) and softening agents for PA6 film (Pebax® Series 13, MP 1878), and as strengthening agents for rubbers.

Pebax® MX 1508 may be used as a process aid for thermoplastic polyurethanes (TPU). Used in combination with fluorinated polymers, Pebax® can also be used as a process aid for polyolefins.

Pebax® breathable resins can be used in a variety of applications requiring films that are highly breathable to water vapor and where waterproof properties are vital. These resins can be used as pure films or applied onto a variety of textiles, including natural or synthetic wovens or non-wovens, by lamination with adhesives or by direct bonding onto the backing (coating). Pebax® waterproof breathable films are suitable for a wide range of markets, including medical/hygiene (surgical garments, surgical sheeting/mattress cover dressings, adult incontinence articles/babies diapers), textile (films for textile lamination for sports, leisure and work wear, footwear, gloves), construction (breathable roofing membranes, outer wall membranes), and post-harvest packaging of fruits, vegetables, and mushrooms. They have also industrial applications where high water vapor breathability is required.

3.2. *Vestamid® engineering thermoplastic elastomers*

The High Performance Polymers Business Unit of the Degussa AG manufactures several polyamide 12 (PA12), polyamide 612 (PA612) and polyamide elastomer compounds, and it sells them under the trade name *Vestamid®* [11]. PA12 elastomers, the most important subgroup of polyamide elastomers, belong to the increasingly important material class of TPE. Because of their excellent properties, they are indispensable in many applications. PA12 elastomers are block copolymers consisting of PA12 segments and polyether segments. PA12-rich products have the major properties of PA12, while the elastomer characteristics become more apparent with increasing polyether content. That is, the polymers become more flexible, with a higher impact strength at low temperatures. Compared to other TPE, PA12 elastomers are distinguished by low density, high resistance to chemicals and solvents, they are easy to process, color, and overmold, can be decorated easily by means of heat transfer printing. Hardness and flexibility can be varied over a wide range, high elasticity and good recovery are inherent properties of these materials, the mechanical properties are only slightly temperature dependent, PA12 elastomers are free of volatile or migrating plasticizers.

The PA12-based elastomer compounds offered are suitable both for precision injection molding and for high-performance extrusion processing, *e.g.*, for making tubing and films. PA12 elastomer compounds and their typical applications are listed in Table 11.

3.3. *D-RIM Nyrim® engineering thermoplastic elastomer*

D-RIM is a completely new plastic material developed by Pucast Oy [12] in order to fulfill, *e.g.*, the electrical requirements of chemical and electronic industries in electrostatic discharge (ESD)-protected areas; its surface resistance is 10^6–10^8 Ohm. The material is based on nylon 6 and is therefore very strong

Table 11. PA12 elastomer compounds and their typical applications

Vestamid® grade	Stabilized against	Shore D hardness	Typical applications
E40-S3	Heat and light	40	Noiseless gears, seals, functional elements of sports shoes, process aids in the extrusion of thermoplastic polyurethanes, films
E47-S1 E47-S3	Heat and light	47	Sports shoe soles, packaging films, non-skid surfaces, sports glasses, protective goggles
E55-S1 E55-S3	Heat and light	55	Alpine ski boot components, sports shoe soles, pneumatic lines, rolls, technical films
E62-S3	Heat and light	62	Alpine ski boots, noiseless gears, conveyor belts
EX9200	Heat and light	68	Decorative and protective films for sports articles and interior/exterior designs on automobiles
E50-R2	Heat (also light from conductive carbon black)	50	Permanently antistatic articles, *e.g.*, conveyor belts, housings, paint spray hoses

mechanically. This enables a very wide range of use. Based on D-RIM, it is easy to machine on-off jigs, machine parts and small-scale series from the standard sheets. For batches of over 100 pieces it is usually better to make a mold. The mold expenses are cost-efficient because D-RIM can be processed with a low-pressure technique. When the smallest details are feasible by machining, a cost-efficient means of manufacturing is achieved. The heat resistance in tough mechanical stress is approximately 80°C, and 155°C under lower stress.

Nyrim® is a thermoplastic material based on nylon 6, produced by using the reactive injection molding (RIM) method. It has been developed by the Dutch DSM Rim Nylon VOF. Nyrim® is a block copolymer based on PA6 where elastomer chains have been chemically bonded to nylon 6. The PA6 component imparts strength, rigidity, heat resistance, and chemical resistance to the material. On the other hand, the elastomer contributes to the rubber-like properties, *i.e.*, toughness and flexibility. The content of the rubber-like component can be varied and in this way the material properties can be tailored to fit the customer requirements. So, Nyrim® is not a single material, but a group of mixtures, which have different properties. Easy recycling (see also Chapter 19), non-toxic and safe end products, and an environmentally friendly manufacturing method will be the characteristics of the structural materials of tomorrow. Nyrim® meets these requirements. When the above aspects are combined with versatile material properties, as well as with manufacturing methods allowing various designs, it is obvious that one has an extremely interesting, new structural material at his disposal. Nyrim® is often used to replace multishaped metal structures in those fields of application where machinery and production costs are high. The plastic structures are prepared by molding and in this way many stages of operation in the metallic manufacture are replaced, *e.g.*, building up the structure of separate pieces, perforating, machining of levels and shapes, corrosion protection, and painting. It is worth noting that a Nyrim® product is lightweight, which means savings in the processing and transportation. Also, when it comes to noise reduction and damping of vibrations, the Nyrim® products reveal properties that are of importance in many fields. The RIM method is usually an option for the injection molding of plastic products, if the latter are either large in volume or they require a large wall thickness, or if the production run is either small or medium. The Nyrim® products can be found in many fields of application in different industries, and their introduction can serve as an example of cost savings. Examples for Nyrim® applications are in: (i) housing, chambers and boxes in places where impact, toughness and strength are expected under varying conditions, (ii) parts that belong to the manufacturing equipment of various materials and articles and where wear resistance, chemical resistance and low noise level are required, (iii) machine components used for power transmission, *e.g.*, tooth wheels and pulleys; fan blades and chambers of the ventilators and the blowers replacing metal structures that can be difficult to manufacture, (iv) various attachment and installation flanges, (v) parts of pumps and of various hydrocyclones where good wear resistance combined with chemical resistance and high working temperatures are required, and (vi) products in the electrical industry where it is possible to reach better impact toughness and lighter structures.

3.4. *Grilon® ELX and Grilamid® ELY engineering thermoplastic elastomers*

Grilon® ELX is a PA6-based elastomer and *Grilamid®* ELY is a PA12-based elastomer from EMS-Chemie [13]. On the basis of an aliphatic polyamide with both rigid and elastomeric components, a unique property profile is achieved.

Grilon® ELX is a general purpose formulation, easily processed on conventional injection molding or extrusion equipments. It is flexible without the addition of a plasticizer, has a favorable specific weight, good chemical resistance, flexibility at low temperatures, easy processing, balanced mechanical properties, versatile coloration.

Grades of Grilon® and Grilamid® are listed in Table 12. Typical applications include soft-faced hammers, bumper strips for ball racquets, ski boots, training shoes, hiking boots, and other applications requiring high flexibility and impact resistance.

4. Polyurethane-based addition thermoplastic elastomers (TPE-U)

Like all thermoplastic elastomers, thermoplastic polyurethane (TPU) is elastic and melt-processable. Further, it can be processed on extrusion as well as on injection, blow and compression molding equipments. It can be vacuum-formed or solution-casted and is well suited for a wide variety of fabrication methodologies. TPU can even be colored through a number of processes. But more

Table 12. Grades of Grilon® ELX and Grilamid® ELY

Grilon®	Shore D hardness	Grilamid®	Shore D hardness
BRZ 323 ELX	50	ELY 20 NZ	52
ELX 40H NZ	58	ELY 42	40
ELX 50H NZ	50	ELY 60	64
ELX 23 NZ	50	ELY 2475	57
		ELY 2694	69
		ELY 2702	68

so than any other thermoplastic elastomer, TPU is fully thermoplastic and can provide a considerable number of physical property combinations, making it an extremely flexible material adaptable to many uses. This is partly because TPU is a linear segmented block copolymer composed of hard and soft segments. The hard segment can be either aromatic or aliphatic. When these isocyanates are combined with short-chain diols, they become the hard block. Normally, the latter is aromatic, but when color or clarity retention by sunlight exposure is a priority, an aliphatic hard segment is often used. The soft segment can be of either polyether or polyester type, depending on the application. For example, wet environments generally require a polyether-based TPU while oil- and hydrocarbon-resistance often demand a polyester-based TPU. For even greater utility, the molecular weight, ratio, and chemical type of the hard and soft segments can be varied. This versatility results from the unique structure of TPU, imparting high resilience, good compression set, and resistance to impacts, abrasions, tears, weather, and even hydrocarbons. TPU offers flexibility without the use of plasticizers, as well as a hardness broad range and high elasticity. In fact, TPU bridges the material gap between rubbers and plastics. Its physical properties enable its use as both a hard rubber and a soft engineering thermoplastic. TPU can be sterilized, welded, easily processed, colored, painted, printed, die-cut, and slitted. It has low temperature flexibility and, in some grades, exhibits biocompatibility, hydrolytic stability, optical clarity, as well as flame-retardant and anti-static properties (the TPU properties are also discussed in Chapters 12 and 13).

These properties make TPU extremely useful for hundreds of products. Some TPU applications are in architectural glass lamination, auto-body side molding, automotive lumbar supports, caster wheels, cattle tags, constant velocity boots (automotive), drive belts, film and sheet, fire hose liner, flexible tubing, food processing equipment, footwear (sport shoe soles), hydraulic hoses, hydraulic seals, inflatable rafts, in-line skates, magnetic media, medical tubing/ biomedical apparatuses, mining screens, sporting goods, swim fins and goggles, coated fabrics, wire and cable coatings.

The trade names of TPE-U from different producers [14] are presented Table 13. Because the properties and applications of TPE-U are more or less

Table 13. Trade names of TPE-U from different producers

Producer	Trade name	Producer	Trade name
Bayer	Desmopan® Texin®	Dow	Isoplast® Pellethane®
Noveon Inc.	Estane® Stat-Rite®	Hunstman	Avalon® Irogran®
CardioTech International	Chronoflex® Chronothane™ Hydromed D640™ Hydroslip C™ Hydrothane™	Merquinsa	Pearlthane® Pearlstick®, Pearlcoat® Pearlbond®
PolyOne	Gravi-Tech® Elastamax®	Coim SPA	Laripur® Urexter®
A. Schulman, Inc.	Polypur™	GLS Corporation	Versollan™
SK Chemicals	Skythane®	RTP Company	RTP® 2300 A
Elastogran	Elastollan®	LATI	Lastane®
Toyobo Company	Toyobo Polyurethane		

similar, only information about some typical TPU of different producers is provided below. In Figure 1 [9], the mechanical properties of TPU are compared to those of other thermoplastic polymers.

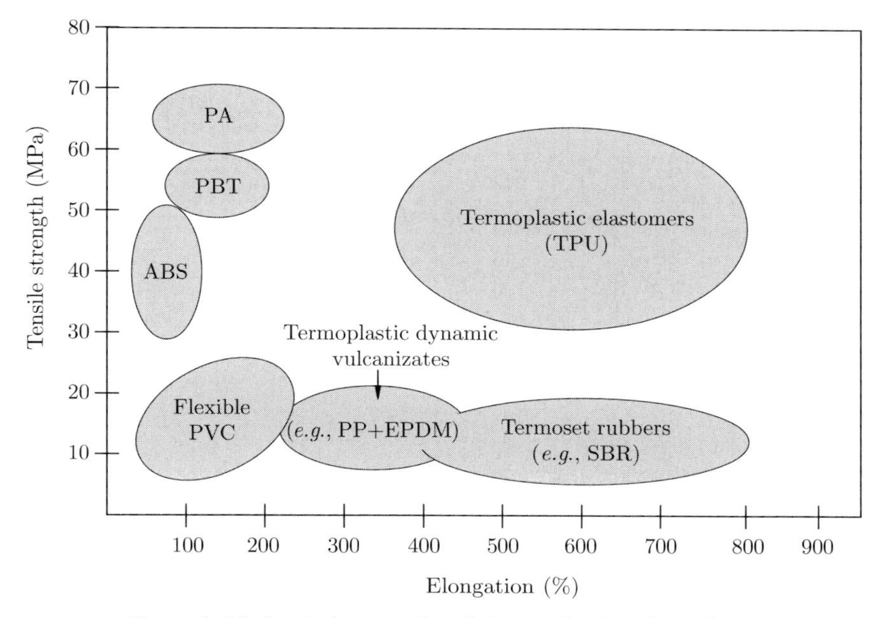

Figure 1. Mechanical properties of thermoplastic polyurethanes

4.1. *Skythane® engineering thermoplastic elastomer*

SK Chemicals produces a thermoplastic polyurethane (TPU) [9] under the trade name *Skythane®*. Being the bridge between rubbers and plastics, this TPU elastomer has innovative potentials for uses requiring special characteristics, such as transparency, elasticity, durability, and resistance to chemical attack, in addition to excellent physical and mechanical properties; Table 14 lists grades and properties of Skythane®.

Table 14. Grades and properties of Skytane®

Skythane®	Type	Characteristic
T Series	Polyester, polyether	Low hardness grade
S100 Series	Polyester	General grade
S300 Series	Polyester	Highly transparent grade
X500 Series	Polyester	Film/sheet grade
L100 Series	Polycaprolactone	Oil-resistant grade
R100 Series	Polyether	Hydrolysis-resistant grade
C Series	Polyester, polyether	Calendering grade
Hot-melt grades	Polyester	Hot-melt
Special grades	Polyester, polyether	Special purpose

When processed by injections molding, shoe outsoles and accessories, as well as side moldings and automobile bumpers can be produced, together with parts for the machine building, *e.g.*, packings, drums. Extrusion is often applied for some industrial parts, such as pneumatic tubes, hoses, belts, electric wires and cables, ABS sensor cables, fire hoses. Automobile bellows and R&P boots as well as shoe air cushion bags are produced by blow molding, whereas calendering is used for the production of films, sheets, conveyor belts, and oil fences.

4.2. *Isoplast® and Pellethane® engineering thermoplastic elastomers*

Isoplast® engineering thermoplastic polyurethane resins of Dow [15] combine the toughness and dimensional stability of amorphous resins with the superior performance and chemical resistance associated with semicrystalline resins. Isoplast® is available in impact-modified, clear and glass-reinforced resins, with a range of options to meet the most demanding applications. Table 15 gives descriptions and applications of Isoplast® TPE-U products.

Pellethane® thermoplastic polyurethane elastomers range from hard to soft and can be fabricated by a variety of methods, from injection molding to extrusion and blow molding. These elastomers offer a combination of properties rarely seen in an engineering thermoplastic. Table 16 gives descriptions and applications of Pellethane® TPE-U products.

Table 15. Description and applications of Isoplast® TPE-U

Product Isoplast®	Applications	Description
	Impact-modified resins	
101	Impact resistant housings, valves, water filter caps, geophysical spacers, electrical transformers	Low heat resistance, excellent room temperature impact, good low temperature impact
202 EZ	Impact resistant housings, valves, water filter caps, geophysical cable spacers, electrical transformers	High heat resistance, very good room temperature impact, good low temperature impact, good for thin walled parts
2510	Medical, catheter, catheter hubs, I.V. connectors, surgical instruments	USP class VI, low heat resistance, excellent room temperature impact, good low temperature impact
	Clear resins	
301	Water filter housings, fuel filter bowls, toothbrush handles	Transparent, medium heat resistance, practical toughness
302	Industrial water filter housings, fuel filter bowls	Excellent for thin-walled parts, high heat resistance, transparent, faster M_w build up, practical toughness
2530	Catheters, valves, needleless syringes, surgical instruments	USP Class VI, low heat resistance, transparent, practical toughness
2531	Catheters, valves, needleless syringes, surgical instruments	USP Class VI, medium heat resistance, transparent, practical toughness

USP Class VI means very good biocompatibility (ISO 10993)

4.3. *Estane® engineering thermoplastic elastomer*

Estane® thermoplastic polyurethane from Noveon [16] is replacing traditional materials, such as rubber, thermoplastic polyolefins and other plastic materials, and its grades are listed in Table 17.

Estane® 58213 is processed by means of injection molding and general extrusion techniques. Estane 58213 NAT is a polyester-based thermoplastic polyurethane compound. The material can be used by itself or as a modifier of other polymers, such as PVC. Estane® T-4057 black (Estane® T4057 BLK 281) is a conductive, polyether-based thermoplastic polyurethane compound intended for extrusion applications. Estane® T-4086 black (Estane® T4086 BLK 281) is a conductive, polyester-based thermoplastic polyurethane.

Estane® polyether polyurethanes are recommended for applications requiring low temperature flexibility, good weathering properties and resistance to wet environments and/or fungus growth. Estane polyester type polyurethanes are generally recommended for applications requiring mechanical strength, resistance to chemical attack, and heat resistance.

Table 16. Applications and description of Pellethane®

Product Pellethane®	Applications	Description
2101 Series	Seals, gaskets, animal tags, molding applications, hose jackets	Ether-ether hybrid; good low temperature properties; fast cycle time with good hydrolytic stability
2102 Series	Seals, gaskets, caster wheels, belting, animal ID tags, other fabricated products, office furniture, cell phones and PDA overmolding	Polycaprolactone-based ester; excellent resistance to fuels and oils; best hydrolytic stability of ester TPUs
2103 Series	Film, tubing and belting, geophysical wire and cable jacketing, cell phones and PDA overmolding	Polyether-based; excellent hydrolytic stability, excellent resistance to microorganisms
2104 Series	Caster wheels, seals, gaskets, animal tags, hose jacketing, ski boots, automotive	Ether-ester hybrid; excellent low temperature properties and cycle time, good resistance to fuels and oils
2355 Series	Film, sheet, tubing, recreation and sporting equipment, industrial hoses, fabric lamination, sewer pipe linings, footwear	Polyester-based; excellent oil and chemical resistance
2363 Series	Medical, catheters, tubing, cuffs, transdermal patches	Polyether-based; excellent hydrolytic stability; excellent resistance to fungi and microorganisms

Table 17. Grades of Estane®

Hardness 64A–90A		Hardness >92A	
Estane® 54660	64A	Estane® GP92AE	92A
Estane® GP65AE	65A	Estane® GP92AT	92A
Estane® GP70AE	70A	Estane® 58130	93A
Estane® GP75AE	75A	Estane® 58212	94A
Estane® GP80AE	80A	Estane® GP95AE	95A
Estane® 58300	80A	Estane® 58141	96A
Estane® 58881	80A	Estane® 58149	96A
Estane® GP85AE	85A	Estane® GP52DE	52D
Estane® 58214	85A	Estane® GP55DT	55D
Estane® 58311	85A	Estane® 58133	55D
Estane® 58134	88A	Estane® GP60DE	60D
Estane® 58887	90A	Estane® 58144	60D
		Estane® GP65DE	65D
		Estane® 58137	70D

Estane® thermoplastic polyurethane provides a binder that is critical to make a good finish on nylon, polyester, and polyurethane fabrics, as well as on leather materials. The Estane® thermoplastic polyurethane binder can be processed into lacquers or into inks with the addition of pigments. Lacquers or inks made of Estane® thermoplastic polyurethane binder can either be printed or spray-coated onto various substrates to give a perfect finish.

Films and sheets produced from this TPU are characterized by an outstanding low temperature flexibility, excellent tear and abrasion resistance, excellent clarity, colorability and printability, and are high water vapor transmission products.

The respective adhesives produced on the basis of Estane® show good chemical, UV, and hydrolysis resistance, low temperature flexibility, high tensile strength, high elongation, excellent flexural fatigue resistance, and a natural affinity to a variety of substrates.

4.4. *Texin® and Desmopan® engineering thermoplastic elastomers*

Texin® and *Desmopan®* resins are thermoplastic polyurethanes of Bayer [17] based on polyesters, polyethers, special copolymers, and polyurethane/polycarbonate blends. These materials exhibit high elasticity and resilience, excellent abrasion resistance, high tensile and tear strength, and are available in unreinforced general-purpose grades of various hardness and durability. Texin® and Desmopan® have excellent resistance to fuels, oils, and greases, superior toughness, and excellent low temperature flexibility and flexural fatigue resistance; their Shore hardness ranges from 60 A to 75 D. Grades of Texin® and Desmopan® are listed in Table 18.

Texin® and Desmopan® can be used in a broad range of applications, including instrument panels, caster wheels, power tools, sporting goods, medical devices, and a variety of extruded films, sheets and profiles. Texin® and Desmopan® resins also readily bond to a variety of substrates to provide a soft-touch feel.

4.5. *Avalon® and Irogran® engineering thermoplastic elastomers*

Avalon® DUO is a new soling material combining the advantages of design durability and comfort. This premium composite soling material from Huntsman Polyurethanes [18] reveals the outstanding wear and flexural properties of Avalon® soft thermoplastic polyurethane (cast or injection molded) for the outsole and the comfort and lightness of polyurethane for the midsole. It also provides greater efficiency as a result of the one-stage production process through casting or injection molding and ensures perfect bonding to the upper. The versatile Avalon® DUO offers a range of benefits, including perfect surface definition, matt finish, shiny and glossy end effects, slip and abrasion resistance, shock absorption, cold flexural properties, and thermal insulation. Offering footwear manufacturers high durability and improved processing, the Huntsman Polyurethanes' footwear team launched the Avalon® ABX range of thermoplastic polyurethane at Simac 2004. This new series of high performance TPU/rub-

Table 18. Grades of Texin® and Desmopan®

Texin®	Shore hardness	Desmopan®	Shore hardness
185	87A	385 E	32D
245	45D	385 S	32D
250	52D	445	45D
255	55D	453	53D
260	60D	5377A	77A
270	70D	5080A	80A
285	85A	DP 9370A	70A
390	88A	W DP 85085A	85A
4210	70D	W DP 89043D	43D
4215	75D	W DP 8905 ID	51D
945U	50D	W DP 89056D	56D
950U	50D		
970U	70D		
985U	85A		
990U	90A		
990R	90A		
DP7-1049	45D		
DP7-3007	55D		

ber compounds is available in a variety of hardnesses, from Shore A 60 to 90, responding to the demands for soft applications in casual and athletic shoes, as well as harder sole types for cleated sports plates. Manufacturers can benefit from the improved processing in terms of faster cycle times, lower mold clamping requirements, and enhanced production robustness. The durability of the Avalon® ABX range is achieved through excellent abrasion and hydrolysis resistance. It has a rubber-like smell and offers improved adhesion to polyurethane midsoles, making it ideal for dual density applications and for heels, top-pieces, kippers, and over-molded designs

Irogran® from Huntsman is a trade name of thermoplastic polyurethane elastomers, which are easily processed by injection molding or extrusion techniques. It is used when a maximum of mechanical strength is a basic requirments. Irogran® materials combine an extreme wear resistance with high tear strength and mechanical reliability. Table 19 shows their typical properties.

4.6. *Elastollan® engineering thermoplastic elastomer*

Elastollan® is a versatile material of Elastogran [19], offering highest innovation potential. Elastollan® is characterized by an extraordinary property profile

Table 19. Typical properties of Irorgan® TPE

Irogran®	Shore A hardness	Tensile strength (MPa)	Elongation (%)	Modulus 20% (MPa)	Modulus 100% (MPa)	Modulus 300% (MPa)	Tear strength (N/mm)	Density (g/cm^3)
A78P4740 8-4740	78	35	670	1.4	4.2	7.8	45	1.12
A80P4699 8-4699	80	35	700	1.5	4.3	8.0	45	1.10
A85P4240 8-4240	87	40	550	3.2	5.5	12	60	1.12
A85P4434 4-4434	86	40	600	3.0	6.0	12	58	1.12
A85P4380 4-4380	87	35	550	3.0	6.0	9.0	55	1.15
A5P4441 8-4441	87	40	550	4.0	7.0	11	60	1.12
A85P4394 8-4394	85	45	550	3.0	6.0	12	70	1.12
A85P4350 4-4350	87	40	550	3.0	6.0	11	60	1.15
A92P4637 8-4637	92	40	500	6.0	10	20	65	1.14

Table 20. Typical properties and applications of Elastollan® TPE

Product range	Shore hardness	Short description	Applications
1100 Polyether-polyurethane	75 A to 74 D	Hydrolysis resistant, resistant to oil and chemicals, flexible at low temperatures, resistant to microbes	Cables, tubing, animal identity tags, railway pads, film
C Polyether-polyurethane	60 A to 74 D	High mechanical properties, resistant to oil and chemicals, resistant to wear, heat-stabilized for highest mechanical requirements	Tubing, round belting, technical parts
B Polyether-polyurethane	85 A to 64 D	Flexible at low temperatures, resistant to wear, good mechanical properties, wide range of physical properties	Shoe soles and accessories, tubing, technical parts
S Polyether-polyurethane	60 A to 74 D	Resistant to wear, good mechanical properties, tear and abrasion resistant, UV-stable	Shoe soles, tubing, castor tyres
600 Polyether-polyurethane	85 A to 60 D	Transparent, good mechanical properties	Tubing, ski tips, decorative parts for sports shoes
500 Polyether-polyurethane	90 A to 64 D	Resistant to wear, tear, and abrasion, used for tyres	Shoe heels, wear parts
Extra products	80 A to 74 D	Applied formulations	Film, several applications

combining elasticity with resistance over a wide range of temperatures. It is abrasion and impact resistant, has a good elastic recovery, is tear and kink-resistant, as well as oil and grease resistant. Moreover, Elastollan® can be modified for specific applications, *e.g.*, by glass-fiber reinforcement. Table 20 shows the typical properties of Elastollan® TPE.

4.7. *Laripur® engineering thermoplastic elastomers*

Laripur® is a trade name of Coim SPA [20]. Laripur® are thermoplastic polyurethanes and, consequently, combine the working technology of thermoplastics together with the well known features of polyurethanes, *e.g.*, excellent abrasion resistance, great flexibility with constancy at temperature variations, good compression resistance, good water and light resistance, good resistance to oils, fats, and to many types of solvents.

Laripur® thermoplastic polyurethanes cover a wide range of hardness, from Shore A 60 to Shore D 75. A total absence of plasticizers is the most important property of these products, with the exception of Series 15 and 18. Also, with very few exceptions, all grades of Laripur® TPU have the EU and the Food and Drug Administration (FDA, USA) food contact approval. The Series 20 standard ester products are based on saturated polyesters. They exhibit strength and durability with excellent tear and abrasion resistance, good hydrolytic stability, resistance to solvents, oxidation, and degradation by UV light. The Series 25 special ester products are based on saturated polyesters. Their features are similar to those of the previous series, but they are characterized by a higher resistance to hydrolysis and improved flexibility at low temperatures. The Series 2102 ester products are based on polycaprolactones; they have the same features as TPU based on quality esters. Compared to Series 25, they show a higher resistance to hydrolysis. The Series 50 modified ester products are obtained by homogeneous incorporation of technical polymers for the achievement of special properties, such as cold impact resistance. The Series 60 and E 2103 ether products are based on quality polyethers. Compared to the ester series, they show a higher resistance to hydrolysis, microbiological attacks, higher cold flexibility, but a lower resistance to oxidation. Specially developed TPU are involved in the manufacturing of hydraulic gaskets and for any other application where high compression set, also in the presence of oils and/or at high temperatures, is required. Two products based on polycaprolactone, E 2202-95A and E 107-93A, and one based on an ether, E 2203-93A, belong to this group to satisfy the various application needs. The Series 15 and 18 products are plasticized TPU having a Shore A hardness between 60 and 70 and the advantage of a reduced processing time. They are used for several technical articles, mainly in footwear, as outsole in combination with a PU microcellular midsole. Alloys are also prepared as homogenous mixtures of other technical polymers and TPU in variable ratios, and their characteristics can vary considerably according to the nature of the binder.

4.8. *Lastane*® *engineering thermoplastic elastomers*

Lastane® thermoplastics are polyurethane products of LATI [21]. They exhibit good wear and impact resistance, even at low temperatures, noise inertness, and excellent mold release. These properties make the Lastane® very suitable for wheels, gears, couplings, parts of textile machinery, shoes, and others. Reinforced Lastane® thermoplastics are used when good rigidity and high dimensional stability are required.

4.9. *Versollan*® *engineering thermoplastic elastomers*

GLS [22] has created a new class of high performance TPU alloys that combines the strength and fluid resistance of TPU with the soft feel, matte appearance and processing advantages of other TPE. The first grades to be introduced are the *Versollan*® RU 2204 and RU 2205 products, with a Shore A hardness of 55 and 65, respectively.

Other products in the Versollan® line include OM 1262 and OM 1255. These products are ideal for long, thin, complex mold geometries. They were formulated specifically for overmolding onto various engineering resin substrates and are characterized by a tactile, rubbery feel, dull, matte finish, exceptional oil resistance, excellent adhesion to PC, ABS, and PC/ABS blends, very good abrasion resistance, improved flexibility (low flexural modulus), fast set-up rates during processing, *i.e.*, reduced cycle times, and spiral flow similar to styrenic TPE.

Versollan® TPU are second-generation high-performance rubberized TPU elastomer alloys that rank as some of the softest in their class at 45 Shore A. Made of specialty TPU from BASF Corp., Wyandotte, these alloys serve as handles and grips for hand and power tools, lawn and garden equipment, and recreational gears. They offer a balance of softness and high performance, and fast set-up rates relative to non-alloyed TPU.

4.10. *Pearlstick*®, *Pearlthane*®, *Pearlcoat*®, *and Pearlbond*® *engineering thermoplastic elastomers*

These tepmoplastic polyurethanes are produced by Merquinsa [23].

Pearlstick® TPU are supplied in pellet form for the adhesives industry and for the production of toe puffs and counters. They are mainly crystalline polymers. Their structure is characterized by a small proportion of hard segments, as compared to their content of soft segments. It is their polyol component (the part traditionally known as the soft segment or amorphous part) that crystallizes and imparts such crystalline characteristics. The resulting polymers show outstandingly high tenacity and stress-strain power. Another interesting feature of these TPU is the low melting point of their crystalline zones (50–60°C), which allows the polymer to be softened or melted for the adhesion process at relatively low temperatures, with sufficient thermoplasticity and surface tack to ensure correct union between substrates of different materials.

The *Pearlthane*® line of TPU is supplied in pellet form and covers a full range of thermoplastic polyurethane elastomers for extrusion, injection molding, and compounding applications. This range has been designed to meet the increasingly demanding needs of a wide variety of technical applications, such as highly flexible films, abrasion resistant hoses, soft touch injection molded parts or special compounds. Table 21 shows the typical properties of some Pearlthane® TPU grades.

Table 21. Typical properties of Pearlthane® TPU grades

Pearlthane®	Shore A hardness	Density (g/cm^3)	Tensile strength (MPa)	Elongation at break (%)	Abrasion loss (mm^3)	Glass transition (°C)	Melting range (°C)
D12K85	85	1.19	35	600	26	−27	160–170
D12K85A	85	1.20	35	600	20	−30	160–170
D12K92	94	1.19	40	470	25	−26	165–175
D12K92A	94	1.19	40	470	25	−26	165–175
D12C75	78	1.19	25	735	24	−32	125–135
D12F75	78	1.19	30	620	20	−30	135–145
D15N70	72	1.16	35	700	23	−53	168–178
15N8O	80	1.13	45	650	25	−54	170–180
15N85	85	1.13	40	630	20	−52	175–185
D15N85 UV	85	1.13	40	630	20	−52	175–185
D15N92	92	1.12	40	520	30	−53	185–195
D15N95UV	95	1.11	40	500	25	−49	192–202
D16N80	81	1.10	35	660	20	−46	160–170
D16N85	85	1.08	40	620	20	−48	168–178
D16N92	92	1.12	40	520	25	−37	179–189

The 11T Series consists of polycaprolactone-based TPU offering outstanding abrasion and wear resistance and a fast set-up in the mold. The 11T series outperforms conventional adipate ester-based TPU in hydrolysis resistance and low temperature flexibility, as well as adhesion strength to other polar plastics, such as ABS, PA, and PVC.

The 12C Series of low melting polyester-based TPU is currently being developed for a wide range of specialty engineering plastics. The interesting thermal and rheological properties of this series provide enhanced compatibility with a wide variety of other thermoplastics.

Based on Merquinsa's expertise in both high-melting film grade TPU and adhesive and coating applications, the new 12F Series is especially being designed for the adhesive film markets in which film quality and outstanding adhesion performance are the key properties.

The polyester-based TPU of the 12K Series are very clear, gel free, low color, and easy to extrude film-forming materials of both high and medium melting range. All grades of this series are FDA approved products. Specific high tack grades are also available in film grade quality.

The 15N is a unique series of TPU grades based on polyether-polycaprolactone copolymers offering excellent hydrolysis resistance in combination with good heat resistance and mechanical properties. Additionally, the 15N Series excels for its outstanding low temperature performance and elastic properties.

The polyether-based TPU of the 16N Series are used when extremely good hydrolysis and microbial resistance are required, *e.g.*, in underground applications, such as certain cables. These polyether products comply with FDA regulations.

Under the trade name *Pearlcoat*®, Merquinsa offers a complete range of thermoplastic polyurethane elastomers for fabric coating applications, supplied in pellet form. The Pearlcoat® range accounts for an important share in all industrial fields where the melt coating process is used, among which the following ones are especially relevant: conveyor belts, inflatable boats, garments, life jackets and many specific applications in other industrial coated fabrics. Grades and typical properties of Pearlcoat® TPU are given in Table 22.

Pearlbond® is supplied in pellet form and has a high molecular weight, high thermoplasticity and very high crystallization rate. Pearlbond® can be

Table 22. Typical properties of Pearlcoat® TPU grades

Pearlcoat®	Shore A hardness	Density (g/cm^3)	Tensile strength (MPa)	Elongation at break (%)	Abrasion loss (mm^3)	Melting temp. (°C)
			Aromatic polyester-based TPU			
125 K	85	1.20	30	500	25	140
D125KB	85	1.22	25	610	30	140
D125 KG	85	1.22	25	620	30	140
125 KW	85	1.23	25	520	40	145
126 K	94	1.20	35	420	40	160
126 KW	94	1.23	30	450	60	160
127 K	92	1.18	30	410	30	150
			Aromatic polyether copolymer-based TPU			
D152K	82	1.11	25	550	25	150
			Aromatic polyether-based TPU			
D162K	82	1.122	30	550	25	150
D163K	85	1.11	30	650	25	170
			Aliphatic polyether-based TPU			
D191K	87	1.03	25	850	35	133

added to the formulation of reactive hot-melts in amounts of 5 to 10%, enabling the improvement of some properties, such as green strength (increased crystallization speed), cohesive strength and specific adhesion to some substrates (polar groups). The Pearlbond® grades must be cryogenically ground into powder so as to be applied by the usual methods, such as scatter coating, paste coating, powder point coating and double dot coating. They are suitable for applications in heat sealable fabrics.

5. Outlook

As emphasized in the introduction, the aim of this chapter is to consider the versatile properties and applications of TPE produced by polycondensation and poyaddition reactions and by no means to compare them.

The future development of TPE will likely be governed by the following aspects: (i) use of polymers from renewable resources, (ii) checking the feasibility of methods of microbiological synthesis, and (iii) development of *in situ* polymerization technologies, especially for composite applications.

"Green chemistry" is pushed forward by concerns about depleting crude oil resources, which will have an impact also on TPE. Unfortunately, no life cycle assessment studies are available in the open literature on this topic. Such information would be, however, most helpful to select/forecast future R&D activities.

References

1. http://plastics.dupont.com, 30.03.2004.
2. http://www.eastman.com, 30.03.2004.
3. http://www.rtpcompany.com, 31.03.2004.
4. http://www.ikolon.com, 31.03.2004.
5. http://www.dsm.com, 25.06.2004.
6. http://www.lgchem.com, 25.06.2004.
7. http://www.p-group.de, 25.06.2004.
8. http://www.ticona.com, 28.06.2004.
9. http://www.skchemicals.com, 28.06.2004.
10. http://www.atofinachemicals.com, 29.07.2004.
11. http://www.degussa-hpp.com, 29.07.2004.
12. http://www.pucast.com, 29.07.2004.
13. http://www.emsgrivory.com, 29.07.2004.
14. http://www.matweb.com, 30.07.2004.
15. http://www.dow.com, 30.07.2004.
16. http://www.noveoninc.com, 30.07.2004.
17. http://www.bayermaterialsciencenafta.com, 02.08.2004.
18. http://www.huntsman.com, 30.07.2004.
19. http://www.elastogran.de, 29.07.2004.
20. http://www.coimgroup.com, 30.07.2004.
21. http://www.lati.com, 02.08.2004.
22. http://www.glscorp.com, 02.08.2004.
23. http://www.merquinsa.com, 02.08.2004.

Chapter 18

Shape Memory Effects of Multiblock Thermoplastic Elastomers

B. K. Kim, S. H. Lee, M. Furukawa

1. Introduction

The *shape memory polymers* (SMP) belong to the group of intelligent or smart materials having the capability to change their shape on exposure to external stimuli, typically temperature [1–3]. Consequently, thermally induced shape memory effects find broad applications in temperature sensing devices, such as actuators, fire alarms, pipe linings, and sporting goods [4–8]. SMP are also useful when high shape retention during shaping or high shape recovery during repeated loading is desired, as in back-counters of footwear. Figure 1 illustrates a simple application in pipe lining, where an SMP pipe should be tightly fitted inside a metal tube. The original SMP pipe is heated and drawn to a diameter small enough for easy insertion into the metal tube, followed by heating to recover the original diameter, which fits the metal tube. The same procedure can be applied to expand a damaged blood vessel, where the recovery temperature of SMP should be close to body temperature [5]. Recently, the substantial change of the thermal expansion coefficient at the transition

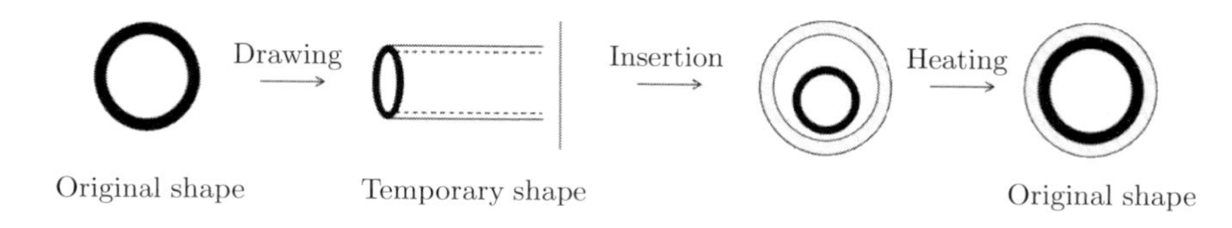

Figure 1. Shape memory polymer applied for pipe lining

temperature *i.e.,* a property inherent to SMP, has been extensively used for the production of water proof/vapor permeable coatings [9–12].

The *shape memory* (SM) effect in polymers is based on the large difference between the glassy state modulus, E_G, and the rubbery state modulus, E_R. That is, the elastic energy stored during deformation in the rubbery state is not high enough to drive the reverse deformation in the glassy state when the load is removed. As a result, the deformation can be frozen in the glassy state. However, ordinary polymers cannot completely restore their deformation upon reheating to the rubbery state. On the contrary, SMP can recover almost all the deformation.

1.1. *Molecular structure requirement*

The shape memory effect is not a specific material property and it results from the combination of polymer structure, microphase morphology, and processing program [1]. These polymers generally consist of two phases, *i.e.*, frozen phase and reversible phase. The frozen phase remains hard during the second shaping process, which normally is performed at a temperature higher than the shape recovery temperature. The reversible phase is subject to softening and hardening upon heating and cooling. Accordingly, excellent shape memory effects have often been observed with styrene-butadiene copolymers (SB) and segmented polyurethanes (PU) [3]. In these materials, the crystalline soft domains (crystalline polybutadiene segments of SB and crystalline soft segments of PU) form the reversible phase and hard domains (styrene blocks in SB and hard segments of PU) become the frozen phase. A shape memory effect has also been obtained with a mixed-phase morphology where a second element acting as the frozen phase should be introduced. This is typically done by crosslinking in *trans*-polyisoprene (TPI) or intensive chain entanglements of high molecular weight in polynorbornene [3].

1.2. *Shape memory programming*

SMP elements are often subject to cyclic deformation, such as in actuators and footwear parts and hence the cyclic characteristics are of practical importance in evaluating their durability. The operating principle is described by a thermomechanical cycle based on an elasticity modulus-temperature relationship, as shown in Figure 2 with a segmented SMP having crystalline *soft* and *hard segments*, their respective melting temperatures being T_{ms} and T_{mh}. The shape memory polymer is first heated above T_{ms} and below T_{mh} to a loading temperature, T_l, and drawn to a predetermined maximum strain, ε_m, at a constant rate. While maintaining the strain at ε_m, the SMP is cooled below T_{ms} to an unloading temperature, T_u. The loading and unloading temperature correspond to the rubbery and glassy state of the polymer, respectively. Upon unloading, part of the strain ($\varepsilon_m - \varepsilon_u$) is instantaneously recovered, leaving an unload strain, ε_u. Finally, the thermomechanical cycle is completed by heating the sample to the loading temperature; the strain is recovered, leaving a permanent strain,

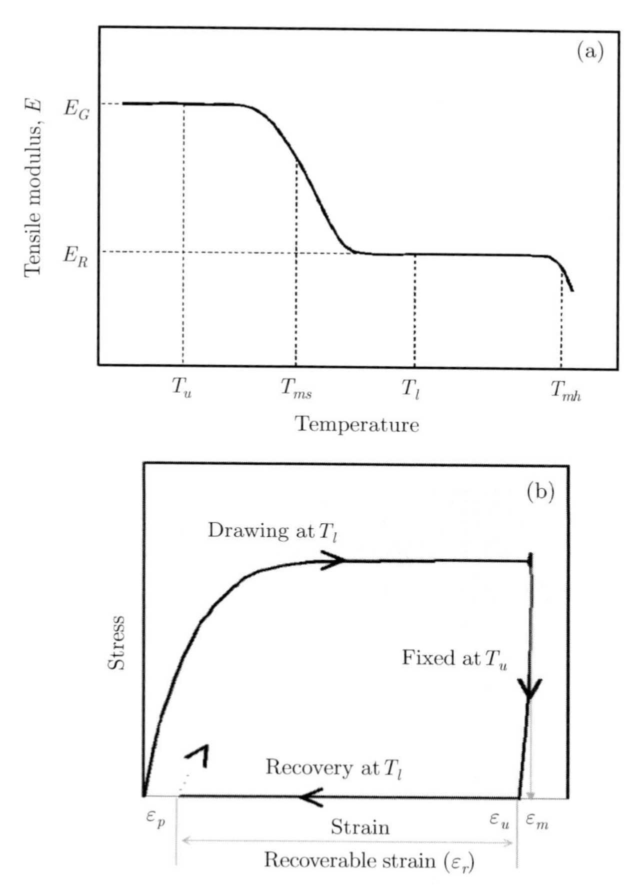

Figure 2. Temperature dependent modulus (a) and a thermomechanical cyclic test evaluating the SMP properties (b)

ε_p. The thermomechanical cyclic behavior of SMP has been modeled on the basis of viscoelastic constitutive equations and the reader is referred to the corresponding references for further details [13,14].

Deformation can be performed either above or below T_{ms} [15,16]. When the sample is deformed at $T < T_{ms}$, only the inelastic component of the strain is simultaneously retained. On the other hand, when the sample is deformed at $T_{ms} < T\ (< T_{mh})$, the deformed shape is frozen upon subsequent cooling below T_{ms}. However, in both types of shaping, the original shape is recovered at $T > T_{ms}$. Deformation at high temperatures is much easier due to the low rubbery state modulus making chain orientations more feasible, although more of the orientations are relaxed before their structures are frozen during the subsequent cooling. Deformation at low temperatures is much more difficult due to the higher glassy state modulus and much less deformation is retained because there is no obvious mechanism (other than the dissipative one) to freeze it.

1.3. *Morphology change*

Typical morphology changes during a thermomechanical cycle of an SMP composed of crystalline hard segment domains, crystalline soft segment domains, and an amorphous phase are schematically shown in Figure 3. A rectangular die-formed SMP sheet is first heated to a loading temperature, T_l, which is higher than T_{ms} and lower than T_{mh}. At this temperature, crystalline domains

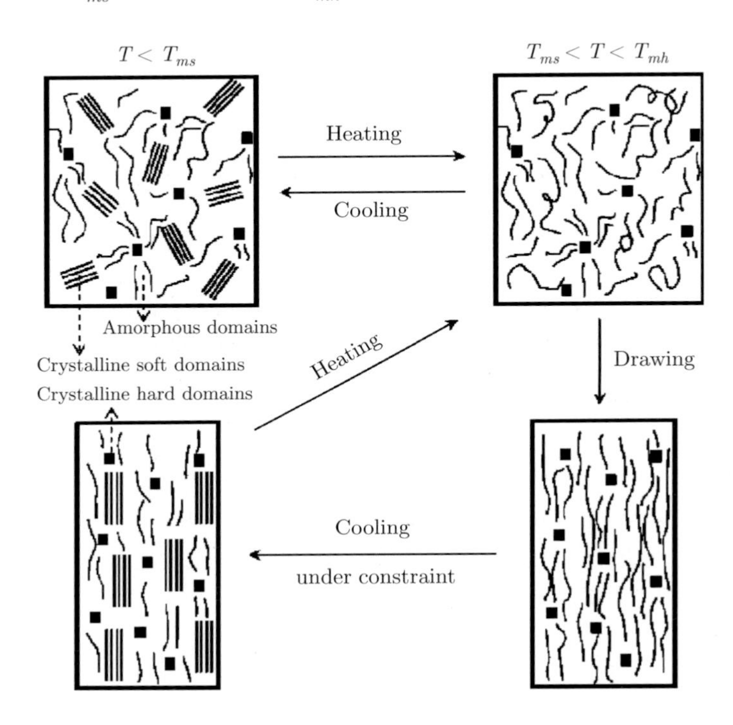

Figure 3. Morphology changes in SMP during a thermomechanical cyclic test

of soft segments are melted while the hard segment domains keep their glassy state. Then the sample is drawn to a preset strain, ε_m, and the soft segments are oriented along the draw direction. Subsequently, SMP is cooled to an unloading temperature, T_u, which is lower than T_{ms} while keeping the strain at ε_m. During cooling, the soft segments are crystallized under stress, retaining more or less the induced orientations. This process is accompanied by a substantial volume shrinkage owing to the density change upon solidification and crystallization. This temporary shape is the one to be achieved in a shape memory article. The highly oriented morphology of this frozen state has a much lower entropy, as compared to the undrawn state. Upon heating above T_{ms}, chain orientations are quickly relaxed due mainly to the entropy elasticity, which keeps the hard segment domains close. So, the amount of recoverable strain depends on the retractive force acting between the frozen phases of the two-phase morphology.

1.4. *Elastic energy balance and entropy elasticity*

Assuming linear elasticity in Figure 2, the strain energy stored per unit mass during stretching to ε_m in the rubbery state, T_l, is given by [17]:

$$\frac{1}{2} E_R \varepsilon_m \tag{1}$$

The strain energy released during shape recovery at T_l is obtained as:

$$\frac{1}{2} E_R(\varepsilon_u - \varepsilon_p) \tag{2}$$

Although it is small, some strain energy is also released upon removal of the strain at T_u:

$$\frac{1}{2} E_G(\varepsilon_m - \varepsilon_u) \tag{3}$$

The strain energy increase during cooling is balanced by the elasticity decrease during subsequent heating, whereas the thermal energy loss during cooling is balanced by the thermal energy gain during heating. The strain energy balance will take the following form:

$$E_R \varepsilon_m = E_R(\varepsilon_u - \varepsilon_p) + E_G(\varepsilon_m - \varepsilon_u) \tag{4}$$

Two important properties of SMP, *i.e.*, shape retention and shape recovery, are obtained as follow:

Shape retention

$$\frac{1}{\varepsilon_m - \varepsilon_u} = \frac{R}{\varepsilon_m - \varepsilon_r} \tag{5}$$

Shape recovery (recoverable srain)

$$\varepsilon_u - \varepsilon_p = \varepsilon_m - R(\varepsilon_m - \varepsilon_u) \geq 0 \tag{6}$$

where $R = E_G/E_R$ (modulus ratio), and $\varepsilon_r = \varepsilon_u - \varepsilon_p$ (recoverable strain). The above simple analysis shows that a high glassy modulus (high R) is desired for high shape retention whereas a high rubbery modulus (low R) is desired for high shape recovery. This seems reasonable since the shape is retained in the glassy state and recovered in the rubbery state.

In the SMP rubbery state, the chains are effectively pinned together by the reversible phase of soft segments entering the rubbery domains and originating from the frozen phase of hard segment domains, thus forming physical crosslinks. Such a morphological texture where hard domains correspond to the tie points in a crosslinked polymer provides entropy elasticity. During stretching, the entropy, S, is decreased due to the decreased distance between the frozen phases and this is the origin of a retractive force, F, according to:

$$F = -T\left(\frac{dS}{dL}\right)_{T,V} \tag{7}$$

In the rubbery state, a change in length, L, does not cause a change in internal energy ($\Delta U = 0$). Therefore, the first law of thermodynamics states that heat, Q, is evolved during stretching:

$$Q = W \tag{8}$$

The entropy force is weak. However, it is strong enough to cause a retraction in the liquid state when the applied stress is removed [18]. When the work done by the internal forces is balanced by the work done by the external forces producing the deformation, the total work, W, done for uniaxial drawing is obtained as:

$$W = \frac{VG_R}{2}\left(\lambda^2 + \frac{2}{\lambda} - 3\right) \tag{9}$$

$$G_R = N\mathrm{k}T \tag{10}$$

where λ, N and k are the draw ratio, the density of subchains (soft segments) bridging the tie points (frozen phases) and the Boltzmann constant, respectively. A high rubbery state modulus will provide the SMP with a great retractive force. Therefore, the molecular design of SMP should consider such molecular parameters as crosslinking density, either chemical or physical, soft and hard segment length and content, in addition to the basic requirement of a two-phase structure.

1.5. *Shape memory polymers vs. alloys*

Among shape memory materials, shape memory alloys and bimetals are well known. Compared to these metallic compounds, SMP have a lower density, high shape recoverability, easy processability, and lower cost. Similar to conventional thermoplastics, SMP can be easily molded by the common methods, such as injection, extrusion, compression, and casting. In addition, their shape recovery temperature can be set at any value in the range room temperature ±50°C, which allows a wide variety of applications. Also, SMP can be colored if desired because they are transparent. However, since the retractive force of the polymer is based on the small entropy elasticity, the SMP applications differ from those of metallic alloys. The basic differences between shape memory polymers and alloys are listed in Table 1.

Among polymers, polynorbornene, TPI, and SB copolymers were the first to exhibit a significant SM effect. However, the commercial application of these polymers has been very much limited by their narrow range of glass transition temperature, T_g, and poor processability. Segmented polyurethanes are basically block copolymers of soft and hard segments [19]. The soft segments are polyols of typical molecular weight 1000–2000, whereas the hard segments are built up of diisocyanates and chain extenders. Depending on the type and composition of soft and hard segments, and the preparation procedure, the structure-

Table 1. Basic differences between shape memory polymers and shape memory alloys

Characteristics	SMP	SM alloy
Density (g/cm^3)	0.96~1.10	6~8
Tensile strength (MPa)	10~35	700~1100
Elongation at break (%)	250~1000	20~60
Recoverable strain (%)	250~800	6~7
Recovery stimulus	Heating	Heating
Recovery temperature (°C)	25~90	−100~100
Deformable stress (MPa)	1~5	50~200
Retractive force (MPa)	1~3	150~300

property relationships of thermoplastic polyurethanes (TPU) are extremely diverse and easily controlled (Figure 4), and hence the shape recovery temperature can be tailored at any temperature between −30 and 70°C, allowing

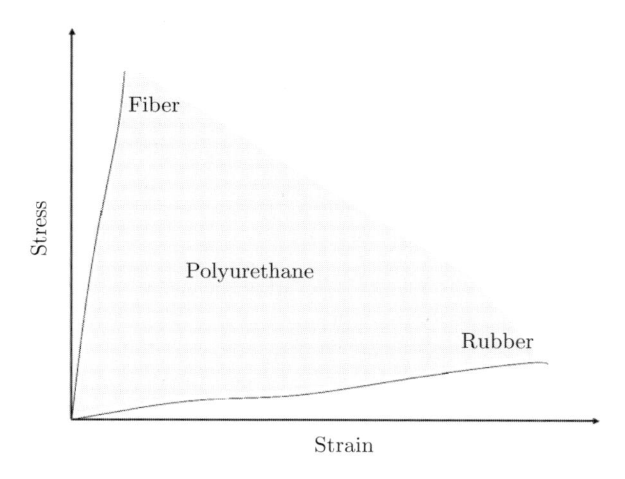

Figure 4. Stress-strain behavior of polyurethanes

a broad range of applications. Similar to the conventional thermoplastic elastomers (TPE), they can be processed by all conventional techniques, which allow versatility of article geometry [20,21]. So, an ample application of TPU as shape memory polymers is expected.

2. Crystalline polyester TPU

2.1. *Basic considerations*

The soft segments of TPU can be designed as crystalline or amorphous depending on the type and length of the polyols being incorporated. Typically, ester and lactone types of polyols with molecular weights over at least 1000 are crystallizable. Among ether type polyols, poly(tetramethylene ether) glycol (PTMG) of high molecular weight can crystallize due to its symmetric structure. By soft segment crystallization, the soft and hard segments are phase-separated, and the melting temperature of the soft segments becomes the "switching temperature" of SMP [22]. On the other hand, when the soft segments are amorphous and miscible with the hard segments, a single T_g of the mixed phase is observed. However, depending on the extent of phase mixing, especially when the latter is poor, the glass transition occurs within a broad temperature range, which reduces the thermal sensitivity of SMP [23,24]. On the contrary, the melting transition is generally much sharper than the glass transition and hence highly temperature-sensitive SMP with crystalline "switching elements" are obtained.

Below, we first consider the shape memory effect, as well as basic structure-property relationships of TPU having crystalline soft segments, *e.g.*, poly(tetramethylene adipate) glycol (PTAd, $M_n = 2000$), and hard segments composed of 4,4′-diphenylmethane diisocyanate (MDI) and ethylene glycol (EG) or ethylene diamine (EDA) [25].

2.2. *Molecular design and synthesis of TPU*

Formulations to prepare the shape memory TPU are shown in Table 2. The OH-terminated polyols (OTP) in the table were obtained by extending the PTAd with hexamethylene diisocyanate (HDI). PH corresponds to a PU obtained from equimolar amounts of PTAd and HDI. The soft segment length was varied by changing the molar (n) ratio of PTAd ($n + 1$) to HDI (n) in OTP.

Table 2. Formulations of crystalline TPUs and their differential scanning calorimetry (DSC) data [25]

Sample	PTAd (moles)	HDI (moles)	OTP[a] (moles)	MDI (moles)	Chain extender (moles)	Soft segment content (wt%)	T_c^b (°C)	T_m^b (°C)	ΔH_f^b (J/g$_{(PTAd)}$)
PTAd	1					100	29.3	50.3	78.1
PH	1	1				100	13.6	53.5	59.2
PHMED1021	1	0	1	2	1(EDA)	78.12	−21.5	30	19.23
PHMED2132	2	1	1	3	2(EDA)	82.73			
PHMED3243	3	2	1	4	3(EDA)	84.3	16.5	41.2	43.2
PHMED4354	4	3	1	5	4(EDA)	85	19	43.6	55.5
PHMEG3243	3	2	1	4	3(EG)	84.23	1.4	48.7	45.7
PHMEG3254	3	2	1	5	4(EG)	80.88	−0.5	47.2	42.3
PHMEG3265	3	2	1	6	5(EG)	77.78	−1.8	46.1	41.9

[a] OH-terminated polyol
[b] Soft segment properties

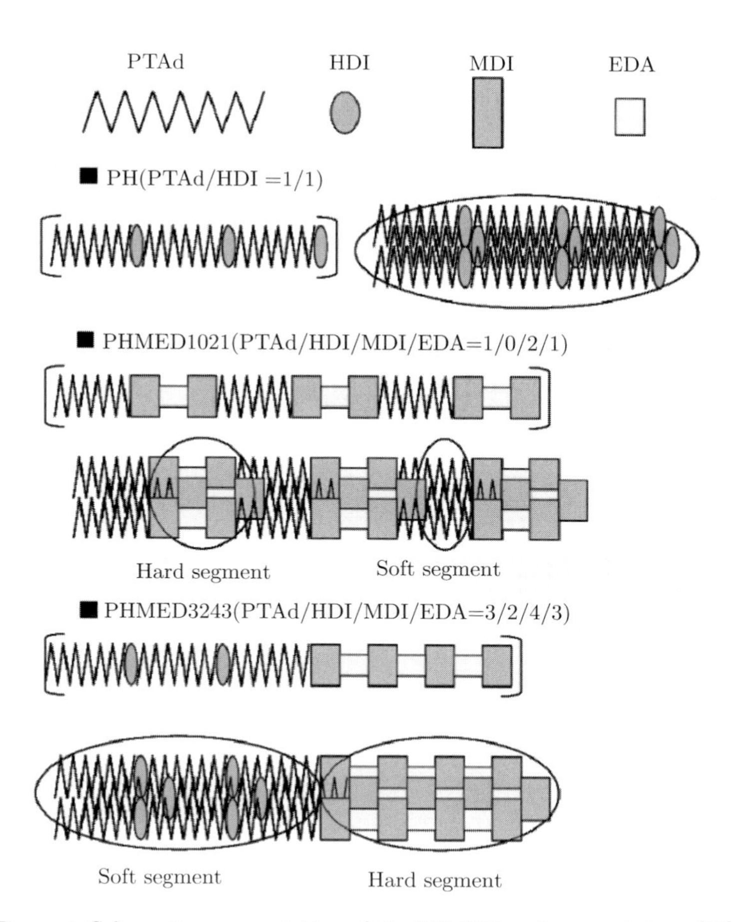

Figure 5. Schematic representation of the PHMED series morphology [25]

To one mole of OTP, various amounts of MDI ($n + 1$) and chain extender (n) were added to perform the synthesis of TPU. For instance, in PHMED3243, the soft segment is composed of three moles of PTAd and two moles of HDI, whereas the hard segment contains four moles of MDI and three moles of EDA. A procedure to prepare the structured TPUs is shown in Scheme 1 and a schematic representation of the morphology is depicted in Figure 5 for the PHMED series.

2.3. *Morphology and thermal properties*

When films are cast at room temperature, PTAd crystallizes in the shape of spherulites upon cooling, with a crystallization temperature, T_c, of 29.3°C and crystalline melting temperature, T_m, of 50.3°C (Figure 6 and Table 2). By extending the PTAd with HDI, the spherulites increase in size, leading to an increase in T_m by about 3°C. This, together with a large drop in T_c by about

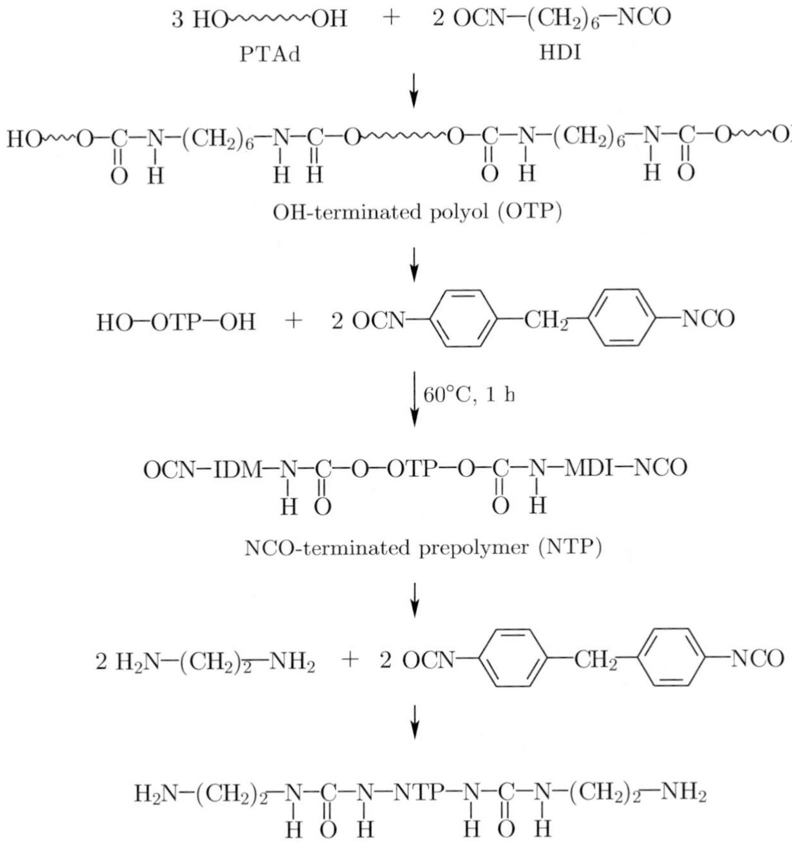

$$3\ \mathrm{HO{\sim\sim\sim}OH}\quad + \quad 2\ \mathrm{OCN{-}(CH_2)_6{-}NCO}$$

PTAd HDI

$$\downarrow$$

$$\mathrm{HO{\sim}O{-}\underset{\underset{O}{\|}}{C}{-}\underset{\underset{H}{|}}{N}{-}(CH_2)_6{-}\underset{\underset{H}{|}}{N}{-}\underset{\underset{H}{|}}{C}{-}O{\sim\sim\sim}O{-}\underset{\underset{O}{\|}}{C}{-}\underset{\underset{H}{|}}{N}{-}(CH_2)_6{-}\underset{\underset{H}{|}}{N}{-}\underset{\underset{O}{\|}}{C}{-}O{\sim}OH}$$

OH-terminated polyol (OTP)

$$\downarrow$$

$$\mathrm{HO{-}OTP{-}OH}\quad + \quad 2\ \mathrm{OCN{-}C_6H_4{-}CH_2{-}C_6H_4{-}NCO}$$

$$\downarrow\ 60°\mathrm{C},\ 1\ \mathrm{h}$$

$$\mathrm{OCN{-}IDM{-}\underset{\underset{H}{|}}{N}{-}\underset{\underset{O}{\|}}{C}{-}O{-}OTP{-}O{-}\underset{\underset{O}{\|}}{C}{-}\underset{\underset{H}{|}}{N}{-}MDI{-}NCO}$$

NCO-terminated prepolymer (NTP)

$$\downarrow$$

$$2\ \mathrm{H_2N{-}(CH_2)_2{-}NH_2}\quad + \quad 2\ \mathrm{OCN{-}C_6H_4{-}CH_2{-}C_6H_4{-}NCO}$$

$$\downarrow$$

$$\mathrm{H_2N{-}(CH_2)_2{-}\underset{\underset{H}{|}}{N}{-}\underset{\underset{O}{\|}}{C}{-}\underset{\underset{H}{|}}{N}{-}NTP{-}\underset{\underset{H}{|}}{N}{-}\underset{\underset{O}{\|}}{C}{-}\underset{\underset{H}{|}}{N}{-}(CH_2)_2{-}NH_2}$$

PHMED3243

Scheme 1. Synthetic route to the preparation of a structured TPU [25]

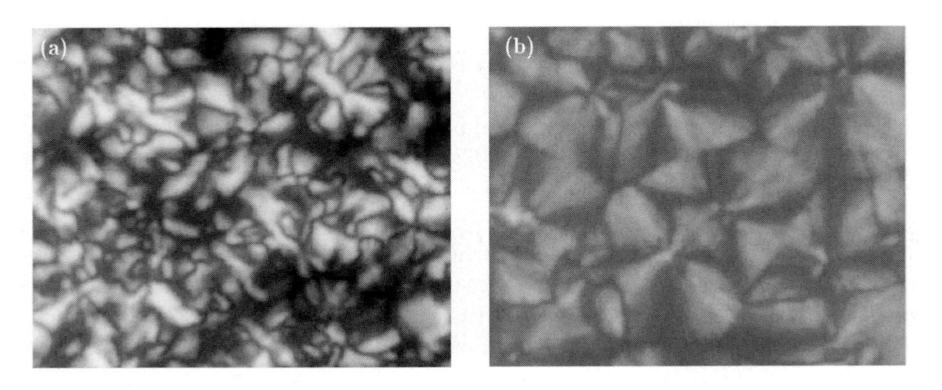

Figure 6. Polarized optical micrographs of PTAd (a) and PH (b) [25]

16°C, results in an increase in the degree of supercooling ($\Delta T = T_m - T_c$), which is indicative of decreased crystallization rate and crystallinity. In the PHMED series, the lengths of soft and hard segments have been simultaneously increased by increasing the number of repeat units, with a net effect of an increased soft segment fraction. When compared with PTAd and PH, the addition of hard segments (PHMED series) gave much smaller spherulites [24]. As the soft segment content increases from 78.12% (PHMED1021) to 85% (PHMED4354), T_c increases from –21.5 to 19°C, and T_m from 30 to 43.6°C (Table 2 and Figure 7). This leads to a decreased degree of supercooling and increased crystallinity.

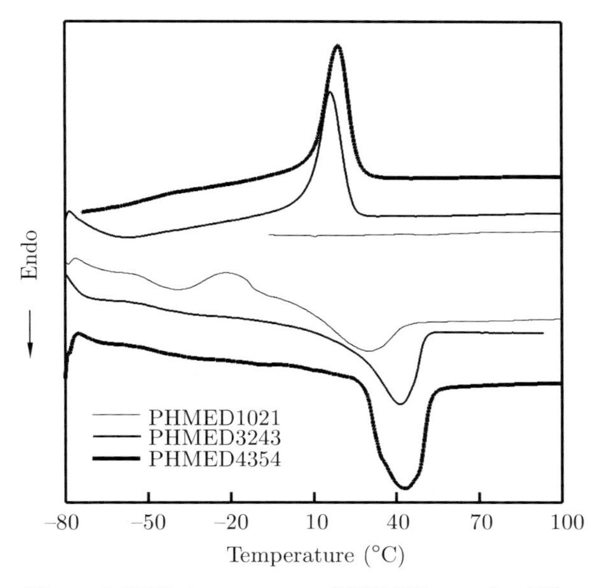

Figure 7. DSC thermograms of PHMED samples [25]

The effect of the hard segment content was studied in the PHMEG series at a constant soft segment length. As the hard segment content increased, T_c decreased and T_m increased, leading to an increase in the degree of supercooling and to a decreased crystallinity. It seems that the hard segments have a slight dilution effect on the soft segments. The influence of copolymerization on the crystallization behavior of the PTAd segments can result in a decrease of the T_c and T_m, with a net effect of increasing the degree of supercooling and decreasing the crystallinity. Essentially identical results have also been reported for TPUs from poly(carprolactone) diol (PCL)/4,4′-diphenylmethane diisocyanate/1,4-butanediol(1,4-BD) [26]. It should be mentioned that no endothermic peak appeared above 200°C (not shown in the figure), which may imply that hard segments are amorphous or partially crystalline. It was generally observed that there is a certain limit in hard segment length and content, above which hard segments get aggregated to form domains and crystalline structures. This minimum hard segment contents were about 30% for polytetramethylene ether glycol (PTMG, MW = 1800)/MDI/1,4-BD, and 18.75% for well annealed PCL (MW = 7000)/MDI/1,4-BD [26,27]. The hard segment crystallization is favored by high molecular weight polyols since long chain segments augment self-aggregation and phase separation. The highest hard segment content in our formulations was 22.32% (PTMEG3265), which seems insufficient to make significantly crystallizable domains separated from the PTAd soft segments of molecular weight 2000. However, an indirect indication of partial hard segment crystallization was found by carrying out dynamic mechanical measurements.

2.4. *Dynamic mechanical properties*

It is generally accepted that microphase separation between soft and hard segments in TPUs is far from complete. In Figure 8, an inflection point is seen

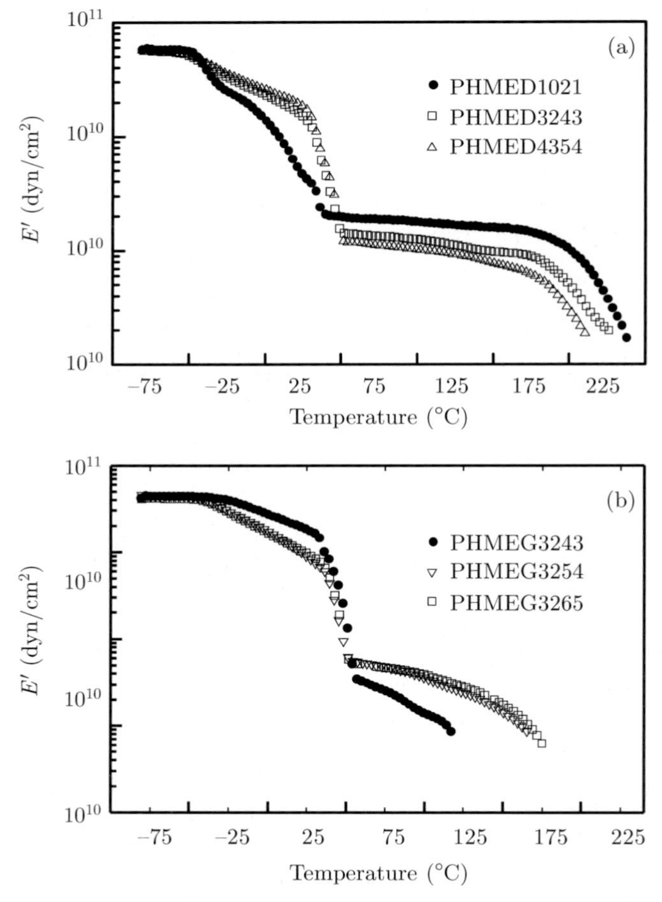

Figure 8. Dynamic storage modulus of the PHMED (a) and PHMEG series (b) [25]

for both PHMED and PHMEG series at about –40°C, *i.e.*, at a temperture higher by about 10°C than the T_g of PTAd and corresponding to the glass transition temperature of the soft segments. It should be noted that T_{gs} has been influenced by the dissolved hard segments. As the soft segment length and content increase (Figure 8a), T_{gs} becomes less significant owing to the high crystallinity of soft segments while their melting transition, T_{ms} (observed at about 40°C as the major transition), becomes more significant. Thus, a TPU with high soft segment content shows a much higher modulus below T_{ms} owing to the high soft segment crystallization and a much lower modulus above T_{ms} owing to decreased hard segment content. On the other hand, as the hard segment content increases, the soft segment length remaining constant (Figure 8b), the glass transition becomes more significant while the melting transition becomes less pronounced, resulting in a lower glassy state modulus and higher rubbery state modulus. Above T_{ms}, the hard segments totally carry the load, so this result can be expected. When different types of chain extenders are compared (PHMED3243 *vs.* PHMEG3243), the EDA extension leads to a much higher and longer rubbery plateau than the EG extension, which is indicative of the higher cohesive energy density and thermal resistance of the urea group, as compared to the urethane group.

The existence of hard domains is indirectly confirmed by the existence of rubbery domains. There is a lower limit of hard segment content and length. Only above this limit do TPUs have enough hard segment domains acting as physical crosslinks at temperatures above the T_{ms} [26]. These physical crosslinks of hard domains are the origin of elastic recoil upon heating and allow shape recovery. When the hard segment length and content are compared, the high content (PHMED1021) results in a higher rubbery plateau extended to a higher temperature, indicating the existence of large hard segment aggregates.

2.5. *Loading in the rubbery state*

Figure 9 shows the typical thermomechanical behavior of PHMEDs subjected to five cycles ($N = 5$) and loaded in the rubbery state ($T = 60$°C) to a maximum strain of 200%, followed by cooling and unloading in the glassy state (20°C). Regardless of the soft segment content, the maximum stress is lower than 7 MPa due to the lower rubbery modulus, resulting in an easy shaping of the SMPs.

The *hysterisis*, that is the area difference between two consecutive cycles, is mostly pronounced for the first pair of cycles. This is an important property for many applications. Thus, the SMP elements are desired to go over the first cycle at high temperature prior to their final shaping as SMP devices [15,16]. It should be noted that the shape retention, *i.e.*, the strain retained after unloading, ε_u, increases with the rise in soft segment content, which is in line with the increasing glassy state modulus (Figure 8) or the high R value (modulus ratio, Eq. (5)). Since the shape is fixed upon cooling and unloading in the glassy state, the slope of the unloading curve should correspond to the glassy state modulus, E_G. From the molecular point of view, soft segments, which are melted

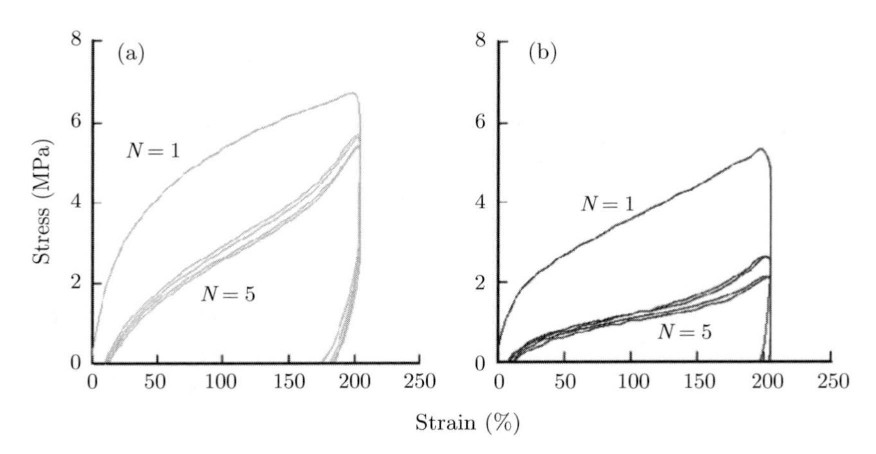

Figure 9. Thermomechanical cyclic test of PHMED1021 (a) and PHMED4354 (b) loaded in the rubbery state ($T_l = T_r = 60$°C, $T_u = 20$°C) [25]

and stretched during loading, crystallize during cooling while maintaining their orientation. The effect of stretching should be more pronounced with the increase of crystallinity, giving rise to a high glassy state modulus and high shape retention, as well. With increased unload strain, the recoverable strain upon heating increases since the residual strain, ε_p, is virtually insensitive to the soft segment content. On the other hand, the yield stress and maximum stress decrease with the rise in soft segment content, these results being consistent with the rubbery state modulus (Figure 9).

2.6. *Loading in the glassy state*

Loading in the glassy state ($T_l = 20$°C) (Figure 10) requires much higher stress to deform and much smaller shape retention is achieved, as compared with high-temperature loading. The high deformation stress (over 10 MPa) is simply caused by the high glassy state modulus, E_G, as shown by a positive yield. The shape retention is much lower than that at the high-temperature loading since there is no mechanism to freeze the deformation, except by the inelastic part of strain. At this small unload strain, ε_u, the recoverable strain, $\varepsilon_u - \varepsilon_p$, upon heating is also small, as compared with high-temperature loading. With regard to the structure effects in TPUs, the shape retention and recoverable strain increase with the rise in soft segment content upon high-temperature loading. In addition, a positive yield (with the stress value increasing with soft segment content), followed by necking and strain hardening are observed due to the crystalline regions in the materials.

With the rise in hard segment length and content, ε_u decreased, which is in line with the decreased glassy state modulus, and upon heating less residual strain was left [25]. That is, the shape is less fixable and more recoverable at high hard segment contents.

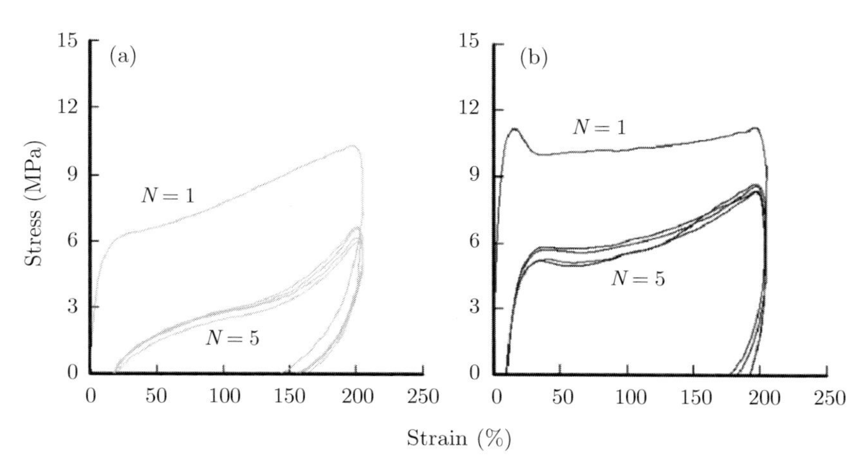

Figure 10. Thermomechanical cyclic test of PHMED1021 (a) and PHMED4354 (b) loaded in the glassy state ($T_l = 20°C$, $T_u = 20°C$, $T_r = 60°C$) [25]

3. Amorphous TPU with allophanate crosslinks

3.1. *Basic considerations*

For the molecular design of SMPs, it is important to tailor the transition temperature as desired. However, for polyurethanes with crystallizable soft segments, the degree of freedom to design the shape transition temperature is limited because the crystalline melting temperature of soft segments becomes the shape transition temperature. For mixed-phase TPUs, one has to control the glass transition temperature of the mixed phase by the proper choice of the soft and hard segment type and content. In such mixed-phase systems, the thermal sensitivity of SMPs is mainly governed by the degree of phase mixing.

In this section, mixed-phase polyurethanes prepared from 1,3-butane diol (1,3-BD), HDI, and MDI are reviewed [28]. When the TPU soft and hard segments are phase-mixed, a second element acting as a fixed phase should be introduced. This second element is typically crosslinks intoduced either by an allophanate reaction or by a multifunctional polyol. Shape memory properties, melt viscosities, dynamic mechanical and thermal properties, and stress relaxations in the glassy and rubbery states are discussed.

3.2. *Synthetic route*

1,3-BD, HDI, and 0.03 wt% DBTDL were reacted in DMF to give OH-terminated prepolymers (first step, $M_n = 7500$, hereafter called HDI blocks), which were subsequently extended with an additional amount 1,3-BD and MDI (second step, MDI blocks). The HDI and MDI blocks do not crystallize due to the geometric asymmetry of 1,3-BD. In order to introduce the allophanate crosslinks between the free isocyanate and urethane groups, MDI (5 wt%) was added to the above system, followed by solution casting at 130°C for 24 h. Formulations and the reaction scheme for the introduction of allophanate crosslinks are given in Table 3 and Scheme 2, respectively.

Table 3. Formulations of amorphous TPUs with allophanate crosslinks [28]

Sample	Soft segment			Hard segment		
	1,3-BD (moles)	HDI (moles)	HTP[a] (moles)	1,3-BD (moles)	MDI (moles)	HSC[b] (wt%)
S75H05	29.7	28.7	1.0	4.0	5.0	22.2
S75H07	29.7	28.7	1.0	6.0	7.0	27.5
S75H09	29.7	28.7	1.0	8.0	9.0	32.0
S75H11	29.7	28.7	1.0	10.0	11.0	36.1
S75H13	29.7	28.7	1.0	12.0	13.0	39.6

[a] OH-terminated prepolymer
[b] Hard segment content

$n+1$ $HO-(CH_2)_2-CH(OH)-CH_3$ (1,3-BD) + n $OCN-(CH_2)_6-NCO$ (HDI)

↓

$H_3C-CH(OH)-(CH_2)_2-O-C(=O)-NH-(CH_2)_6-NH-C(=O)-O-(CH_2)_2-CH(OH)-CH_3$

OH-terminated prepolymer (OTP)

$HO-(CH_2)_2-CH(OH)-CH_3$ (1,3-BD) + ↓ + $OCN-C_6H_4-CH_2-C_6H_4-NCO$ (HDI)

$\left[O-OTP-OC(=O)NH-C_6H_4-CH_2-C_6H_4-NHCO-O-(CH_2)_2-CH(CH_3)-OC(=O)NH-C_6H_4-CH_2-C_6H_4-NHC(=O) \right]_m$

↓ + Extra MDI (5% excess)
casting (130°C)

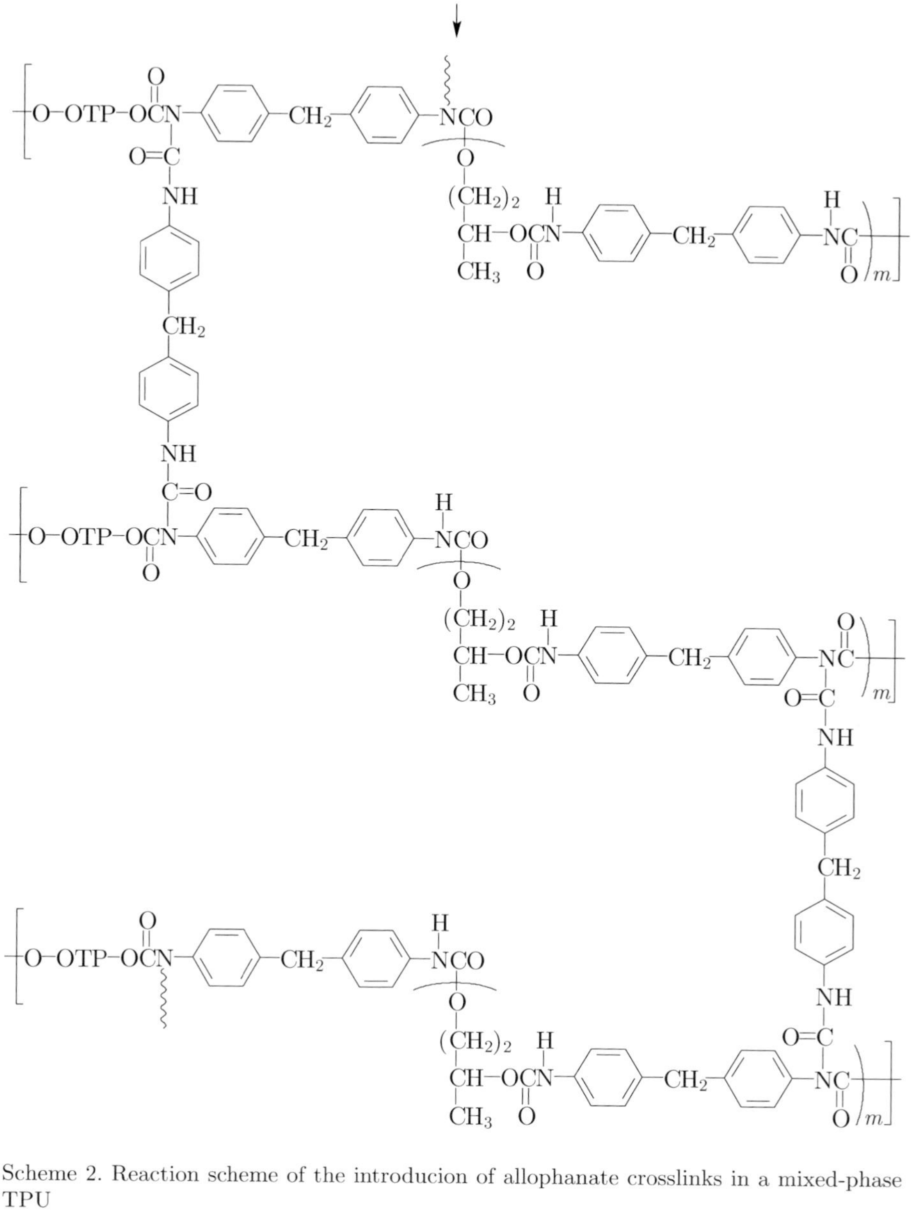

Scheme 2. Reaction scheme of the introducion of allophanate crosslinks in a mixed-phase TPU

3.3. *Dynamic mechanical properties*

The dynamic mechanical properties of polyurethanes as a function of the MDI block content are shown in Figure 11. Regardless of the MDI block content, a

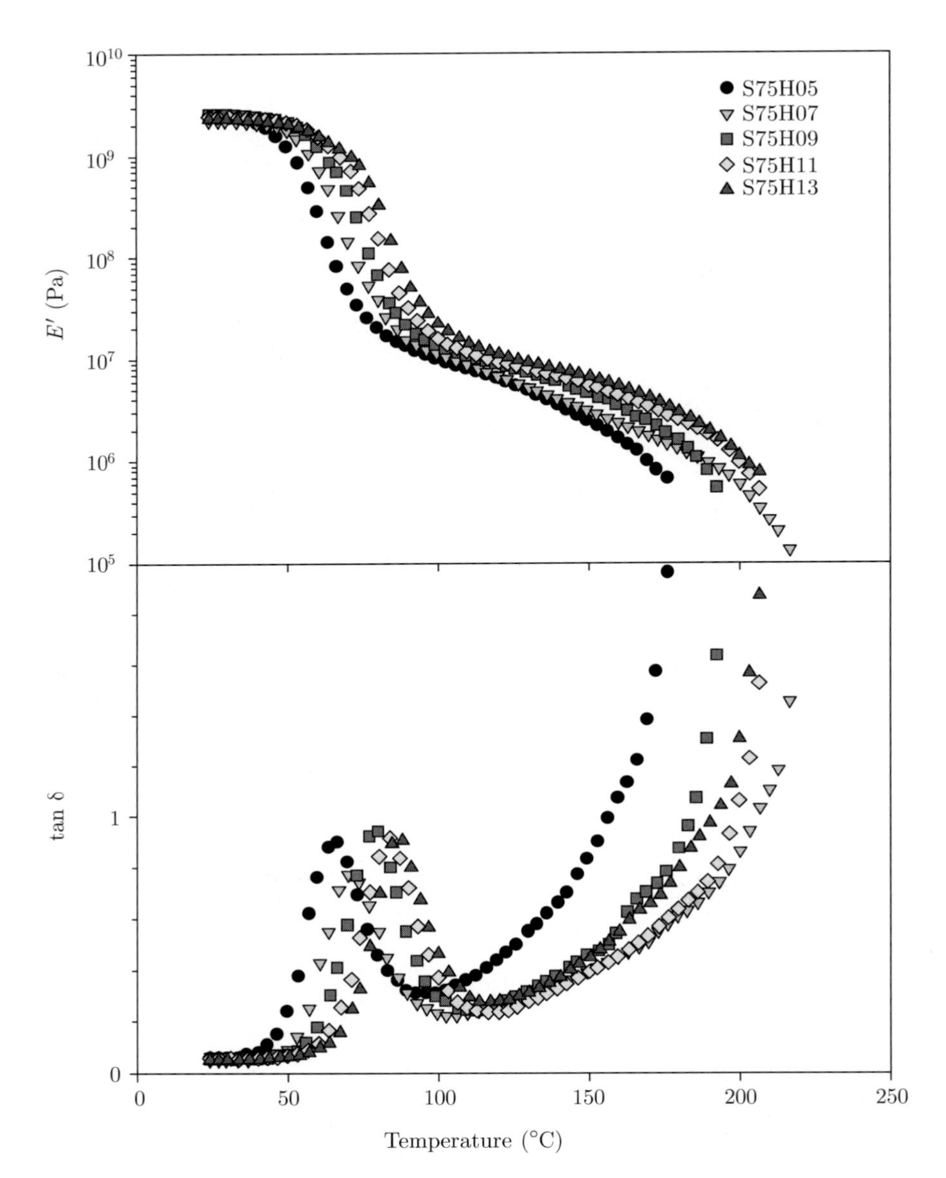

Figure 11. Dynamic mechanical properties of polyurethanes as a function of the MDI block content [28]

single $\tan\delta$ peak is obtained. This implies that the two different blocks of segmented polyurethanes are well phase-mixed. It is also noted that the peak temperature monotonically increases with increasing MDI block content, *i.e.*, from about 30°C to 70°C. This is a great advantage of amorphous polyurethane, which allows to set the shape transition at any desired temperature depending on the specific applications. The miscibility of HDI and MDI blocks is expected from 1,3-BD and urethane groups, which are present in both segments. It is seen that the half height widths of the $\tan\delta$ peaks are within 20°C, indicating a high temperature sensitivity of the materials, which is possibly due to the intensive mixings of the two blocks. It should also be noted that the rubbery state modulus increases with the rise in MDI block content.

3.4. *Melt viscosities*

Figure 12 shows typical complex viscosities, η^*, of the TPUs for S75H07 and S75H11 as a function of the oscillating frequency, ω. The melt viscosities of shape memory polyurethane elastomers are important since these materials are often extruded or injection molded. The viscosity functions of these materials

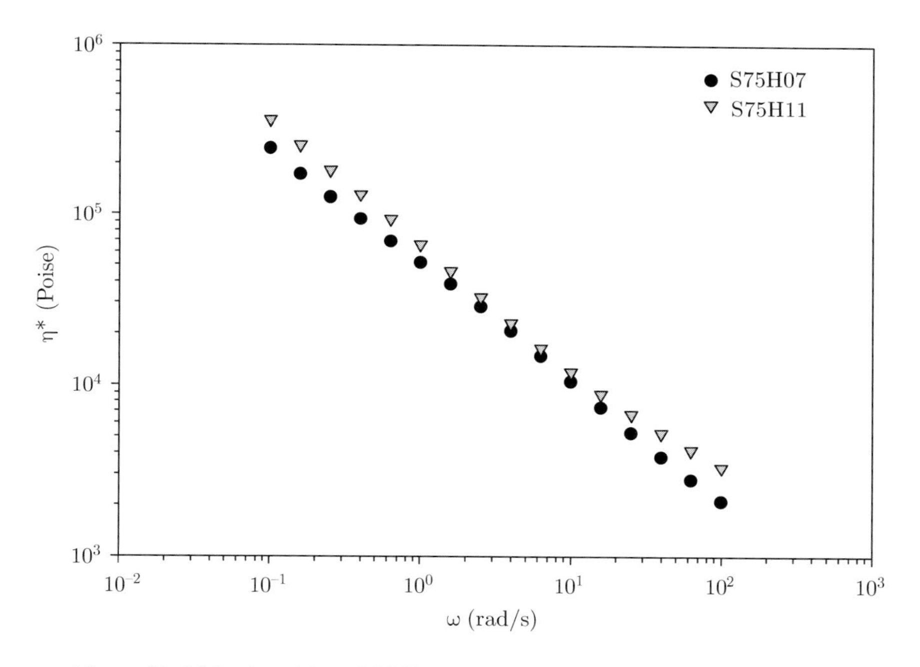

Figure 12. Melt viscosities of TPUs with allophanate crosslinks at 200°C [28]

are almost linear. This is a typical behavior of rubbery materials where crosslinks dominate the melt properties. The sample with a higher MDI block content (S75H11) shows higher melt viscosities at high and low frequencies. However, at intermediate frequencies, the melt viscosities of the two samples are almost the same. Thus, a viscosity upturn is established for the TPU with higher MDI content. Viscosity upturns at low frequencies are often observed with immiscible polymer blends where the dispersed phases are connected in three-dimensional networks [29].

3.5. *Thermal properties*

Differential scanning calorimetry (DSC) measurements [28] showed that the rise in hard segment content results in an increase in T_g from 37.6°C (22.2%) to 61.0°C (39.6%) and this result is consistent with the dynamic mechanical thermal measurements (Figure 11). No crystallization peak was observed because the bulky pendant methyl groups of 1,3-BD disturb the stacking of molecular chains. T_g *vs.* hard segment content is plotted in Figure 13, together

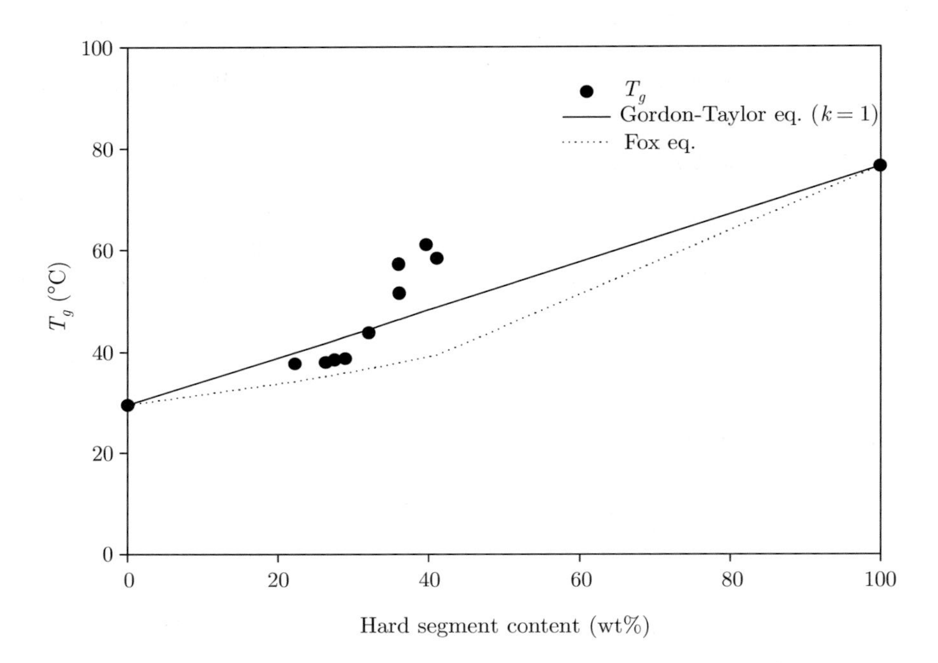

Figure 13. T_g *vs.* hard segment content [28]

with two additive rules of mixing, *i.e.*, Fox equation (Eq. (11)) and Gordon-Taylor equation (Eq. (12)).

$$\frac{1}{T_g} = \frac{W_s}{T_{gs}} + \frac{W_h}{T_{gh}} \tag{11}$$

$$T_g = \frac{W_s T_{gs} + k(1 - W_s) T_{gh}}{W_s + k(1 - W_s)} \tag{12}$$

$$k = \frac{(\alpha_{Rh} - \alpha_{Gh})}{(\alpha_{Rs} - \alpha_{Gs})} \tag{13}$$

$$T_g = W_s T_{gs} + W_h T_{gh} \ \text{(for } k = 1.0\text{)} \tag{14}$$

In these equations, T_g is the glass transition temperature of the mixed phase, T_{gs} and T_{gh} are those of soft and hard segments, and W_s and W_h are the weight fractions of soft and hard segments. The parameter k, *i.e.*, the ratio of the difference in thermal expansion coefficients between the rubbery and glassy state of hard segments to that of soft segments should be unit in copolymer and polymer blends [30]. This leads to a simple additivity given by Eq. (14). It was observed that T_g increased rapidly around the MDI block content of 35 wt% (S75H11, S75H13). This indicates that with the rise in MDI block content, three-dimensional networks are formed *via* allophanate crosslinks and result in an abrupt increase in T_g. The network formation is also suggested by the upturn of the viscosity function at low frequencies, especially at a high hard segment contents, as described above.

3.6. *Thermomechanical properties*

A typical cyclic loading test performed with a shape memory polyurethane cast film is shown in Figure 14. The shape retention and recovery are over 90% for the first cycle, $N = 1$. The shape retention increases and the shape

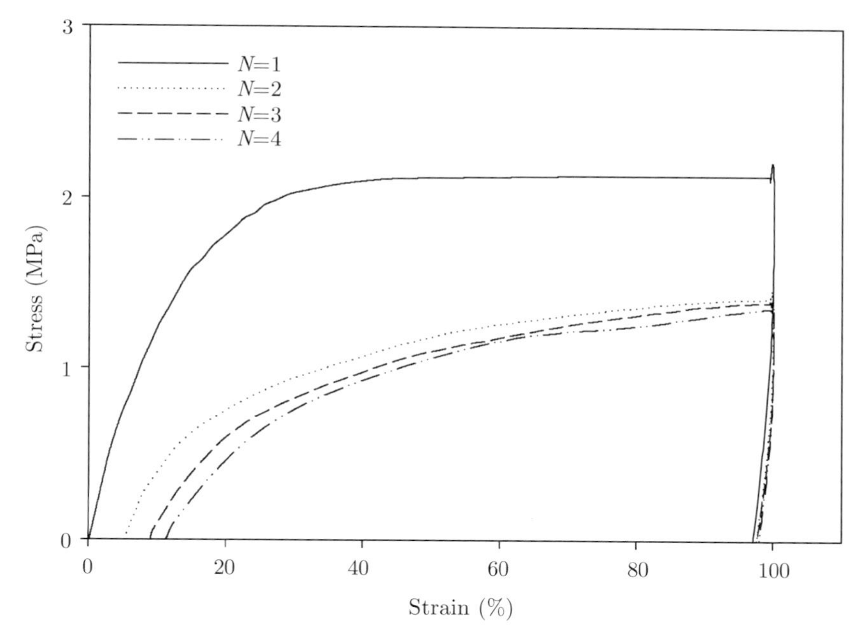

Figure 14. Thermomechanical cyclic test of TPUs with allophanate crosslinks [28]

recovery decreases with the increasing number of cycles. The irreversible parts of the extended molecular chains might cause hystereses in the thermo-mechanical cycles. Table 4 summarizes the shape retention and recovery of all

Table 4. Thermomechanical properties of TPUs with allophanate crosslinks [28]

Sample	Shape retention[a] (%)			Shape recovery[b] (%)		
	$N=1$	$N=2$	$N=3$	$N=1$	$N=2$	$N=3$
S75H05	91.95	95.65	96.85	81.00	75.95	75.45
S75H07	94.95	97.35	97.40	89.80	86.70	84.40
S75H09	96.45	97.50	97.75	91.75	88.45	86.45
S75H11	93.50	96.50	97.60	92.75	88.40	84.00
S75H13	94.80	95.35	95.75	90.20	86.30	85.60

[a] Shape retention = $(\varepsilon_u/\varepsilon_m)\times 100$
[b] Shape recovery = $(\varepsilon_r/\varepsilon_m)\times 100$

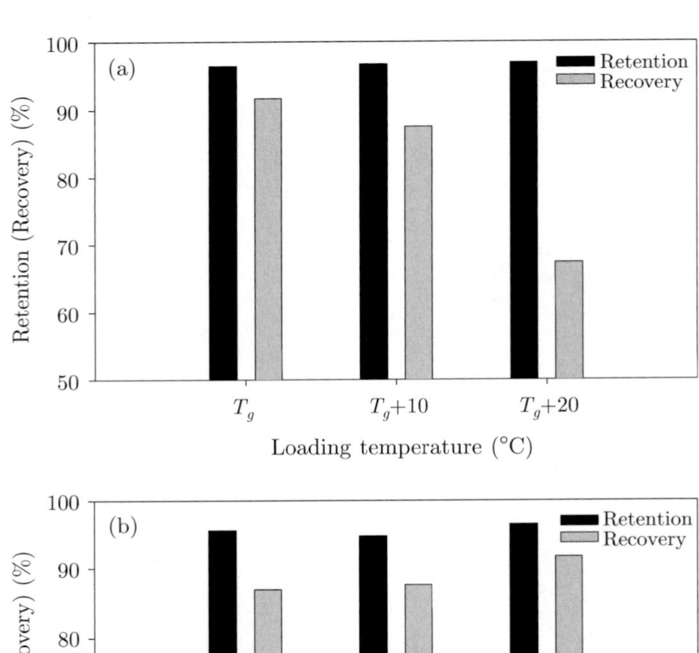

Figure 15. Shape memory properties *vs.* loading temperature [28]

samples up to $N = 3$. It is seen that shape retention and recovery increase with the rise in MDI block content. This is mainly due to the higher glassy state modulus (shape retention) and rubbery state modulus (shape recovery) at a higher MDI block content. However, the longer relaxation time of high MDI block content samples (see below) should also contribute to the shape retention and recovery.

Figure 15 shows that the shape recovery decreases at higher loading temperatures. This means that much more extended chains are relaxed during loading at higher temperatures. On the other hand, high recovery temperatures enhance the shape recovery, which is in agreement with the ideal rubber theory. The stress relaxation in the glassy and rubbery states (Figure 16) support the above results. Based on Maxwell's model [31], the relaxation time is defined as the time for stress decay to 1/e, and it is of the order of 10 s in the figure. So, extensive recoiling as well as slip of extended molecules are expected to occur during loading and cooling. Also, high hard segment content TPUs have longer relaxation times and this result is consistent with Table 4.

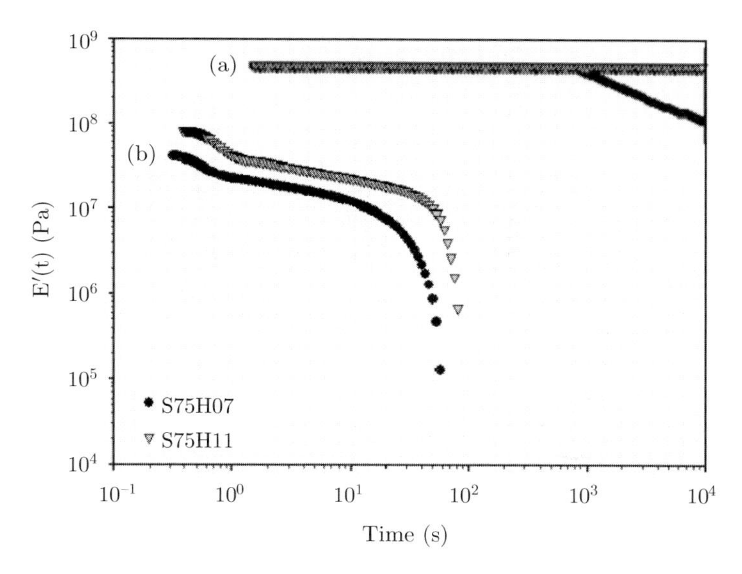

Figure 16. Stress relaxation in the glassy (a) and rubbery state (b) [28]

Shape memory amorphous TPUs with allophanate crosslinking based on PCL diol ($M_n = 500$) or PTAd diol ($M_n = 500$) as soft segments and 1,3-BD/MDI as hard segments were also studied [24]. Dynamic mechanical measurements revealed that low molecular weight polyols had a mixed-phase morphology. However, the extent of phase mixing was greater at a lower crosslink density, a higher hard segment content, as well as in the PTAd-based TPU.

4. Amorphous TPUs based on trifunctional polyols

4.1. *Synthetic route*

When crosslinks are introduced by an allophanate reaction, a quantitative control of their density is not feasible since the reaction occurs reversibly between free isocyanate groups and main-chain urethane groups, and hence the transition temperature is not closely controlled. Thus, when a reproducible close control of the shape recovery temperature is desired, multifunctional polyols or isocyanates can be used to provide TPUs with well controlled crosslink density.

Scheme 3 gives the reaction route to the preparation of the shape memory TPUs crosslinked by trifunctional polyols. PTMG (M_n=250) and a molar excess of MDI were first reacted to obtain NCO-terminated prepolymers (NTP), which were

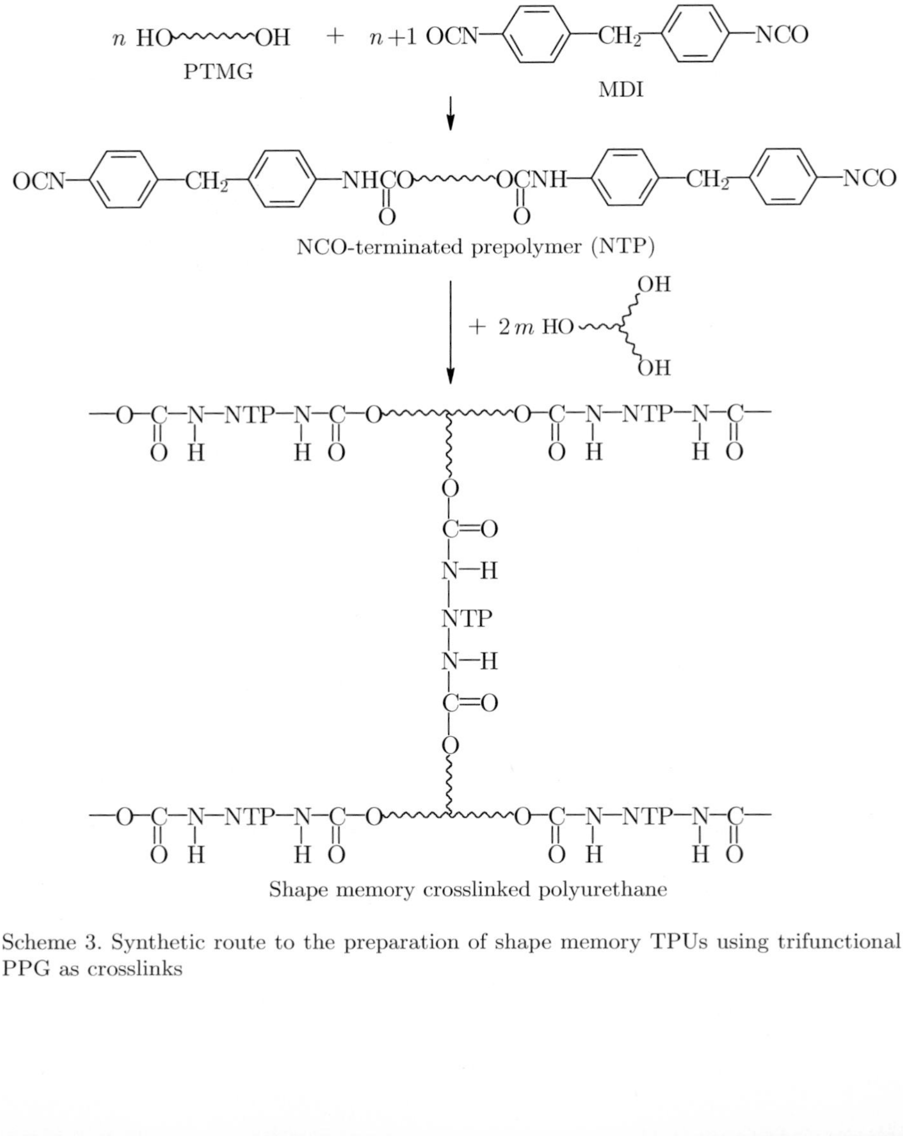

n HO~~~~OH + n+1 OCN–C$_6$H$_4$–CH$_2$–C$_6$H$_4$–NCO

PTMG MDI

↓

OCN–C$_6$H$_4$–CH$_2$–C$_6$H$_4$–NHC(=O)O~~~~OC(=O)NH–C$_6$H$_4$–CH$_2$–C$_6$H$_4$–NCO

NCO-terminated prepolymer (NTP)

↓ + 2m HO~~~(OH)(OH)

Shape memory crosslinked polyurethane

Scheme 3. Synthetic route to the preparation of shape memory TPUs using trifunctional PPG as crosslinks subsequently extended by trifunctional PPG ($M_n = 500$). Thus, the molecular weight of NTP corresponds to the prepolymer molecular weight, M_p, and that between the crosslinks, M_c, the latter taking the values of 3000, 5000, and 7000.

4.2. *Dynamic and thermomechanical properties*

The X-ray diffraction (XRD) profiles of the TPUs show an amorphous shoulder at about $2\theta = 20°$ (Figure 17) and a very sharp glass–rubber transition at about 60°C is obtained from dynamic mechanical measurements (Figure 18). It is seen

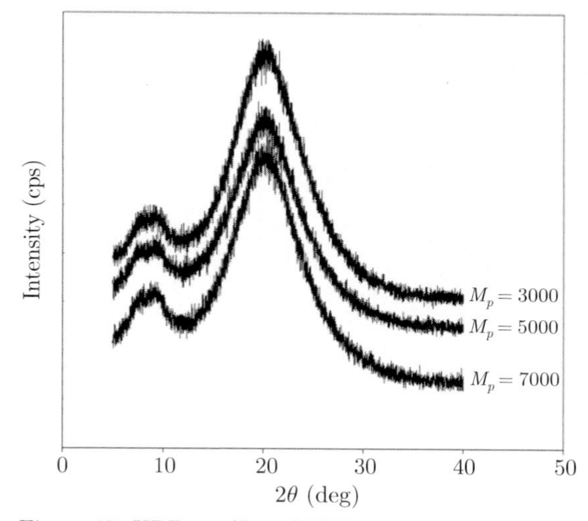

Figure 17. XRD profiles of TPUs with different M_p

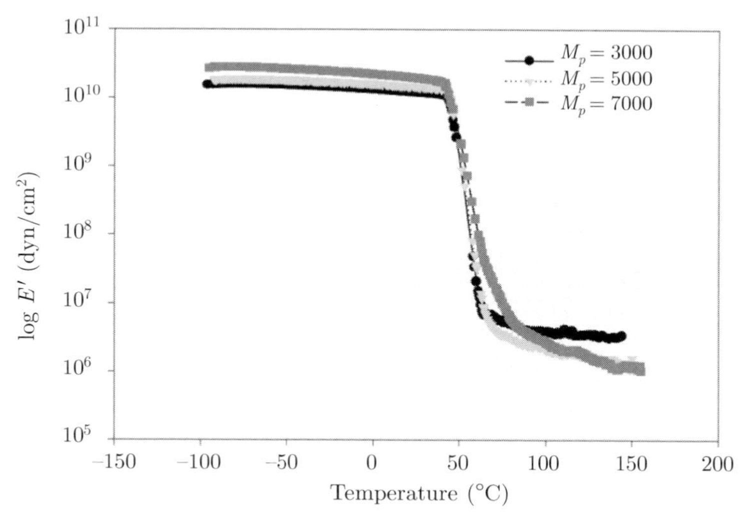

Figure 18. Dynamic storage modulus of TPUs with different M_p

that the transition becomes sharper as the molecular weight of NTP decreases, indicating an improved segment mixing at high crosslink density. The modulus value varies by more than three orders of magnitude within the range of *ca.* 10°C and this is a very important shape memory property related to the temperature sensitivity.

Typical thermomechanical cycles, with loading at T_g+20°C and unloading at T_g–20°C, are shown in Figure 19 for $M_p = 3000$, and the corresponding shape memory data are listed in Table 5, together with those for other TPUs. It is seen that shape retention over 99% and shape recovery over 97% are obtained with $M_p = 3000$ in the first cycle ($N = 1$) and the shape retention remains over 98% until the fifth cycle ($N = 5$). The latter implies that the material responds almost reversibly to the loading in the rubbery state, mainly due to the type of crosslinking. Crosslinks formed by urethane groups should be thermally stronger than those formed by allophanate linkages, which are reversible. The shape reten-

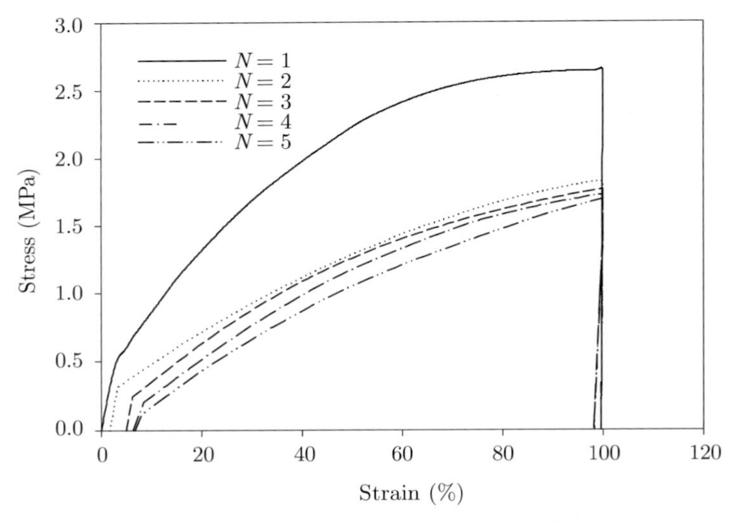

Figure 19. Thermomechanical cyclic test of TPUs (M_p = 3000)

Table 5. Thermomechanical properties of TPUs with trifuctional polyols. M_p is the prepolymer molecular weight

	Shape retention (%)			Shape recovery (%)		
	M_p 3000	M_p 5000	M_p 7000	M_p 3000	M_p 5000	M_p 7000
$N = 1$	99.65	98.60	97.65	97.95	91.70	89.80
$N = 2$	99.65	98.55	96.55	94.80	90.80	86.25
$N = 3$	99.60	98.25	96.30	93.30	86.95	84.80
$N = 4$	98.25	97.80	94.90	91.55	85.95	81.00
$N = 5$	98.20	97.20	92.05	91.15	85.15	77.95

tion and, to a greater extent, the shape recovery decrease with the rise in molecular weight between the crosslinks. This is an indication that the shape recovery is mainly driven by the entropy elasticity, suggesting that the rubbery state modulus is inversely proportional to the molecular weight between the crosslinks.

5. Polyurethanes containing mesogenic moieties

5.1. *Basic considerations*

It has been reported that TPUs with mesogenic moieties in their hard segments have higher values of the rubbery state modulus than those with conventional non-mesogenic chain extenders, even at a low content of hard segments when their liquid crystallinity cannot be manifested. It is then expected that mesogenic hard segments would impart a high shape recoverability to the TPUs and this was tested with three types of mesogenic chain extenders, *i.e.*, 4,4′-dihydroxy biphenyl (DHBP), 4,4′-bis-(2-hydroxyethoxy)biphenyl (BEBP), and 4,4′ bis-(6-hydroxyhexoxy)biphenyl (BHBP) (Scheme 4) to prepare TPUs from PCL diol (M_n = 4000)/HDI/DHBP, PCL diol/MDI/BEBP, and PCL diol/MDI/BHBP [32,33] and the results were compared with conventional TPUs, which were extended by non-mesogenic 1,4-BD [34,35].

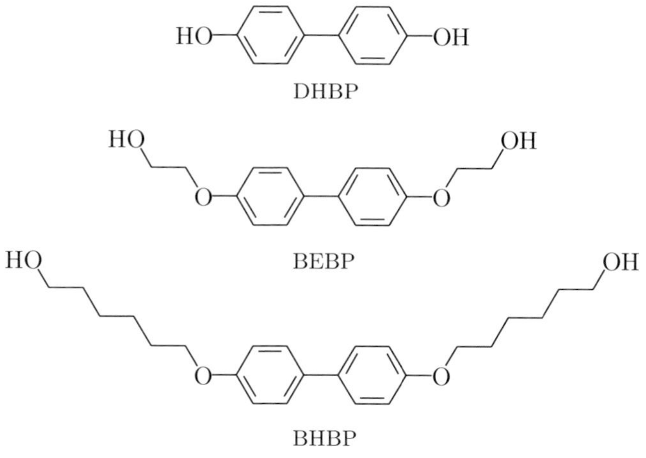

HO—⟨biphenyl⟩—OH

DHBP

HO–CH₂CH₂–O—⟨biphenyl⟩—O–CH₂CH₂–OH

BEBP

HO–(CH₂)₆–O—⟨biphenyl⟩—O–(CH₂)₆–OH

BHBP

Scheme 4. Mesogenic chain extenders

5.2. *Thermal and thermomechanical properties*

The TPUs prepared from PCL/HDI/DHBR showed an enantiotropic mesophase in the hard domains only when the hard segment content was 40% or higher. The tensile moduli of the crystalline TPUs were higher than those of PCL/MDI/1,4-BD based TPUs at temperatures below and above the crystalline melting temperature of the PCL soft segments, T_{ms} [33]. The latter and the heat of fusion at T_m, ΔH_{ms}, generally decreased with the rise in the liquid crystalline hard segment content (Table 6).This might be caused by the hindered crystallization and the reduced crystal perfection and size of the PCL segments by the restricted chain mobility in the presence of rigid hard segments.

Table 6. Thermal properties of TPUs with mesogenic hard segments: ΔH_{ms} – heat of fusion at T_m; ΔH_c – total exothermic heat of transitions at T_{cm} and T_{ci}; T_m – crystal-smectic mesophase transition; T_i – mesophase-isotropic transition; ΔH_h – total endothermic heat of trasitions at T_m and T_i. Number following TPE designates hard segment content (%) [33]

Sample	T_{ms} [a] (°C)	ΔH_{ms} [a] (J/g soft segment)	T_m [b] (°C)	T_i [b] (°C)	ΔH_h [a] (J/g hard segment)	T_{cm} [b] (°C)	T_{ci} [b] (°C)	ΔH_c [a] (J/g hard segment)
TPU-4	59	78	–	–	–	–	–	–
TPU-20	57	71	–	–	24	–	–	–
TPU-30	50	69	–	–	32	–	–	20
TPU-40	47	67	163	178	45	110	125	29
TPU-50	45	45	160	180	47	113	127	33
TPU-60	42	36	163	185	70	112	131	42
TPU-70	41	22	167	188	76	113	135	55
TPU-80	38	11	168	187	77	120	140	53
TPU-100	–	–	170	192	71	126	143	58

a) Measured by DSC
b) Observed by polarizing microscope

In Table 6, T_m, T_i, T_{cm}, T_{ci}, the total endothermic heat of transition at T_m and T_i (ΔH_h), as well as the total exothermic heat of transition at T_{cm} and T_{ci} (ΔH_c) generally decrease when the content of the PCL soft segments increases. At the same time, the average hard segment length decreases, and this seems to cause the drop in transition temperature. The reduced ΔH_h and ΔH_c values support the assumption of diminishing liquid crystallinity and hard segment crystallinity at a higher PCL soft segment content. The shape memory effect of these TPUs having mesogenic hard segments was similar to that of PCL/MDI/1,4-BD [34]. A typical thermomechanical cyclic loading behavior of a shape memory TPU containing mesogenic hard segments is shown in Figure 20 [33].

6. Other thermoplastic elastomers showing shape memory effects

6.1. *Polymer networks with crystalline segments*

SMP should be composed of two phases, *i.e.*, a reversible and a fixed phase. Typically, the transition temperature of the reversible phase becomes the shape recovery temperature and hence its range and sensitivity are of practical significance for their processability. Therefore, any thermoplastic material,

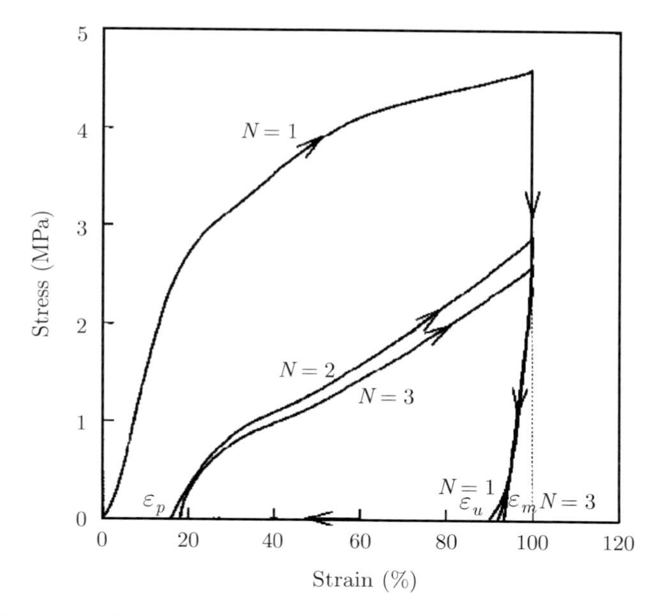

Figure 20. Thermomechanical cyclic test of TPU-20 (loaded at 65°C and unloaded at 25°C) [33]

which satisfies these basic requirements can be considered as SMP. Based on such a phase-separated morphology, several block copolymers were designed and synthesized to show shape memory effect. These include segmented networks composed of poly(octadecyl vinyl ether) (PODVE) and poly(butyl acrylate) (PBA) [36], polycaprolactone-polyamide block copolymer [37], and poly(ethylene terephthalate)-poly(ethylene glycol) copolymer [38]. The segmented networks (Scheme 5) show a high degree of phase separation over a wide range of compositions, which allows the crystallization of the PODVE segments. Then the crystalline PODVE segments having a melting temperature of 40–42°C (depending on the composition) become the reversible phase. By changing the segment length of PODVE and the type of comonomer, the SMP properties can be tailored.

6.2. *Polycaprolactone-polyamide block copolymers*

6.2.1. *Synthetic route*

The diamine-terminated nylon 6/66 copolyamide oligomers (CPA, Scheme 6), having the M_n values shown in Table 7 were synthesized by the melt polycondensation reaction of ε-caprolactam (CA), adipic acid (AA), and hexamethylene diamine (HA) [37,39]. An excess of HA over AA was used to obtain CPA with terminal amine groups and the molecular weight was controlled by a stoichiometric imbalance of reactants, *i.e.*, by varying the AA/HA feed molar ratio at a fixed CA/AA feed molar ratio (Table 7). The diamine-terminated nylon 6

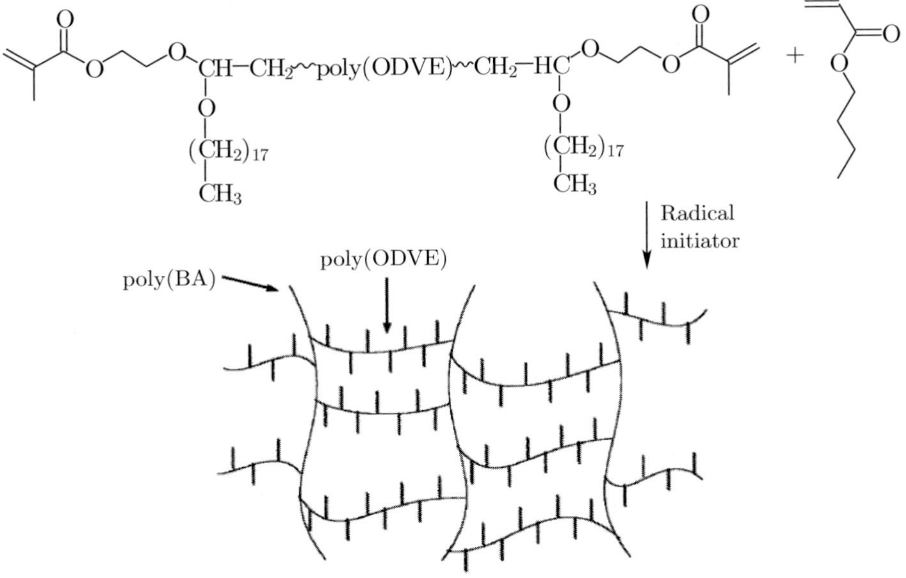

CH–CH$_2$~poly(ODVE)~CH$_2$–HC + O
O O
(CH$_2$)$_{17}$ (CH$_2$)$_{17}$
CH$_3$ CH$_3$
Radical initiator
poly(BA)
poly(ODVE)

Scheme 5. Net-poly(ODVE-*co*-BA) [36]

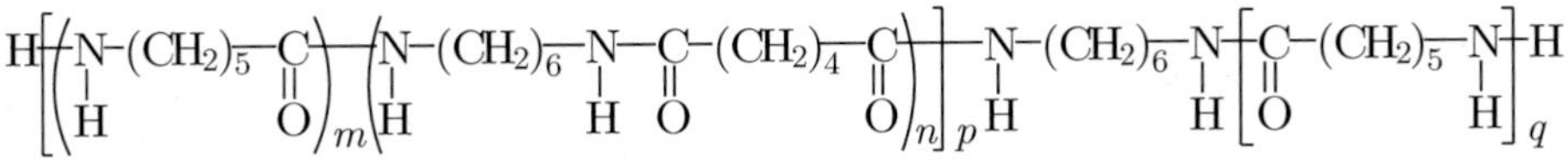

$$H\left[\left(\underset{H}{N}-(CH_2)_5-\underset{O}{C}\right)_m\left(\underset{H}{N}-(CH_2)_6-\underset{H}{N}-\underset{O}{C}-(CH_2)_4-\underset{O}{C}\right)_n\right]_p\underset{H}{N}-(CH_2)_6-\underset{H}{N}\left[\underset{O}{C}-(CH_2)_5-\underset{H}{N}\right]_q H$$

Scheme 6. Chemical structure of a copolyamide oligomer (CPA)

Table 7. Charaterisistics of polyamide oligomers [37]

Designation	Feed molar ratio of CA/AA/HA	M_n NMR	M_n Titration	T_m (°C)	ΔH_m at T_m (J/g)
CPA1	2.00/1.00/2.60	1180	1320	177	81
CPA2	2.00/1.00/1.49	2150	2500	168	53
CPA3	2.00/1.00/1.30	–	3760	174	55
HPA1	2.00/ – /0.29	–	1340	181	111

homopolyamide oligomer (HPA) was similarly synthesized in the absence of AA (Table 7).

PCL-polyamide multiblock copolymers were synthesized by linear chain extension of PCL diol and diamine-terminated polyamide oligomer with HDI in DMF (Scheme 7). The sample designation in Table 8 indicates the molecular weight and the content of PCL block, as well as the type of the polyamide block

Table 8. Charaterisistics of PCL-polyamide block copolymers [37]

Sample	M_n of PCL diol	Polyamide oligomer	Composition in feed (wt%)			Composition in polymer (wt%)			$[\eta]$
			PCL diol	Polyamide oligomer	HDI	PCL diol	Polyamide oligomer	HDI	dL/g
4P1HN73	4000	HPA1	70.00	23.54	6.46	71.2	20.8	8.0	0.89
4P1CN73	4000	CPA1	70.00	23.30	6.70	72.1	20.7	7.2	1.02
4P2CN73	4000	CPA2	70.00	25.15	4.85	70.3	24.0	5.7	1.71
4P3CN73	4000	CPA3	70.00	25.71	4.29	72.0	22.8	5.2	1.39
4P1CN64	4000	CPA1	60.00	32.61	7.39	62.7	28.5	8.8	1.02
4P1CN82	4000	CPA1	80.00	14.47	5.53	81.7	11.3	7.0	1.19
2P1CN73	2000	CPA1	70.00	20.98	9.02	69.2	19.3	11.5	1.35

Table 9. Thermal properties of PCL-polyamide block copolymers [37]

Sample	T_{gs} (°C)	T_{ms} (°C)	ΔH_{ms} (J/g PCL segment)	T_{cs} (°C)	ΔH_{cs} (J/g PCL segment)	T_{mh} (°C)	ΔH_{mh} (J/g polyamide segment)	T_{ch} (°C)	ΔH_{ch} (J/g polyamide segment)
4P1HN73	–58.5	52.8	68.7	11.5	61.2	181.6	77.2	–	–
4P1CN73	–58.8	52.8	65.0	11.8	59.1	183.2	57.8	–	–
4P2CN73	–59.0	51.8	65.2	12.4	58.7	167.7	43.0	–	–
4P3CN73	–62.4	52.9	70.0	18.1	63.7	167.5	38.4	122.2	31.5
4P1CN64	–58.8	51.9	54.7	9.2	45.2	182.3	63.7	–	–
4P1CN82	–60.1	52.5	68.6	16.2	60.4	181.4	62.8	–	–
2P1CN73	–56.8	42.3	47.4	–3.8	44.0	181.5	61.6	–	–

$$\text{HO-PCL-OH} + xx+yz\ \text{OCN-(CH}_2)_6\text{-NCO} + yz\ \text{H}_2\text{N-polyamide-NH}_2$$

PCL diol — HDI — Diamine-terminated polyamide oligomer

↓

$$\left[\left(\text{O-PCL-O-}\underset{\text{O}}{\underset{\|}{\text{C}}}\text{NH-(CH}_2)_6\text{-NH}\underset{\text{O}}{\underset{\|}{\text{C}}}\right)_x\left(\text{NH-polyamide-NH}\underset{\text{O}}{\underset{\|}{\text{C}}}\text{NH-(CH}_2)_6\text{-NH}\underset{\text{O}}{\underset{\|}{\text{C}}}\right)_y\right]_z$$

Scheme 7. Reaction scheme of the obtaining of a PCL-polyamide multiblock copolymer

in the PCL-polyamide block copolymer. For instance, 4P1HN73 indicates that this PCL-polyamide block copolymer was synthesized with PCL diol of M_n 4000 and HPA1 (Table 7), and the content of PCL segments is about 70 wt%. A TPU elastomer, 4TPE73 (Runs 4–70 in [33]), was synthesized from PCL diol (M_n = 4000), MDI, and 1,4-BD. The content of PCL segments in 4TPE73 was 70 wt%, and the detailed synthetic method was described in [32].

6.2.2. *Thermal properties*

The DSC data listed in Table 9 were determined as follows. After melting for 3 min at 20°C above the melting temperature, T_{mh}, of the polyamide segments in the DSC apparatus, the crystallization temperature, T_{ch}, and the heat of crystallization, ΔH_{ch}, of the polyamide segments, as well as the crystallization temperature, T_{cs}, and the heat of crystallization, ΔH_{cs}, of the PCL segments were measured in a cooling scan to –140°C. In the subsequent heating scan, after holding for 1 min at –140°C, the glass transition temperature, T_{gs}, the melting temperature, T_{ms}, and the heat of fusion, ΔH_{ms}, of the PCL segments, as well as the crystallization temperature, T_{ch}, T_{mh}, and the heat of fusion, ΔH_{mh}, of the polyamide segments were determined.

All the T_{gs} values shown in Table 9 are within the temperature range of –63°C ~ –56°C. In TPEs, T_g of the soft segments generally increases as some of the hard segments are dissolved into the soft domains due to phase mixing. Since the T_g of pure PCL is of about –60°C [40], the results of T_{gs} in Table 9 suggest that the dissolution of polyamide segments into PCL domains is not considerable. Compared to other PCL-polyamide block copolymers, 4P3CN73 and 4P1CN82 generally reveal lower T_{gs}, higher ΔH_{ms} and ΔH_{cs}, and smaller supercoolings necessary for crystallization upon cooling ($T_{ms}-T_{cs}$). These results show that in these two polymers the dissolution of polyamide segments into PCL domains and the crystallization of PCL segments are less intense and the crystallization of PCL segments is relatively easier than in other PCL-polyamide block copolymers [37,39,40], due to the higher molecular weight or lower polyamide segment content. 2P1CN73 has a higher T_{gs}, lower ΔH_{ms} and ΔH_{cs}, and a higher value of $T_{ms}-T_{cs}$ than the other block copolymers. These results show that the dissolution of polyamide segments into PCL domains, *i.e.*, phase mixing, is enhanced when the molecular weight of the PCL segments is low. In our pervious papers [34,35], we compared the properties of PCL diol/MDI/1,4-BD based TPE non-ionomers with those of PCL diol/MDI/1,4-BD/dimethylol propionic acid (DMPA) based TPE ionomers. A lower residual strain after shape recovery was observed with TPE ionomers due to the enhanced microphase separation between soft and hard segments and the higher rubbery plateau modulus. When the molecular weight and content of the PCL segments were 4000 and 70 wt%, respectively, the ΔH_{ms} values of the TPE ionomer and non-ionomer were 64 Jg^{-1} and 57 Jg^{-1}, respectively, and the somewhat higher value for the TPE ionomer was explained to be due to enhanced microphase separation. In Table 9, it is seen that ΔH_{ms} is higher than 64 Jg^{-1} at higher molecular weights and contents of the PCL segments.

6.2.3. *Shape memory behavior*

Typical cyclic behaviors of a PCL-polyamide block copolymer and 4TPE73 are shown in Figure 21. The samples were elongated at 65°C to 100% strain, ε_m, at a constant draw rate of 500 mm min^{-1}. While maintaining the strain at ε_m, the samples were cooled to 25°C and unloaded. Upon removing the stress at 25°C, a small recovery of strain to ε_u occurred. The samples were subsequently

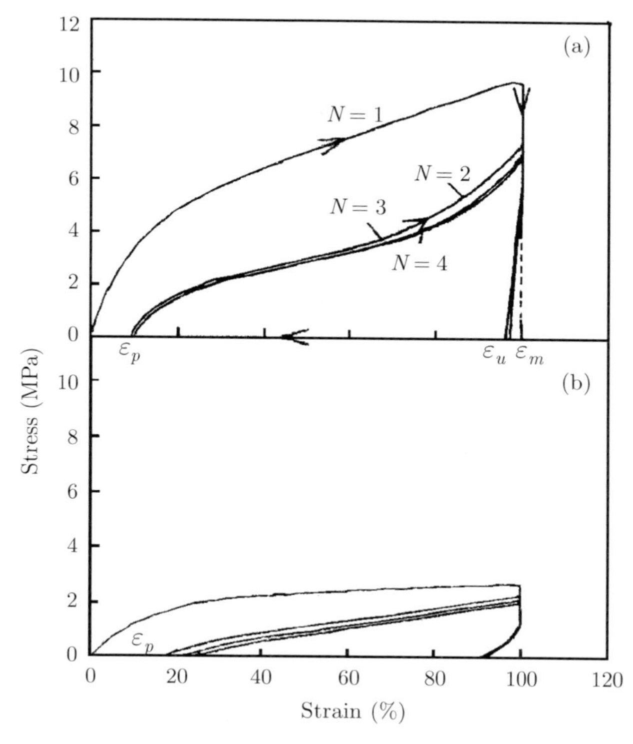

Figure 21. Cyclic tensile behavior of 4P1CN64 (a) and 4TPE73 (b) [37]

heated to 65°C within 5 min, and held at this temperature for the next 10 min, allowing the strain to recover. Thus, the thermomechanical cycle ($N = 1$) was completed, leaving a residual strain, ε_p, at the start of the next cycle ($N = 2$).

It is seen that the shape of the PCL-polyamide block copolymer remains almost the same after the first cycle. Table 10 shows small variations in the ε_p values in all three thermomechanical cycles and they are close to 10%. Even after twenty thermomechanical cycles with 4P1CN73 and 4P1CN82, the results were almost the same. However, in the case of TPEs, the shape of the curves changed and the values of ε_p after the first cycle (*ca.* 20%) strongly increased during the following thermomechanical cycles (see, *e.g.*, Figure 21b). Because fatigue by repeated deformations in TPEs originates from soft-hard phase mixing and orientation, these results show that the changes in phase mixing or segmental orientation by repeated deformations are less considerable in PCL-polyamide block copolymers.

The variations of ε_p (Table 10) in the PCL-polyamide block copolymer samples obey the following orders, depending on: the PCL block content — 4P1CN64 > 4P1CN73 > 4P1CN82, the PCL block length — 4P1CN73 > 4P1CN82, and the polyamide block length — 4P3CN73 > 4P2CN73 > 4P1CN73. The deformation stress of 4P1CN73 is higher than that of 4TPE73, suggesting that the PCL-polyamide block copolymer will have a higher retraction force which is necessary for the practical application of shape memory materials.

Table 10. Residual strain after a thermomechanical cyclic test of PCL-polyamide block copolymers and 4TPE73 [37]

Sample	ε_p (%) after		
	$N = 1$	$N = 2$	$N = 3$
4P1HN73	7.3	8.4	8.5
4P1CN73	7.8	8.6	8.6
4P2CN73	8.1	9.2	10.6
4P3CN73	9.4	11.1	11.4
4P1CN64	9.1	10.0	9.9
4P1CN82	5.6	6.4	6.3
2P1CN73	6.5	7.1	7.3
4TPE73	17.0	20.0	23.0

7. Shape memory blends and composites

7.1. *Basic considerations*

The shape memory effect has also been observed for amorphous non-crosslinked homopolymers, such as polynorbornene, poly(methyl methacrylate) (PMMA), and polycaprolactone [3] where the physical entanglements can play the role of a frozen phase. When they are deformed in the rubbery state and subsequently quenched to the glassy state under constant strain, the deformed shape is retained because the micro-Brownian motions are frozen. Upon reheating to the rubbery state, the original shape is recovered by the elastic energy stored during the deformation. However, when the frozen phase is not rigid enough, the hysteresis increases by repeated deformation-recovery cycles. The hysteresis has been minimized by increasing the physical entanglements with a high molecular weight polymer in the case of polynorbornene [3] and by introducing crosslinking agents in the cases of polyethylene and PVC [41]. In order to control the shape recovery temperature and design the various physical properties of shape memory polymers, miscible blends can be utilized because their T_g and physical properties vary smoothly with composition. However, reports on the shape memory effects of polymer blends are scarce in the open literature [42,43].

We considered blends of TPUs with PVC [44] and phenoxy [45]. It is well known that several aliphatic polyesters are miscible with PVC [46] and phenoxy resins [47]. In the blends of PVC or phenoxy with TPUs based on aliphatic polyester soft segments, it is expected that the PVC (or phenoxy)/aliphatic polyester domains are smooth and the phase-separated hard segment domains act as physical crosslinks [48,49].

7.2. *PVC/TPU blends*

TPUs were prepared from polycaprolactone diol (PCL, $M_n = 2000, 4000$), HDI, and DHBP. Formulations are given in Table 11, together with the feed ratios, hard segment contents, and intrinsic viscosities. In the sample designation, the numbers show the M_n of PCL and the hard segment content, respectively, *e.g.*, 2TPU4 is based on PCL with M_n of 2000 and hard segment content of 40%.

Table 11. Formulations of PVC/TPU blends [44]

Sample	Feed (moles)				Hard segment content (wt%)		$[\eta]$ (dL/g)
	Hard segment		Soft segment				
	HDI	DHBP	PCL2000	PCL4000	In feed	In polymer	
2TPU4	8.38	6.38	2.00	–	40	38	0.53
2TPU6	17.56	15.56	2.00	–	60	62	0.59
4TPU4	7.86	6.86	–	1.00	40	41	0.51
4TPU6	17.06	16.06	–	1.00	60	63	0.57

DSC thermograms for TPUs (Figure 22) show a glass transition of the soft PCL segments, T_{gs}, as well as melting endotherms of soft, T_{ms}, and hard segments, T_{mh}. It is seen that T_{ms}, T_{mh}, the heats of fusion at T_{ms} (ΔH_{ms}) and T_{mh} (ΔH_{mh}) all decrease as the PCL segments decrease in length at a constant

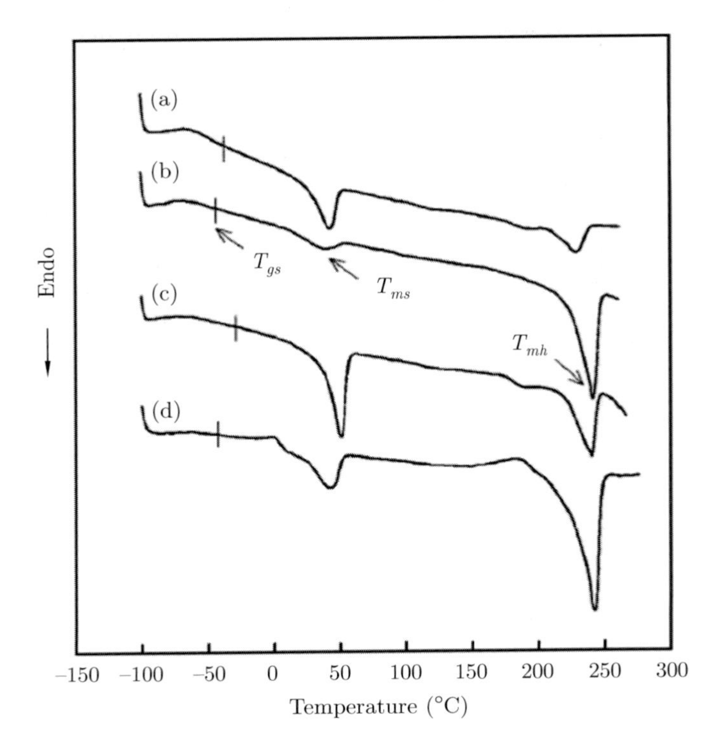

Figure 22. DSC thermograms of TPUs: 2TPU4 (a), 2TPU6 (b), 4TPU4 (c), and 4TPU6 (d) [44]

hard segment content. This implies that phase mixing between soft and hard segments is favored by short PCL segments.

Typical DSC thermograms are shown for PVC/2TPU4 blends in Figure 23, where a single T_g shifting smoothly with composition is seen. The PCL segments do not reveal an endothermic melting peak for PCL segmens up to 200°C where thermal degradation of PVC occurs. This implies that PVC and PCL segments are miscible and the crystallization of PCL segments is inhibited. Other PVC/TPU blends showed essentially identical results, which are summarized in Table 12.

Typical dynamic mechanical properties of the blends are given for 2TPU4 (Figure 24) where the tan δ peak decreases and the rubbery state modulus increases smoothly with increasing amount of TPU. The former confirms the miscibility and the latter implies that hard segment domains do play the role of physical crosslinks or reinforcing filler in the rubbery state. With the rise in hard segment content, the increase in the rubbery state modulus was more pronounced. Similar results were also obtained for TPUs from PCL 2000.

For a thermomechanical cyclic test, the sample was drawn at $T_g + 30$°C to 100% strain, ε_m, at a constant rate of 10 mm min^{-1}. While maintaining the strain at ε_m, the sample was cooled to 20°C and unloaded. As a result, a small

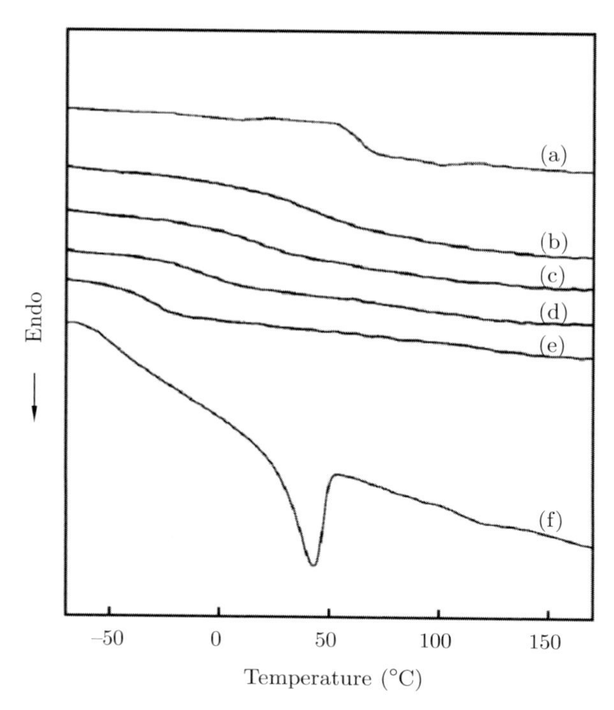

Figure 23. DSC thermograms of PVC/2TPU4 blends: 10/0 (a), 8/2 (b), 6/4 (c), 4/6 (d), 2/8 (e), and 0/10 (f) by wt [44]

Table 12. Thermal properties of PVC/TPU blends [44]

Sample	T_g (°C) of PVC/TPU blends at the weight ratio of					
	10/0	8/2	6/4	4/6	2/8	0/10
PVC blends with						
2TPU4	66.1	46.1	18.7	–7.3	–27.0	–40.7
2TPU6	66.1	47.7	44.2	7.7	–14.1	–47.5
4TPU4	66.1	42.8	17.4	–18.1	–34.8	–37.8
4TPU6	66.1	55.2	48.3	7.4	–13.9	–47.0

instantaneous recovery of strain to ε_u took place. The sample was subsequently heated to T_g+30°C within 5 min, and held at this temperature for the next 10 min, thus allowing a recovery of the strain and completing the first thermomechanical cycle ($N = 1$). The second cycle ($N = 2$) started at a residual strain, ε_p. Figure 25 shows that PVC deforms at a low stress of less than 2 MPa, and that hysterisis develops, *i.e.*, ε_p increases and the stress causing deformation decreases as the thermomechanical cycles are repeated. This shows that the physical entanglement of PVC chains is not rigid enough to sustain its original

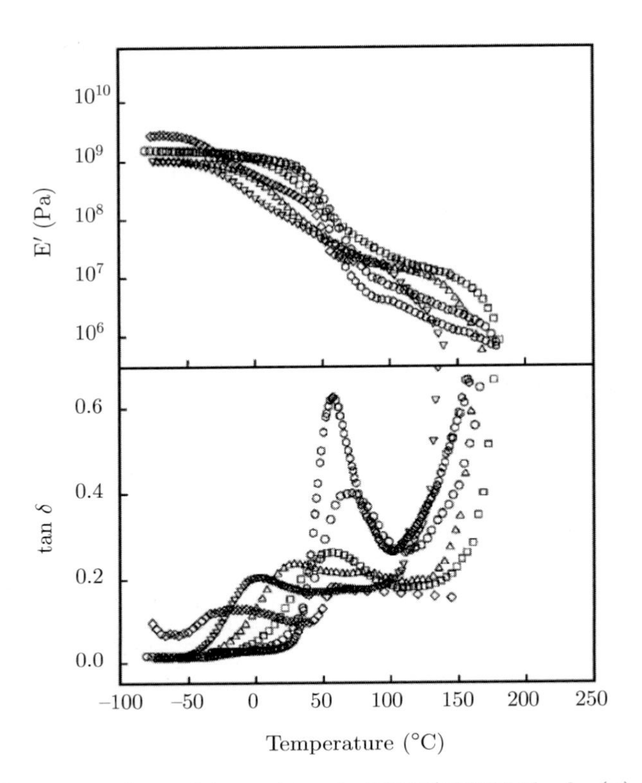

Figure 24. Storage tensile modulus and tan δ of PVC/2TPU4 blends: (⬠) 10/0, (□) 8/2, (○) 6/4, (△) 4/6, (▽) 2/8, and (◇) 0/10 by wt [44]

shape when deformation–recovery is repeated in the rubbery state. When PVC is blended with TPUs, the stress causing deformation increases. However, a high value of ε_p is observed even after the first cycle. This shows that when the glassy/rubbery modulus ratio is smaller than two orders of magnitude (Figure 24), the permanent deformation in the hard segment domains is liable to occur together with the recoverable deformation of the reversible phase of miscible PVC/PCL segments even in the rubbery state.

7.3. *Phenoxy/TPU blend*

TPUs were prepared from polycaprolactone diol (PCL, $M_n = 2000, 4000$), HDI, 1,4-BD, and DHBP. The formulations are given in Table 13, together with the feed ratios, hard segment contents, and intrinsic viscosities. The sample designation is the same as for the PVC/TPU blends described above. It was found that the PCL segments in TPU and phenoxy resin formed a miscible blend, as concluded from its single T_g, which varied smoothly with the relative content of PCL segments and phenoxy resin (Figure 26 and Table 14). A single T_g lying between the two T_g's of the pure components is a widely accepted criterion

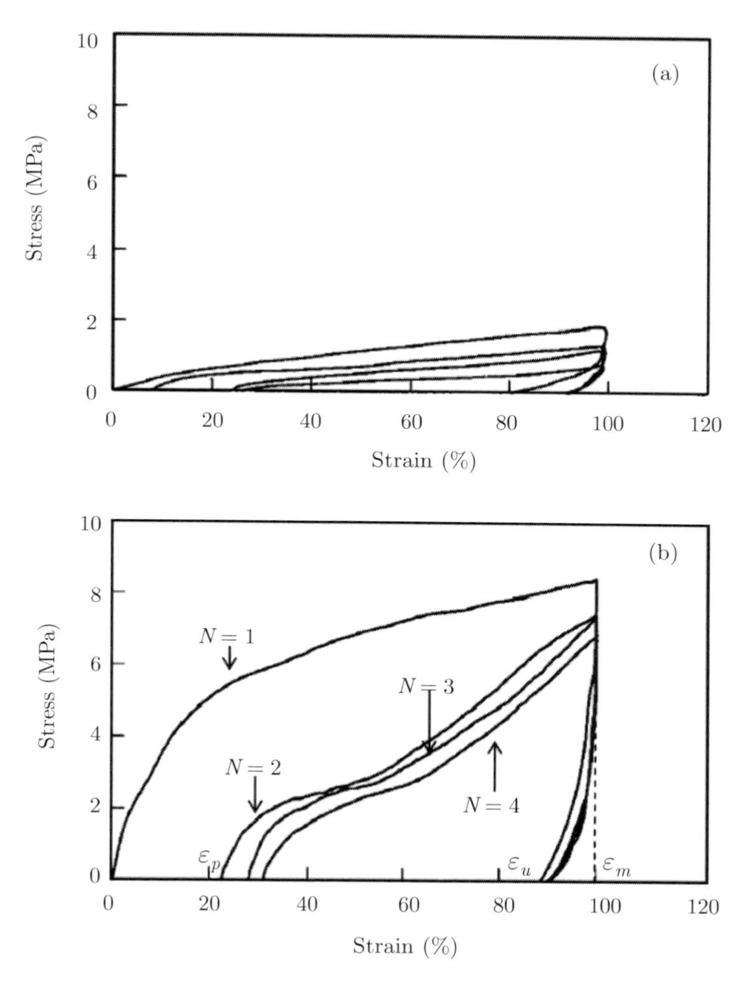

Figure 25. Cyclic tensile behavior of PVC (a), PVC/2TPU4 blend (8/2 by wt) (b) [44]

Table 13. Formulations of phenoxy/TPU blends [45]

Sample	Feed (moles)						Soft content in feed (wt%)	$[\eta]$ (dL/g)
	Soft segment		Hard segment					
	PCL2000	PCL4000	HDI	BD	DHBP	M_n		
2TPU6	1.0	–	5.2	3.5	0.7	1340	60	0.65
2TPU7	1.0	–	3.4	2.0	0.4	840	70	0.65
4TPU6	–	1.0	10.0	7.6	1.4	2630	60	0.65
4TPU7	–	1.0	6.6	4.7	0.9	1720	70	0.65

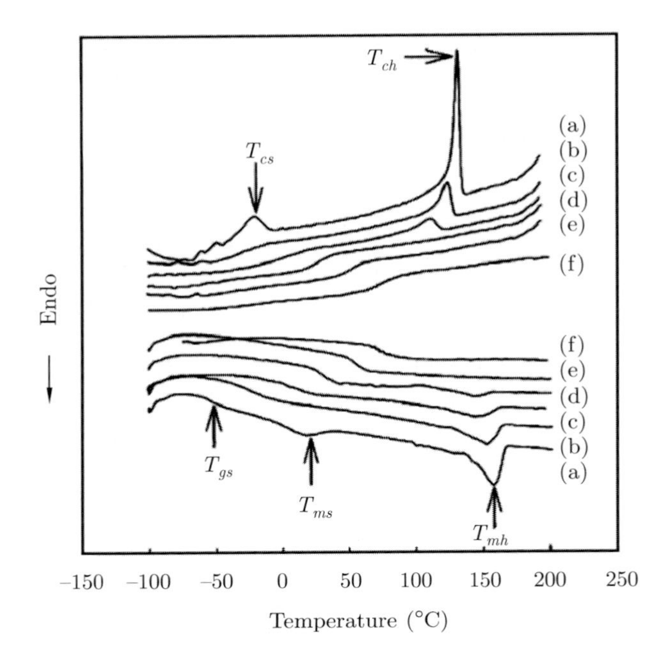

Figure 26. DSC thermograms of 2TPU6/phenoxy resin blends: 10/0 (a), 8/2 (b), 6/4 (c), 4/6 (d), 2/8 (e), and 0/10 (f) by wt [45]

of miscibility. Similar to PVC/TPU blends, crystallization and melting peaks of the PCL segments were not observed, indicating that the crystallization of PCL segments has been severely disturbed in the miscible blends. It was also noted that T_{mh} and ΔH_{mh} of pure TPU decreased in the blends. In crystalline polymers, the decreased free energy change results in decreased equilibrium melting temperature [50]. Even in the absence of equilibrium, the decrease in T_{mh} is an indication of the partial dissolution of PCL segments or phenoxy resin into the hard segment domains.

The phenoxy/TPU blends showed a shape memory effect when the miscible PCL segment/phenoxy resin domains formed the reversible phase and the hard segment domains played the role of the frozen phase. The shape recovery could be designed at any temperature. Shape retention and recovery over 90% were generally obtained (Table 15). The hysteresis in shape recovery decreased as the hard segment content in the blend increased and the block length of TPU decreased.

7.4. *Shape memory composites*

By hybridization or incorporation of inorganic materials in SMP, smart composites can be produced, combining the unique function or property of each component [51]. The morphology and mechanical properties of such composites have been reported recently [52]. The addition of conducting carbon black (CB)

Table 14. Thermal properties of phenoxy/TPU blends [45]: T_g – glass transition temperature; T_{cs} and T_{ms} – crystallization and melting temperatures of soft PCL segments; T_{ch} and T_{mh} – crystallization and melting temperatures of hard segments

Blend	Cooling scan				Heating scan				
(by wt)	T_{cs} (°C)	ΔH_{cs} (J/g)	T_{ch} (°C)	ΔH_{ch} (J/g)	T_g (°C)	T_{ms} (°C)	ΔH_{ms} (J/g)	T_{mh} (°C)	ΔH_{mh} (J/g)
				2TPU6/Phenoxy resin					
10/0	–21.3	8.2	132.0	23.9	–62.4	8.0	5.8	158.4	19.6
8/2	–	–	124.0	12.9	6.5	–	–	152.8	14.9
6/4	–	–	112.2	5.7	36.7	–	–	144.7	9.0
4/6	–	–	–	–	60.6	–	–	128.2	3.9
2/8	–	–	–	–	79.9	–	–	–	–
0/10	–	–	–	–	85.4	–	–	–	–
				2TPU7/Phenoxy resin					
10/0	–17.6	9.5	118.0	7.0	–56.5	22.6	11.8	143.3	13.0
8/2	–	–	104.6	3.3	–6.3	–	–	123.3	9.0
6/4	–	–	107.8	3.2	25.8	–	–	129.8	4.7
4/6	–	–	–	–	49.6	–	–	–	–
2/8	–	–	–	–	73.5	–	–	–	–
0/10	–	–	–	–	85.4	–	–	–	–
				4TPU6/Phenoxy resin					
10/0	3.1	21.3	137.7	22.8	–59.2	42.5	21.4	165.1	23.3
8/2	–	–	99.4	6.9	–1.5	–	–	142.3	11.3
6/4	–	–	93.8	2.5	30.9	–	–	137.6	7.3
4/6	–	–	–	–	53.6	–	–	143.5	3.8
2/8	–	–	–	–	70.6	–	–	–	–
0/10	–	–	–	–	85.4	–	–	–	–
				4TPU7/Phenoxy resin					
10/0	5.6	25.3	132.5	13.5	–57.4	44.4	24.7	159.9	13.4
8/2	–	–	105.4	2.8	8.4	–22.7	2.9	141.4	6.9
6/4	–	–	97.2	1.3	21.4	–	–	135.5	4.5
4/6	–	–	–	–	48.6	–	–	131.2	1.6
2/8	–	–	–	–	64.6	–	–	–	–
0/10	–	–	–	–	85.4	–	–	–	–

Table 15. Residual strain, ε_p, after a thermomechanical cyclic test for phenoxy/TPU blends [45]

Sample (wt ratio)	Temperature (°C)			ε_p (%) after		
	Drawn	Quenched	Recovered	$N=1$	$N=2$	$N=3$
		2TPU6/phenoxy resin				
6/4	50	25	50	9.4	12.6	15.5
4/6	75	50	75	9.8	13.5	15.8
		2TPU7/phenoxy resin				
6/4	40	15	40	13.9	19.5	21.1
4/6	65	40	65	14.2	20.3	21.4
		4TPU6/phenoxy resin				
6/4	45	20	45	14.2	17.0	17.7
4/6	70	45	70	16.5	22.1	22.7
		4TPU7/phenoxy resin				
6/4	35	10	35	15.4	19.9	22.4
4/6	65	40	65	20.3	21.4	26.8

to a TPU based on PCL/MDI/1,4-BD showed a combination of shape memory properties and electric conductivity [53]. The CB content above the percolation threshold has not influenced the crystallization and melting behavior of PCL soft segments, which is the basic requirement for shape memory behavior. However, the addition of CB increased the bulk viscosity, which resulted in a slow strain recovery. Hybridization of PTMG/MDI/1,4-BD based TPU with tetraethoxysilane (TEOS) by a sol-gel process also resulted in good shape retention and shape recovery over 80%, as well as in improved mechanical properties, with a maximum reinforcement at 10% TEOS [54]. A number of contributions before 1998 are available in an extensive review on shape memory hybrid composites [55].

8. Conclusions and outlook

The shape memory effect in polymers is based on a large difference in the glassy state modulus, E_G, and the rubbery state modulus, E_R. Thus, most of the deformations created in the rubbery state during shaping are frozen upon cooling to the glassy state and recovered in the rubbery state upon exposure to high temperature. Consequently, high shape retention in the glassy state and high retractive force in the rubbery state are of prime importance among the properties of SMP. A simple analysis based on elastic energy balance showed that high shape retention is obtained with a high glassy state modulus or high E_G/E_R ratio. Physically, this means that the shape is being retained in the glassy state where the slope of the cooling curve during a thermomechanical cycle is E_G. On the other hand, high retractive force in the rubbery state is essential for shape memory performance since the shape is recovered by the retractive forces acting between the frozen phases.

With crystalline soft segments, high shape retention was obtained with high soft segment content and high shape recovery with high hard segment content, which were caused by the high glassy state modulus and high rubbery state modulus, respectively. It has been observed that the glassy state modulus can be effectively controlled by the amount of crystallizable soft segments.

On the other hand, with amorphous soft segments, shape retention as well as shape recovery increased with increasing hard segment content due to the increased glassy state and rubbery state modulus. Among the two types of crosslinks, *i.e.*, allophanate crosslinks and trifunctional polyols, the frozen phase introduced by the trifuntional polyol showed a much higher shape retention over 99% and shape recovery over 97% depending on the crosslink density. This implies that the quantitative control of crosslink position and density can tailor the shape memory performance with high temperature sensitivity. It was noted that loading at higher temperatures gave poorer shape recovery, implying that a significant amounts of extended chains are relaxed during loading. Shape recovery in the rubbery state seems to follow the ideal rubber theory based on the fact that high recovery temperatures result in good shape recovery. This confirms that a shape memory effect can only be obtained with two-phase structures where the frozen phases are tied up by the reversible phases.

The shape memory behavior of an SMP can further be tailored by blending with a second polymer, which is miscible with the reversible phase of the SMP. Also, by hybridization or incorporation of inorganic materials in SMP, smart composites can be produced, combining the unique function or property of each component. Regarding the molecular design of SMPs, high rubbery state modulus as well as high E_G/E_R ratio polymers would greatly enhance their performance.

Acknowledgements

The authors express their sincere gratitude to Prof. H. M. Jeong (University of Ulsan) for his valuable discussion and help, to Prof. B. C. Chun (University of Suwon) and Prof. J. H. Cho (Konkuk University) for their excellent contributions to the field of SMP. The support from Pusan National University is also gratefully acknowledged.

References

1. Lendlein A and Kelch S (2002) Shape memory effect, *Angew Chem Int Ed* **41**:2034–2057.
2. Wei Z G, Sandstrom R and Miyazaki S (1998) Shape memory materials and hybrid composites for smart systems, Part 1 Shape memory materials, *J Mater Sci* **33**:3743–3762.

3. Takei S (1989) What is shape memory polymer, in *Development and Application of Shape Memory Polymers* (Ed. Irie M) CMC, Tokyo, pp. 11–80 (in Japanese).
4. Monkman G J (2000) Advances in shape memory polymer actuation, *Mechatronics* **10**:489–498.
5. Akaba N (1989) Application for blood vessel surgery, in *Development and Application of Shape Memory Polymers* (Ed. Irie M) CMC, Tokyo, pp. 111–117 (in Japanese).
6. Nakasima A (1989) Application for dental materials, in *Development and Application of Shape Memory Polymers* (Ed. Irie M) CMC, Tokyo, pp. 118–126 (in Japanese).
7. Ueno K and Ota S (1989) Application for cable coatings, in *Development and Application of Shape Memory Polymers* (Ed. Irie M) CMC, Tokyo, pp. 127–143 (in Japanese).
8. Tashiki S (1989) Application to a self-controlable heater, in *Development and Application of Shape Memory Polymers* (Ed. Irie M) CMC, Tokyo, pp. 144–152 (in Japanese).
9. Hayashi S, Kondo S, Kapadia P and Ushioda E (1995) Room-temperature-functional shape-memory polymers, *Plastics Eng,* pp. 29–31.
10. Jeong H M, Ahn B K and Kim B K (2000) Temperature sensitive water vapour permeability and shape memory effect of polyurethane with crystalline reversible phase and hydrophilic segments, *Polym Int* **49**:1714–1721.
11. Jeong H M, Cho S M, Ahn B K and Kim B K (2000) Water vapor permeability of shape memory polyurethanes with amorphous reversible phase, *J Polym Sci Polym Phys* **38**:3009–3017.
12. Cho J W, Jung Y C, Chun B C and Chung Y C (2004) Water vapor permeability and mechanical properties of fabrics coated with shape-memory polyurethane, *J Appl Polym Sci* **92**:2812–2816.
13. Tobushi H, Okumira K, Hayashi S and Ito N (2001) Thermomechanical constitutive model of shape memory polymer, *Mech Mater* **33**:545–554.
14. Lin J R and Chen I W (1999) Shape memorized crosslinked ester type polyurethane and its mechanical viscoelastic model, *J Appl Polym Sci* **73**:1305–1319.
15. Tobushi H, Hayashi S and Kojoma S (1992) Mechanical properties of shape memory polymer of polyurethane series, *JSME Int J* **35**:296–302.
16. Hayashi S, Kondo S and Kawamura K (1992) Structures and properties of shape memory polyurethane, *34th Ann Polyurethane Technical/Marketing Conf*, Dearborn, Michigan, pp. 21–24.
17. Hertzberg R W (1976) *Deformation and Fracture Mechanics of Engineering Materials*, Wiley, pp. 269–324.
18. McCrum N G, Buckley C P and Bucknall C B (1997) *Principles of Polymer Engineering*, Oxford Science, Guildford, pp. 84–116.
19. Smith R N (1985) Introduction to polyurethane chemistry, in *PMA's Reference Guide*, Polyurethane Manufacturers Association, Glen Ellyn, pp. 1–43.
20. Birley A W, Haworth B and Batchelor J (1991) *Physics of Plastics*, Hanser, New York, pp. 163–250.
21. Baird D G and Collias D I (1998) *Polymer Processing*, Wiley, New York, pp. 1–8.
22. Li F, Zhang X, Hou J, Xu M, Luo X, Ma D and Kim B K (1997) Studies on the thermally simulated shape memory effect of segmented polyurethane, *J Appl Polym Sci* **64**:1511–1516.
23. Kim B K, Shin Y J, Cho S M and Jeong H M (2000) Shape memory behavior of segmented polyurthanes with an amorphous reversible phase: The effect of block length and content, *J Appl Polym Sci* **38**:2652–2657.
24. Kim B K, Lee J S, Lee Y M, Shin J H and Paik S H(2001) Shape memory behavior of amorphous polyurethane *J Mater Sci* **B40**:1179–1181
25. Lee J S (1999) Molecular design, synthesis, and properties of high performance polyurethane, PhD Thesis, Pusan National University, Korea.

26. Li F, Hou J, Zhu W, Zhang X, Xu M, Lou X, Ma D and Kim B K (1996) Crystallinity and morphology of segmented polyurethanes with different soft segment lengths, *J Appl Polym Sci* **62**:631–638.
27. Lee B S, Chun B C, Chung Y C, Sul K I and Cho J W (2001) Structure and thermomechanical properties of polyurethane block copolymers with shape memory effect, *Macromolecules* **34**:6431–6437.
28. Park S H, Kim J W, Lee S H and Kim B K (2004) Temperature sensitive amorphous polyurethanes, *J Macromol Sci-Phys* **B43**:447–458.
29. Ahn T O, Kim C K, Kim B K, Jeong H M and Huh J D (1990) Binary blends of nylons with ethylene vinyl alcohol copolymers, *Polym Eng Sci* **30**:15–23.
30. Olabisi O, Robeson L M and Shaw M T (1979) *Polymer-Polymer Miscibility*, Academic Press, New York, pp. 277–279.
31. Morrison F A (2001) *Understanding Rheology*, Oxford University Press, New York, pp. 255–276.
32. Jeong H M, Lee J B, Lee S Y and Kim B K (2000) Shape memory polyurethane containing mesogenic moity, *J Mater Sci* **35**:279–283.
33. Jeong H M, Kim B K and Choi Y J (2000) Synthesis and properties of thermotropic liquid crystalline polyurethane elastomers, *Polymer* **41**:1849–1855.
34. Kim B K, Lee S Y and Xu M (1996) Polyurethane having shape memory effects, *Polymer* **37**:5781–5793.
35. Kim B K, Lee S Y, Baek S H, Choi Y J and Xu M (1998) Polyurethane inomers having shape memory effect, *Polymer* **39**:5949–5959.
36. Reyntjens W G, Du F E and Goethals E J (1999) Polymer networks containing crystallizable poly(octadecyl vinyl ether) segments for shape memory materials, *Macromol Chem Rapid Commun* **20**:251–255.
37. Lee H Y, Jeong H M, Lee J S and Kim B K (1999) Study on the shape memory polyamides. Synthesis and thermomechanical properties of polycaprolactone-polyamide block copolymer, *Polym J* **32**:23–28.
38. Park C, Lee J Y, Chun B C, Chung Y C and Cho J W (2004) Shape memory effect of poly(ethylene terephthalate) and poly(ethylene glycol) copolymer cross-linked with glycerol and sulfoisophthalate group and its application to impact-absorbing composite material, *J Appl Polym Sci* **94**:308–316.
39. Ahn T O, Choi I S, Lee S W and Jeong H M (1994) Thermal and mechanical properties of poly(ether urethane) modified by copolyamide segments, *Macromol Chem Phys* **195**:2559–2567.
40. Ahn T O, Oh M H, Yoo K S and Jeong H M (1995) Thermal and mechanical properties of poly(ester urethane) modified by copolyamide segments of various molecular weight, *Polym Int* **36**:239–245.
41. Skákalová V, Lukeš V and Breza M (1997) Shape memory effect of dehydrochlorinated crosslinked poly(vinyl chloride), *Macromol Chem Phys* **198**:3161–3172.
42. Seefried Jr C G, Koleske J V and Critchfield F E (1975) Thermoplastic urethane elastomers. I. Effects of soft-segment variations, *J Appl Polym Sci* **19**:2493–2502.
43. Kusy R P and Whitley J Q (1994) Thermal characterization of shape memory polymer blends for biomedical implanations, *Thermochim Acta* **243**:253–263.
44. Jeong H M, Song J H, Lee S Y and Kim B K (2001) Miscibility and shape memory property of poly(vinyl chloride)/thermoplastic polyurethane blends, *J Mater Sci* **36**:5457–5463.
45. Jeong H M, Ahn B K and Kim B K (2001) Miscibility and shape memory effect of thermoplastic polyurethane blends with phenoxy resin, *Eur Polym J* **37**:2245–2252.
46. Woo E M, Barlow J W and Paul D R (1985) Thermodynamics of the phase behaviour of poly(vinyl chloride)/aliphatic polyester blends, *Polymer* **26**:763–773.

47. de Juana R and Cortazar M (1993) Study of the melting and crystallization behavior of binary poly(ε-caprolactone)/poly(hydroxyl ether of bisphenol A) blends, *Macromolecules* **26**:1170–1176.
48. Xiao F, Shen D, Zhang X, Hu S and Xu M (1987) Studies on the morphology of blends of poly(vinyl chloride) and segmented polyurethanes, *Polymer* **28**:2335–2345.
49. Ahn T O, Han K T, Jeong H M and Lee S W (1992) Miscibility of thermoplastic polyurethane elastomers with chlorine-containing polymers, *Polymer* **33**:115–120.
50. Flory P J (1953) *Principles of Polymer Chemistry*, Cornell University Press, Ithaca.
51. Jeon H G, Mather P T and Haddad T S (2000) Shape memory and nanostructure in poly(norbornyl-poss) copolymers, *Polym Int* **49**:453–457.
52. Varghese S, Gatos K G, Apostlov A A and Karger-Kocsis J (2004) Morphology and mechanical properties of layered silicate reinforced natural and polyurethane rubber blends produced by latex compounding, *J Appl Polym Sci* **92**:543–551.
53. Li F, Qi L, Yang J, Xu M, Luo X and Ma D (1999) Polyurethane/conducting carbon black composite: structure, electric conductivity, strain recovery behavior, and their relationships, *J Appl Polym Sci* **75**:68–77.
54. Cho J W and Lee S H (2004) Influence of silica on shape memory effect and mechanical properties of polyurethane-silica hybrids, *Eur Polym J* **26**:1–6.
55. Wei Z G, Sandstrom R and Miyazaki S (1998) Shape memory materials and hybrid composites for smart systems, Part 2. Shape memory hybrid composites, *J Mater Sci* **33**:3763–3783.www

Chapter 19

Condensation and Addition Thermoplastic Elastomers: Recycling Aspects

T. Spychaj, M. Kacperski, A. Kozlowska

1. Recycling opportunities for step-growth polymers

Step-growth polymers can be recycled by numerous methods, which can be divided into physical and chemical (and sometimes also physicochemical). Material recycling *via* melt processing is a simple way of the reuse of the first group; a wide range of chemical recycling options may be also realized, resulting in feedstocks for virgin or other polymer/resin syntheses, as well as in additives for various plastics [1]. Other common methods of polymer recycling, *e.g.*, energy recovery, are of minor importance in relation to polycondensation and polyaddition polymers (provided that particular types of polymers are to be collected separately).

Thermoplastic elastomers (TPEs) are block copolymers with macromolecules containing stiff (rigid) parts with relatively high melting, T_m, and glass transition, T_g, temperatures, and elastic (soft) parts characterized in that their T_m and T_g values are much lower than the polymer normal service temperatures [2,3]. These rigid and soft parts of the macromolecules form phase-separated domains, which determine an intrinsic feature of thermoplastic elastomers, *i.e.*, a broad temperature plateau of the strain-stress relationship.

In order to define the scope of this chapter as involving the recycling of step-growth TPEs, the following issues should be mentioned here: (i) the literature concerning the reuse and recycling of block polycondensation/polyaddition copolymers is rather limited, (ii) the description of the relevant recycling methods is most often based on the ester and amide (homo)polymers or polyurethanes, (iii) material recycling and melt blending are recommended for TPEs as the primary recycling methods, whereas feedstock/chemical options complement them as an offer for step-growth homopolymers, (iv) a higher number of specific recycling techniques can be used for waste polyurethanes.

It should be also mentioned that thermoplastic block copolymers composed of chemically different blocks can be considered as potential compatibilizers for the production of some polymer blends, including those of reactive polymers.

Unlike commodity plastics, the upgrading and recycling of TPEs (*e.g.*, poly(ethylene terephthalate) (PET), poly(butylene terephthalate) (PBT), polyamides (PA), polyurethanes (PUR), as well as the copolymers with ester, amide and urethane blocks) or polycarbonates is economically attractive. A reason for this is the beneficial relationship between material costs and the retained performance value after recycling (Figure 1).

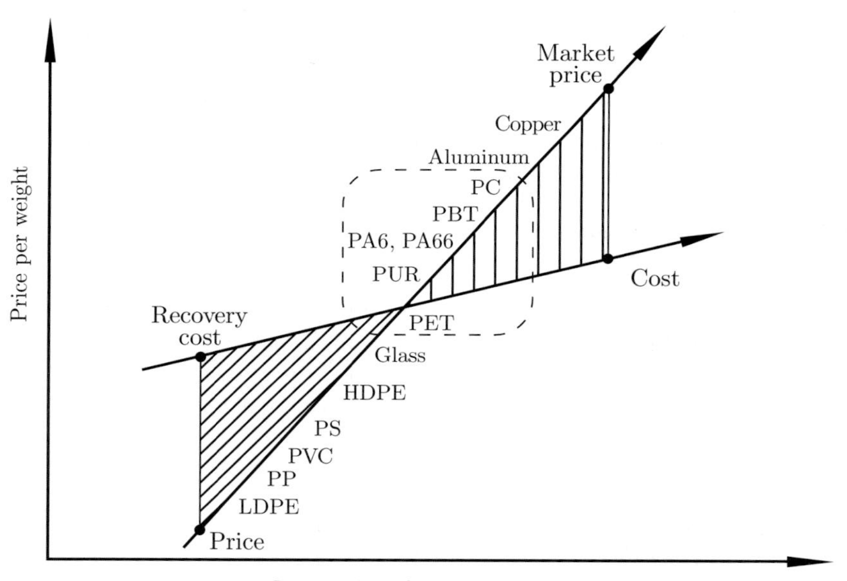

Figure 1. Material costs *vs.* retained performance value after recycling (adapted from [4]): ▨ – recycling cost exceeds the market value of the material; ▯▯ – recycling is economically beneficial

Step-growth (co)polymers bearing ester or amide bonds are synthesized *via* reversible reactions (see also Chapters 2, 3, and 10). It is relatively easy to convert them back to their monomers or oligomers (or value-added chemicals) by solvolytic reactions, such as hydrolysis, alcoholysis (including glycolysis), and other transesterification reactions, as well as by ammonolysis, aminolysis, acidolysis, or transamidation [1].

$$\underset{\text{Diacid}}{HO-\underset{\underset{O}{\|}}{C}-R-\underset{\underset{O}{\|}}{C}-OH} + \underset{\text{Glycol or diamine}}{HZ-(CH_2)_2-ZH} \underset{\text{Solvolysis}}{\overset{\text{Polymerization}}{\rightleftharpoons}}$$

$$\underset{\text{Polycondensation polymer}}{HO\left[\underset{\underset{O}{\|}}{C}-R-\underset{\underset{O}{\|}}{C}-Z-(CH_2)_2-Z\right]_n H} + (2n-1)\ H_2O \qquad (1)$$

where Z is O– in polyesterification and HN– in polyamidation; R is alkylene or arylene

Polyurethanes are formed by irreversible, nonequilibrium reactions of diols and diisocyanates.

$$OCN-R-NCO + HO\sim\sim OH \rightarrow OCN\left[R-NH\underset{\underset{O}{\|}}{C}O\right]_n\sim\sim OH \qquad (2)$$

The solvolytic depolymerization of polyurethanes (PUR) can also be performed, mainly by hydrolysis, alcoholysis/glycolysis or ammonolysis/aminolysis processes (the chemical recycling of PUR is discussed in Section 3.3.5).

2. Cleavage of ester, amide and urethane bonds

The solvolytic reactions of the considered polymers containing heteroatoms in their backbone consist in the cleavage of C–O or C–N bonds of the chain. The polymer chains are degraded according to Schemes 3 to 5, respectively.

$$\underset{\text{Polyester}}{-\underset{\underset{O}{\|}}{C}-O-\overset{|}{\underset{|}{C}}-} + HZ \rightarrow -\underset{\underset{O}{\|}}{C}-Z- + HO-\overset{|}{\underset{|}{C}}- \qquad (3)$$

$$\underset{\text{Polyamide}}{-NH-\underset{\underset{O}{\|}}{C}-\overset{|}{\underset{|}{C}}-} + HZ \rightarrow -NH_2 + -Z-\underset{\underset{O}{\|}}{C}-\overset{|}{\underset{|}{C}}- \qquad (4)$$

$$\underset{\text{Polyurethane}}{-NH-\underset{\underset{O}{\|}}{C}-O-\overset{|}{\underset{|}{C}}-} + HZ \rightarrow -NH-\underset{\underset{O}{\|}}{C}-Z- + HO-\overset{|}{\underset{|}{C}}- \qquad (5)$$

$$-NH-\underset{\underset{O}{\|}}{C}-Z- \xrightarrow{-\ CO_2\ (\text{if } Z = -OH)} -NH_2$$

where Z is –OH (water), –OR or –OROH (monoalcohol or glycol), –OCOR (acid), –OMe (alkali), $-NH_2$, –NHR or $-NR_2$ (ammonia, amine (primary or secondary))

From the point of view of the application of ester, amide, and urethane polymers, hydrolysis, alcoholysis/glycolysis and ammonolysis/aminolysis reactions are the most important. These processes are often used for recycling to feedstocks and offer raw materials, mainly for virgin polymer/resin syntheses. The schemes of the respective reactions with brief comments on the practical importance of the particular processes are presented in Figures 2 to 4. A description of the important solvolytic degradation processes for ester, amide, and urethane polymers is also given in this chapter.

It should be mentioned here that the practical use of the solvolytic methods for the recycling of TPE block copolymers (*i.e.*, composed of polyester and/or polyamide and/or polyurethane blocks) is quite limited. A reason for this is the rather complex composition of the reaction products, limiting the potential field of their application. On the basis of Figures 2 to 4, it may be concluded that only glycolysis of these copolymers seems to offer products with a clear destination — for polyurethane synthesis.

3. Recycling of thermoplastic elastomers

3.1. *Recycling of polyester-based (co)polymers*

Poly(ethylene terephthalate) is the most important polyester; *ca.* 95% of the total polyester consumption belongs to PET [14]. The annual world consumption of PET is higher than 25 million tons, with a further increasing tendency of *ca.* 10% per year. This is an important factor determining the position of polyesters among other polymers.

The characterization of the recycling methods as related to (co)polyesters is focused here on physical or physicochemical processes (*i.e.*, the processing basically does not involve chain cleavage) and melt blending (including reactive melt blending). Solvolytic degradation methods of recycling to feedstocks are briefly described, with emphasis on the latest approaches, because the state of the art of PET chemical recycling has been recently presented in [1,14].

3.1.1. *Physical and physicochemical methods*

The immiscible blends of thermoplastic polymers are assumed to be compatible if the presence of a finely dispersed phase and the resistance to bulk phase separation/segregation result in satisfactory physical and mechanical properties. Compatibility is usually facilitated by copolymers, especially block or graft copolymers, with segments capable of specific interactions at the interface, thus resulting in a decrease of the interfacial tension and promoting mechanical interlocking by the interpenetration and entanglements between the blend components.

TPE (co)polymers from step-growth polymerization processes can be used as compatibilizers for melt blending of the respective polymers, *e.g.*, polyesters with polyamides or polycarbonate. The interfacial chemical reactions between

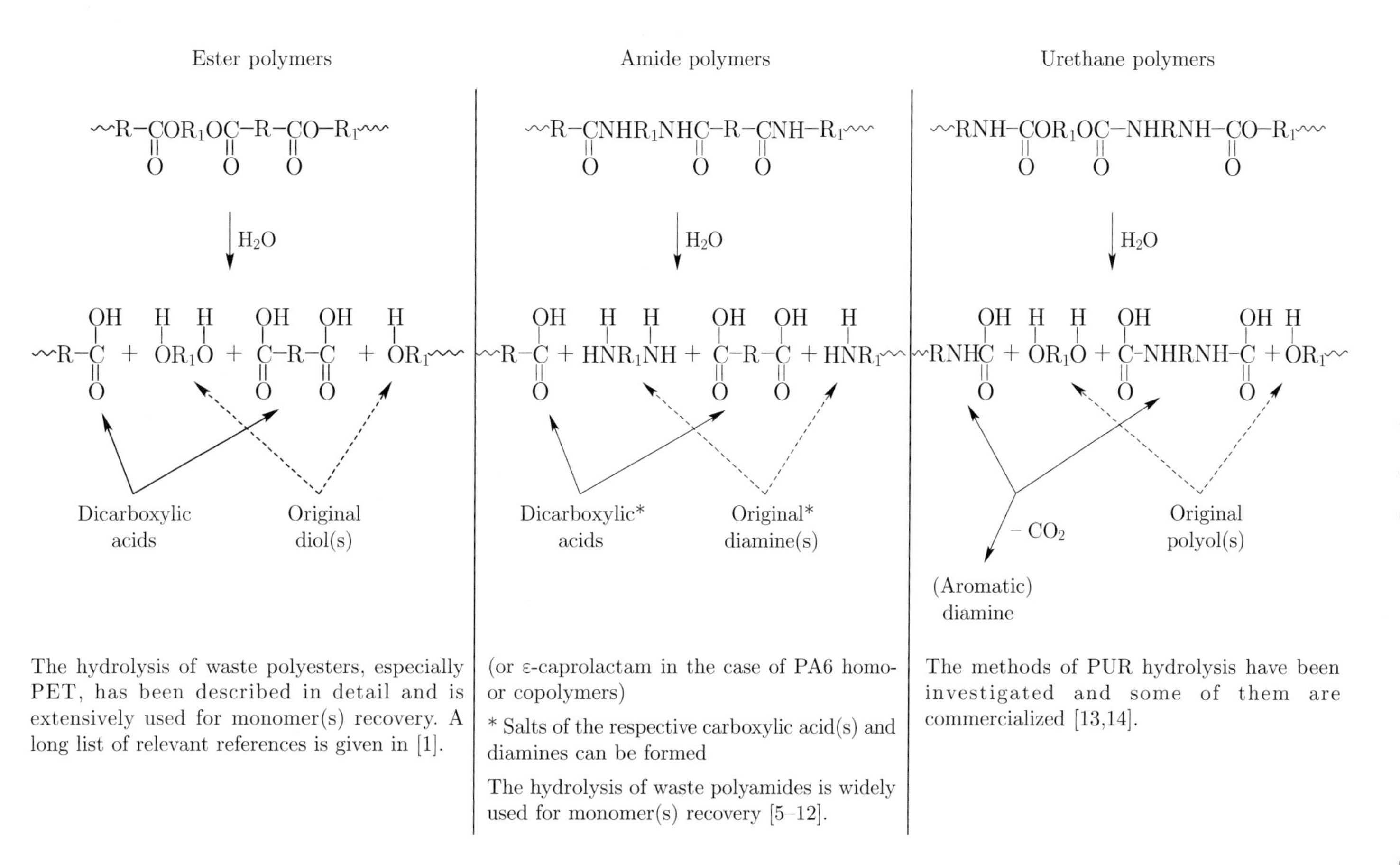

Figure 2. Hydrolysis reactions

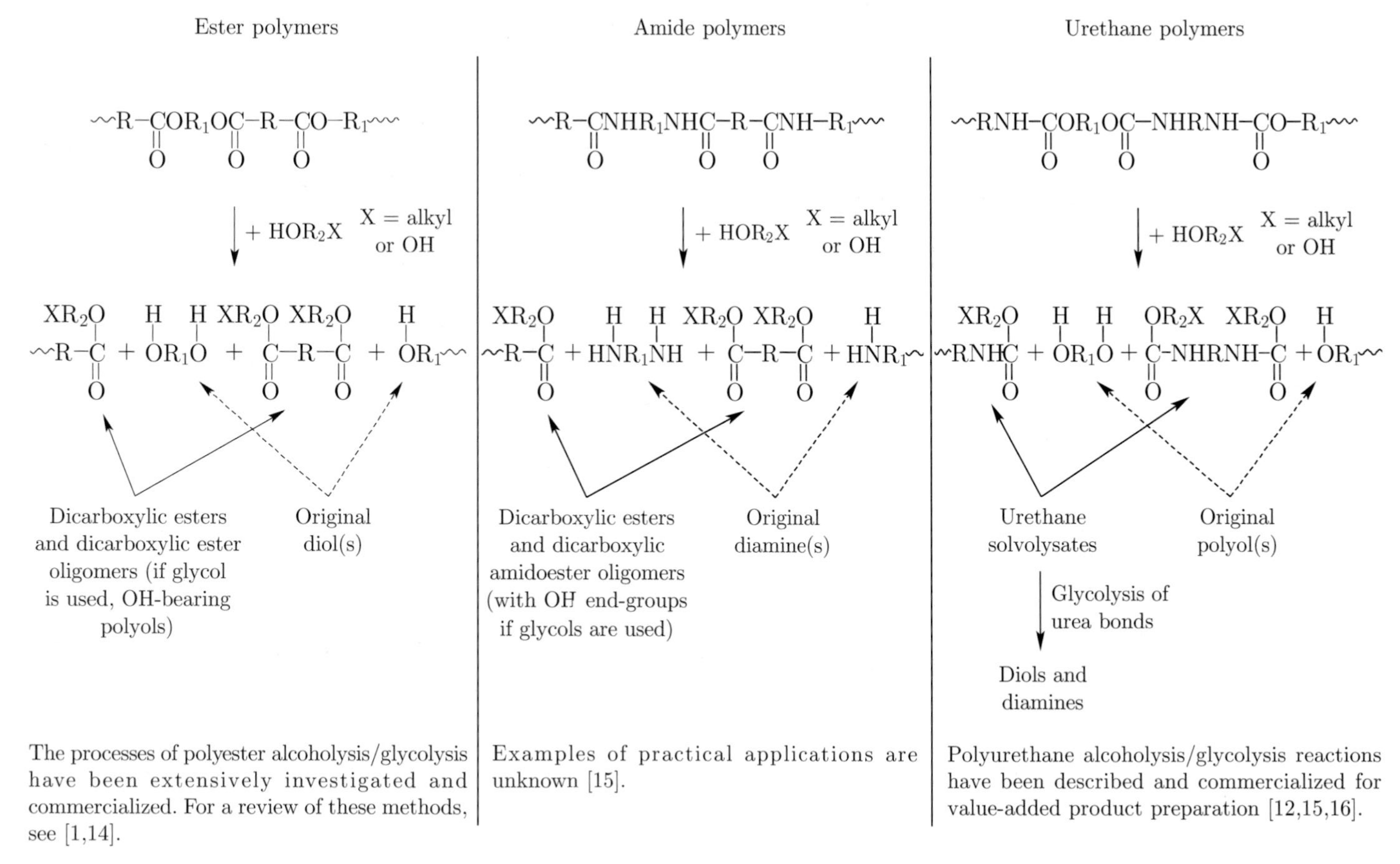

Figure 3. Alcoholysis/glycolysis reactions

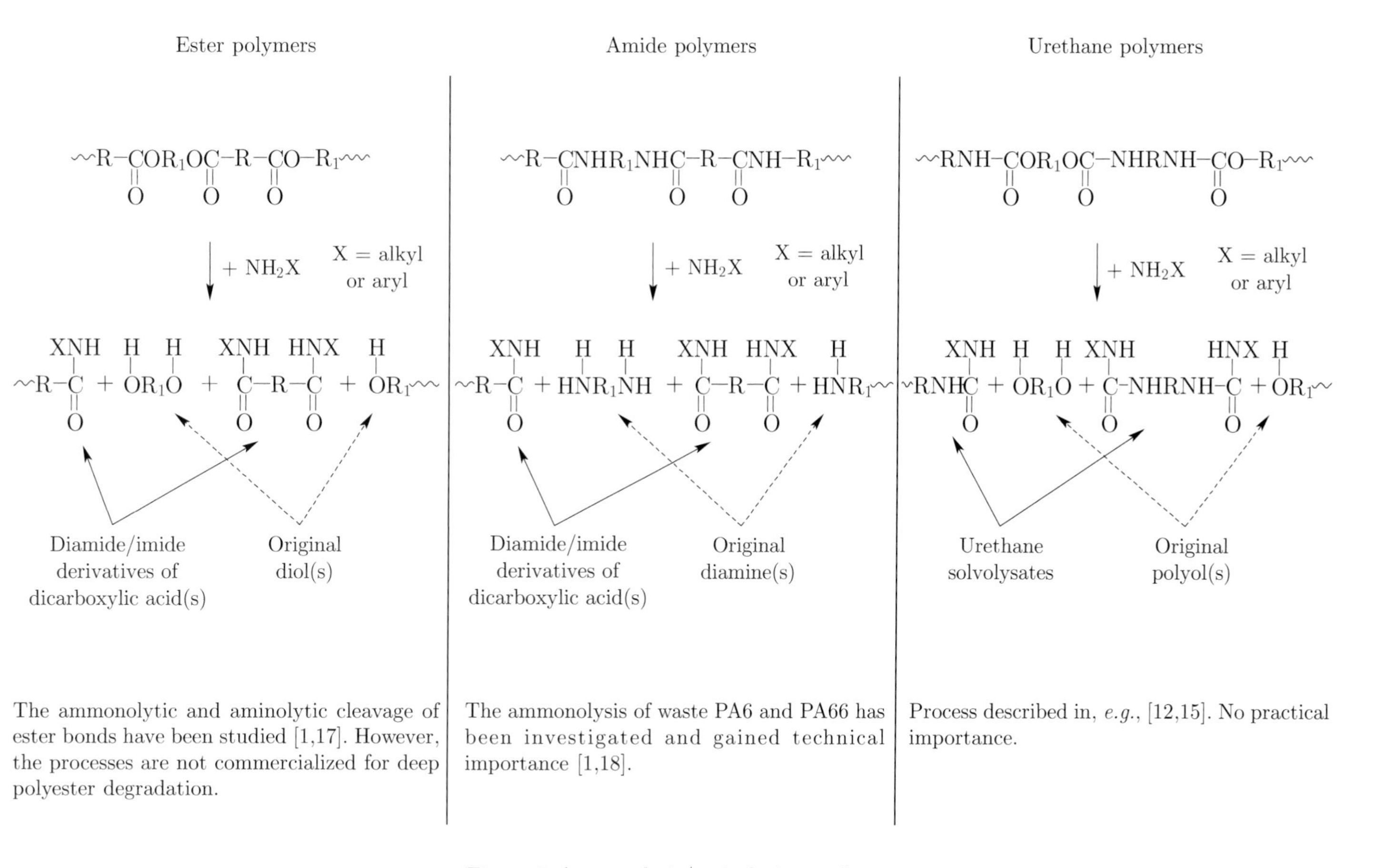

Figure 4. Ammonolysis/aminolysis reactions

the functional groups of the melt blended polymers may be practically employed to obtain polymer blends with useful mechanical properties. Table 1 lists the reactions of the functional groups occurring in polyester, polyamide, and polyurethane chains or blocks.

Table 1. Reactive melt blending of ester, amide, and urethane linear polymers

Polymer end functional groups	Polymers and possible reactions of the functional groups
	Polyesters
	1. Hydroxyl (PET) + anhydride or acid (modifier)
	2. Carboxyl (PET) + epoxy (modifier)
	3. Carboxyl (PET) + oxazoline (modifier)
HO~~~COOH	4. Carboxyl/hydroxyl (PET) + isocyanate (modifier)
	5. Carboxyl (PET) + nitrile (modifier)
	6. Transesterification reactions (between PET and usually glycol type modifier)
	Polyamides
	1. Amine (PA) + anhydride or acid (modifier)
	2. Amine/carboxyl (PA) + epoxy (modifier)
	3. Amine/carboxyl (PA) + isocyanate (modifier)
H_2N~~~COOH	4. Carboxyl (PA) + oxazoline (modifier)
	5. Carboxyl (PA) + nitrile (modifier)
	6. Transamidation reactions (between PA and usually diamine type modifier)
	Polyurethanes
	1. Hydroxyl (PUR) + anhydride or acid (modifier)
	2. Hydroxyl (PUR) + isocyanate (modifier)
HO~~~NCO	3. Isocyanate (PUR) + hydroxyl (modifier)
	4. Isocyanate (PUR) + epoxy (modifier)
	5. Isocyanate (PUR) + amine (modifier)

Since step-growth TPEs have functional groups (*i.e.*, carboxyl, hydroxyl, amine, or isocyanate) at the chain ends, *in situ* or reactive compatibilization is an effective approach for developing engineering thermoplastic blends from these materials. A reactive coupler, such as an epoxide moiety, is sometimes used to compatibilize the blend components [19].

Blending of thermoplastic polyesters with themselves or with other polymers is a significant activity of some polymer/resin suppliers as well as consumers, such as car manufacturers. The polymers blended with polyesters in order to prepare high performance blends include other polyesters, crystalline engineering thermoplastics (*e.g.*, polyamides) and amorphous engineering thermoplastics (*e.g.*, polycarbonate (PC)). A variety of such blends are commercially available to provide competitive products advantageous to the end users (*e.g.*, PET/PBT, PET/PA6, PET/PA66, PBT/PA6, PBT/PA66, PC/PET, PBT/PC; see also [19]). A considerable growth of demand is predicted for the polyester blends during this decade because of the growing market for traditional and new polyester blends and the impact of recycling issues caused by environmental regulations and society expectations (*e.g.*, by EU Directive on "end-of-life" vehicles).

A more detailed description of some aspects of polyester blends may be found in the excellent chapters of Nadkarni and Rath [19], Xanthos [20], and Karger-Kocsis [21] in the handbook edited by Fakirov (for reference data follow, *e.g.*, [1]) as well as in the book of Scheirs (for reference data follow, *e.g.*, [4]).

3.1.2. *Polyester solvolysis: the latest approaches*

The solvolysis of waste polyesters, especially PET, has been a subject of permanent interest from both the scientific and the applied viewpoint. As it was already mentioned, a brief description of the latest polyester solvolytic processes will be presented as a supplement to the recent review chapter [1].

Farahat and Nikles [22,23] described a method of synthesis of acrylic coating resins for magnetic tapes starting from a PET glycolysate. The product of PET glycolysis with diethylene glycol served as a substrate for UV-curable solventless acrylic oligoesters. A basic catalyst and acryloyl or methacryloyl chloride were used in the second stage of the resins' syntheses [22,23].

The aminoglycolysis of PET was developed in an interesting way. The PET aminoglycolysates with triethanolamine were proposed as polyol components for rigid polyurethane foam synthesis [24] or epoxy resin hardeners [25,26]. The two papers concerning PET aminoglycolysis and the product applications appeared very recently [27,28]. Atta [27] has investigated diethanolamine and triethanolamine as solvolytic agents for PET. In a second step, the obtained aminoglycolysates were applied as raw materials for solid multifunctional epoxy resin syntheses *via* a reaction with epichlorohydrine. Epoxy resins based on PET aminoglycolysates hardened with *p*-phenylenediamine or diaminediphenylether were tested as coating materials on blasted carbon steel panels. It was found that the synthesized epoxy resins are thermally stable up to 470°C, whereas cured coatings show a great corrosion resistance (646 h) in salt spray tests. It is worth noting, on the basis of Atta's [27] and our research results [25,26,28] that the PET aminoglycolysate with triethanolamine can play a double role: a raw material for epoxy resin synthesis and a resin hardener by itself. It was also found that in addition to the PET/triethanolamine chemical degradation products, some other aminoglycolysates can also serve as hardeners for epoxy resins at elevated temperatures [28]. In the latter study, it was shown that the molar volume of the nitrogen-containing substitutent(s) is the key factor determining the access of OH groups to the ester bonds of the polymer backbone and the catalytic influence of tertiary nitrogen atom(s) on the C–O bond cleavage. This factor also influences the product reactivity towards epoxy resins. Among eleven tested PET/tertiary alkanolamine aminoglycolysates, the most active and appropriate epoxy resin hardeners are the terephthalic derivatives of N-methyl and N-ethyl diethanolamine, triethanolamine and bis(2-hydroxyethyl)piperazine.

Another polyester type, *i.e.*, polycarbonate, is rather seldom solvolytically depolymerized because of losing (at least partially) carbonate groups/bonds obtained *via* phosgene route synthesis. Troev *et al.* [29] recently described PC chemical degradation with dialkyl phosphonates (dimethyl or diethyl) or triethyl phosphate. The products, oligomeric carbonates containing phosphorus atoms, can be considered as precursors for the modification of various polymers by improving their flame-retardant properties, thermal stability, and adhesion.

Simultaneous dual solvolytic processes of PET with potential technical importance have been developed very recently [30,31]. Guclu *et al.* [30] have described simultaneous glycolysis and neutral hydrolysis of PET waste in the presence of xylene. The organic solvent enables the use of lower temperature and pressure, in contrast to the known hydrolysis methods yielding the intermediates suitable for PET or other polymeric materials. A water soluble crystallizable fraction with high purity, consisting of mono-2-hydroxyethyl ester of terephthalic acid (monohydroxy terephthalate, MMT) monomer has been obtained with a significant yield. It was found by multiple heating/cooling cycles in the differential scanning calorimeter that MMT is supperior to the conventional bis(2-hydroxyethyl)terephthalate (BHET) substrate in the polymer formation due to the catalytic effect of the carboxyl groups [30]. Spychaj *et al.* [31] have developed another simultaneous solvolytic process consisting of aminolysis and alkaline hydrolysis of PET waste. The process is performed in an autoclave and provides equimolar salts of terephthalic acid and the respective diamine (Scheme 6).

$$\left.\begin{array}{c} HO\!\left[\!-\overset{\displaystyle O}{\overset{\|}{C}}-C_6H_4-\overset{\displaystyle O}{\overset{\|}{C}}-OCH_2CH_2O-\right]_n\!H \\ HO\!-\!H \\ \downarrow \\ H_2N(CH_2)_6NH_2 \end{array}\right\} \xrightarrow[-n\ HOCH_2CH_2OH]{} \begin{array}{c} {}^{\ominus}O-\overset{\displaystyle O}{\overset{\|}{C}}-C_6H_4-\overset{\displaystyle O}{\overset{\|}{C}}-O^{\ominus} \\ \overset{\oplus}{H_3N}-(CH_2)_6-\overset{\oplus}{NH_3} \\ \text{6T nylon salt} \end{array} \qquad (6)$$

As a result of accelerated hydrolysis in a slightly alkaline aqueous solution of diamine followed by an instantaneous neutralization of the terephthalic acid formed while maintaining the molar ratio of PET repeat units/diamine = 1:1, the equimolar salt is formed (nylon salt). The nylon salts of terephthalic acid prepared *via* PET aminohydrolysis are obtained in an elegant way and can be applied for copolyamide syntheses or polyamide modification [31,32].

3.2. *Recycling of polyamide-based (co)polymers*

Polyamides and copolymers with amide groups are widely used, most often for fibers and injection molded plastics. However, special methods of copolyamide recycling have not been developed so far. The routes to their reclamation are analogous to polyamide recycling. The most important and also the largest source of polyamides utilized for reprocessing (mainly PA6 and PA66) is waste carpet covering. The consumption of polyamides for the manufacture of carpets and carpet covering constitutes 30–40% of their total global production. The waste carpets together with municipal waste are usually disposed into landfills. With regard to their structure and high resistance to biodegradation, they constitute a serious ecological problem [18,33–36].

A scheme of a typical carpet layout is given in Figure 5 [37]. The multilayer carpet structure has the following composition: a layer of face fibers (most often PA6, PA66, rarely PET, PAN, or wool), two textile layers (carrier and underlay

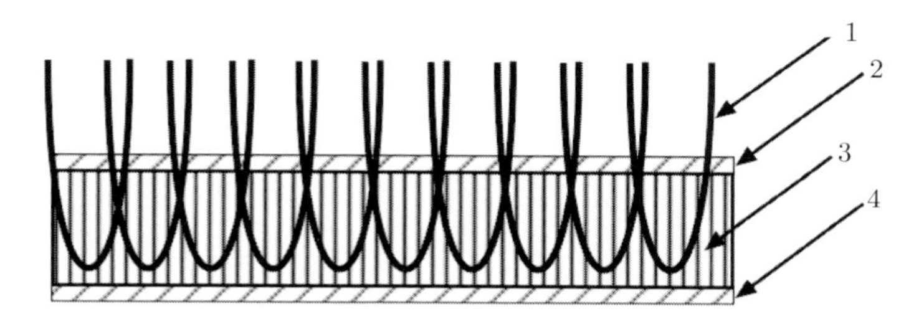

Figure 5. Scheme of a typical carpet floor covering layout: 1 – layer of face fibers; 2 – textile carrier layer; 3 – adhesive layer (latex) with fillers; 4 – textile underlay (cushioning), backing [37]

(PP, sometimes PET, jute, cellulosic fibers)), and a latex gluing layer (rubbers of the SBR type with fillers, most often $CaCO_3$) [18]. A rough composition of the different layers in a typical carpet covering is shown in Table 2 [18].

The first step in the recycling of used carpets and carpet coverings is the sorting and the identification of fibers. The procedure for the recovery of polyamides is represented schematically in Figure 6, whereas the methods of

Table 2. Composition of typical carpet floor covering [18,37]

Structural components of floor covering	Material	Fraction (wt%)
Face layer	Polyamide fibers	46–49
Carrier and backing layer textile	Polypropylene fibers	10–11
Gluing layer with filler	Latex + $CaCO_3$	40–44

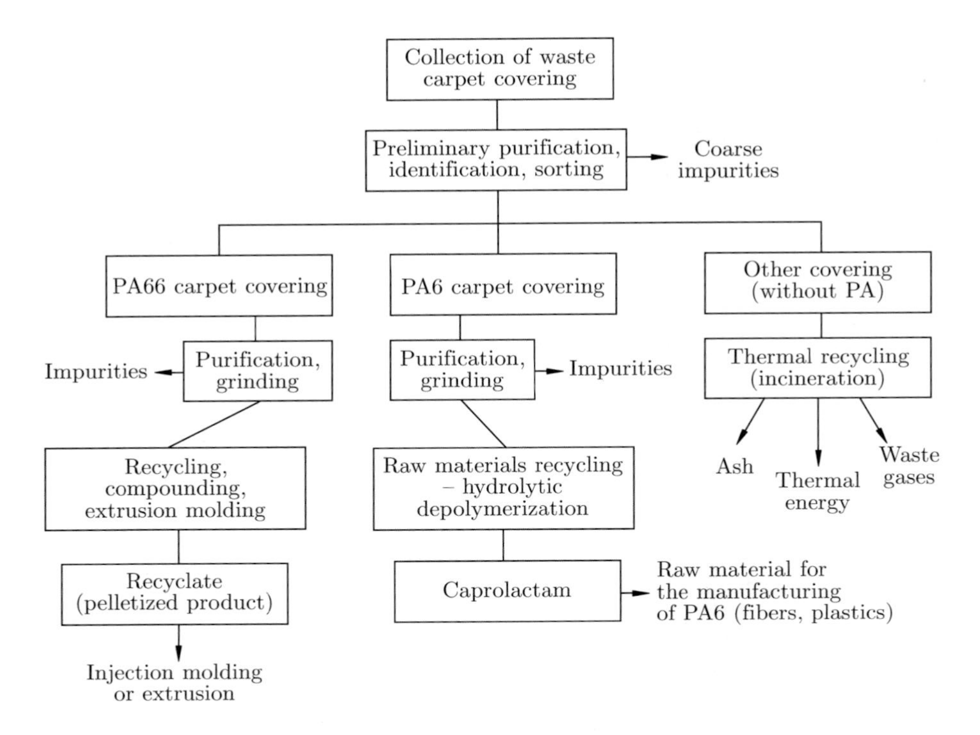

Figure 6. Schematic representation of the procedure for polyamide recovery

identification of the polymeric fibers used for the manufacture of carpets are shown in Figure 7 [37]. With regard to the analysis, price, and time, the identification is performed most often by melting point, solubility, and density determinations, as well as by near infrared spectoscopy (NIR) [18].

An essential method of PA6 and PA66 separation is based on selective dissolution [38,39]. An aliphatic carboxylic acid aqueous solution (*e.g.*, 85% solution of formic acid) is used as a selective solvent of PA6. Precipitation of PA6 takes place after the addition of water. Another method is based on the dissolution of both polyamides followed by a selective precipitation of PA6 or PA66.

The three types of recycling methods of waste products and used carpet covering containing polyamide fibers are chemical or feedstock recycling, material (mechanical) recycling, and energy recovery.

3.2.1. *Chemical/feedstock recycling*

Among the various modes of chemical recycling of the amide polymers, hydrolysis, ammonolysis, and depolymerization under reduced pressure are most important.

The hydrolysis process is the major method of PA6 recycling. It was first patented by Allied Chemical Co. in 1965 [5], and then improved and further

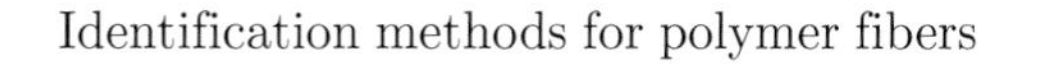

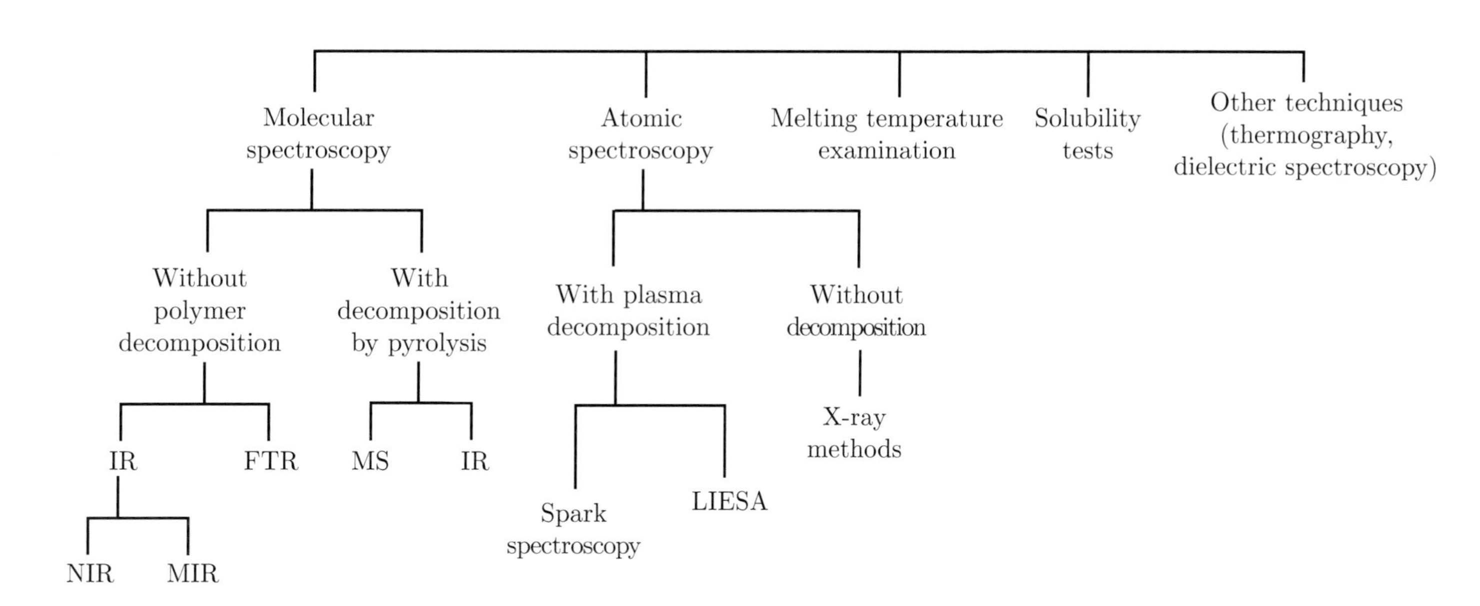

Figure 7. Scheme of the identification methods of polymeric fibers in covering carpets: NIR – near infrared spectroscopy; MIR – middle infrared spectroscopy (14 000÷4000 cm^{-1}); FTR – Fourier-transform Raman spectroscopy (4000÷200 cm^{-1}), MS – mass spectroscopy, LIESA – laser-induced emission atomic spectroscopy [37]

developed [6–8]. PA6 is hydrolytically depolymerized to ε-caprolactam according to the Schemes 7–9.

$$\sim\!\!\sim\!\!\sim \underset{\mathrm{H}}{\underset{|}{\mathrm{N}}}-\underset{\mathrm{O}}{\underset{\|}{\mathrm{C}}}-(\mathrm{CH_2})_5-\mathrm{NH_2} + \mathrm{H_2O} \rightleftharpoons \sim\!\!\sim\!\!\sim \mathrm{NH_2} + \mathrm{HO}-\underset{\mathrm{O}}{\underset{\|}{\mathrm{C}}}-(\mathrm{CH_2})_5-\mathrm{NH_2} \quad (7)$$

$$\sim\!\!\sim\!\!\sim \underset{\mathrm{H}}{\underset{|}{\mathrm{N}}}-\underset{\mathrm{O}}{\underset{\|}{\mathrm{C}}}-(\mathrm{CH_2})_5-\mathrm{COOH} + \mathrm{H_2O} \rightleftharpoons \sim\!\!\sim\!\!\sim \mathrm{NH_2} + \mathrm{HOOC}-(\mathrm{CH_2})_5-\mathrm{COOH} \quad (8)$$

$$\mathrm{HO}-\underset{\mathrm{O}}{\underset{\|}{\mathrm{C}}}-(\mathrm{CH_2})_5-\mathrm{NH_2} \rightleftharpoons \text{caprolactam (ring C=O, N–H)} + \mathrm{H_2O} \quad (9)$$

The first step of the process takes place in superheated steam at a temperature of 175–180°C and pressure of 963–997 kPa, whereas a temperature of 350°C and a pressure of 790 kPa are used in the second step. The process yield amounts to about 94% [5].

The hydrolysis in acidic medium was patented by BASF in the second half of the 1990s [9,10]. The inorganic acids used are most often orthophosphoric and boric acid, whereas formic and benzoic acid are among the organic ones. The acid hydrolysis of PA6 takes place in superheated steam at 250–350°C (Scheme 10).

$$\sim\!\!\sim \underset{\mathrm{O}}{\underset{\|}{\mathrm{C}}}-\mathrm{N}(\ldots\mathrm{HOOC}) \longrightarrow \text{caprolactam (ring C=O, N–H)} + \sim\!\!\sim\!\!\sim \mathrm{COOH} \quad (10)$$

Regardless of the process simplicity, its application is restricted for the following reasons: (i) the fillers or reinforcing fibers in the polymers may react with the acid and decrease the process yield and (ii) the disposal of by-products and waste water is associated with high costs. These restrictions can be avoided by a precise separation of fillers and fibers from the polymers [7,8,18].

In 1994, DuPont [18] patented the ammonolysis process, in which PA6 and PA66 react with ammonia in the presence of a catalyst (ammonium phosphate) and as a result the corresponding monomers can be obtained. A high purity of the waste polyamides (98%) is required. The process conditions are 300–350°C and pressure of 68.9 MPa [18,40,41].

$$\sim\!\!\sim \underset{\mathrm{H}}{\underset{|}{\mathrm{N}}}-\mathrm{C}(\ldots\mathrm{H_2N}) \xrightarrow[\text{Phosphate catalyst}]{\mathrm{NH_3}} \text{caprolactam (ring C=O, N–H)} + \sim\!\!\sim\!\!\sim \mathrm{NH_2} \quad (11)$$

The polyamide is mixed with gaseous ammonia and the catalyst in the ammonolysis reactor. The main reaction products are caprolactam, hexamethylene diamine, aminocapronitrile, and adiponitrile. In addition, there are also other products and the majority of them can be hydrogenated to hexamethylene diamine.

PA6 can be also successfully depolymerized under "vacuum" [7,18]. This method is particularly attractive with regard to the high purity caprolactam obtained in high yields, short reaction times and, unlike other methods, there is no need water to be distilled off. The presence of a catalyst prevents the formation of cyclic olefins and azides. The best catalyst was found to be K_2CO_3 [7,18]. The disadvantages of this method consist in the necessity to use a catalyst and to maintain homogeneity of the reaction mixture. Filler residues may disturb the course of the reaction; moreover, the organic residues of this process cannot be recycled.

Among the other depolymerization methods of amide polymers, it is noteworthy to mention: (i) application of microwaves for the PA6 depolymerization (for the time being, on a laboratory scale) [42]. For instance, under the action of microwaves (200 W) for 12–23 min, aminocapronic acid was obtained in a 90% yield, together with some oligomers and cyclic compounds; orthophosphoric acid was used as catalyst, (ii) pyrolysis with or without the use of catalysts [43,44], (iii) radiation method [45].

3.2.2. *Material recycling*

The material (mechanical) recycling methods of polyamide wastes are based first of all on the separation of the particular components of the carpet covering. The separated components are processed to the appropriate products, which can then compete with the products manufactured from virgin polymers. Such type of recycling brings some difficulties due to the high absorption of water by the polymers.

With regard to the structure of the carpet covering, the wastes mainly contain the two immiscible plastics, polyamide and polypropylene. This immiscibility is the reason for the unsatisfactory mechanical properties of the polymer blends, hence the necessity of the application of compatabilizers. United Recycling Inc. was the first to introduce to the market the two polymeric blends originating from carpet waste (URI 20-001 and URI 10-001) [46]. These blends contained 60 and 75 wt% of PA6; 15 and 10 wt% of PP; 10 and 15 wt% of other polymers, respectively, and 15 wt% of inorganic fillers. They were used for extrusion of carpet adhesive tapes [46].

Monsanto patented in 1994 a method of processing of all the carpet components without a previous separation [47,48]. The product could be extruded in the twin-screw extruder; it contained 35–67 wt% of PA, 8–21 wt% of PP, 5–29 wt% of SBR, and 10–40 wt% of inorganic filler. In the case when compatibilizers were not introduced into the blends, the mechanical properties were close to those of polystyrene, but they were much poorer than those of virgin PA66. Considerably better results were achieved after the introduction into the blends of compatilizers of the following types: Polybond (maleic anhydride grafted onto PP (PP-g-MA)), Kraton (styrene-ethylene/butadiene-styrene

(SEBS)) and poly(ethylene-*co*-vinyl acetate). The recyclate properties could compete with those of the virgin polymers [49–51].

In order to improve the properties of the processed waste carpet covering, glass fibers were also introduced [49]. An example can be the application of these fibers (15 or 30 wt%) in the polymeric blends obtained from the carpet recycling, with PP-g-MA as a compatibilizer in the extrusion process. The average weight ratio of such a composite is PA/$CaCO_3$/PP/SBR = 50/30/15/5. The addition of glass fibers (30 wt%) improves the tensile strength of the composite by about 180%. The competitive properties of such filled recyclates are similar to those of many plastics.

BASF is involved in the recycling of tanks of transportation vehicle radiators made of glass fiber-reinforced PA66 [18]. The tanks are remanufactured from the recyclate after blending with virgin material. Bayer, jointly with Mercedes Benz, utilizes waste touring bus seats made of elastomer-modified and glass fiber-reinforced PA6 [18]. The recyclate properties are very close to those of the initial material. The recyclate is processed into engine fan lids in Mercedes touring buses. It is worth noting that the utilization of recyclate from used parts for a new vehicle was planned right from the beginning of the respective project. A similar action was undertaken by Ford, jointly with DuPont [18]. They used blends containing 25 wt% of recyclate from carpet waste for the manufacture of air-supplying vehicle ducts.

Another utilization of PA wastes originating from carpets is the application of face fibres and textiles as fibrous fillers for the manufacture of composites and laminates [52,53]. Trimmed carpets were soaked with urea-formaldehyde resins, and subsequently pressed at 150–200°C and 3.4 MPa. The products are characterized by good mechanical properties (tensile modulus of 2.4–2.8 GPa).

3.2.3. *Energy recovery*

This recycling technique is based on the incineration of unsorted wastes or of their fraction after sorting and eventually separation of the components. A comparison of the energy required for the obtaining of 1 kg of polymer and that released by its combustion is presented in Table 3 [54]. It is assumed that the energy recovery is profitable when the energy released as a result of incineration is not less than 1/2 of that required for polymer synthesis [44,49].

In the case of polyamides, this recycling technique is unfavorable (Table 3). Additionally, polyamides exhibit the excellent feature that, even after reprocessing, their properties are still relatively high, making them attractive for use as engineering plastics [18]. However, one should bear in mind that incineration is the simplest way of waste utilization, and the combustion energies of plastics are comparable to or higher than that of coal.

3.3. *Recycling of urethane polymers*

The basic raw materials for polyurethane synthesis are isocyanates and polyols. The polyols are usually oligomeric compounds with molecular weight ranging

Table 3. Energy content and combustion energy of polymers used in carpets and of various combustible materials

Material	Energy content (MJ/kg)	Combustion energy (MJ/kg)
PA6	156	28.7
PA66	154	28.7
PA6/30% glass fiber	134	20.1
PA66/30% glass fiber	133	20.1
PP	73	44.0
PET	84	31.4
Wood	Unknown	18.5
Brown coal	Unknown	19.6
Hard coal	Unknown	28.4
Wood coal	Unknown	29.1
Coke	Unknown	29.3
Peat	Unknown	15.9
Petroleum	Unknown	42.2

from a few hundred to several thousand Daltons. Hence, polyurethanes can be considered as block copolymers composed of rigid and soft segments. Polyurethanes represent a specific group of plastics since they occur in many different forms, such as solid or foamed, rigid or flexible, thermosetting or thermoplastic. They found a wide application as flexible or rigid foams, paints and lacquers, elastomers, adhesives, fibers, artificial leather, *etc.* The polyurethanes are mainly manufactured in the form of polymeric foams, primarily flexible foams. During the manufacture of such materials, an exceptionally large quantity of waste (even up to 20 wt%) is generated. This fact, together with the necessity of reusing the post-consumer foams stimulated the recycling of polyurethane waste products. A substantial problem associated with rigid PUR foam recycling is the emission of freons, which are used as physical porofors. In order to avoid the release of these compounds into the atmosphere, their recovery is performed during size reduction/grinding. The methods described in a further section of this chapter find application mainly in the recycling process of foamed polyurethanes, which, however, does not exclude their utilization in the recovery of solid (non-porous) polyurethanes. In practice, for the reuse of polymers bearing urethane groups, the most important methods involve material and chemical/feedstock recycling. The known methods of polyurethane recycling are given in Figure 8.

3.3.1. *Regrinding*

This recycling technique is based on grinding of PUR foams into the form of fine powders and utilization of the obtained material. The PUR grinding processes

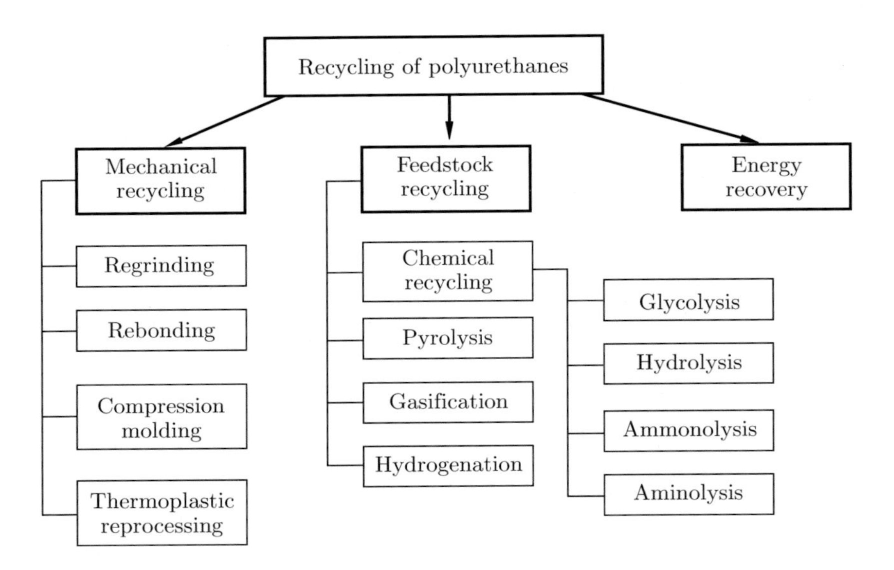

Figure 8. Methods of polyurethane recycling

are particularly difficult to perform in the case of flexible polyurethanes; a possibility of cryogenic grinding (with the use of liquid nitrogen) exists, which significantly improves the quality of obtained materials, but also increases the process costs.

The most important methods of polyurethane waste grinding are [55–57]: (i) precision knife cutting, (ii) two-roll milling, (iii) pellet milling, (iv) impact disc milling, and (v) extruder milling. One of the most frequently used methods of PUR waste grinding is the two-roll milling (Figure 9). A fine powder with grain size generally smaller than 250 μm is formed as a result of these processes. This powder may find application as oil absorber and as a filler for thermo-insulating concretes or plastics.

Oil absorbers: powdered polyurethane foams possess excellent hydrophobic properties and can absorb various types of mineral oils (1 kg of powder can absorb even up to 3.43 dm^3 of oil [58]). Therefore, such reprocessed foams are used for the treatment of scenes of road accidents, in plants involved in the production and trade of petroleum derivatives, and as barriers removing pollution of inland and sea waters.

Fillers for thermo-insulating concretes: PUR particles with a size of up to 1 cm are used in this case. Such concretes are mainly composed of polyurethane (about 90%), special modifiers of cement, and additives. The obtained material is characterized by a low thermal conductivity coefficient (about 60 mW/m K), low density (about 400 kg/m^3), and good acoustic insulation [59].

Fillers for plastics: PUR powders are also used for filling new products, mainly polyurethane ones (including both foams and products manufactured by

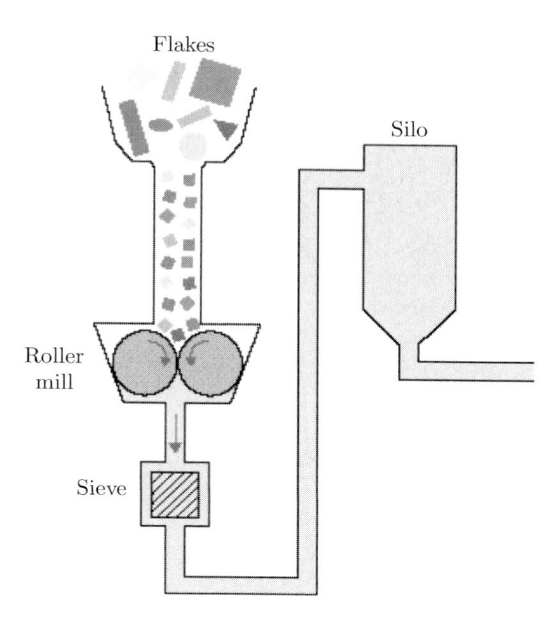

Figure 9. Schematic representation of the two-roll milling method [55]

a technique of reactive injection molding (RIM)). The process is most frequently realized by means of the so-called "three stream process" and consists of the three essential stages: (i) PUR grinding into a very fine powder, (ii) blending of the powder with some amount of polyols, (iii) blending in a mixing head of the three components, *i.e.*, polyol containing the PUR powder, polyol with appropriate additives, and isocyanate. The formation of products takes place directly after mixing. This method is mainly used for the manufacturing of products by the RIM method and for flexible foams. The critical factors determining the good properties of the products are adequately selected quality of the PUR powder, *i.e.*, appropriate and uniform degree of grinding, and a very small water content in the system. The PUR powder may be also used as a filler for other plastics. For instance, the application of powders from flexible PUR foams as fillers for polyester molding compounds of the type of bulk molding compounds (BMC) was studied by the authors' team [60]. An improvement in mechanical properties (especially the impact resistance) was observed at a powder content of several weight percent. Similar results were achieved with the application of grinded RIM waste as a filler for BMC [61]. Other authors showed that grinded PUR foams can be applied in the rubber manufacturing [62]. It was found that the addition of up to 25 wt% of PUR powder does not change significantly the rubber behavior (mechanical properties, vulcanization time, structure). Literature reports concerning the application of grinded PUR waste for filling of thermoplastics, *e.g.* PS and PVC, are also known [63]. An adequately prepared filler was added to the granulates of these polymers and this composition was processed by injection molding. Materials with mechanical properties significantly better than those of the non-modified thermoplastics were obtained.

3.3.2. *Rebonding*

This process was utilized almost from the beginning of PUR flexible foam manufacturing, first for the post-production and then for the post-consumer waste management. Rebonding usually consists of three stages [15,56,57]: (i) grinding of foams into particles with dimensions most often from 5 to 25 mm (such grinded flexible foams can be used in packages as a shock-protecting material and for filling of pillows and toys), (ii) blending of the obtained particles with a binding agent (*e.g.*, polymeric methylene-bisphenyl-diisocyanate and other isocyanate prepolymers), (iii) pressing in a hydraulic or conveyor press using steam as a heating medium (Figure 10).

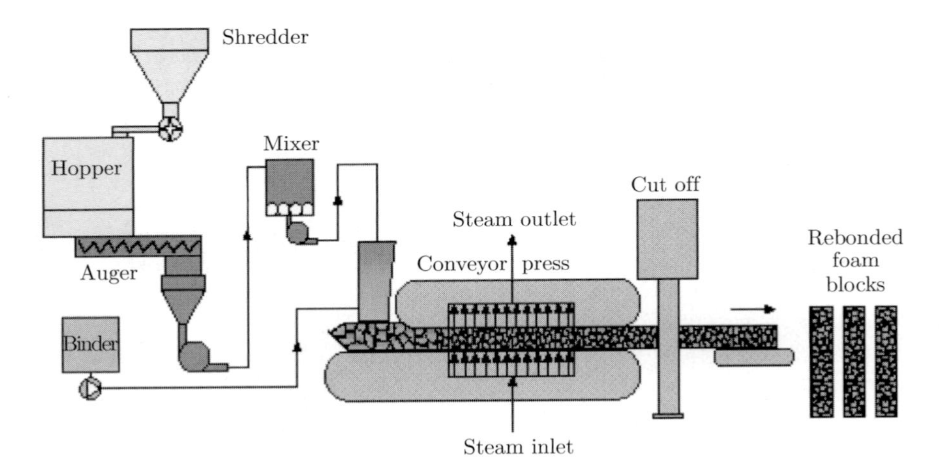

Figure 10. Continuous rebonding equipment [55]

The obtained products find numerous applications, *e.g.*, as sport mattresses and armrests, plates for floor heating assembly, or materials for vibration damping (washers under mechanical devices, railway tracks, *etc.*).

The blending of PUR particles with the binding agent is usually carried out in a mixer, in which spraying of the foams with binding agents takes place for 10–20 min with simultaneous very vigorous stirring. The wetted foam is placed in a mold and molded most often at 100–240°C and 2–20 MPa for about 10–20 min. Steam heated to 110–130°C is simultaneously introduced into the mold interior in order to accelerate the reaction with the prepolymer. During the compression molding, a volume reduction by a factor of 2 to 10 takes place. The products obtained, after demolding and seasoning for 12 to 24 h, are mechanically finished. Examples of accessible rebonding methods may be the REBOTEC and REMOTEC technologies of Hennecke Co. [57].

3.3.3. *Compression molding*

Compression molding is a very interesting technique of material recycling because a 100% content of the recyclate can be achieved in the new product [64]. The material is pressed at high temperature and pressure in a hydraulic press without the use of any binding agent. Monolithic products with good mechanical properties are obtained. The method found application in the recycling of flexible and rigid foams, and of the products molded by the RIM method [57,65]. First, the material is ground into 0.5 to 3 mm pellets, the latter being subsequently preheated at about 150°C for 1–12 min. Then, the pellets are placed in a hydraulic press mold and are compression molded at 180–200°C for 1–4 min under a pressure of above 35 MPa. A typical compression molding process is shown in Figure 11.

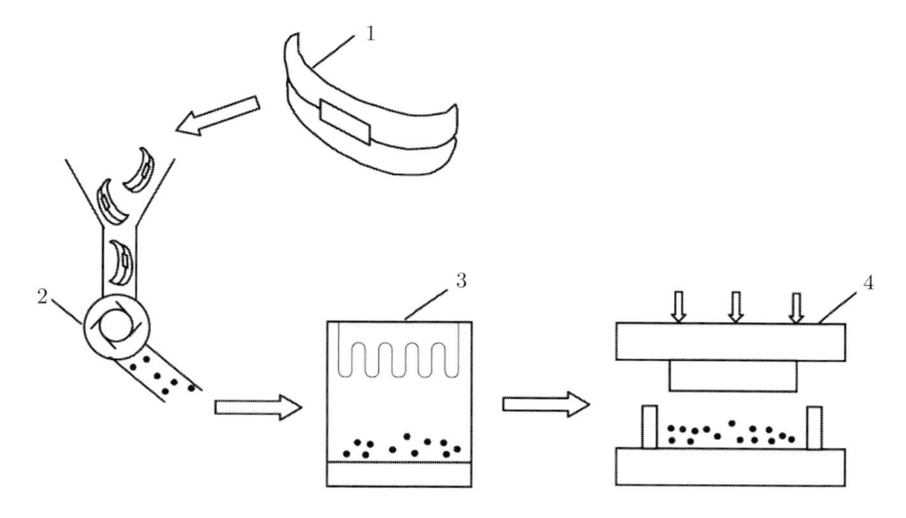

Figure 11. Schematic representation of a compression molding process [57]: 1 – used car parts (*e.g.*, bumpers), 2 – pelletizing, 3 – preheating, 4 – forming of new automotive parts (*e.g.*, battery trays)

The method currently finds application, among others, in the manufacture of various parts for the automotive industry, such as, *e.g.*, headlamps, buckets, seat backs, and air ducts.

3.3.4. *Thermoplastic reprocessing*

Some polyurethane materials (*e.g.*, flexible foams with a small degree of crosslinking or monolithic thermoplastic polyurethanes) can be processed by methods typical of thermoplastic polymers. Compression molding in a hydraulic press is most often used to this purpose. At high temperature and pressure, the crosslinks are broken and the material flows in the mold. In this way, different types of monolithic products with good mechanical properties can be prepared [15].

3.3.5. *Methods of chemical recycling*

The glycolysis of PUR products is the most important and popular method of chemical recycling [15,66]. This method is used both on a small scale (several tons per month) and on a very large scale (thousands of tons per year). In the urethane polymer chain fragments, most susceptible to glycolysis are the ester bonds. The urethane bonds are definitely less reactive, whereas the ether bonds are the least reactive [67].

The glycolysis of polyurethanes has been extensively studied, *e.g.*, by Simioni *et al.* [68–70]. These authors investigated the chemical solvolysis of PUR rigid foams and polyurethane waste from the reinforced reactive injection molding (R-RIM) by means of diethylene (DEG) and dipropylene (DPG) glycols. Potassium hydroxide and potassium acetate were used as catalysts. The obtained polyols have a hydroxyl number of 470–812 mg KOH/g and viscosity of 180–9000 mPa·s. They were used in a mixture with standard oligoetherol at the weight ratio of 1:1 for the preparation of foams and, after distilling off the free glycol, as additives for the PUR system intended for processing by the R-RIM method. The obtained polyols behaved in the foaming process in the same way as the standard ones. PUR foams obtained with the participation of the degradation products of PUR have better mechanical properties, lower flammability and lower thermal conductivity than those synthesized with the application of solvolysates prepared with the use of DPG. A 4 m^3 industrial reactor with a capacity of 1200 kg of PUR foam waste per one shift (8 h), in which the above described process can be carried out is described in [70].

At present, the glycolysis process of PUR foams is well known, although investigations are still ongoing in this field. In [71], detailed studies of the glycolysis of PUR flexible foams from car seats are reported. The optimum reaction conditions, such as reaction time, relative quantities of reagents (PUR, DEG, catalyst), and their influence on the selected properties of the products have been determined.

Polyurethane adhesives can be produced from polyurethane elastomer waste by the application of glycolysis [72,73]. In the process of adhesive manufacturing, both the upper layer of the post-reaction mixture (mainly consisting of polyols appropriate for the elastomer synthesis) and the bottom layer (consisting of ethylene glycol, ethanolamines and their derivatives, as well as the remaining compounds useful in the elastomer synthesis) were utilized. The obtained adhesives were characterized by good mechanical properties.

In [74], a method of reclaiming of PUR flexible foam waste by grinding of foam scraps and heating in the presence of mono- and diethanolamine at a temperature of 130–150°C is described. The reaction process was considerably faster when monoethanolamine was used at elevated temperature. After the completion of the process, the degradation product underwent separation into two layers. The upper layer contained oligoether, whereas the bottom one was formed by amine derivatives of MDI. It was found that alcoholysis is the main reaction in the process due to the formation by the amine groups of adducts with the carbonyl groups of the urethane bond. The degradation products originating from this process can find application as raw materials for the manufacture of PUR foams.

In [75,76], a method of decomposition of polyurethane rigid foams is described, using EG, DEG, and polyoxypropylenated pentaerythritol (with a hydroxyl number of 170 mg KOH/g) mixed with ethanolamines (mono-, di- and triethanolamine). Various salts of metals, such as tin, iron, sodium, potassium, zinc, calcium, copper, and cobalt were used as catalysts. The process products can be used as raw materials for the production of polyurethane plastics.

Nowadays, many companies use the glycolysis method on a commercial scale (Table 4). The process of split phase glycolysis is an interesting method developed by ICI [77]; glycolysis is carried out by means of DEG at 200°C. The post-reaction mixture separates into two layers. The upper layer can be utilized for the production of flexible foams, whereas the bottom layer is subjected to propoxylation and finds application in the manufacturing of PUR rigid foams. A detailed description of this method and of some other commercial ones is given in [15].

Table 4. Examples of European companies using the solvolysis processes [16]

Company	Chemolysis method	Capacity (tons/year)	Application
Getzner	Glycolysis	max 1000	Elastomers, rigid foams
Regra	Glycolysis, Acidolysis	max 1000	Shoe soles, rigid foams, flexible foams
Stankiewicz	Acidolysis	max 1000	Flexible foams
Aprithane	Glycolysis	max 2000	Rigid foams
ICI/du Vergier	Split phase hydroglycolysis	max 500	Rigid foams, flexible foams
Lögstör	Glycolysis	max 3000	Piping insulating foams

The polyurethane hydrolysis is interesting, although it is not used on a larger scale as a method of polyurethane recycling. As a rule, the process is carried out at temperatures varying from 200 to 350°C and elevated pressure. The products include polyols and various amine derivatives. The polyols obtained in this way can be used in the manufacturing of new polyurethane articles.

The degradation of elastomers and PUR flexible foams can be also performed by the use of ammonia in the supercritical state [78]. The process was carried out at 139°C and 14 MPa for 120 min. After the process completion, the ammonia was evaporated. Pure polyols remain in the reactor after extraction with water in order to recover diamines and the chain extender. The diamines can be subjected to phosgenization in order to achieve the recovery of isocyanates. In this way, almost all raw materials necessary for the production of new polyurethanes can be obtained.

A two-stage method of waste polyurethane degradation is described in [79]. In the first stage, scission of the polyurethane chain takes place at a temperature of 120°C in the presence of dialkanolamine and metal hydroxide (*e.g.*, KOH). Under these conditions, the reaction products include polyols, aromatic amines, short-chain ureas, and urea derivatives. The second stage is based on the alcoxylation of the hydroxyl groups and the primary and secondary amine groups, *e.g.*, by the use of propylene oxide. In this way, polyols with a hydroxyl number of 156–271 mg KOH/g and viscosity within the range of 1950–57 000 mPa·s can be obtained. The flexible foams prepared from recycled polyols revealed favorable mechanical properties.

Troev *et al.* (see, *e.g.*, [80,81]) degraded polyurethanes (microporous elastomers or flexible foams) *via* exchange reactions between urethane groups and alkoxy groups of phosphonic or phosphoric acid derivatives using dialkyl phosphonates or trialkyl phosphates. The potential application of the oligomeric liquid products is similar to that of the polycarbonate degradation products [29], *i.e.*, for flame retardant formulations and polymer modification.

3.3.6. *Other techniques*

Pyrolysis and gasification processes are also utilized, most often for the recycling of polymer blends. The respective facilities are usually parts of the installations in large petrochemical plants. A more comprehensive discussion on the thermal recycling of different polymers is given in [82].

Incineration with energy recovery is recommended in the case when a material is so worn out, that its other utilization is impossible and when polyurethanes occur as blends with other plastics and municipal waste. This process is currently utilized on a large scale in many countries.

4. Life cycle analysis

The basic recycling options, *i.e.*, physical or physicochemical methods, chemical processes, and energy recovery have certain advantages and disadvantages. In their recent book [82], Azapagic *et al.* pointed out the following features of the above mentioned recycling processes: (i) material recycling — simplicity, necessity to identify and separate the individual plastic components causing labor and energy consumption, (ii) chemical methods — recycling of mixed or soiled waste plastics possible, high capital costs necessary, (iii) incineration — causes energy recovery, but generates dangerous solids (ashes) and gaseous products. The former two recycling methods preserve petrochemical or, more generally, carbon-based resources.

It should be mentioned here that the recycling of waste plastics is not necessarily more sustainable than their disposal to landfill after being used only once. In other words, recycling is not totally impact-free. Recycling processes consume energy and materials, and usually generate additional air and/or water emissions and solid waste. Other aspects connected directly or indirectly with transportation steps should also be taken into account. From the environmental viewpoint, the polymer recycling process is not sustainable if it consumes more resources and energy than those necessary for the production of the respective polymer.

A detailed examination of the advantages and disadvantages of polymer recycling by considering several life cycle analysis case studies is given in [82]. Here, a short description of the relevant life cycle analysis of flexible polyurethane foam wastes is presented, with emphasis on the relation between the start and the end of the product's life, *i.e.*, including process and product design.

In their case study, Markovic and Hicks [83] used the life cycle design principles to develop a novel recyclable polyurethane furniture cushioning material (mattress) called "Waterlily". The study has been divided into two stages; in the first one, identification of the main stages in the life cycle had to be targeted for maximum improvements with respect to environmental performance, while the second stage has aimed to identify the most appropriate end-of-life options for polyurethane foams necessary to redesign the existing product for improved recyclability.

The results of life cycle analysis for the energy consumption of the existing products have revealed that nearly 80% of the total energy is consumed from the extraction of oil from the ground to the manufacturing of the principal raw materials (mainly diisocyanates and polyols). The remaining 20% of the total energy is necessary for the product fabrication (*ca.* 5%), packaging and transportation (5%), and waste management (10%).

For the raw material manufacturing stage, the main aim has been to develop a simpler formulation (tolylene 2,4-diisocyanate (TDI) with higher volatility has been replaced by the less volatile 4,4′-methylene-bis(phenyl isocyanate) (MDI)) and, doing so, to improve the working environment conditions.

In addition to the minimization of isocyanate vapors at the working place, some other factors have environmental impact, *i.e.*, the use of fluoroorganic blowing agents should be avoided and the possibility of autocombustion should be eliminated. The packaging material should be recyclable and transportation costs of low density polyurethane foams should be minimized. The mattress product should exhibit reduced flammability (without the use of halogenated flame-retardants). Finally, the waste management stage should enable physical and chemical recycling or energy recovery, as appropriate. Several recycling options have been considered [83]. The final decision has been chemical recycling, giving feedstocks for polyurethane synthesis, *i.e.*, for the same application. Compacted pellets of the used "Waterlily" mattresses have been subjected to the so-called split-phase glycolysis [77] with diethylene glycol. In such a way, a new "Waterlily" process for the production of mattresses, which incorporates all the above environmental improvements, has been developed. An important feature of this process is its closed-loop character.

Comparison of the various considered recycling options in the described example has shown that chemical recycling can save up to 50% of the energy required to manufacture the mattress from virgin materials. Energy recovery and mechanical recycling would result in a saving of up to 15% and 10%, respectively. Another recycling option, *i.e.*, rebonding foam chips for carpet underlay consumes *ca.* 16% more energy than the mattress manufacturing from virgin materials.

5. Summary and conclusions

Thermoplastic elastomers synthesized by polycondensation and polyaddition processes can be recycled mainly by the conventional methods used for plastics, *i.e.*, melt processing, chemical/feedstock degradation and, sometimes, by energy recovery (incineration).

Undoubtedly, the most important method of TPE recycling is melt processing, which makes use of the important and reproducible TPE feature, *i.e.*, thermoplastic flow and elastomeric properties after solidification. Melt blending of TPE copolymers (which can play the role of compatibilizers) with other recycled polymers, including reactive blending, broadens the application of this recycling technique.

Solvolytic methods, as related to the step-growth copolymers, seem to be of minor significance because of the rather complex mixture of the chemical degradation products. However, these methods have some importance for waste polyurethane elastomers and some polycondensation homopolymers. Incineration for energy recovery seems to be of minor importance as a recycling option for TPEs, as compared to some commodity waste polymers.

Life cycle assessment should be adopted as a necessary and valuable tool when considering the selection of the recycling methods on a case-by-case basis.

References

1. Spychaj T (2002) Chemical recycling of PET: methods and products, in *Handbook of Thermoplastic Polyesters* (Ed. Fakirov S) Wiley-VCH, Weinheim, pp. 1251–1290.
2. Roslaniec Z and Pietkiewicz D (2002) Synthesis and characteristics of polyester-based thermoplastic elastomers: chemical aspects, in *Handbook of Thermoplastic Polyesters* (Ed. Fakirov S) Wiley-VCH, Weinheim, pp. 582–658.
3. Aharoni S M (2002) Probable future trends in various classes of thermoplastic polyesters, in *Handbook of Thermoplastic Polyesters* (Ed. Fakirov S) Wiley-VCH, Weinheim, pp. 1321–1338.
4. Scheirs R (1998) Recycling of engineering thermoplastics, in *Polymer Recycling: Science, Technology and Application* (Ed. Scheirs R) J. Wiley & Sons, Chichester, pp. 303–338.
5. Bonfield J H, Hecker R C, Snider O E and Apostle B G (1965) Regeneration of ε-caprolactam from polyamides, US Patent 3,182,055.
6. Lazarus S D, Twilley I C and Snider O E (1967) Simultaneous depolymerisation of polycaproamide and polyester with recovery of caprolactam, US Patent 3,317,519.
7. Disselhoff R (1995) Chemical recycling of polyamide, *Proc 17th Annual Int Conf on Advances Stabilization and Degradation of Polymers*, Luzern, Switzerland.
8. Braun D, Disselhoff R, Guckel C and Illing G (2001) Chemical (tertiary) recycling of fiber glass reinforced polyamide-6, *Chem-Ing-Tech* **73**:183–190.

9. Bassler P and Kopietz M (1996) Process for producing caprolactam through hydrolytic cleavage of molten polycaprolactam, US Patent 5,495,015.
10. Corbin T, Handermann A, Kotek R, Porter W, Dellinger J and Davis E (1996) Reclaiming epsilon-caprolactam from nylon 6 carpet, US Patent 5,977,193.
11. Fisher M M (2003) Plastics recycling, in *Plastics and the Environment* (Ed. Andrady A L), J. Wiley & Sons, New Jersey, pp. 563–627.
12. Polk M B (2003) Depolymerization and recycling, in *Synthetic Methods in Step-Growth Polymers* (Eds. Roger M E and Long T E) J. Wiley & Sons, New Jersey, pp.527–574.
13. Meluch W C, Campbell G A (1976) Recovering amines by the hydrolytic decomposition of polyurethanes, US Patent 3,978,128.
14. Nadkarni V M (2002) Recycling of polyesters, in *Handbook of Thermoplastic Polyesters* (Ed. Fakirov S) Wiley-VCH, Weinheim, pp. 1223–1249.
15. Scheirs R (1998) Recycling of polyurethanes, in *Polymer Recycling: Science, Technology and Applications* (Ed. Scheirs R) J. Wiley & Sons, Chichester, pp. 340–377.
16. Rasshofer W (2002) Chemical recycling of PU, www.plastics-in-elv.org
17. Paszun D and Spychaj T (1997) Chemical recycling of poly(ethylene terephthalate), *Ind Eng Chem Res* **36**:1373–1383.
18. Scheirs (1998) Nylon recycling, in *Polymer Recycling: Science, Technology and Application* (Ed. Scheirs R) J. Wiley & Sons, Chichester, pp. 287–302.
19. Nadkarni V M and Rath A K (2002) Blends of thermoplastic polyesters, in *Handbook of Thermoplastic Polyesters* (Ed. Fakirov S) Wiley-VCH, Weinheim, pp. 835–893.
20. Xanthos M (2002) Reactive modification/compatibilization of polyesters, in *Handbook of Thermoplastic Polyesters* (Ed. Fakirov S) Wiley-VCH, Weinheim, pp. 815–833.
21. Karger-Kocsis J (2002) Recycling options for post-consumer PET and PET containing waste by melt blending, in *Handbook of Thermoplastic Polyesters* (Ed. Fakirov S) Wiley-VCH, Weinheim, pp. 1291–1318.
22. Farahat M S and Nikles D E (2001) On the UV curability and mechanical properties of novel binder systems derived from polyethylene terephthalate waste for solventless magnetic tape manufacturing. 1. Acrylated oligoesters, *Macromol Mater Eng* **286**:695–704.
23. Farahat M S and Nikles D E (2002) On the UV curability and mechanical properties of novel binder systems derived from polyethylene terephthalate waste for solventless magnetic tape manufacturing. 2. Methacrylated oligoesters, *Macromol Mater Eng* **287**:353–362.
24. Kacperski M and Spychaj T (1999) Rigid polyurethane foam with poly(ethylene terephthalate)/triethanolamine recycling products, *Polym Adv Technol* **10**:620–624.
25. Fabrycy E, Leistner A and Spychaj T (2000) New epoxy resin hardeners from PET scrap, *Adhesion* **44**:35–39.
26. Spychaj T, Fabrycy E, Spychaj S and Kacperski M (2001) Aminolysis and aminoglycolysis of waste poly(ethylene terephthalate), *J Mater Cycles & Waste Managment* **3**:24–31.
27. Atta A M (2003) Epoxy resin based on poly(ethylene terephthalate) waste: synthesis and characterization, *Progr Rubber Plastics Recycl Technol* **19**:17–40.
28. Spychaj T, Pilawka R, Spychaj S and Bartkowiak A (2004) Tertiary alkanolamines as solvolytic agents for poly(ethylene terephthalate). Evaluation of the products as epoxy resin hardeners, *Ind Eng Chem Res* **43**:862–874.
29. Troev K, Tsevi R and Gitsov I (2000) A novel depolymerization route to phosphorus-containing oligocarbonates, *Polymer* **42**:39–42.

30. Guclu G, Yalcinyuva T, Ozyumus S and Orbay M (2003) Simultaneous glycolysis and hydrolysis of poly(ethylene terephthalate) and characterization of products by differential scanning calorimetry, *Polymer* **44**:7609–7616.
31. Spychaj T, Malkiewicz E and Ukielski R (2003) Method of the aromatic dicarboxylic acid and diamine salts preparation, Polish Patent Appl P 360 748.
32. Spychaj T, Pilawka R and Ukielski R (2004) New possibilities of chemical recycling methods in synthesis and/or modification of step-growth polymers, *Proc 3rd Int Conf Polymer Modification, Degradation and Stabilization MODEST*, Lyon, CDROM abstract.
33. Wang Y, Zang Y, Polk M B, Kumar S and Muzzy J D (2003) Recycling of carpet and textile fibres, in *Plastics and the Environment* (Ed. Andrady A L), John Wiley & Sons, Hoboken, NJ, pp. 697–725.
34. Realff M J, Ammons J C and Newton D (1999) Carpet recycling: determining the reverse production system design, *Polym-Plast Technol* **38**:547–567.
35. Braun M, Levy A B, Sifniades S (1999) Recycling nylon 6 carpet to caprolactam, *Polym-Plast Technol* **38**:471–484.
36. Dahlhoff G, Niederer J P M and Hoelderich W F (2001) Epsilon-caprolactam: new by-product free synthesis routes, *Catal Rev* **43**:381–441.
37. Krolikowski W and Bledzki A (1998) Methods of identification and sorting of the waste carpet coverings, *Ekoplast* **14**:55–60 (in Polish).
38. Moran J (1994) Separation of nylon-6 from mixtures with nylon-66, US Patent 5,280,105.
39. Berthold H and Hagen R (1994) Recovery of polyamides, European Patent 603,434.
40. Bodrero S, Canivenc E and Cansell F (1998) Chemical recycling of polyamide 6,6 and polyamide 6 through a two step AMI-/ammonolysis process, *Abstr Pap ACS* **16**:395-Poly Part 3 Aug 23.
41. Smith R A and Gracon B E (1995) Polyamide 66 and 6 chemical recycling, *Proc Recycle'95*, Davos, Switzerland, paper 5–2.1.
42. Klun U and Krzan A (2000) Rapid microwave induced depolymerisation of polyamide-6, *Polymer* **41**:4361–4365.
43. Czernik S, Elam C C, Evans R J, Meglen R R, Moens L and Tatsumoto K (1998) Catalytic pyrolysis of nylon-6 to recover caprolactam, *J Anal Appl Pyrol* **46**:51–64.
44. Bockhorn H, Donner S, Gernsbeck M, Hornung A and Hornung U (2001) Pyrolysis of polyamide 6 under catalytic conditions and its application to reutilization of carpets, *J Anal Appl Pyrol* **58**:79–94.
45. Burillo G, Clough R L, Czvikovszky T, Guven O, Le Moel A, Liu W W, Singh A, Yang J T and Zaharescu T (2002) Polymer recycling: potential application of radiation technology, *Radiat Phys Chem* **64**:41–51.
46. Schut J H (1993) A recycling first: carpets!, *Plastic Tech*, April, 22–25.
47. David D J, Dickerson J L and Sincock T F (1994) Thermoplastic composition and method for producing thermoplastic composition by melt blending carpet, US Patent 5,294,384.
48. Hagberg C G and Dickerson J L (1997) Recycling nylon carpet *via* reactive extrusion, *Plastic Eng* **53**(4):41–43.
49. Datta R J, Polk M B and Kumar S (1995) Reactive compatibilization of polypropylene and nylon, *Polym Plast Technol Eng* **34**:551–560.
50. La Mantia F P and Capizzi L (2001) Recycling of compatibilized and uncompatibilized nylon/polypropylene blends, *Polym Degrad Stabil* **71**:285–291.
51. Young D, Chlystek S, Malloy R and Rios I (1998) Recycling of carpet scrap, US Patent 5,719,198.

52. Kotlair A M, Fountain D P (1997) Synthetic wood from waste fibrous waste products, US Patent 5,626,939.
53. Kotlair A M and Michielsen S (1999) Utilization of waste fibers in laminates, US Patent 5,912,062.
54. Kindler H and Nikles A (1980) Energy requirements in the production of materials – fundamentals and energy equivalent values for plastics, *Kunststoffe* **70**:802–807.
55. www.isopa.org.
56. Stone H, Villwock R and Martel B (2000) Recent technical advances in recycling of scrap polyurethane foam as finely ground powder in flexible foam, *Proc Polyurethane Conf*, May, www.mobiustechnologies.com.
57. Neuray M P, Sulzbach H M and Wirth J, Methods for the recycling of polyurethane and polyurethane composites, www.henneckemachinery.com.
58. www.seg-online.de.
59. www.isola.be.
60. Nowaczek W and Kacperski M, unpublished results.
61. Hulme A J and Goodhead T C (2003) Cost effective reprocessing of polyurethane by hot compression moulding, *J Mater Proc Technol* **139**:322–326.
62. Sims G L A and Sombatsompop N (1996) Pulverised flexible polyurethane foam particles as a filler in natural rubber vulkanisates, *Cellular Polymers* **15**:90–104.
63. Shutov F, Arastoopour H and Ivanov G (1993) Recycling of polyurethane waste foams using new pulverisation principle: Solid State Shear Extrusion (SSSE), *Cellular Polymers 2nd Int Conf*, Edinburgh, Paper 18.
64. Kardasz D (1999) Possibilities of materials recycling of rigid polyurethane foams, PhD Thesis, Technical University of Szczecin, Szczecin (in Polish).
65. Rasshofer W (2002) Mechanical recycling of PU foam, www.plastics-in-elv.org.
66. Kacperski M and Spychaj T (1999) Chemical recycling of waste ester and urethane polymers as a source of raw materials for production of polyurethane, *Polimery* (Warsaw) **44**:1–5 (in Polish).
67. Wirpsza Z (1991) *Polyurethanes: Chemistry, Technology, Application*, WNT, Warsaw (in Polish).
68. Simioni F, Bisello S and Tavan M (1983) Polyol recovery from rigid polyurethane waste, *Cellular Polymers* **2**:281–293.
69. Rienzi S A, Simioni F and Modesti M (1993) Glycolysis of polyurethane and polyurea polymers, *Cellular Polymers 2nd Int Conf*, Edinburgh, Paper 14.
70. Simioni F, Modesti M and Rienzi S A (1991) Recycling of polyurethane waste, *Cellular Polymer 1st Int Conf*, London.
71. Wu C-H, Chang C-Y, Cheng C-M and Huang H-C (2003) Glycolysis of waste flexible polyurethane foam, *Polym Degrad Stabil* **80**:103–111.
72. Borda J, Pasztor G and Zsuga M (2000) Glycolysis of polyurethane foams and elastomers, *Polym Degrad Stabil* **68**:419–422.
73. Borda J, Racz A and Zsuga M (2002) Recycled polyurethane elastomers: a universal adhesive, *J Adhesion Sci Technol* **16**:1225–1234.
74. Kanaya K and Tokahoshi S (1994) Decomposition of polyurethane foam by alkanolamines, *J Appl Polym Sci* **51**:675–682.
75. Maslowski H, Kozlowski K and Czuprynski B (1977) Method of thermal degradation of waste rigid polyurethane foams to obtain raw materials for foam synthesis, Polish Patent 94 455.
76. Czuprynski B, Paciorek-Sadowska J, Liszkowska J and Czuprynska J (2002) The utilization of rigid polyurethane-polyisocyanurate foams by the combined alcoholysis-aminolysis process, *Polimery* (Warsaw) **47**:104–109.

77. Leaflet of ICI Company (1999).
78. Lentz H and Mormann W (1992) Chemical recycling of polyurethanes and separation of the components by supercritical ammonia, *Macromol Symp* **57**:305–310.
79. Van Der Wal H R (1994) New chemical recycling process for polyurethanes, *J Reinf Plast Comp* **13**:87–96.
80. A. Troev K, Atanasov V I, Tsevi R, Grancharov G and Tsekova A (2000) Chemical degradation of polyurethanes. Degradation of microporous polyurethane elastomer by dimethyl phosphonate, *Polym Degrad Stabil* **67**:159–165.
81. Troev K, Grancharov G, Tsevi R and Tsekova A (2000) A novel approach to recycling of polyurethanes: chemical degradation of flexible polyurethane foam by triethyl phosphate, *Polymer* **41**:7017–7022.
82. Azapagic A, Emsley A and Hamerton I (2003) Environmental impacts of recycling, in *Polymers, the Environment, Sustainable Development* (Ed. Hamerton I) J Wiley & Sons, Chichester, pp. 155–172.
83. Markovic V and Hicks D A (1997) Design for chemical recycling, *Phil Trans Royal Soc* London Ser A **355**:1415–1424.

List of Acknowledgements

The authors gratefully acknowledge permissions to reproduce copyrighted materials from a number of sources. Every effort has been made to trace copyright ownership and to give accurate and complete credit to copyright owners, but if, inadvertently, any mistake or omission has occurred, full apologies are herewith tendered

Chapter 1. Creation and Development of Thermoplastic Elastomers, and Their Position Among Organic Materials

1. Figure 1 reprinted from *Rubber Technology - Compounding and Testing for Performance,* 2001, Editor: Dick J, Author: Rader Ch R, Chapter 10, p. 265, Copyright (2004), with permission from Hanser Publishers, Munich.
2. Figure 2 reprinted from *Polymer J*, Vol. 36, 2004, Authors: Yokozawa T and Yokoyama A, Title: Chain-growth polycondensation: polymerization nature in polycondensation and approach to condensation polymer architecture, pp. 65–83, Copyright (2004), with permission from Yokoyama T.

Chapter 2. Polycondensation Reactions in Thermoplastic Elastomer Chemistry: State of the Art, Trends, and Future Developments

1. Figure 1 reprinted from *Adv Biochem Eng*, Vol. 71, 2001, Authors: Kim Y B and Lenz R W, Title: Polyesters from microorganisms, pp. 51–79, Copyright (2004), with permission from Lenz R W and Springer-Verlag GmbH&Co.KG.

Chapter 3. Polyester Thermoplastic Elastomers: Synthesis, Properties, and Some Applications

1. Table 2 reprinted from *Polimery*, Vol. 33, 1988, Authors: Ukielski R and Wojcikiewicz H, Title: Poly(ether ester) block copolymers modified by pentaerythritol, pp. 9–12, Copyright (2004), with permission from Industrial Chemistry Research Institute, Warsaw.

2. Figures 2 and 3 reprinted from *Sci Papers Tech Uni Szczecin*, Vol. 479, 1992, Author: Slonecki J, Title: Morphology and some properties of poly(ether ester)s, Copyright (2004), with permission from Szczecin Technical University Press.
3. Figure 6 reprinted from *J Polym Sci Part A Polym Chem*, Vol. 37, 1999, Authors: Cheng X, Luo X L and Ma D Z, Title: Proton NMR characterization of chain structure in butylene terephthalate-EPSILON-caprolactone copolymers, pp. 3770–3777, Copyright (2004), with permission from John Wiley & Sons, Inc.
4. Figures 9 and 10 reprinted from *Elastomery*, Vol. 1, 1997, Authors: El Fray M, Kozlowska A and Slonecki J, Title: Influence of the oligoamide's soft segments mass concentration on some selected properties of copoly(ester amide)s, pp. 12–20, Copyright (2004), with permission from Institute of Rubber Industry, Piastow.
5. Figures 13 and 14 reprinted from *Polymer*, Vol. 35, 1994, Authors: Walch E and Gaymans R J, Title: Synthesis and properties of poly(butylene terephthalate)-b-polyisobutylene segmented block copolymers, pp. 636–641,Copyright (2004), with permission from Elsevier.
6. Figures 15 and 16 reprinted from *Sci Papers Inst Org Polym Technol Wroclaw Uni Technol*, Vol. 52, 2003, Authors: Kwiatkowski K and Roslaniec Z, Title: Preparation of thermoplastic elastomers by the reactive modification of poly(ethylene terephthalate), pp. 500–504, Copyright (2004), with permission from Publishing House of the Wroclaw University of Technology.
7. Figures 17 and 18 reprinted from *Nanostructured Elastomeric Biomaterials for Soft Tissue Reconstruction*, 2003, Author: El Fray M, Copyright (2004), with permission from Publishing House of the Warsaw University of Technology.

Chapter 4. Terpoly(Ester-*b*-Ether-*b*-Amide) Thermoplastic Elastomers: Synthesis, Structure, and Properties

1. Figures 1–9 and 11–14 reprinted from *Sci Papers Techn Uni Szczecin*, Vol. 556, 2000, Author: Ukielski R, Title: Multiblock terpoly(ester-*b*-ether-*b*-amide) elastomers: synthesis, structure, properties, pp. 41,58, 63, 65–67, 69, 70, 96, 109, 110, and 113, with permission from "ZAPOL" Technical University of Szczecin.
2. Figure 10 reprinted from *Angew Makromol Chem*, Vol. 271, 1999, Authors: Ukielski R, Lembicz F and Majszczyk J, Title: New multiblock terpoly(ester-ether-amide) thermoplastic elastomers, p. 57, Copyright (2004), with permission from Wiley VCH.

Chapter 5. High Performance Thermoplastic Aramid Elastomers: Synthesis, Properties, and Applications

1. Figure 8 reprinted from *J Polym Sci, Part A Polym Chem,* Vol. 42, 2004, Authors: Rabani G, Rosair G M and Kraft A, Title: Low-temperature route to thermoplastic polyamide elastomers, pp. 1449–1460, Copyright (2004), with permission from John Wiley & Sons, Inc.
2. Table 6 reproduced from Thesis, University of Twente, Enschede, The Netherlands, 2000, Author: Niesten M C E J, Title: Polyether based segmented copolymers with uniform aramid units, pp. 1–167, Copyright (2004), with permission from Niesten M C E J.
3. Table 7 reproduced from Thesis, University of Twente, Enschede, The Netherlands, 2000, Author: Niesten M C E J, Title: Polyether based segmented copolymers with uniform aramid units, pp. 1–167, Copyright (2004), with permission from Niesten M C E J.

Chapter 6. Poly(Ether Ester) Thermoplastic Elastomers: Phase and Deformation Behavior on the Nano- and Microlevel

1. Figures 1–3 reprinted from *J Macromol Sci-Phys,* Vol. B27, 1990, Authors: Fakirov S, Apostolov A A, Boeseke P and Zachmann H G, Title: Structure of segmented poly(ether ester)s as revealed by synchrotron radiation, pp. 379–395, Copyright (2004), with permission from Marcel Dekker, Inc.
2. Figures 4 and 5 reprinted from *J Macromol Sci-Phys,* Vol. B31, 1992, Authors: Apostolov A A and Fakirov S, Title: Effect of the block length on the deformation behavior of poly(ether ester)s as revealed by SAXS, pp. 329–355, Copyright (2004), with permission from Marcel Dekker, Inc.

Chapter 8. Dielectric Relaxation of Polyester-Based Thermoplastic Elastomers

1. Figure 3 reprinted from *Macromolecules,* Vol. 22, 1989, Authors: Runt J, Du L, Martynowicz L M, Brezny D M and Mayo M, Title: Dielectric properties and cocrystallization of mixtures of poly(butylene terephthalate) and poly(ester ether) segmented block copolymers, pp. 3908–3913, Copyright (2004), with permission from the American Chemical Society.

Chapter 9. Thermoplastic Poly(Ether-*b*-Amide) Elastomers: Synthesis

1. Figures 7–9 and Table 3 reprinted from *Eur Polym J,* Vol. 13, 1977, Authors: Deleens G, Foy P and Maréchal E, Title: Synthesis and characterization of copolycondensatesequences poly(amide-*seq*-ether) - II - Polycondensation of polyamide 11, ω,ω'-diacid or diester oligomers with polyether ω,ω'-dihydroxy oligomers, pp. 343–360, Copyright (2004), with permission from Elsevier.

Chapter 11. Semicrystalline Segmented Poly(Ether-*b*-Amide) Copolymers: Overview of Solid-State Structure-Property Relationships and Uniaxial Deformation Behavior

1. Figure 8 reprinted from *Progr Colloid Polym Sci*, Vol. 72, 1986, Authors: Goldbach G, Kita M, Meyer K and Richter K P, Title: Structure and properties of polyamide 12 alloys, pp. 83–96, Copyright (2004), with permission from Springer-Verlag GmbH & Co. KG.
2. Figure 10 (AFM) reprinted from *J Polym Sci Part B Polym Phys*, Vol. 37, 1999, Authors: McLean R S and Sauer B B, Title: Nano-deformation of crystalline domains during tensile stretching studied by atomic force microscopy, pp. 859–866, Copyright (2003), with permission from Wiley Periodicals Inc.
3. Figure 10 (SAXS) reprinted from *J Polym Sci Part B Polym Phys*, Vol. 40, 2002, Authors: Sauer B B, McLean R S, Brill D J and Londono D J, Title: Morphology and orientation during the deformation of segmented elastomers studied via small-angle x-ray scattering and atomic force microscopy, pp. 1727–1740, Copyright (2003), with permission from Wiley Periodicals Inc.
4. Figure 11 reprinted from *J Elastomers and Plastics*, Vol. 22, 1990, Author: Warner S, Title: Strain-induced crystallization and melting behavior of polyetheramide block copolymer, pp.166–173, Copyright (2004), with permission from SAGE Publications.
5. Figure 15 reprinted from *Polymer*, Vol. 41, 2000, Authors: Niesten M C E J, Feijen J and Gaymans R J, Title: Synthesis and properties of segmented copolymers having aramid units of uniform length, pp. 8487–8500, Copyright (2003), with permission from Elsevier Ltd.
6. Figure 16 reprinted from *Polymer*, Vol. 42, 2001, Authors: Niesten M C E J and Gaymans R J, Title: Tensile and elastic properties of segmented copolyetheresteramides with uniform aramid units, pp. 6199–6207, Copyright (2003), with permission from Elsevier Ltd.
7. Figure 17 reprinted from *Polymer*, Vol. 42, 2001, Authors: Niesten M C E J and Gaymans R J, Title: Tensile and elastic properties of segmented copolyetheresteramides with uniform aramid units, pp. 6199–6207, Copyright (2003), with permission from Elsevier Ltd.
8. Figure 18 reprinted from *Polymer*, Vol. 42, 2001, Authors: Niesten M C E J, Harkema S, van der Heide E and Gaymans R J, Title: Structural changes of segmented copolyetheresteramides with uniform aramid units induced by melting and deformation, pp. 1131–1142, Copyright (2003), with permission from Elsevier Ltd.

Chapter 12. Thermoplastic Polyurethane Elastomers in Interpenetrating Polymer Networks

1. Table 1 reprinted from *Thermochimica Acta*, Vol. 371, 2001, Authors: Vatalis A, Delides C, Georgoussis G, Kyritsis A, Grigoryeva O, Sergeeva L, Brovko A, Zimich O, Shtompel V, Neagu E and Pissis P,

Title: Characterization of thermoplastic interpenetrating polymer networks by various thermal analysis techniques, pp. 87–93, Copyright (2004) with permission from Elsevier.

Chapter 14. Molecular dynamics and ionic conductivity studies in polyurethane thermoplastic elastomers

1. Figures 4, 5, 7, and 8, Scheme 1 and Tables 2, 3, and 4 reprinted from *e-Polymers*, No. 042, 2004, Authors: Roussos M, Konstantopoulou A, Kalogeras I M, Kanapitsas A, Pissis P, Savelyev Yu and Vassilikou-Dova A, Title: Comparative dielectric studies of segmental mobility in novel polyurethanes, Copyright (2004), with permission from Dr. B. Jung, Editor.
2. Figures 9 and 10, and Tables 5, 6 and 7 reprinted from *e-Polymers*, No. 043, 2004, Authors Raftopoulos K, Zegkinoglou I, Kanapitsas A, Kripotou S, Christakis I, Vassilikou-Dova A, Pissis P and Savelyev Yu, Title: Dielectric and hydration properties of segmental polyurethanes, Copyright (2004),with permission from Dr. B. Jung, Editor.
3. Figures 11, 12, 13, and 14, Scheme 2 and Table 8 reprinted from *Eur Polym J*, Vol. 39, 2003, Authors: Charnetskaya A G, Polizos G, Shtompel V I, Privalko E G, Kercha Yu Yu and Pissis P, Title: Phase morphology and molecular dynamics of a polyurethane ionomer reinforced with a liquid crystalline filler, pp. 2167–2174, Copyright (2004), with permission from Elsevier Science Ltd.
4. Figure 16 reprinted from *Solid State Ionics*, Vol. 125, 1999, Authors: Pissis P, Kyritsis A and Shilov V V, Title: Molecular mobility and protonic conductivity in polymers: hydrogels and ionomers, pp. 203–212, Copyright (2004), with permission from Elsevier Science Ltd.
5. Figures 18 and 20, and Table 11 reprinted from *Solid State Ionics*, Vol. 136-137, 2000, Authors: Pissis P, Kyritsis A, Georgoussis G, Shilov V V and Shevchenko V V, Title: Structure-property relationships in proton conductors based on polyurethanes, pp. 255–260, Copyright (2004),with permission from Elsevier Science Ltd.
6. Figures 22 and 23, Scheme 5 and Table 14 reprinted from *Solid State Ionics*, Vol. 145, 2001, Authors: Polizos G, Shilov V V and Pissis P, Title: Molecular mobility and protonic conductivity studies in telechelics based on poly(ethylene oxide) capped with hydroxyl groups at both ends, pp. 93–100, Copyright (2004), with permission from Elsevier Science Ltd.
7. Figures 24 and 25 reprinted from *J Non-Cryst Solids*, Vol. 305, 2002, Authors Polizos G, Shilov V V and Pissis P, Title: Temperature and pressure effects on molecular mobility and ionic conductivity in telechelics based on poly(ethylene oxide) capped with hydroxyl groups at both ends, pp. 212–217,Copyright (2004), with permission from Elasvier Science Ltd.

8. Scheme 3 and Tables 9 and 10 reprinted from *Solid State Ionics*, Vol. 120, 1999, Authors: Shilov V V, Shevchenko V V, Pissis P, Kyritsis A, Gomza Yu P, Nesin S D and Klimenko N S, Title: Morphology and protonic conductivity of the polyurethanes with acid groups in the flexible segment, pp. 43–50, Copyright (2004), with permission from Elsevier Science Ltd.
9. Scheme 4 reprinted from *Solid State Ionics*, Vol. 136–137, 2000, Authors: Polizos G, Kyritsis A, Pissis P, Shilov V V and Shevchenko V V, Title: Structure and molecular mobility studies in novel polyurethane ionomers based on poly(ethylene oxide), pp. 1139–1146, Copyright (2004), with permission from Elsevier Science Ltd.

Chapter 18. Shape Memory Effects of Multiblock Thermoplastic Elastomers

1. Tables 3 and 4, and Figures 11, 12, 13, 14, 15 and 16 reprinted from *J Macromol Sci Polym Phys*, Vol. 43, 2004, Authors: Park S H, Kim J W, Lee S H and Kim B K, Title: Temperature sensitive amorphous polyurethanes, pp. 447–458, Copyright (2004), with permission from Marcel Dekker.
2. Table 6 and Figure 20 reprinted from *Polymer*, Vol. 41, 2000, Authors: Jeong H M, Kim B K and Choi Y J, Title: Synthesis and properties of thermotropic liquid crystalline polyurethane elastomers, pp. 1849–1855, Copyright (2004), with permission from Elsevier Science.
3. Tables 7, 8, 9 and 10, and Figure 21 reprinted from *Polymer J*,Vol. 32, 2000, Authors: Lee H Y, Jeong H M, Lee J S and Kim B K, Title: Study on the shape memory polyamides. Synthesis and thermomechanical properties of polycaprolactone-polyamide block copolymers, pp. 23–28, Copyright (2004), with permission from the Society of Polymer Science, Japan.
4. Tables 11 and 12, and Figures 22, 23, 24 and 25 reprinted from *J Mater Sci*, Vol. 36, 2001, Authors: Jeong H M, Song J H, Lee S Y and Kim B K, Title: Miscibility and shape memory property of poly(vinyl chloride)/thermoplastic polyurethane blends, pp. 5457–5463, Copyright (2004), with permission from Kluwer Academic.
5. Tables 13, 14 and 15, and Figure 26, reprinted from *Eur Polym J*, Vol. 37, 2001, Authors: Jeong H M, Ahn B K and Kim B K, Title: Miscibility and shape memory effect of thermoplastic polyurethane blends with phenoxy resin, pp. 2245–2252, Copyright (2004), with permission from Elsevier Science.

Chapter 19. Condensation and Addition Thermoplastic Elastomers: Recyclyng aspects

1. Figure 1 reprinted from *Polymer Recycling. Science, Technology and Application,* 1998, Author: Schreis R, Title: Recycling of engineering thermoplastics, pp. 303–338, Copyright (1998), with permission from J. Wiley & Sons, Inc.

2. Figures 5, 6, 7 and Table 2 reprinted form *Ekoplast*, Vol. 14, 1998, Authors: Krolikowski W and Bledzki A, Title: Methods of identification and sorting of the waste carpet covering, pp. 55–60, Copyright (1998), with permission from Technical University of Szczecin Publishing House.
3. Figures 9 and 10 reprinted from www.isopa.org.
4. Figure 11 reprinted from www.henneckemachinery.com, Authors: Neuray M P, Sulzbach H M and Wirth J, Title: Methods for the recycling of polyurethane composites.
5. Table 4 reprinted from www.plastics-in-elv.org, 2002, Author: Rasshofer W, Title: Chemical recycling of PU foam.
6. Part of the text reprinted from *Polymers, the Environment, Sustainable Development*, 2003, Authors: Azapagic A, Emsley A and Hamerton I, Title: Life cycle product design for chemical recycling: "Waterlily" cushioning, pp. 169–171, Copyright (2003), with permission from J. Wiley & Sons, Inc.

Author Index

Subject Index